# Representations of Hecke Algebras at Roots of Unity

For further volumes:
www.springer.com/series/6253

# Algebra and Applications

Volume 15



*Algebra and Applications* aims to publish well written and carefully refereed monographs with up-to-date information about progress in all fields of algebra, its classical impact on commutative and noncommutative algebraic and differential geometry, K-theory and algebraic topology, as well as applications in related domains, such as number theory, homotopy and (co)homology theory, physics and discrete mathematics.

Particular emphasis will be put on state-of-the-art topics such as rings of differential operators, Lie algebras and super-algebras, group rings and algebras, $C^*$-algebras, Kac-Moody theory, arithmetic algebraic geometry, Hopf algebras and quantum groups, as well as their applications. In addition, Algebra and Applications will also publish monographs dedicated to computational aspects of these topics as well as algebraic and geometric methods in computer science.

Meinolf Geck · Nicolas Jacon

# Representations of Hecke Algebras at Roots of Unity

Springer

Meinolf Geck
University of Aberdeen
Institute of Mathematics
Aberdeen AB24 3UE
UK
m.geck@abdn.ac.uk

Nicolas Jacon
Université de Franche-Comté
UFR Sciences et Techniques
Route de Gray 16
Besancon 25030
France
njacon@univ-fcomte.fr

ISBN 978-0-85729-715-0 e-ISBN 978-0-85729-716-7
DOI 10.1007/978-0-85729-716-7
Springer London Dordrecht Heidelberg New York

British Library Cataloguing in Publication Data
A catalogue record for this book is available from the British Library

Library of Congress Control Number: 2011929782

Mathematics Subject Classification (2010): 20C08, 20C20, 20C33, 20F55, 20G42, 17B37



*Cover design*: VTeX UAB, Lithuania

Printed on acid-free paper

Springer is part of Springer Science+Business Media (www.springer.com)

# Preface

One of the major open problems in the representation theory of finite groups is the determination of the irreducible representations of the symmetric group $\mathfrak{S}_n$ over a field of characteristic $p > 0$. Thanks to the work of James [179] in the 1970s, we do have a natural parametrisation of the irreducible representations in the framework of the theory of Specht modules, but explicit combinatorial formulae for their dimensions are not known in general! Note that the analogous problem in characteristic 0 has been solved for a long time, by the work of Frobenius around 1900.

In a wider context, this problem is a special case of the problem of determining the irreducible representations of Iwahori–Hecke algebras. These algebras arise naturally in the representation theory of finite groups of Lie type, but they can also be defined abstractly as certain deformations of group algebras of finite Coxeter groups, where the deformation depends on one or several parameters. For the purposes of this introduction, let us assume that all the parameters are integral powers of a fixed element in the base field. If this base parameter has infinite order and the base field has characteristic 0, then we are in the "generic case" where the algebras are semisimple; this case is quite well understood [132], [231]. Also note that, both for historical reasons and as far as applications are concerned, the case where all parameters are equal is particularly important.

The main focus in this text will be on the "modular case" where the algebras are non-semisimple. This situation typically occurs over fields of positive characteristic (a familiar phenomenon from the representation theory of finite groups), but it also occurs over fields of characteristic 0 when the base parameter is a root of unity. While leading to a highly interesting and rich theory in its own right, it turns out that the study of the characteristic 0 situation also provides a crucial step for understanding the positive characteristic case, which is most important for applications to finite groups of Lie type.

Over the last two decades, there has been considerable progress on the characteristic 0 situation. One of the most spectacular advances is the "LLT conjecture" [208] (where "LLT" stands for Lascoux, Leclerc, Thibon) and its proof by Ariki [7], [10]. This brings deep geometric methods and the combinatorics of crystal/canonical bases of quantum groups into the picture, opening the way for a variety of new theoretical connections and practical applications. Combined with sophisticated computational methods, the theory has now reached the following state:

- The classical theory of "Specht modules" has been generalised to Iwahori–Hecke algebras associated to arbitrary finite Coxeter groups, giving rise to natural parametrisations of the irreducible representations.
- Explicit descriptions of these parametrisations are now known in terms of so-called "canonical basic sets". Also, the dimensions of the irreducible representations are known, either by purely combinatorial algorithms (for the classical types) or in the form of explicit tables (for the exceptional types).

These results remain valid over fields of characteristic $p > 0$, as soon as $p$ is larger than some bound depending on the type of the algebra. As far as the parametrisation of the irreducible representations is concerned, the bound is very mild. For example, in the equal-parameter case, it will turn out that it is sufficient to assume that the characteristic is "good" in the sense of the theory of algebraic groups. However, as far as the dimensions of the irreducible representations are concerned, no explicit bound on $p$ is known at the present state of knowledge.

But there is a general conjecture – first formulated by James [181] in type $A$ – specifying such a bound. This conjecture has been verified in a number of cases, including algebras of type $A_n$ for $n \leqslant 9$ (see [181]) and all algebras of exceptional type (see Geck, Lux, and Müller [94], [126], [129]). If true, this conjecture would also yield explicit results about the dimensions of the irreducible representations of the symmetric group $\mathfrak{S}_n$ in characteristic $p > 0$ where $p$ is such that $p^2 > n$.

The purpose of this book is to develop the general theory along the above lines and to show how it is transformed into explicit results. In a sense, this book tries to do for representations of Iwahori–Hecke algebras at roots of unity what the book by Geck and Pfeiffer [132] did for the "generic case". However, while [132] was essentially self-contained, the situation is more complex here. In fact, in order to obtain our main results, we rely on the following sources:

- Ariki's proof [7] of the LLT conjecture.
- Certain deep properties of Kazhdan–Lusztig cells [222], [231] which do not seem to be accessible by elementary methods.
- The existence and basic properties of "canonical bases" and "crystals" for the Fock space representations of certain quantised enveloping algebras.

The first two ultimately rely on deep geometric theories, an exposition of which would go far beyond the scope of this text. Fortunately, this material is now more readily accessible through a number of books; for example, Kirwan [201], Chriss and Ginzburg [50], Hotta et al. [159], Kiehl and Weissauer [197]. Also note that the geometry only plays a role in the proofs, but not in the formulation of the results! (It is not completely impossible that, some day, more direct and purely algebraic proofs will be found.) Much of what we need about crystal and canonical bases can be found in Ariki's book [10]; see also Jantzen [185], Kashiwara [191], Lusztig [230]. Our general policy regarding these topics is that we shall introduce the required notation to state the results that we need, but we will not endeavour to give the proofs. In this way, we can keep the size of this text within reasonable limits, and yet present some substantial results and applications.

The origin of the theory of Iwahori–Hecke algebras lies in the representation theory of finite groups of Lie type, where these algebras arise as endomorphism algebras of certain induced representations. Via some natural functors, a well-defined part of the representation theory of a finite group of Lie type is controlled by the representation theory of Iwahori–Hecke algebras. Thus, the theory and the results that we are going to present in this book form a contribution to the general project of determining the irreducible representations of all non-abelian finite simple groups. Note that such a group is either an alternating group of degree at least 5, or a simple group of Lie type, or one of 26 sporadic simple groups; see Gorenstein et al. [142].

A rough outline of the contents of this book now follows.

Chapters 1 and 2 provide a general introduction to the representation theory of Iwahori–Hecke algebras and, thus, may be of some independent interest. The discussion will be based on the Kazhdan–Lusztig theory of "cells" [195], [219]. In Lusztig's work [220] on characters of reductive groups over finite fields, a crucial role is played by the "**a**-function", which associates with every irreducible representation $E$ of a finite Coxeter group a numerical invariant $\mathbf{a}_E$. One of the main themes of this book will be to show that these invariants play a similarly important role for "modular" representations. In Theorem 2.6.12, this culminates in the construction of a "cell datum" in the sense of Graham and Lehrer [144], giving rise to a general theory of "Specht modules" for Iwahori–Hecke algebras. (These results originally appeared in [111], [112].) Thus, we now see that the original Specht module theory in type $A$, due to Dipper and James [62] and Murphy [256], [257] (see also the exposition by Mathas [245]), is the prototype of a picture which applies to all Iwahori–Hecke algebras associated with finite Coxeter groups.

In our exposition, we pay a particular attention to treating Iwahori–Hecke algebras of type $A$ as a model case. The required results on Kazhdan–Lusztig cells will be established in a complete and self-contained manner, where no use of geometry is required; see Section 2.8. This treatment of type $A$ is new and entirely independent of the original approach by Dipper, James, and Murphy.

In Chapter 3, we study non-semisimple Iwahori–Hecke algebras in the spirit of Brauer's classical "modular representation theory" involving, in particular, blocks and decomposition numbers. We shall assume that the reader has some familiarity which the basic features of this theory (for a general finite-dimensional associative algebra); this is readily accessible in standard reference texts, like Curtis and Reiner [53] and Feit [83]. In this setting, we define the key concept of a "canonical basic set" in Section 3.2. This concept is independent of the existence of a Graham–Lehrer cell datum, but, in a sense, it captures precisely those features of a cell datum which can be seen by looking only at the decomposition matrix of the algebra. Again, we treat Iwahori–Hecke algebras of type $A$ as a model case. In Section 3.5 we give a new proof of the classification of the modular irreducible representations of these algebras. For this purpose, we have found it convenient to introduce the formal concept of an "abstract Fock datum" in Section 3.4. In another direction, we present a factorisation result for decomposition matrices and formulate a general version of

James's conjecture. The exposition in Section 3.7 unifies the original formulation of James [181] with the further developments in [92], [98], [129], [133].

In Chapter 4 we explain the fundamental connection between Iwahori–Hecke algebras and representations of a finite group of Lie type $G(\mathbb{F}_q)$ (where $q$ is a power of a prime number and $\mathbb{F}_q$ denotes a finite field with $q$ elements). We begin with a self-contained discussion of the Schur functor and its variations, where we combine the original approach of Dipper [58] with later developments by Cline et al. [51] and Schubert [279]. Following [109], we then show in Theorem 4.4.1 how our results on "cell data" and "canonical basic sets" lead to a natural parametrisation of the modular irreducible representations of $G(\mathbb{F}_q)$ which admit non-zero vectors fixed under a Borel subgroup. This generalises classical results from the characteristic 0 situation (due to Bourbaki, Iwahori, Tits, ...) to positive characteristic. We also explain how this fits into a (conjectural) classification of all irreducible representations of $G(\mathbb{F}_q)$ in the "non-defining characteristic case".

The determination of canonical basic sets for the classical types $B_n$ and $D_n$ has turned out to be an extremely difficult problem. At the end of Chapter 4 we shall discuss some cases that can be dealt with by elementary methods, based on the work of Dipper, James, and Murphy [66], [68]. The solution in the general case requires completely new methods; this will be achieved as a consequence of the results presented in Chapters 5 and 6.

For this purpose, it will be convenient to work in the framework of the theory of *Ariki–Koike algebras*, which are generalisations of Iwahori–Hecke algebras of type $B_n$. The main idea of Chapter 5 is to try to generalise as much as possible the combinatorial constructions involved in the discussion of type $A$ in Chapter 3. This leads us to consider in Section 5.7 certain special choices of the parameters which arise from the combinatorics of "FLOTW multipartitions" (where FLOTW stands for Foda, Leclerc, Okado, Thibon, Welsh [88]); these special choices cover, in particular, the equal parameter case for Iwahori–Hecke algebras of type $B_n$ and $D_n$. As a consequence, in Theorem 5.8.2, we can state the main result concerning the determination of canonical basic sets for this choice of parameters. The methods in Chapter 5 do not allow us to complete the proof of this theorem. The missing piece is a result about the number of irreducible representations of Ariki–Koike algebras which is due to Ariki and Mathas [15] and which relies on the deep work of Ariki [7] on the proof of the LLT conjecture. This will be discussed in Chapter 6.

The idea that FLOTW multipartitions are relevant in the modular representation theory of Iwahori–Hecke algebras of classical type first appeared in the work of Jacon [172], [173], [174]. Originally, the base field for the algebras was assumed to be of characteristic 0. The new approach developed in Chapter 5 shows that these results also hold for fields of positive characteristic.

In Chapter 6 we introduce the quantised enveloping algebra $\mathcal{U}_q(\widehat{\mathfrak{sl}_e})$ and study the canonical bases of certain Fock space representations. The associated "crystals" carry some rich combinatorial structure which will be discussed in detail. We can state (without proof) Ariki's theorem [7] which links the canonical bases of the Fock space representations to the irreducible representations of Ariki–Koike algebras at roots of unity. This allows us to complete the proofs of the main results of the previous chapter; see Section 6.3. Since this only covers certain choices of the parameters for Ariki–Koike algebras, we then go further and present some deep results of Uglov [291] on the canonical bases of the Fock space representations. We show how this leads to an explicit description of the "canonical basic sets" for Ariki–Koike algebras at roots of unity – and, hence, of Iwahori–Hecke algebras of classical type – for any choice of the parameters, assuming that the base field is of characteristic 0; see Theorem 6.7.2. We also derive purely combinatorial algorithms for computing decomposition numbers and the dimensions of the irreducible representations (in characterictic 0).

Finally, Chapter 7 contains explicit results concerning Iwahori–Hecke algebras of exceptional type $H_3$, $H_4$, $F_4$, $E_6$, $E_7$, $E_8$. We also explain some basic algorithmic methods, including Parker's MeatAxe. The project of computing the decomposition matrices for these algebras (over fields of characteristic 0) was started almost 20 years ago in [126] and finally completed in [129]; the matrices for type $E_8$ appear here for the first time in print. From these matrices, one can simply read off the corresponding "canonical basic sets".

*Acknowledgments.* While the idea of writing this book has been around for some time, the actual work began in 2008. Much of the writing of the first chapters was done while the first author enjoyed the hospitality of the Newton Institute (Cambridge, UK) during the special semester on algebraic Lie theory (January to June 2009). We thank the anonymous referees for detailed comments on early versions of the manuscript which, we hope, led to considerable improvements. We are indebted to Gunter Malle, who carefully read all chapters and made numerous and detailed comments, and Cédric Lecouvey for his comments on Chapters 5 and 6.

Finally, we wish to thank Springer Verlag for their support and their patiently accepting our repeated excuses for delaying the delivery of the final manuscript.

Aberdeen and Besançon, *Meinolf Geck*
March 2011 *Nicolas Jacon*

# Contents


**1 Generic Iwahori–Hecke Algebras** ..... 1
1.1 Coxeter Groups and Weight Functions ..... 2
1.2 Representations of **H** ..... 7
1.3 Lusztig's **a**-Invariants ..... 13
1.4 Balanced Representations ..... 20
1.5 The Asymptotic Algebra ..... 28
1.6 Introducing Cells ..... 36
1.7 Examples of Cells ..... 43
1.8 Cells and Leading Coefficients ..... 52

**2 Kazhdan–Lusztig Cells and Cellular Bases** ..... 59
2.1 The Kazhdan–Lusztig Basis ..... 60
2.2 A Pre-order Relation on $\mathrm{Irr}(W)$ ..... 69
2.3 On Lusztig's Conjectures, I ..... 78
2.4 On Lusztig's Conjectures, II ..... 86
2.5 On Lusztig's Conjectures, III ..... 96
2.6 A Cellular Basis for **H** ..... 105
2.7 Further Properties of the Cellular Basis of **H** ..... 114
2.8 The Case of the Symmetric Group ..... 123

**3 Specialisations and Decomposition Maps** ..... 133
3.1 Grothendieck Groups and Decomposition Maps ..... 134
3.2 Canonical Basic Sets ..... 143
3.3 Principal Specialisations and Blocks of Defect 1 ..... 153
3.4 Towards Canonical Basic Sets for Classical Types ..... 162
3.5 A Canonical Basic Set for the Symmetric Group ..... 171
3.6 Factorisation of Decomposition Maps ..... 181
3.7 The General Version of James's Conjecture ..... 190
3.8 Blocks and Bad Specialisations ..... 198

**4 Hecke Algebras and Finite Groups of Lie Type** ... 207
4.1 The Schur Functor and its Variations ... 208
4.2 Hom Functors and Harish-Chandra Series ... 215
4.3 Unipotent Principal Series Representations, I ... 223
4.4 Unipotent Principal Series Representations, II ... 232
4.5 Examples and Conjectures ... 240
4.6 A First Approach to Type $B_n$ ... 250

**5 Representation Theory of Ariki–Koike Algebras** ... 261
5.1 The Complex Reflection Group of Type $G(l,1,n)$ ... 262
5.2 Basic Properties of Ariki–Koike Algebras ... 266
5.3 Ariki–Koike Algebras as Cellular Algebras ... 270
5.4 Decomposition Maps for Ariki–Koike Algebras ... 274
5.5 Cyclotomic Ariki–Koike Algebras ... 276
5.6 A Fock Datum for Ariki–Koike Algebras ... 284
5.7 FLOTW Multipartitions ... 289
5.8 On Basic Sets for Ariki–Koike Algebras ... 302

**6 Canonical Bases in Affine Type $A$ and Ariki's Theorem** ... 309
6.1 The Quantum Affine Algebra $\mathcal{U}_q(\widehat{\mathfrak{sl}_e})$ ... 310
6.2 The Fock Space and Ariki's Theorem ... 316
6.3 Crystal and Canonical Bases for Highest Weight Modules ... 325
6.4 Computing Decomposition Matrices of Ariki–Koike Algebras ... 328
6.5 Uglov's Theory of Fock Spaces ... 334
6.6 Canonical Bases for Fock Spaces ... 347
6.7 Computation of Canonical Basic Sets for Iwahori–Hecke Algebras . 354
6.8 Recent Developments and Conjectures ... 359

**7 Decomposition Numbers for Exceptional Types** ... 363
7.1 Algorithmic Methods ... 364
7.2 Decomposition Matrices for $W$ of Dihedral Type ... 371
7.3 Decomposition Matrices for $W$ of Type $F_4$ ... 374
7.4 Decomposition Matrices for Types $H_3$, $H_4$, $E_6$, $E_7$, $E_8$ ... 379

**References** ... 387

**Index** ... 399

# Chapter 1
# Generic Iwahori–Hecke Algebras

In this chapter we introduce the main objects of our study: Finite Coxeter groups, generic Iwahori–Hecke algebras, and their representations. The groups and algebras are defined in a purely algebraic way, in terms of generators and defining relations. Thus, a generic Iwahori–Hecke algebra $\mathbf{H}$ is seen to be a deformation of the group algebra of a Coxeter group $W$, where the deformation depends on one or several parameters. We recall the relevant definitions and basic results in Sections 1.1 and 1.2, but we tacitly assume that the reader already has some familiarity with them.

Following Lusztig [219], [231], we specify the parameters of $\mathbf{H}$ by a "weight function" $L: W \to \Gamma$, where $\Gamma$ is a totally ordered abelian group. The typical and most familiar example is the case where $\Gamma = \mathbb{Z}$, with its natural order. More general choices for $\Gamma$ are useful for several reasons: first of all, this provides the greatest level of generality and flexibility; furthermore, it brings out more clearly the role that is played by the given total order on $\Gamma$.

In Section 1.3, the total order is used to define Lusztig's $\mathbf{a}$-function, which associates with every irreducible representation $E \in \mathrm{Irr}_{\mathbb{C}}(W)$ an element $\mathbf{a}_E \in \Gamma$. The construction relies in an essential way on the generic algebra $\mathbf{H}$ and the known connection (via Tits's deformation theorem) between the irreducible representations of $W$ and those of $\mathbf{H}$. The explicit results summarized at the end of Section 1.3 show the remarkable dependence of the function $E \mapsto \mathbf{a}_E$ on the total order of $\Gamma$.

The study of the $\mathbf{a}$-function, and its subtle relation with the Kazhdan–Lusztig basis of $\mathbf{H}$, will be one of the main themes of this book. As a first step we will introduce in Section 1.5 an "asymptotic" version of $\mathbf{H}$. Our construction, following [112], is logically independent of Lusztig's construction of the asymptotic ring $\mathbf{J}$ in [223], [231, Chap. 18], but it is, of course, motivated by it. The advantage of our approach is that it does not rely on certain deep properties of the Kazhdan–Lusztig basis of $\mathbf{H}$ which are not (yet) known to hold in the general multiparameter case. Instead, we rely on properties of the "balanced representations" in Section 1.4.

In Section 1.6, using the asymptotic version of $\mathbf{H}$, we can then give a first definition of the partition of $W$ into left, right, and two-sided "cells" and establish some basic properties of them. This is followed by the discussion of a number of examples and further results in Sections 1.7 and 1.8.

M. Geck, N. Jacon, *Representations of Hecke Algebras at Roots of Unity*,
Algebra and Applications 15, DOI 10.1007/978-0-85729-716-7_1,

## 1.1 Coxeter Groups and Weight Functions

We briefly recall the basic definitions concerning Coxeter groups and Iwahori–Hecke algebras. More details and references can be found in [29], [132], [231].

**1.1.1.** Let $S$ be a finite set and $M = (m_{st})_{s,t \in S}$ be a matrix satisfying $m_{ss} = 1$ for all $s \in S$, and $m_{st} = m_{ts} \in \{2,3,4,5,\ldots\} \cup \{\infty\}$ for all $s \neq t$ in $S$. Such a matrix is called a *Coxeter matrix*. Then we define a group $W = W(M)$ by a presentation with generators $S$ and defining relations as follows:

- $s^2 = 1$ for all $s \in S$;
- $(st)^{m_{st}} = 1$ for all $s \neq t$ in $S$ with $m_{st} < \infty$.

The pair $(W,S)$ is called a *Coxeter system* and $W$ is called a *Coxeter group*.

We encode the above presentation in a graph, called the *Coxeter graph* of $W$. It has vertices labelled by the elements of $S$, and two vertices labelled by $s \neq t$ are joined by an edge if $m_{st} \geqslant 3$. Moreover, if $m_{st} \geqslant 4$, we label the edge by $m_{st}$. If the graph is connected, we say that $W$ is an *irreducible Coxeter group*. If this is not the case, we have a direct product decomposition $W = W_1 \times \cdots \times W_d$, where $W_i = \langle S_i \rangle$ and each subset $S_i \subseteq S$ corresponds to the vertices in a connected component of the Coxeter graph; furthermore, each $W_i$ is a Coxeter group with generating set $S_i$. The groups $W_i$ will be called the *irreducible components* of $W$.

**Table 1.1** Coxeter graphs of irreducible finite Coxeter groups

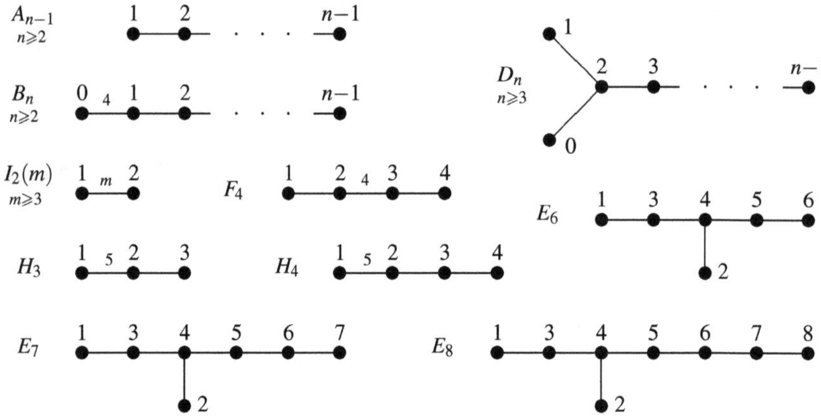


(The numbers on the vertices correspond to a chosen labelling of the elements of $S$.)

| Type | $A_{n-1}$ | $B_n$ | $D_n$ | $I_2(m)$ |
|---|---|---|---|---|
| Order | $n!$ | $2^n n!$ | $2^{n-1} n!$ | $2m$ |

| Type | $H_3$ | $H_4$ | $F_4$ | $E_6$ | $E_7$ | $E_8$ |
|---|---|---|---|---|---|---|
| Order | 120 | 14400 | 1152 | 51840 | 2903040 | 696729600 |

There is a complete classification of the finite Coxeter groups. The graphs corresponding to the irreducible finite Coxeter groups, and the group orders, are given in Table 1.1. We say that $W$ is a finite *Weyl group* or is of *crystallographic type* if $m_{st} \in \{2,3,4,6\}$. These are precisely the finite Coxeter groups which arise, for example, in the theory of finite-dimensional semisimple complex Lie algebras, or in the theory of connnected reductive algebraic groups (see also Chapter 4).

The standard example is type $A_{n-1}$, where $W$ can be identified with the symmetric group $\mathfrak{S}_n$, generated by the basic transpositions $s_i = (i, i+1)$ for $1 \leqslant i \leqslant n-1$. This is the Weyl group for the simple Lie algebra $\mathfrak{sl}_n(\mathbb{C})$ of $n \times n$-matrices with trace 0, or for the simple algebraic group $\mathrm{SL}_n(k)$ of $n \times n$-matrices with determinant 1 over any algebraically closed field $k$.

The groups of type $H_3$, $H_4$ or $I_2(m)$ ($m = 5$ or $m > 7$) are non-crystallographic.

**1.1.2.** Let $k$ be any commutative ring (with 1) and $\{\xi_s \mid s \in S\} \subseteq k^\times$ be a collection of elements such that $\xi_s = \xi_t$ whenever $s,t \in S$ are conjugate in $W$. Then, by Bourbaki [29, Chap. IV, §2, Exc. 23], we have a corresponding *Iwahori–Hecke algebra*

$$H_k = H_k(W, S, \{\xi_s\}).$$

This is an associative algebra over $k$ which is free as a $k$-module, with basis $\{T_w \mid w \in W\}$; the multiplication is uniquely determined by the rule

$$T_s T_w = \begin{cases} T_{sw} & \text{if } l(sw) > l(w), \\ T_{sw} + (\xi_s - \xi_s^{-1}) T_w & \text{if } l(sw) < l(w), \end{cases}$$

where $s \in S$ and $w \in W$. Here, $l \colon W \to \mathbb{Z}_{\geqslant 0}$ is the length function on $W$. Recall that, given $w \in W$, we can write $w = s_1 \cdots s_p$, where $s_i \in S$. If $p$ is minimal with this property, we say that this is a *reduced expression* for $w$; then $l(w) = p$ is called the length of $w$. In this case, the above rules imply that $T_w = T_{s_1} \cdots T_{s_p}$.

We note that $T_1$ is the identity element of $H_k$. The elements $\{\xi_s\}$ are called the *parameters* of $H_k$. We also remark that if $W = W_1 \times \cdots \times W_d$ is the decomposition into irreducible components (where $W_i = \langle S_i \rangle$ as above), then we have

$$H_k \cong H_k(W_1, S_1, \{\xi_s\}_{s \in S_1}) \otimes_k \cdots \otimes_k H_k(W_d, S_d, \{\xi_s\}_{s \in S_d});$$

see [132, Exc. 8.4]. In this way, many questions about Iwahori–Hecke algebras in general can be reduced to the case where $(W,S)$ is irreducible.

**Example 1.1.3.** (a) Assume that $\xi_s = 1$ for all $s \in S$. Then the map $w \mapsto T_w$ defines an isomorphism of $k$-algebras from $kW$ (the group algebra of $W$ over $k$) onto $H_k$.

(b) Assume that $\xi_s = \xi_t$ for all $s,t \in S$; this case will be referred to as the *equal-parameter case*. We are automatically in this case when $W$ is of type $A_{n-1}$, $D_n$, $I_2(m)$ ($m$ odd), $H_3$, $H_4$, $E_6$, $E_7$ or $E_8$ (since all generators in $S$ are conjugate in $W$).

(c) Assume that $W$ is finite and irreducible. Then unequal parameters can only occur in types $B_n$, $F_4$ or $I_2(m)$ ($m$ even). In these cases, the set $S$ falls into two classes under conjugation by $W$; see also Example 1.1.11(b) below.

**1.1.4.** The purpose of this book is to address the following problem.

**Fundamental Problem.** *Assume that $W$ is finite and $k$ is a field. Then determine the irreducible representations of $H_k(W,S,\{\xi_s\})$.*

By Example 1.1.3, this includes the problem of determining the irreducible representations of the symmetric group $\mathfrak{S}_n \cong W(A_{n-1})$ over fields of positive characteristic. Note that there are many open questions even in this special case; in particular, the dimensions of the irreducible representations are not known!

If $H_k(W,S,\{\xi_s\})$ is semisimple, then the above problem is essentially solved; see [132], [231]. So the main focus in this text will be on the non-semisimple situation. The first step consists of noting that any algebra $H_k$ as above can be obtained from a suitable "generic" Iwahori–Hecke algebra by a process of specialisation. For this purpose, we introduce the following notion where, following a suggestion of Bonnafé [21], we combine the two settings in [219], [231].

**Definition 1.1.5 (Lusztig).** Let $\Gamma$ be an abelian group (written additively). We say that a function $L\colon W \to \Gamma$ is a *weight function* if the following condition holds:

$$L(ww') = L(w) + L(w') \text{ for all } w,w' \in W \text{ such that } l(ww') = l(w) + l(w').$$

Note that $L$ is uniquely determined by the values $\{L(s) \mid s \in S\}$. Furthermore, if $\{c_s \mid s \in S\}$ is a collection of elements in $\Gamma$ such that $c_s = c_t$ whenever $s,t \in S$ are conjugate in $W$, then there is a (unique) weight function $L\colon W \to \Gamma$ such that $L(s) = c_s$ for all $s \in S$. (This follows from Matsumoto's lemma; see [132, §1.2].)

**1.1.6.** Let us assume that a weight function $L\colon W \to \Gamma$ has been fixed. Let $R \subseteq \mathbb{C}$ be a subring and $A = R[\Gamma]$ be the free $R$-module with basis $\{\varepsilon^g \mid g \in \Gamma\}$. There is a well-defined ring structure on $A$ such that $\varepsilon^g \varepsilon^{g'} = \varepsilon^{g+g'}$ for all $g, g' \in \Gamma$. We write $1 = \varepsilon^0 \in A$. Given $a \in A$ we denote by $a_g$ the coefficient of $\varepsilon^g$, so that $a = \sum_{g \in \Gamma} a_g \varepsilon^g$.

We apply the general construction in 1.1.2 to the ring $A$ and the collection of elements $\{v_s \mid s \in S\}$ where $v_s := \varepsilon^{L(s)}$ for $s \in S$. The corresponding algebra will be denoted by $\mathbf{H} = \mathbf{H}_A(W,S,L)$ and called the *generic Iwahori–Hecke algebra* associated with $W, L$. Thus, $\mathbf{H}$ is an associative algebra which is free as an $A$-module, with basis $\{T_w \mid w \in W\}$; the multiplication is given by

$$T_sT_w = \begin{cases} T_{sw} & \text{if } l(sw) > l(w), \\ T_{sw} + (v_s - v_s^{-1})T_w & \text{if } l(sw) < l(w), \end{cases}$$

where $s \in S$ and $w \in W$. The element $T_1$ is the identity element.

In the setting of 1.1.2, assume that there is a ring homomorphism $\theta : A \to k$ such that $\theta(v_s) = \xi_s$ for all $s \in S$. Then we can regard $k$ as an $A$-module (via $\theta$), and we find that $H_k$ is obtained by extension of scalars from $\mathbf{H}$:

$$H_k(W,S,\{\xi_s\}) \cong k \otimes_A \mathbf{H}_A(W,S,L).$$

In this situation, we say that $\theta\colon A \to k$ is a *specialisation* and that $H_k(W,S,\{\xi_s\})$ is obtained from $\mathbf{H}$ by *specialisation* (via $\theta$). For example, if $\theta_1 : A \to k$ is a ring homomorphism such that $\theta_1(\varepsilon^g) = 1$ for all $g \in \Gamma$, then $k \otimes_A \mathbf{H}_A(W,S,L) \cong kW$.

**Example 1.1.7.** Let $\Gamma = \mathbb{Z}$. Then $A = R[v, v^{-1}]$ is the ring of Laurent polynomials over $R$ in one indeterminate $v := \varepsilon$. Let $L\colon W \to \mathbb{Z}$ be a weight function and set $c_s = L(s)$ for $s \in S$. Then the relations in $\mathbf{H}$ read as follows, where $s \in S$ and $w \in W$:

$$T_sT_w = \begin{cases} T_{sw} & \text{if } l(sw) > l(w), \\ T_{sw} + (v^{c_s} - v^{-c_s})T_w & \text{if } l(sw) < l(w). \end{cases}$$

This is the setting of Lusztig [231]; it is particularly relevant for applications to the representation theory of reductive groups over finite fields; see Section 4.2.

*Remark 1.1.8.* For various applications, it will be convenient to set $\dot{T}_w := \varepsilon^{L(w)} T_w$ for all $w \in W$ and $u_s := v_s^2$ for all $s \in S$. Then, clearly, $\{\dot{T}_w \mid w \in W\}$ also is an $A$-basis of $\mathbf{H}$; furthermore, we have the multiplication rules:

$$\dot{T}_s\dot{T}_w = \begin{cases} \dot{T}_{sw} & \text{if } l(sw) > l(w), \\ u_s\dot{T}_{sw} + (u_s - 1)\dot{T}_w & \text{if } l(sw) < l(w), \end{cases}$$

where $s \in S$ and $w \in W$. Thus, the introduction of the basis $\{\dot{T}_w \mid w \in W\}$ shows that $\mathbf{H}$ is already defined over the subring $\mathbb{Z}[u_s \mid s \in S] \subseteq A$; that is, all structure constants with respect to this basis lie in $\mathbb{Z}[u_s \mid s \in S]$.

**Example 1.1.9.** A "universal" weight function is given as follows. For $s, t \in S$, we write $s \sim t$ if $s, t$ are conjugate in $W$. Let $S' \subseteq S$ be a set of representatives for the equivalence classes of $S$ under this relation. Let $\Gamma_0$ be the group of all tuples $(n_s)_{s \in S'}$ where $n_s \in \mathbb{Z}$ for all $s \in S'$. (The addition is defined componentwise.) Let $L_0\colon W \to \Gamma_0$ be the weight function given by sending $s \in S$ to the tuple $(n_t)_{t \in S'}$, where $n_t = 1$ if $t$ is conjugate to $s$ and $n_t = 0$ otherwise. Let $A_0 = R[\Gamma_0]$ and $\mathbf{H}_0 = \mathbf{H}_{A_0}(W, S, L_0)$ be the associated Iwahori–Hecke algebra; we denote the parameters by $\{v_s^\circ \mid s \in S\}$ in this case. Note that $A_0$ is nothing but the ring of Laurent polynomials in $\{v_s^\circ \mid s \in S'\}$ (and these elements are algebraically independent).

Any algebra $H_k(W, S, \{\xi_s\})$ as above is obtained by specialisation from $\mathbf{H}_0$ (where we take $R = \mathbb{Z}$). Indeed, since $\{v_s^\circ \mid s \in S'\}$ are algebraically independent, we can certainly find a unital ring homomorphism $\theta_0\colon A_0 \to k$ such that $\theta_0(v_s^\circ) = \xi_s$ for all $s \in S$. Thus, $H_k(W, S, \{\xi_s\}) \cong k \otimes_{A_0} \mathbf{H}_0$ (via $\theta_0$).

**1.1.10.** As in [219], we shall assume that $\Gamma$ admits a total ordering $\leqslant$ which is compatible with the group structure; that is, whenever $g, g' \in \Gamma$ are such that $g \leqslant g'$, we have $g + h \leqslant g' + h$ for all $h \in \Gamma$. Such an order will be called a *monomial order*. We usually assume that $L(s) \geqslant 0$ for all $s \in S$. (We will see in Lemma 1.1.12 below that this is no severe restriction.) The existence of a monomial order on $\Gamma$ implies that $\Gamma$ is torsion free. Furthermore, we can write any $0 \neq a \in A$ uniquely in the form

$$a = a_1\varepsilon^{g_1} + \cdots + a_d\varepsilon^{g_d} \quad \text{where} \quad 0 \neq a_i \in R,\ g_i \in \Gamma \text{ and } g_1 < \ldots < g_d.$$

We denote $\mathrm{lt}(a) := a_1\varepsilon^{g_1}$ and call this the *leading term* of $a$. We also set $\mathrm{lt}(0) := 0$. Then one easily checks that $\mathrm{lt}(aa') = \mathrm{lt}(a)\mathrm{lt}(a')$ for any $a, a' \in A$. In particular, if $a \neq 0$ and $a' \neq 0$, then $\mathrm{lt}(aa') \neq 0$ and so $aa' \neq 0$; hence, $A$ is an integral domain.

Finally, since $S$ is a finite set, it is usually sufficient to consider the case where $\Gamma$ is finitely generated. Consequently, in this case, we have $\Gamma \cong \mathbb{Z}^r$ for some $r \geqslant 1$, which means that $A$ is a ring of Laurent polynomials in $r$ variables. Then specifying a monomial order $\leqslant$ on $\Gamma$ amounts to specifying a total order on the monomials in $A$ (compatible with the multiplication).

**Example 1.1.11.** (a) In the set-up of Example 1.1.7, there is a natural monomial order on $\Gamma = \mathbb{Z}$, and we will usually assume that $c_s \geqslant 0$ for all $s \in S$.

(b) Assume that $W$ is of type $B_n$, $F_4$ or $I_2(m)$ ($m$ even). Then, in general, $L$ depends on two values $a, b \in \Gamma$, which are attached to the generators in $S$:

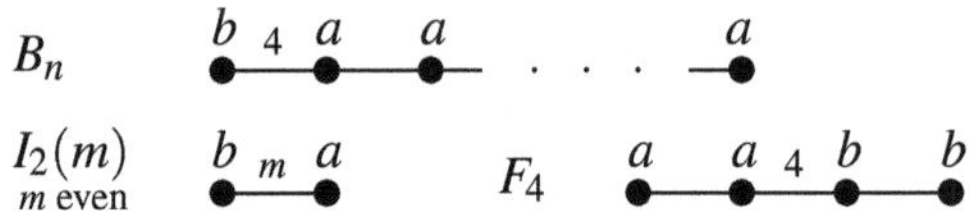


The possible choices of monomial orders that are available here can best be seen by taking $L = L_0$ to be the "universal" weight function in Example 1.1.9, where $\Gamma_0 = \mathbb{Z}^2$, $b = (1,0)$ and $a = (0,1)$. Then $A_0 = R[V^{\pm 1}, v^{\pm 1}]$ is the ring of Laurent polynomials in the indeterminates $V := \varepsilon^{(1,0)}$ and $v := \varepsilon^{(0,1)}$. A familiar monomial order is the *pure lexicographic order* given by

$$(i,j) \leqslant_{\text{lex}} (i',j') \quad \stackrel{\text{def}}{\Leftrightarrow} \quad i \lessdot i' \quad \text{or} \quad i = i' \text{ and } j \leqslant j' \qquad (i,i',j,j' \in \mathbb{Z}).$$

More generally, for any $\alpha \in \mathbb{R}$, we have a monomial ordering $\leqslant_\alpha$ given by

$$(i,j) \leqslant_\alpha (i',j') \quad \stackrel{\text{def}}{\Leftrightarrow} \quad \begin{array}{l} i + \alpha j < i' + \alpha j' \text{ or} \\ i + \alpha j = i' + \alpha j' \text{ and } j \leqslant j'. \end{array}$$

In particular, we see that there are infinitely many monomial orders on $\Gamma_0$. For a classification of all orderings on $\Gamma_0$, see Tutorial 10 in [206, §1.4] and also [271].

(c) Now assume that $W$ is any finite Coxeter group and $L\colon W \to \Gamma$ is a weight function. Let $\leqslant$ be a monomial order on $\Gamma$. By analogy to Bonnafé and Iancu [26], we say that we are in the *asymptotic case* if $L(s) > 0$ for all $s \in S$ and if, on any irreducible component of type $B_n$, $F_4$ or $I_2(m)$ ($m$ even), where $L$ takes values $a, b \in \Gamma$ as above, we have $b > ra > 0$ for all $r \in \mathbb{Z}_{\geqslant 1}$.

As already mentioned above, we will usually assume that $L(s) \geqslant 0$ for all $s \in S$. This is justified by the following result, observed by Bonnafé [22, Cor. 5.8].

**Lemma 1.1.12.** *Let $\leqslant$ be a monomial order on $\Gamma$. For $s \in S$, set $\delta_s = 1$ if $L(s) \geqslant 0$, and $\delta_s = -1$ if $L(s) < 0$. Then there is a well-defined weight function $L'\colon W \to \Gamma$ such that $L'(s) = \delta_s L(s)$ ($s \in S$); note that $L'(s) \geqslant 0$ for all $s \in S$.*

*Let $\mathbf{H}'$ be the generic Iwahori–Hecke algebra associated with $W, L'$ and let $\{T'_w \mid w \in W\}$ be the standard basis of $\mathbf{H}'$. Then there is a unique $A$-algebra isomorphism $\mathbf{H} \to \mathbf{H}'$ such that $T_s \mapsto \delta_s T'_s$ for all $s \in S$.*

*Proof.* To show that there is a weight function $L'$ as above, we need to check that $\delta_s = \delta_t$ whenever $s, t \in S$ are conjugate in $W$. But, if $s, t \in S$ are conjugate, then

$L(s) = L(t)$ and so $\delta_s$, $\delta_t$ will either both be $+1$ or both be $-1$. Now let us show that there is an algebra isomorphism $\mathbf{H} \to \mathbf{H}'$ as above. By [132, 4.4.5], the algebra $\mathbf{H}$ has a presentation with generators $\{T_s \mid s \in S\}$ and defining relations

$$\begin{aligned} T_s^2 &= T_1 + (\varepsilon^{L(s)} - \varepsilon^{-L(s)})T_s && (s \in S), \\ \underbrace{T_sT_tT_s\cdots}_{m_{st}\text{ terms}} &= \underbrace{T_tT_sT_t\cdots}_{m_{st}\text{ terms}} && (s,t \in S, s \neq t, m_{st} < \infty). \end{aligned}$$

Hence, all we need to check is whether the elements $\{\delta_s T_s' \mid s \in S\}$ satisfy the above relations. Now, we have

$$(\delta_s T_s')^2 = T_{s'}^2 = T_1' + (\varepsilon^{L'(s)} - \varepsilon^{-L'(s)})T_s' = T_1' + (\delta_s\varepsilon^{L'(s)} - \delta_s\varepsilon^{-L'(s)})(\delta_s T_s').$$

It remains to note that if $\delta_s = -1$, then $\delta_s\varepsilon^{L'(s)} - \delta_s\varepsilon^{-L'(s)} = -\varepsilon^{-L(s)} + \varepsilon^{L(s)}$, as required. Now let us check the second type of relations. Let $s \neq t$ in $S$ be such that $m_{st} < \infty$. Then the verification reduces to proving that $\delta_s\delta_t\delta_s\cdots = \delta_t\delta_s\delta_t\cdots$ (with $m_{st}$ factors on both sides). If $\delta_s = \delta_t$, this is clear. Now assume that $\delta_s \neq \delta_t$. In particular, $L(s) \neq L(t)$ in this case and so $m_{st}$ must be even. But then, on both sides of the above identity, we have $m_{st}/2$ factors corresponding to $s$ and $m_{st}/2$ factors corresponding to $t$. Hence, we get the same result on both sides. □

*Remark 1.1.13.* Let $\theta\colon A \to k$ be a specialisation into a field $k$ and consider the specialised algebra $\mathbf{H}_k$. Then, in the setting of Lemma 1.1.12, the algebras $\mathbf{H}_k$ and $\mathbf{H}_k'$ are isomorphic. Hence, if we have solved our "fundamental problem" in 1.1.4 for $\mathbf{H}_k'$, then this problem is automatically solved for $\mathbf{H}_k$ as well. Thus, indeed, it will be sufficient to consider weight functions such that $L(s) \geqslant 0$ for all $s \in S$.

## 1.2 Representations of H

We will assume from now on that $W$ is finite. Let $L\colon W \to \Gamma$ be a weight function and $\mathbf{H}$ be the associated generic Iwahori–Hecke algebra. Let us now turn to the representation theory of $W$ and of $\mathbf{H}$.

**1.2.1.** We set $\mathbb{Z}_W := \mathbb{Z}[2\cos(2\pi/m_{st}) \mid s,t \in S] \subseteq \mathbb{R}$. For example, $\mathbb{Z}_W = \mathbb{Z}$ if $W$ is a finite Weyl group. We shall always assume that $\mathbb{Z}_W \subseteq R$ (where $R \subseteq \mathbb{C}$ is the subring used to define $A$). Then the field of fractions of $R$, which will be denoted by $\mathbb{K}$, is a splitting field for $W$; see [132, Theorem 6.3.8]. Throughout, we use the following notation for the irreducible representations of $W$ (up to isomorphism):

$$\mathrm{Irr}_{\mathbb{K}}(W) = \{E^\lambda \mid \lambda \in \Lambda\},$$

where $\Lambda$ is a finite indexing set and each $E^\lambda$ is a $\mathbb{K}$-vector space with a given $\mathbb{K}W$-module structure. We also use the notation

$$d_\lambda = \dim E^\lambda, \qquad M(\lambda) = \text{ an indexing set for a basis of } E^\lambda$$

(where $M(\lambda)$ is ordered in some way so that it makes sense to write down matrices with rows and columns indexed by $M(\lambda)$).

Finally, we assume that $\Gamma$ admits a monomial ordering as in 1.1.10. As we have seen, this implies that $A$ is an integral domain. Let $K$ be the field of fractions of $A$; by extension of scalars, we obtain a finite-dimensional $K$-algebra $\mathbf{H}_K = K \otimes_A \mathbf{H}$ which is known to be split semisimple; see [132, Theorem 9.3.5]. Let $\mathrm{Irr}(\mathbf{H}_K)$ denote the set of irreducible representations of $\mathbf{H}_K$ (up to isomorphism). By *Tits's deformation theorem*, there is a bijection between this set and $\mathrm{Irr}_{\mathbb{K}}(W)$; see [132, 8.1.7] and also Exercises 26 and 27 of Bourbaki [29, Chap. IV, §2]. Thus, we can write

$$\mathrm{Irr}(\mathbf{H}_K) = \{E_\varepsilon^\lambda \mid \lambda \in \Lambda\} \qquad (d_\lambda = \dim E_\varepsilon^\lambda),$$

where each $E_\varepsilon^\lambda$ is a $K$-vector space with a given $\mathbf{H}_K$-module structure. The correspondence $E^\lambda \leftrightarrow E_\varepsilon^\lambda$ is uniquely determined by the condition

$$\mathrm{trace}(w, E^\lambda) = \theta_1\big(\mathrm{trace}(T_w, E_\varepsilon^\lambda)\big) \qquad \text{for all } w \in W,$$

where $\theta_1 \colon A \to R$ is the unique $R$-linear ring homomorphism such that $\theta_1(\varepsilon^g) = 1$ for all $g \in \Gamma$. Note that, by [132, Theorem 9.3.5], we have

$$\varepsilon^{L(w)}\,\mathrm{trace}(T_w, E_\varepsilon^\lambda) \in \mathbb{Z}_W[v_s \mid s \in S] \qquad \text{for all } w \in W;$$

in particular, these traces lie in $A$ and so it makes sense to apply $\theta_1$ to them. It also follows that, for *any* $\mathbf{H}_K$-module $V$, we have $\mathrm{trace}(T_w, V) \in A$ for all $w \in W$.

*Remark 1.2.2.* The proofs of the statements summarized in 1.2.1, especially the statements concerning splitting fields for $W$ and $\mathbf{H}_K$, are by no means easy. In fact, various chapters of [132] are concerned with these questions, where case-by-case arguments (according to the classification of finite Coxeter groups) are required. A number of authors have contributed to the establishment of these results, over an extended period of time; see the bibliographic comments in [132, §5.7 and §9.5].

If one is mainly interested in finite Weyl groups and the equal-parameter case, then more conceptual arguments are available via the geometry of an associated algebraic group; see Springer [282] and Lusztig [216] (see also Example 2.5.7).

**Example 1.2.3.** Let $\mathrm{Cl}(W)$ be the set of conjugacy classes of $W$. For $C \in \mathrm{Cl}(W)$, let $w_C \in C$ be a representative which has minimal length in $C$. Then the matrix

(a) $$X(\mathbf{H}) := \big(\mathrm{trace}(\dot{T}_{w_C}, E_\varepsilon^\lambda)\big)_{\lambda \in \Lambda, C \in \mathrm{Cl}(W)}$$

is called the *character table* of $\mathbf{H}$, where we define $\dot{T}_w := \varepsilon^{L(w)}\,T_w$ for any $w \in W$, as in Remark 1.1.8. By a result due to Geck and Pfeiffer [132, 8.2.9], $X(\mathbf{H})$ does not depend on the choice of the representatives $\{w_C \mid C \in \mathrm{Cl}(W)\}$; furthermore, there is a unique set of polynomials $\{f_{w,C} \mid w \in W, C \in \mathrm{Cl}(W)\} \subseteq \mathbb{Z}[v_s^2 \mid s \in S]$ such that

(b) $$\mathrm{trace}(\dot{T}_w, E_\varepsilon^\lambda) = \sum_{C \in \mathrm{Cl}(W)} f_{w,C}\,\mathrm{trace}(\dot{T}_{w_C}, E_\varepsilon^\lambda) \qquad \text{for any } \lambda \in \Lambda \text{ and } w \in W.$$

The tables $X(\mathbf{H})$ are explicitly known for all $W, L$; see [132, Chap. 10 and 11] and the references there. When we apply the specialisation homomorphism $\theta_1 : A \to R$ to the entries of $X(\mathbf{H})$, we obtain the classical character table of the finite group $W$.

**Example 1.2.4.** We shall frequently apply the following "specialisation argument" for representations. Let $V$ be an $\mathbf{H}_K$-module and $V'$ be a $\mathbb{K}W$-module such that

$$(*) \qquad \operatorname{trace}(w, V') = \theta_1\big(\operatorname{trace}(T_w, V)\big) \qquad \text{for all } w \in W,$$

where $\theta_1$ is defined as above; recall that $\mathbb{K}W \cong \mathbb{K} \otimes_A \mathbf{H}$, where $\mathbb{K}$ is regarded as an $A$-module via $\theta_1$. For example, $(*)$ will certainly hold when $V \cong K \otimes_A M$ and $V' \cong \mathbb{K} \otimes_A M$, where $M$ is an $\mathbf{H}$-module which is finitely generated and free over $A$.

For any $\lambda \in \Lambda$, denote by $m(V, \lambda)$ the multiplicity of $E_\varepsilon^\lambda$ as an irreducible constituent of $V$, and denote by $m(V', \lambda)$ the multiplicity of $E^\lambda$ as an irreducible constituent of $V'$. Thus, we have

$$\operatorname{trace}(T_w, V) = \sum_{\lambda \in \Lambda} m(V, \lambda) \operatorname{trace}(T_w, E_\varepsilon^\lambda) \qquad \text{for all } w \in W,$$
$$\operatorname{trace}(w, V') = \sum_{\lambda \in \Lambda} m(V', \lambda) \operatorname{trace}(w, E^\lambda) \qquad \text{for all } w \in W.$$

Applying $\theta_1$ and using Tits's deformation theorem, we obtain that

$$\sum_{\lambda \in \Lambda} m(V, \lambda) \operatorname{trace}(w, E^\lambda) = \sum_{\lambda \in \Lambda} m(V', \lambda) \operatorname{trace}(w, E^\lambda) \qquad \text{for all } w \in W.$$

Since the trace functions associated with the irreducible representations of $W$ are linearly independent, we deduce that $m(V, \lambda) = m(V', \lambda)$ for all $\lambda \in \Lambda$.

**Example 1.2.5.** It is known that every $w \in W$ is conjugate to its inverse; see [132, 3.2.14]. Hence, we have $\operatorname{trace}(w, E^\lambda) = \operatorname{trace}(w^{-1}, E^\lambda)$ for all $\lambda \in \Lambda$. A similar property holds on the level of $\mathbf{H}_K$; that is, we have

$$\text{(a)} \qquad \operatorname{trace}(T_w, E_\varepsilon^\lambda) = \operatorname{trace}(T_{w^{-1}}, E_\varepsilon^\lambda) \qquad \text{for all } w \in W.$$

This is seen as follows. It is easily checked that the $A$-linear map $h \mapsto h^\flat$ defined by $T_w^\flat = T_{w^{-1}}$ $(w \in W)$ is an anti-involution of $\mathbf{H}$. So we can define the *contragredient module* $\hat{E}_\varepsilon^\lambda := \operatorname{Hom}_K(E_\varepsilon^\lambda, K)$ where $T_w$ acts via $T_w \colon \varphi \mapsto \varphi \circ T_{w^{-1}}$ for $\varphi \in \hat{E}_\varepsilon^\lambda$.

For any $w \in W$, we have $\operatorname{trace}(T_w, \hat{E}_\varepsilon^\lambda) = \operatorname{trace}(T_{w^{-1}}, E_\varepsilon^\lambda)$ and, hence,

$$\theta_1\big(\operatorname{trace}(T_w, \hat{E}_\varepsilon^\lambda)\big) = \operatorname{trace}(w^{-1}, E^\lambda) = \operatorname{trace}(w, E^\lambda) = \theta_1\big(\operatorname{trace}(T_w, E_\varepsilon^\lambda)\big).$$

By Tits's deformation theorem, this implies that $\hat{E}_\varepsilon^\lambda \cong E_\varepsilon^\lambda$ and so (a) holds.

**Example 1.2.6.** Let sgn denote the sign representation of $W$, which is given by the group homomorphism sending each $w \in W$ to $(-1)^{l(w)}$. Via tensoring with sgn, we obtain a bijection $\lambda \mapsto \lambda^\dagger$ of $\Lambda$ such that

$$\text{(a)} \qquad E^{\lambda^\dagger} \cong E^\lambda \otimes \operatorname{sgn} \qquad \text{for all } \lambda \in \Lambda.$$

This operation can be lifted to representations of $\mathbf{H}_K$. Namely, there is a unique $A$-algebra automorphism $\dagger \colon \mathbf{H} \to \mathbf{H}$ such that $T_s^\dagger = -T_s^{-1}$ for all $s \in S$; see [132, Exc. 8.2]. By extension of scalars, this induces a $K$-algebra automorphism of $\mathbf{H}_K$, which we denote by the same symbol. Given any finite-dimensional $\mathbf{H}_K$-module $V$, denote by $V^\dagger$ the $\mathbf{H}_K$-module with the same underlying vector space $V$, but where $h \in \mathbf{H}_K$ acts via $h^\dagger$. Then, by [132, Prop. 9.4.1], we have

(b) $$(E^{\lambda^\dagger})_\varepsilon \cong (E_\varepsilon^\lambda)^\dagger \qquad \text{for all } \lambda \in \Lambda.$$

The trace of $T_w$ on $E_\varepsilon^{\lambda^\dagger}$ is determined as follows. There is a unique $R$-linear ring homomorphism $A \to A$, $a \mapsto \bar{a}$, such that $\overline{\varepsilon^g} = \varepsilon^{-g}$ for all $g \in \Gamma$. Then we have

(c) $$\operatorname{trace}(T_w, E_\varepsilon^{\lambda^\dagger}) = (-1)^{l(w)}\overline{\operatorname{trace}(T_w, E_\varepsilon^\lambda)} \qquad \text{for all } w \in W;$$

see [132, Prop. 9.4.1].

**Example 1.2.7.** Let $w_0 \in W$ be the longest element. Then $T_{w_0}^2$ lies in the centre of $\mathbf{H}_K$ and, hence, acts by a scalar in every irreducible representation of $\mathbf{H}_K$. This scalar can be explicitly described, as follows. Let $T := \{wsw^{-1} \mid s \in S, w \in W\}$ be the set of all reflections in $W$. Let $S' \subseteq S$ be a set of representatives of the conjugacy classes of $W$ which are contained in $T$. For $s \in S'$, let $N_s$ be the cardinality of the conjugacy class of $s$; thus, $|T| = \sum_{s \in S'} N_s$. Let $\rho^\lambda : \mathbf{H}_K \to M_{d_\lambda}(K)$ be a representation afforded by $E_\varepsilon^\lambda$. Then, by an argument due to Springer (see [132, Theorem 9.2.2]), we have

(a) $$\rho^\lambda(T_{w_0}^2) = \varepsilon^{2N_\lambda} I_{d_\lambda} \quad \text{where} \quad N_\lambda := \sum_{s \in S'} \Big(\frac{N_s \operatorname{trace}(s, E^\lambda)}{\dim E^\lambda}\Big) L(s) \in \Gamma$$

and $I_{d_\lambda}$ denotes the identity matrix of size $d_\lambda$. Note that, since every $s \in S'$ has order 2, we have $\operatorname{trace}(s, E^\lambda) \in \mathbb{Z}$ and so, by a well-known result in the character theory of finite groups, the quantity $N_s\operatorname{trace}(s, E^\lambda)/\dim E^\lambda$ also is an integer. Thus, the expression defining $N_\lambda$ is a well-defined element of $\Gamma$.

Now let us set $P := \varepsilon^{-N_\lambda}\rho^\lambda(T_{w_0})$. Then $P^2 = I_{d_\lambda}$ and so $P$ is a diagonalisable matrix with eigenvalues $\pm 1$; in particular, $m := \operatorname{trace}(P) \in \mathbb{Z}$. Thus, we obtain that $\operatorname{trace}(T_{w_0}, E_\varepsilon^\lambda) = m\varepsilon^{N_\lambda}$. Applying the specialisation homomorphism $\theta_1 : A \to R$, we conclude that $m = \operatorname{trace}(w_0, E^\lambda)$ and, hence,

(b) $$\operatorname{trace}(T_{w_0}, E_\varepsilon^\lambda) = \operatorname{trace}(w_0, E^\lambda)\, \varepsilon^{N_\lambda}.$$

Thus, $\operatorname{trace}(T_{w_0}, E_\varepsilon^\lambda)$ is explicitly described in terms of character values of $W$.

*Remark 1.2.8.* We have already remarked in 1.2.1 that $\operatorname{trace}(T_w, E_\varepsilon^\lambda) \in A$ for all $w \in W$. So it is natural to ask if it is even possible to find a representation $\rho^\lambda$ afforded by $E_\varepsilon^\lambda$ such that $\rho^\lambda(T_w) \in M_{d_\lambda}(A)$ for all $w \in W$. This is indeed the case, but the proof requires some deep results on the Kazhdan–Lusztig basis of $\mathbf{H}$ and a case-by-case analysis; see [132, 9.3.8] and the references there. (We will recover this result in the context of cellular algebras in Corollary 2.7.14.) However, we can establish a weak version of this statement by a general argument, and this will be useful in our discussion of cells in Section 1.6. This relies on the following ring-theoretic result.

**Lemma 1.2.9 (Rouquier [272]).** *Assume that $\Gamma = \mathbb{Z}$, so that A is the ring of Laurent polynomials in one indeterminate $v = \varepsilon$. Then the subring*

$$\mathscr{R} := \left\{ \frac{f}{g} \;\middle|\; f \in \mathbb{Z}[v, v^{-1}], g \in \mathbb{Z}[v] \text{ and } g \text{ has constant term } 1 \right\} \subseteq K.$$

*is a principal ideal domain.*

*Proof.* We need some standard results from commutative ring theory; a suitable reference is Matsumura [248]. Following Broué and Kim [33, §2.B], the first and crucial step is to show that $\mathscr{R}$ is a Dedekind domain. For this purpose, by [248, Theorem 11.6], it is enough to show that $\mathscr{R}$ is a one-dimensional noetherian domain which is integrally closed in $\mathbb{Q}(v)$ (its field of fractions). Now note that $\mathscr{R} \subseteq \mathbb{Q}(v)$ is the localisation of $\mathbb{Z}[v, v^{-1}]$ with respect to the multiplicatively closed set

$$M = \{g \in \mathbb{Z}[v] \mid g \text{ has constant term } 1\}.$$

But the ring $\mathbb{Z}[v, v^{-1}]$ is known to be noetherian and integrally closed in $\mathbb{Q}(v)$. These properties pass on to localisations and, hence, $\mathscr{R}$ is noetherian and integrally closed in $\mathbb{Q}(v)$. In order to show that $\mathscr{R}$ is one-dimensional, we must show that every non-zero prime ideal of $\mathscr{R}$ is maximal. So let $\mathfrak{p}$ be a non-zero prime ideal of $\mathscr{R}$. By [248, Theorem 4.1], $\mathfrak{p}$ is generated by a prime ideal $I \subseteq \mathbb{Z}[v, v^{-1}]$ such that $I \cap M = \varnothing$. The prime ideals in $\mathbb{Z}[v, v^{-1}]$ are explicitly known (see, for example, [132, Exc. 7.9]). Thus, $I$ is either principal (generated by a prime number in $\mathbb{Z}$ or by an irreducible polynomial in $\mathbb{Z}[v]$) or generated by two elements $\ell$ and $f$ where $\ell > 0$ is a prime number and $f \in \mathbb{Z}[v]$ is a monic polynomial whose reduction modulo $\ell$ is an irreducible polynomial in $\mathbb{F}_l[v]$. But ideals of the latter type have non-empty intersection with $M$. Indeed, every irreducible polynomial in $\mathbb{F}_l[v]$ divides $v^{l^m} - v$ for some $m \geqslant 1$. Hence, we have $v^{l^m} - v \in I$ for some $m \geqslant 1$. Since $v$ is a unit in $\mathbb{Z}[v, v^{-1}]$, this implies that $1 - v^{l^m - 1} \in I \cap M$ and so $I \cap M$ is non-empty, as claimed.

Consequently, all the non-zero prime ideals in $\mathscr{R}$ are principal. In particular, this shows that every non-zero prime ideal in $\mathscr{R}$ is maximal and, hence, $\mathscr{R}$ is a Dedekind domain. Finally, in a Dedekind domain, every non-zero ideal is a product of a finite number of prime ideals. We have just seen that every prime ideal in $\mathscr{R}$ is principal. Hence, every ideal in $\mathscr{R}$ is principal. □

**Corollary 1.2.10.** *Assume that W is a Weyl group and $\Gamma = \mathbb{Z}$. Then, for each $\lambda \in \Lambda$, there exists a representation $\pi^\lambda : \mathbf{H}_K \to M_{d_\lambda}(K)$ afforded by $E_\varepsilon^\lambda$ such that $\pi^\lambda(T_w) \in M_{d_\lambda}(\mathscr{R})$ for all $w \in W$.*

*Proof.* Since $W$ is a Weyl group, we have $\mathbb{Z}_W = \mathbb{Z}$ and so we can take $R = \mathbb{Z}$. Hence, $K = \mathbb{Q}(v)$ is the field of fractions of the ring $\mathscr{R}$, which is a principal ideal domain by Lemma 1.2.9. But then a standard argument (see, for example, [132, 7.3.7]) shows that every irreducible representation of $\mathbf{H}_K$ can be realised over $\mathscr{R}$. □

**1.2.11.** We define an $A$-linear map $\tau\colon \mathbf{H} \to A$ by $\tau(T_1) = 1$ and $\tau(T_w) = 0$ for $1 \neq w \in W$. Then we have $\tau(T_yT_x) = \tau(T_xT_y) = \delta_{x^{-1}y}$ (Kronecker delta); see [132, 8.1.1]. So $\tau$ defines a non-degenerate symmetric bilinear form on $\mathbf{H}$, where $\{T_w \mid w \in W\}$ and $\{T_{w^{-1}} \mid w \in W\}$ form a pair of dual bases. Thus, $\mathbf{H}$ is a *symmetric algebra*, with trace form $\tau$. Let $\tau_K$ be the canonical extension to a trace form on $\mathbf{H}_K$.

For $\lambda \in \Lambda$, let $\chi^\lambda$ denote the *character* of $E_\varepsilon^\lambda$; we have $\chi^\lambda(h) = \operatorname{trace}(h, E_\varepsilon^\lambda) \in A$ for all $h \in \mathbf{H}$. Since $\mathbf{H}_K$ is split semisimple, the characters $\{\chi^\lambda \mid \lambda \in \Lambda\}$ form a basis of the vector space of all trace functions on $\mathbf{H}_K$. Hence, by the general theory of symmetric algebras, there is a unique expression (see [132, 7.2.6]):

(a) $$\tau_K = \sum_{\lambda \in \Lambda} \mathbf{c}_\lambda^{-1} \chi^\lambda \qquad \text{where} \qquad 0 \neq \mathbf{c}_\lambda \in K.$$

The elements $\mathbf{c}_\lambda$ were called *Schur elements* in [96]. By [132, Theorem 9.3.5], we have the following important integrality property:

(b) $$\varepsilon^{L(w_0)} \mathbf{c}_\lambda \in \mathbb{Z}_W[v_s \mid s \in S] \qquad \text{for all } \lambda \in \Lambda;$$

In particular, $\mathbf{c}_\lambda$ lies in $A$. The *generic degree* corresponding to $E^\lambda$ is defined by

(c) $$\delta_\lambda := \mathbf{c}_\lambda^{-1} P_{W,L} \in K \qquad \text{where} \qquad P_{W,L} := \sum_{w \in W} \varepsilon^{2L(w)},$$

but note that this may no longer be an element of $A$.

Choosing a basis of $E_\varepsilon^\lambda$, indexed by $M(\lambda)$ as above, we obtain a matrix representation $\rho^\lambda\colon \mathbf{H}_K \to M_{d_\lambda}(K)$. Given $h \in \mathbf{H}_K$ and $\mathfrak{s}, \mathfrak{t} \in M(\lambda)$, we denote by $\rho_{\mathfrak{st}}^\lambda(h)$ the $(\mathfrak{s}, \mathfrak{t})$-entry of $\rho^\lambda(h)$. We now have the following *Schur relations*. In particular, these yield an alternative characterisation of the elements $\mathbf{c}_\lambda$.

**Proposition 1.2.12 (Schur relations; cf. [132, 7.2.2]).** *Let $\{B_w \mid w \in W\}$ be any basis of $\mathbf{H}$ and $\{B_w^\vee \mid w \in W\}$ the corresponding dual basis, such that $\tau(B_xB_y^\vee) = \delta_{xy}$ for all $x, y \in W$. Given $\lambda, \mu \in \Lambda$, let $\mathfrak{s}, \mathfrak{t} \in M(\lambda)$ and $\mathfrak{u}, \mathfrak{v} \in M(\mu)$. Then*

$$\sum_{w \in W} \rho_{\mathfrak{st}}^\lambda(B_w)\rho_{\mathfrak{uv}}^\mu(B_w^\vee) = \begin{cases} \mathbf{c}_\lambda & \text{if } \lambda = \mu,\ \mathfrak{s} = \mathfrak{v},\ \mathfrak{t} = \mathfrak{u}, \\ 0 & \text{otherwise.} \end{cases}$$

*In particular, this implies the orthogonality relations*

$$\sum_{w \in W} \chi^\lambda(B_w)\chi^\mu(B_w^\vee) = \begin{cases} d_\lambda \mathbf{c}_\lambda & \text{if } \lambda = \mu, \\ 0 & \text{otherwise.} \end{cases}$$

The origin of the definition of the elements $\mathbf{c}_\lambda$ and of the generic degrees $\delta_\lambda$ lies in the representation theory of finite groups of Lie type. Without going into much detail at this stage, let us briefly describe this connection.

**1.2.13.** Let us assume that $W$ is of crystallographic type and arises as the Weyl group of a family of finite groups of Lie type

$$\mathscr{S} = \{G(q) \mid q \text{ any prime power}\}.$$

Assume that all groups $G(q)$ are of "split type". Some examples are given by

$$\begin{array}{lll} W = W(A_{n-1}) & \text{where} & G(q) = \mathrm{GL}_n(\mathbb{F}_q) \text{ for all } q, \\ W = W(B_n) & \text{where} & G(q) = \mathrm{SO}_{2n+1}(\mathbb{F}_q) \text{ for all } q, \\ W = W(E_8) & \text{where} & G(q) = E_8(\mathbb{F}_q) \text{ for all } q. \end{array}$$

Let $\Gamma = \mathbb{Z}$ and the weight function $L$ be such that $L(s) = 1$ for all $s \in S$. Then $A = R[v, v^{-1}]$ is the ring of Laurent polynomials in one indeterminate $v = \varepsilon$. In this situation, it is known that $\mathbf{c}_\lambda$ divides $P_{W,L}$ in $\mathbb{K}[v, v^{-1}]$ and that there is a polynomial $D_\lambda \in \mathbb{Q}[u]$ (where $u$ is an indeterminate) such that $\delta_\lambda = D_\lambda(v^2) \in \mathbb{Q}[v]$; see [132, 9.3.6]. This polynomial $D_\lambda$ has the following interpretation.

Given a prime power $q$, let us consider the complex irreducible representations of $G(q)$. Let $B(q) \subseteq G(q)$ be a Borel subgroup and define

$$\mathrm{Irr}_{\mathbb{C}}(G(q), B(q)) := \left\{ \rho \in \mathrm{Irr}_{\mathbb{C}}(G(q)) \,\middle|\, \begin{array}{l} \rho \text{ occurs in the permutation} \\ \text{representation } \mathbb{C}[G(q)/B(q)] \end{array} \right\},$$

the set of (unipotent) *principal series representations*. Then, by classical results due to Iwahori, Tits, Benson and Curtis, there exists a bijection

$$\mathrm{Irr}_{\mathbb{C}}(W) \xrightarrow{\sim} \mathrm{Irr}_{\mathbb{C}}(G(q), B(q)), \qquad E^\lambda \mapsto \rho_q^\lambda,$$

such that $\dim \rho_q^\lambda = D_\lambda(q)$ for all $\lambda \in \Lambda$. (See [53, §68], [132, §8.4] or Section 4.3 in this book.) Thus, $D_\lambda(q)$ is the dimension of an irreducible representation of $G(q)$.

## 1.3 Lusztig's a-Invariants

In Lusztig's work [220] on characters of reductive groups over finite fields, a crucial role is played by a certain function which attaches to each irreducible representation $E^\lambda$ of $W$ an invariant $\mathbf{a}_\lambda \in \mathbb{Z}_{\geqslant 0}$. In the situation of 1.2.13, this is defined by

$$\mathbf{a}_\lambda := \max\{i \geqslant 0 \mid u^i \text{ divides } D_\lambda\} \qquad \text{for any } \lambda \in \Lambda.$$

(See also 2.2.12 and 4.3.12 for further interpretations of $\mathbf{a}_\lambda$.) One of the main themes of this book will be to show that these **a**-*invariants* play a similarly important role for "modular" representations.

Throughout this section, let $W$ be a finite Coxeter group and $L\colon W \to \Gamma$ be a weight function. Let $\leqslant$ be a monomial order on $\Gamma$ such that $L(s) \geqslant 0$ for all $s \in S$. Let $\Gamma_{\geqslant 0} := \{g \in \Gamma \mid g \geqslant 0\}$ and denote by $R[\Gamma_{\geqslant 0}]$ the set of all $R$-linear combinations of terms $\varepsilon^g$ where $g \geqslant 0$. The notations $R[\Gamma_{>0}]$, $R[\Gamma_{\leqslant 0}]$, $R[\Gamma_{<0}]$ have a similar meaning. We are now ready to introduce the invariants $\mathbf{a}_\lambda$ in general.

**Proposition 1.3.1 (Lusztig).** *Let $\lambda \in \Lambda$. Then there exist some $\mathbf{a}_\lambda \in \Gamma_{\geqslant 0}$ and a strictly positive real number $f_\lambda \in \mathbb{Z}_W$ such that*

$$\varepsilon^{2\mathbf{a}_\lambda} \mathbf{c}_\lambda \in R[\Gamma_{\geqslant 0}] \quad \textit{and} \quad \varepsilon^{2\mathbf{a}_\lambda} \mathbf{c}_\lambda \equiv f_\lambda \mod R[\Gamma_{>0}].$$

*Note that $\mathbf{a}_\lambda$ and $f_\lambda$ are uniquely determined by these conditions. We have*

$$\mathbf{a}_\lambda = \min\{g \in \Gamma_{\geqslant 0} \mid \varepsilon^g \chi^\lambda(T_w) \in \mathbb{K}[\Gamma_{\geqslant 0}] \textit{ for all } w \in W\}.$$

*Proof.* (Cf. [217, 1.9].) For the following discussion, it will be useful to assume that $\mathbb{K} \subseteq \mathbb{R}$. By Example 1.2.5(a), we have $\chi^\lambda(T_w) = \chi^\lambda(T_{w^{-1}})$ for all $w \in W$. Hence, using the orthogonality relations in Proposition 1.2.12 we obtain

$$\sum_{w \in W} \chi^\lambda(T_w)^2 = \sum_{w \in W} \chi^\lambda(T_w)\chi^\lambda(T_{w^{-1}}) = d_\lambda \mathbf{c}_\lambda.$$

Now we set $\mathbf{a}_\lambda := \min\{g \in \Gamma_{\geqslant 0} \mid \varepsilon^g \chi^\lambda(T_w) \in \mathbb{K}[\Gamma_{\geqslant 0}]$ for all $w \in W\}$. For each $w \in W$, denote by $c_{w,\lambda} \in \mathbb{K}$ the constant term of $\varepsilon^{\mathbf{a}_\lambda}\chi^\lambda(T_w)$; note that $c_{w,\lambda} \neq 0$ for at least one $w \in W$. Let $f'_\lambda := \sum_{w \in W} c_{w,\lambda}^2$. Since all $c_{w,\lambda}$ are real numbers, not all of which are zero, we conclude that $f'_\lambda$ is a strictly positive real number. Now we obtain

$$\varepsilon^{2\mathbf{a}_\lambda} \sum_{w \in W} \chi^\lambda(T_w)^2 \equiv \sum_{w \in W} \left(\varepsilon^{\mathbf{a}_\lambda}\chi^\lambda(T_w)\right)^2 \equiv \sum_{w \in W} c_{w,\lambda}^2 \equiv f'_\lambda \mod \mathbb{K}[\Gamma_{>0}].$$

Hence, setting $f_\lambda = f'_\lambda/d_\lambda$, we see that $\varepsilon^{2\mathbf{a}_\lambda}\mathbf{c}_\lambda \in f_\lambda + \mathbb{K}[\Gamma_{>0}]$. Finally, since $\mathbf{c}_\lambda \in \mathbb{Z}_W[\Gamma]$, we must have $f_\lambda \in \mathbb{Z}_W$. □

*Remark 1.3.2.* Let $\lambda \in \Lambda$ and $w \in W$. By Proposition 1.3.1, we can write

$$\varepsilon^{\mathbf{a}_\lambda}\chi^\lambda(T_w) = c_{w,\lambda} + \mathbb{K}\text{-linear combination of terms } \varepsilon^g \text{ where } g > 0,$$

where $c_{w,\lambda} \in \mathbb{K}$. These are the "leading coefficients of character values" considered by Lusztig [220, Chap. 5], [225]. From the orthogonality relations in Proposition 1.2.12 (see also [132, Exc. 9.8]), we immediately deduce that

$$\sum_{w \in W} c_{w,\lambda}\, c_{w^{-1},\mu} = \begin{cases} f_\lambda d_\lambda & \text{if } \lambda = \mu, \\ 0 & \text{otherwise.} \end{cases}$$

Thus, the coefficients $c_{w,\lambda}$ behave as if they were the character values of an algebra with a basis indexed by the elements of $W$; see Section 1.5 for a further discussion.

**Example 1.3.3.** Assume that $L(s) > 0$ for all $s \in S$. Let $\lambda \in \Lambda$.

(a) If $E^\lambda$ is the unit representation, then $\mathbf{a}_\lambda = 0$ and $f_\lambda = 1$.
(b) If $E^\lambda$ is the sign representation, then $\mathbf{a}_\lambda = L(w_0)$ and $f_\lambda = 1$.
(c) If $E^\lambda$ is neither the unit nor the sign representation, then $0 < \mathbf{a}_\lambda < L(w_0)$.

Here, $w_0 \in W$ denotes the longest element. (Note that these statements fail if $L(s) = 0$ for some $s \in S$; see Example 1.3.7 below.) First note that the unit and the sign representation of $W$ correspond to the following representations of $\mathbf{H}_K$, respectively:

$$\begin{aligned} \operatorname{ind}_\varepsilon \colon \mathbf{H}_K \to K, \qquad & T_w \mapsto \varepsilon^{L(w)}, \\ \operatorname{sgn}_\varepsilon \colon \mathbf{H}_K \to K, \qquad & T_w \mapsto (-1)^{l(w)} \varepsilon^{-L(w)}. \end{aligned}$$

The condition that $L(s) > 0$ for all $s \in S$ implies that $0 < L(w) < L(w_0)$ for all $w \in W \setminus \{1, w_0\}$. This immediately yields (a) and (b). To prove (c), first note that $\operatorname{ind}_\varepsilon(T_w) \in R[\Gamma_{>0}]$ for $w \neq 1$. This yields that

$$\varepsilon^{\mathbf{a}_\lambda} \sum_{w \in W} \chi^\lambda(T_w) \operatorname{ind}_\varepsilon(T_{w^{-1}}) \equiv \varepsilon^{\mathbf{a}_\lambda} \chi^\lambda(T_1) \bmod R[\Gamma_{>0}].$$

Since $E_\varepsilon^\lambda \not\cong \operatorname{ind}_\varepsilon$, the sum on the left-hand side must be zero by Proposition 1.2.12. Hence, $\mathbf{a}_\lambda > 0$, as required. On the other hand, by 1.2.1, we have $\varepsilon^{L(w)} \chi^\lambda(T_w) \in R[\Gamma_{\geqslant 0}]$ for all $w \in W$. This already shows that $\mathbf{a}_\lambda \leqslant L(w_0)$. Now assume, if possible, that $\mathbf{a}_\lambda = L(w_0)$. Then $\varepsilon^{L(w_0)} \chi^\lambda(T_w) \in R[\Gamma_{>0}]$ for all $w \neq w_0$ and $\varepsilon^{L(w_0)} \chi^\lambda(T_{w_0})$ has a non-zero constant term. Using (b), it follows that

$$\varepsilon^{2L(w_0)} \sum_{w \in W} \chi^\lambda(T_w) \operatorname{sgn}_\varepsilon(T_{w^{-1}}) \equiv \pm \varepsilon^{L(w_0)} \chi^\lambda(T_{w_0}) \not\equiv 0 \bmod R[\Gamma_{>0}].$$

But, since $E_\varepsilon^\lambda \not\cong \operatorname{sgn}_\varepsilon$, we also deduce from Proposition 1.2.12 that the sum on the left-hand side is zero, which is a contradiction. So we have $\mathbf{a}_\lambda < L(w_0)$, as required.

**Example 1.3.4.** Recall from Example 1.2.6 that we have a "duality" operation $\lambda \mapsto \lambda^\dagger$ on $\Lambda$ such that $E^{\lambda^\dagger} \cong E^\lambda \otimes \operatorname{sgn}$. By [132, Prop. 9.4.3], we have

$$\mathbf{c}_{\lambda^\dagger} = \bar{\mathbf{c}}_\lambda = \varepsilon^{-2N_\lambda} \mathbf{c}_\lambda \quad \text{and} \quad \mathbf{a}_{\lambda^\dagger} - \mathbf{a}_\lambda = N_\lambda, \quad \text{with } N_\lambda \text{ as in Example 1.2.7(a).}$$

Here, the map $A \to A$, $a \mapsto \bar{a}$, is defined as in Example 1.2.6.

*Remark 1.3.5.* Let $W = W_1 \times \cdots \times W_d$ be the decomposition into irreducible components. Correspondingly, we have

$$\operatorname{Irr}_{\mathbb{K}}(W) = \{E^{\lambda_1} \boxtimes \cdots \boxtimes E^{\lambda_d} \mid \lambda_i \in \Lambda_i\}, \quad \text{where} \quad \operatorname{Irr}_{\mathbb{K}}(W_i) = \{E^{\lambda_i} \mid \lambda_i \in \Lambda_i\}.$$

Thus, we can identify $\Lambda = \Lambda_1 \times \cdots \times \Lambda_d$. As already noted in 1.1.2, we have a tensor product decomposition $\mathbf{H} \cong \mathbf{H}_1 \otimes_A \cdots \otimes_A \mathbf{H}_d$, where $\mathbf{H}_i$ is the generic algebra associated with $W_i$ and the restriction of $L$ to $W_i$. Hence, we also have

$$\operatorname{Irr}(\mathbf{H}_K) = \{E_\varepsilon^{\lambda_1} \boxtimes \cdots \boxtimes E_\varepsilon^{\lambda_d} \mid \lambda_i \in \Lambda_i\}.$$

By [132, Exc. 8.5], this yields that $\mathbf{c}_\lambda = \mathbf{c}_{\lambda_1} \cdots \mathbf{c}_{\lambda_d}$, where $\lambda = (\lambda_1, \ldots, \lambda_d)$. Consequently, we have $\mathbf{a}_\lambda = \mathbf{a}_{\lambda_1} + \cdots + \mathbf{a}_{\lambda_d}$ and $f_\lambda = f_{\lambda_1} \cdots f_{\lambda_d}$. Thus, the determination of $\mathbf{a}_\lambda$ and $f_\lambda$ can be reduced to the case where $(W,S)$ is irreducible.

*Remark 1.3.6.* The elements $\mathbf{c}_\lambda$ are explicitly known for all types of $W$; see [132, Chap. 10 and 11] and the references there. It turns out that they have a quite special

**Table 1.2** The invariants $f_\lambda$ and $\mathbf{a}_\lambda$ for type $F_4$

| | $b>2a>0$ | | $b=2a>0$ | | $2a>b>a>0$ | | $b=a>0$ | | $b>a=0$ | |
|---|---|---|---|---|---|---|---|---|---|---|
| $E^\lambda$ | $f_\lambda$ | $\mathbf{a}_\lambda$ | $f_\lambda$ | $\mathbf{a}_\lambda$ | $f_\lambda$ | $\mathbf{a}_\lambda$ | $f_\lambda$ | $\mathbf{a}_\lambda$ | $f_\lambda$ | $\mathbf{a}_\lambda$ |
| $1_1$ | 1 | 0 | 1 | 0 | 1 | 0 | 1 | 0 | 6 | 0 |
| $1_2$ | 1 | $12b-9a$ | 2 | $15a$ | 1 | $11b-7a$ | 8 | $4a$ | 6 | $12b$ |
| $1_3$ | 1 | $3a$ | 2 | $3a$ | 1 | $-b+5a$ | 8 | $4a$ | 6 | 0 |
| $1_4$ | 1 | $12b+12a$ | 1 | $36a$ | 1 | $12b+12a$ | 1 | $24a$ | 6 | $12b$ |
| $2_1$ | 1 | $3b-3a$ | 2 | $3a$ | 1 | $2b-a$ | 2 | $a$ | 12 | $3b$ |
| $2_2$ | 1 | $3b+9a$ | 2 | $15a$ | 1 | $2b+11a$ | 2 | $13a$ | 12 | $3b$ |
| $2_3$ | 1 | $a$ | 1 | $a$ | 1 | $a$ | 2 | $a$ | 3 | 0 |
| $2_4$ | 1 | $12b+a$ | 1 | $25a$ | 1 | $12b+a$ | 2 | $13a$ | 3 | $12b$ |
| $4_1$ | 2 | $3b+a$ | 2 | $7a$ | 2 | $3b+a$ | 8 | $4a$ | 6 | $3b$ |
| $9_1$ | 1 | $2b-a$ | 2 | $3a$ | 1 | $b+a$ | 1 | $2a$ | 2 | $2b$ |
| $9_2$ | 1 | $6b-2a$ | 1 | $10a$ | 1 | $6b-2a$ | 8 | $4a$ | 2 | $6b$ |
| $9_3$ | 1 | $2b+2a$ | 1 | $6a$ | 1 | $2b+2a$ | 8 | $4a$ | 2 | $2b$ |
| $9_4$ | 1 | $6b+3a$ | 2 | $15a$ | 1 | $5b+5a$ | 1 | $10a$ | 2 | $6b$ |
| $6_1$ | 3 | $3b+a$ | 3 | $7a$ | 3 | $3b+a$ | 3 | $4a$ | 12 | $3b$ |
| $6_2$ | 3 | $3b+a$ | 3 | $7a$ | 3 | $3b+a$ | 12 | $4a$ | 12 | $3b$ |
| $12_1$ | 6 | $3b+a$ | 6 | $7a$ | 6 | $3b+a$ | 24 | $4a$ | 6 | $3b$ |
| $4_2$ | 1 | $b$ | 1 | $2a$ | 1 | $b$ | 2 | $a$ | 6 | $b$ |
| $4_3$ | 1 | $7b-3a$ | 1 | $11a$ | 1 | $7b-3a$ | 4 | $4a$ | 6 | $7b$ |
| $4_4$ | 1 | $b+3a$ | 1 | $5a$ | 1 | $b+3a$ | 4 | $4a$ | 6 | $b$ |
| $4_5$ | 1 | $7b+6a$ | 1 | $20a$ | 1 | $7b+6a$ | 2 | $13a$ | 6 | $7b$ |
| $8_1$ | 1 | $3b$ | 1 | $6a$ | 1 | $3b$ | 1 | $3a$ | 12 | $3b$ |
| $8_2$ | 1 | $3b+6a$ | 1 | $12a$ | 1 | $3b+6a$ | 1 | $9a$ | 12 | $3b$ |
| $8_3$ | 1 | $b+a$ | 2 | $3a$ | 1 | $3a$ | 1 | $3a$ | 3 | $b$ |
| $8_4$ | 1 | $7b+a$ | 2 | $15a$ | 1 | $6b+3a$ | 1 | $9a$ | 3 | $7b$ |
| $16_1$ | 2 | $3b+a$ | 2 | $7a$ | 2 | $3b+a$ | 4 | $4a$ | 6 | $3b$ |

The notation for $\mathrm{Irr}_{\mathbb{K}}(W)$ is defined in [132, Appendix C.3].
Here, $a := L(s_1) = L(s_2)$ and $b := L(s_3) = L(s_4)$, cf. Table 1.1.

form. Indeed, one checks that there is a family $\{\Phi_d \mid d \in I\}$ of monic polynomials in one variable over $\mathbb{Z}_W$ such that

(a) $\mathbf{c}_\lambda = f_\lambda \varepsilon^{\gamma_\lambda} \prod_{d\in I} \Phi_d\big(\varepsilon^{\gamma_{\lambda,d}}\big)^{n_{\lambda,d}}$, where $\gamma_\lambda \in \Gamma$, $n_{\lambda,d} \geqslant 0$ and $0 \neq \gamma_{\lambda,d} \in \Gamma$;

(b) all the (complex) roots of $\Phi_d$ are roots of unity;

(c) the product of all $\mathbf{c}_\lambda$ (for $\lambda \in \Lambda$) lies in $\mathbb{Z}[u_s^{\pm 1} \mid s \in S]$, where $u_s := v_s^2$ $(s \in S)$.

Note that the monomials $\gamma_\lambda$ and $\gamma_{\lambda,i}$ are not uniquely determined. In fact, depending on the monomial order $\leqslant$ on $\Gamma$, the terms involving those monomials are rearranged so as to produce the relation $\varepsilon^{2\mathbf{a}_\lambda}\mathbf{c}_\lambda \equiv f_\lambda \bmod R[\Gamma_{>0}]$ in Proposition 1.3.1.

From the explicit knowledge of $\mathbf{c}_\lambda$ one can deduce explicit formulae for the invariants $\mathbf{a}_\lambda$ and $f_\lambda$. If $L(s) = 0$ for all $s \in S$, then $\mathbf{c}_\lambda = |W|/d_\lambda$. Hence, $\mathbf{a}_\lambda = 0$ and $f_\lambda = |W|/d_\lambda$ for all $\lambda \in \Lambda$ in this case. Now assume that $L(s) > 0$ for at least some $s \in S$. For $W$ of exceptional type $H_3$, $H_4$, $E_6$, $E_7$, $E_8$ (where we are automatically in the equal-parameter case), see the tables in [220, Chap. 4] and in the Appendices C and E in [132]. For type $F_4$, see Table 1.2 (p. 16); note that, by the symmetry of the diagram, we can assume without loss generality that $L(s_1) = L(s_2) \leqslant L(s_3) = L(s_4)$. For the types $I_2(m)$, $A_{n-1}$, $B_n$ and $D_n$, see the examples below.

**Table 1.3** The irreducible representations of $\mathbf{H}_K$ in type $I_2(m)$

$$\begin{array}{llll}
1_W^\varepsilon: & T_{s_1} \mapsto v_{s_1}, & & T_{s_2} \mapsto v_{s_2}, \\
\mathrm{sgn}^\varepsilon: & T_{s_1} \mapsto -v_{s_1}^{-1}, & & T_{s_2} \mapsto -v_{s_2}^{-1}, \\
\mathrm{sgn}_1^\varepsilon: & T_{s_1} \mapsto v_{s_1}, & & T_{s_2} \mapsto -v_{s_2}^{-1}, \\
\mathrm{sgn}_2^\varepsilon: & T_{s_1} \mapsto -v_{s_1}^{-1}, & & T_{s_2} \mapsto v_{s_2}, \\
\sigma_j^\varepsilon: & T_{s_1} \mapsto \begin{pmatrix} -v_{s_1}^{-1} & 0 \\ \mu_j & v_{s_1} \end{pmatrix}, & & T_{s_2} \mapsto \begin{pmatrix} v_{s_2} & 1 \\ 0 & -v_{s_2}^{-1} \end{pmatrix}
\end{array}$$

| $\lambda$ | $1_W$ | $(\mathrm{sgn}_1)$ | $\sigma_j$ | $(\mathrm{sgn}_2)$ | $\mathrm{sgn}$ | |
|---|---|---|---|---|---|---|
| $\mathbf{a}_\lambda$ | 0 | $a$ | $a$ | $a$ | $ma$ | $(b=a>0)$ |
| $f_\lambda$ | 1 | $\frac{m}{2}$ | $\frac{m}{2-\zeta^j-\zeta^{-j}}$ | $\frac{m}{2}$ | 1 | $(b=a>0)$ |
| $\mathbf{a}_\lambda$ | 0 | $a$ | $b$ | $\frac{m}{2}(b-a)+a$ | $\frac{m}{2}(a+b)$ | $(b>a\geqslant 0)$ |
| $f_\lambda$ | 1 | 1 | $\frac{m}{2-\zeta^{2j}-\zeta^{-2j}}$ | 1 | 1 | $(b>a>0)$ |
| $f_\lambda$ | 2 | 2 | $\frac{m}{2-\zeta^{2j}-\zeta^{-2j}}$ | 2 | 2 | $(b>a=0)$ |

(where $b := L(s_1)$, $a := L(s_2)$ and $\mu_j := v_{s_1}v_{s_2}^{-1} + \zeta^j + \zeta^{-j} + v_{s_1}^{-1}v_{s_2}$)

**Example 1.3.7.** Let $W$ be of type $I_2(m)$ $(m \geqslant 3)$; that is, $W = \langle s_1, s_2 \rangle$, where $s_1^2 = s_2^2 = (s_1s_2)^m = 1$. In this case, $\mathbb{Z}_W = \mathbb{Z}[\zeta + \zeta^{-1}]$, where $\zeta \in \mathbb{C}$ is a root of unity of order $m$, chosen such that $\zeta + \zeta^{-1} = 2\cos(2\pi/m)$. By [132, §5.4], we have

$$\mathrm{Irr}_{\mathbb{K}}(W) = \begin{cases} \{1_W, \mathrm{sgn}, \sigma_1, \sigma_2, \ldots, \sigma_{(m-1)/2}\} & \text{if } m \text{ is odd}, \\ \{1_W, \mathrm{sgn}, \sigma_1, \sigma_2, \ldots, \sigma_{(m-2)/2}, \mathrm{sgn}_1, \mathrm{sgn}_2\} & \text{if } m \text{ is even}, \end{cases}$$

where $1_W$ is the unit and sgn is the sign representation, all $\sigma_j$ are two-dimensional, and $\mathrm{sgn}_1, \mathrm{sgn}_2$ are two further one-dimensional representations when $m$ is even, in which case we fix the notation such that $s_1$ acts as $+1$ in $\mathrm{sgn}_1$ and as $-1$ in $\mathrm{sgn}_2$. By [132, §8.3], explicit realisations of the corresponding representations of $\mathbf{H}_K$ are known; see Table 1.3. Using the formulae for $\mathbf{c}_\lambda$ in [132, Theorem 8.3.4], one obtains the invariants $\mathbf{a}_\lambda$ and $f_\lambda$. In Table 1.3, the columns are ordered such that the invariants $\mathbf{a}_\lambda$ are increasing from left to right; the columns corresponding to $\mathrm{sgn}_1$, $\mathrm{sgn}_2$ must be deleted if $m$ is odd.

Let us give a concrete example to see how $\mathbf{a}_\lambda$ and $f_\lambda$ are computed as a function of the monomial order $\leqslant$. Let $m = 4$ so that $W$ is the dihedral group of order 8. Let $\zeta \in \mathbb{C}$ be a fourth root of unity; then, by the general formula in [132, 8.3.4], we have

$$\begin{aligned}\mathbf{c}_{\sigma_1} &= 4\frac{(v_{s_1}^2 v_{s_2}^2 - (\zeta+\zeta^{-1})v_{s_1}v_{s_2}+1)(v_{s_1}^2 + (\zeta+\zeta^{-1})v_{s_1}v_{s_2}+v_{s_2}^2)}{v_{s_1}^2 v_{s_2}^2(2-(\zeta^2+\zeta^{-2}))}\\ &= v_{s_1}^{-2}v_{s_2}^{-2}(v_{s_1}^2 v_{s_2}^2+1)(v_{s_1}^2+v_{s_2}^2)\\ &= \varepsilon^{-2a}+\varepsilon^{-2b}+\varepsilon^{2a}+\varepsilon^{2b},\end{aligned}$$

where $b := L(s_1)$ and $a := L(s_2)$. Hence, if $b > a > 0$, then all four terms are different, $f_{\sigma_1} = 1$ and the "lowest" term is $\varepsilon^{-2b}$; so $\mathbf{a}_{\sigma_1} = b$. Similarly, if $a > b > 0$, then $f_{\sigma_1} = 1$ and $\mathbf{a}_{\sigma_1} = a$. If $b > a = 0$, then $\mathbf{c}_{\sigma_1} = \varepsilon^{-2b}+2+\varepsilon^{2b}$ and so $f_{\sigma_1} = 1$, $\mathbf{a}_{\sigma_1} = b$. Finally, if $a = b > 0$, then $\mathbf{c}_{\sigma_1} = 2(\varepsilon^{-2a}+\varepsilon^{2a})$ and so $f_{\sigma_1} = 2$, $\mathbf{a}_{\sigma_1} = a$.

In the next three examples, we describe the invariants $\mathbf{a}_\lambda$ for $W$ of type $A_{n-1}$, $B_n$, $D_n$. In these cases, $\mathrm{Irr}(W)$ is parametrised by suitable partitions or pairs of partitions, where we follow the notational conventions of [220, Chap. 4], [132, Chap. 5].

Recall that a *partition* is a finite, weakly decreasing sequence of non-negative integers; we often write this in the form $\lambda = (\lambda_1 \geqslant \lambda_2 \geqslant \ldots \geqslant \lambda_N \geqslant 0)$. We say that $\lambda$ is a partition of $n$, and write $\lambda \vdash n$, if $|\lambda| := \lambda_1 + \cdots + \lambda_N$ equals $n$; the numbers $\lambda_i$ which are non-zero are called the *parts* of $\lambda$. As is usual, we will not distinguish between two partitions which have identical parts.

**Example 1.3.8.** Let $W$ be of type $A_{n-1}$, where $W \cong \mathfrak{S}_n$. We are automatically in the equal-parameter case; write $a := L(s) > 0$ for $s \in S$. There is a standard labelling

$$\mathrm{Irr}_{\mathbb{K}}(W) = \{E^\lambda \mid \lambda \in \Lambda\}, \quad \text{where} \quad \Lambda = \{\text{set of all partitions of } n\}.$$

For example, the unit and the sign representation are labelled by $(n)$ and $(1^n)$, respectively; see [132, §5.4]. (Here, $(1^n)$ denotes the partition which has $n$ parts equal to 1.) By [132, Prop. 9.4.5], given a partition $\lambda$ of $n$, we have

$$f_\lambda = 1 \qquad \text{and} \qquad \mathbf{a}_\lambda = n(\lambda)a, \quad \text{where} \quad n(\lambda) := \sum_{1\leqslant i\leqslant N}(i-1)\lambda_i;$$

here, we write $\lambda = (\lambda_1 \geqslant \lambda_2 \geqslant \ldots \geqslant \lambda_N \geqslant 0)$ for some $N \geqslant 1$. We have

$$\lambda^\dagger = \lambda^* \qquad \text{and} \qquad \mathbf{a}_{\lambda^*} = a\sum_{1\leqslant i\leqslant N}\frac{1}{2}\lambda_i(\lambda_i-1),$$

where $\lambda^*$ denotes the conjugate (or transpose) partition; see [132, 5.4.3, 5.4.9].

**Example 1.3.9.** Let $W$ be of type $B_n$ with weight function $L\colon W \to \Gamma$ given by

$B_n$ 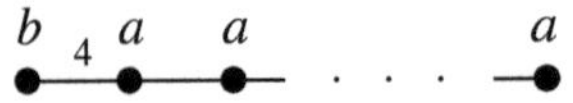 where $a, b \geqslant 0$


There is a standard labelling $\mathrm{Irr}_{\mathbb{K}}(W) = \{E^\lambda \mid \lambda \in \Lambda\}$, where

$$\Lambda = \{\text{set of all pairs of partitions } (\lambda, \mu) \text{ such that } |\lambda| + |\mu| = n\}.$$

For example, the unit and the sign representation are labelled by $((n), \varnothing)$ and $(\varnothing, (1^n))$ respectively; see [132, §5.5]. We have $(\lambda, \mu)^\dagger = (\mu^*, \lambda^*)$; see [132, 5.5.6].

The elements $\mathbf{c}_\lambda$ have been determined explicitly by Hoefsmit [157]; see also Lusztig [212, 9.6], and Iancu [169] for alternative proofs. The case where $a=b=0$ has already been dealt with in Remark 1.3.6. Now we distinguish three further cases:

*Case 1.* We have $a=0$ and $b>0$. Then we have

$$\mathbf{a}_{(\lambda,\mu)} = b\,|\mu| \qquad \text{and} \qquad f_{(\lambda,\mu)} \in \{1,2,\ldots,n\}.$$

This directly follows from Hoefsmit's original formulas for $\mathbf{c}_\lambda$.

*Case 2.* We have $a>0$ and $b \geqslant ra$ for all $r \in \mathbb{Z}_{\geqslant 0}$. By [103, Remark 5.1], we have

$$\mathbf{a}_{(\lambda,\mu)} = \big(n(\lambda)+2n(\mu)-n(\mu^*)\big)\,a+|\mu|\,b \qquad \text{and} \qquad f_{(\lambda,\mu)}=1,$$

where $n(\lambda)$, $n(\mu)$, $n(\mu^*)$ are defined using the formula in Example 1.3.8. These statements actually hold as soon as $b>(n-1)a>0$; see [121, Example 3.6].

*Case 3.* There exists an integer $r \geqslant 0$ (which is then unique) such that $ra \leqslant b < (r+1)a$. In this case, we set $b' := b-ra \in \Gamma_{\geqslant 0}$. Following Lusztig [231, 22.10], we choose a sufficiently large integer $N \geqslant 0$ such that we can write

$$\lambda = (\lambda_1 \geqslant \lambda_2 \geqslant \ldots \geqslant \lambda_{N+r} \geqslant 0) \quad \text{and} \quad \mu = (\mu_1 \geqslant \mu_2 \geqslant \ldots \geqslant \mu_N \geqslant 0).$$

Then we define $Z_{a,b}^N(\lambda,\mu)$ to be the multiset formed by the $2N+r$ entries

$$\begin{array}{ll} (\lambda_i+N+r-i)a+b' & (1 \leqslant i \leqslant N+r), \\ (\mu_j+N-j)a & (1 \leqslant j \leqslant N). \end{array}$$

Furthermore, let $Z^0$ be the multiset whose entries are

$$0,\ a,\ 2a,\ \ldots,\ (N-1)a,\ b',\ a+b',\ 2a+b',\ \ldots,\ (N+r-1)a+b'.$$

Then, by Lusztig [231, Prop. 22.14], we have

$$\mathbf{a}_{(\lambda,\mu)} = \sum_{1 \leqslant i \leqslant 2N+r} (i-1)z_i - \sum_{1 \leqslant i \leqslant 2N+r} (i-1)z_i^0,$$

where $z_1, z_2, \ldots, z_{2N+r}$ are the entries of $Z$ and $z_1^0, z_2^0, \ldots, z_{2N+r}^0$ are the entries of $Z^0$, both arranged in decreasing order. Note that the second sum only depends on $a,b,N$ and the whole expression does not depend on $N$. (It is precisely the extra term arising from $Z^0$ which guarantees this independence of $N$.) Furthermore, we have $f_{(\lambda,\mu)} = 2^c$ for some $c \geqslant 0$ where $c=0$ if $b'>0$.

**Example 1.3.10.** Let $n \geqslant 2$ and $W$ be of type $B_n$, as above. Let us now assume that $b=0$ and $a>0$. Consider the reflection subgroup $\tilde{W} := \langle s_0s_1s_0, s_1, \ldots, s_{n-1}\rangle \subseteq W$, where the generators of $W$ are labelled as in Table 1.1. Let $\tilde{L}$ denote the restriction of $L$ to $\tilde{W}_n$. Then $\tilde{W}$ is a Coxeter group of type $D_n$, with diagram as given below, and $\tilde{L}$ is a multiple of the length function on $\tilde{W}$; see [132, §1.4].

$D_n$

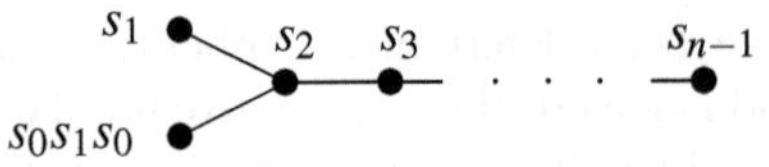


The irreducible representations of $\tilde{W}$ are classified as follows. Given a pair of partitions $(\lambda,\mu)$ such that $|\lambda|+|\mu|=n$, we denote by $E^{[\lambda,\mu]}$ the restriction of $E^{(\lambda,\mu)} \in \mathrm{Irr}(W)$ to $\tilde{W}$. Then the following hold (see [132, 5.6.1]).

- If $\lambda \neq \mu$, then $E^{[\lambda,\mu]} = E^{[\mu,\lambda]}$ is an irreducible representation of $\mathrm{Irr}(\tilde{W})$.
- If $\lambda = \mu$, then $E^{[\lambda,\lambda]} = E^{[\lambda,+]} \oplus E^{[\lambda,-]}$, where $E^{[\lambda,+]}$, $E^{[\lambda,-]}$ are non-isomorphic irreducible representations of $\tilde{W}$. (This can only occur if $n$ is even.)

Furthermore, all irreducible representations of $\tilde{W}$ arise in this way. The **a**-invariants of these representations are given as follows. Let $\tilde{E} \in \mathrm{Irr}(\tilde{W})$ and let $(\lambda,\mu)$ be a pair of partitions as above such that $\tilde{E}$ is a constituent of $E^{[\lambda,\mu]}$. Then, by [132, §10.5], the **a**-invariant of $\tilde{E}$ is $\mathbf{a}_{(\lambda,\mu)}$, where the latter is given by the formula in Case 3 of Example 1.3.9 (with $r=0$, $b'=0$); furthermore, each $f_{\tilde{E}}$ is a power of 2.

*Remark 1.3.11.* The invariants $\mathbf{a}_\lambda$ are inductively determined purely in terms of the characters of $W$ and the parabolic subgroups of $W$. Indeed, if $W=\{1\}$, then $\mathrm{Irr}_{\mathbb{K}}(W)$ only consists of the unit representation and this has **a**-invariant 0. Now assume that $W \neq \{1\}$ and that the **a**-invariants have already been determined for all proper parabolic subgroups of $W$. For any subset $J \subsetneq S$, let $W_J \subseteq W$ denote the corresponding parabolic subgroup and write $\mathrm{Irr}_{\mathbb{K}}(W_J) = \{M^\mu \mid \mu \in \Lambda_J\}$. Then, for any $E^\lambda \in \mathrm{Irr}_{\mathbb{K}}(W)$, we can define

$$\mathbf{a}'_\lambda := \max\{\mathbf{a}_\mu \mid \mu \in \Lambda_J \text{ where } J \subsetneq S \text{ and } M^\mu \uparrow E^\lambda\}.$$

Here, we write $M^\mu \uparrow E^\lambda$ if $E^\lambda$ is an irreducible constituent of the representation obtained by inducing $M^\mu$ from $W_J$ to $W$. Then, finally, we have

$$\mathbf{a}_\lambda = \begin{cases} \mathbf{a}'_\lambda & \text{if } \mathbf{a}'_{\lambda^\dagger} - \mathbf{a}'_\lambda \leqslant N_\lambda, \\ \mathbf{a}'_{\lambda^\dagger} - N_\lambda & \text{otherwise,} \end{cases}$$

where $N_\lambda \in \Gamma$ is given by the formula in Example 1.2.7(a). In the equal-parameter case, this was observed in [132, 6.5.6]; see [115, §4] for the general case.

## 1.4 Balanced Representations

In the previous section we have attached to each irreducible representation $E^\lambda$ of $W$ an invariant $\mathbf{a}_\lambda \in \Gamma_{\geqslant 0}$. In Proposition 1.3.1, we have seen that $\varepsilon^{\mathbf{a}_\lambda}\chi^\lambda(T_w) \in \mathbb{K}[\Gamma_{\geqslant 0}]$ for all $w \in W$, and that $\mathbf{a}_\lambda$ is minimal with this property. One might ask if something like this is true not only for the character, but also for the individual entries of a matrix representation afforded by $E^\lambda_\varepsilon$. It will turn out that this is the case in a very strong sense, as a result of the construction of a cellular basis of $\mathbf{H}$ (see Section 2.7). In this section, we provide the foundations for this construction. Throughout, we assume that $\leqslant$ is a monomial order on $\Gamma$ such that $L(s) \geqslant 0$ for all $s \in S$.

**1.4.1.** We will consider a certain valuation ring $\mathscr{O}_0$ in $K$. Let us denote by $\mathbb{K}[\Gamma_{>0}]$ the set of all $\mathbb{K}$-linear combinations of terms $\varepsilon^g$, where $g > 0$; the notation $\mathbb{K}[\Gamma_{\geqslant 0}]$ has a similar meaning. Note that $1+\mathbb{K}[\Gamma_{>0}]$ is multiplicatively closed. Furthermore, every element $x \in K$ can be written in the form

$$x = r_x \varepsilon^{g_x} \frac{1+p}{1+q}, \qquad \text{where } r_x \in \mathbb{K},\ g_x \in \Gamma \text{ and } p,q \in \mathbb{K}[\Gamma_{>0}];$$

note that, if $x \neq 0$, then $r_x$ and $g_x$ indeed are *uniquely determined* by $x$; if $x = 0$, we have $r_0 = 0$ and we set $g_0 := +\infty$ by convention. We set

$$\mathscr{O}_0 := \{x \in K \mid g_x \geqslant 0\} \qquad \text{and} \qquad \mathfrak{m} := \{x \in K \mid g_x > 0\}.$$

One easily checks that $\mathscr{O}_0$ is a valuation ring in $K$, with maximal ideal $\mathfrak{m}$; note that

$$\mathscr{O}_0 \cap \mathbb{K}[\Gamma] = \mathbb{K}[\Gamma_{\geqslant 0}] \qquad \text{and} \qquad \mathfrak{m} \cap \mathbb{K}[\Gamma] = \mathbb{K}[\Gamma_{>0}].$$

We have a well-defined $\mathbb{K}$-linear ring homomorphism $\mathscr{O}_0 \to \mathbb{K}$ with kernel $\mathfrak{m}$. The image of $x \in \mathscr{O}_0$ in $\mathbb{K}$ is called the *constant term* of $x$. Thus, the constant term of $x$ is 0 if $x \in \mathfrak{m}$; the constant term equals $r_x$ if $x \in \mathscr{O}_0^\times$.

**Warning:** Note that the algebra $\mathbf{H}$ is not defined over $\mathscr{O}_0$! For example, the coefficients in the quadratic relations for $T_s$ ($s \in S$, $L(s) > 0$) do not lie in $\mathscr{O}_0$.

**Example 1.4.2.** Let $\Gamma = \mathbb{Z}$ so that $A = R[v, v^{-1}]$ is the ring of Laurent polynomials in one indeterminate $v = \varepsilon$. Then $K = \mathbb{K}(v)$ is the field of rational functions in $v$. In this case (which is the most important one for many applications), $\mathscr{O}_0$ is just the localisation of $A$ in the prime ideal generated by $v$:

$$\mathscr{O}_0 = \{f/g \mid f, g \in \mathbb{K}[v] \text{ such that } g(0) \neq 0\} \subseteq \mathbb{K}(v).$$

Here, the canonical map $\mathscr{O}_0 \to \mathbb{K}$ is given by evaluating $f/g$ at $v = 0$.

**Definition 1.4.3.** Let $\lambda \in \Lambda$ and consider a matrix representation $\rho^\lambda : \mathbf{H}_K \to M_{d_\lambda}(K)$ afforded by $E_\varepsilon^\lambda$. We say that $\rho^\lambda$ is a *balanced representation* if

$$\varepsilon^{\mathbf{a}_\lambda} \rho^\lambda(T_w) \in M_{d_\lambda}(\mathscr{O}_0) \qquad \text{for all } w \in W.$$

In this case, we denote by $c_{w,\lambda}^{\mathfrak{st}} \in \mathbb{K}$ the constant term of $\varepsilon^{\mathbf{a}_\lambda} \rho_{\mathfrak{st}}^\lambda(T_w)$, for any $w \in W$ and $\mathfrak{s}, \mathfrak{t} \in M(\lambda)$. The coefficients $c_{w,\lambda}^{\mathfrak{st}}$ will be called the *leading matrix coefficients* of $\mathbf{H}$. (They will play a major role in our discussion of "cell data" in Section 1.6.)

*Remark 1.4.4.* (a) Consider the leading coefficients of character values $c_{w,\lambda}$ in Remark 1.3.2. If $\rho^\lambda$ is a balanced representation afforded by $E_\varepsilon^\lambda$, then

$$c_{w,\lambda} = \sum_{\mathfrak{s} \in M(\lambda)} c_{w,\lambda}^{\mathfrak{ss}} \qquad \text{for all } w \in W.$$

Thus, the leading matrix coefficients in Definition 1.4.3 are refinements of Lusztig's leading coefficients of character values.

(b) Balanced representations and, hence, the corresponding leading matrix coefficients are not uniquely determined (say, by the characters $\chi^\lambda$). For example, given a balanced representation $\rho^\lambda : \mathbf{H}_K \to M_{d_\lambda}(K)$, one can just conjugate $\rho^\lambda$ by an invertible matrix with coefficients in $\mathbb{K}$ and the resulting new representation will still be balanced. However, we will see that the main constructions derived from balanced representations (for example, the "asymptotic algebra" in Section 1.5 or the relation "$\leftrightsquigarrow_L$" in 1.6.10) are independent of any choices.

The condition in Definition 1.4.3 is hard to verify, even in concrete examples. Following [103], [112], we will now establish a basic result which provides an efficient criterion for verifying that a given representation is balanced.

**Proposition 1.4.5 (Cf. [103, §4], [112, §4]).** *Assume that $\mathbb{K} \subseteq \mathbb{R}$. Let $\lambda \in \Lambda$ and $\rho^\lambda : \mathbf{H}_K \to M_{d_\lambda}(K)$ be any matrix representation afforded by $E_\varepsilon^\lambda$. Assume that there exists a symmetric matrix $\Omega^\lambda \in M_{d_\lambda}(K)$ such that the following two conditions hold:*

*(a) we have $\Omega^\lambda \rho^\lambda(T_{w^{-1}}) = \rho^\lambda(T_w)^{\mathrm{tr}} \Omega^\lambda$ for all $w \in W$;*
*(b) all entries of $\Omega^\lambda$ lie in $\mathscr{O}_0$ and we have $\Omega^\lambda \equiv D^\lambda \bmod \mathfrak{m}$, where $D^\lambda$ is a diagonal matrix with diagonal coefficients $\{b_\mathfrak{s} \mid \mathfrak{s} \in M(\lambda)\} \subseteq \mathbb{K}^\times$.*

*Then, $\rho^\lambda$ is balanced and the corresponding leading matrix coefficients satisfy*

$$b_\mathfrak{t}\, c_{w^{-1},\lambda}^{\mathfrak{ts}} = c_{w,\lambda}^{\mathfrak{st}}\, b_\mathfrak{s} \quad \text{for all } w \in W \text{ and } \mathfrak{s},\mathfrak{t} \in M(\lambda).$$

*Proof.* Define $\gamma := \min\{g \in \Gamma \mid \varepsilon^g \rho_{\mathfrak{st}}^\lambda(T_w) \in \mathscr{O}_0$ for all $w \in W$ and all $\mathfrak{s},\mathfrak{t} \in M(\lambda)\}$. In order to show that $\rho^\lambda$ is balanced, we must show that $\mathbf{a}_\lambda \geqslant \gamma$. First note that, by the Schur relations (see Proposition 1.2.12), we have for any $\mathfrak{s},\mathfrak{t} \in M(\lambda)$

$$\varepsilon^{2\gamma} \mathbf{c}_\lambda = \sum_{w \in W} \left(\varepsilon^\gamma \rho_{\mathfrak{st}}^\lambda(T_w)\right)\left(\varepsilon^\gamma \rho_{\mathfrak{ts}}^\lambda(T_{w^{-1}})\right) \in \mathscr{O}_0.$$

By Proposition 1.3.1, we also know that $\varepsilon^{2\mathbf{a}_\lambda} \mathbf{c}_\lambda$ lies in $\mathscr{O}_0$ and has a non-zero constant term. Thus, it remains to show that $\varepsilon^{2\gamma} \mathbf{c}_\lambda$ has a non-zero constant term. (This will actually imply that $\mathbf{a}_\lambda = \gamma$.) For this purpose, let $\hat{c}_{w,\lambda}^{\mathfrak{st}} \in \mathbb{R}$ be the constant term of $\varepsilon^\gamma \rho_{\mathfrak{st}}^\lambda(T_w)$. Now we multiply the relation $\Omega^\lambda \rho^\lambda(T_{w^{-1}}) = \rho^\lambda(T_w)^{\mathrm{tr}} \Omega^\lambda$ by $\varepsilon^\gamma$ and consider constant terms. Using (b), we obtain that

$$b_\mathfrak{t}\, \hat{c}_{w^{-1},\lambda}^{\mathfrak{ts}} = \hat{c}_{w,\lambda}^{\mathfrak{st}}\, b_\mathfrak{s} \qquad \text{for all } w \in W \text{ and all } \mathfrak{s},\mathfrak{t} \in M(\lambda).$$

Now choose $\mathfrak{s},\mathfrak{t} \in M(\lambda)$ such that $\hat{c}_{y,\lambda}^{\mathfrak{st}} \neq 0$ for some $y \in W$. Then

$$\varepsilon^{2\gamma} \mathbf{c}_\lambda \equiv \sum_{w \in W} \hat{c}_{w,\lambda}^{\mathfrak{st}}\, \hat{c}_{w^{-1},\lambda}^{\mathfrak{ts}} \equiv b_\mathfrak{s} b_\mathfrak{t}^{-1} \sum_{w \in W} \left(\hat{c}_{w,\lambda}^{\mathfrak{st}}\right)^2 \bmod \mathfrak{m}.$$

The sum on the right-hand side is a non-zero real number since $\hat{c}_{y,\lambda}^{\mathfrak{st}} \neq 0$ for some $y \in W$. Thus, $\varepsilon^{2\gamma} \mathbf{c}_\lambda$ has a non-zero constant term, as desired. As already mentioned above, this implies that $\mathbf{a}_\lambda = \gamma$. Once this is established, it also follows that $c_{w,\lambda}^{\mathfrak{st}} = \hat{c}_{w,\lambda}^{\mathfrak{st}}$ for all $w \in W$ and all $\mathfrak{s}, \mathfrak{t} \in M(\lambda)$, and this yields the remaining assertions. □

Note that the relation in (a) will hold for all $w \in W$ if it holds for all $s \in S$. This remark, although almost trivial, is nevertheless useful in dealing with examples.

**Example 1.4.6.** Let $W$ be of type $I_2(m)$ ($m \geqslant 3$), with generators $s_1, s_2$ such that $(s_1 s_2)^m = 1$. Assume that $L(s_1) \geqslant L(s_2) \geqslant 0$. We have $\mathbb{Z}_W = \mathbb{Z}[\zeta + \zeta^{-1}]$, where $\zeta \in \mathbb{C}$ is a root of unity of order $m$. We show that the matrix representations in Example 1.3.7 are balanced. Since these representations are realised over the ring $\mathbb{Z}_W[v_{s_1}^{\pm 1}, v_{s_2}^{\pm 1}]$, it follows that the corresponding leading matrix coefficients lie in $\mathbb{Z}_W$.

Note that the above statements are clear for one-dimensional representations. Now consider a two-dimensional representation $\sigma_j^\varepsilon$. Let

$$\Omega_j = \begin{pmatrix} v_{s_1}\mu_j(v_{s_2} + v_{s_2}^{-1}) & v_{s_1}\mu_j \\ v_{s_1}\mu_j & v_{s_1}^2 + 1 \end{pmatrix} \in M_2(\mathbb{Z}_W[v_{s_1}^{\pm 1}, v_{s_2}^{\pm 1}]),$$

where $\mu_j = v_{s_1} v_{s_2}^{-1} + \zeta^j + \zeta^{-j} + v_{s_1}^{-1} v_{s_2}$. Then $\Omega_j$ is a symmetric matrix and one immediately checks that

$$\Omega_j \sigma_j^\varepsilon(T_{w^{-1}}) = \sigma_j^\varepsilon(T_w)^{\mathrm{tr}} \Omega_j \qquad \text{for all } w \in W.$$

(Note that it is enough to do this for $w \in \{s_1, s_2\}$.) Note that all entries of $\Omega_j$ lie in $\mathcal{O}_0$. This is clear for the entry $v_{s_1}^2 + 1$. Furthermore, we have

$$v_{s_1}\mu_j = v_{s_1}^2 v_{s_2}^{-1} + (\zeta^j + \zeta^{-j}) v_{s_1} + v_{s_2},$$

which lies in $\mathcal{O}_0$ since $L(s_1) \geqslant L(s_2) \geqslant 0$. Finally,

$$v_{s_1}\mu_j(v_{s_2} + v_{s_2}^{-1}) = v_{s_1}\mu_j v_{s_2} + v_{s_1}^2 v_{s_2}^{-2} + (\zeta^j + \zeta^{-j}) v_{s_1} v_{s_2}^{-1} + 1,$$

which is still in $\mathcal{O}_0$. Using the above formulae, we obtain that

$$\Omega_j \equiv \begin{pmatrix} 1 & 0 \\ 0 & 1 \end{pmatrix} \bmod \mathfrak{m} \qquad \text{if } L(s_1) > L(s_2) > 0,$$

$$\Omega_j \equiv \begin{pmatrix} 2 + \zeta^j + \zeta^{-j} & 0 \\ 0 & 1 \end{pmatrix} \bmod \mathfrak{m} \qquad \text{if } L(s_1) = L(s_2) > 0.$$

Hence, in these two cases, the conditions in Proposition 1.4.5 are satisfied and so $\sigma_j^\varepsilon$ is balanced. Now assume that $L(s_2) = 0$; that is, $v_{s_2} = 1$. Then we perform a slight transformation as follows. We set

$$\hat{\sigma}_j^\varepsilon(T_w) := P^{-1} \sigma_j^\varepsilon(T_w) P \quad (w \in W), \qquad \text{where} \qquad P := \begin{pmatrix} 1 & -\frac{1}{2} \\ 0 & 1 \end{pmatrix}.$$

Then $\hat{\Omega}_j \hat{\sigma}_j^\varepsilon(T_{w^{-1}}) = \hat{\sigma}_j^\varepsilon(T_w)^{\mathrm{tr}} \hat{\Omega}_j$ for all $w \in W$, where $\hat{\Omega}_j := P^{\mathrm{tr}} \Omega_j P$. Furthermore,

$$\hat{\Omega}_j \equiv \begin{pmatrix} 2 & 0 \\ 0 & \frac{1}{2} \end{pmatrix} \mod \mathfrak{m} \qquad \text{if } L(s_1) > L(s_2) = 0,$$

$$\hat{\Omega}_j \equiv \begin{pmatrix} 2(2+\zeta^j+\zeta^{-j}) & 0 \\ 0 & \frac{1}{2}(2-\zeta^j-\zeta^{-j}) \end{pmatrix} \mod \mathfrak{m} \qquad \text{if } L(s_1) = L(s_2) = 0.$$

Hence, Proposition 1.4.5 applies again and so $\hat{\sigma}_j^\varepsilon$ is balanced. But then $\sigma_j^\varepsilon$ must also be balanced, since the transforming matrix $P$ has all its entries in $\mathbb{K}$.

We now turn to the existence of balanced representations in general. This will be based on the following result.

**Lemma 1.4.7 (Cf. Lusztig [217, 1.7]).** *Assume that $\mathbb{K} \subseteq \mathbb{R}$. For each $\lambda \in \Lambda$, there exists a symmetric bilinear form $\langle\ ,\ \rangle_\lambda : E_\varepsilon^\lambda \times E_\varepsilon^\lambda \to K$ such that the following hold.*

*(a) We have $\langle T_w.e, e'\rangle_\lambda = \langle e, T_{w^{-1}}.e'\rangle_\lambda$ for all $e, e' \in E_\varepsilon^\lambda$ and $w \in W$.*
*(b) For every $0 \neq e \in E_\varepsilon^\lambda$, there exist some $g \in \Gamma$ and some strictly positive real number $b \in \mathbb{K}$ such that $\varepsilon^{2g}\langle e, e\rangle_\lambda \in b + \mathfrak{m}$.*

*In particular, the form $\langle\ ,\ \rangle_\lambda$ is non-degenerate.*

*Proof.* Let $(\ ,\ )$ be any symmetric bilinear form on $E_\varepsilon^\lambda$ which admits an orthonormal basis, $\{e_\mathfrak{s} \mid \mathfrak{s} \in M(\lambda)\}$ say. Then we define

$$\langle e, e'\rangle_\lambda := \sum_{w\in W} (T_w.e, T_w.e') \qquad \text{for any } e, e' \in E_\varepsilon^\lambda.$$

We claim that $\langle T_s.e, e'\rangle_\lambda = \langle e, T_s.e'\rangle_\lambda$ for all $s \in S$. Indeed, let $s \in S$ and split the above sum over $w \in W$ into parts according to whether multiplication by $s$ increases or decreases $l(w)$. This yields

$$\langle T_s.e, e'\rangle_\lambda = \sum_{w\in W} (T_{ws}.e, T_w.e') + (v_s - v_s^{-1}) \sum_{w\in W : l(ws)<l(w)} (T_w.e, T_w.e'),$$
$$\langle e, T_s.e'\rangle_\lambda = \sum_{w\in W} (T_w.e, T_{ws}.e') + (v_s - v_s^{-1}) \sum_{w\in W : l(ws)<l(w)} (T_w.e, T_w.e').$$

Since the first two sums are the same in each case, we conclude that $\langle T_s.e, e'\rangle = \langle e, T_s.e'\rangle_\lambda$, as claimed. Consequently, (a) holds.

Now let $0 \neq e \in E_\varepsilon^\lambda$. Writing each term $T_w.e$ in the defining formula for $\langle e, e\rangle_\lambda$ as a linear combination of the original basis $\{e_\mathfrak{s}\}$, one readily sees that

$$\langle e, e\rangle_\lambda = \sum_{x\in N_e} x^2, \quad \text{where} \quad N_e \subseteq K^\times,\ N_e \neq \varnothing.$$

Now write each $x \in N_e$ as $x = r_x \varepsilon^{g_x} u_x$, where $0 \neq r_x \in \mathbb{K}$, $g_x \in \Gamma$ and $u_x \in 1 + \mathfrak{m}$. Let $g := -\min\{g_x \mid x \in N_\mathfrak{t}\}$. Then $\varepsilon^{2g}\langle e, e\rangle_\lambda$ lies in $\mathscr{O}_0$ and has constant term $b := \sum_{x\in M : g_x = g} r_x^2 \in \mathbb{K}$. Since all $r_x$ are real numbers, there are no cancellations in this sum and so $b > 0$. Thus, (b) holds. □

**Theorem 1.4.8 (Cf. [103, §4]).** *For each $\lambda \in \Lambda$, there exists a matrix representation $\rho^\lambda : \mathbf{H}_K \to M_{d_\lambda}(K)$ afforded by $E^\lambda_\varepsilon$ which is balanced. This representation $\rho^\lambda$ can be chosen such that the conditions in Proposition 1.4.5 are satisfied, where $b_\mathfrak{s}$ is a positive real number in $\mathbb{K}$ for all $\mathfrak{s} \in M(\lambda)$. The corresponding leading matrix coefficients have the following properties:*

*(a) We have $c^{\mathfrak{st}}_{w,\lambda} \in \mathbb{K} \cap \mathbb{R}$ for all $w \in W$ and $\mathfrak{s}, \mathfrak{t} \in M(\lambda)$.*

*(b) We have $b_\mathfrak{t} c^{\mathfrak{ts}}_{w^{-1},\lambda} = c^{\mathfrak{st}}_{w,\lambda} b_\mathfrak{s}$ for all $w \in W$ and $\mathfrak{s}, \mathfrak{t} \in M(\lambda)$. In particular, $c^{\mathfrak{st}}_{w,\lambda} \neq 0$ if and only if $c^{\mathfrak{ts}}_{w^{-1},\lambda} \neq 0$.*

*Proof.* We can assume that $\mathbb{K} \subseteq \mathbb{R}$; see 1.2.1. Let $\langle\ ,\ \rangle_\lambda$ be as in Lemma 1.4.7. Since we are working in characteristic 0, the vector space $E^\lambda_\varepsilon$ has an orthogonal basis, $\{h_\mathfrak{s} \mid \mathfrak{s} \in M(\lambda)\}$ say, with respect to this form. For each $\mathfrak{s} \in M(\lambda)$, we have

$$\varepsilon^{2g_\mathfrak{s}} \langle h_\mathfrak{s}, h_\mathfrak{s} \rangle_\lambda \in b_\mathfrak{s} + \mathfrak{m}, \qquad \text{where } g_\mathfrak{s} \in \Gamma \text{ and } b_\mathfrak{s} \in \mathbb{K},\ b_\mathfrak{s} > 0.$$

Now set $f_\mathfrak{s} := \varepsilon^{g_\mathfrak{s}} h_\mathfrak{s}$ for $\mathfrak{s} \in M(\lambda)$. Then $\{f_\mathfrak{s} \mid \mathfrak{s} \in M(\lambda)\}$ still is an orthogonal basis, and we have $\langle f_\mathfrak{s}, f_\mathfrak{s} \rangle_\lambda \in b_\mathfrak{s} + \mathfrak{m}$ for all $\mathfrak{s} \in M(\lambda)$. Let $\Omega^\lambda$ be the Gram matrix of $\langle\ ,\ \rangle_\lambda$ with respect to the basis $\{f_\mathfrak{s}\}$ and $D^\lambda$ be the diagonal matrix with diagonal coefficients $b_\mathfrak{s}$ ($\mathfrak{s} \in M(\lambda)$). Then the conditions in Proposition 1.4.5 are satisfied for the matrix representation $\rho^\lambda$ defined with respect to the basis $\{f_\mathfrak{s}\}$. □

**1.4.9.** We summarize the procedure for constructing balanced representations as follows, where we express everything in matrix terms.

Let $\lambda \in \Lambda$ and $\sigma^\lambda : \mathbf{H}_K \to M_{d_\lambda}(K)$ be a representation afforded by $E^\lambda_\varepsilon$. Then set

$$\Omega^\lambda_0 := \sum_{w \in W} \sigma^\lambda(T_w)^{\mathrm{tr}}.\sigma^\lambda(T_w) \in M_{d_\lambda}(K).$$

Clearly, $\Omega^\lambda_0$ is a symmetric matrix. By Lemma 1.4.7, we have

$$\det(\Omega^\lambda_0) \neq 0 \qquad \text{and} \qquad \Omega^\lambda_0\, \sigma^\lambda(T_{w^{-1}}) = \sigma^\lambda(T_w)^{\mathrm{tr}}\, \Omega^\lambda_0 \quad \text{for all } w \in W.$$

In fact, as shown in Lemma 1.4.7, $\Omega^\lambda_0$ is the Gram matrix of a bilinear form which has no non-zero isotropic vectors. Thus, we can apply the Gram–Schmidt orthogonalisation procedure, which yields an upper triangular matrix $P^\lambda$ with 1 on the diagonal and such that

$$(P^\lambda)^{\mathrm{tr}}\, \Omega^\lambda_0\, P^\lambda = \Omega^\lambda_1,$$

where $\Omega^\lambda_1$ is a diagonal matrix with the following property. There exist strictly positive real numbers $\{b_\mathfrak{s} \mid \mathfrak{s} \in M(\lambda)\}$ and elements $\{g_\mathfrak{s} \mid \mathfrak{s} \in M(\lambda)\} \subseteq \Gamma$ such that the diagonal coefficient of $\Omega^\lambda_1$ at position $\mathfrak{s} \in M(\lambda)$ has the form

$$b_\mathfrak{s} \varepsilon^{-2g_\mathfrak{s}} + \text{combination of terms } \varepsilon^g \text{ where } g > -2g_\mathfrak{s}.$$

Let $E^\lambda$ be the diagonal matrix with diagonal coefficients $\{\varepsilon^{g_\mathfrak{s}} \mid \mathfrak{s} \in M(\lambda)\}$. We set

$$\Omega^\lambda := E^\lambda\, \Omega_1^\lambda\, E^\lambda \in M_{d_\lambda}(\mathscr{O}_0),$$
$$\rho^\lambda(T_w) := (E^\lambda)^{-1}(P^\lambda)^{-1}\sigma^\lambda(T_w)\,P^\lambda\, E^\lambda \qquad \text{for all } w \in W.$$

Then $\Omega^\lambda\rho^\lambda(T_{w^{-1}}) = \rho^\lambda(T_w)^{\mathrm{tr}}\,\Omega^\lambda$ for all $w \in W$ and $\Omega^\lambda$ is a diagonal matrix where the diagonal coefficient at position $\mathfrak{s} \in M(\lambda)$ is congruent to $b_\mathfrak{s}$ modulo $\mathfrak{m}$. Hence, by Proposition 1.4.5, $\rho^\lambda$ is a balanced representation afforded by $E_\varepsilon^\lambda$.

**Proposition 1.4.10.** *For every $\lambda \in \Lambda$, let $\rho^\lambda : \mathbf{H}_K \to M_{d_\lambda}(K)$ be a matrix representation afforded by $E_\varepsilon^\lambda$ which is balanced. Then the corresponding leading matrix coefficients have the following properties.*

*(a) Given $\lambda, \mu \in \Lambda$, let $\mathfrak{s}, \mathfrak{t} \in M(\lambda)$ and $\mathfrak{u}, \mathfrak{v} \in M(\mu)$. Then*

$$\sum_{w\in W} c_{w,\lambda}^{\mathfrak{st}}\, c_{w^{-1},\mu}^{\mathfrak{uv}} = \begin{cases} f_\lambda & \text{if } \lambda = \mu,\ \mathfrak{s} = \mathfrak{v},\ \mathfrak{t} = \mathfrak{u}, \\ 0 & \text{otherwise.} \end{cases}$$

*(b) Let $x, y \in W$. Then*

$$\sum_{\lambda\in\Lambda}\sum_{\mathfrak{s},\mathfrak{t}\in M(\lambda)} f_\lambda^{-1}\, c_{x,\lambda}^{\mathfrak{st}}\, c_{y^{-1},\lambda}^{\mathfrak{ts}} = \begin{cases} 1 & \text{if } x = y, \\ 0 & \text{otherwise.} \end{cases}$$

*(c) Given $\lambda \in \Lambda$ and $\mathfrak{s}, \mathfrak{t} \in M(\lambda)$, there exists some $w \in W$ such that $c_{w,\lambda}^{\mathfrak{st}} \neq 0$. Conversely, given $w \in W$, there exist some $\lambda, \mathfrak{s}, \mathfrak{t}$ such that $c_{w,\lambda}^{\mathfrak{st}} \neq 0$. In particular,*

$$\mathbf{a}_\lambda = \min\{g \in \Gamma_{\geqslant 0} \mid \varepsilon^g \rho_{\mathfrak{st}}^\lambda(T_w) \in \mathscr{O}_0 \text{ for all } w \in W \text{ and } \mathfrak{s}, \mathfrak{t} \in M(\lambda)\}.$$

*Proof.* The Schur relations in Proposition 1.2.12 yield

$$\delta_{\mathfrak{sv}}\delta_{\mathfrak{ut}}\delta_{\lambda\mu}\varepsilon^{\mathbf{a}_\lambda+\mathbf{a}_\mu}\mathbf{c}_\lambda = \sum_{w\in W}\left(\varepsilon^{\mathbf{a}_\lambda}\rho_{\mathfrak{st}}^\lambda(T_w)\right)\left(\varepsilon^{\mathbf{a}_\mu}\rho_{\mathfrak{uv}}^\mu(T_{w^{-1}})\right).$$

By Definition 1.4.3, the right-hand side is congruent to $\sum_{w\in W} c_{w,\lambda}^{\mathfrak{st}}\, c_{w^{-1},\lambda}^{\mathfrak{uv}}$ modulo $\mathfrak{m}$. If $\lambda \neq \mu$, the left-hand side is zero. If $\lambda = \mu$, then the left-hand side is congruent to $\delta_{\mathfrak{sv}}\delta_{\mathfrak{ut}}f_\lambda$; see Proposition 1.3.1. This yields (a).

Now (b) is a formal consequence of (a). Indeed, by the Artin–Wedderburn theorem, the set of all triples $(\lambda, \mathfrak{s}, \mathfrak{t})$ has cardinality $|W|$. Arranging these triples, and the elements of $W$, in some order, we obtain $|W| \times |W|$-matrices

$$A := \left(c_{w,\lambda}^{\mathfrak{st}}\right) \qquad \text{and} \qquad B := \left(f_\lambda^{-1}\, c_{w^{-1},\lambda}^{\mathfrak{st}}\right),$$

where the triples $(\lambda, \mathfrak{s}, \mathfrak{t})$ label the rows and the elements of $W$ label the columns. Then (a) means that the product $A.B^{\mathrm{tr}}$ is the identity matrix. So $B^{\mathrm{tr}}.A$ also is the identity matrix. Writing out the coefficients of $B^{\mathrm{tr}}.A$ yields (b). Finally, (c) clearly is a formal consequence of (a) and (b). □

We now introduce a particular class of matrix representations of $\mathbf{H}$. These are defined using the concept of a $W$-graph which first appeared in the work of Kazh-

dan and Lusztig [195] on "cells" in the equal-parameter case. (We shall say more on these "cells" in Chapter 2.) The following definition works for general weight functions. (In [132, Def. 11.1.1] we made a first attempt to generalise the original definition of [195], but we did not take into account a monomial order on $\Gamma$.)

**Definition 1.4.11.** Assume that $L(s) > 0$ for all $s \in S$. A $W$-*graph* for $\mathbf{H}$ (with respect to $\leqslant$) consists of the following data:

(a) a set $X$ together with a map $I$ which assigns to each $x \in X$ a subset $I(x) \subseteq S$;
(b) a collection of elements $\{m_{x,y}^s\}$ in $A$, where $x,y \in X$ and $s \in S$ are such that $s \in I(x)$ and $s \notin I(y)$.

These data are subject to the following requirements. First, we require that

$$v_s m_{x,y}^s \in R[\Gamma_{>0}] \quad \text{and} \quad \overline{m_{x,y}^s} = m_{x,y}^s \quad \text{for all } x,y \in X,\ s \in I(x) \setminus I(y),$$

where $a \mapsto \bar{a}$ ($a \in A$) is as in Example 1.2.6. Furthermore, let $V$ be a free $A$-module with a basis $\{e_y \mid y \in X\}$. For $s \in S$, define an $A$-linear map $\rho_s \colon V \to V$ by

$$\rho_s(e_y) = \begin{cases} v_s e_y + \sum\limits_{x \in X : s \in I(x)} m_{x,y}^s e_x & \text{if } s \notin I(y), \\ -v_s^{-1} e_y & \text{if } s \in I(y). \end{cases}$$

Then we require that the assignment $T_s \mapsto \rho_s$ defines a representation of $\mathbf{H}$.

**Example 1.4.12.** (a) Assume that $W$ is of type $I_2(m)$, with generators $S = \{s_1, s_2\}$ such that $L(s_1) \geqslant L(s_2) > 0$. The irreducible representations of $\mathbf{H}_K$ are described in Example 1.3.7. The one-dimensional representations certainly are given by $W$-graphs. Now let $\sigma_j$ be one of the two-dimensional representations and define a new representation $\tilde{\sigma}_j$ by sending $\sigma_j(T_{s_i})$ to $\sigma_j(T_{s_i})^{\text{tr}}$ for $i = 1, 2$. Then $\tilde{\sigma}_j$ is given in terms of a $W$-graph. (Note that $\tilde{\sigma}_j$ is equivalent to $\sigma_j$.)

(b) Let $W$ be of type $F_4$ with generators labelled as in Table 1.1 (p. 2). The weight function $L \colon W \to \Gamma$ is determined by $a := L(s_1) = L(s_2)$ and $b := L(s_3) = L(s_4)$. If $\leqslant$ is a monomial order on $\Gamma$ such that $b > 3a > 0$, then the graphs in [132, Fig. 11.3 (pp. 377)] are $W$-graphs with respect to $\leqslant$.

**Example 1.4.13.** Assume that we are in the equal-parameter case where $\Gamma = \mathbb{Z}$ and $L(s) = 1$ for all $s \in S$, as originally considered by Kazhdan and Lusztig [195]. Then the conditions in Definition 1.4.11 imply that $m_{x,y}^s \in R$. In this case, Gyoja [150] and Lusztig [224] have shown that every irreducible representation of $\mathbf{H}_K$ can be realised in terms of a $W$-graph. (We shall discuss this general existence result in Section 2.7.) For $W$ of exceptional type, such $W$-graphs have been constructed explicitly:

| | |
|---|---|
| $H_3$ | Lusztig [216, §5]; |
| $H_4$ | Alvis–Lusztig [3]; |
| $F_4, E_6$ | Naruse (personal communication January and July 1998); |
| $E_7, E_8$ | Howlett and Yin [164], [297] (and personal communication 2004). |

(The $W$-graphs for types $H_3$ and $H_4$ are also printed in [132, §11.4].)

Experiments with the above examples suggest the following connection between $W$-graphs and balanced representations:

**Conjecture 1.4.14.** *Let $\lambda \in \Lambda$ and assume that a corresponding matrix representation $\rho^\lambda : \mathbf{H}_K \to M_{d_\lambda}(K)$ is realised in terms of a $W$-graph. Then $\rho^\lambda$ is balanced.*

In Section 2.7 we will see that, under some hypothesis on $W, L$, for any $\lambda \in \Lambda$ there always exists a balanced representation $\rho^\lambda$ which is afforded by a $W$-graph.

## 1.5 The Asymptotic Algebra

We shall assume from now on that, for every $\lambda \in \Lambda$, we are given a matrix representation $\rho^\lambda$ afforded by $E_\varepsilon^\lambda$ which is balanced; this is possible by Theorem 1.4.8. Let $c_{w,\lambda}^{\mathfrak{st}}$ be the corresponding leading matrix coefficients. We now use these coefficients, following [112], to construct a $\mathbb{K}$-algebra. (This construction is independent of Lusztig's construction of the asymptotic ring $\mathbf{J}$ in [223], [231, Chap. 18]; we will see in Corollary 2.3.16 that, up to the change of a sign of basis elements, the two constructions give the same result under some hypothesis on $W, L$.)

**Definition 1.5.1 (See [112]).** Let $\tilde{\mathbf{J}}$ be a $\mathbb{K}$-vector space with a basis $\{t_w \mid w \in W\}$. We define an element of $\tilde{\mathbf{J}}$ and a bilinear product on $\tilde{\mathbf{J}}$ by

$$1_{\tilde{\mathbf{J}}} := \sum_{w \in \tilde{\mathscr{D}}} \tilde{n}_w t_w \qquad \text{and} \qquad t_x t_y := \sum_{z \in W} \tilde{\gamma}_{x,y,z} t_{z^{-1}} \quad (x, y \in W),$$

where

$$\tilde{\gamma}_{x,y,z} := \sum_{\lambda \in \Lambda} \sum_{\mathfrak{s},\mathfrak{t},\mathfrak{u} \in M(\lambda)} f_\lambda^{-1} c_{x,\lambda}^{\mathfrak{st}} c_{y,\lambda}^{\mathfrak{tu}} c_{z,\lambda}^{\mathfrak{us}} \quad (x, y, z \in W),$$

$$\tilde{n}_w := \sum_{\lambda \in \Lambda} \sum_{\mathfrak{s} \in M(\lambda)} f_\lambda^{-1} c_{w^{-1},\lambda}^{\mathfrak{ss}} \quad (w \in W),$$

$$\tilde{\mathscr{D}} := \{w \in W \mid \tilde{n}_w \neq 0\}.$$

We will show in several steps that $\tilde{\mathbf{J}}$ is a symmetric algebra where $1_{\tilde{\mathbf{J}}}$ is the identity. First, we show that the ingredients in the definition of $\tilde{\mathbf{J}}$ only depend on $\mathbf{H}$.

**Lemma 1.5.2.** *The constants $\tilde{\gamma}_{x,y,z}$ and $\tilde{n}_w$ do not depend on the choice of the balanced representations $\rho^\lambda$. More precisely, $\tilde{\gamma}_{x,y,z}$ and $\tilde{n}_w$ are equal, respectively, to the constant terms of*

$$\sum_{\lambda \in \Lambda} f_\lambda^{-1} \varepsilon^{3\mathbf{a}_\lambda} \chi^\lambda(T_x T_y T_z) \in \mathscr{O}_0 \qquad \textit{and} \qquad \sum_{\lambda \in \Lambda} f_\lambda^{-1} \varepsilon^{\mathbf{a}_\lambda} \chi^\lambda(T_{w^{-1}}) \in \mathscr{O}_0.$$

*Furthermore, $\tilde{\gamma}_{x,y,z}$ and $\tilde{n}_w$ lie in the subring $\mathbb{Z}_W[f_\lambda^{-1} \mid \lambda \in \Lambda] \subseteq \mathbb{C}$.*

*Proof.* Given $\lambda \in \Lambda$, consider the expression

$$\begin{aligned}\varepsilon^{3\mathbf{a}_\lambda}\chi^\lambda(T_xT_yT_z) &= \sum_{\mathfrak{s}\in M(\lambda)} \varepsilon^{3\mathbf{a}_\lambda}\rho^\lambda_{\mathfrak{s}\mathfrak{s}}(T_xT_yT_z)\\ &= \sum_{\mathfrak{s},\mathfrak{t},\mathfrak{u}\in M(\lambda)} \big(\varepsilon^{\mathbf{a}_\lambda}\rho^\lambda_{\mathfrak{s}\mathfrak{t}}(T_x)\big)\big(\varepsilon^{\mathbf{a}_\lambda}\rho^\lambda_{\mathfrak{t}\mathfrak{u}}(T_y)\big)\big(\varepsilon^{\mathbf{a}_\lambda}\rho^\lambda_{\mathfrak{u}\mathfrak{s}}(T_z)\big).\end{aligned}$$

By the definition of balanced representations, the terms $\varepsilon^{\mathbf{a}_\lambda}\rho^\lambda_{\mathfrak{s}\mathfrak{t}}(T_x)$, $\varepsilon^{\mathbf{a}_\lambda}\rho^\lambda_{\mathfrak{t}\mathfrak{u}}(T_y)$, $\varepsilon^{\mathbf{a}_\lambda}\rho^\lambda_{\mathfrak{u}\mathfrak{s}}(T_z)$ all lie in $\mathscr{O}_0$. So the whole sum lies in $\mathscr{O}_0$ and its constant term can be computed term by term. Hence,

$$\sum_{\mathfrak{s},\mathfrak{t},\mathfrak{u}\in M(\lambda)} f_\lambda^{-1} c^{\mathfrak{s}\mathfrak{t}}_{x,\lambda}\, c^{\mathfrak{t}\mathfrak{u}}_{y,\lambda}\, c^{\mathfrak{u}\mathfrak{s}}_{z,\lambda} \equiv f_\lambda^{-1}\varepsilon^{3\mathbf{a}_\lambda}\chi^\lambda(T_xT_yT_z) \bmod \mathfrak{m}.$$

This yields the required interpretation of $\tilde{\gamma}_{x,y,z}$. In particular, it shows that $\tilde{\gamma}_{x,y,z}$ only depends on the character of $\rho^\lambda$, not on the choice of $\rho^\lambda$ itself. A similar argument applies to the defining formula for $\tilde{n}_w$: We have

$$\sum_{\mathfrak{s}\in M(\lambda)} f_\lambda^{-1} c^{\mathfrak{s}\mathfrak{s}}_{w^{-1},\lambda} \equiv f_\lambda^{-1}\varepsilon^{\mathbf{a}_\lambda}\chi^\lambda(T_{w^{-1}}) \bmod \mathfrak{m} \qquad \text{for all } \lambda \in \Lambda.$$

Finally, recall from 1.2.1 that $\chi^\lambda(T_w) \in \mathbb{Z}_W[\Gamma]$ for all $w \in W$. Furthermore, the structure constants of $\mathbf{H}$ with respect to the basis $\{T_w\}$ lie in $\mathbb{Z}[\Gamma]$. This immediately yields the statement about the ring in which the constants $\tilde{\gamma}_{x,y,z}$ and $\tilde{n}_w$ lie. □

**Lemma 1.5.3.** *We have the following relations:*

(a) $\tilde{\gamma}_{x,y,z} = \tilde{\gamma}_{y,z,x}$ *for all* $x,y,z \in W$,

(b) $\sum_{w\in W} \tilde{\gamma}_{x^{-1},y,w}\,\tilde{n}_w = \delta_{xy}$ *for all* $x,y \in W$,

(c) $\tilde{n}_w = \tilde{n}_{w^{-1}}$ *and* $\tilde{\gamma}_{x,y,z} = \tilde{\gamma}_{y^{-1},x^{-1},z^{-1}}$ *for all* $w,x,y,z \in W$.

*Proof.* (a) Just note that the defining formula for $\tilde{\gamma}_{x,y,z}$ is symmetrical under cyclic permutations of $x,y,z$.

(b) Using the defining formulae for $\tilde{\gamma}_{x,y,z}$ and $\tilde{n}_w$, the left-hand side evaluates to

$$\begin{aligned}&\sum_{w\in W}\Big(\sum_{\lambda\in\Lambda}\sum_{\mathfrak{s},\mathfrak{t},\mathfrak{u}\in M(\lambda)} f_\lambda^{-1} c^{\mathfrak{s}\mathfrak{t}}_{x^{-1},\lambda}\, c^{\mathfrak{t}\mathfrak{u}}_{y,\lambda}\, c^{\mathfrak{u}\mathfrak{s}}_{w,\lambda}\Big)\Big(\sum_{\mu\in\Lambda}\sum_{\mathfrak{v}\in M(\mu)} f_\mu^{-1} c^{\mathfrak{v}\mathfrak{v}}_{w^{-1},\mu}\Big)\\ &\quad= \sum_{\lambda,\mu\in\Lambda}\sum_{\mathfrak{s},\mathfrak{t},\mathfrak{u}\in M(\lambda)}\sum_{\mathfrak{v}\in M(\mu)} f_\lambda^{-1} f_\mu^{-1} c^{\mathfrak{s}\mathfrak{t}}_{x^{-1},\lambda}\, c^{\mathfrak{t}\mathfrak{u}}_{y,\lambda}\Big(\sum_{w\in W} c^{\mathfrak{u}\mathfrak{s}}_{w,\lambda}\, c^{\mathfrak{v}\mathfrak{v}}_{w^{-1},\mu}\Big).\end{aligned}$$

By Proposition 1.4.10(a), the parenthesised sum evaluates to $\delta_{\mathfrak{u}\mathfrak{v}}\delta_{\mathfrak{s}\mathfrak{v}}\delta_{\lambda\mu}f_\lambda$. Inserting this into the above expression yields $\sum_{\lambda\in\Lambda}\sum_{\mathfrak{s},\mathfrak{t}\in M(\lambda)} f_\lambda^{-1} c^{\mathfrak{s}\mathfrak{t}}_{x^{-1},\lambda}\, c^{\mathfrak{t}\mathfrak{s}}_{y,\lambda} = \delta_{xy}$, where the last equality holds by Proposition 1.4.10(b).

(c) Let $\lambda \in \Lambda$. By Lemma 1.5.2, we can assume without loss of generality that $\rho^\lambda$ has been chosen such that the additional conditions in Theorem 1.4.8 are satisfied.

Thus, we have $b_\mathfrak{t}\, c^{\mathfrak{ts}}_{w^{-1},\lambda} = c^{\mathfrak{st}}_{w,\lambda}\, b_\mathfrak{s}$, where $b_\mathfrak{s}, b_\mathfrak{t}$ are positive real numbers (depending on $\lambda$). This immediately shows that $c^{\mathfrak{ss}}_{w,\lambda} = c^{\mathfrak{ss}}_{w^{-1},\lambda}$ and so $\tilde{n}_w = \tilde{n}_{w^{-1}}$. Similarly,

$$c^{\mathfrak{st}}_{y^{-1},\lambda}\, c^{\mathfrak{tu}}_{x^{-1},\lambda}\, c^{\mathfrak{us}}_{z^{-1},\lambda} = \left(b_\mathfrak{s}^{-1} b_\mathfrak{t}\, c^{\mathfrak{ts}}_{y,\lambda}\right)\left(b_\mathfrak{t}^{-1} b_\mathfrak{u}\, c^{\mathfrak{ut}}_{x,\lambda}\right)\left(b_\mathfrak{u}^{-1} b_\mathfrak{s}\, c^{\mathfrak{su}}_{z,\lambda}\right) = c^{\mathfrak{ut}}_{x,\lambda}\, c^{\mathfrak{ts}}_{y,\lambda}\, c^{\mathfrak{su}}_{z,\lambda}.$$

Summing over all $\lambda \in \Lambda$ and $\mathfrak{s},\mathfrak{t},\mathfrak{u} \in M(\lambda)$, this yields $\tilde{\gamma}_{y^{-1},x^{-1},z^{-1}} = \tilde{\gamma}_{x,y,z}$. □

In general, the constants $\tilde{\gamma}_{x,y,z}$ are hard to compute. In the following example, we can at least deal with some special case.

**Example 1.5.4.** Assume that $L(s) > 0$ for all $s \in S$ and that $1 \in \{x,y,z\}$. Then

$$\tilde{\gamma}_{x,y,z} = \begin{cases} 1 & \text{if } x=y=z=1, \\ 0 & \text{otherwise.} \end{cases}$$

Indeed, by Lemma 1.5.3(a), we can choose notation such that $z = 1$. Now let $\lambda_0 \in \Lambda$ be such that $E^{\lambda_0}$ is the unit representation of $W$. Then $\rho^{\lambda_0}(T_w) = (\varepsilon^{L(w)})$ for all $w \in W$. Let $M(\lambda_0) = \{1\}$. Since $\mathbf{a}_{\lambda_0} = 0$, it follows that $c^{11}_{w,\lambda_0} = 0$ for $w \neq 1$, and $c^{11}_{1,\lambda_0} = 1$. Now let $\lambda \in \Lambda$ be such that $\lambda \neq \lambda_0$. By Example 1.3.3, we have $\mathbf{a}_\lambda > 0$. Since $\rho^\lambda(T_1)$ is the identity matrix, we conclude that $c^{\mathfrak{st}}_{1,\lambda} = 0$ for all $\mathfrak{s},\mathfrak{t} \in M(\lambda)$. Hence, in the defining formula for $\tilde{\gamma}_{x,y,1}$, the terms corresponding to $\lambda$ are zero. So we obtain $\tilde{\gamma}_{x,y,1} = c^{11}_{x,\lambda_0} c^{11}_{y,\lambda_0}$ and this yields the above formula.

**Proposition 1.5.5.** *$\tilde{\mathbf{J}}$ is an associative algebra with identity element $1_{\tilde{\mathbf{J}}}$.*

*Proof.* Let $x,y,z \in W$. We must check that $(t_x t_y)t_z = t_x(t_y t_z)$, which is equivalent to

$$\sum_{u\in W} \tilde{\gamma}_{x,y,u^{-1}}\, \tilde{\gamma}_{u,z,w^{-1}} = \sum_{u\in W} \tilde{\gamma}_{x,u,w^{-1}}\, \tilde{\gamma}_{y,z,u^{-1}} \qquad \text{for all } w \in W.$$

Using the defining formula, the left-hand side evaluates to

$$\sum_{u\in W}\Bigg(\sum_{\substack{\lambda\in\Lambda\\ \mathfrak{s},\mathfrak{t},\mathfrak{u}\in M(\lambda)}} f_\lambda^{-1}\, c^{\mathfrak{st}}_{x,\lambda}\, c^{\mathfrak{tu}}_{y,\lambda}\, c^{\mathfrak{us}}_{u^{-1},\lambda}\Bigg)\Bigg(\sum_{\substack{\mu\in\Lambda\\ \mathfrak{s}',\mathfrak{t}',\mathfrak{u}'\in M(\mu)}} f_\mu^{-1}\, c^{\mathfrak{s}'\mathfrak{t}'}_{u,\mu}\, c^{\mathfrak{t}'\mathfrak{u}'}_{z,\mu}\, c^{\mathfrak{u}'\mathfrak{s}'}_{w^{-1},\mu}\Bigg)$$

$$= \sum_{\lambda,\mu\in\Lambda}\ \sum_{\substack{\mathfrak{s},\mathfrak{t},\mathfrak{u}\in M(\lambda)\\ \mathfrak{s}',\mathfrak{t}',\mathfrak{u}'\in M(\mu)}} f_\lambda^{-1} f_\mu^{-1}\, c^{\mathfrak{st}}_{x,\lambda}\, c^{\mathfrak{tu}}_{y,\lambda}\, c^{\mathfrak{t}'\mathfrak{u}'}_{z,\mu}\, c^{\mathfrak{u}'\mathfrak{s}'}_{w^{-1},\mu}\Bigg(\sum_{u\in W} c^{\mathfrak{us}}_{u^{-1},\lambda}\, c^{\mathfrak{s}'\mathfrak{t}'}_{u,\mu}\Bigg).$$

By Proposition 1.4.10(a), the parenthesised sum evaluates to $\delta_{\mathfrak{ut}'}\delta_{\mathfrak{ss}'}\delta_{\lambda\mu} f_\lambda$. Hence, the above expression equals

$$\sum_{\lambda\in\Lambda}\ \sum_{\mathfrak{s},\mathfrak{t},\mathfrak{u},\mathfrak{u}'\in M(\lambda)} f_\lambda^{-1}\, c^{\mathfrak{st}}_{x,\lambda}\, c^{\mathfrak{tu}}_{y,\lambda}\, c^{\mathfrak{uu}'}_{z,\lambda}\, c^{\mathfrak{u}'\mathfrak{s}}_{w^{-1},\lambda}.$$

By a similar computation, the right-hand side evaluates to

$$\sum_{u\in W}\Big(\sum_{\substack{\lambda\in\Lambda\\ \mathfrak{s},\mathfrak{t},\mathfrak{u}\in M(\lambda)}} f_\lambda^{-1}\, c_{x,\lambda}^{\mathfrak{st}}\, c_{u,\lambda}^{\mathfrak{tu}}\, c_{w^{-1},\lambda}^{\mathfrak{us}}\Big)\Big(\sum_{\substack{\mu\in\Lambda\\ \mathfrak{s}',\mathfrak{t}',\mathfrak{u}'\in M(\mu)}} f_\mu^{-1}\, c_{y,\mu}^{\mathfrak{s}'\mathfrak{t}'}\, c_{z,\mu}^{\mathfrak{t}'\mathfrak{u}'}\, c_{u^{-1},\mu}^{\mathfrak{u}'\mathfrak{s}'}\Big)$$
$$= \sum_{\lambda,\mu\in\Lambda}\sum_{\substack{\mathfrak{s},\mathfrak{t},\mathfrak{u}\in M(\lambda)\\ \mathfrak{s}',\mathfrak{t}',\mathfrak{u}'\in M(\mu)}} f_\lambda^{-1} f_\mu^{-1}\, c_{x,\lambda}^{\mathfrak{st}}\, c_{w^{-1},\lambda}^{\mathfrak{us}}\, c_{y,\mu}^{\mathfrak{s}'\mathfrak{t}'}\, c_{z,\mu}^{\mathfrak{t}'\mathfrak{u}'}\Big(\sum_{u\in W} c_{u,\lambda}^{\mathfrak{tu}}\, c_{u^{-1},\mu}^{\mathfrak{u}'\mathfrak{s}'}\Big)$$
$$= \sum_{\lambda\in\Lambda}\sum_{\mathfrak{s},\mathfrak{t},\mathfrak{t}',\mathfrak{u}\in M(\lambda)} f_\lambda^{-1}\, c_{x,\lambda}^{\mathfrak{st}}\, c_{y,\lambda}^{\mathfrak{tt}'}\, c_{z,\lambda}^{\mathfrak{t}'\mathfrak{u}}\, c_{w^{-1},\lambda}^{\mathfrak{us}}.$$

We see that both sides are equal, hence $\tilde{\mathbf{J}}$ is associative. To show that $1_{\tilde{\mathbf{J}}}$ is the identity element of $\tilde{\mathbf{J}}$, we let $x \in W$ and note that

$$t_x 1_{\tilde{\mathbf{J}}} = \sum_{w\in W} \tilde{n}_w\, t_x t_w = \sum_{y\in W}\Big(\sum_{w\in W} \tilde{n}_w\, \tilde{\gamma}_{x,w,y^{-1}}\Big) t_y = \sum_{y\in W}\Big(\sum_{w\in W} \tilde{n}_w\, \tilde{\gamma}_{y^{-1},x,w}\Big) t_y = t_x$$

by Lemma 1.5.3(a) and (b). A similar argument shows that $1_{\tilde{\mathbf{J}}} t_x = t_x$. Thus, $1_{\tilde{\mathbf{J}}}$ is the identity element of $\tilde{\mathbf{J}}$. □

**Proposition 1.5.6.** *The linear map* $\bar{\tau}\colon \tilde{\mathbf{J}} \to \mathbb{K}$ *defined by* $\bar{\tau}(t_w) = \tilde{n}_{w^{-1}}$ *is a symmetrising trace such that* $\bar{\tau}(t_x t_{y^{-1}}) = \delta_{xy}$ *for all* $x, y \in W$.

*Proof.* Let $x, y \in W$. Then, using Lemma 1.5.3(b), we obtain

$$\bar{\tau}(t_{x^{-1}} t_y) = \sum_{w\in W} \tilde{\gamma}_{x^{-1},y,w^{-1}}\, \bar{\tau}(t_w) = \sum_{w\in W} \tilde{\gamma}_{x^{-1},y,w^{-1}}\, \tilde{n}_{w^{-1}} = \delta_{xy}.$$

This implies that $\bar{\tau}(t_x t_y) = \bar{\tau}(t_y t_x)$ for all $x, y \in W$; hence, $\bar{\tau}$ is a trace function. We also see that $\{t_w \mid w \in W\}$ and $\{t_{w^{-1}} \mid w \in W\}$ form a pair of dual bases; hence, $\bar{\tau}$ is non-degenerate. □

**Proposition 1.5.7.** *For* $\lambda \in \Lambda$, *define a linear map*

$$\bar{\rho}^\lambda : \tilde{\mathbf{J}} \to M_{d_\lambda}(\mathbb{K}), \qquad t_w \mapsto \big(c_{w,\lambda}^{\mathfrak{st}}\big)_{\mathfrak{s},\mathfrak{t}\in M(\lambda)}.$$

*Then* $\bar{\rho}^\lambda$ *is an absolutely irreducible representation of* $\tilde{\mathbf{J}}$, *and all irreducible representations of* $\tilde{\mathbf{J}}$ *(up to equivalence) arise in this way. In particular,* $\tilde{\mathbf{J}}$ *is a split semisimple algebra. (Recall that* $\mathbb{K}$ *is any field containing* $\mathbb{Z}_W$.*) We have*

$$\operatorname{trace}(t_w, \bar{\rho}^\lambda) \equiv \sum_{\mathfrak{s}\in M(\lambda)} c_{w,\lambda}^{\mathfrak{ss}} \equiv \varepsilon^{\mathbf{a}_\lambda} \chi^\lambda(T_w) \mod \mathfrak{m} \qquad \text{for all } w \in W.$$

*Hence, the character of* $\bar{\rho}^\lambda$ *is uniquely determined by* $\chi^\lambda$ *(see also Remark 1.4.4).*

*Proof.* We must show that $\bar{\rho}^\lambda(t_x t_y) = \bar{\rho}^\lambda(t_x)\bar{\rho}^\lambda(t_y)$ for all $x, y \in W$. Now, by the definition of $\tilde{\gamma}_{x,y,z}$, we have

$$\bar{\rho}^\lambda_{\mathfrak{st}}(t_x t_y) = \sum_{z\in W} \tilde{\gamma}_{x,y,z^{-1}}\, c_{z,\lambda}^{\mathfrak{st}} = \sum_{z\in W}\Big(\sum_{\substack{\mu\in\Lambda\\ \mathfrak{s}',\mathfrak{t}',\mathfrak{u}'\in M(\mu)}} f_\mu^{-1}\, c_{x,\mu}^{\mathfrak{s}'\mathfrak{t}'}\, c_{y,\mu}^{\mathfrak{t}'\mathfrak{u}'}\, c_{z^{-1},\mu}^{\mathfrak{u}'\mathfrak{s}'}\Big) c_{z,\lambda}^{\mathfrak{st}}.$$

Using Proposition 1.4.10(a), the right-hand side evaluates to

$$\sum_{\substack{\mu\in\Lambda\\ \mathfrak{s}',\mathfrak{t}',\mathfrak{u}'\in M(\mu)}} f_\mu^{-1}\, c_{x,\mu}^{\mathfrak{s}'\mathfrak{t}'}\, c_{y,\mu}^{\mathfrak{t}'\mathfrak{u}'}\, \delta_{\mathfrak{u}'\mathfrak{t}}\, \delta_{\mathfrak{s}'\mathfrak{s}}\, \delta_{\lambda\mu} f_\lambda = \sum_{\mathfrak{t}'\in M(\lambda)} c_{x,\mu}^{\mathfrak{s}\mathfrak{t}'}\, c_{y,\mu}^{\mathfrak{t}'\mathfrak{t}} = \big(\bar\rho^\lambda(t_x)\bar\rho^\lambda(t_y)\big)_{\mathfrak{s}\mathfrak{t}},$$

as required. To show that $\bar\rho^\lambda$ is absolutely irreducible, we argue as follows. By Proposition 1.5.6, we have a symmetrising trace where $\{t_w \mid w\in W\}$ and $\{t_{w^{-1}} \mid w\in W\}$ form a pair of dual bases. Consequently, the relations in Proposition 1.4.10 can be interpreted as "Schur relations" for the representations $\bar\rho^\lambda$. Thus, we have

$$\sum_{w\in W} \bar\rho^\lambda_{\mathfrak{s}\mathfrak{t}}(t_w)\,\bar\rho^\lambda_{\mathfrak{u}\mathfrak{v}}(t_{w^{-1}}) = \delta_{\mathfrak{s}\mathfrak{v}}\,\delta_{\mathfrak{t}\mathfrak{u}}\, f_\lambda \qquad \text{for all } \mathfrak{s},\mathfrak{t},\mathfrak{u},\mathfrak{v}\in M(\lambda).$$

By a general result about symmetric algebras (see [132, Remark 7.2.3]), the validity of these relations implies that $\bar\rho^\lambda$ is absolutely irreducible. Finally, if $\lambda\neq\mu$ in $\Lambda$, then we also have the relations $\sum_{w\in W}\bar\rho^\lambda_{\mathfrak{s}\mathfrak{t}}(t_w)\,\bar\rho^\mu_{\mathfrak{u}\mathfrak{v}}(t_{w^{-1}})=0$. In particular, this implies that $\bar\rho^\lambda$ and $\bar\rho^\mu$ are not equivalent.

Since $\dim\tilde{\mathbf{J}} = |W| = \sum_{\lambda\in\Lambda} d_\lambda^2$, we conclude that $\tilde{\mathbf{J}}$ is split semisimple, and that $\{\bar\rho^\lambda \mid \lambda\in\Lambda\}$ are the irreducible representations of $\tilde{\mathbf{J}}$ (up to equivalence). □

*Remark 1.5.8.* Let $W = W_1\times\cdots\times W_d$ be the decomposition into irreducible components. As already noted in Remark 1.3.5, we have a tensor product decomposition $\mathbf{H}\cong\mathbf{H}_1\otimes_A\cdots\otimes_A\mathbf{H}_d$, where $\mathbf{H}_i$ is the generic algebra associated with $W_i$ and the restriction of $L$ to $W_i$. Furthermore,

$$\operatorname{Irr}(\mathbf{H}_K) = \{E_\varepsilon^{\lambda_1}\boxtimes\cdots\boxtimes E_\varepsilon^{\lambda_d} \mid \lambda_i\in\Lambda_i\}, \quad \text{where} \quad \operatorname{Irr}_{\mathbb{K}}(W_i) = \{E^{\lambda_i} \mid \lambda_i\in\Lambda_i\}.$$

For each $i$ and $\lambda_i\in\Lambda_i$, let $\rho^{\lambda_i}$ be a balanced representation afforded by $E_\varepsilon^{\lambda_i}$. Then, using the compatibility of the invariants $\mathbf{a}_\lambda$ with the above decomposition (see Remark 1.3.5), we conclude that $\rho^{\lambda_1}\boxtimes\cdots\boxtimes\rho^{\lambda_d}$ (where $\lambda_i\in\Lambda_i$ for all $i$) is a balanced representation for $\mathbf{H}_K$. Consequently, we also have a tensor product decomposition

$$\tilde{\mathbf{J}}\cong\tilde{\mathbf{J}}_1\otimes_{\mathbb{K}}\cdots\otimes_{\mathbb{K}}\tilde{\mathbf{J}}_d,$$

where $\tilde{\mathbf{J}}_i$ is the asymptotic algebra defined with respect to $\mathbf{H}_i$ and the balanced representations $\{\rho^{\lambda_i} \mid \lambda_i\in\Lambda_i\}$. Thus, to describe the structure of $\tilde{\mathbf{J}}$ in general, it will be sufficient to do this assuming that $(W,S)$ is irreducible.

**Definition 1.5.9.** Recall that $\mathbb{Z}_W = \mathbb{Z}[2\cos(2\pi/m_{st}) \mid s,t\in S]$ and $R\subseteq\mathbb{C}$ is a subring such that $\mathbb{Z}_W\subseteq R$. A prime ideal $\mathfrak{p}$ of $R$ is called *L-bad* if $f_\lambda\in\mathfrak{p}$ for some $\lambda\in\Lambda$. We say that $R$ is *L-good* if there are no $L$-bad prime ideals in $R$. Thus, we have

$$R \text{ is } L\text{-good} \qquad\Leftrightarrow\qquad f_\lambda\in R^\times \text{ for all } \lambda\in\Lambda$$

(where $R^\times$ denotes the multiplicative group of invertible elements in $R$). We say that a commutative ring $k$ (with 1) is $L$-good if there is a unital ring homomorphism $R\to k$ where $R$ is $L$-good.

**Table 1.4** Conditions for $R$ to be $L$-good

| Type | Condition on $R$ (where $L(s)>0$ for some $s\in S$) |
|---|---|
| $A_{n-1}$ | no condition |
| $B_n$ | $\begin{cases} 2\in R^\times \text{ if } L(s_0)=rL(s_1) \text{ for some } r\in\{0,1,\ldots,n-1\} \\ n!\in R^\times \text{ if } L(s_1)=0 \\ \text{no condition, otherwise}\end{cases}$ |
| $D_n$ | $2\in R^\times$ |
| $F_4,E_6,E_7$ | $2,3\in R^\times$ |
| $E_8$ | $2,3,5\in R^\times$ |
| $I_2(m)$ $(m\geqslant 5)$ | $\begin{cases} 2\in R^\times \text{ if } m=6 \text{ and } L(s_1)\neq L(s_2) \\ m\in R^\times \text{ and } 2\cos(2\pi/m)\in R, \text{ otherwise}\end{cases}$ |
| $H_3$ | $2,5\in R^\times$ and $2\cos(2\pi/5)=\frac{1}{2}(-1+\sqrt{5})\in R$ |
| $H_4$ | $2,3,5\in R^\times$ and $2\cos(2\pi/5)=\frac{1}{2}(-1+\sqrt{5})\in R$ |

(The generators in $S$ are labelled as in Table 1.1)

**Example 1.5.10.** Assume that $L(s)=0$ for all $s\in S$. Then $f_\lambda=|W|/d_\lambda$ for all $\lambda$; see Remark 1.3.6. Hence, in this case, $R$ is $L$-good if and only if $\mathbb{Z}_W\subseteq R$ and $|W|\in R^\times$. (In fact, in this case, one easily sees that the map $w\mapsto t_w$ defines an algebra isomorphism $\mathbb{K}W\cong\tilde{\mathbf{J}}$.)

Now assume that $W$ is irreducible and that $L(s)>0$ for at least some $s\in S$. Then, the conditions for $R$ to be $L$-good are given by Table 1.4. This information is extracted from the knowledge of $\mathbf{c}_\lambda$ in all cases; see Remark 1.3.6 and the examples following it. To deal with type $I_2(m)$, note the identities

| $m$ even | $m$ odd |
|---|---|
| $\displaystyle\prod_{1\leqslant j\leqslant(m-2)/2}(2-\zeta^{2j}-\zeta^{-2j})=\frac{m^2}{4}$ | $\displaystyle\prod_{1\leqslant j\leqslant(m-1)/2}(2-\zeta^{j}-\zeta^{-j})=m$ |
| $\displaystyle\prod_{1\leqslant j\leqslant(m-2)/2}(2\pm(\zeta^{j}+\zeta^{-j}))=\frac{m}{2}$ | $\displaystyle\prod_{1\leqslant j\leqslant(m-1)/2}(2+\zeta^{j}+\zeta^{-j})=1$ |

These are used as follows. Assume, for example, that $m$ is even and $L(s_1)>L(s_2)>0$. Then, using the formulae in Example 1.3.7 and the above identities, we obtain

$$\prod_{\lambda\in\Lambda}f_\lambda=\prod_{1\leqslant j\leqslant(m-2)/2}\frac{m}{2-\zeta^{2j}-\zeta^{-2j}}=4m^{(m-2)/2-2}.$$

Hence, for $m\neq 6$, we need to require that $m\in R^\times$. On the other hand, for $m=6$, we only need to require that $2\in R^\times$. The other cases are treated similarly.

**Proposition 1.5.11 (Cf. [111, 2.6], [112, 4.10]).** *Let $R\subseteq\mathbb{C}$ be a subring such that $\mathbb{Z}_W\subseteq R$ and $\tilde{\gamma}_{x,y,z}\in R$, $\tilde{n}_w\in R$ for all $x,y,z,w\in W$. (For example, by Lemma 1.5.2, any $L$-good subring $R\subseteq\mathbb{C}$ has these properties.) Then, for each $\lambda\in\Lambda$, the balanced representation $\rho^\lambda$ can be chosen such that the following two conditions hold.*

*(a) We have $\bar{\rho}^\lambda(t_w)=(c_{w,\lambda}^{\mathfrak{st}})\in M_{d_\lambda}(R)$ for all $w\in W$.*

*(b) There exists a symmetric matrix $B^\lambda \in M_{d_\lambda}(R)$ such that*

$$\det(B^\lambda) \neq 0 \qquad \textit{and} \qquad B^\lambda \bar{\rho}^\lambda(t_{w^{-1}}) = \bar{\rho}^\lambda(t_w)^{\text{tr}} B^\lambda \quad \textit{for all } w \in W;$$

*furthermore, if $\mathfrak{p}$ is a prime ideal in $R$ such that $\det(B^\lambda) \in \mathfrak{p}$, then $\mathfrak{p}$ is $L$-bad. In particular, if $R$ is $L$-good, then $\det(B^\lambda) \in R^\times$.*

*Proof.* By Remark 1.5.8, we can assume that $W$ is irreducible; furthermore, we may assume that $R$ is the subring of $\mathbb{C}$ generated by $\mathbb{Z}_W$ and all constants $\tilde{\gamma}_{x,y,z}$, $\tilde{n}_w$ for $x, y, z, w \in W$. In particular, $R \subseteq \mathbb{R}$. Recall that: if $W$ is a Weyl group, then $\mathbb{Z}_W = \mathbb{Z}$; if $W$ is of type $H_3$ or $H_4$, then $\mathbb{Z}_W = \mathbb{Z}[\frac{1}{2}(-1+\sqrt{5})]$; if $W$ is of type $I_2(m)$, then $\mathbb{Z}_W = \mathbb{Z}[\zeta + \zeta^{-1}]$, where $\zeta \in \mathbb{C}$ is a root of unity of order $m$. Thus, except possibly in type $I_2(m)$, the ring $\mathbb{Z}_W$ is a principal ideal domain and, hence, so is $R$.

So let us first deal separately with $W$ of type $I_2(m)$. We can assume that $L(s_1) \geqslant L(s_2) \geqslant 0$. Note that (a) and (b) are clear for one-dimensional representations, where we can just take $B^\lambda = (1)$. Now consider a two-dimensional representation $\sigma_j^\varepsilon$. In Example 1.4.6, we have seen that $\sigma_j^\varepsilon$ is balanced and that the corresponding leading matrix coefficients lie in $\mathbb{Z}_W$. Hence, (a) holds. Furthermore, let $\Omega_j$ be as in Example 1.4.6. Then all entries of $\Omega_j$ lie in $\mathbb{Z}_W[\Gamma]$. Let $B_j$ be the matrix obtained by taking the constant terms of the entries of $\Omega_j$. Then the invariance property in (b) holds. Furthermore, if $L(s_1) > L(s_2) \geqslant 0$, then $\det(B_j) = 1$. If $L(s_1) = L(s_2) \geqslant 0$, then $\det(B_j)$ is non-zero and divides $m$. So the identities mentioned in Example 1.5.10 show that the additional requirements in (b) hold.

We can now assume that $W$ is not of type $I_2(m)$. Then, as remarked above, $R$ is a principal ideal domain. Let $\tilde{\mathbf{J}}_R$ be the $R$-span of $\{t_w \mid w \in W\}$. Since the structure constants of $\tilde{\mathbf{J}}$ lie in $R$, we see that $\tilde{\mathbf{J}}_R$ is an $R$-subalgebra of $\tilde{\mathbf{J}}$ and $\tilde{\mathbf{J}} = \mathbb{K} \otimes_R \tilde{\mathbf{J}}_R$. Then a standard argument (e.g. see [132, Remark 7.3.7]) implies that $\bar{\rho}^\lambda$ is equivalent to a representation which is realized over $R$; that is, there exists an invertible matrix $P_\lambda \in M_{d_\lambda}(\mathbb{K})$ such that

$$P_\lambda^{-1} \bar{\rho}^\lambda(t_w) P_\lambda \in M_{d_\lambda}(R) \qquad \text{for all } w \in W.$$

Now use $P_\lambda$ to transform the original balanced representation $\rho^\lambda$ to a representation

$$\tilde{\rho}^\lambda \colon \mathbf{H}_K \to M_{d_\lambda}(K), \qquad T_w \mapsto P_\lambda^{-1} \rho^\lambda(T_w) P_\lambda.$$

Since all entries of $P_\lambda$ lie in $\mathbb{K}$, it is clear that $\tilde{\rho}^\lambda$ is also balanced. The corresponding leading matrix coefficients are given by the entries of the matrices $P_\lambda^{-1} \bar{\rho}^\lambda(t_w) P_\lambda$ and, hence, lie in $R$. Thus, replacing $\rho^\lambda$ by $\tilde{\rho}^\lambda$, we see that (a) holds. To prove (b), let

$$B_1^\lambda = \sum_{y \in W} \bar{\rho}^\lambda(t_y)^{\text{tr}} . \bar{\rho}^\lambda(t_y) \in M_{d_\lambda}(R).$$

This matrix clearly is symmetric. Let $0 \neq e = (e_1, \ldots, e_{d_\lambda}) \in R^d$. Since $\bar{\rho}^\lambda$ is irreducible, there exists some $y \in W$ such that $\bar{\rho}^\lambda(t_y)^{\text{tr}} \neq 0$ and, hence, the standard scalar product of this vector with itself will be strictly positive. (Recall that $R \subseteq \mathbb{R}$.)

Consequently, we have $eB_1^\lambda e^{\mathrm{tr}} > 0$. Thus, $B_1^\lambda$ is positive-definite and, in particular, $\det(B_1^\lambda) \neq 0$. For any $x \in W$, we have

$$B_1^\lambda.\bar{\rho}^\lambda(t_{x^{-1}}) = \sum_{y\in W} \bar{\rho}^\lambda(t_y)^{\mathrm{tr}}.\bar{\rho}^\lambda(t_y t_{x^{-1}}) = \sum_{y,z\in W} \tilde{\gamma}_{y,x^{-1},z^{-1}} \bar{\rho}^\lambda(t_y)^{\mathrm{tr}}.\bar{\rho}^\lambda(t_z).$$

Now $\tilde{\gamma}_{y,x^{-1},z^{-1}} = \tilde{\gamma}_{x^{-1},z^{-1},y} = \tilde{\gamma}_{z,x,y^{-1}}$ by Lemma 1.5.3. Hence, the right-hand side of the above identity equals

$$\sum_{y,z\in W} \tilde{\gamma}_{z,x,y^{-1}} \bar{\rho}^\lambda(t_y)^{\mathrm{tr}}.\bar{\rho}^\lambda(t_z) = \sum_{z\in W} \bar{\rho}^\lambda(t_z t_x)^{\mathrm{tr}}.\bar{\rho}^\lambda(t_z) = \bar{\rho}^\lambda(t_x)^{\mathrm{tr}}.B_1^\lambda.$$

Let $0 \neq n \in R$ be a greatest common divisor of all non-zero coefficients of $B_1^\lambda$ and $B^\lambda = n^{-1}B_1^\lambda \in M_d(R)$. Then $B^\lambda$ is a non-singular symmetric matrix such that the invariance property in (b) holds. It remains to prove the statement about $\det(B^\lambda)$. This follows by a standard argument on symmetric algebras. Let $\mathfrak{p}$ be a prime ideal of $R$ such that $f_\lambda \notin \mathfrak{p}$. Note that $B^\lambda \not\equiv 0 \bmod \mathfrak{p}$, since 1 is a greatest common divisior of all non-zero coefficients of $B^\lambda$. Let $k$ be the field of fractions of $R/\mathfrak{p}$ and denote by $B_k^\lambda$ the matrix obtained by reducing all coefficents modulo $\mathfrak{p}$. By reduction modulo $\mathfrak{p}$, we also obtain a $k$-algebra $\tilde{\mathbf{J}}_k = k \otimes_R \tilde{\mathbf{J}}_R$ and a corresponding matrix representation $\bar{\rho}_k^\lambda : \tilde{\mathbf{J}}_k \to M_{d_\lambda}(k)$. Hence, we have

$$B_k^\lambda \neq 0 \qquad \text{and} \qquad B_k^\lambda.\bar{\rho}_k^\lambda(t_w) = \bar{\rho}_k^\lambda(t_{w^{-1}})^{\mathrm{tr}}.B_k^\lambda \quad \text{for all } w \in W.$$

Now let us regard $k^{d_\lambda}$ as a $\tilde{\mathbf{J}}_k$-module via $\bar{\rho}_k^\lambda$. Then the above relations imply that the nullspace of $B_k^\lambda$ is a proper $\tilde{\mathbf{J}}_k$-invariant subspace. Hence, if we can show that $\bar{\rho}_k^\lambda$ is irreducible, then that nullspace must be zero and so $B_k^\lambda$ is invertible; that is, $\det(B^\lambda) \notin \mathfrak{p}$, as required.

To show that $\bar{\rho}_k^\lambda$ is irreducible, we argue as follows. By Proposition 1.5.6, we know that $\tilde{\mathbf{J}}$ is symmetric and so we have the Schur relations for the matrix coefficients of $\bar{\rho}^\lambda$. Reducing these relations modulo $\mathfrak{p}$, we obtain

$$\sum_{w\in W} \bar{\rho}_{k,\mathfrak{st}}^\lambda(t_w)\bar{\rho}_{k,\mathfrak{uv}}^\lambda(t_{w^{-1}}) = \begin{cases} f_\lambda \bmod \mathfrak{p} & \text{if } \mathfrak{s}=\mathfrak{v}, \mathfrak{t}=\mathfrak{u}, \\ 0 & \text{otherwise.} \end{cases}$$

Since $f_\lambda \not\equiv 0 \bmod \mathfrak{p}$, one easily deduces from this that $\bar{\rho}_k^\lambda$ is (absolutely) irreducible; see [132, Remark 7.2.3]. □

**Conjecture 1.5.12.** *(a) We have $\tilde{\gamma}_{x,y,z} \in \mathbb{Z}$ for all $x,y,z \in W$ and $\tilde{n}_w \in \mathbb{Z}$ for all $w \in W$.*

*(b) The elements $\{\tilde{n}_d t_d \mid d \in \mathscr{D}\}$ are orthogonal idempotents in $\tilde{\mathbf{J}}$. (Recall that $1_{\tilde{\mathbf{J}}} = \sum_{d\in\mathscr{D}} \tilde{n}_d t_d$.)*

We will see in Section 2.5 that this conjecture holds under certain assumptions on $W, L$ which are satisfied, for example, in the equal-parameter case. In Theorem 1.7.10, we will see that it also holds for certain choices of unequal parameters in type $B_n$. It would be very interesting and useful to find a general proof of this conjecture which works without any assumptions on $W, L$.

## 1.6 Introducing Cells

Lusztig's work [214, §8] (see also [220, Chap. 4]) on representations of finite groups of Lie type showed that there is a natural partition of $\mathrm{Irr}_{\mathbb{K}}(W)$ into "cells" or "families", where $W$ is a finite Weyl group and $L$ is the length function on $W$. In a completely different context, "cells" appeared in the work of Joseph on primitive ideals in the enveloping algebra of a complex semisimple Lie algebra; see the survey [187].

Then Kazhdan and Lusztig [195], [219] (see also Chapter 2) constructed a new basis $\{C_w \mid w \in W\}$ of $\mathbf{H}$ (for general $W, L$) and used this basis to define partitions of $W$ into left, right, and two-sided "cells" (depending on $L$); furthermore, each "cell" in this sense is shown to carry a representation of $\mathbf{H}$, with a "canonical" basis indexed by the elements in the cell. In 1.6.1 and 1.6.2 we shall give axiomatic versions of these constructions and then apply them to the algebra $\tilde{\mathbf{J}}$ with its basis $\{t_w \mid w \in W\}$. Following [112], this yields a first approach to the theory of "cells".

**1.6.1.** Let $k$ be a commutative ring (with 1) and $\underline{H}$ any associative algebra (with identity) which is free over $k$ with a basis indexed by the elements of our finite Coxeter group $W$. Let us denote this basis by $\{c_w \mid w \in W\}$ and write

$$c_x c_y = \sum_{z \in W} \underline{h}_{x,y,z}\, c_z, \qquad \text{where} \qquad \underline{h}_{x,y,z} \in k.$$

Also assume that the $k$-linear map $\underline{H} \to \underline{H}$ defined by $c_w \mapsto c_{w^{-1}}$ is an anti-automorphism of $\underline{H}$. By analogy with Kazhdan and Lusztig [195], [219], we introduce the following relations on $W$.

Let $y, z \in W$. We write $z \leftarrow_{\mathscr{L}} y$ if there exists some $x \in W$ such that $\underline{h}_{x,y,z} \neq 0$; that is, $c_z$ occurs with non-zero coefficient in $c_x c_y$ (when expressed in the $c$-basis). Let $\leqslant_{\mathscr{L}}$ be the *pre-order*[1] relation on $W$ generated by $\leftarrow_{\mathscr{L}}$; that is, we have $z \leqslant_{\mathscr{L}} y$ if there exist elements $z = y_0, y_1, \ldots, y_m = y$ in $W$ such that $y_{i-1} \leftarrow_{\mathscr{L}} y_i$ for $1 \leqslant i \leqslant m$. Let $\sim_{\mathscr{L}}$ denote the associated equivalence relation; the corresponding equivalence classes are called the *left cells* of $W$.

Similarly, we can define a pre-order $\leqslant_{\mathscr{R}}$ by considering multiplication by $c_x$ on the right. The equivalence relation associated with $\leqslant_{\mathscr{R}}$ will be denoted by $\sim_{\mathscr{R}}$ and the corresponding equivalence classes are called the *right cells* of $W$. We have

$$z \leqslant_{\mathscr{R}} y \quad \Leftrightarrow \quad z^{-1} \leqslant_{\mathscr{L}} y^{-1}.$$

This follows by using the anti-homomorphism $c_w \mapsto c_{w^{-1}}$. Thus, any statement concerning the left pre-order relation $\leqslant_{\mathscr{L}}$ has an equivalent version for the right pre-order relation $\leqslant_{\mathscr{R}}$, via the map $c_w \mapsto c_{w^{-1}}$.

Finally, we define a pre-order $\leqslant_{\mathscr{LR}}$ by the condition that $z \leqslant_{\mathscr{LR}} y$ if there exists a sequence $z = y_0, y_1, \ldots, y_m = y$ such that, for each $i \in \{1, \ldots, m\}$, we have $y_{i-1} \leqslant_{\mathscr{L}} y_i$

[1] A pre-order on a set is, by definition, just a transitive reflexive relation (but we do not necessarily have anti-symmetry). Thus, a partial order is a pre-order which is also anti-symmetric. If $\leqslant$ is a pre-order, we have an associated equivalence relation given by $x \sim y$ if and only if $x \leqslant y$ and $y \leqslant x$. Consequently, $\leqslant$ induces a partial order on the equivalence classes under $\sim$.

or $y_{i-1} \leqslant_{\mathscr{R}} y_i$. The equivalence relation associated with $\leqslant_{\mathscr{L}\mathscr{R}}$ will be denoted by $\sim_{\mathscr{L}\mathscr{R}}$ and the corresponding equivalence classes are called the *two-sided cells* of $W$. Thus, we have defined partitions

$$W = \bigsqcup\{\text{left cells}\} = \bigsqcup\{\text{right cells}\} = \bigsqcup\{\text{two-sided cells}\}.$$

Obviously, these notions heavily depend on $\underline{H}$ and on the choice of the basis $\{c_w\}$.

**1.6.2.** Let $\mathfrak{C}$ be a left cell of $W$ (with respect to $\underline{H}$). Then $\mathfrak{C}$ gives rise to a representation of $\underline{H}$. Following Kazhdan and Lusztig [195], [219], this is constructed as follows. Let $y \in W$ and define the following two $k$-submodules of $\underline{H}$:

$$\begin{aligned}
\mathfrak{I}_y^{\mathscr{L}} &:= \langle c_x \mid x \in W \text{ and } x \leqslant_{\mathscr{L}} y \rangle_k,\\
\hat{\mathfrak{I}}_y^{\mathscr{L}} &:= \langle c_x \mid x \in W \text{ and } x \leqslant_{\mathscr{L}} y \text{ but } x \not\sim_{\mathscr{L}} y \rangle_k.
\end{aligned}$$

By the definition of the relations $\leqslant_{\mathscr{L}}$ and $\sim_{\mathscr{L}}$, it is clear that $\mathfrak{I}_y^{\mathscr{L}}$ and $\hat{\mathfrak{I}}_y^{\mathscr{L}}$ are left ideals of $\underline{H}$; furthermore, we have $\hat{\mathfrak{I}}_y^{\mathscr{L}} \subseteq \mathfrak{I}_y^{\mathscr{L}}$. Hence, the quotient $\mathfrak{I}_y^{\mathscr{L}}/\hat{\mathfrak{I}}_y^{\mathscr{L}}$ is a left $\underline{H}$-module. Note that

$$\mathfrak{I}_y^{\mathscr{L}} = \mathfrak{I}_{y'}^{\mathscr{L}} \quad \text{and} \quad \hat{\mathfrak{I}}_y^{\mathscr{L}} = \hat{\mathfrak{I}}_{y'}^{\mathscr{L}} \qquad \text{if } y \sim_{\mathscr{L}} y'.$$

Hence, if $\mathfrak{C}$ is the left cell containing $y$, then the above quotient only depends on $\mathfrak{C}$ and we shall write

$$[\mathfrak{C}]_k := \mathfrak{I}_y^{\mathscr{L}}/\hat{\mathfrak{I}}_y^{\mathscr{L}} \quad \text{and} \quad e_x := \text{residue class of } c_x \text{ in } [\mathfrak{C}]_k \text{ (where } x \in \mathfrak{C}).$$

Thus, $[\mathfrak{C}]_k$ is a left $\underline{H}$-module with a "canonical" basis $\{e_x \mid x \in \mathfrak{C}\}$. Explicitly, the action of $c_w$ is given by the formula

$$c_w.e_x := \sum_{y \in \mathfrak{C}} \underline{h}_{w,x,y}\, e_y, \qquad \text{where } w \in W \text{ and } x \in \mathfrak{C}.$$

Similarly, we define right ideals $\mathfrak{I}_y^{\mathscr{R}}$, $\hat{\mathfrak{I}}_y^{\mathscr{R}}$ giving rise to right $\underline{H}$-modules associated with the various right cells of $W$. Finally, we can define two-sided ideals $\mathfrak{I}_y^{\mathscr{L}\mathscr{R}}$, $\hat{\mathfrak{I}}_y^{\mathscr{L}\mathscr{R}}$ giving rise to $(\underline{H},\underline{H})$-bimodules associated with the various two-sided cells of $W$.

Finally, assume that $k$ is a field and $\underline{H}$ is semisimple. Then, since the left cells form a partition of $W$, we obtain a corresponding direct sum decomposition of $\underline{H}$:

$$\underline{H} \cong \bigoplus_{\mathfrak{C} \text{ left cell}} [\mathfrak{C}]_k \qquad \text{(isomorphism of left } \underline{H}\text{-modules).}$$

We obtain analogous decompositions with respect to right cells or two-sided cells.

**Example 1.6.3.** Let $\underline{H} = \mathbb{Z}W$, the group algebra of $W$ where $c_w = w$ for all $w \in W$. Then each basis element $w$ is invertible in $\mathbb{Z}W$ and, hence, there is only one left cell, only one right cell and only one two-sided cell, namely the whole of $W$. So these notions are not very interesting in this case.

The same thing happens when we take $\underline{H} = \mathbf{H}$ where $c_w = T_w$ for all $w \in W$. Just note that each basis element $T_w$ is invertible in $\mathbf{H}$.

In Chapter 2 we shall consider again the case where $\underline{H} = \mathbf{H}$, but work with the Kazhdan–Lusztig basis of $\mathbf{H}$. For the remainder of this section, let us assume that

$$\boxed{\underline{H} = \tilde{\mathbf{J}} \qquad \text{and} \qquad c_w = t_w \text{ for all } w \in W.}$$

Note that, in this case, we have $\underline{h}_{x,y,z} = \tilde{\gamma}_{x,y,z^{-1}}$ for all $x,y,z \in W$. Also note that, by Lemma 1.5.3(c), the map $t_w \mapsto t_{w^{-1}}$ is an anti-automorphism of $\tilde{\mathbf{J}}$.

**Definition 1.6.4.** The left, right or two-sided cells defined by taking $\underline{H} = \tilde{\mathbf{J}}$ (with basis $\{t_w \mid w \in W\}$) are called the left, right or two-sided *$\tilde{\mathbf{J}}$-cells* of $W$ respectively.

**Lemma 1.6.5.** *Let $x,y,z \in W$ be such that $\tilde{\gamma}_{x,y,z} \neq 0$. Then $x \sim_{\mathscr{L}} y^{-1}$, $y \sim_{\mathscr{L}} z^{-1}$, $z \sim_{\mathscr{L}} x^{-1}$.*

*Proof.* By the definition of $\leqslant_{\mathscr{L}}$, we have the implication:

(1) If $\tilde{\gamma}_{x,y,z} \neq 0$, then $z^{-1} \leqslant_{\mathscr{L}} y$.

Now, by Lemma 1.5.3(a), we have $\tilde{\gamma}_{x,y,z} = \tilde{\gamma}_{z,x,y} = \tilde{\gamma}_{y,z,x}$. Hence, (1) implies:

(2) If $\tilde{\gamma}_{x,y,z} \neq 0$, then $z^{-1} \leqslant_{\mathscr{L}} y$, $y^{-1} \leqslant_{\mathscr{L}} x$ and $x^{-1} \leqslant_{\mathscr{L}} z$.

By Lemma 1.5.3(c), we also have $\tilde{\gamma}_{x,y,z} = \tilde{\gamma}_{y^{-1},x^{-1},z^{-1}}$. Hence, (1), (2) imply:

(3) If $\tilde{\gamma}_{x,y,z} \neq 0$, then $z \leqslant_{\mathscr{L}} x^{-1}$, $x \leqslant_{\mathscr{L}} y^{-1}$ and $y \leqslant_{\mathscr{L}} z^{-1}$.

Combining now (2) and (3), we deduce that $x \sim_{\mathscr{L}} y^{-1}$, $y \sim_{\mathscr{L}} z^{-1}$, $z \sim_{\mathscr{L}} x^{-1}$. □

**Lemma 1.6.6.** *If $x,y \in W$ are such that $x \leqslant_{\mathscr{L}} y$, then $x \sim_{\mathscr{L}} y$. (Analogous statements hold for $\leqslant_{\mathscr{R}}$ and $\leqslant_{\mathscr{LR}}$.) Furthermore, we have $w \sim_{\mathscr{LR}} w^{-1}$ for any $w \in W$.*

*Proof.* It is sufficient to prove the first statement for an elementary step in the definition of $\leqslant_{\mathscr{L}}$; that is, we can assume that $x \leftarrow_{\mathscr{L}} y$. This means that $\tilde{\gamma}_{w,y,x^{-1}} \neq 0$ for some $w \in W$ and so Lemma 1.6.5 shows that $x \sim_{\mathscr{L}} y$, as required. Now fix $w \in W$. By Lemma 1.5.3, we have $\sum_{x \in W} \tilde{\gamma}_{w^{-1},w,x} \tilde{n}_x = 1$ and so $\tilde{\gamma}_{w^{-1},w,x} \neq 0$ for some $x \in W$. This implies that $x^{-1} \leqslant_{\mathscr{L}} w$ and so $x^{-1} \sim_{\mathscr{L}} w$. On the other hand, by Lemma 1.5.3(c), we have $\tilde{\gamma}_{w^{-1},w,x^{-1}} = \tilde{\gamma}_{w^{-1},w,x} \neq 0$. So we can also deduce that $x \leqslant_{\mathscr{L}} w$ and $x \sim_{\mathscr{L}} w$. Thus, we have shown that $w \sim_{\mathscr{L}} x \sim_{\mathscr{R}} w^{-1}$; in particular, $w \sim_{\mathscr{LR}} w^{-1}$. □

**Corollary 1.6.7.** *Let $x,y,z \in W$ be such that $\tilde{\gamma}_{x,y,z} \neq 0$. Then the elements $x^{\pm 1}$, $y^{\pm 1}$, $z^{\pm 1}$ all lie in the same two-sided $\tilde{\mathbf{J}}$-cell.*

*Proof.* This is an immediate consequence of Lemmas 1.6.5 and 1.6.6. □

*Remark 1.6.8.* Let $x,y \in W$. By Lemma 1.6.6, we have $x \sim_{\mathscr{LR}} y$ if and only if there is a sequence of elements $x = x_0, x_1, \ldots, x_m = y$ in $W$ such that, for each $i \in \{1,\ldots,m\}$, we have $x_{i-1} \sim_{\mathscr{L}} x_i$ or $x_{i-1} \sim_{\mathscr{R}} x_i$. It follows that the two-sided $\tilde{\mathbf{J}}$-cells of $W$ are the smallest subsets of $W$ which are at the same time unions of left cells and unions of right cells. Since, furthermore, $x \sim_{\mathscr{R}} y$ if and only if $x^{-1} \sim_{\mathscr{L}} y^{-1}$, we conclude that the whole cell structure of $W$ is determined by the knowledge of the left $\tilde{\mathbf{J}}$-cells.

Indeed, by Lemma 1.5.3(b), we have $\sum_{d'\in\tilde{\mathscr{D}}} \tilde{\gamma}_{x^{-1},y,d'}\tilde{n}_{d'} = \delta_{xy}$. Now, if $\tilde{\gamma}_{x^{-1},y,d'} \neq 0$, then $x \sim_{\mathscr{L}} d'$ by Lemma 1.6.5. So, by Conjecture 1.6.18, the sum reduces to $\tilde{\gamma}_{x^{-1},y,d}\tilde{n}_d$, where $d \in \tilde{\mathscr{D}}$ is the unique element such that $x \sim_{\mathscr{L}} d$. Thus, we have $\tilde{\gamma}_{x^{-1},y,d}\,\tilde{n}_d = \delta_{xy}$. Now let $x = y$. Then $\tilde{\gamma}_{x^{-1},x,d} \neq 0$ and so we also have $x \sim_{\mathscr{L}} d^{-1}$ by Lemma 1.6.5. On the other hand, by Lemma 1.5.3(c), we already know that $d^{-1} \in \tilde{\mathscr{D}}$. Hence, Conjecture 1.6.18 implies that $d = d^{-1}$. Thus, (a) is proved.

Now let $d, d' \in \tilde{\mathscr{D}}$. Then $t_d t_{d'} = \sum_{z\in W} \tilde{\gamma}_{d,d',z^{-1}}\, t_z = \sum_{z\in W} \tilde{\gamma}_{z^{-1},d,d'}\, t_z$, where the second equality holds by Lemma 1.5.3(a). Hence, (a) yields that $t_d^2 = \tilde{n}_d^{-1} t_d$ and $t_d t_{d'} = 0$ if $d \neq d'$. Thus, the elements $\{\tilde{n}_d t_d \mid d \in \tilde{\mathscr{D}}\}$ are orthogonal idempotents, as claimed.

## 1.7 Examples of Cells

We consider a number of examples where the two-sided $\tilde{\mathbf{J}}$-cells and the corresponding partition of $\mathrm{Irr}_{\mathbb{K}}(W)$ (see Definition 1.6.12) can be explicitly determined. A considerable reduction of the required computations is achieved by the following general result, a first version of which appeared in [103, Theorem 4.10].

**Proposition 1.7.1.** *Let $\lambda \in \Lambda$ and assume that*

$$f_\lambda = 1 \qquad \text{and} \qquad c^{\mathfrak{st}}_{w^{-1},\lambda} = c^{\mathfrak{ts}}_{w,\lambda} \in \mathbb{Z} \quad \text{for all } w \in W \text{ and } \mathfrak{s},\mathfrak{t} \in M(\lambda).$$

*Then the following hold.*

*(a) For each pair $(\mathfrak{s},\mathfrak{t}) \in M(\lambda) \times M(\lambda)$, there exists a unique $w \in W$ such that $c^{\mathfrak{st}}_{w,\lambda} \neq 0$; we have in fact $c^{\mathfrak{st}}_{w,\lambda} = \pm 1$. Let us denote this element by $w_\lambda(\mathfrak{s},\mathfrak{t})$.*
*(b) The map $M(\lambda) \times M(\lambda) \to W$, $(\mathfrak{s},\mathfrak{t}) \mapsto w_\lambda(\mathfrak{s},\mathfrak{t})$, is injective.*
*(c) The two-sided $\tilde{\mathbf{J}}$-cell associated to $E^\lambda$ (see Proposition 1.6.11) is given by $\mathscr{F}_\lambda = \{w_\lambda(\mathfrak{s},\mathfrak{t}) \mid \mathfrak{s},\mathfrak{t} \in M(\lambda)\}$ and we have $\mathrm{Irr}_{\mathbb{K}}(W \mid \mathscr{F}_\lambda) = \{E^\lambda\}$.*
*(d) For a fixed $\mathfrak{t} \in M(\lambda)$, the set $\{w_\lambda(\mathfrak{s},\mathfrak{t}) \mid \mathfrak{s} \in M(\lambda)\}$ is a left $\tilde{\mathbf{J}}$-cell and, for a fixed $\mathfrak{s} \in M(\lambda)$, the set $\{w_\lambda(\mathfrak{s},\mathfrak{t}) \mid \mathfrak{t} \in M(\lambda)\}$ is a right $\tilde{\mathbf{J}}$-cell.*

*Proof.* (a) Let $\mathfrak{s},\mathfrak{t} \in M(\lambda)$. Using the assumptions on $f_\lambda$ and $c^{\mathfrak{st}}_{w,\lambda}$, the "Schur relations" in Proposition 1.4.10(a) show that

$$1 = f_\lambda = \sum_{w\in W} c^{\mathfrak{st}}_{w,\lambda} c^{\mathfrak{ts}}_{w^{-1},\lambda} = \sum_{w\in W} \left(c^{\mathfrak{st}}_{w,\lambda}\right)^2.$$

Since $c^{\mathfrak{st}}_{w,\lambda} \in \mathbb{Z}$, there will be a unique $w \in W$ such that $c^{\mathfrak{st}}_{w,\lambda} \neq 0$; in fact, $c^{\mathfrak{st}}_{w,\lambda} = \pm 1$.

(b) Let $\mathfrak{s},\mathfrak{t},\mathfrak{s}',\mathfrak{t}' \in M(\lambda)$ be such that $(\mathfrak{s},\mathfrak{t}) \neq (\mathfrak{s}',\mathfrak{t}')$. Assume, if possible, that $w_\lambda(\mathfrak{s},\mathfrak{t}) = w_\lambda(\mathfrak{s}',\mathfrak{t}')$. Then (a) and the relations in Proposition 1.4.10(a) yield that

$$0 = \sum_{w\in W} c^{\mathfrak{st}}_{w,\lambda} c^{\mathfrak{t}'\mathfrak{s}'}_{w^{-1},\lambda} = \sum_{w\in W} c^{\mathfrak{st}}_{w,\lambda} c^{\mathfrak{s}'\mathfrak{t}'}_{w,\lambda} = \pm 1,$$

which is a contradiction.

(c) By (a), we have $E^\lambda \leftrightsquigarrow_L w_\lambda(\mathfrak{s},\mathfrak{t})$ for all $\mathfrak{s},\mathfrak{t} \in M(\lambda)$ and so

$$\mathscr{F}'_\lambda := \{w_\lambda(\mathfrak{s},\mathfrak{t}) \mid \mathfrak{s},\mathfrak{t} \in M(\lambda)\} \subseteq \mathscr{F}_\lambda.$$

Now let us fix $\mathfrak{s},\mathfrak{t} \in M(\lambda)$ and denote $w = w_\lambda(\mathfrak{s},\mathfrak{t})$. Consider the relations in Proposition 1.4.10(b):

$$\sum_{\mu\in\Lambda} \sum_{\mathfrak{u},\mathfrak{v}\in M(\mu)} f_\mu^{-1}\, c_{w,\mu}^{\mathfrak{u}\mathfrak{v}}\, c_{w^{-1},\mu}^{\mathfrak{v}\mathfrak{u}} = 1.$$

The contribution to this sum coming from the term where $\mu$ equals $\lambda$ is given by

$$\sum_{\mathfrak{u},\mathfrak{v}\in M(\lambda)} f_\lambda^{-1}\, c_{w,\lambda}^{\mathfrak{u}\mathfrak{v}}\, c_{w^{-1},\lambda}^{\mathfrak{v}\mathfrak{u}} = \sum_{\mathfrak{u},\mathfrak{v}\in M(\lambda)} \left(c_{w,\lambda}^{\mathfrak{u}\mathfrak{v}}\right)^2 = \left(c_{w,\lambda}^{\mathfrak{s}\mathfrak{t}}\right)^2 = 1,$$

where we used (a) and the injectivity of the map in (b). Hence, we deduce that

$$\sum_{\mu\in\Lambda,\mu\neq\lambda} \sum_{\mathfrak{u},\mathfrak{v}\in M(\mu)} f_\mu^{-1}\, c_{w,\mu}^{\mathfrak{u}\mathfrak{v}}\, c_{w^{-1},\mu}^{\mathfrak{v}\mathfrak{u}} = 0.$$

Now, we can assume that the balanced representations $\rho^\mu$ satisfy the additional conditions in Theorem 1.4.8. So, for each $\mu \in \Lambda$ and $\mathfrak{u},\mathfrak{v} \in M(\mu)$, we have

$$f_\mu^{-1}\, c_{w,\mu}^{\mathfrak{u}\mathfrak{v}}\, c_{w^{-1},\mu}^{\mathfrak{v}\mathfrak{u}} = f_\mu^{-1} b_\mathfrak{v}\, b_\mathfrak{u}^{-1} \left(c_{w,\mu}^{\mathfrak{u}\mathfrak{v}}\right)^2$$

and this is a non-negative real number. Hence, there are no cancellations in the above sum. So, all terms in that sum must be zero; that is, we have

$$(*) \qquad c_{w,\mu}^{\mathfrak{u}\mathfrak{v}} = 0 \qquad \text{for all } \mathfrak{u},\mathfrak{v} \in M(\mu) \text{ where } \mu \neq \lambda.$$

This already shows that $w \notin \mathscr{F}_\mu$ for any $\mu \neq \lambda$. So all that remains is to show that $\mathscr{F}_\lambda \subseteq \mathscr{F}'_\lambda$. For this purpose, let $x, y \in W$ and consider the structure constant $\tilde{\gamma}_{x,y,w}$. In the defining formula, all terms corresponding to $\mu \neq \lambda$ will be zero, thanks to $(*)$. In combination with (a) and Lemma 1.5.3(a) this yields that

$$\tilde{\gamma}_{w,x,y} = \tilde{\gamma}_{y,w,x} = \tilde{\gamma}_{x,y,w} = \sum_{\mathfrak{s},\mathfrak{t}\in M(\lambda)} f_\lambda^{-1}\, c_{x,\lambda}^{\mathfrak{s}\mathfrak{t}}\, c_{y,\lambda}^{\mathfrak{t}\mathfrak{u}}\, c_{w,\lambda}^{\mathfrak{u}\mathfrak{s}} = 0 \quad \text{unless} \quad x, y \in \mathscr{F}'_\lambda.$$

By the definition of $\sim_{\mathscr{LR}}$, this implies that $\mathscr{F}_\lambda \subseteq \mathscr{F}'_\lambda$, as desired.

(d) Let $\mathfrak{t} \in M(\lambda)$ be fixed and $y = w_\lambda(\mathfrak{t},\mathfrak{t})$. Let $\mathfrak{s} \in M(\lambda)$ and $x = w_\lambda(\mathfrak{s},\mathfrak{t})$. We claim that $x \sim_{\mathscr{L}} y$. By Lemma 1.6.6, it is enough to show that $x \leqslant_{\mathscr{L}} y$ and, for this purpose, it is enough to show that $\tilde{\gamma}_{x,y,x^{-1}} \neq 0$. Now, we have

$$\tilde{\gamma}_{x,y,x^{-1}} = \sum_{\mu\in\Lambda} \sum_{\mathfrak{s}',\mathfrak{t}',\mathfrak{u}'\in M(\mu)} f_\mu^{-1}\, c_{x,\mu}^{\mathfrak{s}'\mathfrak{t}'}\, c_{y,\mu}^{\mathfrak{t}'\mathfrak{u}'}\, c_{x^{-1},\mu}^{\mathfrak{u}'\mathfrak{s}'} = \sum_{\mathfrak{s}',\mathfrak{t}',\mathfrak{u}'\in M(\lambda)} c_{x,\lambda}^{\mathfrak{s}'\mathfrak{t}'}\, c_{y,\lambda}^{\mathfrak{t}'\mathfrak{u}'}\, c_{x,\lambda}^{\mathfrak{s}'\mathfrak{u}'},$$

where the second equality holds by $(*)$ and the assumptions on $\lambda$. But then (a) and (b) show that $(\mathfrak{s}',\mathfrak{t}') = (\mathfrak{s},\mathfrak{t})$ and $(\mathfrak{t}',\mathfrak{u}') = (\mathfrak{t},\mathfrak{t})$. Hence, we obtain that $\tilde{\gamma}_{x,y,x^{-1}} =$

$(c^{\mathfrak{st}}_{x,\lambda})^2 c^{\mathfrak{tt}}_{y,\lambda} \neq 0$, as required. Thus, we have shown that the set $\{w_\lambda(\mathfrak{s},\mathfrak{t}) \mid \mathfrak{s} \in M(\lambda)\}$ is contained in a left $\tilde{\mathbf{J}}$-cell.

Conversely, let $z \in W$ and assume that $y \sim_{\mathscr{L}} z$. Then we must show that $z = w_\lambda(\mathfrak{s}',\mathfrak{t})$ for some $\mathfrak{s}' \in M(\lambda)$. For this purpose, it will be sufficient to consider an elementary step in the definition of $\sim_{\mathscr{L}}$; recall that $\sim_{\mathscr{L}}$ is equivalent to $\leqslant_{\mathscr{L}}$ by Lemma 1.6.6. Thus, we can assume that $\tilde{\gamma}_{x,y,z^{-1}} \neq 0$ for some $x \in W$. By the defining formula, there exists some $\mu \in \Lambda$ and $\mathfrak{s}',\mathfrak{t}',\mathfrak{u}' \in M(\mu)$ such that

$$c^{\mathfrak{s}'\mathfrak{t}'}_{x,\mu} c^{\mathfrak{t}',\mathfrak{u}'}_{y,\mu} c^{\mathfrak{u}'\mathfrak{s}'}_{z^{-1},\mu} \neq 0.$$

Using (∗), we conclude that $\mu = \lambda$. By the assumptions on $\lambda$, we have $c^{\mathfrak{u}'\mathfrak{s}'}_{z^{-1},\lambda} = c^{\mathfrak{s}'\mathfrak{u}'}_{z,\lambda}$. Then (a) and (b) show that $x = w_\lambda(\mathfrak{s}',\mathfrak{t}')$, $z = w_\lambda(\mathfrak{s}',\mathfrak{u}')$ and $(\mathfrak{t}',\mathfrak{u}') = (\mathfrak{s},\mathfrak{t})$. Hence, we have $z = w_\lambda(\mathfrak{s}',\mathfrak{t})$ as required. Thus, we have shown that the set $\{w_\lambda(\mathfrak{s},\mathfrak{t}) \mid \mathfrak{s} \in M(\lambda)\}$ is a left $\tilde{\mathbf{J}}$-cell. The argument for right $\tilde{\mathbf{J}}$-cells is entirely analogous. □

**Example 1.7.2.** Let $\lambda \in \Lambda$ be such that $d_\lambda = 1$ and $f_\lambda = 1$. Let us write $M(\lambda) = \{1\}$. For $s \in S$, we have $\rho^\lambda(T_s) = (v_s)$ or $(-v_s^{-1})$. So, for any $w \in W$, we have $\rho^\lambda(T_w) = \rho^\lambda(T_{w^{-1}}) = (\pm\varepsilon^{g_w})$, where $g_w = g_{w^{-1}} \in \Gamma$. In particular, we have $c^{11}_{w,\lambda} = c^{11}_{w^{-1},\lambda} \in \mathbb{Z}$ for all $w \in W$. So we can apply Proposition 1.7.1 and this yields that

$$\mathrm{Irr}_{\mathbb{K}}(W \mid \mathscr{F}_\lambda) = \{E^\lambda\}, \qquad \text{where} \qquad \mathscr{F}_\lambda = \{w_\lambda(1,1)\}.$$

The above conditions are satisfied, for example, for the unit and the sign representation of $W$, where $L(s) > 0$ for all $s \in S$ (see Example 1.3.3). This yields another proof for the fact that $\{1\}$ and $\{w_0\}$ are two-sided $\tilde{\mathbf{J}}$-cells (see Example 1.6.17).

**Table 1.5** Two-sided $\tilde{\mathbf{J}}$-cells $\mathscr{F}$ and the corresponding sets $\mathrm{Irr}_{\mathbb{K}}(W \mid \mathscr{F})$ for type $I_2(m)$

| | |
|---|---|
| $L(s_1) = L(s_2) > 0$ | |
| $\{1_0\}$ | $\{1_W\}$ |
| $W \setminus \{1_0, 1_m\}$ | $\{(\mathrm{sgn}_1), (\mathrm{sgn}_2), \sigma_j^\varepsilon \text{ (all } j)\}$ |
| $\{1_m\}$ | $\{\mathrm{sgn}\}$ |
| (where $\mathrm{sgn}_1$ and $\mathrm{sgn}_2$ have to be omitted if $m$ is odd) | |
| $L(s_1) > L(s_2) > 0$ | |
| $\{1_0\}$ | $\{1_W\}$ |
| $\{2_1\}$ | $\{\mathrm{sgn}_1\}$ |
| $W \setminus \{1_0, 2_1, 1_{m-1}, 1_m\}$ | $\{\sigma_j^\varepsilon \text{ (all } j)\}$ |
| $\{1_{m-1}\}$ | $\{\mathrm{sgn}_2\}$ |
| $\{1_m\}$ | $\{\mathrm{sgn}\}$ |
| $L(s_1) > L(s_2) = 0$ | |
| $\{1_0, 2_1\}$ | $\{1_W, \mathrm{sgn}_1\}$ |
| $W \setminus \{1_0, 2_1, 1_{m-1}, 1_m\}$ | $\{\sigma_j^\varepsilon \text{ (all } j)\}$ |
| $\{1_m, 1_{m-1}\}$ | $\{\mathrm{sgn}, \mathrm{sgn}_2\}$ |

**Example 1.7.3.** Let $W$ be of type $I_2(m)$ $(m \geqslant 3)$, with generators $s_1, s_2$ such that $(s_1s_2)^m = 1$. Assume that $L(s_1) \geqslant L(s_2) \geqslant 0$. Let us determine the two-sided $\tilde{\mathbf{J}}$-cells for $W$. We use the following notation. For any $k \geqslant 0$, write $1_k = s_1s_2s_1\cdots$ ($k$ factors) and $2_k = s_2s_1s_2\cdots$ ($k$ factors); note that $1_m = 2_m$. Then the two-sided $\tilde{\mathbf{J}}$-cells and the corresponding sets $\mathrm{Irr}_{\mathbb{K}}(W \mid \mathscr{F})$ are given in Table 1.5. To prove this, write $b := L(1_1)$ and $a := L(2_1)$, where $b \geqslant a \geqslant 0$. We begin with some general remarks. By Example 1.4.6, the irreducible representations as in Example 1.3.7 are balanced. The representations $\sigma_j^\varepsilon$ all have $\mathbf{a}$-invariant $b$. We find that

$$\varepsilon^b \sigma^{\varepsilon_j}(T_{1_1}) = \begin{pmatrix} -1 & 0 \\ \varepsilon^b \mu_j & \varepsilon^{2b} \end{pmatrix} \qquad \text{and so} \qquad \sigma_j^\varepsilon \leftrightsquigarrow_L 1_1.$$

This already shows that all the representations $\sigma_j^\varepsilon$ belong to $\mathrm{Irr}_{\mathbb{K}}(W \mid \mathscr{F}_1)$, where $\mathscr{F}_1$ is the two-sided cell containing $1_1 = s_1$. Next, Example 1.7.2 shows that if $b \geqslant a > 0$, then $\{1_0\}$ and $\{1_m\}$ are two-sided $\tilde{\mathbf{J}}$-cells, where $\mathrm{Irr}_{\mathbb{K}}(W \mid \{1_0\}) = \{1_W\}$ and $\mathrm{Irr}_{\mathbb{K}}(W \mid \{1_m\}) = \{\mathrm{sgn}\}$. Now we go through the above three cases individually.

First assume that $a = b > 0$. All that remains to show is that if $m$ is even, then the representations $\mathrm{sgn}_1$, $\mathrm{sgn}_2$ also belong to $\mathrm{Irr}_{\mathbb{K}}(W \mid \mathscr{F}_1)$. Now, $\mathrm{sgn}_1$ and $\mathrm{sgn}_2$ both have $\mathbf{a}$-invariant $b$. We have $\varepsilon^b \mathrm{sgn}_1^\varepsilon(T_{2_1}) = (-1)$ and so $\mathrm{sgn}_1 \leftrightsquigarrow_L 2_1$. Furthermore, we have $\varepsilon^b \mathrm{sgn}_2^\varepsilon(T_{1_1}) = (-1)$ and so $\mathrm{sgn}_2 \leftrightsquigarrow_L 1_1$. Also note that

$$\varepsilon^b \sigma_j^\varepsilon(T_{2_1}) = \begin{pmatrix} \varepsilon^{2b} & \varepsilon^b \\ 0 & -1 \end{pmatrix} \qquad \text{and so} \qquad \sigma_j^\varepsilon \leftrightsquigarrow_L 2_1,$$

$$\varepsilon^b \sigma_j^\varepsilon(T_{1_1}) = \begin{pmatrix} -1 & 0 \\ \varepsilon^b \mu_j & \varepsilon^{2b} \end{pmatrix} \qquad \text{and so} \qquad \sigma_j^\varepsilon \leftrightsquigarrow_L 1_1.$$

Hence, $\mathrm{sgn}_1$ and $\mathrm{sgn}_2$ belong to $\mathrm{Irr}_{\mathbb{K}}(W \mid \mathscr{F}_1)$, as required.

Now assume that $b > a > 0$. Then all that remains to do is to determine the two-sided $\tilde{\mathbf{J}}$-cells to which $\mathrm{sgn}_1$ and $\mathrm{sgn}_2$ belong. Now $\mathrm{sgn}_1$ and $\mathrm{sgn}_2$ have $\mathbf{a}$-invariants $a$ and $\frac{m}{2}(b-a)+a$ respectively. We have $\varepsilon^a \mathrm{sgn}_1^\varepsilon(T_{2_1}) = (-1)$ and so $\mathrm{sgn}_1 \leftrightsquigarrow_L 2_1$. Furthermore, we have $\mathrm{sgn}_2^\varepsilon(T_{1_{m-1}}) = \big((-1)^{m/2}\varepsilon^{m(a-b)/2-a}\big)$ and so $\mathrm{sgn}_2 \leftrightsquigarrow_L 1_{m-1}$. Finally, the assumptions in Example 1.7.2 are satisfied and, hence, $\{2_1\}$ and $\{1_{m-1}\}$ are two-sided $\tilde{\mathbf{J}}$-cells such that $\mathrm{Irr}_{\mathbb{K}}(W \mid \{2_1\}) = \{\mathrm{sgn}_1\}$ and $\mathrm{Irr}_{\mathbb{K}}(W \mid \{1_{m-1}\}) = \{\mathrm{sgn}_2\}$, as required.

Finally, assume that $b > a = 0$. We still have $\sigma_j \in \mathrm{Irr}_{\mathbb{K}}(W \mid \mathscr{F}_1)$ for all $j$. It remains to consider what happens with the one-dimensional representations. Both $1_W$ and $\mathrm{sgn}_1$ have $\mathbf{a}$-invariant 0. We find that

$$\{w \in W \mid 1_W \leftrightsquigarrow_L w\} = \{w \in W \mid \mathrm{sgn}_1 \leftrightsquigarrow_L w\} = \{1_0, 2_1\}.$$

Both $\mathrm{sgn}_2$ and $\mathrm{sgn}_1$ have $\mathbf{a}$-invariant $\frac{m}{2}b$. We find that

$$\{w \in W \mid \mathrm{sgn}_2 \leftrightsquigarrow_L w\} = \{w \in W \mid \mathrm{sgn} \leftrightsquigarrow_L w\} = \{1_{m-1}, 1_m\}.$$

Furthermore, since $b > 0$,

$$\varepsilon^b \sigma_j^\varepsilon(T_{1_0}) \equiv \varepsilon^b \sigma_j^\varepsilon(T_{2_1}) \equiv \begin{pmatrix} 0 & 0 \\ 0 & 0 \end{pmatrix} \mod \mathfrak{m}.$$

Then the characterisation of $\sim_{\mathscr{LR}}$ in Lemma 1.6.14 shows that $\{1_0, 2_1\}$ is a two-sided $\tilde{\mathbf{J}}$-cell and that $\mathrm{Irr}_{\mathbb{K}}(W \mid \{1_0, 2_1\}) = \{1_W, \mathrm{sgn}_1\}$. Applying Corollary 1.6.16 yields the two-sided $\tilde{\mathbf{J}}$-cell $\{1_{m-1}, 1_m\}$.

**Example 1.7.4.** Let $W$ be of type $I_2(m)$, where $b := L(s_1)$ and $a := L(s_2)$, as in the previous example. Using similar computations as above, one finds that

$$1_{\tilde{\mathbf{J}}} = \sum_{d \in \tilde{\mathscr{D}}} \tilde{n}_d t_d = \begin{cases} t_{1_0} - t_{1_1} - t_{2_1} + t_{1_m} & \text{if } b = a > 0, \\ t_{1_0} - t_{1_1} - t_{2_1} - t_{2_3} + (-1)^{m/2} t_{1_{m-1}} + t_{1_m} & \text{if } b > a > 0, \\ t_{1_0} - t_{1_1} - t_{2_3} + t_{1_m} & \text{if } b > a = 0. \end{cases}$$

Furthermore, in every irreducible representation of $\tilde{\mathbf{J}}$ (arising from the balanced representations as in Example 1.3.7), each $t_d$ ($d \in \tilde{\mathscr{D}}$) is represented by a diagonal matrix with $0, \pm 1$ on the diagonal. Let us just prove this in the case where $b > a > 0$. (The other cases are treated by similar methods.) We have

$$1_W^\varepsilon(T_{1_0}) = 1, \qquad \varepsilon^a \mathrm{sgn}_1^\varepsilon(T_{2_1}) = -1,$$

$$\varepsilon^{m(a+b)/2} \mathrm{sgn}^\varepsilon(T_{1_m}) = 1, \qquad \varepsilon^{m(b-a)/2+a} \mathrm{sgn}_2^\varepsilon(T_{1_{m-1}}) = (-1)^{m/2},$$

$$\varepsilon^b \sigma_j^\varepsilon(T_{1_1}) \equiv \begin{pmatrix} -1 & 0 \\ 0 & 0 \end{pmatrix} \mod \mathfrak{m}, \qquad \varepsilon^b \sigma_j^\varepsilon(T_{2_3}) \equiv \begin{pmatrix} 0 & 0 \\ 0 & -1 \end{pmatrix} \mod \mathfrak{m}.$$

Now set $h := t_{1_0} - t_{1_1} - t_{2_1} - t_{2_3} + (-1)^{m/2} t_{1_{m-1}} + t_{1_m} \in \tilde{\mathbf{J}}$. Then the above formulae show that, in every irreducible representation of $\tilde{\mathbf{J}}$, the element $h$ is represented by the identity matrix. (Here, we also use the information on the sets $\mathscr{F}_\lambda$ in Example 1.7.3 and the fact that $c_{w,\lambda}^{\mathfrak{st}} = 0$ unless $w \in \mathscr{F}_\lambda$.) Hence, we must have $h = 1_{\tilde{\mathbf{J}}}$.

*Remark 1.7.5.* We notice that the two-sided $\tilde{\mathbf{J}}$-cells for $W$ of type $I_2(m)$, as determined above, coincide with the two-sided cells in the sense of Kazhdan–Lusztig; see Example 2.1.18. (A similar statement can also be shown for left and right cells.) This seems to be a general phenomenon, and will be further discussed in Section 2.5.

**1.7.6.** Let $n \geqslant 1$ and $(W, S)$ be of type $A_{n-1}$ or $B_n$, where $S = \{s_1, \ldots, s_{n-1}\}$ (in type $A_{n-1}$) or $S = \{s_0, s_1, \ldots, s_{n-1}\}$ (in type $B_n$), with the labelling in Table 1.1. For any $m \in \{1, \ldots, n\}$, let $W_m \subseteq W$ be the parabolic subgroup generated by $s_0, s_1, \ldots, s_{m-1}$ (where we set $s_0 = 1$ in type $A_{n-1}$). Then $W_m$ is of type $A_{m-1}$ or $B_m$ respectively. Let $\Lambda_m$ be an indexing set for $\mathrm{Irr}_{\mathbb{K}}(W_m)$. In particular, we have $W_n = W$ and $\Lambda_n = \Lambda$.

A special feature of this case is that, for $n \geqslant 2$, the irreducible representations of $W_n$ restrict multiplicity-freely to $W_{n-1}$. More precisely, for each $\lambda \in \Lambda_n$, we have

(a) $$\mathrm{Res}_{W_{n-1}}^{W_n}(E^\lambda) \cong \bigoplus_{\mu \in \Lambda'(\lambda)} E^\mu \qquad \text{for some subset } \Lambda'(\lambda) \subseteq \Lambda_{n-1};$$

see [132, 6.1.8 and 6.1.9]. We shall consider a weight function $L_n \colon W_n \to \Gamma$ satisfying the following conditions where $a, b \in \Gamma$:

$$L_n(s_1) = \cdots = L_n(s_{n-1}) = a > 0,$$
$$L_n(s_0) = b > 0, \qquad b \notin \{a, 2a, \ldots, (n-1)a\} \quad \text{(in type } B_n\text{)}.$$

Note that, if $n \geqslant 2$, then the restriction of $L_n$ to $W_{n-1}$ is just $L_{n-1}$. Under these conditions, by Examples 1.3.8 and 1.3.9, we have

(b) $$\mathbf{c}_\lambda \in \mathbb{Z}[\Gamma] \qquad \text{and} \qquad f_\lambda = 1 \qquad \text{for all } \lambda \in \Lambda_n.$$

Note that, since $W_n$ is a Weyl group, we have $\mathbb{Z}_{W_n} = \mathbb{Z}$. In particular, we can take $R = \mathbb{Z}$ as our $L_n$-good subring of $\mathbb{C}$; then $\mathbb{K} = \mathbb{Q}$. Let us denote the generic algebra associated with $W_n, L_n$ by $\mathbf{H}_n$ and write $\mathbf{H}_{n,K} := K \otimes_A \mathbf{H}_n$.

**Proposition 1.7.7.** *In the setting of 1.7.6, assume that $\Gamma = \mathbb{Z}$. Then there exists a family of representations*

$$\{\sigma^\lambda : \mathbf{H}_{n,K} \to M_{d_\lambda}(K) \mid \lambda \in \Lambda_n,\ n \geqslant 1\}$$

*such that the following conditions are satisfied for all $n \geqslant 1$ and $\lambda \in \Lambda_n$.*

*(a) The representation $\sigma^\lambda$ is afforded by $E_\varepsilon^\lambda$ and we have $\sigma^\lambda(T_w) \in M_{d_\lambda}(\mathscr{R})$ for all $w \in W_n$, where $\mathscr{R}$ is Rouquier's ring in Lemma 1.2.9.*

*(b) Writing $\Lambda'(\lambda) = \{\mu_1, \ldots, \mu_r\} \subseteq \Lambda_{n-1}$, we have a block diagonal shape*

$$\sigma^\lambda(T_w) = \begin{bmatrix} \sigma^{\mu_1}(T_w) & 0 & \ldots & 0 \\ 0 & \sigma^{\mu_2}(T_w) & \ddots & \vdots \\ \vdots & \ddots & \ddots & 0 \\ 0 & \ldots & 0 & \sigma^{\mu_r}(T_w) \end{bmatrix} \qquad \text{for all } w \in W_{n-1}.$$

*Proof.* By assumption, we have $A = \mathbb{Z}[v, v^{-1}]$, where $v = \varepsilon$ is an indeterminate. We construct the representations $\sigma^\lambda$ of $\mathbf{H}_{n,K}$ inductively, as follows. If $n = 1$, then all irreducible representations of $W_1$ are one-dimensional and there is nothing to prove. Now assume that $n \geqslant 2$ and that $\sigma^\mu$ has already been constructed for every $\mu \in \Lambda_{n-1}$. Let $\lambda \in \Lambda_n$ and write $\Lambda'(\lambda) = \{\mu_1, \ldots, \mu_r\} \subseteq \Lambda_{n-1}$. By a general compatibility property (see [132, 9.1.9]), the relation in 1.7.6(a) also holds for the corresponding representations of $\mathbf{H}_{n-1,K} \subseteq \mathbf{H}_{n,K}$. Hence, denoting by $e_i \in \mathbf{H}_{n-1,K}$ the central primitive idempotent corresponding to $E_\varepsilon^{\mu_i}$ for $1 \leqslant i \leqslant r$, we have

$$E_\varepsilon^\lambda = E_1 \oplus \cdots \oplus E_r, \qquad \text{where} \qquad E_i := e_i E_\varepsilon^\lambda \cong E_\varepsilon^{\mu_i}.$$

We will now show that this decomposition already works over $\mathscr{R}$. Indeed, by Corollary 1.2.10, we can find a basis of $E_\varepsilon^\lambda$ such that the corresponding matrix representation $\pi^\lambda : \mathbf{H}_{n,K} \to M_{d_\lambda}(K)$ satisfies (a). Let $V^\lambda$ be the $\mathscr{R}$-submodule of $E_\varepsilon^\lambda$ spanned by that basis; thus, we can identify $E_\varepsilon^\lambda = K \otimes_{\mathscr{R}} V^\lambda$. By general results on symmetric algebras (see [132, 7.2.7]), the idempotent $e_i$ can be written as

$$e_i = \mathbf{c}_{\mu_i}^{-1} \sum_{w \in W_{n-1}} \operatorname{trace}(T_w, E_\varepsilon^{\mu_i})\, T_{w^{-1}} \in \mathbf{H}_{n-1,K}.$$

Now we use 1.7.6(b). This shows that $f_{\mu_i} = 1$ and, hence, $\mathbf{c}_{\mu_i} \in \mathbb{Z}[v, v^{-1}]$ is a unit in $\mathscr{R}$. It follows that each $e_i$ is an $\mathscr{R}$-linear combination of the basis elements $T_w$ ($w \in W_{n-1}$). Consequently, $e_i V^\lambda$ is an $\mathscr{R}$-submodule of $V^\lambda$ and we have a direct sum decomposition $V^\lambda = V_1 \oplus \cdots \oplus V_r$, where $V_i := e_i V^\lambda$ and $E_i \cong K \otimes_{\mathscr{R}} V_i$. Now consider an $\mathscr{R}$-basis of $V^\lambda$ which is compatible with this decomposition. Transforming the original basis to this new one, we obtain a representation $\tilde{\pi}^\lambda : \mathbf{H}_{n,K} \to M_{d_\lambda}(K)$ such that (a) holds and, in addition, we have

$$\tilde{\pi}^\lambda(T_w) = \begin{bmatrix} \tilde{\sigma}^{\mu_1}(T_w) & 0 & \dots & 0 \\ 0 & \tilde{\sigma}^{\mu_2}(T_w) & \ddots & \vdots \\ \vdots & \ddots & \ddots & 0 \\ 0 & \dots & 0 & \tilde{\sigma}^{\mu_r}(T_w) \end{bmatrix} \qquad \text{for all } w \in W_{n-1},$$

where $\tilde{\sigma}^{\mu_i}$ is afforded by $E_i$ for all $i$. In particular, $\tilde{\sigma}^{\mu_i}$ and $\sigma^{\mu_i}$ are equivalent over $K$. So there exists some invertible matrix $P_i \in M_{d_i}(K)$, where $d_i = \dim E_i$, such that

$$\sigma^{\mu_i}(T_w) = P_i^{-1}\, \tilde{\sigma}^{\mu_i}(T_w)\, P_i \qquad \text{for all } w \in W_{n-1}.$$

Multiplying $P_i$ by a suitable polynomial in $\mathbb{Z}[v]$, we may assume that $P_i \in M_{d_i}(\mathscr{R})$ and that the greatest common divisior of all non-zero entries of $P_i$ is 1. (Recall from Lemma 1.2.9 that $\mathscr{R}$ is a principal ideal domain and, hence, a unique factorisation domain.) Now $\det(P_i)$ is a non-zero element of $\mathscr{R}$. In fact, we claim that $\det(P_i)$ is a unit in $\mathscr{R}$. This follows by a standard argument on symmetric algebras, which we already employed at the end of the proof of Proposition 1.5.11. Let $\mathfrak{p}$ be a maximal ideal in $R$; we must show that $\det(P_i) \notin \mathfrak{p}$. For this purpose, let $k := \mathscr{R}/\mathfrak{p}$ be the residue field and consider the specialisation $\theta : A \to k$ induced by reduction modulo $\mathfrak{p}$. Applying $\theta$, we obtain a specialised algebra $\mathbf{H}_{n-1,k}$ and specialised representations $\sigma_k^{\mu_i}$ and $\tilde{\sigma}_k^{\mu_i}$. Since each $\mathbf{c}_{\mu_i}$ is a unit in $\mathscr{R}$, we have $\theta(\mathbf{c}_{\mu_i}) \neq 0$. As in the proof of Proposition 1.5.11, this implies that $\sigma_k^{\mu_i}$ and $\tilde{\sigma}_k^{\mu_i}$ are irreducible. The reduction of $P_i$ modulo $\mathfrak{p}$ defines a non-zero module homomorphism between these representations. Hence, by Schur's lemma, this homomorphism must be invertible and so $\det(P_i) \notin \mathfrak{p}$, as required. Thus, each $P_i$ is invertible over $\mathscr{R}$. Let $P$ be the block diagonal matrix with diagonal blocks given by $P_1, \dots, P_d$. Then $P$ is invertible over $\mathscr{R}$ and, if we set

$$\sigma^\lambda(T_w) := P^{-1}\, \tilde{\pi}^\lambda(T_w)\, P \qquad \text{for all } w \in W_n,$$

then $\sigma^\lambda$ is a matrix representation of $\mathbf{H}_{n,K}$ which satisfies (a) and (b). □

*Remark 1.7.8.* The existence of a compatible family of representations $\{\sigma^\lambda\}$ as above was known before. Indeed, the "seminormal representations" constructed by Hoesfmit [157] satisfy these properties; see also Ram [268]. Of course, we could have referred to Hoefsmit's proof, but the above argument provides an alternative approach and is quite self-contained and much less computational.

However, Hoefsmit's results are actually stronger: they can be used to drop the asumption that $\Gamma = \mathbb{Z}$. Indeed, in the general case, consider the ring

$$\mathscr{R}(\Gamma) := \left\{ \frac{f}{g} \;\middle|\; f \in \mathbb{Z}[\Gamma],\, g \in 1 + \mathbb{Z}[\Gamma_{>0}] \right\} \subseteq K.$$

Then Hoefsmit's construction [157] yields a family of representations as in Proposition 1.7.7 where, in (a), we have $\sigma^\lambda(T_w) \in M_{d_\lambda}(\mathscr{R}(\Gamma))$ for all $w \in W_n$. (This is immediately seen from the explicit form of the matrices in [157, §2.2].)

**Corollary 1.7.9.** *Assume we are in the setting of 1.7.6. Then, for each $\lambda \in \Lambda_n$, there exists a balanced representation $\rho^\lambda$ afforded by $E_\varepsilon^\lambda$ such that the conditions in Proposition 1.7.1 are satisfied. Thus, we have*

$$c_{w,\lambda}^{\mathfrak{st}} = c_{w^{-1},\lambda}^{\mathfrak{ts}} \in \{0, \pm 1\} \qquad \textit{for all } w \in W_n \textit{ and } \mathfrak{s}, \mathfrak{t} \in M(\lambda).$$

*Proof.* Let $\sigma^\lambda$ be as in Proposition 1.7.7 (or as in Remark 1.7.8 if $\Gamma \neq \mathbb{Z}$). Let $B \in M_{d_\lambda}(K)$ be any matrix such that $B\sigma^\lambda(T_{w^{-1}}) = \sigma^\lambda(T_w)^{\mathrm{tr}} B$ for all $w \in W_n$. Then we claim that $B$ must be diagonal. We prove this by induction on $n$. If $n = 1$, then $\sigma^\lambda$ is one-dimensional and there is nothing to prove. Now assume that $n \geqslant 2$ and consider the restriction of $\sigma^\lambda$ to $\mathbf{H}_{n-1,K}$. This restriction has a block form as described in Proposition 1.7.7(b). Correspondingly, let us also write $B$ in block form:

$$B = \begin{bmatrix} B_{1,1} & B_{1,2} & \dots & B_{1,r} \\ B_{2,1} & B_{2,2} & \ddots & \vdots \\ \vdots & \ddots & \ddots & B_{r-1,r} \\ B_{r,1} & \dots & B_{r,r-1} & B_{r,r} \end{bmatrix}.$$

Then, for any $i, j \in \{1, \dots, r\}$, we have $B_{i,j}\sigma^{\mu_j}(T_{w^{-1}}) = \sigma^{\mu_i}(T_w)^{\mathrm{tr}} B_{i,j}$ for all $w \in W_{n-1}$. This means that $B_{i,j}$ defines an $\mathbf{H}_{n-1,K}$-module homomorphism between $E_\varepsilon^{\mu_j}$ and the contragredient dual of $E_\varepsilon^{\mu_i}$. By Example 1.2.5, the latter is isomorphic to $E_\varepsilon^{\mu_i}$. So we must have $B_{i,j} = 0$ if $i \neq j$. Furthermore, if $i = j$, then $B_{i,i}$ is a diagonal matrix by induction. Hence, $B$ is a diagonal matrix, as claimed.

We now apply the procedure in 1.4.9 and set $\Omega_0^\lambda := \sum_{w \in W_n} \sigma(T_w)^{\mathrm{tr}}.\sigma(T_w)$. Then $\Omega_0^\lambda$ is a symmetric matrix such that $\Omega_0^\lambda \sigma^\lambda(T_{w^{-1}}) = \sigma^\lambda(T_w)^{\mathrm{tr}} \Omega_0^\lambda$ for all $w \in W_n$. We have just seen that this implies that $\Omega_0^\lambda$ is a diagonal matrix. Hence, the Gram–Schmidt orthogonalisation procedure is not required here and we have $\Omega_0^\lambda = \Omega_1^\lambda$ in the notation of 1.4.9. Furthermore, the diagonal coefficient of $\Omega_0^\lambda$ at position $\mathfrak{s} \in M(\lambda)$ has the form

$$b_{\mathfrak{s}} \varepsilon^{-2m_{\mathfrak{s}}} + \text{combination of terms } \varepsilon^g \text{ where } g > -2m_{\mathfrak{s}},$$

where $b_{\mathfrak{s}}$ is a positive real number and $m_{\mathfrak{s}} \in \Gamma$. Finally, let $E^\lambda$ be the diagonal matrix with diagonal entries $\{\varepsilon^{m_{\mathfrak{s}}} \mid \mathfrak{s} \in M(\lambda)\}$ and set

$$\rho^\lambda(T_w) := (E^\lambda)^{-1} \sigma^\lambda(T_w) E^\lambda \qquad \text{for all } w \in W_n.$$

Then $\rho^\lambda$ is a balanced representation and the corresponding leading matrix coefficients satisfy $b_\mathfrak{t} c_{w^{-1},\lambda}^{\mathfrak{ts}} = c_{w,\lambda}^{\mathfrak{st}} b_\mathfrak{s}$ for all $w \in W_n$ and $\mathfrak{s}, \mathfrak{t} \in M(\lambda)$.

Since $\sigma^\lambda$ and, hence, $\rho^\lambda$ are realised over $\mathscr{R}$ (or over $\mathscr{R}(\Gamma)$), we have $c_{w,\lambda}^{\mathfrak{st}} \in \mathbb{Z}$ for all $w \in W_n$ and $\mathfrak{s}, \mathfrak{t} \in M(\lambda)$. So it remains to show that $b_\mathfrak{s} = b_\mathfrak{t}$ for all $\mathfrak{s}, \mathfrak{t} \in M(\lambda)$. Now, given $\mathfrak{s}, \mathfrak{t} \in M(\lambda)$, we have the relation (see Proposition 1.4.10)

$$1 = f_\lambda = \sum_{w \in W_n} c_{w,\lambda}^{\mathfrak{st}} c_{w^{-1},\lambda}^{\mathfrak{ts}}.$$

But, since all leading matrix coefficients are integers, we conclude that

$$c_{w,\lambda}^{\mathfrak{st}} c_{w^{-1},\lambda}^{\mathfrak{ts}} = b_\mathfrak{s} b_\mathfrak{t}^{-1} (c_{w,\lambda}^{\mathfrak{st}})^2 \quad \text{is a non-negative integer for all } w \in W_n.$$

Hence, there is a unique $w \in W_n$ such that $c_{w,\lambda}^{\mathfrak{st}} c_{w^{-1},\lambda}^{\mathfrak{ts}} = b_\mathfrak{s} b_\mathfrak{t}^{-1} (c_{w,\lambda}^{\mathfrak{st}})^2 = 1$. So we must have $c_{w,\lambda}^{\mathfrak{st}} = c_{w^{-1},\lambda}^{\mathfrak{ts}} = \pm 1$ and $b_\mathfrak{s} b_\mathfrak{t}^{-1} = 1$, as required. □

**Theorem 1.7.10 (Cf. [103, §5], [25, §2]).** *Assume that $W = W_n$ is of type $A_{n-1}$ or $B_n$, and that $L = L_n$ is a weight function satisfying the conditions in 1.7.6. Then there exists a bijection*

$$\coprod_{\lambda \in \Lambda} (M(\lambda) \times M(\lambda)) \quad \stackrel{1-1}{\longrightarrow} \quad W, \qquad (\mathfrak{s}, \mathfrak{t}) \mapsto w_\lambda(\mathfrak{s}, \mathfrak{t}),$$

*such that the following hold.*

(a) *Let $w \in W$, $\lambda \in \Lambda$ and $\mathfrak{s}, \mathfrak{t} \in M(\lambda)$. Then $c_{w,\lambda}^{\mathfrak{st}} \neq 0$ if and only if $w = w_\lambda(\mathfrak{s}, \mathfrak{t})$. Also, if $c_{w,\lambda}^{\mathfrak{st}} \neq 0$, then $c_{w,\lambda}^{\mathfrak{st}} = \pm 1$. Furthermore, $w_\lambda(\mathfrak{s}, \mathfrak{t})^{-1} = w_\lambda(\mathfrak{t}, \mathfrak{s})$.*

(b) *The left, right and two-sided $\tilde{\mathbf{J}}$-cells of $W$ are given as described in Proposition 1.7.1. In particular, $\operatorname{Irr}(W \mid \mathscr{F}_\lambda) = \{E^\lambda\}$ for all $\lambda \in \Lambda$.*

(c) *We have $\tilde{\mathscr{D}} = \{d \in W \mid d^2 = 1\} = \{w_\lambda(\mathfrak{s}, \mathfrak{s}) \mid \lambda \in \Lambda, \mathfrak{s} \in M(\lambda)\}$ and $\tilde{n}_d = \pm 1$ for all $d \in \tilde{\mathscr{D}}$. Every left $\tilde{\mathbf{J}}$-cell contains a unique element of $\tilde{\mathscr{D}}$.*

*In particular, Conjectures 1.5.12 and 1.6.18 hold in this case.*

*Proof.* By Corollary 1.7.9, the assumptions of Proposition 1.7.1 are satisfied for every $\lambda \in \Lambda$. This yields the existence of the required bijection satisfying (a) and (b). The statement about $w_\lambda(\mathfrak{s}, \mathfrak{t})^{-1}$ is then clear. In particular, this shows that

$$\{w \in W \mid w^2 = 1\} = \{w_\lambda(\mathfrak{s}, \mathfrak{s}) \mid \lambda \in \Lambda, \mathfrak{s} \in M(\lambda)\}.$$

To prove the remaining assertions in (c), let $w \in W$ and write $w = w_\lambda(\mathfrak{s}, \mathfrak{t})$, where $\lambda \in \Lambda$ and $\mathfrak{s}, \mathfrak{t} \in M(\lambda)$. Using the defining formula for $\tilde{n}_w$ and (a), we obtain

$$\tilde{n}_w = \tilde{n}_{w^{-1}} = \sum_{\mu \in \Lambda} \sum_{\mathfrak{u} \in M(\mu)} c_{w,\lambda}^{\mathfrak{uu}} = \delta_{\mathfrak{st}} c_{w,\mu}^{\mathfrak{st}} = \pm \delta_{\mathfrak{st}}.$$

In particular, $\tilde{\mathscr{D}}$ has the required form and $\tilde{n}_d = \pm 1$ for all $d \in \tilde{\mathscr{D}}$. Using (b), we then see that every left $\tilde{\mathbf{J}}$-cell contains a unique element of $\tilde{\mathscr{D}}$. In particular, Conjecture 1.6.18 holds. By Remark 1.6.19, this implies that Conjecture 1.5.12(b) also holds. Finally, Conjecture 1.5.12(a) is seen to hold directly from the defining formulae for $\tilde{\gamma}_{x,y,z}$ and $\tilde{n}_w$, using that $f_\lambda = 1$ for all $\lambda \in \Lambda$. □

*Remark 1.7.11.* The original treatment in [103], [25] relied on the Dipper–James–Murphy approach [62], [68] to Hoefsmit's construction [157] of "seminormal representations". Here, at least for $\Gamma = \mathbb{Z}$, we have replaced this reference by our Proposition 1.7.7, which essentially relies on the use of Rouquier's ring $\mathscr{R}$.

*Remark 1.7.12.* Let $W$ be irreducible of arbitrary type, and $L\colon W \to \Gamma$ be any weight function such that $L(s) \geqslant 0$ for all $s \in S$. Assume that $f_\lambda = 1$ for all $\lambda \in \Lambda$. Then, using the explicit knowledge of $\mathbf{a}_\lambda$ and $f_\lambda$ in all cases (see Remark 1.3.6), we see that $W$ must be of type $A_{n-1}$ or $B_n$ where $L$ satisfies the conditions in 1.7.6.

*Remark 1.7.13.* Let $W$ be of type $B_n$, with generators $S = \{s_0, s_1, \ldots, s_{n-1}\}$ as above. What happens in the cases which are not covered by the conditions in 1.7.6? To fix some notation, let $r \in \{0, 1, \ldots, n-1\}$, $a \in \Gamma_{>0}$, and consider a weight function $L\colon W \to \Gamma$ such that $L(s_0) = ra$ and $L(s_i) = a$ for $1 \leqslant i \leqslant n-1$. Then one may conjecture that the left, right and two-sided $\tilde{\mathbf{J}}$-cells of $W$ sets are given by the "combinatorial $r$-cells" of Bonnafé et al. [25, §4]. This is supported by the results of Bonnafé [23], Pietraho [264], [265], Taskin [289].

## 1.8 Cells and Leading Coefficients

The main result of this section is a refinement of the "orthogonality relations" for Lusztig's leading coefficients of character values (see Remark 1.3.2) with respect to a given left $\tilde{\mathbf{J}}$-cell. This is modelled on the discussion in [220, Chap. 5].

**Theorem 1.8.1 (Cf. Lusztig [220, 5.8] in the equal-parameter case).** *Let $\mathfrak{C}$ be a left $\tilde{\mathbf{J}}$-cell in $W$ and $\lambda, \mu \in \Lambda$. Then the following hold.*

*(a) If $\lambda \neq \mu$, then for any $\mathfrak{s}, \mathfrak{t} \in M(\lambda)$ and $\mathfrak{u}, \mathfrak{v} \in M(\mu)$, we have*

$$\sum_{w \in \mathfrak{C}} c_{w,\lambda}^{\mathfrak{st}}\, c_{w^{-1},\mu}^{\mathfrak{uv}} = 0 \qquad \textit{and} \qquad \sum_{w \in \mathfrak{C}} c_{w,\lambda}\, c_{w^{-1},\mu} = 0.$$

*(b) If $\lambda = \mu$, then for any $\mathfrak{s} \in M(\lambda)$, we have*

$$\sum_{\mathfrak{t} \in M(\lambda)} \sum_{w \in \mathfrak{C}} c_{w,\lambda}^{\mathfrak{st}}\, c_{w^{-1},\lambda}^{\mathfrak{ts}} = \sum_{w \in \mathfrak{C}} c_{w,\lambda}\, c_{w^{-1},\lambda} = \tilde{m}(\mathfrak{C}, \lambda)\, f_\lambda,$$

*where $\tilde{m}(\mathfrak{C}, \lambda)$ denotes the multiplicity of $\bar{\rho}^\lambda$ as an irreducible constituent of the left cell module $[\mathfrak{C}]_{\mathbb{K}}$ of $\tilde{\mathbf{J}}$.*

*In particular, $\tilde{m}(\mathfrak{C}, \lambda) > 0$ if and only if there exists some $w \in \mathfrak{C}$ such that $c_{w,\lambda} \neq 0$.*

*Proof.* This is, to a large extent, analogous to the general proof of the Schur relations for symmetric algebras; see, for example, [132, §7.2]. Let $P$ be any matrix of size $d_\lambda \times d_\mu$, with coefficients in $\mathbb{K}$, and set

$$I(P) := \sum_{w\in\mathfrak{C}} \bar{\rho}^\lambda(t_w)\, P\, \bar{\rho}^\mu(t_{w^{-1}}).$$

We claim that $\bar{\rho}^\lambda(t_x)I(P) = I(P)\bar{\rho}^\mu(t_x)$ for all $x \in W$. Indeed, we have

$$\bar{\rho}^\lambda(t_x)I(P) = \sum_{w\in\mathfrak{C}} \bar{\rho}^\lambda(t_x t_w)I(P)\bar{\rho}^\mu(t_{w^{-1}}) = \sum_{w\in\mathfrak{C}}\sum_{y\in W} \tilde{\gamma}_{x,w,y^{-1}}\, \bar{\rho}^\lambda(t_y)\, I(P)\, \bar{\rho}^\mu(t_{w^{-1}}).$$

Suppose that, in the above sum, we have $\tilde{\gamma}_{x,w,y^{-1}} \neq 0$. Then, by Lemma 1.6.5, we must have $y \sim_{\mathscr{L}} w$. Hence, we can first let the two sums runs over all $y, w \in \mathfrak{C}$ and then, applying the argument once more, let $w$ run over all elements of $W$ and $y$ run over all elements in $\mathfrak{C}$. Thus, we obtain

$$\bar{\rho}^\lambda(t_x)I(P) = \sum_{y\in\mathfrak{C}} \bar{\rho}^\lambda(t_y)I(P)\Big(\sum_{w\in W} \tilde{\gamma}_{x,w,y^{-1}}\, \bar{\rho}^\mu(t_{w^{-1}})\Big).$$

Since $\tilde{\gamma}_{x,w,y^{-1}} = \tilde{\gamma}_{y^{-1},x,w}$ (see Lemma 1.5.3), we have $\sum_{w\in W} \tilde{\gamma}_{x,w,y^{-1}}\, t_{w^{-1}} = t_{y^{-1}} t_x$. Hence, the expression on the right-hand side equals $I(P)\,\bar{\rho}^\mu(t_x)$, as required.

Now let $\mathfrak{t} \in M(\lambda)$, $\mathfrak{u} \in M(\mu)$ and $P = P_{\mathfrak{tu}}$ be the matrix with $(\mathfrak{t},\mathfrak{u})$-coefficient 1, and coefficient 0 otherwise. Then, for any $\mathfrak{s} \in M(\lambda)$ and $\mathfrak{v} \in M(\mu)$, we have

$$I_{\mathfrak{sv}}(P_{\mathfrak{tu}}) = \sum_{w\in\mathfrak{C}} \bar{\rho}^\lambda_{\mathfrak{st}}(t_w)\,\bar{\rho}^\mu_{\mathfrak{uv}}(t_{w^{-1}}) = \sum_{w\in\mathfrak{C}} c^{\mathfrak{st}}_{w,\lambda}\, c^{\mathfrak{uv}}_{w^{-1},\mu}.$$

If $\lambda \neq \mu$, then $I(P) = 0$ for every matrix $P$ and, hence,

$$\sum_{w\in\mathfrak{C}} c^{\mathfrak{st}}_{w,\lambda}\, c^{\mathfrak{uv}}_{w^{-1},\mu} = 0 \qquad \text{for all } \mathfrak{s},\mathfrak{t} \in M(\lambda) \text{ and } \mathfrak{u},\mathfrak{v} \in M(\mu),$$

as desired. This also implies that

$$\sum_{w\in\mathfrak{C}} c_{w,\lambda}\, c_{w^{-1},\mu} = \sum_{\mathfrak{s}\in M(\lambda)}\sum_{\mathfrak{u}\in M(\mu)}\sum_{w\in\mathfrak{C}} c^{\mathfrak{ss}}_{w,\lambda}\, c^{\mathfrak{uu}}_{w^{-1},\mu} = 0 \qquad \text{if } \lambda \neq \mu.$$

So let us now assume that $\lambda = \mu$. Then $I(P)$ must be a scalar multiple of the identity for every matrix $P$ and, hence,

$$\sum_{w\in\mathfrak{C}} c^{\mathfrak{st}}_{w,\lambda}\, c^{\mathfrak{uv}}_{w^{-1},\lambda} = 0 \qquad \text{if } \mathfrak{s} \neq \mathfrak{v},$$

$$\sum_{w\in\mathfrak{C}} c^{\mathfrak{st}}_{w,\lambda}\, c^{\mathfrak{us}}_{w^{-1},\lambda} \quad \text{does not depend on } \mathfrak{s}.$$

These two properties imply that, for a fixed $\mathfrak{s}_0 \in M(\lambda)$, we have

$$\sum_{\mathfrak{s},\mathfrak{t}\in M(\lambda)} \sum_{w\in\mathfrak{C}} c_{w,\lambda}^{\mathfrak{s}\mathfrak{s}}\, c_{w^{-1},\lambda}^{\mathfrak{t}\mathfrak{t}} = \sum_{\mathfrak{t}\in M(\lambda)} \sum_{w\in\mathfrak{C}} c_{w,\lambda}^{\mathfrak{t}\mathfrak{t}}\, c_{w^{-1},\lambda}^{\mathfrak{t}\mathfrak{t}} = \sum_{\mathfrak{t}\in M(\lambda)} \sum_{w\in\mathfrak{C}} c_{w,\lambda}^{\mathfrak{s}_0\mathfrak{t}}\, c_{w^{-1},\lambda}^{\mathfrak{t}\mathfrak{s}_0}.$$

Now note that the left-hand side equals $\sum_{w\in\mathfrak{C}} c_{w,\lambda}\, c_{w^{-1},\lambda}$; furthermore, the right hand side is the $(\mathfrak{s}_0,\mathfrak{s}_0)$-coefficient of

$$I\Big(\sum_{\mathfrak{t}\in M(\lambda)} P_{\mathfrak{t}\mathfrak{t}}\Big) = \bar{\rho}^\lambda(z), \qquad \text{where} \qquad z := \sum_{w\in\mathfrak{C}} t_w t_{w^{-1}},$$

and this does not depend on $\mathfrak{s}_0$. So, it will be sufficient to prove that

$$\operatorname{trace}\big(\bar{\rho}^\lambda(z)\big) = \tilde{m}(\mathfrak{C},\lambda)\, d_\lambda\, f_\lambda.$$

Now, we have

$$z = \sum_{w\in\mathfrak{C}} t_w t_{w^{-1}} = \sum_{w\in\mathfrak{C}} \sum_{x\in W} \tilde{\gamma}_{w,w^{-1},x}\, t_{x^{-1}}.$$

By Lemma 1.5.3, we have $\tilde{\gamma}_{w,w^{-1},x} = \tilde{\gamma}_{x,w,w^{-1}}$. Hence, we obtain that

$$z = \sum_{x\in W} \Big(\sum_{w\in\mathfrak{C}} \tilde{\gamma}_{x,w,w^{-1}}\Big) t_{x^{-1}} = \sum_{x\in W} \operatorname{trace}(t_x, [\mathfrak{C}]_{\mathbb{K}})\, t_{x^{-1}} = \sum_{x\in W} \sum_{\mu\in\Lambda} \tilde{m}(\mathfrak{C},\mu)\, c_{x,\mu}\, t_{x^{-1}}.$$

Consequently, using the orthogonality relations for Lusztig's leading coefficients of character values, we find that

$$\operatorname{trace}\big(\bar{\rho}^\lambda(z)\big) = \sum_{\mu\in\Lambda} \tilde{m}(\mathfrak{C},\mu) \sum_{x\in W} c_{x,\mu}\, c_{x^{-1},\lambda} = \tilde{m}(\mathfrak{C},\lambda)\, d_\lambda\, f_\lambda,$$

as desired. □

**Example 1.8.2.** Let $\lambda\in\Lambda$ and $\mathfrak{C}$ be a left $\tilde{\mathbf{J}}$-cell of $W$ such that $\tilde{m}(\mathfrak{C},\lambda)>0$. Then we claim that $\mathfrak{C}\subseteq\mathscr{F}_\lambda$. Indeed, by Theorem 1.8.1(b), there exists some $w\in\mathfrak{C}$ such that $c_{w,\lambda}\neq 0$. Hence, we also have $c_{w,\lambda}^{\mathfrak{s}\mathfrak{s}}\neq 0$ for some $\mathfrak{s}\in M(\lambda)$ and so $E^\lambda \leftsquigarrow\!\!\rightsquigarrow_L w$. This shows that $w\in\mathscr{F}_\lambda$ and so $\mathscr{F}_\lambda\cap\mathfrak{C}\neq\varnothing$. Since $\mathscr{F}_\lambda$ is a union of left $\tilde{\mathbf{J}}$-cells, we must have $\mathfrak{C}\subseteq\mathscr{F}_\lambda$, as claimed.

**Example 1.8.3.** Assume that the balanced representations $\{\rho^\lambda \mid \lambda\in\Lambda\}$ are chosen such that the additional conditions in Theorem 1.4.8 are satisfied; that is, we have

$$b_{\mathfrak{t}}\, c_{w^{-1},\lambda}^{\mathfrak{t}\mathfrak{s}} = c_{w,\lambda}^{\mathfrak{s}\mathfrak{t}}\, b_{\mathfrak{s}} \qquad \text{for all } w\in W,\ \lambda\in\Lambda \text{ and } \mathfrak{s},\mathfrak{t}\in M(\lambda),$$

where $\{b_{\mathfrak{s}} \mid \mathfrak{s}\in M(\lambda)\}\subseteq\mathbb{R}_{>0}$. Now let $\mathfrak{C}$ be a left $\tilde{\mathbf{J}}$-cell in $W$. Then the identity in Theorem 1.8.1(b) reads

$$\sum_{\mathfrak{t}\in M(\lambda)} \sum_{w\in\mathfrak{C}} b_{\mathfrak{s}}\, b_{\mathfrak{t}}^{-1} \big(c_{w,\lambda}^{\mathfrak{s}\mathfrak{t}}\big)^2 = \tilde{m}(\mathfrak{C},\lambda)\, f_\lambda \qquad \text{for any } \lambda\in\Lambda \text{ and } \mathfrak{s}\in M(\lambda),$$

and there are no cancellations in the above sum. Hence, we obtain the implication

(a) $\quad c_{w,\lambda}^{\mathfrak{s}\mathfrak{t}}\neq 0$ for some $w\in\mathfrak{C}$, $\lambda\in\Lambda$, $\mathfrak{s},\mathfrak{t}\in M(\lambda) \quad\Rightarrow\quad \tilde{m}(\mathfrak{C},\lambda)>0.$

Conversely, the formula in Theorem 1.8.1(b) shows that if $\tilde{m}(\mathfrak{C},\lambda) > 0$, then there exists some $w \in \mathfrak{C}$ such that $c_{w,\lambda} \neq 0$ and, hence, $c_{w,\lambda}^{\mathfrak{s}\mathfrak{s}} \neq 0$ for some $\mathfrak{s} \in M(\lambda)$.

We now obtain the following alternative description of the sets $\mathrm{Irr}_{\mathbb{K}}(W \mid \mathscr{F})$.

**Lemma 1.8.4.** *Let $\lambda,\mu \in \Lambda$. Then $\mathscr{F}_\lambda = \mathscr{F}_\mu$ if and only if there exist a sequence $\lambda = \lambda_0, \lambda_1, \ldots, \lambda_m = \mu$ in $\Lambda$ and a sequence of left $\tilde{\mathbf{J}}$-cells $\mathfrak{C}_1, \ldots, \mathfrak{C}_m$ such that, for each $i \in \{1,\ldots,m\}$, we have $\tilde{m}(\mathfrak{C}_i,\lambda_{i-1}) > 0$ and $\tilde{m}(\mathfrak{C}_i,\lambda_i) > 0$.*

*Proof.* We choose the balanced representations of $\mathbf{H}_K$ as in Example 1.8.3. Now assume first that $\mathscr{F} := \mathscr{F}_\lambda = \mathscr{F}_\mu$. There exist some $x,y \in \mathscr{F}$ such that $E^\lambda \leftrightsquigarrow_L x$ and $E^\mu \leftrightsquigarrow_L y$. By the definition of $\sim_{\mathscr{LR}}$ and Lemma 1.6.6, there exists a sequence $x = x_0, x_1, \ldots, x_m = y$ in $W$ such that, for each $i \in \{1,\ldots,m\}$, we have $x_{i-1} \sim_{\mathscr{L}} x_i$ or $x_{i-1} \sim_{\mathscr{R}} x_i$. By Proposition 1.4.10(c), there exist $\lambda = \lambda_0, \lambda_1, \ldots, \lambda_m = \lambda'$ in $\Lambda$ such that, for each $i \in \{0,1,\ldots,m\}$, we have

$$c_{x_i,\lambda_i}^{\mathfrak{s}_i\mathfrak{t}_i} \neq 0 \qquad \text{for some } \mathfrak{s}_i,\mathfrak{t}_i \in M(\lambda_i).$$

Note that, by our choice of the balanced representations, we then also have

$$c_{x_i^{-1},\lambda_i}^{\mathfrak{t}_i\mathfrak{s}_i} \neq 0 \qquad \text{for some } \mathfrak{s}_i,\mathfrak{t}_i \in M(\lambda_i).$$

We define a sequence of left $\tilde{\mathbf{J}}$-cells $\mathfrak{C}_1, \ldots, \mathfrak{C}_m$ as follows. If $x_{i-1} \sim_{\mathscr{L}} x_i$, then let $\mathfrak{C}_i$ be the left $\tilde{\mathbf{J}}$-cell containing $x_{i-1}$ and $x_i$. If $x_{i-1} \sim_{\mathscr{R}} x_i$, then $x_{i-1}^{-1} \sim_{\mathscr{L}} x_i^{-1}$ and we let $\mathfrak{C}_i$ be the left $\tilde{\mathbf{J}}$-cell containing $x_{i-1}^{-1}$ and $x_i^{-1}$. In both cases, by Example 1.8.3(a), we have $\tilde{m}(\mathfrak{C}_i,\lambda_{i-1}) > 0$ and $\tilde{m}(\mathfrak{C}_i,\lambda_i) > 0$, as required.

Conversely, assume that $\lambda = \lambda_0, \lambda_1, \ldots, \lambda_m = \lambda'$ and $\mathfrak{C}_1, \ldots, \mathfrak{C}_m$ are sequences with the required properties. For each $i$, let $\mathscr{F}_i$ be the two-sided $\tilde{\mathbf{J}}$-cell containing $\mathfrak{C}_i$. By Theorem 1.8.1(b), there exist elements $x_i, y_i \in \mathfrak{C}_i$ ($1 \leqslant i \leqslant m$) such that

$$c_{x_i,\lambda_i}^{\mathfrak{s}_i\mathfrak{t}_i} \neq 0 \quad \text{and} \quad c_{y_i,\lambda_{i-1}}^{\mathfrak{u}_i\mathfrak{v}_i} \neq 0, \quad \text{where} \quad \mathfrak{s}_i,\mathfrak{t}_i \in M(\lambda_i),\ \mathfrak{u}_i,\mathfrak{v}_i \in M(\lambda_{i-1}).$$

Thus, $E^{\lambda_{i-1}} \leftrightsquigarrow_L y_i$ and $E^{\lambda_i} \leftrightsquigarrow_L x_i$ for $1 \leqslant i \leqslant m$. Since $x_i, y_i \in \mathscr{F}_i$, we deduce that

$$E^{\lambda_{i-1}} \in \mathrm{Irr}_{\mathbb{K}}(W \mid \mathscr{F}_i) \qquad \text{and} \qquad E^{\lambda_i} \in \mathrm{Irr}_{\mathbb{K}}(W \mid \mathscr{F}_i) \qquad \text{for } 1 \leqslant i \leqslant m.$$

But this certainly implies that $\mathscr{F}_\lambda = \mathscr{F}_1 = \mathscr{F}_2 = \cdots = \mathscr{F}_m = \mathscr{F}_\mu$, as required. □

**Example 1.8.5.** Let $\mathfrak{C}$ be a left $\tilde{\mathbf{J}}$-cell. Then we claim that

$$\tilde{m}(\mathfrak{C},\lambda) = \sum_{d \in \tilde{\mathscr{D}} \cap \mathfrak{C}} \tilde{n}_d c_{d,\lambda} \qquad \text{and} \qquad \sum_{\lambda \in \Lambda} f_\lambda^{-1} \tilde{m}(\mathfrak{C},\lambda) = \sum_{d \in \tilde{\mathscr{D}} \cap \mathfrak{C}} \tilde{n}_d^2.$$

Indeed, using the defining formula for $\tilde{n}_w$ (for any $w \in W$) and the fact that $\tilde{n}_w = \tilde{n}_{w^{-1}}$ (see Section 1.5), we obtain that

$$\sum_{d \in \tilde{\mathscr{D}} \cap \mathfrak{C}} \tilde{n}_d c_{d,\lambda} = \sum_{w \in \mathfrak{C}} \tilde{n}_w c_{w,\lambda} = \sum_{\mu \in \Lambda} f_\mu^{-1} \sum_{w \in \mathfrak{C}} c_{w^{-1},\mu} c_{w,\lambda},$$

and it remains to use the refined orthogonality relations in Theorem 1.8.1. The second formula is proved similarly.

**Example 1.8.6.** Assume that Conjectures 1.5.12 and 1.6.18 hold. Then $\tilde{\gamma}_{x,y,z} \in \mathbb{Z}$ and $\tilde{n}_d \in \mathbb{Z}$ for all $x, y, z \in W$ and $d \in \tilde{\mathscr{D}}$. So the relation in Remark 1.6.19(a) shows that $\tilde{n}_d = \pm 1$ for all $d \in \tilde{\mathscr{D}}$. Now let $\mathfrak{C}$ be a left $\tilde{\mathbf{J}}$-cell and $d \in \tilde{\mathscr{D}}$ be the unique element in $\mathfrak{C}$. Then the formulae in Example 1.8.5 reduce to

(a) $$\tilde{m}(\mathfrak{C}, \lambda) = \tilde{n}_d\, c_{d,\lambda} = \pm c_{d,\lambda} \qquad \text{and} \qquad \sum_{\lambda \in \Lambda} f_\lambda^{-1}\, \tilde{m}(\mathfrak{C}, \lambda) = 1.$$

Furthermore, for any $y \in W$, we have $t_y t_d = \sum_{x \in W} \tilde{\gamma}_{y,d,x^{-1}}\, t_x = \sum_{x \in W} \tilde{\gamma}_{x^{-1},y,d}\, t_x$, where the second equality holds by Lemma 1.5.3. Then the formula in Remark 1.6.19(a) shows that $t_y t_d = \tilde{n}_d\, t_y$ if $y \sim_{\mathscr{L}} d$, and $t_y t_d = 0$ if $y \not\sim_{\mathscr{L}} d$. Thus, we have

(b) $$\tilde{\mathbf{J}} t_d = \langle t_y \mid y \in \mathfrak{C} \rangle_{\mathbb{K}} \cong [\mathfrak{C}]_{\mathbb{K}} \qquad \text{(isomorphism of left } \tilde{\mathbf{J}}\text{-modules).}$$

Again, it would be very interesting and useful to find direct and elementary proofs of these statements which do not rely on the above conjectures.

**1.8.7.** Let us assume that $W = W_n$ is of type $A_{n-1}$ or $B_n$ and that $L = L_n$ is a weight function satisfying the conditions in 1.7.6. Then we have a bijection

$$\bigsqcup_{\lambda \in \Lambda} (M(\lambda) \times M(\lambda)) \xrightarrow{1-1} W, \qquad (\mathfrak{s}, \mathfrak{t}) \mapsto w_\lambda(\mathfrak{s}, \mathfrak{t}),$$

with the properties in Theorem 1.7.10. In particular, we have

$$\mathscr{F}_\lambda = \{ w_\lambda(\mathfrak{s}, \mathfrak{t}) \mid \mathfrak{s}, \mathfrak{t} \in M(\lambda) \} \qquad \text{for any } \lambda \in \Lambda;$$

furthermore, the set $\mathfrak{C}_{\mathfrak{t}} := \{ w_\lambda(\mathfrak{s}, \mathfrak{t}) \mid \mathfrak{s} \in M(\lambda) \}$ is a left $\tilde{\mathbf{J}}$-cell for each $\mathfrak{t} \in M(\lambda)$. Let $[\mathfrak{C}_{\mathfrak{t}}]_{\mathbb{K}}$ be the corresponding left cell module, with basis $\{ e_{\mathfrak{s}} \mid \mathfrak{s} \in M(\lambda) \}$, where $e_{\mathfrak{s}} = t_{w_\lambda(\mathfrak{s},\mathfrak{t})}$ for $\mathfrak{s} \in M(\lambda)$. Explicitly, the action of $\tilde{\mathbf{J}}$ on $[\mathfrak{C}_{\mathfrak{t}}]_{\mathbb{K}}$ is given by

$$t_w . e_{\mathfrak{s}'} = \sum_{\mathfrak{s} \in M(\lambda)} \tilde{\gamma}_{w, w_\lambda(\mathfrak{s}', \mathfrak{t}), w_\lambda(\mathfrak{t}, \mathfrak{s})}\, e_{\mathfrak{s}} \qquad \text{for all } w \in W \text{ and } \mathfrak{s}' \in M(\lambda),$$

where we used that $w_\lambda(\mathfrak{s}, \mathfrak{t})^{-1} = w_\lambda(\mathfrak{t}, \mathfrak{s})$. Let $\hat{\rho}^{\mathfrak{t}} \colon \tilde{\mathbf{J}} \to M_{d_\lambda}(\mathbb{K})$ denote the corresponding matrix representation; we have

$$\hat{\rho}^{\mathfrak{t}}_{\mathfrak{s}\mathfrak{s}'}(t_w) = \tilde{\gamma}_{w, w_\lambda(\mathfrak{s}', \mathfrak{t}), w_\lambda(\mathfrak{t}, \mathfrak{s})} \qquad \text{for all } w \in W \text{ and } \mathfrak{s}, \mathfrak{s}' \in M(\lambda).$$

(Here, we assume that $M(\lambda)$ has been ordered in some way so that it makes sense to write down matrices with rows and columns indexed by $M(\lambda)$.)

**Lemma 1.8.8.** *We have $\hat{\rho}^{\mathfrak{t}}_{\mathfrak{s}\mathfrak{s}'}(t_w) = \pm c_{w,\lambda}^{\mathfrak{s}\mathfrak{s}'}$ for all $w \in W$ and all $\mathfrak{s}, \mathfrak{s}' \in M(\lambda)$. In particular, the matrix $\bar{\rho}^\lambda(t_w)$ is zero unless $w \in \mathscr{F}_\lambda$, in which case it has precisely one non-zero entry (which is $\pm 1$).*

*Proof.* If $w \notin \mathscr{F}_\lambda$, then $\hat{\rho}^{\mathfrak{t}}(t_w) = 0$ by Corollary 1.6.7. Now assume that $w \in \mathscr{F}_\lambda$ and write $w = w_\lambda(\mathfrak{u}, \mathfrak{v})$, where $\mathfrak{u}, \mathfrak{v} \in M(\lambda)$. Also write $x = w_\lambda(\mathfrak{s}', \mathfrak{t})$ and $y = w_\lambda(\mathfrak{t}, \mathfrak{s})$. Then the defining equation for $\tilde{\gamma}_{w,x,y}$ in combination with Theorem 1.7.10 yields that

$$\hat{\rho}^{\mathfrak{t}}_{\mathfrak{s}\mathfrak{s}'}(t_w) = \tilde{\gamma}_{w,x,y} = c^{\mathfrak{u}\mathfrak{v}}_{w,\lambda}\, c^{\mathfrak{v}\mathfrak{t}}_{x,\lambda}\, c^{\mathfrak{t}\mathfrak{u}}_{y,\lambda} = \pm\delta_{\mathfrak{u}\mathfrak{s}}\delta_{\mathfrak{v}\mathfrak{s}'}\, c^{\mathfrak{u}\mathfrak{v}}_{w,\lambda},$$

as required. This also shows that $\hat{\rho}^{\mathfrak{t}}_{\mathfrak{s}\mathfrak{s}'}(t_w)$ has precisely one non-zero entry. □

**Proposition 1.8.9.** *In the setting of 1.8.7, let $\lambda \in \Lambda$ and $\mathfrak{t} \in M(\lambda)$.*

*(a) The cell representation $\hat{\rho}^{\mathfrak{t}}$ is irreducible and equivalent to $\bar{\rho}^\lambda$. In fact, we have*

$$\hat{\rho}^{\mathfrak{t}}(t_w) = D_{\mathfrak{t}}\bar{\rho}^\lambda(t_w)D_{\mathfrak{t}}^{-1} \qquad \textit{for all } w \in W,$$

*where $D_{\mathfrak{t}} \in M_{d_\lambda}(\mathbb{K})$ is a diagonal matrix with $\pm 1$ on the diagonal.*

*(b) The balanced representation $\rho^\lambda$ of $\mathbf{H}_K$ can be chosen such that, in addition to the conditions in Proposition 1.5.11, we have $\bar{\rho}^\lambda(t_w) = \hat{\rho}^{\mathfrak{t}}(t_w)$ for all $w \in W$. Then, in particular, each $c^{\mathfrak{s}\mathfrak{t}}_{w,\lambda}$ is equal to $\tilde{\gamma}_{w,x,y}$ for suitable $x, y \in \mathscr{F}_\lambda$.*

*Proof.* Let $\mu \in \Lambda$ be such that $\tilde{m}(\mathfrak{C}_{\mathfrak{t}}, \mu) > 0$. By Example 1.8.2, we have $\mathfrak{C}_{\mathfrak{t}} \subseteq \mathscr{F}_\mu$ and so $\mu = \lambda$. Now $|\mathfrak{C}_{\mathfrak{t}}| = |M(\lambda)| = d_\lambda = \dim E^\lambda$ and so $\hat{\rho}^{\mathfrak{t}}$ must be irreducible and equivalent to $\bar{\rho}^\lambda$. So there exists an invertible matrix $D_{\mathfrak{t}} \in M_{d_\lambda}(\mathbb{K})$ such that $\hat{\rho}^{\mathfrak{t}}(t_w) = D_{\mathfrak{t}}\bar{\rho}^\lambda(t_w)D_{\mathfrak{t}}^{-1}$ for all $w \in W$. Taking into account Lemma 1.8.8, a straightforward computation now shows that $D_{\mathfrak{t}}$ must be a scalar multiple of a diagonal matrix with $\pm 1$ on the diagonal. Thus, (a) is proved.

To prove (b) it is sufficient to note that if $\rho^\lambda$ is a balanced representation satisfying the conditions in Proposition 1.5.11, then the representation defined by $T_w \mapsto D_{\mathfrak{t}}\rho^\lambda(T_w)D_{\mathfrak{t}}^{-1}$ is still balanced and it still satisfies the conditions in Proposition 1.5.11. Hence, we just have to replace $\rho^\lambda$ by this new representation. □

**Conjecture 1.8.10.** *In the setting of 1.8.7, let $\lambda \in \Lambda$ and $\mathfrak{t}, \mathfrak{t}' \in M(\lambda)$. Then we actually have $\hat{\rho}^{\mathfrak{t}}(t_w) = \hat{\rho}^{\mathfrak{t}'}(t_w)$ for all $w \in W$.*

This is known to be true when $W \cong \mathfrak{S}_n$, but the proof requires much more work. (It can be deduced, for example, by combining Lemma 1.8.12 below with the discussion in Remark 2.8.18.) It would be very useful to have a direct proof of Conjecture 1.8.10, valid in all situations described in 1.7.6.

*Remark 1.8.11.* Let $\lambda \in \Lambda$ and $x, x', y, y' \in \mathscr{F}_\lambda$ be such that $x \sim_{\mathscr{R}} x'$, $y \sim_{\mathscr{R}} y'$, $x \sim_{\mathscr{L}} y$, $x' \sim_{\mathscr{L}} y'$. Then Conjecture 1.8.10 is equivalent to

$$(\heartsuit) \qquad \tilde{\gamma}_{w,x,y^{-1}} = \tilde{\gamma}_{w,x',y'^{-1}} \qquad \text{for all } w \in W \text{ and all } x, x', y, y' \in \mathscr{F}_\lambda \text{ as above.}$$

Indeed, by Theorem 1.7.10, we have $x = w_\lambda(\mathfrak{s}, \mathfrak{t})$, $y = w_\lambda(\mathfrak{s}', \mathfrak{t})$, $x' = w_\lambda(\mathfrak{s}, \mathfrak{t}')$, $y' = w_\lambda(\mathfrak{s}', \mathfrak{t}')$, where $\mathfrak{s}, \mathfrak{s}', \mathfrak{t}, \mathfrak{t}' \in M(\lambda)$, and it remains to recall the definition of $\hat{\rho}^{\mathfrak{t}}$, $\hat{\rho}^{\mathfrak{t}'}$.

**Lemma 1.8.12 (Cf. [107, 5.10]).** *Assume that Conjecture 1.8.10 holds. Then there exists a sign $n_\lambda = \pm 1$ (depending only on $\lambda$) such that, for all $x, y, z \in \mathscr{F}_\lambda$, we have*

$$\tilde{\gamma}_{x,y,z} = \begin{cases} n_\lambda & \text{if } x \sim_{\mathscr{L}} y^{-1},\, y \sim_{\mathscr{L}} z^{-1},\, z \sim_{\mathscr{L}} x^{-1}, \\ 0 & \text{otherwise.} \end{cases}$$

*Conversely, if the above formula holds, then Conjecture 1.8.10 is true.*

*Proof.* By Lemma 1.6.5, we have $\tilde{\gamma}_{x,y,z} = 0$ unless $x \sim_{\mathscr{L}} y^{-1}, y \sim_{\mathscr{L}} z^{-1}, z \sim_{\mathscr{L}} x^{-1}$. Now assume that $x, y, z$ are related in this way. Then $x = w_\lambda(\mathfrak{s}, \mathfrak{t})$, $y = w_\lambda(\mathfrak{t}, \mathfrak{u})$, $z = w_\lambda(\mathfrak{u}, \mathfrak{s})$ for some $\mathfrak{s}, \mathfrak{t}, \mathfrak{u} \in M(\lambda)$; see Theorem 1.7.10. Let us also set $d := w_\lambda(\mathfrak{t}, \mathfrak{t})$. Then $d \sim_{\mathscr{L}} d$, $d = d^{-1} \sim_{\mathscr{R}} y$, $x \sim_{\mathscr{R}} z^{-1}$. So we can apply (♡) and this yields

$$\tilde{\gamma}_{x,y,z} = \tilde{\gamma}_{x,d,x^{-1}} = \tilde{\gamma}_{x^{-1},x,d},$$

where we also used Lemma 1.5.3(a). Furthermore, since every left cell contains a unique element in $\tilde{\mathscr{D}}$ (see Theorem 1.7.10(c)), the identity $\sum_{w \in W} \tilde{\gamma}_{x^{-1},x,w} \tilde{n}_w = 1$ in Lemma 1.5.3(b) now reduces to $\tilde{\gamma}_{x^{-1},x,d} \tilde{n}_d = 1$. Hence, we conclude that $\tilde{\gamma}_{x,y,z} = \tilde{n}_d$, where $d = w_\lambda(\mathfrak{t}, \mathfrak{t})$. So it remains to show that $\tilde{n}_d = \tilde{n}_{d'}$ for any $d' \in \tilde{\mathscr{D}} \cap \mathscr{F}_\lambda$. Let $d' = w_\lambda(\mathfrak{t}', \mathfrak{t}')$, where $\mathfrak{t}' \in M(\lambda)$, and consider the element $w = w_\lambda(\mathfrak{t}, \mathfrak{t}')$. Then $w^{-1} = w_\lambda(\mathfrak{t}', \mathfrak{t})$ and so $w \sim_{\mathscr{L}} d'$, $d \sim_{\mathscr{L}} w^{-1}$, $w \sim_{\mathscr{R}} d$, $d' \sim_{\mathscr{R}} w^{-1}$. Again, we can apply (♡); arguing as above, this now yields $\tilde{n}_{d'} = \tilde{\gamma}_{w^{-1},w,d'} = \tilde{\gamma}_{w^{-1},d,w} = \tilde{n}_d$, as required.

Conversely, assume that the above formula for $\tilde{\gamma}_{x,y,z}$ is true. Then all the non-zero entries of $\hat{\rho}^{\mathfrak{t}}(t_w)$ have the same sign, for any $w \in W$ and any $\mathfrak{t} \in M(\lambda)$. Hence, the required identity in Conjecture 1.8.10 is clear by the formula in Lemma 1.8.8. □

# Chapter 2
# Kazhdan–Lusztig Cells and Cellular Bases

The aim of this chapter is to develop a general framework for studying the representation theory of Iwahori–Hecke algebras associated with finite Coxeter groups.

The motivating example is the representation theory of the symmetric group $\mathfrak{S}_n$. Frobenius showed around 1900 that the irreducible representations of $\mathfrak{S}_n$ over a field of characteristic 0 are naturally parametrised by the partitions of $n$. In the 1970s, James [181] developed a "characteristic-free" approach to the representation theory of $\mathfrak{S}_n$, where Specht modules and certain bilinear forms on them play a crucial role. Dipper and James [62] extended this theory to Iwahori–Hecke algebras associated with $\mathfrak{S}_n$. A considerable simplification was then achieved through the powerful new ideas introduced by Murphy [256], [257]. In fact, what we nowadays call the "Murphy basis" is an example of a "cellular basis" in the formal sense defined later by Graham and Lehrer [144].

Here, we shall construct such a "cellular basis" in the sense of Graham and Lehrer, for the generic algebra **H** associated with an arbitrary (finite) Coxeter group $W$. For this purpose, we need two basic ingredients:

(1) a basis of **H** with certain specific multiplicative properties and
(2) a suitable partial ordering on $\mathrm{Irr}(W)$.

Already Graham and Lehrer identified the Kazhdan–Lusztig basis $\{C_w \mid w \in W\}$ (see Section 2.1) of **H** as a natural candidate for (1). However, it is only in some very special examples (in type $A$ or $B$) that $\{C_w\}$ itself has the required multiplicative properties. But in any case, this new basis of **H** provides the necessary tools to define a partial ordering on $\mathrm{Irr}(W)$; see Section 2.2.

In order to proceed, we have to rely on certain deep properties of the basis $\{C_w\}$ for which no elementary proofs are known. Sections 2.3–2.5 are devoted to a discussion of these properties, which appear as conjectures **P1–P15** in Lusztig's book [231]. We can then put all the pieces together and construct, following [111], [112], a "cellular basis" for **H**; see Sections 2.6 and 2.7. In the final section, we present an elementary treatment of the case where $W \cong \mathfrak{S}_n$.

M. Geck, N. Jacon, *Representations of Hecke Algebras at Roots of Unity*,
Algebra and Applications 15, DOI 10.1007/978-0-85729-716-7_2,

## 2.1 The Kazhdan–Lusztig Basis

Let $W$ be a finite Coxeter group and $L\colon W \to \Gamma$ a weight function, where $\Gamma$ is an abelian group admitting a monomial order $\leqslant$ such that $L(s) \geqslant 0$ for all $s \in S$ (as in Chapter 1). Let $\mathbf{H} = \mathbf{H}_A(W,S,L)$ be the corresponding generic Iwahori–Hecke algebra over $A = R[\Gamma]$, where $R \subseteq \mathbb{C}$ is a subring as in 1.2.1. The main purpose of this section is to introduce the *Kazhdan–Lusztig basis* $\{C_w \mid w \in W\}$ of $\mathbf{H}$. This basis first appeared in [195], in the equal-parameter case. Then Lusztig [219] showed that the construction also works in the general multiparameter case. These results are now readily accessible in Lusztig's book [231], so we will outline the main constructions and formulate the main results but refer to [231] for further details.

**2.1.1.** Given elements $y, w \in W$, we write $y \leqslant w$ if $y$ can be obtained by omitting some terms in a reduced expression for $w$. This defines a partial order relation on $W$, called the *Bruhat–Chevalley order*. Here are some properties (see [231, Chap. 2]):

(a) Let $w \in W$ and $s \in S$. Then $sw < w$ if and only if $l(sw) = l(w) - 1$.
(b) Let $y, w \in W$ and $s \in S$ be such that $sw < w$. Then

$$y \leqslant w \qquad \Leftrightarrow \qquad \begin{cases} sy \leqslant sw & \text{if } sy < y, \\ y \leqslant sw & \text{if } sy > y. \end{cases}$$

Note that (b) provides a recursive description of $\leqslant$.

**2.1.2.** Let $w_0 \in W$ be the longest element. For any $w \in W$, we can write uniquely

$$T_w T_{w_0} = \sum_{y \in W} R^*_{y,w} T_{yw_0}, \qquad \text{where } R^*_{y,w} \in \mathbb{Z}[\Gamma].$$

If $w = 1$, then $R^*_{1,1} = 0$ and $R^*_{y,1} = 0$ for all $y \neq 1$. Now assume that $w \neq 1$ and let $s \in S$ be such that $sw < w$. Then one easily checks the following relation:

$$R^*_{y,w} = \begin{cases} R^*_{sy,sw} & \text{if } sy < y, \\ R^*_{sy,sw} + (v_s - v_s^{-1})R^*_{y,sw} & \text{if } sy > y. \end{cases}$$

By using 2.1.1 and the above formulae, we obtain (see also [231, 4.5 and 4.7])

(a) $$R^*_{w,w} = 1 \qquad \text{and} \qquad R^*_{y,w} = 0 \quad \text{unless } y \leqslant w,$$

(b) $$\bar{R}^*_{y,w} = (-1)^{l(y)+l(w)} R^*_{y,w}.$$

(Here, $\bar{a}$ for any $a \in A$ is defined in Example 1.2.6.) The above recursion formulae are the same as those for the elements $r_{y,w}$ in [231, 4.4]. Consequently, we have

(c) $$T_{w^{-1}}^{-1} = \sum_{y \in W} \bar{R}^*_{y,w} T_y \qquad \text{for any } w \in W.$$

(The relation between the expressions for $T_w T_{w_0}$ and $T_{w^{-1}}^{-1}$ already appeared in the remarks following [195, Lemma A.4].)

**2.1.3.** We set $\Gamma_{\geqslant 0} = \{g \in \Gamma \mid g \geqslant 0\}$ and denote by $\mathbb{Z}[\Gamma_{\geqslant 0}]$ the set of all integral linear combinations of terms $\varepsilon^g$, where $g \geqslant 0$. The notations $\mathbb{Z}[\Gamma_{>0}]$, $\mathbb{Z}[\Gamma_{\leqslant 0}]$, $\mathbb{Z}[\Gamma_{<0}]$ have a similar meaning. Then, by the proof of [231, Theorem 5.2] (see also [228, 7.10]), there exists a unique collection of elements $\{P^*_{y,w} \mid y,w \in W\} \subseteq \mathbb{Z}[\Gamma]$ satisfying the following conditions:

(a) $P^*_{w,w} = 1$ and $P^*_{y,w} = 0$ unless $y \leqslant w$; furthermore, $P^*_{y,w} \in \mathbb{Z}[\Gamma_{<0}]$ if $y < w$.
(b) For any $y, w \in W$, we have

$$\bar{P}^*_{y,w} = \sum_{z \in W : y \leqslant z \leqslant w} R^*_{y,z} P^*_{z,w}.$$

Note that $P^*_{y,w}$ can be constructed recursively using (a) and (b); see Example 2.1.5. (Here, the notation is as in [219]; $R^*_{y,w}$, $P^*_{y,w}$ are denoted by $r_{y,w}$, $p_{y,w}$ in [231].)

**Definition 2.1.4 (Kazhdan and Lusztig [195], Lusztig [219]).** For $w \in W$, we set

$$C_w := \sum_{y \in W} (-1)^{l(w)+l(y)} \bar{P}^*_{y,w} T_y \in \mathbf{H},$$

with $P^*_{y,w}$ as in 2.1.3. The elements $\{C_w \mid w \in W\}$ form an $A$-basis of $\mathbf{H}$; to see this just note that, by 2.1.3(a), we have $T_w \in C_w + \sum_{y \in W : y < w} \mathbb{Z}[\Gamma_{>0}] T_y$ for any $w \in W$.

For $x, y \in W$, let us write

$$C_x C_y = \sum_{z \in W} h_{x,y,z} C_z, \qquad \text{where} \qquad h_{x,y,z} \in A.$$

**Example 2.1.5.** The formulae in 2.1.2 yield a straighforward algorithm for computing the polynomials $R^*_{y,w}$. As already mentioned above, the formulae in 2.1.3 can then be used to construct $P^*_{y,w}$ recursively. Indeed, given $y < w$, note that

$$\bar{P}^*_{y,w} - P^*_{y,w} = \sum_{z \in W : z < y \leqslant w} R^*_{y,z} P^*_{z,w}.$$

Proceeding by induction on $l(w) - l(y)$, all terms on the right-hand side are known. Then $P^*_{y,w}$ itself is determined by the additional condition that $P^*_{y,w} \in \mathbb{Z}[\Gamma_{<0}]$.

For example, it is clear that $C_1 = T_1$. Now let $s \in S$. Then $R^*_{1,s} = v_s - v_s^{-1}$ and so

$$\bar{P}^*_{1,s} - P^*_{1,s} = \sum_{z \in W : 1 < z \leqslant s} R^*_{1,z} P^*_{z,s} = R^*_{1,s} P^*_{s,s} = v_s - v_s^{-1}.$$

Since $P^*_{1,s} \in \mathbb{Z}[\Gamma_{<0}]$, it follows that $P^*_{1,s} = 0$ (if $L(s) = 0$) and $P^*_{1,s} = v_s^{-1}$ (if $L(s) > 0$). Thus, we obtain

$$C_s = \begin{cases} T_s & \text{if } L(s) = 0, \\ T_s - v_s T_1 & \text{if } L(s) > 0. \end{cases}$$

In order to see some more complicated polynomials $P^*_{y,w}$, let $s, t \in S$ be such that $m_{st} \geqslant 3$ and assume that $L(t) \geqslant L(s) > 0$. Then the above precedure yields

$$C_{st} = T_{st} - v_t T_s - v_s T_t + v_s v_t T_1$$

and

$$C_{tst} = T_{tst} - v_t T_{st} - v_t T_{ts} + v_t^2 T_s$$
$$+ \begin{cases} v_s v_t T_t - v_s v_t^2 T_1 & \text{if } L(s) = L(t), \\ (v_s v_t - v_s^{-1} v_t) T_t - (v_s v_t^2 - v_s^{-1} v_t^2) T_1 & \text{if } L(t) > L(s). \end{cases}$$

With some more effort, it is possible to write down explicit formulae for all basis elements $C'_w$, where $W$ is of type $I_2(m)$; see [231, Chap. 7], [132, Exc. 11.4].

We can now state the following characterisation of the element $C_w$. This version of the characterisation (which works for finite $W$) is due to Lusztig [232]; the proof is very similar to (but the statement as such is different from) the original one in [195], [219] (which relied on the "bar involution" on $\mathbf{H}$).

**Theorem 2.1.6 (Kazhdan and Lusztig [195], Lusztig [219], [231], [232]).** *For any $w \in W$, the element $C_w$ is uniquely determined by the following two conditions:*

$$C_w \in T_w + \sum_{y \in W} \mathbb{Z}[\Gamma_{>0}] T_y \qquad \textit{and} \qquad C_w T_{w_0} \in \sum_{y \in W} \mathbb{Z}[\Gamma_{\leqslant 0}] T_y.$$

*Proof.* Let us verify that $C_w$ satisfies the above two conditions. The first one is clear by 2.1.3(a); furthermore, using the relations (a), (b) in 2.1.2, we obtain:

$$\begin{aligned} C_w T_{w_0} &= \sum_{y \in W} (-1)^{l(w)+l(y)} \Big( \sum_{z \in W : y \leqslant z \leqslant w} (-1)^{l(y)+l(z)} R^*_{y,z} \bar{P}^*_{z,w} \Big) T_{yw_0} \\ &= \sum_{y \in W} (-1)^{l(w)+l(y)} \Big( \sum_{z \in W : y \leqslant z \leqslant w} \overline{R}^*_{y,z} \overline{P}^*_{z,w} \Big) T_{yw_0} \\ &= \sum_{y \in W} (-1)^{l(w)+l(y)} P^*_{y,w} T_{yw_0}, \end{aligned}$$

where the last equality holds by 2.1.3(b). Thus, we have in fact

$$C_w T_{w_0} \in T_{ww_0} + \sum_{y \in W : y < w} \mathbb{Z}[\Gamma_{<0}] T_{yw_0} \quad \subseteq \quad \sum_{y \in W} \mathbb{Z}[\Gamma_{\leqslant 0}] T_y,$$

as required. Using this expression for $C_w T_{w_0}$, one easily deduces the following statement. Let $h = \sum_{x \in W} a_x C_x \in \mathbf{H}$, where $a_x \in \mathbb{Z}[\Gamma]$ for all $x \in W$. Then we have

$$(*) \qquad hT_{w_0} \in \sum_{y \in W} \mathbb{Z}[\Gamma_{\leqslant 0}] T_y \qquad \Rightarrow \qquad a_x \in \mathbb{Z}[\Gamma_{\leqslant 0}] \quad \text{for all } x \in W.$$

This immediately implies the uniqueness of $C_w$. Indeed, assume that $\tilde{C}_w \in \mathbf{H}$ also satisfies the desired conditions. Let $h := C_w - \tilde{C}_w$. Then we have

$$h \in \sum_{y \in W} \mathbb{Z}[\Gamma_{>0}] T_y \subseteq \sum_{y \in W} \mathbb{Z}[\Gamma_{>0}] C_y \qquad \text{and} \qquad hT_{w_0} \in \sum_{y \in W} \mathbb{Z}[\Gamma_{\leqslant 0}] T_y.$$

Hence, using $(*)$, we conclude that, in an expression of $h$ as a linear combination of basis elements $\{C_x\}$, all coefficients must be zero and so $\tilde{C}_w = C_w$. □

*Remark 2.1.7.* As in [195], [219], we set $C'_w := (-1)^{l(w)} C_w^\dagger$ for all $w \in W$, where $\dagger$ is defined in Example 1.2.6. (The element $C'_w$ is denoted by $c_w$ in [231].) Using the formula $T_w^\dagger = (-1)^{l(w)} T_{w^{-1}}^{-1}$ and the relations in 2.1.2, 2.1.3, we obtain

$$C'_w = \sum_{z\in W} (-1)^{l(z)} \bar{P}^*_{z,w} T_z^\dagger = \sum_{z\in W} \bar{P}^*_{z,w} T_{z^{-1}}^{-1} = \sum_{y\in W} (\bar{P}^*_{z,w}\bar{R}^*_{y,z}) T_y = \sum_{y\in W} P^*_{y,w} T_y.$$

Furthermore, applying $\dagger$ to the relation $C_x C_y = \sum_{z\in W} h_{x,y,z} C_z$, we obtain

$$C'_x C'_y = \sum_{z\in W} (-1)^{l(x)+l(y)+l(z)}\, h_{x,y,z}\, C'_z \qquad \text{for any } x, y \in W.$$

We shall write $h'_{x,y,z} := (-1)^{l(x)+l(y)+l(z)}\, h_{x,y,z}$ for any $x, y, z \in W$.

Thus, any statement about $C_w$ has an equivalent version for $C'_w$ (where typically some signs need to be arranged). For applications to representation theory, it is more convenient to work with $C_w$; see, for example, Remark 2.1.12. In this book, we will systematically work with $C_w$.

**Theorem 2.1.8 (Kazhdan and Lusztig [195], Lusztig [219], [231, Chap. 6]).** *For any $x, y, z \in W$, we have $h_{x,y,z} = \bar{h}_{x,y,z}$. Furthermore, for $s \in S$ and $w \in W$, we have*

$$C_s C_w = \begin{cases} C_{sw} & \text{if } L(s) = 0, \\ -(v_s + v_s^{-1})\, C_w & \text{if } L(s) > 0 \text{ and } sw < w, \\ C_{sw} - \displaystyle\sum_{y\in W: sy<y<w} (-1)^{l(w)+l(y)} \mu^s_{y,w} C_y & \text{if } L(s) > 0 \text{ and } sw > w, \end{cases}$$

*where $\mu^s_{y,w} \in \mathbb{Z}[\Gamma]$ is such that $\bar{\mu}^s_{y,w} = \mu^s_{y,w}$.*

(The analogous formulae for the elements $\{C'_w\}$ are proved in [231, Chap. 6]; then it remains to use the conversion formulae in 2.1.7.)

**2.1.9.** There is a direct recursive algorithm for simultaneously computing

$$\{P^*_{y,w} \mid y, w \in W\} \quad \text{and}$$
$$\{\mu^s_{y,w} \mid s \in S, y, w \in W \text{ such that } L(s) > 0 \text{ and } sy < y < w < sw\},$$

without reference to the polynomials $\{R^*_{y,w}\}$. Recall that, first of all, we have

(a) $\qquad P^*_{w,w} = 1$ for all $w \in W \qquad$ and $\qquad P^*_{y,w} = 0$ unless $y < w$;

see 2.1.3. We shall now list some further properties of these elements. Let $y, w \in W$ be such that $y < w$. Let $t \in S$ be such that $tw < w$. Then we have

(b1) $\quad P^*_{y,w} = P^*_{ty,tw}$ $\qquad$ if $L(t) = 0$,

(b2) $\quad P^*_{y,w} = v_t^{-1} P^*_{ty,w}$ $\qquad$ if $L(t) > 0$, $ty > y$,

(b3) $\quad P^*_{y,w} = v_t P^*_{y,tw} + P^*_{ty,tw} - \displaystyle\sum_{z\in W: y\leqslant z<tw, tz<z} P^*_{y,z} \mu^t_{z,tw}$ $\qquad$ if $L(t) > 0$, $ty < y$.

Furthermore, for any $s \in S$ such that $L(s) > 0$ and $sy < y < w < sw$, we have

(c1) $$\mu^s_{y,w} - v_s P^*_{y,w} + \sum_{z \in W: y<z<w, sz<z} P^*_{y,z}\mu^s_{z,w} \in \mathbb{Z}[\Gamma_{<0}],$$

(c2) $$\overline{\mu}^s_{y,w} = \mu^s_{y,w}.$$

See [231, Chap. 6] and [132, §11.1]. In order to describe a recursion based on the above properties, we need to define an ordering on all pairs of elements $(y,w)$, where $y,w \in W$ and $y \leqslant w$. This is done as follows:

$$(y',w') \sqsubseteq (y,w) \quad \stackrel{\text{def}}{\Leftrightarrow} \quad w' < w \quad \text{or} \quad w' = w \text{ and } y \leqslant y'.$$

The recursion starts with the pair $(y,w) = (1,1)$. We have $P^*_{1,1} = 1$ and there are no $\mu$-polynomials to determine for this pair. Now let $(y,w)$ be such that $w \neq 1$ and $y \leqslant w$. Assume that $P^*_{y',w'}$ and the relevant $\mu$-polynomials are already known for all pairs $(y',w') \sqsubset (y,w)$. Then we proceed as follows.

(1) First we determine $P^*_{y,w}$. If $y = w$, then $P^*_{w,w} = 1$. If $y < w$, then choose some $t \in S$ such that $tw < w$. There are three cases to distinguish:

  (i) If $L(t) = 0$, then $(ty,tw) \sqsubset (y,w)$ and so the right-hand side of (b1) is known by induction.
  (ii) If $L(t) > 0$ and $ty > y$, then $(ty,w) \sqsubset (y,w)$ and so the right-hand side of (b2) is known by induction.
  (iii) If $L(t) > 0$ and $ty < y$, then all terms on the right-hand side of (b3) involve pairs $(y',w')$, where $w' < w$. In particular, $(y',w') \sqsubset (y,w)$ for all such pairs and so, by induction, the right-hand side of (b3) is known.

(2) Now assume that $y < w$. Then we have to determine $\mu^s_{y,w}$ for any $s \in S$ such that $L(s) > 0$ and $sy < y < w < sw$. For this purpose, we set

$$\alpha := v_s P^*_{y,w} - \sum_{z \in W: y<z<w, sz<z} P^*_{y,z}\mu^s_{z,w}.$$

  (i) For all $z$ appearing in the above sum, we have $(y,z) \sqsubset (y,w)$ and $(z,w) \sqsubset (y,w)$ and, hence, the corresponding terms are known by induction. By (1), we also know $P^*_{y,w}$. Thus, $\alpha$ is determined.
  (ii) Write $\alpha = \alpha_+ + \alpha_0 + \alpha_-$, where $\alpha_+ \in \mathbb{Z}[\Gamma_{>0}]$, $\alpha_- \in \mathbb{Z}[\Gamma_{<0}]$ and $\alpha_0 \in \mathbb{Z}$ are uniquely determined. By (c1) and (c2), we have $\mu^s_{y,w} = \alpha_+ + \alpha_0 + \overline{\alpha}_+$. Thus, $\mu^s_{y,w}$ is determined.

For readers with an interest in "computer algebra" we just mention that it is an excellent programming exercise to implement the above recursion on a computer. For further details see, for example, DuCloux [75] and his COXETER system, CHEVIE [105], [118], and the references in these articles.

The above recursion formulae can actually be used to establish some further properties of $P^*_{y,w}$ and $\mu^s_{y,w}$. We illustrate this with a few examples.

**Example 2.1.10.** Let $y, w \in W$ and $s \in S$. Then we claim that

(a) $$v_s \mu^s_{y,w} \in \mathbb{Z}[\Gamma_{>0}], \quad \text{where} \quad L(s) > 0 \text{ and } sy < y < w < sw.$$

Indeed, by 2.1.9(c2), this is equivalent to showing that $v_s^{-1}\mu^s_{y,w} \in \mathbb{Z}[\Gamma_{<0}]$. Multiplying 2.1.9(c1) by $v_s^{-1}$, we obtain

$$v_s^{-1}\mu^s_{y,w} - P^*_{y,w} + \sum_{z \in W: y<z<w, sz<z} P^*_{y,z}\left(v_s^{-1}\mu^s_{z,w}\right) \in \mathbb{Z}[\Gamma_{<0}].$$

By an inductive argument, we can assume that we already know that $v_s^{-1}\mu^s_{z,w} \in \mathbb{Z}[\Gamma_{<0}]$ for all $z$ in the above sum. Hence, we also have $v_s^{-1}\mu^s_{y,w} \in \mathbb{Z}[\Gamma_{<0}]$, as required.

Assume, furthermore, that we are in the equal-parameter case where $\Gamma = \mathbb{Z}$ and $L(s) = 1$ for all $s \in S$. Then $\mathbb{Z}[\Gamma]$ is the ring of Laurent polynomials in one indeterminate $v = \varepsilon$. Let $y, w \in W$ and $s \in S$ be such that $sy < y < w < sw$. We have just seen that $v\mu^s_{y,w} \in v\mathbb{Z}[v]$. Hence, we have $\mu^s_{y,w} \in \mathbb{Z}[v]$. Since $\overline{\mu}^s_{y,w} = \mu^s_{y,w}$, we conclude that $\mu^s_{y,w} \in \mathbb{Z}$. In fact, we have

(b) $$\mu^s_{y,w} = \text{coefficient of } v^{-1} \text{ in } P^*_{y,w} \in v^{-1}\mathbb{Z}[v^{-1}].$$

Indeed, since $\mu^s_{y,w} \in \mathbb{Z}$, the relation in 2.1.9(c1) reduces to the condition that $\mu^s_{y,w} - vP^*_{y,w} \in v^{-1}\mathbb{Z}[v^{-1}]$, which immediately yields the above statement.

**Example 2.1.11.** Let $y, w \in W$ be such that $y \leqslant w$ and set $P_{y,w} := v_w v_y^{-1} P^*_{y,w}$. Then the following holds:

(a) If $L(s) > 0$ for all $s \in S$, then $P_{y,w} \in \mathbb{Z}[\Gamma_{\geqslant 0}]$ is non-zero, with constant term 1.

This is proved as follows (see also [231, Prop. 5.4]). If $y = w$, then $P_{w,w} = 1$ and so (a) holds. Now assume that $y < w$ and choose some $t \in S$ such that $tw < w$. If $ty > y$, then 2.1.9(b2) yields $P_{y,w} = P_{ty,w}$ and so (a) holds by induction. (Note that $y \leqslant tw$ by 2.1.1(b) and, hence, $ty \leqslant t(tw) = w$.) If $ty < y$, then 2.1.9(b3) yields

$$P_{y,w} = v_t^2 P_{y,tw} + P_{ty,tw} - \sum_{z \in W: y \leqslant z < tw, tz<z} P_{y,z} v_{tw} v_z^{-1}\left(v_t \mu^t_{z,tw}\right).$$

By Example 2.1.10 and induction, we have $P_{y,z} \in \mathbb{Z}[\Gamma_{\geqslant 0}]$ and $v_t\mu^t_{z,tw} \in \mathbb{Z}[\Gamma_{>0}]$ for all $z$ in the above sum. Hence, we conclude that $P_{y,w} \equiv P_{ty,tw} \mod \mathbb{Z}[\Gamma_{>0}]$. Since $ty \leqslant tw$ by 2.1.1(b), this yields (a) by induction.

Note that if $L(s) = 0$ for some $s \in S$, then the conclusion in (a) no longer holds. For example, if $L(s) = 0$, then $C'_s = T_s$ and so $P_{1,s} = 0$.

*Remark 2.1.12.* Assume that $L(s) > 0$ for all $s \in S$. Then the basis $\{C_w\}$ gives rise to a $W$-graph structure on $W$. Indeed, let us set $I(w) := \{s \in S \mid sw < w\}$ for $w \in W$. Furthermore, if $y, w \in W$ and $s \in S$ are such that $s \in I(y)$ and $s \notin I(w)$, we set

$$m^s_{y,w} := \begin{cases} 1 & \text{if } y = sw, \\ -(-1)^{l(w)+l(y)}\mu^s_{y,w} & \text{if } y < w, \\ 0 & \text{otherwise.} \end{cases}$$

Then we see that the data $\{I(w)\}$, $\{m^s_{y,w}\}$ give rise to a $W$-graph structure on the set $W$, in the sense of Definition 1.4.11. Note that $v_s m^s_{y,w} \in \mathbb{Z}[\Gamma_{>0}]$ by Example 2.1.10.

**2.1.13.** Recall that $\mathbf{H}$ is a symmetric algebra, where $\{T_w \mid w \in W\}$ and $\{T_{w^{-1}} \mid w \in W\}$ form a pair of dual bases of $\mathbf{H}$. Since each $C_w$ equals $T_w$ plus a $\mathbb{Z}[\Gamma_{>0}]$-linear combination of basis elements $T_z$ ($z \in W$), it easily follows that

(a) $$\tau(C_{x^{-1}}C_y) \in \delta_{xy} + \mathbb{Z}[\Gamma_{>0}] \qquad \text{for all } x, y \in W;$$

(see [220, 5.3.3] where this appeared in the equal-parameter case). Now set

$$D_w := T_w + \sum_{y \in W : w < y} \overline{P}^*_{yw_0, ww_0} T_y \qquad (w \in W),$$

where $w_0 \in W$ is the longest element. Then, $\{C_w \mid w \in W\}$ and $\{D_{w^{-1}} \mid w \in W\}$ form a pair of dual bases; that is, we have

(b) $$\tau(C_x D_{y^{-1}}) = \delta_{xy} \qquad \text{for all } x, y \in W.$$

In particular, $h_{x,y,z} = \tau(C_x C_y D_{z^{-1}})$ for all $x, y, z \in W$, a relation which will be used repeatedly in what follows. The relation (b) follows from the following identity:

$$\sum_{z \in W : y \leqslant z \leqslant w} (-1)^{l(w)+l(y)} P^*_{y,z} P^*_{ww_0, zw_0} = \delta_{yw} \qquad \text{for all } y \leqslant w \text{ in } W,$$

which appeared as [195, Theorem 3.1] in the equal-parameter case; see [231, 10.7 and 11.4] or [103, §2] for the general case. Once the above identity is proved, one also obtains the following relation (see [231, 11.6] or [103, 2.6]):

(c) $$\mu^s_{ww_0, yw_0} = -(-1)^{l(w)+l(y)} \mu^s_{y,w}$$

for any $s \in S$ and $y, w \in W$ such that $sy < y < w < sw$.

**2.1.14.** The $A$-linear map $\mathbf{H} \to \mathbf{H}$, $h \mapsto h^\flat$, defined by $T_w^\flat = T_{w^{-1}}$ ($w \in W$) is an anti-involution of $\mathbf{H}$; see Example 1.2.5. Applying $\flat$ to the relation $T_{w^{-1}}^{-1} = \sum_{y \in W} \bar{R}^*_{y,w} T_y$, we find that $R^*_{y^{-1},w^{-1}} = R^*_{y,w}$ for all $y, w \in W$. Then, using 2.1.3, it also follows that

$$C_w^\flat = C_{w^{-1}} \qquad \text{and} \qquad P^*_{y,w} = P^*_{y^{-1},w^{-1}} \qquad \text{for all } y, w \in W.$$

We can now apply the general definitions concerning "cells" in Section 1.6 to the algebra $\underline{H} = \mathbf{H}$ with its basis $\{C_w \mid w \in W\}$. Thus, we obtain pre-order relations $\leqslant_{\mathscr{L}}$, $\leqslant_{\mathscr{R}}$, $\leqslant_{\mathscr{LR}}$ on $W$. Recall, for example, that $\leqslant_{\mathscr{L}}$ is defined as the transitive closure of the relation $\leftarrow_{\mathscr{L}}$; by the multiplicition formulae in Theorem 2.1.6, we have

$$y \leftarrow_{\mathscr{L}} w \quad \Leftrightarrow \quad \begin{cases} \text{either } y = sw, & \text{where } s \in S \text{ is such that } L(s) = 0 \text{ or } sw > w, \\ \text{or } \mu^s_{y,w} \neq 0, & \text{where } s \in S,\ L(s) > 0 \text{ and } sy < y < w < sw. \end{cases}$$

Furthermore, we have $y \leqslant_{\mathscr{R}} w$ if and only if $y^{-1} \leqslant_{\mathscr{L}} w^{-1}$. And, finally, $\leqslant_{\mathscr{LR}}$ is the union of $\leqslant_{\mathscr{L}}$ and $\leqslant_{\mathscr{R}}$.

**Definition 2.1.15.** The left, right or two-sided cells defined, in the sense of 1.6.1, by taking $\underline{H} = \mathbf{H}$ with its basis $\{C_w \mid w \in W\}$, are called the left, right or two-sided *Kazhdan–Lusztig cells* of $W$ respectively.

From now on, unless explicitly stated otherwise, the symbols $\leqslant_{\mathscr{L}}$, $\leqslant_{\mathscr{R}}$, $\leqslant_{\mathscr{LR}}$, $\sim_{\mathscr{L}}$, $\sim_{\mathscr{R}}$, $\sim_{\mathscr{LR}}$ will always refer to the pre-order relations defined using the Kazhdan–Lusztig basis $\{C_w\}$ of $\mathbf{H}$.

**Lemma 2.1.16 (Lusztig [231, 8.6]).** *Given $w \in W$, define $\mathscr{L}(w) := \{s \in S \mid sw < w$ and $L(s) > 0\}$ and $\mathscr{R}(w) := \mathscr{L}(w^{-1})$. Then the following hold:*

*(a) If $z, y \in W$ are such that $z \leqslant_{\mathscr{L}} y$, then $\mathscr{R}(y) \subseteq \mathscr{R}(z)$.*
*(b) If $z, y \in W$ are such that $z \leqslant_{\mathscr{R}} y$, then $\mathscr{L}(y) \subseteq \mathscr{L}(z)$.*

*In particular, the function $w \mapsto \mathscr{R}(w)$ is constant on left cells and the function $w \mapsto \mathscr{L}(w)$ is constant on right cells.*

*Proof.* Since the formulation in [231, 8.6] does not include the possibility that $L(s) = 0$ for some $s \in S$, let us briefly sketch the argument. To prove (a), we may assume that $z, y$ are related by an elementary step in the definition of $\leqslant_{\mathscr{L}}$; that is, there is some $s \in S$ such that $h'_{s,y,z} \neq 0$. If $L(s) > 0$, then the argument is exactly the same as in [231, 8.6], using the fact that ${}^t\mathbf{H} := \langle C_w \mid w \in W, wt < w\rangle_A \subseteq \mathbf{H}$ is a left ideal for any $t \in S$ such that $L(t) > 0$; see [231, 8.4].

Now assume that $L(s) = 0$. Then $z = sy$ by the multiplication formulae in Theorem 2.1.6. Let $t \in \mathscr{R}(y)$. If $sy > y$, then $\mathscr{R}(y) \subseteq \mathscr{R}(sy)$ and so $t \in \mathscr{R}(z)$, as required. Finally, assume that $sy < y$; then $l(sy) = l(yt)$. If we had $zt > z$, then $l(syt) = l(y)$ and so $syt = y$; see [132, 1.2.6]. Hence, $s, t$ would be conjugate in this case and so $L(s) = L(t)$, which is a contradiction. Thus, we must have $zt < z$, as required.

The proof of (b) is analogous. □

**Example 2.1.17.** Assume that $L(s) > 0$ for all $s \in S$. Let $w \in W$ be such that $w \sim_{\mathscr{L}} 1$. Then Lemma 2.1.16 implies that $\mathscr{R}(w) = \mathscr{R}(1) = \varnothing$ and so $w = 1$. Hence, $\{1\}$ is a left Kazhdan–Lusztig cell.

Similarly, let $w \in W$ be such that $w \sim_{\mathscr{L}} w_0$, where $w_0 \in W$ is the longest element. Then Lemma 2.1.16 implies that $\mathscr{R}(w) = \mathscr{R}(w_0) = S$ and so $w = w_0$. Hence, $\{w_0\}$ is a left Kazhdan–Lusztig cell. We have the following explicit formula:

$$C_{w_0} = \sum_{w \in W} (-1)^{l(w_0)+l(w)} \varepsilon^{L(w_0)-L(w)} T_w.$$

Indeed, if $w \in W$ is such that $w \neq w_0$, then there exists some $s \in S$ such that $sw > w$. Hence, the formula in 2.1.9(b2) yields that $P^*_{w,w_0} = v_s^{-1} P^*_{sw,w_0}$. By a simple downward induction on $l(w)$, we conclude that $P^*_{w,w_0} = \varepsilon^{L(w)-L(w_0)}$, as required.

**Example 2.1.18.** Let $W$ be of type $I_2(m)$ $(m \geqslant 3)$; that is, we have $W = \langle s_1, s_2\rangle$, where $s_1^2 = s_2^2 = (s_1 s_2)^m = 1$. Let $L$ be a weight function where $b := L(s_1) \geqslant 0$ and $a := L(s_2) \geqslant 0$; here, $a = b$ if $m$ is odd.

The relations $\leqslant_{\mathscr{L}}$, $\leqslant_{\mathscr{R}}$ and $\leqslant_{\mathscr{LR}}$ are determined in [231, 8.8]. (See also [132, Exc. 11.4] for the case $a \neq b$.) For any $k \geqslant 0$, write $1_k = s_1 s_2 s_1 \cdots$ ($k$ factors) and $2_k = s_2 s_1 s_2 \cdots$ ($k$ factors); note that $1_m = 2_m$. With this notation, we have:

- If $m$ is odd and $a=b>0$, then the left cells are
$$\{1_0\},\quad \{1_m\},\quad \{2_1,1_2,2_3,\ldots,1_{m-1}\},\quad \{1_1,2_2,1_3,\ldots,2_{m-1}\}.$$
- If $m$ is even and $a=b>0$, then the left cells are
$$\{1_0\},\quad \{1_m\},\quad \{2_1,1_2,2_3,\ldots,2_{m-1}\},\quad \{1_1,2_2,1_3,\ldots,1_{m-1}\}.$$
- If $m$ is even and $b>a>0$, then the left cells are
$$\{1_0\},\ \{2_1\},\ \{1_{m-1}\},\ \{1_m\},\ \{1_1,2_2,1_3,\ldots,2_{m-2}\},\ \{1_2,2_3,1_4,\ldots,2_{m-1}\}.$$
- If $m$ is even and $b>a=0$, then the left cells are
$$\{1_0,2_1\},\ \{1_m,1_{m-1}\},\ \{1_1,2_2,1_3,\ldots,2_{m-2}\},\ \{1_2,2_3,1_4,\ldots,2_{m-1}\}.$$

The two-sided cells and the partial order induced on them are given by

$$\{1_m\} \leqslant_{\mathcal{LR}} W\setminus\{1_0,1_m\} \leqslant_{\mathcal{LR}} \{1_0\} \quad (a=b>0),$$
$$\{1_m\} \leqslant_{\mathcal{LR}} \{1_{m-1}\} \leqslant_{\mathcal{LR}} W\setminus\{1_0,2_1,1_{m-1},1_m\} \leqslant_{\mathcal{LR}} \{2_1\} \leqslant_{\mathcal{LR}} \{1_0\} \quad (b>a>0),$$
$$\{1_m,1_{m-1}\} \leqslant_{\mathcal{LR}} W\setminus\{1_0,2_1,1_{m-1},1_m\} \leqslant_{\mathcal{LR}} \{1_0,2_1\} \quad (b>a=0).$$

Recall that, in Definition 1.6.4, we have introduced the left, right and two-sided $\tilde{\mathbf{J}}$-cells of $W$, using the algebra $\underline{H}=\tilde{\mathbf{J}}$ with its basis $\{t_w \mid w\in W\}$. In the above example where $W$ is of type $I_2(m)$, the two-sided Kazhdan–Lustzig cells are precisely the two-sided $\tilde{\mathbf{J}}$-cells determined in Example 1.7.3. If this was known to be true in general, then our task in this book would be considerably simpler! (We will discuss this in more detail in Section 2.5.) To close this section, we will show by a general argument that, at least, the Kazhdan–Lusztig cells are always unions of $\tilde{\mathbf{J}}$-cells.

**2.1.19.** We will now bring back into the picture the balanced matrix representations $\{\rho^\lambda \mid \lambda\in\Lambda\}$ and the corresponding leading matrix coefficients $c_{w,\lambda}^{\mathfrak{st}}$; see Section 1.4. Recall that, for any $w\in W$, we have

$$\varepsilon^{\mathbf{a}_\lambda}\,\rho^\lambda(T_w)\in M_{d_\lambda}(\mathcal{O}_0) \qquad\text{and}\qquad c_{w,\lambda}^{\mathfrak{st}} \equiv \varepsilon^{\mathbf{a}_\lambda}\,\rho_{\mathfrak{st}}^\lambda(T_w) \bmod \mathfrak{m}$$

for all $\lambda\in\Lambda$ and $\mathfrak{s},\mathfrak{t}\in M(\lambda)$. Now consider the expressions for $C_w$ and $D_w$. Since $\overline{P}_{y,w}^{*}\in\mathbb{Z}[\Gamma_{>0}]$ for all $y\neq w$, we deduce that

$$\varepsilon^{\mathbf{a}_\lambda}\rho_{\mathfrak{st}}^\lambda(C_w)\in\mathcal{O}_0 \qquad\text{and}\qquad \varepsilon^{\mathbf{a}_\lambda}\rho_{\mathfrak{st}}^\lambda(D_w)\in\mathcal{O}_0,$$
$$\varepsilon^{\mathbf{a}_\lambda}\rho_{\mathfrak{st}}^\lambda(T_w)\equiv\varepsilon^{\mathbf{a}_\lambda}\rho_{\mathfrak{st}}^\lambda(C_w)\equiv\varepsilon^{\mathbf{a}_\lambda}\rho_{\mathfrak{st}}^\lambda(D_w)\equiv c_{w,\lambda}^{\mathfrak{st}} \bmod \mathfrak{m},$$

for all $\lambda\in\Lambda$ and $\mathfrak{s},\mathfrak{t}\in M(\lambda)$. Thus, the leading matrix coefficients can be taken with respect to any of the bases $\{T_w\}$, $\{C_w\}$ or $\{D_w\}$.

**Proposition 2.1.20.** *Every left Kazhdan–Lusztig cell of $W$ is a union of left $\tilde{\mathbf{J}}$-cells (see Definition 1.6.4). Analogous statements hold for right and two-sided cells. In particular, if $x,y,z\in W$ are such that $\tilde{\gamma}_{x,y,z}\neq 0$, then the elements $x^{\pm1}$, $y^{\pm1}$, $z^{\pm1}$ all lie in the same two-sided Kazhdan–Lusztig cell.*

*Proof.* Let $y,z\in W$ belong to the same left $\tilde{\mathbf{J}}$-cell. It will be sufficient to consider an elementary step in the definition of this relation; that is, we can assume that

$$\tilde{\gamma}_{x,y,z^{-1}} = \Big(\sum_{\lambda\in\Lambda}\sum_{\mathfrak{s},\mathfrak{t},\mathfrak{u}\in M(\lambda)} f_\lambda^{-1}\, c_{x,\lambda}^{\mathfrak{st}}\, c_{y,\lambda}^{\mathfrak{tu}}\, c_{z,\lambda}^{\mathfrak{us}}\Big) \neq 0 \quad \text{for some } x\in W.$$

We deduce that

$$\sum_{\mathfrak{u}\in M(\lambda)} c_{y,\lambda}^{\mathfrak{tu}}\, c_{z^{-1},\lambda}^{\mathfrak{us}} \neq 0 \qquad \text{for some } \lambda\in\Lambda \text{ and } \mathfrak{s},\mathfrak{t}\in M(\lambda).$$

Using the relations in 2.1.19 we obtain that

$$\varepsilon^{2\mathbf{a}_\lambda}\rho_{\mathfrak{ts}}^\lambda(C_yD_{z^{-1}}) \equiv \sum_{\mathfrak{u}\in M(\lambda)} \big(\varepsilon^{\mathbf{a}_\lambda}\rho_{\mathfrak{tu}}^\lambda(C_y)\big)\big(\varepsilon^{\mathbf{a}_\lambda}\rho_{\mathfrak{us}}^\lambda(D_{z^{-1}})\big) \equiv \sum_{\mathfrak{u}\in M(\lambda)} c_{y,\lambda}^{\mathfrak{tu}}\, c_{z^{-1},\lambda}^{\mathfrak{us}}$$

modulo $\mathfrak{m}$. Since the expression on the right-hand side is non-zero modulo $\mathfrak{m}$, we conclude that $\rho^\lambda(C_yD_{z^{-1}}) \neq 0$ and so $C_yD_{z^{-1}} \neq 0$. Since $\tau$ is non-degenerate, we have $\tau(C_wC_yD_{z^{-1}}) \neq 0$ for some $w\in W$. This yields $h_{w,y,z} = \tau(C_wC_yD_{z^{-1}}) \neq 0$ and so $z \leqslant_{\mathscr{L}} y$ (in the Kazhdan–Lusztig pre-order). Similarly, we find that

$$\varepsilon^{2\mathbf{a}_\lambda}\rho_{\mathfrak{ts}}^\lambda(D_yC_{z^{-1}}) \equiv \sum_{\mathfrak{u}\in M(\lambda)} \big(\varepsilon^{\mathbf{a}_\lambda}\rho_{\mathfrak{tu}}^\lambda(D_y)\big)\big(\varepsilon^{\mathbf{a}_\lambda}\rho_{\mathfrak{us}}^\lambda(C_{z^{-1}})\big) \equiv \sum_{\mathfrak{u}\in M(\lambda)} c_{y,\lambda}^{\mathfrak{tu}}\, c_{z^{-1},\lambda}^{\mathfrak{us}}$$

modulo $\mathfrak{m}$ and, hence, $D_yC_{z^{-1}} \neq 0$. Again, we can find some $w\in W$ such that $\tau(C_wD_yC_{z^{-1}}) \neq 0$. It follows that $h_{z^{-1},w,y^{-1}} = \tau(C_{z^{-1}}C_wD_y) = \tau(C_wD_yC_{z^{-1}}) \neq 0$. Hence, we have $y^{-1} \leqslant_{\mathscr{R}} z^{-1}$ and so $y \leqslant_{\mathscr{L}} z$. Thus, we have shown that $y,z$ belong to the same left Kazhdan–Lusztig cell. The arguments for right and two-sided cells are analogous. The last statement (involving $\tilde{\gamma}_{x,y,z}$) follows from Corollary 1.6.7. □

## 2.2 A Pre-order Relation on Irr$(W)$

We have just seen that the weight function $L\colon W\to\Gamma$ and the monomial order $\leqslant$ on $\Gamma$ give rise to the Kazhdan–Lusztig pre-order relations $\leqslant_{\mathscr{L}}$, $\leqslant_{\mathscr{R}}$, $\leqslant_{\mathscr{LR}}$ on $W$. We will now use the two-sided relation $\leqslant_{\mathscr{LR}}$ to define a pre-order relation on the set $\mathrm{Irr}_{\mathbb{K}}(W) = \{E^\lambda \mid \lambda\in\Lambda\}$. Recall that, in Proposition 1.6.11, we constructed a natural surjective map

$$\mathrm{Irr}_{\mathbb{K}}(W) \to \{\text{set of two-sided } \tilde{\mathbf{J}}\text{-cells of } W\}, \qquad E^\lambda \mapsto \mathscr{F}_\lambda.$$

By Proposition 2.1.20, we also know that each $\mathscr{F}_\lambda$ is contained in a two-sided Kazhdan–Lusztig cell. This leads us to the following definition.

**Definition 2.2.1 (Cf. Lusztig [220, 5.15]).** Let $\lambda,\mu\in\Lambda$. Then we define

$$E^\lambda \preceq_L E^\mu \quad \overset{\text{def}}{\Leftrightarrow} \quad x \leqslant_{\mathscr{LR}} y \quad \text{for all } x\in\mathscr{F}_\lambda \text{ and } y\in\mathscr{F}_\mu,$$

where $\leqslant_{\mathscr{LR}}$ is the Kazhdan–Lusztig pre-order relation on $W$; see 2.1.14. Since each two-sided $\tilde{\mathbf{J}}$-cell is contained in a two-sided Kazhdan–Lusztig cell, we have

$$E^\lambda \preceq_L E^\mu \quad\Leftrightarrow\quad x \leqslant_{\mathcal{LR}} y \quad \text{for } \textit{some } x \in \mathscr{F}_\lambda \text{ and } \textit{some } y \in \mathscr{F}_\mu.$$

Furthermore, we write $E^\lambda \sim_L E^\mu$ if $E^\lambda \preceq_L E^\mu$ and $E^\mu \preceq_L E^\lambda$. Thus, $E^\lambda \sim_L E^\mu$ if and only if $\mathscr{F}_\lambda$, $\mathscr{F}_\mu$ are contained in the same two-sided Kazhdan–Lusztig cell.

We wish to mention right away that the relation $\sim_L$ on $\mathrm{Irr}_{\mathbb{K}}(W)$ is not yet completely understood nor explicitly known in all cases (and even less so the relation $\preceq_L$). In this section we will, therefore, content ourselves with giving some examples and explaining the open questions. Of particular interest for us will be the relation between $\preceq_L$ and the function $E^\lambda \mapsto \mathbf{a}_\lambda$. We will see that even the first example that one might think of, namely the case where $W \cong \mathfrak{S}_n$, requires a considerable amount of work; see Example 2.2.13 and Section 2.8. We begin by showing that $\preceq_L$ and $\sim_L$ can be expressed without reference to the map $E^\lambda \mapsto \mathscr{F}_\lambda$.

**2.2.2.** By the general method described in 1.6.1, each left Kazhdan–Lusztig cell $\mathfrak{C}$ gives rise to a representation of $\mathbf{H}$. This is constructed as follows. Let $[\mathfrak{C}]_A$ be an $A$-module with a free $A$-basis $\{e_w \mid w \in \mathfrak{C}\}$. Then the action of $\mathbf{H}$ on $[\mathfrak{C}]_A$ is given by

(a) $$C_w.e_x := \sum_{y\in\mathfrak{C}} h_{w,x,y}\, e_y, \qquad \text{where } w \in W \text{ and } x \in \mathfrak{C}.$$

(Similarly, right cells give rise to right $\mathbf{H}$-modules and two-sided cells give rise to $\mathbf{H}$-bimodules.) Now let $\theta\colon A \to k$ be a ring homomorphism into a field $k$. Then $[\mathfrak{C}]_k := k \otimes_A [\mathfrak{C}]_A$ is a left module for the specialised algebra $\mathbf{H}_k$. In particular, let $\theta_1 : A \to \mathbb{K}$ be the ring homomorphism such that $\theta_1(\varepsilon^g) = 1$ for all $g \in \Gamma$, as in 1.2.1. Then we obtain a module $[\mathfrak{C}]_1 := \mathbb{K} \otimes_A [\mathfrak{C}]_A$ for $\mathbb{K}W = \mathbb{K} \otimes_A \mathbf{H}$. For any $\lambda \in \Lambda$, denote by $m(\mathfrak{C},\lambda)$ the multiplicity of $E^\lambda$ as an irreducible constituent of $[\mathfrak{C}]_1$. Then the "specialisation argument" in Example 1.2.4 immediately shows that

(b) $\quad m(\mathfrak{C},\lambda) =$ multiplicity of $E_\varepsilon^\lambda$ as an irreducible constituent of $[\mathfrak{C}]_K$,

where $[\mathfrak{C}]_K$ is the $\mathbf{H}_K$-module obtained from $[\mathfrak{C}]_A$ by extending scalars from $A$ to $K$.

*Remark 2.2.3.* Assume that $L(s) > 0$ for all $s \in S$. Let $\mathfrak{C}$ be a left Kazhdan–Lusztig cell and consider the left cell module $[\mathfrak{C}]_A$. As in Remark 2.1.12, we see that the action of $\mathbf{H}$ on $[\mathfrak{C}]_A$ is given by a $W$-graph, where $X = \mathfrak{C}$, $I(x) = \mathscr{L}(x)$ $(x \in \mathfrak{C})$ and

$$m_{x,y}^s = \begin{cases} 1 & \text{if } x = sy > y, \\ -(-1)^{l(y)+l(x)}\mu_{x,y}^s & \text{if } sx < x < y < sy, \\ 0 & \text{otherwise.} \end{cases}$$

**Lemma 2.2.4.** *Let $\lambda \in \Lambda$ and $\mathfrak{C}$ be a left Kazhdan–Lusztig cell such that $m(\mathfrak{C},\lambda) > 0$. Then we have $E^\lambda \leftrightsquigarrow_L w$ for some $w \in \mathfrak{C}$, that is, $w \in \mathfrak{C} \cap \mathscr{F}_\lambda$.*

*Proof.* (Compare with the proof of Theorem 1.8.1.) Consider the identity

$$\sum_{w\in\mathfrak{C}} C_w D_{w^{-1}} = \sum_{x\in W}\sum_{y\in\mathfrak{C}} h_{x,y,y} D_{x^{-1}}.$$

(This is proved by multiplying both sides by $C_z$ for some $z \in W$ and applying the trace form $\tau$.) Now note that $\operatorname{trace}(C_x, [\mathfrak{C}]_A) = \sum_{y \in \mathfrak{C}} h_{x,y,y}$. Taking also into account the formula $[\mathfrak{C}]_K \cong \bigoplus_{\mu \in \Lambda} m(\mathfrak{C}, \mu) E_\varepsilon^\mu$, we obtain

$$\sum_{y \in \mathfrak{C}} h_{x,y,y} = \sum_{\mu \in \Lambda} m(\mathfrak{C}, \mu) \chi^\mu(C_x) \qquad \text{for all } x \in W.$$

Then the orthogonality relations in Proposition 1.2.12 yield that

$$\chi^\lambda\Big(\sum_{w \in \mathfrak{C}} C_w D_{w^{-1}}\Big) = \sum_{\mu \in \Lambda} m(\mathfrak{C}, \mu)\Big(\sum_{x \in W} \chi^\mu(C_x)\chi^\lambda(D_{x^{-1}})\Big) = m(\mathfrak{C}, \lambda)\, d_\lambda\, \mathbf{c}_\lambda.$$

Multiplying this relation by $\varepsilon^{2\mathbf{a}_\lambda}$ and taking constant terms, we deduce that

$$\sum_{\mathfrak{s},\mathfrak{t} \in M(\lambda)} \sum_{w \in \mathfrak{C}} c_{w,\lambda}^{\mathfrak{st}}\, c_{w^{-1},\lambda}^{\mathfrak{ts}} = m(\mathfrak{C}, \lambda)\, d_\lambda\, f_\lambda.$$

Since the right-hand side is non-zero by assumption, we conclude that $c_{w,\lambda}^{\mathfrak{st}} \neq 0$ for some $w \in \mathfrak{C}$ and some $\mathfrak{s}, \mathfrak{t} \in M(\lambda)$, as required. □

**Corollary 2.2.5.** *Let $\lambda, \mu \in \Lambda$ and $\mathfrak{C}, \mathfrak{C}'$ be left Kazhdan–Lusztig cells such that $m(\mathfrak{C}, \lambda) > 0$ and $m(\mathfrak{C}', \mu) > 0$. Then $E^\lambda \preceq_L E^\mu$ if and only if $w \leqslant_{\mathscr{LR}} w'$ for some $w \in \mathfrak{C}$ and some $w' \in \mathfrak{C}'$ (where $\leqslant_{\mathscr{LR}}$ denotes the Kazhdan–Lusztig pre-order relation).*

*In particular, $E^\lambda \sim_L E^\mu$ if and only if $\mathfrak{C}, \mathfrak{C}'$ are contained in the same two-sided Kazhdan–Lusztig cell.*

*Proof.* First assume that $E^\lambda \preceq_L E^\mu$. By definition, this means that $x \leqslant_{\mathscr{LR}} y$ for all $x \in \mathscr{F}_\lambda$ and $y \in \mathscr{F}_\mu$. Now, by Lemma 2.2.4, there exist elements $w \in \mathfrak{C} \cap \mathscr{F}_\lambda$ and $w' \in \mathfrak{C}' \cap \mathscr{F}_\mu$. Hence, we have $w \leqslant_{\mathscr{LR}} w'$, as required.

Conversely, assume that $w \leqslant_{\mathscr{LR}} w'$ for some $w \in \mathfrak{C}$ and some $w' \in \mathfrak{C}'$. Since $m(\mathfrak{C}, \lambda) > 0$, there exists some $w_1 \in \mathfrak{C} \cap \mathscr{F}_\lambda$; see Lemma 2.2.4. Similarly, there exists some $w_1' \in \mathfrak{C}' \cap \mathscr{F}_\mu$. Hence, we have $w_1 \sim_{\mathscr{L}} w \leqslant_{\mathscr{LR}} w' \sim_{\mathscr{L}} w_1'$ and so $w_1 \leqslant_{\mathscr{LR}} w_1'$. As pointed out in Definition 2.2.1, this already implies that $E^\lambda \preceq_L E^\mu$. □

*Remark 2.2.6.* Let $W = W_1 \times \cdots \times W_d$ be the decomposition into irreducible components. Correspondingly, we have

$$\operatorname{Irr}_{\mathbb{K}}(W) = \{E^{\lambda_1} \boxtimes \cdots \boxtimes E^{\lambda_d} \mid \lambda_i \in \Lambda_i\}, \quad \text{where} \quad \operatorname{Irr}_{\mathbb{K}}(W_i) = \{E^{\lambda_i} \mid \lambda_i \in \Lambda_i\}.$$

Thus, as in Remark 1.3.5, we identify $\Lambda = \Lambda_1 \times \cdots \times \Lambda_d$. Furthermore, we have $\mathbf{H} \cong \mathbf{H}_1 \otimes_A \cdots \otimes_A \mathbf{H}_d$, where $\mathbf{H}_i$ is the generic algebra associated with $W_i$, $L|_{W_i}$. The Kazhdan–Lusztig basis of $\mathbf{H}$ behaves well with respect to this decomposition, that is, if $w = w_1 \cdot \ldots \cdot w_d$, where $w_i \in W_i$, then $C_w = C_{w_1} \cdot \ldots \cdot C_{w_d}$. It follows that

$$E^\lambda \preceq_L E^\mu \qquad \Leftrightarrow \qquad E^{\lambda_i} \preceq_{L_i} E^{\mu_i} \quad \text{for } i = 1, \ldots, d,$$

where $\lambda = (\lambda_1, \ldots, \lambda_d)$ and $\mu = (\mu_1, \ldots, \mu_d)$. Thus, the determination of $\preceq_L$ can be reduced to the case where $(W, S)$ is irreducible.

**2.2.7.** Assume that $W, L$ are such that the following data are explicitly available:

- The matrices $\{\rho^\lambda(T_s) \mid s \in S\}$ for all $\lambda \in \Lambda$. (Recall the algorithm in 1.4.9 for turning any given representation into a balanced representation.)
- All the polynomials $\{P^*_{y,w}\}$ and $\{\mu^s_{y,w}\}$. (See the recursive description in 2.1.9.)

Since the invariants $\mathbf{a}_\lambda$ are also known (see Section 1.3), we can then work out all leading matrix coefficients $c^{\mathfrak{st}}_{w,\lambda}$ and the Kazhdan–Lusztig pre-order relations $\leqslant_{\mathcal{L}}$, $\leqslant_{\mathcal{R}}$, $\leqslant_{\mathcal{LR}}$. This, in turn, enables us to explicitly determine the pre-order relation $\preceq_L$ on $\mathrm{Irr}_{\mathbb{K}}(W)$, via the characterisation in Corollary 2.2.5. Now, the above data are available for $W$ of type $I_2(m)$ (any $m \geqslant 3$), $H_3$, $H_4$, $F_4$. We will now go through these examples one by one and describe the relation $\preceq_L$ on $\mathrm{Irr}_{\mathbb{K}}(W)$ in each case.

**Example 2.2.8.** Let $W$ be of type $I_2(m)$ ($m \geqslant 3$); that is, we have $W = \langle s_1, s_2 \rangle$, where $s_1^2 = s_2^2 = (s_1s_2)^m = 1$. Recall from Example 1.3.7 the description of $\mathrm{Irr}_{\mathbb{K}}(W)$. By Example 2.1.18, we know the left and two-sided cells. It it also not difficult to determine the cell modules $[\mathfrak{C}]_1$ and decompose them into irreducibles. Let $\psi$ denote the sum of all the two-dimensional representations.

- If $m$ is odd and $L(s_1) = L(s_2) > 0$, then the left cell $\{1_0\}$ affords $1_W$, $\{2_m\}$ affords sgn, and the two remaining left cells afford $\psi$.
- If $m$ is even and $L(s_1) = L(s_2) > 0$, then $\{1_0\}$ affords $1_W$, $\{2_m\}$ affords sgn, the first of the two remaining left cells affords $\psi \oplus \mathrm{sgn}_2$, and the second affords $\psi \oplus \mathrm{sgn}_1$.
- If $m$ is even and $L(s_1) > L(s_2) > 0$, then $\{1_0\}$ affords $1_W$, $\{2_1\}$ affords $\mathrm{sgn}_1$, $\{1_{m-1}\}$ affords $\mathrm{sgn}_2$, $\{1_m\}$ affords sgn, and the two remaining left cells afford $\psi$.
- If $m$ is even and $L(s_1) > L(s_2) = 0$, then $\{1_0, 2_1\}$ affords $1_W \oplus \mathrm{sgn}_1$, $\{1_m, 1_{m-1}\}$ affords $\mathrm{sgn} \oplus \mathrm{sgn}_2$, and the two remaining left cells afford $\psi$.

Using this information together with the knowledge of $\leqslant_{\mathcal{LR}}$ (see Example 2.1.18) and of $\mathbf{a}_\lambda$ (see Example 1.3.7), we find that the pre-order $\preceq_L$ on $\mathrm{Irr}_{\mathbb{K}}(W)$ is "linear" in the sense that, for any $\lambda, \mu \in \Lambda$, we have

$$E^\lambda \preceq_L E^\mu \quad \Leftrightarrow \quad \mathbf{a}_\mu \leqslant \mathbf{a}_\lambda.$$

In particular, $E^\lambda \sim_L E^\mu$ if and only if $\mathbf{a}_\lambda = \mathbf{a}_\mu$.

**Example 2.2.9.** Let $W$ be of type $H_3$ or $H_4$. Then all generators are conjugate, so we are automatically in the equal-parameter case. Assume that $L(s) > 0$ for $s \in S$. Then Alvis [2] has computed all polynomials $P^*_{y,w}$ and $\mu^s_{y,w}$. In this way, he explicitly determined the relations $\leqslant_{\mathcal{L}}$ and $\leqslant_{\mathcal{LR}}$; he also found the decomposition of the left cell representations into irreducibles (see [216, §5] for type $H_3$). The partial order induced by $\leqslant_{\mathcal{LR}}$ on the set of two-sided cells is, in fact, a total order.[1] The equivalence classes of $\mathrm{Irr}_{\mathbb{K}}(W)$ under $\sim_L$ are explicitly known by [218, §13] and [2, 3.5]. This information, together with the invariants $\mathbf{a}_\lambda$, is listed in [132, Tables C.1 and C.2]. It turns out that, again, the pre-order $\preceq_L$ is "linear" such that

[1] This statement is not contained in [2]; we thank Alvis (personal communication, 2008) for having verified this using his programs for producing the data in [2].

$$E^{\lambda} \preceq_L E^{\mu} \quad \Leftrightarrow \quad \mathbf{a}_{\mu} \leqslant \mathbf{a}_{\lambda}$$

for any $\lambda, \mu \in \Lambda$. In particular, $E^{\lambda} \sim_L E^{\mu}$ if and only if $\mathbf{a}_{\lambda} = \mathbf{a}_{\mu}$.

**Fig. 2.1** Partial order on two-sided cells in type $F_4$

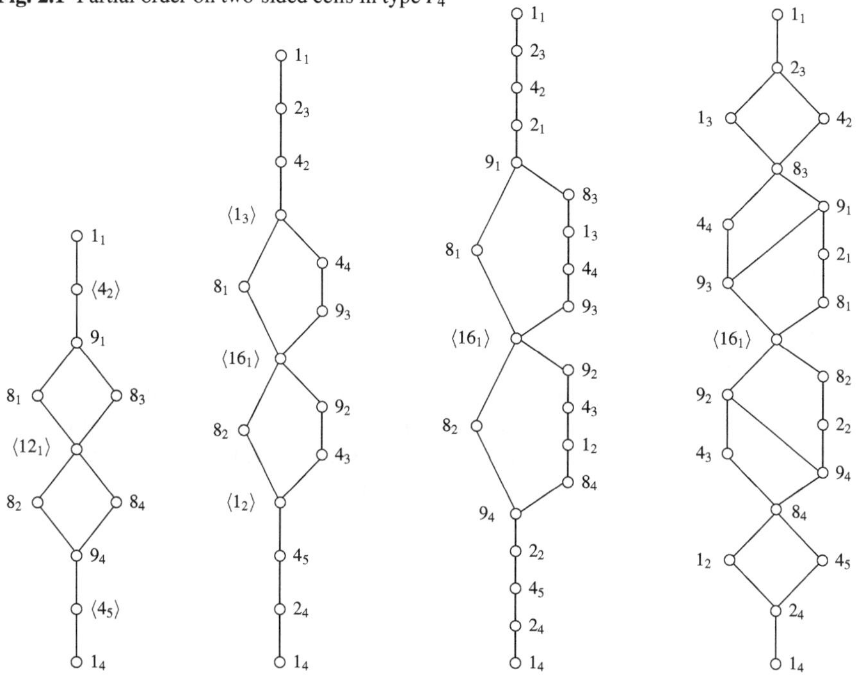


Brackets ⟨ ⟩ indicate a two-sided cell with several irreducible components, given as follows:
$\langle 4_2 \rangle = \{2_1, 2_3, 4_2\}$, $\langle 4_5 \rangle = \{2_2, 2_4, 4_5\}$, $\langle 1_3 \rangle = \{1_3, 2_1, 8_3, 9_1\}$, $\langle 1_2 \rangle = \{1_2, 2_2, 8_4, 9_4\}$,
$\langle 12_1 \rangle = \{1_2, 1_3, 4_1, 4_3, 4_4, 6_1, 6_2, 9_2, 9_3, 12_1, 16_1\}$, $\langle 16_1 \rangle = \{4_1, 6_1, 6_2, 12_1, 16_1\}$.
Otherwise, the two-sided cell contains just one irreducible respresentation.

**Example 2.2.10.** Let $W$ be of type $F_4$ with generators labelled as in Table 1.1 (p. 2). Assume that $a := L(s_1) = L(s_2) > 0$ and $b := L(s_3) = L(s_4) > 0$. (The case where $L(s_i) = 0$ for some $i$ will be considered in Remark 2.4.13.) We may also assume without loss of generality that $b \geqslant a$. The pre-order relations $\leqslant_{\mathcal{L}}$, $\leqslant_{\mathcal{R}}$, $\leqslant_{\mathcal{LR}}$ have been determined in [105], based on an explicit computation of all the polynomials $P^*_{y,w}$ and $\mu^s_{y,w}$ using CHEVIE [118]. The resulting pre-order relations $\preceq_L$ on $\mathrm{Irr}_{\mathbb{K}}(W)$ are given in Figure 2.1. The notation for the irreducible representations is compatible with that in Table 1.2 (p. 16). For example, $1_1$ is the trivial representation, $1_4$ is the sign representation and $4_2$ is the reflection representation. The pre-order $\preceq_L$ is not "linear" in these cases, but by inspection of the tables we notice that, at least, the following property holds:

$$E^\lambda \preceq_L E^\mu \qquad \Rightarrow \qquad \mathbf{a}_\mu \leqslant \mathbf{a}_\lambda \quad (\text{with equality only if } E^\lambda \sim_L E^\mu).$$

In particular, if $E^\lambda \sim_L E^\mu$, then $\mathbf{a}_\lambda = \mathbf{a}_\mu$.

*Remark 2.2.11.* The diagrams in Figure 2.1 have a striking symmetry. This is a general phenomenon. Indeed, recall the definition of the bijection $\lambda \mapsto \lambda^\dagger$ on $\Lambda$ from Example 1.2.6. By Corollary 1.6.16, we have $\mathrm{Irr}_{\mathbb{K}}(W \mid \mathscr{F}w_0) = \mathrm{Irr}_{\mathbb{K}}(W \mid \mathscr{F})^\dagger$ for every two-sided $\tilde{\mathbf{J}}$-cell $\mathscr{F}$ of $W$. Now, 2.1.13(c) implies that if $x, y \in W$ are such that $x \leqslant_{\mathscr{LR}} y$, then $yw_0 \leqslant_{\mathscr{LR}} xw_0$. It follows that, for any $\lambda, \mu \in \Lambda$, we have

$$E^\lambda \preceq_L E^\mu \qquad \Leftrightarrow \qquad E^{\mu^\dagger} \preceq_L E^{\lambda^\dagger}.$$

Thus, the pre-order $\preceq_L$ admits a natural symmetry with respect to $\lambda \mapsto \lambda^\dagger$.

**2.2.12.** Assume that $W$ is of crystallographic type and that we are in the equal-parameter case where $\Gamma = \mathbb{Z}$ and $L(s) = 1$ for all $s \in S$. It has recently been shown [113] that then $\preceq_L$ admits a geometric interpretation, and this actually yields an explicit description of $\preceq_L$. It would be beyond the scope of this book to explain this in detail, but we can at least outline the general idea, assuming some familiarity with the theory of algebraic groups and Lusztig's work [220] (see also Section 4.4).

So let $G$ be a connected reductive algebraic group (over $\mathbb{C}$ or over $\overline{\mathbb{F}}_p$, where $p$ is a large prime), with Weyl group $W$. Then, by the *Springer correspondence* (see [197], [221], [282]), we can naturally associate with every $E^\lambda \in \mathrm{Irr}_{\mathbb{K}}(W)$ a pair consisting of a unipotent class of $G$, which we denote by $O_\lambda$, and a $G$-equivariant irreducible local system on $O_\lambda$. Thus, we obtain a map

$$\mathrm{Irr}_{\mathbb{K}}(W) \to \{\text{set of unipotent classes in } G\}, \quad E^\lambda \mapsto O_\lambda.$$

(The local system on $O_\lambda$ will not play a role for our purposes here.) We now need the concept of a "special" unipotent class. This is defined as follows. Given $\lambda \in \Lambda$, let $\mathbf{b}_\lambda$ be the smallest $i \geqslant 0$ such that $E^\lambda$ is a constituent of the $i$th symmetric power of the natural reflection representation of $W$. Lusztig [215] observed that we always have $\mathbf{a}_\lambda \leqslant \mathbf{b}_\lambda$. We say that $E^\lambda$ is a *special representation* if $\mathbf{a}_\lambda = \mathbf{b}_\lambda$; let

$$\mathscr{S}(W) := \{\lambda \in \Lambda \mid \mathbf{a}_\lambda = \mathbf{b}_\lambda\}.$$

Following Lusztig [215], the classes $\{O_\lambda \mid \lambda \in \mathscr{S}(W)\}$ are called the *special unipotent classes* of $G$ (although the word "special" only appeared in later references; see also 4.3.13). By [220, 13.1.1], we have

$$\mathbf{a}_\lambda = \dim \mathscr{B}_u \qquad \text{for any } \lambda \in \mathscr{S}(W),$$

where $u \in O_\lambda$ and $\mathscr{B}_u$ is the variety of Borel subgroups in $G$ containing $u$. Now [113, Cor. 5.6] shows that, for any $\lambda, \mu \in \mathscr{S}(W)$, we have

$$(*)\qquad E^{\lambda} \preceq_L E^{\mu} \qquad \Leftrightarrow \qquad O_{\lambda} \subseteq \overline{O}_{\mu} := \text{Zariski closure of } O_{\mu}.$$

The map $E^{\lambda} \mapsto O_{\lambda}$ is explicitly known in all cases; see Carter [45, §13.3] and the references there. Also, the Zariski closure relation among the special unipotent classes of $G$ is explicitly known in all cases; see Carter [45, §13.4] and Spaltenstein [280]. Hence, $(*)$ provides an explicit description of the pre-order $\preceq_L$ for special representations. On the other hand, given any $\lambda \in \Lambda$, we have

$$E^{\lambda} \sim_L E^{\lambda_0} \quad \text{for a unique } \lambda_0 \in \mathscr{S}(W);$$

see [220, 4.14 and 5.25]. Hence, since the equivalence relation $\sim_L$ is explicitly known by Lusztig [220, 4.4–4.13 and 5.25], the relation $\preceq_L$ is determined once we know it for special representations. Finally, by [220, 5.27], the function $\lambda \mapsto \mathbf{a}_{\lambda}$ is known to be constant on the equivalence classes under $\sim_L$. Hence, by the above characterisation of $\mathbf{a}_{\lambda}$ for $\lambda \in \mathscr{S}(W)$, we also find that, for any $\lambda, \mu \in \Lambda$, we have

$$E^{\lambda} \preceq_L E^{\mu} \quad \Rightarrow \quad \mathbf{a}_{\mu} \leqslant \mathbf{a}_{\lambda} \quad (\text{with equality only if } E^{\lambda} \sim_L E^{\mu}).$$

(In the next two sections, we will say more about the proof of this implication.)

**Example 2.2.13.** In the setting of 2.2.12, let $W$ be of type $A_{n-1}$. Then $W \cong \mathfrak{S}_n$ and $\Lambda$ is the set of all partitions of $n$; see Example 1.3.8. By [220, 4.4], all irreducible representations of $W$ are special. Now $W$ is the Weyl group of $G = \mathrm{GL}_n$ (over $\mathbb{C}$ or over $\overline{\mathbb{F}}_p$, where $p$ is a large prime). Let $\lambda = (\lambda_1 \geqslant \lambda_2 \geqslant \ldots \geqslant 0) \in \Lambda$. Then the Springer correspondence associates with $E^{\lambda}$ the unipotent class $O_{\lambda}$ consisting of all unipotent matrices in $G$ whose Jordan normal form has blocks of size $\lambda_1, \lambda_2, \ldots$; see Springer [282, p. 293], [45, §13.3]. By [104, §2.6] and 2.2.12$(*)$, we have

$$E^{\lambda} \preceq_L E^{\mu} \qquad \Leftrightarrow \qquad O_{\lambda} \subseteq \overline{O}_{\mu} \qquad \Leftrightarrow \qquad \lambda \trianglelefteq \mu,$$

where $\trianglelefteq$ denotes the *dominance order*, which is defined as follows. Write $\lambda = (\lambda_1 \geqslant \lambda_2 \geqslant \cdots \geqslant 0)$ and $\mu = (\mu_1 \geqslant \mu_2 \geqslant \cdots \geqslant 0)$. Then

$$\lambda \trianglelefteq \mu \quad \stackrel{\text{def}}{\Leftrightarrow} \quad \sum_{1 \leqslant i \leqslant d} \lambda_i \leqslant \sum_{1 \leqslant i \leqslant d} \mu_i \quad (\text{for all } d \geqslant 1).$$

It then follows by a completely elementary argument that we have the implication

$$\lambda \trianglelefteq \mu \qquad \Rightarrow \qquad \lambda = \mu \quad \text{or} \quad \mathbf{a}_{\lambda} > \mathbf{a}_{\mu};$$

see, for example, [132, Exc. 5.6]. See Corollary 2.8.14 for a much more direct and elementary proof of the above characterisation of $\preceq_L$ (following [107]).

**Example 2.2.14.** In the setting of 2.2.12, let $W$ be of type $B_n$. Then $\Lambda$ is the set of all pairs of partitions $(\lambda, \mu)$ such that $|\lambda| + |\mu| = n$; see Example 1.3.9. For any $(\lambda, \mu) \in \Lambda$ and $(\lambda', \mu') \in \Lambda$, let us define

$$(\lambda,\mu) \preceq (\lambda',\mu') \quad \overset{\text{def}}{\Leftrightarrow} \quad \begin{cases} \sum_{1\leqslant i\leqslant d} (\lambda_i+\mu_i) \leqslant \sum_{1\leqslant i\leqslant d} (\lambda_i'+\mu_i') \\ \lambda_d + \sum_{1\leqslant i<d} (\lambda_i+\mu_i) \leqslant \lambda_d' + \sum_{1\leqslant i<d} (\lambda_i'+\mu_i'), \\ \text{(for all } d \geqslant 1) \end{cases}$$

where $\lambda = (\lambda_1 \geqslant \lambda_2 \geqslant \ldots \geqslant 0)$, $\lambda' = (\lambda_1' \geqslant \lambda_2' \geqslant \ldots \geqslant 0)$, $\mu = (\mu_1 \geqslant \mu_2 \geqslant \ldots \geqslant 0)$ and $\mu' = (\mu_1' \geqslant \mu_2' \geqslant \ldots \geqslant 0)$. By [220, 4.5], we have:

$$(\lambda,\mu) \in \mathscr{S}(W) \quad \Leftrightarrow \quad \lambda_i + 1 \geqslant \mu_i \geqslant \lambda_{i+1} \quad \text{(for all } i \geqslant 1).$$

Now $W$ is the Weyl group of $G = \mathrm{SO}_{2n+1}$ (over $\mathbb{C}$ or over $\overline{\mathbb{F}}_p$, where $p$ is a large prime). Then, by Spaltenstein [281, §4] and 2.2.12(∗), we have

$$E^{(\lambda,\mu)} \preceq_L E^{(\lambda',\mu')} \quad \Leftrightarrow \quad O_{(\lambda,\mu)} \subseteq \overline{O}_{(\lambda',\mu')} \quad \Leftrightarrow \quad (\lambda,\mu) \preceq (\lambda',\mu')$$

for $(\lambda,\mu) \in \mathscr{S}(W)$ and $(\lambda',\mu') \in \mathscr{S}(W)$. See [122, §5] for further details.

**Example 2.2.15.** Let again $W$ be of type $B_n$ and consider the reflection subgroup $\tilde{W} \subseteq W$ in Example 1.3.10. Then $\tilde{W}$ is of type $D_n$ and this is the Weyl group of $G = \mathrm{SO}_{2n}$ (over $\mathbb{C}$ or over $\overline{\mathbb{F}}_p$, where $p$ is a large prime). To be consistent with the notation in Example 1.3.10, the equal-parameter weight function on $\tilde{W}$ will be denoted by $\tilde{L}$. Now $\mathrm{Irr}_{\mathbb{K}}(\tilde{W})$ is described in terms of the restrictions of the irreducible representations of $W$ to $\tilde{W}$. Given $\tilde{E} \in \mathrm{Irr}_{\mathbb{K}}(\tilde{W})$ and $(\lambda,\mu) \in \Lambda$, we write

$$\tilde{E} \mid E^{(\lambda,\mu)} \quad \overset{\text{def}}{\Leftrightarrow} \quad \tilde{E} \text{ is a constituent of the restriction of } E^{(\lambda,\mu)} \text{ to } \tilde{W}.$$

To characterise the special representations in $\mathrm{Irr}_{\mathbb{K}}(\tilde{W})$, it is convenient to define

$$\tilde{\mathscr{S}}(W) := \{(\lambda,\mu) \in \Lambda) \mid \lambda_i \geqslant \mu_i \geqslant \lambda_{i+1} - 1 \text{ for all } i \geqslant 1\},$$

where we write $\lambda = (\lambda_1 \geqslant \lambda_2 \geqslant \ldots \geqslant 0)$ and $\mu = (\mu_1 \geqslant \mu_2 \geqslant \ldots \geqslant 0)$. Now let $\tilde{E} \in \mathrm{Irr}_{\mathbb{K}}(\tilde{W})$. Then, by [220, 4.6], we have

$$\tilde{E} \text{ is special} \quad \Leftrightarrow \quad \tilde{E} \mid E^{(\lambda,\mu)} \text{ for some } (\lambda,\mu) \in \tilde{\mathscr{S}}(W).$$

Note that if $\tilde{E}$ is special, then $(\lambda,\mu)$ on the right-hand side is uniquely determined: just observe that if both $(\lambda,\mu)$ and $(\mu,\lambda)$ belong to $\tilde{\mathscr{S}}(W)$, then $\lambda = \mu$.

Now, by Spaltenstein [281, §4] and 2.2.12(∗), we obtain the following result. Let $\tilde{E}, \tilde{E}' \in \mathrm{Irr}_{\mathbb{K}}(\tilde{W})$ be special representations. Let $(\lambda,\mu) \in \tilde{S}(W)$ and $(\lambda',\mu') \in \tilde{\mathscr{S}}(W)$ be such that $\tilde{E} \mid E^{(\lambda,\mu)}$ and $\tilde{E}' \mid E^{(\lambda',\mu')}$. Then

$$\tilde{E} \preceq_{\tilde{L}} \tilde{E}' \quad \Leftrightarrow \quad \begin{cases} \tilde{E} = \tilde{E}' & \text{if } \lambda = \lambda' = \mu = \mu', \\ (\lambda,\mu) \preceq (\lambda',\mu') & \text{otherwise,} \end{cases}$$

where $(\lambda,\mu) \preceq (\lambda',\mu')$ is defined in Example 2.2.15. See [122, §5] for further details.

**Example 2.2.16.** Assume that $W$ is of type $E_6$, $E_7$ or $E_8$. The equivalence classes of $\mathrm{Irr}_{\mathbb{K}}(W)$ under $\sim_L$ are listed in Lusztig [220, 4.11–4.13]; see also [132, App. C]. The Springer correspondence is explicitly described in the tables in [45, 13.3].

($E_6$) We have $|\mathrm{Irr}_{\mathbb{K}}(W)| = 25$ and there are 17 equivalence classes under $\sim_L$. The partially ordered set of special unipotent classes is printed in [45, p. 441].
($E_7$) We have $|\mathrm{Irr}_{\mathbb{K}}(W)| = 60$ and there are 35 equivalence classes under $\sim_L$. The partially ordered set of special unipotent classes is printed in [45, p. 443].
($E_8$) We have $|\mathrm{Irr}_{\mathbb{K}}(W)| = 112$ and there are 46 equivalence classes under $\sim_L$. The partially ordered set of special unipotent classes is printed in [45, p. 445].

**Example 2.2.17.** Let $W$ be of type $B_n$ and $L\colon W \to \Gamma$ be a weight function given by

$B_n$ 

where $b > (n-1)a > 0$.

Recall that $\Lambda$ is the set of all pairs of partitions $(\lambda,\mu)$ such that $|\lambda|+|\mu| = n$; see Example 1.3.9. By [122, Example 5.1] (which relies on the series of papers by Bonnafé, Geck, Iancu [21], [26], [108], [114], [121]), we have

(a) $$E^{(\lambda,\mu)} \preceq_L E^{(\lambda',\mu')} \quad\Leftrightarrow\quad (\lambda,\mu) \trianglelefteq (\lambda',\mu').$$

Here, $\trianglelefteq$ denotes the *dominance order* on pairs of partitions, which is defined by

(b) $$(\lambda,\mu) \trianglelefteq (\lambda',\mu') \quad\overset{\text{def}}{\Leftrightarrow}\quad \begin{cases} \displaystyle\sum_{1\leqslant i\leqslant d} \lambda_i \leqslant \sum_{1\leqslant i\leqslant d} \lambda_i' \\ \displaystyle|\lambda| + \sum_{1\leqslant i\leqslant d} \mu_i \leqslant |\lambda'| + \sum_{1\leqslant i\leqslant d} \mu_i', \\ \text{(for all } d \geqslant 1) \end{cases}$$

where $\lambda = (\lambda_1 \geqslant \lambda_2 \geqslant \ldots \geqslant 0)$, $\lambda' = (\lambda_1' \geqslant \lambda_2' \geqslant \ldots \geqslant 0)$, $\mu = (\mu_1 \geqslant \mu_2 \geqslant \ldots \geqslant 0)$ and $\mu' = (\mu_1' \geqslant \mu_2' \geqslant \ldots \geqslant 0)$. Furthermore, by [121, Cor. 5.5], we have

(c) $$E^{(\lambda,\mu)} \preceq_L E^{(\lambda',\mu')} \quad\Rightarrow\quad \mathbf{a}_{(\lambda',\mu')} \leqslant \mathbf{a}_{(\lambda,\mu)},$$

with equality only if $(\lambda,\mu) = (\lambda',\mu')$.

For the (infinitely many) remaining open cases in type $B_n$, at least a conjecture is formulated in [122, 4.11].

*Remark 2.2.18.* Lusztig's definition [220, 5.15] of a pre-order relation on $\mathrm{Irr}_{\mathbb{K}}(W)$ looks somewhat different from that in Definition 2.2.1, but it is really the same. Let us briefly indicate why this is the case. By the Artin–Wedderburn theorem, the split semisimple algebra $\mathbf{H}_K$ decomposes as a direct sum of simple two-sided ideals $\mathbf{H}_K = \bigoplus_{\lambda\in\Lambda} \mathbf{H}_K(\lambda)$, where $\mathbf{H}_K(\lambda)$ is the sum of all left ideals in $\mathbf{H}_K$ which are isomorphic to $E_\varepsilon^\lambda$ (as left $\mathbf{H}_K$-modules). On the other hand, for any $y \in W$, we have the two-sided ideals $\mathfrak{I}_y^{\mathscr{LR}}$ and $\hat{\mathfrak{I}}_y^{\mathscr{LR}}$ defined by the general procedure in 1.6.2, with respect to the basis $\{C_w \mid w \in W\}$ of $\mathbf{H}_K$. Now let $\mathfrak{T}$ be a two-sided Kazhdan–Lusztig cell and $\lambda \in \Lambda$. Then we claim that the following two statements are equivalent:

(a) $E^\lambda \leftrightsquigarrow_L w$ for some $w \in \mathfrak{T}$.
(b) $\mathbf{H}_K(\lambda) \subseteq \mathfrak{I}_y^{\mathscr{LR}}$ and $\mathbf{H}_K(\lambda) \not\subseteq \hat{\mathfrak{I}}_y^{\mathscr{LR}}$ for some $y \in \mathfrak{T}$.

Indeed, if (a) holds, then $c_{w,\lambda}^{\mathfrak{st}} \neq 0$ and $c_{w^{-1},\lambda}^{\mathfrak{uv}} \neq 0$ for some $\mathfrak{s},\mathfrak{t},\mathfrak{u},\mathfrak{v} \in M(\lambda)$; see 1.6.10. Hence, by 2.1.19, we have $\rho^\lambda(C_w) \neq 0$ and $\rho^\lambda(D_{w^{-1}}) \neq 0$. This yields $C_w.\mathbf{H}_K(\lambda) \neq \{0\}$ and $D_{w^{-1}}.\mathbf{H}_K(\lambda) \neq \{0\}$. Then the argument in the proof of [220, Lemma 5.2] shows that (b) holds. Conversely, assume that (b) holds. Then the inclusion $\mathbf{H}_K(\lambda) \subseteq \mathfrak{I}_y^{\mathscr{LR}}$ induces an $(\mathbf{H}_K,\mathbf{H}_K)$-bimodule homomorphism $\varphi\colon \mathbf{H}_K(\lambda) \to \mathfrak{I}_y^{\mathscr{LR}}/\hat{\mathfrak{I}}_y^{\mathscr{LR}}$, which is non-zero since $\mathbf{H}_K(\lambda) \not\subseteq \hat{\mathfrak{I}}_y^{\mathscr{LR}}$. Now let us just consider the left $\mathbf{H}_K$-module structure. Since $\mathfrak{T}$ is a union of left cells, the left $\mathbf{H}_K$-module $\mathfrak{I}_y^{\mathscr{LR}}/\hat{\mathfrak{I}}_y^{\mathscr{LR}}$ has a filtration by left cell modules $[\mathfrak{C}_i]_K$, where each $\mathfrak{C}_i$ is a left cell contained in $\mathfrak{T}$. Hence, there exists a non-zero $\mathbf{H}_K$-module homomorphism $E_\varepsilon^\lambda \to [\mathfrak{C}_i]_K$ for some $i$. Then $m(\mathfrak{C}_i,\lambda) > 0$ and so there exists some $w \in \mathfrak{C}_i$ such that $E^\lambda \leftrightsquigarrow_L w$; see Lemma 2.2.4. Thus, the equivalence of (a) and (b) is proved. Once this is established, we can conclude that

$$\text{(c)} \quad E^\lambda \preceq_L E^\mu \quad \Leftrightarrow \quad \begin{cases} \text{there exists some } w \in W \text{ such that} \\ \mathbf{H}_K(\lambda) \subseteq \mathfrak{I}_w^{\mathscr{LR}},\ \mathbf{H}_K(\mu) \subseteq \mathfrak{I}_w^{\mathscr{LR}},\ \mathbf{H}_K(\mu) \not\subseteq \hat{\mathfrak{I}}_w^{\mathscr{LR}}. \end{cases}$$

The condition on the right-hand side is the one used by Lusztig [220, 5.15].

## 2.3 On Lusztig's Conjectures, I

In the previous section, we have defined a pre-order relation $\preceq_L$ on $\mathrm{Irr}_{\mathbb{K}}(W)$ and we have seen that, in many examples, the following implication holds for any $\lambda,\mu \in \Lambda$:

$$(\clubsuit) \qquad E^\lambda \preceq_L E^\mu \quad \Rightarrow \quad \mathbf{a}_\mu \leqslant \mathbf{a}_\lambda \quad \text{(with equality only if } E^\lambda \sim_L E^\mu).$$

This property will turn out to be *the* key to our main results on representations of Hecke algebras at roots of unity. The somewhat weaker implication

$$(\clubsuit') \qquad E^\lambda \sim_L E^\mu \quad \Rightarrow \quad \mathbf{a}_\mu = \mathbf{a}_\lambda$$

was a key ingredient in Lusztig's work [220] on characters of reductive groups over finite fields. Now, a general proof of these apparently simple statements is not yet known. And in those situations where ($\clubsuit$) and ($\clubsuit'$) are known to hold, the proofs rely on deep results from algebraic geometry, or explicit computations. It is the purpose of this and the following two sections to discuss this in some more detail.

Assume first that $W$ is a Weyl group and that we are in the equal-parameter case. Then the proof of ($\clubsuit'$) in [220, Chap. 5] relies on the theory of primitive ideals in enveloping algebras. Subsequently, Lusztig [225] found a new proof which uses a geometric interpretation of $\{C_w\}$ and the results in [222], [223]. This interpretation implies, for example, that all coefficients of the polynomials $P_{y,w}^*$ are non-negative integers. In the general multiparameter case, such a geometric interpretation is not known – and the coefficients of $P^*_{y,w}$ may be strictly negative; see Example 2.1.5!

In his book [231, Chap. 14], Lusztig has extended the known situation in the equal-parameter case and stated 15 conjectural properties **P1–P15** of the basis $\{C_w\}$ which should hold for any Coxeter group (finite or infinite) and in the general multiparameter case. In [231, Chap. 20], Lusztig shows that (♣) and (♣′) are formal consequences of **P1–P15**. Thus, **P1–P15** appear to provide the appropriate framework for establishing substantial results concerning the representation theory of **H**. (See 2.4.1 for a summary of the cases where **P1–P15** are known to hold.)

**2.3.1.** For the convenience of the reader, we state here Lusztig's conjectures **P1–P15** in [231, Chap. 14] in the general framework involving a totally ordered abelian group $\Gamma$, and taking into account the possibility that $L(s) = 0$ for some $s \in S$. Also note that these properties are formulated in [231] with respect to the basis $\{C'_w\}$, but, using the formulae in Remark 2.1.7, it is a straightforward matter to switch back and forth between $C_w$ and $C'_w$. The following definitions originally appeared in [222], in the equal-parameter case. For a fixed $z \in W$, we set

$$\mathbf{a}(z) := \min\{g \in \Gamma_{\geqslant 0} \mid \varepsilon^g h_{x,y,z} \in \mathbb{Z}[\Gamma_{\geqslant 0}] \text{ for all } x,y \in W\}.$$

Given $x,y,z \in W$, we define $c_{x,y,z^{-1}} \in \mathbb{Z}$ by

$$c_{x,y,z^{-1}} := \text{ constant term of } \varepsilon^{\mathbf{a}(z)} h_{x,y,z} \in \mathbb{Z}[\Gamma_{\geqslant 0}].$$

Furthermore, if $z \in W$ is such that $P^*_{1,z} \neq 0$, we define an element $\Delta(z) \in \Gamma_{\geqslant 0}$ and an integer $0 \neq n_z \in \mathbb{Z}$ by the condition

$$\varepsilon^{\Delta(z)} P^*_{1,z} \equiv n_z \mod \mathbb{Z}[\Gamma_{<0}]; \qquad \text{see [231, 14.1].}$$

(Note that we can only have $P^*_{1,z} = 0$ if $L(s) = 0$ for some $s \in S$; see Example 2.1.11; this is the only place where we explicitly have to mention if $L(s)$ equals zero or not.) Finally, we set $\mathscr{D} := \{z \in W \mid P^*_{1,z} \neq 0 \text{ and } \mathbf{a}(z) = \Delta(z)\}$.

**Conjecture 2.3.2 (Lusztig [231, 14.2]).** *The following properties hold.*

**P1.** *For any $z \in W$ such that $P^*_{1,z} \neq 0$, we have $\mathbf{a}(z) \leqslant \Delta(z)$.*
**P2.** *If $d \in \mathscr{D}$ and $x,y \in W$ satisfy $c_{x,y,d} \neq 0$, then $x = y^{-1}$.*
**P3.** *If $y \in W$, there exists a unique $d \in \mathscr{D}$ such that $c_{y^{-1},y,d} \neq 0$.*
**P4.** *If $x,y \in W$ are such that $x \leqslant_{\mathscr{LR}} y$, then $\mathbf{a}(y) \geqslant \mathbf{a}(x)$. In particular, if $x \sim_{\mathscr{LR}} y$, then $\mathbf{a}(x) = \mathbf{a}(y)$.*
**P5.** *If $d \in \mathscr{D}$, $y \in W$, $c_{y^{-1},y,d} \neq 0$, then $c_{y^{-1},y,d} n_d = (-1)^{l(d)}$.*
**P6.** *If $d \in \mathscr{D}$, then $d^2 = 1$.*
**P7.** *For any $x,y,z \in W$, we have $c_{x,y,z} = c_{y,z,x}$.*
**P8.** *Let $x,y,z \in W$ be such that $c_{x,y,z} \neq 0$. Then $x \sim_{\mathscr{L}} y^{-1}$, $y \sim_{\mathscr{L}} z^{-1}$, $z \sim_{\mathscr{L}} x^{-1}$.*
**P9.** *If $x \leqslant_{\mathscr{L}} y$ and $\mathbf{a}(x) = \mathbf{a}(y)$, then $x \sim_{\mathscr{L}} y$.*
**P10.** *If $x \leqslant_{\mathscr{R}} y$ and $\mathbf{a}(x) = \mathbf{a}(y)$, then $x \sim_{\mathscr{R}} y$.*
**P11.** *If $x \leqslant_{\mathscr{LR}} y$ and $\mathbf{a}(x) = \mathbf{a}(y)$, then $x \sim_{\mathscr{LR}} y$.*

**P12.** *Let $I \subseteq S$ and $W_I$ be the parabolic subgroup generated by $I$. If $y \in W_I$, then $\mathbf{a}(y)$ computed in terms of $W_I$ is equal to $\mathbf{a}(y)$ computed in terms of $W$.*

**P13.** *Any left cell $\mathfrak{C}$ of $W$ contains a unique element $d \in \mathscr{D}$. We have $c_{x^{-1},x,d} \neq 0$ for all $x \in \mathfrak{C}$.*

**P14.** *For any $z \in W$, we have $z \sim_{\mathscr{LR}} z^{-1}$.*

**P15.** *If $w, w', x, y \in W$ are such that $\mathbf{a}(x) = \mathbf{a}(y)$, then*

$$\sum_{z \in W} h_{x,w',z} \otimes_{\mathbb{Z}} h_{w,z,y} = \sum_{z \in W} h_{z,w',y} \otimes_{\mathbb{Z}} h_{w,x,z} \quad \textit{in } \mathbb{Z}[\Gamma] \otimes_{\mathbb{Z}} \mathbb{Z}[\Gamma].$$

We just remark (a) that there are some logical dependences between these properties (for example, "**P1** + **P3** $\Rightarrow$ **P5**" by [231, 14.5]) and (b) that some of these properties seem to be more crucial than others (for example, **P4** will appear almost everywhere while **P6** will not be needed in the whole discussion below).

*Remark 2.3.3.* In 2.1.14, we have seen that $C_w^\flat = C_{w^{-1}}$ for all $w \in W$, where $h \mapsto h^\flat$ is the anti-automorphism of $\mathbf{H}$ defined by $T_w^\flat = T_{w^{-1}}$. This immediately implies that

$$h_{x,y,z} = h_{y^{-1},x^{-1},z^{-1}}, \qquad \mathbf{a}(z) = \mathbf{a}(z^{-1}) \qquad \text{and} \qquad c_{x,y,z} = c_{y^{-1},x^{-1},z^{-1}}$$

for all $x, y, z \in W$. Furthermore, we have $n_z = n_{z^{-1}}$, $\Delta(z) = \Delta(z^{-1})$, $\mathscr{D} = \mathscr{D}^{-1}$.

*Remark 2.3.4.* **P14** holds for finite $W$ by Lemma 1.6.6 and Proposition 2.1.20. (See [220, 5.2] for the equal-parameter case.) The reason why it appears in the above list is that Conjecture 2.3.2 is stated in [231] for arbitrary (possibly infinite) Coxeter groups satisfying a certain boundedness condition.

*Remark 2.3.5.* Assume that we are in the equal-parameter case where $\Gamma = \mathbb{Z}$ and $L(s) = 1$ for all $s \in S$. Now $A$ is the ring of Laurent polynomials in one indeterminate $v = \varepsilon$. One easily checks that there is a well-defined ring homomorphism $\alpha \colon \mathbf{H} \to \mathbf{H}$ such that $\alpha(v) = -v$, $\alpha(r) = r$ for all $r \in R$ and $\alpha(T_w) = (-1)^{l(w)} T_w$ for all $w \in W$. Hence, by the characterisation in Theorem 2.1.6, we must have $\alpha(C_w) = (-1)^{l(w)} C_w$. Applying $\alpha$ to the relation $C_x C_y = \sum_{z \in W} h_{x,y,z} C_z$, we deduce that

(a) $$h_{x,y,z}(-v) = (-1)^{l(x)+l(y)+l(z)}\, h_{x,y,z}(v) \qquad \text{for all } x, y, z \in W.$$

This also implies that

(b) $$(-1)^{l(x)+l(y)+l(z)} c_{x,y,z} = (-1)^{\mathbf{a}(z)} c_{x,y,z} \qquad \text{for all } x, y, z \in W.$$

(These observations are due to Lusztig [222, 3.2].)

*Remark 2.3.6.* Let $\mathbf{J}$ be the free $\mathbb{Z}$-module with basis $\{t_w \mid w \in W\}$. We define an element of $\mathbf{J}$ by $1_{\mathbf{J}} := \sum_{d \in \mathscr{D}} n_d t_d$. We define a bilinear product on $\mathbf{J}$ by

$$t_x t_y = \sum_{z \in W} \gamma_{x,y,z} t_{z^{-1}}, \qquad \text{where} \qquad \gamma_{x,y,z} := (-1)^{l(x)+l(y)+l(z)} c_{x,y,z^{-1}}.$$

(Note that this agrees with the notation in [231, 13.6].) Using **P1**, **P4**, one deduces that $\mathbf{J}$ is an associative ring, where $1_{\mathbf{J}}$ is the identity. Since we will not need this construction here, we refer to Lusztig [231, Chap. 18] for further details. For the identification with our algebra $\tilde{\mathbf{J}}$, see Proposition 2.3.16 below.

*Remark 2.3.7.* We note that **P15** really is a statement about a certain bimodule structure (which appeared in [216], [222, 9.2]). To see this, consider the ring $\mathscr{A} = R[\Gamma] \otimes_R R[\Gamma]$ and let $\mathscr{E}$ be a free $\mathscr{A}$-module with basis $\{e_z \mid z \in W\}$. Let

$$\mathbf{H}_1 = \mathscr{A} \otimes_A \mathbf{H}, \quad \text{where } A \text{ is embedded into } \mathscr{A} \text{ via } a \mapsto 1 \otimes a,$$
$$\mathbf{H}_2 = \mathscr{A} \otimes_A \mathbf{H}, \quad \text{where } A \text{ is embedded into } \mathscr{A} \text{ via } a \mapsto a \otimes 1.$$

By **P4** and the definition of the Kazhdan–Lusztig pre-order relation $\leqslant_{\mathscr{LR}}$, there is a left action of $\mathbf{H}_1$ on $\mathscr{E}$ via

$$C_w.e_x = \sum_{z \in W:\, \mathbf{a}(z)=\mathbf{a}(x)} (1 \otimes h_{w,x,z})\, e_z \qquad \text{for } x, w \in W.$$

Similarly, there is a right action of $\mathbf{H}_2$ on $\mathscr{E}$ via

$$e_x.C_w = \sum_{z \in W:\, \mathbf{a}(z)=\mathbf{a}(x)} (h_{x,w,z} \otimes 1)\, e_z \qquad \text{for } x, w \in W.$$

Now let $x, w, w' \in W$. Then

$$C_w.(e_x.C_{w'}) = \sum_{\substack{z \in W \\ \mathbf{a}(z)=\mathbf{a}(x)}} (h_{x,w',z} \otimes 1)\, C_w.e_z = \sum_{\substack{y,z \in W \\ \mathbf{a}(y)=\mathbf{a}(z)=\mathbf{a}(x)}} (h_{x,w',z} \otimes h_{w,z,y})\, e_y.$$

Here, we recognise the terms appearing on the left-hand side of **P15**. Similarly, when we expand $(C_w.e_x).C_{w'}$, we will recognise the terms appearing on the right-hand side of **P15**. Thus, we conclude that **P15** holds if and only if $\mathscr{E}$ is an $(\mathbf{H}_1, \mathbf{H}_2)$-bimodule.

As already mentioned in the introduction to this section, Lusztig [231, Chap. 20] has shown that (♣′) formally follows from **P1**–**P15**. We will now give a somewhat streamlined exposition of this deduction which, eventually, only requires **P1**, **P4**. (The stronger property (♣) will be considered in the next section.) For this purpose, we need to relate the functions $\mathbf{a}(z)$ and $\mathbf{a}_\lambda$. The following result (which first appeared in [114]) seems to be the only known connection between these two functions which can be proved without assuming any of the properties **P1**–**P15**.

**Lemma 2.3.8.** *Let $\lambda \in \Lambda$ and $w \in W$ be such that $E^\lambda \rightsquigarrow_L w$. Then $\mathbf{a}(w) \geqslant \mathbf{a}_\lambda$.*

*Proof.* By assumption, there exist some $\mathfrak{s}, \mathfrak{t} \in M(\lambda)$ such that $c_{w,\lambda}^{\mathfrak{st}} \neq 0$. Furthermore, by 2.1.19, we have $\varepsilon^{\mathbf{a}_\lambda} \rho_{\mathfrak{st}}^\lambda(D_w) \equiv c_{w,\lambda}^{\mathfrak{st}} \mod \mathfrak{m}$. Now we claim that

(a) $$\rho_{\mathfrak{st}}^\lambda(D_w) = \sum_{x,y \in W} \mathbf{c}_\lambda^{-1}\, \rho_{\mathfrak{st}}^\lambda(D_{x^{-1}})\, \rho_{\mathfrak{ss}}^\lambda(D_{y^{-1}})\, h_{x,y,w^{-1}}.$$

This is seen as follows. Let $x,y \in W$. Then $h_{x,y,w^{-1}} = \tau(C_xC_yD_w)$; see 2.1.13. Furthermore, $\tau = \sum_{\mu\in\Lambda} \mathbf{c}_\mu^{-1}\chi^\mu$ and so $h_{x,y,w^{-1}} = \sum_{\mu\in\Lambda} \mathbf{c}_\mu^{-1}\chi^\mu(C_xC_yD_w)$. This yields

$$h_{x,y,w^{-1}} = \sum_{\mu\in\Lambda} \sum_{\mathfrak{u},\mathfrak{u}',\mathfrak{v}\in M(\mu)} \mathbf{c}_\mu^{-1}\,\rho^\mu_{\mathfrak{u}\mathfrak{u}'}(C_x)\,\rho^\mu_{\mathfrak{u}'\mathfrak{v}}(C_y)\,\rho^\mu_{\mathfrak{v}\mathfrak{u}}(D_w).$$

Now multiply on both sides by $\rho^\lambda_{\mathfrak{s}\mathfrak{t}}(D_{x^{-1}})\,\rho^\lambda_{\mathfrak{s}\mathfrak{s}}(D_{y^{-1}})$ and sum over all $x,y \in W$. Using the Schur relations in Proposition 1.2.12, a straightforward computation yields (a).

Now note that $\mathbf{c}_\lambda^{-1} = f_\lambda^{-1}\varepsilon^{2\mathbf{a}_\lambda}/(1+g_\lambda)$, where $g_\lambda \in F[\Gamma_{>0}]$. Hence, we obtain

$$\varepsilon^{\mathbf{a}(w)}\rho^\lambda_{\mathfrak{s}\mathfrak{t}}(D_w) = \sum_{x,y\in W} \frac{f_\lambda^{-1}}{1+g_\lambda}\big(\varepsilon^{\mathbf{a}_\lambda}\rho^\lambda_{\mathfrak{s}\mathfrak{t}}(D_{x^{-1}})\big)\big(\varepsilon^{\mathbf{a}_\lambda}\rho^\lambda_{\mathfrak{s}\mathfrak{s}}(D_{y^{-1}})\big)\big(\varepsilon^{\mathbf{a}(w)}h_{x,y,w^{-1}}\big).$$

All terms in the above sum lie in $\mathscr{O}_0$; see 2.1.19 and also note that $\mathbf{a}(w) = \mathbf{a}(w^{-1})$ by Remark 2.3.3. Hence the whole sum will lie in $\mathscr{O}_0$ and so $\varepsilon^{\mathbf{a}(w)}\rho^\lambda_{\mathfrak{s}\mathfrak{t}}(\mathbf{D}_w) \in \mathscr{O}_0$. Since $\varepsilon^{\mathbf{a}_\lambda}\rho^\lambda_{\mathfrak{s}\mathfrak{t}}(D_w) \not\equiv 0 \bmod \mathfrak{m}$, we conclude that $\mathbf{a}(w) \geqslant \mathbf{a}_\lambda$, as claimed. □

**Lemma 2.3.9.** *Let $\mathfrak{C}$ be a left Kazhdan–Lusztig cell and $\lambda \in \Lambda$ be such that $m(\mathfrak{C},\lambda) > 0$.*

*(a) If $y \in W$ is such that $\rho^\lambda(C_y) \neq 0$, then $y' \leqslant_{\mathscr{R}} y$ for some $y' \in \mathfrak{C}$.*
*(b) If $z \in W$ is such that $\rho^\lambda(D_{z^{-1}}) \neq 0$, then $z \leqslant_{\mathscr{R}} z'$ for some $z' \in \mathfrak{C}$.*

*Proof.* Since $m(\mathfrak{C},\lambda) > 0$, we have that $E^\lambda_\varepsilon$ is an irreducible constituent of $[\mathfrak{C}]_K$; see 2.2.2. Now assume that $\rho^\lambda(C_y) \neq 0$; that is, $C_y$ does not act as zero on $E^\lambda_\varepsilon$. Then $C_y$ cannot act as zero on $[\mathfrak{C}]_K$ either. By the definition of this action, there exist some $x,y' \in \mathfrak{C}$ such that $h_{y,x,y'} \neq 0$. In particular, $y' \leqslant_{\mathscr{R}} y$. Thus, (a) is proved. Now assume that $\rho^\lambda(D_{z^{-1}}) \neq 0$. Then $D_{z^{-1}}$ cannot act as zero on $[\mathfrak{C}]_K$. So, by the definition of this action, there exists some $z' \in \mathfrak{C}$ such that $D_{z^{-1}}C_{z'} \neq 0$. Since $\tau$ is non-degenerate, there exists some $x \in W$ such that $\tau(C_xD_{z^{-1}}C_{z'}) \neq 0$. This yields $h_{z',x,z} = \tau(C_{z'}C_xD_{z^{-1}}) = \tau(C_xD_{z^{-1}}C_{z'}) \neq 0$ and so $z \leqslant_{\mathscr{R}} z'$, as required. □

**Lemma 2.3.10.** *Assume that* **P4** *holds. Let $x,y,z \in W$. Then*

$$c_{x,y,z} = \sum_{\lambda} \sum_{\mathfrak{s},\mathfrak{t},\mathfrak{u}\in M(\lambda)} f_\lambda^{-1}\,c^{\mathfrak{s}\mathfrak{t}}_{x,\lambda}\,c^{\mathfrak{t}\mathfrak{u}}_{y,\lambda}\,c^{\mathfrak{u}\mathfrak{s}}_{z,\lambda},$$

*where the first sum runs over all $\lambda \in \Lambda$ such that $\mathbf{a}_\lambda = \mathbf{a}(z)$.*

*Proof.* We have $h_{x,y,z^{-1}} = \tau(C_xC_yD_z)$ and $\tau = \sum_{\lambda\in\Lambda} \mathbf{c}_\lambda^{-1}\chi^\lambda$. This yields

$$\begin{aligned} h_{x,y,z^{-1}} &= \sum_{\lambda\in\Lambda} \mathbf{c}_\lambda^{-1}\operatorname{trace}\big(\rho^\lambda(C_x)\rho^\lambda(C_y)\rho^\lambda(D_z)\big) \\ &= \sum_{\lambda\in\Lambda} \sum_{\mathfrak{s},\mathfrak{t},\mathfrak{u}\in M(\lambda)} \mathbf{c}_\lambda^{-1}\,\rho^\lambda_{\mathfrak{s}\mathfrak{t}}(C_x)\,\rho^\lambda_{\mathfrak{t}\mathfrak{u}}(C_y)\,\rho^\lambda_{\mathfrak{u}\mathfrak{s}}(D_z). \end{aligned}$$

Now note that $\mathbf{c}_\lambda^{-1} = f_\lambda^{-1}\varepsilon^{2\mathbf{a}_\lambda}/(1+g_\lambda)$, where $g_\lambda \in F[\Gamma_{>0}]$. Hence, we obtain

$$\varepsilon^{\mathbf{a}(z)} h_{x,y,z^{-1}} = \sum_{\lambda \in \Lambda} \sum_{\mathfrak{s},\mathfrak{t},\mathfrak{u} \in M(\lambda)} \frac{f_\lambda^{-1}}{1+g_\lambda} \left(\varepsilon^{\mathbf{a}_\lambda} \rho^\lambda_{\mathfrak{st}}(C_x)\right)\left(\varepsilon^{\mathbf{a}_\lambda} \rho^\lambda_{\mathfrak{tu}}(C_y)\right)\left(\varepsilon^{\mathbf{a}(z)} \rho^\lambda_{\mathfrak{us}}(D_z)\right).$$

By 2.1.19, the terms $\varepsilon^{\mathbf{a}_\lambda} \rho^\lambda_{\mathfrak{st}}(C_x)$, $\varepsilon^{\mathbf{a}_\lambda} \rho^\lambda_{\mathfrak{tu}}(C_y)$ and $\varepsilon^{\mathbf{a}_\lambda} \rho^\lambda_{\mathfrak{us}}(D_z)$ lie in $\mathscr{O}_0$. Let $\lambda \in \Lambda$ be such that $\rho^\lambda_{\mathfrak{us}}(D_z) \neq 0$. Let $\mathfrak{C}$ be a left Kazhdan–Lusztig cell such that $m(\mathfrak{C}, \lambda) > 0$. Then, by Lemma 2.3.9(b), there exists some $w \in \mathfrak{C}$ such that $z^{-1} \leqslant_{\mathscr{R}} w$. By **P4**, we must have $\mathbf{a}(w) \leqslant \mathbf{a}(z^{-1}) = \mathbf{a}(z)$. Furthermore, by Lemma 2.2.4, there exists some $w'' \in \mathfrak{C} \cap \mathscr{F}_\lambda$. Hence, by **P4** and Lemma 2.3.8, we have $\mathbf{a}_\lambda \leqslant \mathbf{a}(w'') = \mathbf{a}(w) \leqslant \mathbf{a}(z)$. But if $\mathbf{a}_\lambda < \mathbf{a}(z)$, then $\varepsilon^{\mathbf{a}(z)} \rho^\lambda_{\mathfrak{us}}(D_z) \in \mathfrak{m}$ for all $\mathfrak{s}, \mathfrak{u} \in M(\lambda)$ and so these terms do not contribute to $\varepsilon^{\mathbf{a}(z)} h_{x,y,z^{-1}} \bmod \mathbb{Z}[\Gamma_{>0}]$. We conclude that

$$\varepsilon^{\mathbf{a}(z)} h_{x,y,z^{-1}} \equiv \Big( \sum_{\substack{\lambda \in \Lambda \\ \mathbf{a}_\lambda = \mathbf{a}(z)}} \sum_{\mathfrak{s},\mathfrak{t},\mathfrak{u} \in M(\lambda)} f_\lambda^{-1} c^{\mathfrak{st}}_{x,\lambda} c^{\mathfrak{tu}}_{y,\lambda} c^{\mathfrak{us}}_{z,\lambda} \Big) \bmod \mathfrak{m}.$$

This yields the desired formula for $c_{x,y,z}$. □

**Corollary 2.3.11.** *Assume that* **P4** *holds. Then* $\{\mathbf{a}(z) \mid z \in W\} \subseteq \{\mathbf{a}_\lambda \mid \lambda \in \Lambda\}$.

*Proof.* Given $z \in W$, let $x, y \in W$ be such that $c_{x,y,z} \neq 0$. Then Lemma 2.3.10 shows that there exists some $\lambda \in \Lambda$ such that $\mathbf{a}_\lambda = \mathbf{a}(z)$. □

**Lemma 2.3.12 (Lusztig [222, 6.1]).** *Assume that* **P4** *holds. Then* **P7** *also holds. Furthermore, if* $c_{x,y,z} \neq 0$, *then* $\mathbf{a}(x) = \mathbf{a}(y) = \mathbf{a}(z)$.

*Proof.* We first show that, for any $x', y', z \in W$, we have

$$(*) \qquad c_{x',y',z} = \text{constant term of } \varepsilon^{\mathbf{a}(z)} \tau(T_{x'} T_{y'} D_z) \in \mathbb{Z}[\Gamma_{\geqslant 0}].$$

Indeed, as already noted in Definition 2.1.4, we have

$$T_w = C_w + \sum_{w' \in W : w' < w} \alpha_{w,w'} C_{w'}, \qquad \text{where} \qquad \alpha_{w,w'} \in \mathbb{Z}[\Gamma_{>0}].$$

Since $\mathbf{a}(z) = \mathbf{a}(z^{-1})$ and $c_{x',y',z} \equiv \varepsilon^{\mathbf{a}(z)} \tau(C_{x'} C_{y'} D_z) \bmod \mathbb{Z}[\Gamma_{>0}]$, this shows that

$$c_{x',y',z} \equiv \varepsilon^{\mathbf{a}(z)} \tau(T_{x'} T_{y'} D_z) + \sum_{x'' < x'} \sum_{y'' < y'} \alpha_{x',x''} \alpha_{y',y''} \varepsilon^{\mathbf{a}(z)} \tau(C_{x''} C_{y''} D_z) \bmod \mathbb{Z}[\Gamma_{>0}].$$

Since $\varepsilon^{\mathbf{a}(z)} \tau(C_{x''} C_{y''} D_z) \in \mathbb{Z}[\Gamma_{\geqslant 0}]$, we see that $(*)$ holds.

Now we argue as follows. Let $x, y, z \in W$ and set $c := c_{x,y,z}$. Assume first that $c \neq 0$. Hence, by $(*)$, we have that $\varepsilon^{\mathbf{a}(z)} \tau(T_x T_y D_z) \in \mathbb{Z}[\Gamma_{\geqslant 0}]$ has constant term $c$. Now, writing $D_x$ in terms of the $T$-basis (see 2.1.13) and using $(*)$, we see that $\varepsilon^{\mathbf{a}(z)} \tau(D_x T_y D_z) \in \mathbb{Z}[\Gamma_{\geqslant 0}]$ has constant term $c$ and, hence, $\varepsilon^{\mathbf{a}(z)} \tau(T_y D_z D_x) \in \mathbb{Z}[\Gamma_{\geqslant 0}]$ has constant term $c$. Using once more the expression of $D_z$ in 2.1.13, we obtain

$$\varepsilon^{\mathbf{a}(z)} \tau(T_y T_z D_x) = \varepsilon^{\mathbf{a}(z)} \tau(T_y D_z D_x) - \sum_{w \in W : z < w} \overline{P}^*_{ww_0, zw_0} \varepsilon^{\mathbf{a}(z)} \tau(T_y T_w D_x).$$

Now, since $c = c_{x,y,z} \neq 0$, we have $h_{x,y,z^{-1}} \neq 0$ and, hence, $z^{-1} \leqslant_{\mathscr{R}} x$. So, by **P4**, we have $\mathbf{a}(x) \leqslant \mathbf{a}(z^{-1}) = \mathbf{a}(z)$. Combining this with $(*)$, we deduce that $\varepsilon^{\mathbf{a}(z)}\tau(T_yT_wD_x) \in \mathbb{Z}[\Gamma_{\geqslant 0}]$ for all $y,w \in W$. Consequently,

$$\varepsilon^{\mathbf{a}(z)}\tau(T_yT_zD_x) \equiv \varepsilon^{\mathbf{a}(z)}\tau(T_yD_zD_x) \equiv c \mod \mathbb{Z}[\Gamma_{\geqslant 0}].$$

Using $(*)$, this shows that $\mathbf{a}(x) \geqslant \mathbf{a}(z)$. Since we also have $\mathbf{a}(z) \geqslant \mathbf{a}(x)$, we conclude that $\mathbf{a}(x) = \mathbf{a}(z)$ and, hence, $c_{y,z,x} = c$. Since $c \neq 0$, we can repeat the whole argument with $c_{y,z,x}$ and find that $c_{z,x,y} = c$; furthermore, $\mathbf{a}(z) = \mathbf{a}(y)$.

Thus, if one of the numbers $c_{x,y,z}$, $c_{y,z,x}$, $c_{z,x,y}$ is non-zero, then these three numbers are equal to each other and we have $\mathbf{a}(x) = \mathbf{a}(y) = \mathbf{a}(z)$. If all three numbers are zero, they are again equal. □

**Lemma 2.3.13 (Lusztig [231, 14.5]).** *Assume that* **P1** *holds. Then*

$$\sum_{d\in\mathscr{D}} (-1)^{l(d)}\, n_d\, c_{x^{-1},y,d} = \delta_{xy} \qquad \textit{for any } x,y \in W.$$

*Proof.* Since $C_{y^{-1}}C_x = \sum_{z\in W} h_{y^{-1},x,z}C_z$ and $\tau(C_z) = (-1)^{l(z)}\overline{P}^*_{1,z}$, we have

$$\begin{aligned}\tau(C_{y^{-1}}C_x) &= \sum_{z\in W} (-1)^{l(z)} h_{y^{-1},x,z}\overline{P}^*_{1,z}\\ &= \sum_{z\in W:\,P^*_{1,z}\neq 0} \left(\varepsilon^{\mathbf{a}(z)}h_{y^{-1},x,z}\right)\left((-1)^{l(z)}\varepsilon^{-\mathbf{a}(z)}\overline{P}^*_{1,z}\right).\end{aligned}$$

By 2.1.13(a), the left-hand side is congruent to $\delta_{xy}$ modulo $\mathbb{Z}[\Gamma_{>0}]$. Now consider the right-hand side. By the definition of $\Delta(z)$, we have $\varepsilon^{-\Delta(z)}\overline{P}^*_{1,z} \equiv n_z \bmod \mathbb{Z}[\Gamma_{>0}]$. Since **P1** is assumed to hold, we have $\mathbf{a}(z) \leqslant \Delta(z)$. This yields that

$$\varepsilon^{-\mathbf{a}(z)}\overline{P}^*_{1,z} \equiv \begin{cases} n_z \bmod \mathbb{Z}[\Gamma_{>0}] & \text{if } z \in \mathscr{D},\\ 0 \bmod \mathbb{Z}[\Gamma_{>0}] & \text{otherwise.}\end{cases}$$

Hence, we obtain $\delta_{xy} \equiv \tau(C_{y^{-1}}C_x) \equiv \sum_{d\in\mathscr{D}}(-1)^{l(d)}c_{y^{-1},x,d^{-1}}n_d \bmod \mathbb{Z}[\Gamma_{>0}]$. Finally, by Remark 2.3.3, we have $c_{y^{-1},x,d^{-1}} = c_{x^{-1},y,d}$, which yields the desired formula. □

**Proposition 2.3.14.** *Assume that* **P1**, **P4** *hold. If $\lambda \in \Lambda$ and $w \in W$ are such that $E^\lambda \leftrightsquigarrow_L w$, then $\mathbf{a}(w) = \mathbf{a}_\lambda$. In particular, (♣′) holds.*

*Proof.* By Lemma 2.3.8, we already know that $\mathbf{a}(w) \geqslant \mathbf{a}_\lambda$. So it will now be sufficient to prove that $\mathbf{a}(w) \leqslant \mathbf{a}_\lambda$. For this purpose, we consider the identity

$$\sum_{d\in\mathscr{D}} (-1)^{l(d)}n_dC_wC_d = \sum_{d\in\mathscr{D},y\in W} (-1)^{l(d)}n_dh_{w,d,y}C_y.$$

Applying $\rho^\lambda$ and multiplying by $\varepsilon^{\mathbf{a}_\lambda+\mathbf{a}(w)}$, we obtain

$$\sum_{d\in\mathscr{D}} (-1)^{l(d)}n_d\varepsilon^{\mathbf{a}_\lambda+\mathbf{a}(w)}\rho^\lambda_{\mathfrak{st}}(C_wC_d) = \sum_{d\in\mathscr{D},y\in W} (-1)^{l(d)}n_d\left(\varepsilon^{\mathbf{a}(w)}h_{w,d,y}\right)\left(\varepsilon^{\mathbf{a}_\lambda}\rho^\lambda_{\mathfrak{st}}(C_y)\right).$$

Assume that the terms corresponding to $d \in \mathcal{D}$, $y \in W$ give a non-zero contribution to the sum on the right-hand side; that is, $h_{w,d,y} \neq 0$ and $\rho^{\lambda}_{\mathfrak{st}}(C_y) \neq 0$. Let $\mathfrak{C}$ be a left Kazhdan–Lusztig cell such that $m(\mathfrak{C},\lambda) > 0$. By Lemma 2.3.9(a), there exists some $z' \in \mathfrak{C}$ such that $z' \leqslant_{\mathcal{R}} y$. On the other hand, since $h_{w,d,y} \neq 0$, we have $y \leqslant_{\mathcal{R}} w$. Furthermore, by Lemma 2.2.4, there exists some $w' \in \mathfrak{C} \cap \mathcal{F}_\lambda$. Thus, we obtain

$$w, w' \in \mathcal{F}_\lambda, \qquad w', z' \in \mathfrak{C}, \qquad z' \leqslant_{\mathcal{R}} y \leqslant_{\mathcal{R}} w.$$

Since every two-sided Kazhdan–Lusztig cell is a union of two-sided $\tilde{\mathbf{J}}$-cells (see Proposition 2.1.20) and also a union of left Kazhdan–Lusztig cells, we conclude that $w, w', y, z'$ all lie in the same two-sided Kazhdan–Lusztig cell. In particular, since **P4** holds, we have $\mathbf{a}(y) = \mathbf{a}(w)$. Hence, using 2.1.19, the right-hand side of the above identity can be rewritten as

$$\begin{aligned}\sum_{d\in\mathcal{D}, y\in W} (-1)^{l(d)} n_d \big(\varepsilon^{\mathbf{a}(y)} h_{w,d,y}\big)\big(\varepsilon^{\mathbf{a}_\lambda} \rho^{\lambda}_{\mathfrak{st}}(C_y)\big) \\ \equiv \sum_{d\in\mathcal{D}, y\in W} (-1)^{l(d)} n_d\, \gamma_{w,d,y^{-1}}\, c^{\mathfrak{st}}_{y,\lambda} \qquad \bmod \mathfrak{m}.\end{aligned}$$

By Lemma 2.3.12, we have $\gamma_{w,d,y^{-1}} = \gamma_{y^{-1},w,d}$. Hence, Lemma 2.3.13 yields that

$$\sum_{d\in\mathcal{D}, y\in W} (-1)^{l(d)} n_d\, \gamma_{w,d,y^{-1}}\, c^{\mathfrak{st}}_{y,\lambda} \equiv c^{\mathfrak{st}}_{w,\lambda} \qquad \bmod \mathfrak{m}.$$

Since $c^{\mathfrak{st}}_{w,\lambda} \neq 0$, we can go back to the left-hand side of the original identity above and conclude that $\sum_{d\in\mathcal{D}} (-1)^{l(d)} n_d \varepsilon^{\mathbf{a}_\lambda + \mathbf{a}(w)} \rho^{\lambda}_{\mathfrak{st}}(C_w C_d) \not\equiv 0 \bmod \mathfrak{m}$. Thus, we have

$$\sum_{d\in\mathcal{D}} \sum_{\mathfrak{u}\in M(\lambda)} (-1)^{l(d)} n_d \big(\varepsilon^{\mathbf{a}(w)} \rho^{\lambda}_{\mathfrak{su}}(C_w)\big)\big(\varepsilon^{\mathbf{a}_\lambda} \rho^{\lambda}_{\mathfrak{ut}}(C_d)\big) \not\equiv 0 \bmod \mathfrak{m}.$$

So there must be some $d \in \mathcal{D}$ and some $\mathfrak{u} \in M(\lambda)$ such that $\varepsilon^{\mathbf{a}(w)} \rho^{\lambda}_{\mathfrak{ut}}(C_d) \not\in \mathfrak{m}$. Consequently, by Proposition 1.4.10(c) and 2.1.19, we have $\mathbf{a}_\lambda \geqslant \mathbf{a}(w)$. Thus, we have shown that $\mathbf{a}(w) = \mathbf{a}_\lambda$ if $E^\lambda \leftrightsquigarrow_L w$.

Now assume that $\lambda, \mu \in \Lambda$ are such that $E^\lambda \sim_L E^\mu$. By definition, this means that $w \sim_{\mathcal{LR}} w'$, where $w, w' \in W$ are such that $E^\lambda \leftrightsquigarrow_L w$ and $E^\mu \leftrightsquigarrow_L w'$. Using **P4**, we obtain $\mathbf{a}_\lambda = \mathbf{a}(w) = \mathbf{a}(w') = \mathbf{a}_\mu$; that is, (♣′) holds. □

*Remark 2.3.15.* In the above discussion, we have not found it necessary to use any of the properties **P2**, **P3**, **P5**, **P6**, **P13** in Lusztig's list. All of these express properties of the elements in $\mathcal{D}$. It seems that these are logically independent of **P1**, **P4**, **P15**.

Finally, we show that our algebra $\tilde{\mathbf{J}}$ constructed in Section 1.5 really is an incarnation of Lusztig's asymptotic ring $\mathbf{J}$ (see Remark 2.3.6).

**Proposition 2.3.16 (Cf. [114, §3]).** *Assume that* **P1**, **P4** *hold. Then*

$$\tilde{\gamma}_{x,y,z} = c_{x,y,z} \qquad \textit{and} \qquad \tilde{n}_w = \begin{cases} (-1)^{l(w)} n_w & \textit{if } w \in \mathcal{D}, \\ 0 & \textit{otherwise}, \end{cases}$$

*for all $x, y, z, w \in W$. In particular, Conjecture 1.5.12(a) holds; that is, $\tilde{\gamma}_{x,y,z}$ and $\tilde{n}_w$ are integers. Furthermore, we have $\tilde{\mathscr{D}} = \mathscr{D}$.*

Thus, the map $t_w \mapsto (-1)^{l(w)} t_w$ defines an algebra isomorphism $\tilde{\mathbf{J}} \xrightarrow{\sim} \mathbb{K} \otimes_{\mathbb{Z}} \mathbf{J}$, where $\mathbf{J}$ is Lusztig's asymptotic ring; see Remark 2.3.6.

*Proof.* Let $x, y, z \in W$. By **P4** and Proposition 2.1.20, we have $\tilde{\gamma}_{x,y,z} = 0$ unless $\mathbf{a}(x) = \mathbf{a}(y) = \mathbf{a}(z)$. The analogous statement also holds for $c_{x,y,z}$ by Lemma 2.3.12. Thus, in order to show that $\tilde{\gamma}_{x,y,z} = c_{x,y,z}$, we can assume without loss of generality that $\mathbf{a}(x) = \mathbf{a}(y) = \mathbf{a}(z)$. But, in this case, we have

$$c_{x,y,z} = \sum_{\lambda} \sum_{\mathfrak{s},\mathfrak{t},\mathfrak{u} \in M(\lambda)} f_\lambda^{-1} c_{x,\lambda}^{\mathfrak{st}} c_{y,\lambda}^{\mathfrak{tu}} c_{z,\lambda}^{\mathfrak{us}}$$

by Lemma 2.3.10, where the first sum runs over all $\lambda \in \Lambda$ such that $\mathbf{a}_\lambda = \mathbf{a}(x) = \mathbf{a}(y) = \mathbf{a}(z)$. On the other hand, we have

$$\tilde{\gamma}_{x,y,z} = \sum_{\lambda \in \Lambda} \sum_{\mathfrak{s},\mathfrak{t},\mathfrak{u} \in M(\lambda)} f_\lambda^{-1} c_{x,\lambda}^{\mathfrak{st}} c_{y,\lambda}^{\mathfrak{tu}} c_{z,\lambda}^{\mathfrak{us}}.$$

But, by Proposition 2.3.14, the leading matrix coefficients appearing in the above expression are zero unless $\mathbf{a}_\lambda = \mathbf{a}(x) = \mathbf{a}(y) = \mathbf{a}(z)$. Thus, the desired identity between $\tilde{\gamma}_{x,y,z}$ and $c_{x,y,z}$ is proved.

The identity in Lemma 2.3.13 and the fact that **P7** holds (see Lemma 2.3.12) now imply that $\sum_{d \in \mathscr{D}} (-1)^{l(d)} n_d t_d \in \tilde{\mathbf{J}}$ is an identity element in $\tilde{\mathbf{J}}$; see the analogous argument in the proof of Proposition 1.5.5. Since the identity element of $\tilde{\mathbf{J}}$ is uniquely determined, we obtain the desired statement about $\tilde{n}_w$. □

## 2.4 On Lusztig's Conjectures, II

The conjectural properties **P1–P15** are known to hold in a number of situations (including the equal-parameter case), but a general proof is still missing. There does not even seem to be a general idea of how to prove one of the crucial properties **P1**, **P4**, **P15** for an arbitrary weight function $L$. In this section, we first give a summary about the present state of knowledge concerning the validity of **P1–P15**. This will be followed by a detailed discussion of the case where $L(s) = 0$ for some $s \in S$.

**2.4.1.** Here is a summary of the cases where **P1–P15** are known to hold.

(a) **P1–P15** hold for any finite $W$, assuming that we are in the equal-parameter case where $\Gamma = \mathbb{Z}$ and there is some $a > 0$ such that $L(s) = a$ for all $s \in S$. (Here, $A$ is the ring of Laurent polynomials in one indeterminate $v = \varepsilon$.) Indeed, as already mentioned, Lusztig [231, Chap. 15] deduces **P1–P15** from the following "positivity" properties:

$$P_{y,w}^* \in \mathbb{Z}_{\geqslant 0}[v, v^{-1}] \quad \text{and} \quad h'_{x,y,z} \in \mathbb{Z}_{\geqslant 0}[v, v^{-1}] \quad \text{for all } x, y, z, w \in W.$$

(Recall that $h'_{x,y,z} = (-1)^{l(x)+l(y)+l(z)} h_{x,y,z}$; see also Remark 2.3.5.) If $W$ is a Weyl group, then these "positivity" properties follow from a geometric interpretation; see Kazhdan and Lusztig [196], Lusztig and Vogan [233] and Springer [283]. If $W$ is of type $I_2(m)$ (any $m \geqslant 2$), $H_3$ or $H_4$, they follow by explicit computations; see Alvis [2] and DuCloux [76].

(b) **P1–P15** have been checked by explicit computations for $W$ of type $I_2(m)$ (any $m \geqslant 3$) and any weight function such that $L(s) > 0$ for $s \in S$; see [76], [114, §5].

(c) **P1–P15** have been checked by explicit computations (with the help of a computer and CHEVIE [118]) for $W$ of type $F_4$ and any weight function such that $L(s) > 0$ for $s \in S$; see [105], [114, §5].

(d) **P1–P15** hold for $W$ of type $B_n$ and any weight function $L\colon W \to \Gamma$ given by

$B_n$ 

where $b > (n-1)a > 0$.

See the series of papers by Bonnafé, Geck, Iancu [21], [26], [108], [114], [121].

(e) **P1–P15** hold if $(W,S)$ is irreducible, $\Gamma = \mathbb{Z}$ and $L(s) = 0$ for some $s \in S$; see 2.4.8 below. (Here, we are essentially reduced to the equal-parameter case; see also Lusztig [224], [225], where a more general setting is considered.)

It is beyond the scope of this book to discuss the proofs of (a)–(d) in any more detail. An elementary proof of **P1–P15** for $W \cong \mathfrak{S}_n$ is given in [107]; see also Section 2.8. The geometric arguments used in (a) can be extended to the so-called *quasi-split case*, in which some choices of unequal parameters occur; see Table 4.1 (p. 227). (The proofs are sketched in [219] and [231, Chap. 16].) Of course, it would be highly desirable to find general proofs (at least for **P1, P4, P15**) which uniformly work for any $W, L$. The above results imply the following general statement:

**Corollary 2.4.2.** *Let $W$ be any finite Coxeter group and $L_0\colon W \to \Gamma_0$ be the universal weight function in Example 1.1.9. Let $\leqslant$ be a monomial order on $\Gamma_0$ such that we are in the "asymptotic case" as in Example 1.1.11(c). Then* **P1–P15** *hold for $W, L_0$.*

For the remainder of this section, we address in some more detail the question of what happens when $W$ is a finite Coxeter group and $L(s) = 0$ for some $s \in S$.

**2.4.3.** Let $\Omega \subseteq W$ be the parabolic subgroup generated by all $t \in S$ such that $L(t) = 0$. Then we can break down the structure of $W$ as follows. Let $W_1 \subseteq W$ be the subgroup generated by $S_1 := \{\omega s \omega^{-1} \mid \omega \in \Omega, s \in S \text{ where } L(s) > 0\}$. Then, by Bonnafé and Dyer [24, Theorem 1.1], $W_1$ is a normal subgroup of $W$ such that $W_1 \cap \Omega = \{1\}$; furthermore, we have a semidirect product decomposition

(a) $W = \Omega \ltimes W_1$ and $(W_1, S_1)$ is a Coxeter system.

Given $w \in W$, let $w = s_1 \cdots s_p$ ($s_i \in S$) be a reduced expression. We denote by $l_\Omega(w)$ the number of $i \in \{1, \ldots, p\}$ such that $L(s_i) = 0$, and by $l_1(w)$ the number of $i \in \{1, \ldots, p\}$ such that $L(s_i) > 0$. (Note that these two numbers do not depend on the choice of the reduced expression.) By [24, Cor. 1.3], we have

(b) $l(w) = l_\Omega(w) + l_1(w)$ and $l_1|_{W_1}$ is the length function for $(W_1, S_1)$.

(c) $l_1(\omega w \omega^{-1}) = l_1(w)$ for all $w \in W_1$ and $\omega \in \Omega$.

Now let $\tilde{s} \in S_1$ and write $\tilde{s} = \omega s \omega^{-1}$, where $\omega \in \Omega$ and $s \in S$ is such that $L(s) > 0$. Then one readily checks that $L(\tilde{s}) = L(s)$; moreover, we have

(d) The restriction $L|_{W_1} : W_1 \to \Gamma$ is a weight function, which we denote by $L_1$.

Indeed, let $w \in W_1$ and let $w = \tilde{s}_1 \cdots \tilde{s}_p$ ($\tilde{s}_i \in S_1$) be a reduced expression for $w$ with respect to $S_1$. For each $i$, let $s_i \in S$ and $\omega_i \in \Omega$ be such that $\tilde{s}_i = \omega_i s_i \omega_i^{-1}$ and $L(s_i) > 0$. Writing each $\omega_i$ as a product of generators $t \in S$ such that $L(t) = 0$, we obtain an expression for $w$ in terms of $S$ which is not necessarily reduced. But we can extract a reduced expression from it, and this reduced expression will contain the factors $s_1, \ldots, s_p$ and various generators $t \in S$ such that $L(t) = 0$. (By (b), all the factors $s_1, \ldots, s_p$ must occur since $l_1(w) = p$.) Thus, we have $L(w) = L(s_1) + \cdots + L(s_p)$. The argument also shows that $L(\tilde{s}_i) = L(s_i)$ for all $i$ and so $L(w) = L(\tilde{s}_1) + \cdots + L(\tilde{s}_p)$. Hence, $L|_{W_1} : W_1 \to \Gamma$ is a weight function, as required.

**Example 2.4.4.** Assume that $(W,S)$ is irreducible and that $\{1\} \neq \Omega \neq W$. According to the classification in Table 1.1, we are in one of the following cases.

(a) Let $(W,S)$ be of type $B_n$, where the generators are labelled as in the diagram below. Let $L(s_0) = 0$ and $L(s_1) = \cdots = L(s_{n-1}) > 0$. Then $\Omega = \{1, s_0\}$ and $S_1 = \{s_0 s_1 s_0, s_1, s_2, \ldots, s_{n-1}\}$. The Coxeter system $(W_1, S_1)$ is of type $D_n$:

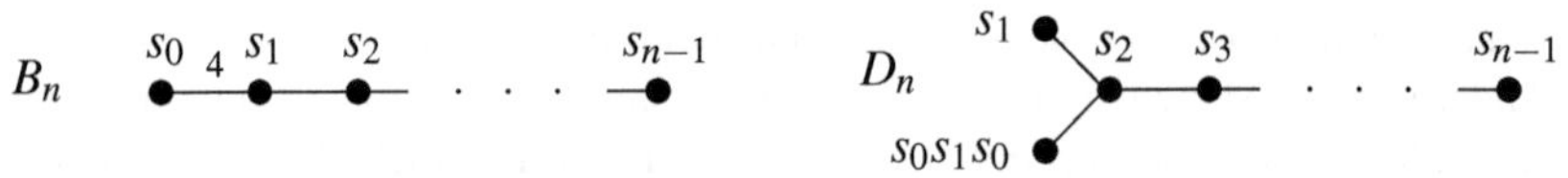


(b) Let $(W,S)$ be of type $B_n$ (with generators labelled as above), but now let $L(s_0) > 0$ and $L(s_1) = \cdots = L(s_{n-1}) = 0$. Then $\Omega = \langle s_1, s_2, \ldots, s_{n-1} \rangle \cong \mathfrak{S}_n$ and $S_1 = \{t_1, t_2, \ldots, t_n\}$, where $t_1 = s_0$ and $t_i = s_{i-1} t_{i-1} s_{i-1}$ for $2 \leqslant i \leqslant n$. One easily checks that all the $t_i$ commute with each other and so $(W_1, S_1)$ is of type $A_1 \times \cdots \times A_1$ ($n$ factors).

(c) Let $(W,S)$ be of type $F_4$, where the generators are labelled as in Table 1.1. Let $L(s_1) = L(s_2) = 0$ and $L(s_3) = L(s_4) > 0$. Then $\Omega = \langle s_1, s_2 \rangle \cong \mathfrak{S}_3$ and $S_1 = \{s_3, s_4, s_2 s_3 s_2, s_1 s_2 s_3 s_2 s_1\}$. One easily checks that $(W_1, S_1)$ is of type $D_4$, where $s_3$, $s_2 s_3 s_2$ and $s_1 s_2 s_3 s_2 s_1$ commute with each other,

(d) Let $(W,S)$ be of type $I_2(m)$, where $m \geqslant 4$ is even. Write $S = \{s_1, s_2\}$ and let $L(s_1) > 0$ and $L(s_2) = 0$. Then $\Omega = \{1, s_2\rangle$ and $S_1 = \{s_1, s_2 s_1 s_2\}$. One easily checks that $(W_1, S_1)$ is of type $I_2(m/2)$.

Note that, in all of the above cases, $L_1$ is a multiple of the length function of $W_1$.

On the level of **H**, we have the following result.

**Proposition 2.4.5 (Cf. Bonnafé [22, §2.E]).** *Assume we are in the setting of 2.4.3.*

*(a) For all $\omega, \omega' \in \Omega$ and $w \in W$, we have*

$$T_\omega T_w = T_{\omega w}, \qquad T_w T_\omega = T_{w\omega}, \qquad T_\omega T_{\omega'} = T_{\omega\omega'}, \qquad T_{\omega^{-1}} = T_\omega^{-1}.$$

(b) *For any* $\tilde{s} \in S_1$ *and* $w \in W_1$, *we have*

$$T_{\tilde{s}}T_w = \begin{cases} T_{\tilde{s}w} & \text{if } l_1(\tilde{s}w) > l_1(w), \\ T_{\tilde{s}w} + (\varepsilon^{L(\tilde{s})} - \varepsilon^{-L(\tilde{s})})T_w & \text{if } l_1(\tilde{s}w) < l_1(w). \end{cases}$$

*In particular,* $\mathbf{H}_1 := \langle T_w \mid w \in W_1 \rangle_A \subseteq \mathbf{H}$ *is a subalgebra, and this is the generic Iwahori–Hecke algebra associated with* $(W_1, S_1)$ *and* $L_1 \colon W_1 \to \Gamma$.

(c) *Let* $\{C_w \mid w \in W\}$ *be the Kazhdan–Lusztig basis of* $\mathbf{H}$. *Then*

$$C_\omega C_w = C_{\omega w}, \qquad C_w C_\omega = C_{w\omega}, \qquad C_\omega = T_\omega$$

*for all* $\omega \in \Omega$ *and* $w \in W$. *Furthermore, if* $w \in W_1$, *then* $C_w \in \mathbf{H}_1$ *and this is the Kazhdan–Lusztig basis element constructed within* $\mathbf{H}_1$.

*Proof.* (a) Let $t \in S$ be such that $L(t) = 0$. Then $T_t T_w = T_{tw}$ and $T_w T_t = T_{wt}$ for all $w \in W$ (independently of whether $l(tw) > l(w)$ or $l(tw) < l(w)$). This yields that $T_\omega T_w = T_{\omega w}$ and $T_w T_\omega = T_{w\omega}$ for all $w \in W$ and $\omega \in \Omega$. Hence, (a) follows.

(b) Let $\tilde{s} \in S_1$ and write $\tilde{s} = \omega s \omega^{-1}$ where $\omega \in \Omega$ and $s \in S$ is such that $L(s) > 0$. Let $w \in W_1$ and set $w' = \omega^{-1} w \omega$. Using 2.4.3(b), we obtain

$$l_1(\tilde{s}w) = l_1(\omega^{-1}\tilde{s}w\omega) = l_1(s\omega^{-1}w\omega) = l_1(sw') = l(sw') - l_\Omega(sw').$$

We certainly have $l_\Omega(sw') = l_\Omega(w')$. Using 2.4.3(b), this yields that

$$l_1(\tilde{s}w) - l_1(w) = l(sw') - l(w').$$

Now assume that $l_1(\tilde{s}w) > l_1(w)$. Then the above relation implies that $l(sw') > l(w')$ and so $T_s T_{w'} = T_{sw'}$. By (a), we have $T_{\tilde{s}} = T_\omega T_s T_\omega^{-1}$ and $T_w = T_\omega T_{w'} T_\omega^{-1}$. This yields

$$T_{\tilde{s}} T_w = T_\omega (T_s T_{w'}) T_\omega^{-1} = T_\omega T_{sw'} T_\omega^{-1} = T_{\omega s w' \omega^{-1}} = T_{\tilde{s}w},$$

as required. Similarly, if $l_1(\tilde{s}w) < l_1(w)$, then $l(sw') < l(w')$ and so $T_s T_{w'} = T_{sw} + (\varepsilon^{L(s)} - \varepsilon^{-L(s)})T_{w'}$. Using (a), we deduce that

$$\begin{aligned} T_{\tilde{s}} T_w &= T_\omega (T_s T_{w'}) T_\omega^{-1} = T_\omega \big(T_{sw'} + (\varepsilon^{L(s)} - \varepsilon^{-L(s)})T_{w'}\big) T_\omega^{-1} \\ &= \big(T_{\omega s w' \omega^{-1}} + (\varepsilon^{L(s)} - \varepsilon^{-L(s)}) T_{\omega w' \omega^{-1}}\big) = T_{\tilde{s}w} + (\varepsilon^{L(s)} - \varepsilon^{-L(s)}) T_w, \end{aligned}$$

as required; note that $L(\tilde{s}) = L(s)$. Once these relations are established, we see that $T_{\tilde{s}} T_w \in \mathbf{H}_1$ for all $\tilde{s} \in S_1$ and $w \in W_1$. It follows that $\mathbf{H}_1 \subseteq \mathbf{H}$ is a subalgebra. The relations then show that $\mathbf{H}_1 \cong \mathbf{H}_A(W_1, S_1, L_1)$, as required.

(c) By the formulae in Theorem 2.1.8, Example 2.1.5 and Remark 2.3.3, we have $C_t = T_t$, $C_t C_w = C_{tw}$ and $C_w C_t = C_{wt}$ for $w \in W$ and $t \in S$ such that $L(t) = 0$. This immediately yields the formulae for $C_\omega$, $C_\omega C_w$ and $C_w C_\omega$, where $\omega \in \Omega$. Now let $w \in W_1$ and denote by $\tilde{C}_w$ the Kazhdan–Lusztig basis element constructed inside $\mathbf{H}_1$. In order to show that $\tilde{C}_w = C_w$, we verify that $\tilde{C}_w$ satisfies the two conditions in Theorem 2.1.6 (with respect to $W$). We have $\tilde{C}_w \in T_w + \sum_{y \in W_1} \mathbb{Z}[\Gamma_{>0}] T_y$ and so the first condition is satisfied. Now let $\tilde{w}_0 \in W_1$ be the longest element (with respect to

$S_1$). Then $\tilde{w}_0^{-1}w_0 \in \Omega$ and so we can write $w_0 = \tilde{w}_0\omega_0$, where $\omega_0 \in \Omega$. By Theorem 2.1.6 (applied to $W_1$), we have $\tilde{C}_w T_{\tilde{w}_0} \in \sum_{y\in W_1} \mathbb{Z}[\Gamma_{\leqslant 0}]T_y$. Since $T_y T_{\omega_0} = T_{y\omega_0}$ for all $y \in W_1$, we deduce that $\tilde{C}_w T_{w_0} = \tilde{C}_w T_{\tilde{w}_0} T_{\omega_0} \in \sum_{y\in W_1} \mathbb{Z}[\Gamma_{\leqslant 0}]T_{y\omega_0}$ and so the second condition also holds. Hence, we must have $\tilde{C}_w = C_w$, as required. □

*Remark 2.4.6.* Define $\mathbf{H}_\omega := T_\omega.\mathbf{H}_1 = \mathbf{H}_1.T_\omega$ for $\omega \in \Omega$. Then Proposition 2.4.5 shows that

$$\mathbf{H} = \bigoplus_{\omega\in\Omega} \mathbf{H}_\omega \qquad \text{and} \qquad \mathbf{H}_\omega.\mathbf{H}_{\omega'} = \mathbf{H}_{\omega\omega'} \quad \text{for all } \omega, \omega' \in \Omega.$$

Thus, $\mathbf{H}$ is an *extended Iwahori–Hecke algebra* and the subspaces $\{\mathbf{H}_\omega \mid \omega \in \Omega\}$ form an $\Omega$-graded *Clifford system* in $\mathbf{H}$, in the sense of [53, Def. 11.12].

**2.4.7.** Let $w \in W$ and write $w = \omega w_1$, where $\omega \in \Omega$ and $w_1 \in W_1$. By Proposition 2.4.5(c), we have $T_w = T_\omega T_{w_1}$, $C_\omega = T_\omega$ and $C_w = C_\omega C_{w_1}$, where $C_{w_1}$ is the Kazhdan–Lusztig basis element defined within $\mathbf{H}_1$. Hence, we obtain

$$\sum_{y\in W}(-1)^{l(\omega w_1)+l(y)}\bar{P}^*_{y,\omega w_1}T_y = C_{\omega w_1} = C_\omega C_{w_1} = \sum_{y_1\in W_1}(-1)^{l(w_1)+l(y_1)}\bar{P}^*_{y_1,w_1}T_{\omega y_1},$$

where $P^*_{y_1,w_1}$ is defined by the element $C'_{w_1} \in \mathbf{H}_1$. Thus, given any $y \in W$ and writing $y = \omega' y_1$, where $\omega' \in \Omega$, $y_1 \in W_1$, we have

(a) $$P^*_{y,w} = \begin{cases} P^*_{y_1,w_1} & \text{if } \omega = \omega', \\ 0 & \text{otherwise.} \end{cases}$$

A similar relation can be established for the structure constants $h_{x,y,z}$. By definition, given $x_1, x_2 \in W_1$ and $\omega_1, \omega_2 \in \Omega$, we have

$$C_{\omega_1 x_1} C_{\omega_2 x_2} = \sum_{\omega_3\in\Omega, x_3\in W_1} h_{\omega_1 x_1, \omega_2 x_2, \omega_3 x_3} C_{\omega_3 x_3}.$$

Note that, if $\omega_1 = \omega_2 = \omega_3 = 1$, then $h_{x_1,x_2,x_3}$ is a structure constant with respect to the Kazhdan–Lusztig basis in $\mathbf{H}_1$. By the relations in Proposition 2.4.5(c), we have

(b) $$C_\omega^{-1} = C_{\omega^{-1}} \qquad \text{and} \qquad C_{\omega^{-1}} C_{w_1} C_\omega = C_{\omega^{-1} w_1 \omega}$$

for any $\omega \in \Omega$ and $w_1 \in W_1$. Using these relations, we obtain

$$\begin{aligned} C_{\omega_1 x_1} C_{\omega_2 x_2} &= C_{\omega_1} C_{x_1} C_{\omega_2} C_{x_2} = C_{\omega_1\omega_2} C_{\omega_2^{-1} x_1 \omega_2} C_{x_2} \\ &= \sum_{x_3\in W_1} h_{\omega_2^{-1} x_1 \omega_2, x_2, x_3} C_{\omega_1\omega_2 x_3}. \end{aligned}$$

Thus, for any $x_3 \in W_1$ and $\omega_3 \in \Omega$, we have

(c) $$h_{\omega_1 x_1, \omega_2 x_2, \omega_3 x_3} = \begin{cases} h_{\omega_2^{-1} x_1 \omega_2, x_2, x_3} & \text{if } \omega_3 = \omega_1\omega_2, \\ 0 & \text{otherwise.} \end{cases}$$

We see that the structure constants for the Kazhdan–Lusztig basis in $\mathbf{H}$ are completely determined by the structure constants inside $\mathbf{H}_1$.

**2.4.8.** Let $\mathbf{a}(z)$, $\Delta(z)$ ($z \in W$) and $\mathscr{D}$ be defined as in 2.3.1, with respect to the weight function $L\colon W \to \Gamma$. Define $\mathbf{a}_1(z_1)$, $\Delta_1(z_1)$ ($z_1 \in W_1$) and $\mathscr{D}_1$ analogously, with respect to the weight function $L_1 : W_1 \to \Gamma$. Then 2.4.7(c) shows that

$$\text{(a)} \qquad \mathbf{a}(\omega z_1) = \mathbf{a}_1(z_1) \qquad \text{for all } \omega \in \Omega \text{ and } z_1 \in W_1.$$

Furthermore, since $L_1(\tilde{s}) > 0$ for all $\tilde{s} \in S_1$, we have $P^*_{1,z_1} \neq 0$ for all $z_1 \in W_1$. Then 2.4.7(a) shows that

$$\text{(b)} \qquad \Delta(\omega z_1) = \Delta_1(z_1) \quad (\text{if } \omega = 1) \qquad \text{and} \qquad \mathscr{D} = \mathscr{D}_1.$$

Assume now that $\Gamma = \mathbb{Z}$ and that $L_1$ is a multiple of the length function of $W_1$. In particular, $A$ is the ring of Laurent polynomials in one indeterminate $v = \varepsilon$. Then the "positivity" properties in 2.4.1(a) hold for $W_1, L_1$. Using Remark 2.1.7 and the formulae in 2.4.7, we conclude that these "positivity" properties also hold for $W, L$:

$$P^*_{y,w} \in \mathbb{Z}_{\geqslant 0}[v, v^{-1}] \quad \text{and} \quad h'_{x,y,z} \in \mathbb{Z}_{\geqslant 0}[v, v^{-1}] \quad \text{for all } x, y, z, w \in W.$$

Taking into account (a) and (b), we can now follow Lusztig's arguments in [231, Chap. 15] to conclude that **P1–P15** hold for $W, L$. In particular, we see that **P1–P15** hold for $W, L$ in all situations described in Example 2.4.4 (where $\Gamma = \mathbb{Z}$).

**Proposition 2.4.9 (Bonnafé [22, §2.E]).**

(a) *Let $x_1, x_2 \in W_1$ and $\omega_1, \omega_2 \in \Omega$. Then $\omega_1 x_1 \leqslant_{\mathscr{L}} \omega_2 x_2$ (with respect to $L$) if and only if $x_1 \leqslant_{\mathscr{L}} x_2$ (with respect to $L_1$). Similarly, $x_1\omega_1 \leqslant_{\mathscr{R}} x_2\omega_2$ (with respect to $L$) if and only if $x_1 \leqslant_{\mathscr{R}} x_2$ (with respect to $L_1$).*

(b) *The left cells of $W$ (with respect to $L$) are of the form $\Omega.\mathfrak{C}_1$ where $\mathfrak{C}_1$ is a left cell of $W_1$ (with respect to $L_1$). The left cell module $[\Omega.\mathfrak{C}_1]_A$ is isomorphic to the induced module $\operatorname{Ind}_{\mathbf{H}_1}^{\mathbf{H}}([\mathfrak{C}_1]_A) := \mathbf{H} \otimes_{\mathbf{H}_1} [\mathfrak{C}_1]_A$.*

(c) *Let $x_1, x_2 \in W_1$ and $\omega_1, \omega_2 \in \Omega$. Then $\omega_1 x_1 \leqslant_{\mathscr{LR}} \omega_2 x_2$ (with respect to $L$) if and only if there exists some $\omega \in \Omega$ such that $x_1 \leqslant_{\mathscr{LR}} \omega x_2 \omega^{-1}$ (with respect to $L_1$).*

(d) *The two-sided cells of $W$ (with respect to $L$) are of the form $\Omega.\mathscr{F}_1.\Omega$, where $\mathscr{F}_1$ is a two-sided cell of $W_1$ (with respect to $L_1$).*

*Proof.* (a) Assume first that $\omega_1 x_1 \leqslant_{\mathscr{L}} \omega_2 x_2$ (with respect to $L$). It is enough to consider the case where $\omega_1 x_1 \leftarrow_{\mathscr{L}} \omega_2 x_2$; that is, there exists some $w \in W$ such that $h_{w,\omega_2 x_2,\omega_1 x_1} \neq 0$. By 2.4.7(c), this structure constant equals $h_{w_1,x_2,x_1}$ (for some $w_1 \in W_1$). Consequently, we have $x_1 \leqslant_{\mathscr{L}} x_2$ (with respect to $L_1$). Conversely, assume that $x_1 \leftarrow_{\mathscr{L}} x_2$ (with respect to $L_1$). Thus, $h_{w_1,x_2,x_1} \neq 0$ for some $w_1 \in W_1$. By 2.4.7(c), $h_{w_1,x_2,x_1}$ also is a structure constant for the Kazhdan–Lusztig basis in $\mathbf{H}$ and so $x_1 \leqslant_{\mathscr{L}} x_2$ (with respect to $L$). Furthermore, $\omega_1 x_1 \sim_{\mathscr{L}} x_1$ and $\omega_2 x_2 \sim_{\mathscr{L}} x_2$. (This immediately follows from the fact that $C_\omega C_w = C_{\omega w}$ for all $\omega \in \Omega$ and $w \in W$; see Proposition 2.4.5(c).) Hence, we also have $\omega_1 x_1 \leqslant_{\mathscr{L}} \omega_2 x_2$ (with respect to $L$). The statement about the relation $\leqslant_{\mathscr{R}}$ is proved using the fact that $x \leqslant_{\mathscr{L}} y \Leftrightarrow x^{-1} \leqslant_{\mathscr{R}} y^{-1}$.

(b) The statement about the left cells is an immediate consequence of (a). Now consider the left cell module $[\Omega.\mathfrak{C}_1]_A$. This module has a basis $\{e_{\omega_1 x_1} \mid \omega_1 \in \Omega, x_1 \in \mathfrak{C}_1\}$, where the action of $C_{\omega w_1}$ ($\omega \in \Omega$, $w_1 \in W_1$) is given by

$$C_{\omega w_1}.e_{\omega_1 x_1} = \sum_{\omega_2 \in \Omega, x_2 \in \mathfrak{C}_1} h_{\omega w_1, \omega_1 x_1, \omega_2 x_2}\, e_{\omega_2 x_2}.$$

Using 2.4.7(c), we obtain that

$$C_{\omega w_1}.e_{\omega_1 x_1} = \sum_{x_2 \in \mathfrak{C}_1} h_{\omega_1^{-1} w_1 \omega_1, x_1, x_2}\, e_{\omega\omega_1 x_2}.$$

On the other hand, by definition, $\mathrm{Ind}_{\mathbf{H}_1}^{\mathbf{H}}([\mathfrak{C}_1]_A)$ has a basis $\{\omega_1 \otimes e_{x_1} \mid \omega_1 \in \Omega, x_1 \in \mathfrak{C}_1\}$, where the action of $C_{\omega w_1}$ ($\omega \in \Omega$, $w_1 \in W_1$) is given by

$$C_{\omega w_1}.(\omega_1 \otimes e_{x_1}) = \sum_{x_2 \in \mathfrak{C}_1} h_{\omega_1^{-1} w_1 \omega_1, x_1, x_2}\, (\omega\omega_1 \otimes e_{x_2}).$$

Hence, $[\Omega.\mathfrak{C}_1]_A \rightarrow \mathrm{Ind}_{\mathbf{H}_1}^{\mathbf{H}}([\mathfrak{C}_1]_A)$, $e_{\omega_1 x_1} \mapsto \omega_1 \otimes e_{x_1}$, is an $\mathbf{H}$-module isomorphism

(c) For any $\omega \in \Omega$, the map $w_1 \mapsto \omega w_1 \omega^{-1}$ is a Coxeter group automorphism of $(W_1, S_1)$. Furthermore, by 2.4.7(b), we have $C_\omega^{-1} = C_{\omega^{-1}}$ and $C_{\omega w_1 \omega^{-1}} = C_\omega C_{w_1} C_{\omega^{-1}}$ for all $w_1 \in W_1$. Hence, for any $x_1, x_2 \in W_1$, we have

$$(*) \qquad x_1 \leqslant_{\mathscr{L}} x_2 \quad \Leftrightarrow \quad \omega x_1 \omega^{-1} \leqslant_{\mathscr{L}} \omega x_2 \omega^{-1} \qquad \text{(with respect to } L_1\text{)}.$$

Now let $x_1, x_2 \in W_1$ and $\omega_1, \omega_2 \in \Omega$. Assume first that $\omega_1 x_1 \leqslant_{\mathscr{LR}} \omega_2 x_2$ (with respect to $L$). It is enough to consider the case where $\omega_1 x_1 \leqslant_{\mathscr{L}} \omega_2 x_2$ or $\omega_1 x_1 \leqslant_{\mathscr{R}} \omega_2 x_2$ (with respect to $L$). Note that, in the latter case, we have $(\omega_1 x_1 \omega_1^{-1})\omega_1 \leqslant_{\mathscr{R}} (\omega_2 x_2 \omega_2^{-1})\omega_2$. Hence, using (a), we conclude that $x_1 \leqslant_{\mathscr{L}} x_2$ or $\omega_1 x_1 \omega_1^{-1} \leqslant_{\mathscr{R}} \omega_2 x_2 \omega_2^{-1}$ (with respect to $L_1$). Setting $\omega = 1$ or $\omega = \omega_1^{-1}\omega_2$ according to these two cases, and using $(*)$, we obtain $x_1 \leqslant_{\mathscr{LR}} \omega x_2 \omega^{-1}$ (with respect to $L_1$), as required. Conversely, assume that there is some $\omega \in \Omega$ such that $x_1 \leqslant_{\mathscr{LR}} \omega x_2 \omega^{-1}$ (with respect to $L_1$). But then, by (a), we have $\omega x_2 \omega^{-1} \sim_{\mathscr{L}} x_2 \omega^{-1} \sim_{\mathscr{R}} x_2$ and so $x_1 \leqslant_{\mathscr{LR}} x_2$ (with respect to $L$).

(d) This immediately follows from (c). □

*Remark 2.4.10.* Let $\mathscr{F}$ be a two-sided cell of $W$ (with respect to $L$). By Proposition 2.4.9(d), we have $\mathscr{F} \cap W_1 \neq \varnothing$. We claim that the following implication holds.

(a) If $x_1, x_2 \in \mathscr{F} \cap W_1$ are such that $x_1 \leqslant_{\mathscr{LR}} x_2$ (with respect to $L_1$), then we have $x_1 \sim_{\mathscr{LR}} x_2$ (with respect to $L_1$).

This is seen as follows. By Proposition 2.4.9(c), since $x_2 \leqslant_{\mathscr{LR}} x_1$ (with respect to $L$), there exists some $\omega \in \Omega$ such that $x_2 \leqslant_{\mathscr{LR}} \omega x_1 \omega^{-1}$ (with respect to $L_1$). Hence, since $x_1 \leqslant_{\mathscr{LR}} x_2$ (with respect to $L_1$), we have $x_1 \leqslant_{\mathscr{LR}} \omega x_1 \omega^{-1}$ (with respect to $L_1$). Relation $(*)$ in the proof of Proposition 2.4.9 shows that then we also have $\omega x_1 \omega^{-1} \leqslant_{\mathscr{LR}} \omega^2 x_1 \omega^{-2}$ (with respect to $L_1$). Repeating this argument, we obtain that $\omega^{i-1} x_1 \omega^{-(i-1)} \leqslant_{\mathscr{LR}} \omega^i x_1 \omega^{-i}$ (with respect to $L_1$), for all $i \geqslant 1$. But $\Omega$ has finite order, and so $\omega^i = 1$ for some $i \geqslant 1$. We conclude that $\omega x_1 \omega^{-1} \leqslant_{\mathscr{LR}} x_1$ and, hence, $x_2 \leqslant_{\mathscr{LR}} x_1$ (with respect to $L_1$). Thus, (a) is proved.

**2.4.11.** Assume that $\mathbb{K} \subseteq \mathbb{C}$ is a splitting field for both $W_1$ and $W$. Then we write

$$\mathrm{Irr}_{\mathbb{K}}(W_1) = \{E^{\lambda_1} \mid \lambda_1 \in \Lambda_1\} \qquad \text{and} \qquad \mathrm{Irr}_{\mathbb{K}}(W) = \{E^{\lambda} \mid \lambda \in \Lambda\}.$$

By the argument in Example 1.2.4, we have the following compatibility between specialisation and restriction, where $\lambda \in \Lambda$ and $\lambda_1 \in \Lambda_1$:

$$\begin{aligned} &\text{multiplicity of } E^{\lambda_1} \text{ in the restriction of } E^{\lambda} \text{ to } W_1 \\ &= \text{multiplicity of } E_\varepsilon^{\lambda_1} \text{ in the restriction of } E_\varepsilon^{\lambda} \text{ to } \mathbf{H}_{1,K} := K \otimes_A \mathbf{H}_1. \end{aligned}$$

Now, the group $\Omega$ acts on $W_1$ and, hence, on $\mathrm{Irr}_{\mathbb{K}}(W_1)$. Thus, there is an action of $\Omega$ on $\Lambda_1$ (which we write as $\lambda_1 \mapsto \omega.\lambda_1$) such that

(a) $\quad \mathrm{trace}(w_1, E^{\omega.\lambda_1}) = \mathrm{trace}(\omega^{-1} w_1 \omega, E^{\lambda_1}) \qquad$ for all $\lambda_1 \in \Lambda_1$ and $w_1 \in W_1$.

Using this notation, *Clifford's theorem* ([53, 11.1]) states the following: Let $\lambda \in \Lambda$ and $\lambda_1 \in \Lambda_1$ be such that $E^{\lambda_1}$ is a constituent of the restriction of $E^{\lambda}$ to $W_1$. Then this restriction is a direct sum of simple modules of the form $E^{\omega.\lambda_1}$, for various $\omega \in \Omega$. Since we have an $\Omega$-graded Clifford system as in Remark 2.4.6, there is also a version of Clifford's theorem on the level of $\mathbf{H}$ (see [53, (11.16)]):

(b) The restriction of $E_\varepsilon^{\lambda} \in \mathrm{Irr}(\mathbf{H}_K)$ is a direct sum of simple $\mathbf{H}_{1,K}$-modules of the form $E_\varepsilon^{\omega.\lambda_1}$, for various $\omega \in \Omega$.

Now let $\mathfrak{C}_1$ be a left cell of $W_1$ (with respect to $L_1$) and $\omega \in \Omega$. Then, by relation $(*)$ in the proof of Proposition 2.4.9(c), the set $\omega \mathfrak{C}_1 \omega^{-1}$ also is a left cell of $W_1$ (with respect to $L_1$). Now, using the formulae in 2.4.7, one sees that

$$h_{x_1,x_2,x_2} = h_{\omega x_1 \omega^{-1}, \omega x_2 \omega^{-1}, \omega x_3 \omega^{-1}} \qquad \text{for all } x_1, x_2, x_3 \in W_1.$$

Hence, the action of $C_{w_1}$ ($w_1 \in W_1$) on $\omega \mathfrak{C}_1 \omega^{-1}$ is the same as the action of $C_{\omega^{-1} w_1 \omega}$ on $\mathfrak{C}_1$. Combining this with (a), we conclude that

(c) $\qquad m(\mathfrak{C}_1, \mu_1) = m(\omega \mathfrak{C}_1 \omega^{-1}, \omega.\mu_1) \qquad$ for all $\omega \in \Omega$.

With these preparations, we obtain the following corollary.

**Corollary 2.4.12.** *Let $\lambda, \mu \in \Lambda$ and $\lambda_1, \mu_1 \in \Lambda_1$ be such that $E^{\lambda_1}$ appears in the restriction of $E^{\lambda}$ to $W_1$ and $E^{\mu_1}$ appears in the restriction of $E^{\mu}$ to $W_1$. Then $\mathscr{F}_{\lambda_1} \subseteq \mathscr{F}_{\lambda}$ and $\mathscr{F}_{\mu_1} \subseteq \mathscr{F}_{\mu}$. Furthermore, we have*

$$E^{\lambda} \preceq_L E^{\mu} \qquad \Leftrightarrow \qquad E^{\lambda_1} \preceq_{L_1} E^{\omega.\mu_1} \quad \textit{for some } \omega \in \Omega.$$

*Proof.* Let $\mathfrak{C}_1$, $\mathfrak{C}_1'$ be left cells of $W_1$ such that $m(\mathfrak{C}_1, \lambda_1) > 0$ and $m(\mathfrak{C}_1', \mu_1) > 0$. By Proposition 2.4.9(b), $\mathfrak{C} := \Omega.\mathfrak{C}_1$ and $\mathfrak{C}' := \Omega.\mathfrak{C}_1'$ are left cells of $W$; furthermore, we have $[\mathfrak{C}]_A \cong \mathrm{Ind}_{\mathbf{H}_1}^{\mathbf{H}}([\mathfrak{C}_1]_A)$ and $[\mathfrak{C}']_A \cong \mathrm{Ind}_{\mathbf{H}_1}^{\mathbf{H}}([\mathfrak{C}'_1]_A)$. Hence, by Frobenius reciprocity, we have $m(\mathfrak{C},\lambda) > 0$ and $m(\mathfrak{C}',\mu) > 0$. Then Corollary 2.2.5 also shows that $\mathscr{F}_{\lambda_1} \subseteq \mathscr{F}_\lambda$ (since $\mathfrak{C}_1 \subseteq \mathfrak{C}$) and $\mathscr{F}_{\mu_1} \subseteq \mathscr{F}_\mu$ (since $\mathfrak{C}'_1 \subseteq \mathfrak{C}'$).

Now assume that $E^\lambda \preceq_L E^\mu$. By Corollary 2.2.5, this implies that $w \leqslant_{\mathscr{LR}} w'$ (with respect to $L$), for some $w \in \mathfrak{C}$ and some $w' \in \mathfrak{C}'$. Let us write $w = \omega_1 w_1$ and $w' = \omega'_1 w'_1$ where $w_1 \in \mathfrak{C}_1$, $w'_1 \in \mathfrak{C}'_1$ and $\omega_1, \omega'_1 \in \Omega$. Then, by Proposition 2.4.9(c), we have $w_1 \leqslant_{\mathscr{LR}} \omega w'_1 \omega^{-1}$ (with respect to $L_1$) for some $\omega \in \Omega$. Using the formula in 2.4.11(c), we conclude that $E^{\lambda_1} \preceq_{L_1} E^{\omega.\mu_1}$, as required.

Conversely, assume that $E^{\lambda_1} \preceq_{L_1} E^{\omega.\mu_1}$, where $\omega \in \Omega$. By 2.4.11(c), we have $m(\omega\mathfrak{C}'_1\omega^{-1}, \omega.\mu_1) = m(\mathfrak{C}'_1, \mu_1) > 0$. So Corollary 2.2.5 implies $w_1 \leqslant_{\mathscr{LR}} \omega w'_1 \omega^{-1}$ (with respect to $L_1$), for some $w_1 \in \mathfrak{C}_1$ and some $w'_1 \in \mathfrak{C}'_1$. Hence, by Proposition 2.4.9(c), we also have $w_1 \leqslant_{\mathscr{LR}} w'_1$ (with respect to $L$). Since $w_1 \in \mathfrak{C}$ and $w'_1 \in \mathfrak{C}'_1$, we can use once more Corollary 2.2.5 and conclude that $E^\lambda \preceq_L E^\mu$. □

**Fig. 2.2** Two-sided cells in type $F_4$ with parameters $0,0,a,a$; see Example 2.4.4(c)

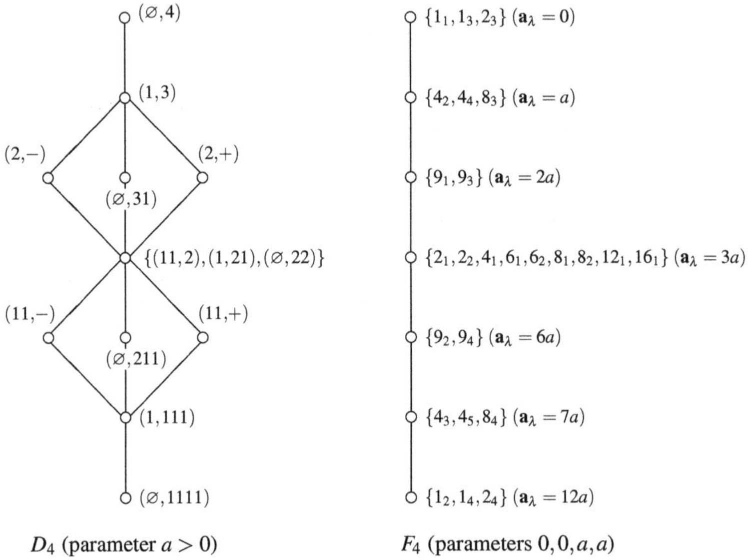


*Remark 2.4.13.* The above result shows that the relation $\preceq_L$ on $\mathrm{Irr}_{\mathbb{K}}(W)$ is completely determined by the relation $\preceq_{L_1}$ on $\mathrm{Irr}_{\mathbb{K}}(W_1)$; an example is given in Figure 2.2. Note that the converse is not true, at least not in any straightforward way. For example, using the above notation, assume that $\lambda = \mu$ and $\omega \in \Omega$ is such that $\lambda_1 = \omega.\mu_1 \neq \mu_1$. Then both sides of the equivalence in Corollary 2.4.3 are trivially true, but we cannot tell whether it is true that $E^{\lambda_1} \preceq_{L_1} E^{\mu_1}$ or not. One can show that, in general, we have:

(a) $$E^{\lambda_1} \preceq_{L_1} E^{\mu_1} \quad \Leftrightarrow \quad \begin{cases} \lambda_1 = \mu_1 & \text{if } \lambda = \mu, \\ E^\lambda \preceq_L E^\mu & \text{otherwise.} \end{cases}$$

(See Example 2.2.15 for the case where $W_1$ is of type $D_n$ and $W$ is of type $B_n$; see Figure 2.2, where $W_1$ is of type $D_4$ and $W$ is of type $F_4$. The remaining cases are much easier to deal with; we omit further details.)

**Proposition 2.4.14 (Cf. [132, 10.5.6], [101, 4.6]).** *Let $\lambda \in \Lambda$ and $\lambda_1 \in \Lambda_1$ be such that $E^{\lambda_1}$ appears in the restriction of $E^\lambda$ to $W_1$. Then*

$$\mathbf{c}_\lambda \dim E^\lambda = |\Omega|\, \mathbf{c}_{\lambda_1} \dim E^{\lambda_1}, \quad \mathbf{a}_\lambda = \mathbf{a}_{\lambda_1}, \quad f_\lambda \dim E^\lambda = |\Omega|\, f_{\lambda_1} \dim E^{\lambda_1}.$$

*Proof.* Let $d_\lambda = \dim E^\lambda$ and denote by $I_{d_\lambda}$ the identity matrix of size $d_\lambda$. Considering a matrix representation $\rho^\lambda$ afforded by $E_\varepsilon^\lambda$, we have that

$$d_\lambda \mathbf{c}_\lambda I_{d_\lambda} = \sum_{w \in W} \rho^\lambda(T_w)\rho^\lambda(T_{w^{-1}}).$$

(Indeed, the $(\mathfrak{s},\mathfrak{t})$-coefficient of the expression on the right-hand side equals

$$\sum_{w \in W} \sum_{\mathfrak{u} \in M(\lambda)} \rho^\lambda_{\mathfrak{s},\mathfrak{u}}(T_w)\rho^\lambda_{\mathfrak{u},\mathfrak{t}}(T_{w^{-1}}).$$

By the Schur relations in Proposition 1.2.12, this evaluates to $\delta_{\mathfrak{s}\mathfrak{t}} d_\lambda \mathbf{c}_\lambda$, as required.) Now let us write $W = \{w_1\omega \mid w_1 \in W_1, \omega \in \Omega\}$. By Proposition 2.4.5(a), we have $T_{w_1\omega} = T_{w_1}T_\omega$ and $T_\omega^{-1} = T_{\omega^{-1}}$. This yields that

$$\begin{aligned} d_\lambda \mathbf{c}_\lambda I_{d_\lambda} &= \sum_{w_1 \in W_1} \sum_{\omega \in \Omega} \rho^\lambda(T_{w_1})\rho^\lambda(T_\omega)\rho^\lambda(T_\omega^{-1})\rho^\lambda(T_{w_1^{-1}}) \\ &= |\Omega| \sum_{w_1 \in W_1} \rho^\lambda(T_{w_1})\rho^\lambda(T_{w_1^{-1}}). \end{aligned}$$

Since $E^{\lambda_1}$ appears in the restriction of $E^\lambda$ from $W$ to $W_1$, a specialisation argument (see Example 1.2.4) shows that $E_\varepsilon^{\lambda_1}$ appears in the restriction of $E_\varepsilon^\lambda$ from $\mathbf{H}_K$ to $\mathbf{H}_{1,K}$. Thus, choosing a suitable basis of $E_\varepsilon^\lambda$ we can assume that, for each $w_1 \in W_1$, the matrix $\rho^\lambda(T_{w_1})$ has a block diagonal shape, where one of the blocks equals $\rho^{\lambda_1}(T_{w_1})$. Let $d_{\lambda_1} = \dim E^{\lambda_1}$. Considering the corresponding block in the above identity arising from the Schur relations, we obtain

$$d_\lambda \mathbf{c}_\lambda I_{d_{\lambda_1}} = |\Omega| \sum_{w_1 \in W_1} \rho^{\lambda_1}(T_{w_1})\rho^{\lambda_1}(T_{w_1^{-1}}).$$

But then the sum on the right-hand side can be evaluated using the Schur relations for $\mathbf{H}_{1,K}$. This yields the desired identity $d_\lambda \mathbf{c}_\lambda = |\Omega| d_{\lambda_1} \mathbf{c}_{\lambda_1}$. Once this is established, the identities concerning $\mathbf{a}_\lambda$ and $f_\lambda$ are immediate consequences. □

## 2.5 On Lusztig's Conjectures, III

Our aim now is to formulate a version of the properties **P1–P15** purely in terms of the invariants $\mathbf{a}_\lambda$ and our algebra $\tilde{\mathbf{J}}$ constructed in Section 1.5.

**Proposition 2.5.1 (Cf. Lusztig [231, 18.9(b)]).** *Assume that* **P1, P4, P15** *hold.*

*(a) Let $w, w', x, y \in W$ and assume that $\mathbf{a}(x) = \mathbf{a}(y)$. Then*

$$\sum_{z \in W} \tilde{\gamma}_{x,w',z^{-1}} h_{w,z,y} = \sum_{z \in W} h_{w,x,z} \tilde{\gamma}_{z,w',y^{-1}}.$$

*(b) Let $w, w', y \in W$ and assume that $\mathbf{a}(w') = \mathbf{a}(y)$. Then*

$$h_{w,w',y} = \sum_{\substack{z \in W, d \in \tilde{\mathscr{D}} \\ \mathbf{a}(z) = \mathbf{a}(d)}} \tilde{n}_d h_{w,d,z} \tilde{\gamma}_{z,w',y^{-1}}.$$

*Proof.* (a) Consider the identity **P15**; by **P4**, we can assume that on both sides the sum is over all $z \in W$ such that $a := \mathbf{a}(z) = \mathbf{a}(x) = \mathbf{a}(y)$. Now, we can write

$$\varepsilon^a h_{x,w',z} = c_{x,w',z^{-1}} + g_{x,w',z}, \qquad \text{where } g_{x,w',z} \in \mathbb{Z}[\Gamma_{>0}],$$
$$\varepsilon^a h_{z,w',y} = c_{z,w',y^{-1}} + g_{z,w',y}, \qquad \text{where } g_{z,w',y} \in \mathbb{Z}[\Gamma_{>0}].$$

Hence, multiplying both sides of **P15** by $\varepsilon^a \otimes 1$, we obtain

$$1 \otimes \Big( \sum_{z \in W : \mathbf{a}(z) = a} c_{x,w',z^{-1}} h_{w,z,y} \Big) + \sum_{z \in W : \mathbf{a}(z) = a} g_{x,w',z} \otimes h_{w,z,y}$$
$$= 1 \otimes \Big( \sum_{z \in W : \mathbf{a}(z) = a} c_{z,w',y^{-1}} h_{w,x,z} \Big) + \sum_{z \in W : \mathbf{a}(z) = a} g_{z,w',y} \otimes h_{w,x,z}.$$

Finally, $\mathbb{Z}[\Gamma] \otimes_{\mathbb{Z}} \mathbb{Z}[\Gamma]$ is a free $\mathbb{Z}$-module with basis $\{\varepsilon^g \otimes \varepsilon^{g'} \mid g, g' \in \Gamma\}$. Comparing the coefficients of $1 \otimes \varepsilon^{g'}$ on both sides, we obtain the identity

$$\sum_{z \in W} c_{x,w',z^{-1}} h_{w,z,y} = \sum_{z \in W} h_{w,x,z} c_{z,w',y^{-1}}.$$

The desired identity in (a) now follows from Proposition 2.3.16.

(b) Taking $x = d \in \tilde{\mathscr{D}}$, we multiply both sides of the identity in (a) by $\tilde{n}_d$ and then sum over all $d \in \tilde{\mathscr{D}}$ such that $\mathbf{a}(d) = \mathbf{a}(y)$. This yields

$$\sum_{z \in W} \sum_{\substack{d \in \tilde{\mathscr{D}} \\ \mathbf{a}(y) = \mathbf{a}(d)}} \tilde{n}_d \tilde{\gamma}_{d,w',z^{-1}} h_{w,z,y} = \sum_{z \in W} \sum_{\substack{d \in \tilde{\mathscr{D}} \\ \mathbf{a}(y) = \mathbf{a}(d)}} \tilde{n}_d h_{w,d,z} \tilde{\gamma}_{z,w',y^{-1}}.$$

On the right-hand side, we can replace the condition "$\mathbf{a}(y) = \mathbf{a}(d)$" by the condition "$\mathbf{a}(z) = \mathbf{a}(d)$", since $\tilde{\gamma}_{z,w',y^{-1}} \neq 0$ implies $\mathbf{a}(z) = \mathbf{a}(y)$ by **P4** and Proposition 2.1.20. On the other hand, by Lemma 1.5.3(a), we have $\tilde{\gamma}_{d,w',z^{-1}} = \tilde{\gamma}_{w',z^{-1},d}$. Hence, the left-hand side of the above identity equals

$$\sum_{z\in W}\sum_{\substack{d\in\tilde{\mathscr{D}}\\ \mathbf{a}(y)=\mathbf{a}(d)}} \tilde{n}_d\,\tilde{\gamma}_{w',z^{-1},d}\,h_{w,z,y} = \sum_{z\in W}\Big(\sum_{\substack{d\in\tilde{\mathscr{D}}\\ \mathbf{a}(y)=\mathbf{a}(d)}} \tilde{\gamma}_{w',z^{-1},d}\,\tilde{n}_d\Big)h_{w,z,y}.$$

Now, if $\tilde{\gamma}_{w',z^{-1},d}\neq 0$, then $\mathbf{a}(d)=\mathbf{a}(w')$ by **P4** and Proposition 2.1.20. Since $\mathbf{a}(y)=\mathbf{a}(w')$, we can omit the condition "$\mathbf{a}(y)=\mathbf{a}(d)$" in the above sum. So Lemma 1.5.3(b) yields that the above sum evaluates to $h_{w,w',y}$, as required. □

**Corollary 2.5.2.** *Assume that* **P1**, **P4**, **P15** *hold. Then* **P9**, **P10**, **P11** *also hold.*

*Proof.* To prove **P9**, let $y,w\in W$ be such that $y\leqslant_{\mathscr{L}} w$ and $\mathbf{a}(y)=\mathbf{a}(w)$. We must show that $y\sim_{\mathscr{L}} w$. It is enough to consider the case where $y,w$ are related by an elementary step of the relation $\leqslant_{\mathscr{L}}$; that is, we have $h_{x,y,w}\neq 0$ for some $x\in W$. But then Proposition 2.5.1(b) shows that there exist some $z\in W$ and $d\in\tilde{\mathscr{D}}$ such that $\mathbf{a}(z)=\mathbf{a}(d)$, $h_{x,d,z}\neq 0$ and $\tilde{\gamma}_{z,y,w^{-1}}\neq 0$. In particular, by Lemma 1.6.5 and Proposition 2.1.20, $y$ and $w$ belong to the same Kazhdan–Lusztig left cell, as required. Once **P9** is established, **P10** and **P11** easily follow as well; see [231, 14.10, 14.11]. Indeed, to obtain **P10**, just note that $\mathbf{a}(z)=\mathbf{a}(z^{-1})$ and that $y\leqslant_{\mathscr{R}} w$ if and only if $y^{-1}\leqslant_{\mathscr{L}} w^{-1}$. Finally, to prove **P11**, let $y\leqslant_{\mathscr{LR}} w$ be such that $\mathbf{a}(y)=\mathbf{a}(w)$. By definition, there is a sequence $y=y_0,y_1,\ldots,y_m=w$ such that, for each $i\in\{1,\ldots,m\}$, we have $y_{i-1}\leqslant_{\mathscr{L}} y_i$ or $y_{i-1}\leqslant_{\mathscr{R}} y_i$. By **P4**, we have $\mathbf{a}(w)=\mathbf{a}(y_m)\geqslant\mathbf{a}(y_{m-1})\geqslant\ldots\geqslant\mathbf{a}(y_1)\geqslant\mathbf{a}(y_0)=\mathbf{a}(y)$. Since $\mathbf{a}(y)=\mathbf{a}(w)$, we have $\mathbf{a}(y)=\mathbf{a}(y_0)=\mathbf{a}(y_1)=\ldots=\mathbf{a}(y_m)=\mathbf{a}(w)$. Applying **P9** or **P10** to $y_{i-1},y_i$, we obtain $y_{i-1}\sim_{\mathscr{L}} y_i$ or $y_{i-1}\sim_{\mathscr{R}} y_i$. Hence, $y\sim_{\mathscr{LR}} w$. □

**2.5.3.** Let us consider the following three statements (♣), (♠), (♦). These should be regarded as our adaptation of Lusztig's properties **P1**–**P15** in Conjecture 2.3.2 for the purposes of this book. Note that (♣), (♠), (♦) do not refer to the function $\mathbf{a}(z)$ or to $\gamma_{x,y,z}$, as defined in 2.3.1; these have only played an auxiliary role.

(♣) Let $\lambda,\mu\in\Lambda$. If $E^\lambda\preceq_L E^\mu$, then $\mathbf{a}_\mu\leqslant\mathbf{a}_\lambda$. In particular, if $E^\lambda\sim_L E^\mu$, then $\mathbf{a}_\lambda=\mathbf{a}_\mu$. Furthermore, if $E^\lambda\preceq_L E^\mu$ and $\mathbf{a}_\mu=\mathbf{a}_\lambda$, then $E^\lambda\sim_L E^\mu$.

(♠) Let $w,w',x,y\in W$ be such that $x\sim_{\mathscr{LR}} y$. Then

$$\sum_{z\in W}\tilde{\gamma}_{x,w',z^{-1}}\,h_{w,z,y}=\sum_{z\in W}h_{w,x,z}\,\tilde{\gamma}_{z,w',y^{-1}}.$$

(Here, $\sim_{\mathscr{LR}}$ refers to the two-sided Kazhdan–Lusztig relation.)

(♦) Every Kazhdan–Lusztig left cell contains a unique element of $\tilde{\mathscr{D}}$.

Let us briefly recall how the first two statements are deduced from **P1**, **P4**, **P15**. To prove (♣), let $x\in\mathscr{F}_\lambda$ and $y\in\mathscr{F}_\mu$. By Proposition 2.3.14, we have $\mathbf{a}(x)=\mathbf{a}_\lambda$ and $\mathbf{a}(y)=\mathbf{a}_\mu$. So, if $E^\lambda\preceq_L E^\mu$, then $x\leqslant_{\mathscr{LR}} y$ and so $\mathbf{a}_\mu=\mathbf{a}(y)\leqslant\mathbf{a}(x)=\mathbf{a}_\lambda$, using **P4**. If $E^\lambda\preceq_L E^\mu$ and $\mathbf{a}_\lambda=\mathbf{a}_\mu$, then $x\leqslant_{\mathscr{LR}} y$ and $\mathbf{a}(x)=\mathbf{a}(y)$. By Corollary 2.5.2, **P11** holds and so $x\sim_{\mathscr{LR}} y$; hence, $E^\lambda\sim_{\mathscr{LR}} E^\mu$, as required. Finally, if $w,w',x,y\in W$

are as in (♠), then $\mathbf{a}(x) = \mathbf{a}(y)$ by **P4** and so the desired identity holds by Proposition 2.5.1. (One can show that, conversely, (♣) and (♠) imply **P1, P4, P15**; see [114, 3.8, 4.7].)

Finally, if **P1, P4, P13** hold, then (♦) holds, since $\tilde{\mathscr{D}} = \mathscr{D}$ by Proposition 2.3.16.

*Remark 2.5.4.* As in Remark 2.3.7, we note that (♠) really is a statement about a certain bimodule structure. Indeed, let $R \subseteq \mathbb{C}$ be an $L$-good subring and consider the algebra $\tilde{\mathbf{J}}$; see Section 1.5. Then $\tilde{\mathbf{J}}_R := \langle t_w \mid w \in W \rangle_R \subseteq \tilde{\mathbf{J}}$ is an $R$-subalgebra of $\tilde{\mathbf{J}}$ and $\tilde{\mathbf{J}} = \mathbb{K} \otimes_R \tilde{\mathbf{J}}_R$. By the identification $C_w \leftrightarrow t_w$, the natural left $\mathbf{H}$-module structure on $\mathbf{H}$ (given by left multiplication) can be transported to a left $\mathbf{H}$-module structure on $\tilde{\mathbf{J}}_A := A \otimes_R \tilde{\mathbf{J}}_R$. Explicitly, the action is given by

$$C_w * t_x = \sum_{z \in W} h_{w,x,z} t_z \qquad \text{for all } x, w \in W.$$

By the definition of the Kazhdan–Lusztig pre-order $\leqslant_{\mathscr{LR}}$, we can define a left $\mathbf{H}$-module structure on $\tilde{\mathbf{J}}_A$ by the formula

$$C_w \diamond t_x = \sum_{z \in W : z \sim_{\mathscr{LR}} x} h_{w,x,z} t_z \qquad \text{for all } x, w \in W.$$

For any $h \in \mathbf{H}$ and $x \in W$, the difference $h * t_x - h \diamond t_x$ is an $A$-linear combination of terms $t_y$, where $y \leqslant_{\mathscr{LR}} w$ and $y \not\sim_{\mathscr{LR}} w$ (in the Kazhdan–Lusztig pre-order). On the other hand, we have a natural right $\tilde{\mathbf{J}}_A$-module structure on $\tilde{\mathbf{J}}_A$ (given by right multiplication). Then these two actions commute if and only if

$$C_w \diamond (t_x t_{w'}) = (C_w \diamond t_x) t_{w'} \qquad \text{for all } x, w, w' \in W.$$

Writing this out using the defining equations, the above identity is equivalent to

$$\sum_{z \in W : z \sim_{\mathscr{LR}} y} \tilde{\gamma}_{x,w',z^{-1}} h_{w,z,y} = \sum_{z \in W : z \sim_{\mathscr{LR}} x} h_{w,x,z} \tilde{\gamma}_{z,w',y^{-1}} \qquad \text{for all } y \in W.$$

Now, by Proposition 2.1.20, we can assume that $z \sim_{\mathscr{LR}} x$ for all $z$ on the left-hand side and $z \sim_{\mathscr{LR}} y$ for all $z$ on the right-hand side. Thus, if $x \not\sim_{\mathscr{LR}} y$, then both sides of the above identity are zero. Hence, the above identity holds if and only if (♠) holds. So we conclude

(a) (♠) holds if and only if $\tilde{\mathbf{J}}_A$ is an $(\mathbf{H}, \tilde{\mathbf{J}}_A)$-bimodule (with the above actions).

Since the algebra $\mathbf{H}$ is generated by $\{C_s \mid s \in S\} \cup \{T_1\}$, we also conclude

(b) in order to verify (♠), it is sufficient to do this assuming that $w = s \in S$.

The following result was proved by Lusztig [223] in the equal-parameter case and in [231, 18.9 and 18.10] in general, assuming that **P1–P15** hold. Here, we follow the proof given in [112], which is much less "computational" than that in [223], [231].

**Theorem 2.5.5 (Lusztig [231, 18.9]; see also [112, §5]).** *Assume that property (♠) in 2.5.3 holds. Then there is a unique unital A-algebra homomorphism* $\phi\colon \mathbf{H} \to \tilde{\mathbf{J}}_A$

*such that, for any $h \in \mathbf{H}$ and $w \in W$, the difference $\phi(h)t_w - h * t_w$ is an $A$-linear combination of terms $t_y$, where $y \leqslant_{\mathscr{LR}} w$ and $y \not\sim_{\mathscr{LR}} w$. Explicitly, $\phi$ is given by*

$$\phi(C_w) = \sum_{\substack{z \in W, d \in \tilde{\mathscr{D}} \\ z \sim_{\mathscr{LR}} d}} h_{w,d,z} \tilde{n}_d t_z \qquad (w \in W).$$

*Proof.* In the setting of Remark 2.5.4, the left $\mathbf{H}$-module structure on $\tilde{\mathbf{J}}_A$ gives rise to an $A$-algebra homomorphism

$$\psi \colon \mathbf{H} \to \mathrm{End}_A(\tilde{\mathbf{J}}_A) \quad \text{such that} \quad \psi(h)(t_w) = h \diamond t_w.$$

Since the left action of $\mathbf{H}$ on $\tilde{\mathbf{J}}_A$ commutes with the right action of $\tilde{\mathbf{J}}_A$, the image of $\psi$ lies in $\mathrm{End}_{\tilde{\mathbf{J}}_A}(\tilde{\mathbf{J}}_A^{\mathrm{r}})$, where the superscript "r" indicates that we consider the right action of $\tilde{\mathbf{J}}_A$ on itself. Now, we have a natural $A$-algebra isomorphism

$$\eta \colon \mathrm{End}_{\tilde{\mathbf{J}}_A}(\tilde{\mathbf{J}}_A^{\mathrm{r}}) \to \tilde{\mathbf{J}}_A, \qquad f \mapsto f(1_{\tilde{\mathbf{J}}_A}).$$

We define $\phi = \eta \circ \psi \colon \mathbf{H} \to \tilde{\mathbf{J}}_A$. Then $\phi$ is an $A$-algebra homomorphism such that

$$\phi(h) = \psi(h)(1_{\tilde{\mathbf{J}}_A}) = h \diamond 1_{\tilde{\mathbf{J}}_A} \qquad \text{for all } h \in \mathbf{H}.$$

This yields $\phi(h)t_w = (h \diamond 1_{\tilde{\mathbf{J}}_A})t_w = h \diamond 1_{\tilde{\mathbf{J}}_A} t_w = h \diamond t_w$ or, in other words, the difference $\phi(h)t_w - h * t_w$ is an $A$-linear combination of terms $t_y$, where $y \leqslant_{\mathscr{LR}} w$ and $y \not\sim_{\mathscr{LR}} w$, as required. Furthermore, we immediately obtain the formula

$$\phi(C_w) = C_w \diamond 1_{\tilde{\mathbf{J}}_A} = \sum_{d \in \tilde{\mathscr{D}}} \tilde{n}_d C_w \diamond t_d = \sum_{z \in W, d \in \tilde{\mathscr{D}} : z \sim_{\mathscr{LR}} d} h_{w,d,z} \tilde{n}_d t_z.$$

Since $h_{1,d,z} = \delta_{d,z}$, this yields $\phi(C_1) = 1_{\tilde{\mathbf{J}}_A}$; hence, $\phi$ is unital.

Finally, assume that $\phi' \colon \mathbf{H} \to \tilde{\mathbf{J}}_A$ is another homomorphism satisfying the required conditions. But these imply that $\phi'(h)t_w = h \diamond t_w$ for all $w \in W$ and, hence, $\phi'(h) = \phi'(h)1_{\tilde{\mathbf{J}}_A} = h \diamond 1_{\tilde{\mathbf{J}}_A}$ for all $h \in \mathbf{H}$. So we have $\phi' = \phi$ as required. □

*Remark 2.5.6.* Once Theorem 2.5.5 is established, the further theory of $\tilde{\mathbf{J}}$ and $\mathbf{H}$ can be developed as in [231, Chap. 18–20], with essentially the same proofs. We just single out the following statement; cf. Lusztig [231, 18.11]:

(a) Let $\theta \colon A \to k$ be a specialisation, where $k$ is a commutative ring with 1. Let $\phi_k \colon \mathbf{H}_k \to \tilde{\mathbf{J}}_k$ be the induced map. Then $\ker(\phi_k)$ is a nilpotent ideal of $\mathbf{H}_k$.

*Proof.* Let $\mathscr{F}_1, \ldots, \mathscr{F}_N$ be the two-sided Kazhdan–Lusztig cells of $W$, where the labelling is such that if $x \leqslant_{\mathscr{LR}} y$ for all $x \in \mathscr{F}_i$ and $y \in \mathscr{F}_j$, then $i \leqslant j$. Consequently, each $\mathbf{H}_{k,\leqslant i} := \langle C_w \mid w \in \mathscr{F}_j,\ 1 \leqslant j \leqslant i \rangle_k$ is a two-sided ideal of $\mathbf{H}_k$. Now let $h \in \ker(\phi_k)$. Then, by Theorem 2.5.5, $h * t_w$ is a $k$-linear combination of terms $t_y$, where $y \leqslant_{\mathscr{LR}} w$ and $y \not\sim_{\mathscr{LR}} w$. Recalling the definition of the $*$-action, we deduce that $h\mathbf{H}_{k,\leqslant i} \subseteq \mathbf{H}_{k,\leqslant i-1}$ for all $i$, where we set $\mathbf{H}_{k,\leqslant 0} = \{0\}$. Hence, given $N$ elements $h_1, \ldots, h_N \in \ker(\phi_k)$, then we have $h_1 \cdots h_N \in \mathbf{H}_{k,\leqslant 0} = \{0\}$. Thus, (a) is proved. □

**Example 2.5.7 (Cf. Lusztig [231, 20.1]).** The inclusion $A \subseteq K$ induces an algebra homomorphism $\phi\colon \mathbf{H}_K \to \tilde{\mathbf{J}}_K$. Since $\mathbf{H}_K$ is semisimple, Remark 2.5.6(a) shows that $\phi$ is an isomorphism. Next, consider the specialisation $\theta_1 : A \to \mathbb{K}$ such that $\theta_1(\varepsilon^g) = 1$ for all $g \in \Gamma$. Let $\phi_1 \colon \mathbb{K}W \to \tilde{\mathbf{J}}$ be the induced map. Since $\mathbb{K}W$ is semisimple, Remark 2.5.6(a) shows that $\phi_1$ is an isomorphism. Thus, $\mathbb{K}W \cong \tilde{\mathbf{J}}$ as $\mathbb{K}$-algebras. Finally, the inclusion $\mathbb{K} \subseteq K$ induces an algebra isomorphism $(\phi_1)_K \colon KW \to \tilde{\mathbf{J}}_K$. Hence, the composition

$$\psi = (\phi_1)_K^{-1} \circ \phi_K \colon \mathbf{H}_K \to KW \quad \text{is an algebra isomorphism.}$$

(This first appeared in [216] in the equal-parameter case.) Thus, using $\psi$, one obtains a more natural explanation for the correspondence $\mathrm{Irr}_{\mathbb{K}}(W) \leftrightarrow \mathrm{Irr}(\mathbf{H}_K)$ in 1.2.1. But note that the results in 1.2.1 do not rely on the assumption that (♠) holds!

**Lemma 2.5.8.** *Assume that (♠) holds. Let $x, y, w \in W$ be such that $y \sim_{\mathcal{LR}} w$. Then*

$$h_{x,w,y} = \sum_{\substack{z \in W, d \in \tilde{\mathcal{D}} \\ z \sim_{\mathcal{LR}} d}} \tilde{n}_d \, h_{x,d,z} \, \tilde{\gamma}_{z,w,y^{-1}}.$$

*Proof.* The left-hand side of the above identity is the coefficient of $t_y$ in the expansion of $C_x * t_w$, and the right-hand side is the coefficient of $t_y$ in the expansion of $\phi(C_x)t_w$. By Theorem 2.5.5, these two coefficients must be the same. □

**Lemma 2.5.9.** *Assume that (♠) holds. Then the left $\tilde{\mathbf{J}}$-cells are precisely the left Kazhdan–Lusztig cells. (Analogous statements hold for right and two-sided cells.) Furthermore, the following implication holds for any $y, w \in W$:*

$$y \leqslant_{\mathcal{L}} w \quad \text{and} \quad y \sim_{\mathcal{LR}} w \qquad \Rightarrow \qquad y \sim_{\mathcal{L}} w,$$

*where $\leqslant_{\mathcal{L}}$, $\sim_{\mathcal{L}}$ and $\sim_{\mathcal{LR}}$ refer to the Kazhdan–Lusztig relations.*

*Proof.* Recall that, by Proposition 2.1.20, every left (or right or two-sided) $\tilde{\mathbf{J}}$-cell is contained in a left (or right or two-sided respectively) Kazhdan–Lusztig cell. To prove the reverse implications, we begin by showing the following two statements:

(a) Let $y, w \in W$ be such that $y \leqslant_{\mathcal{L}} w$ and $y \sim_{\mathcal{LR}} w$ (with respect to the Kazhdan–Lusztig relations). Then $y, w$ belong to the same left $\tilde{\mathbf{J}}$-cell (and, hence, $y \sim_{\mathcal{L}} w$).
(b) Let $y, w \in W$ be such that $y \leqslant_{\mathcal{R}} w$ and $y \sim_{\mathcal{LR}} w$ (with respect to the Kazhdan–Lusztig relations). Then $y, w$ belong to the same right $\tilde{\mathbf{J}}$-cell.

To prove (a), we may assume that $y, w$ are related by an elementary step in the Kazhdan–Lusztig pre-order relation $\leqslant_{\mathcal{L}}$; that is, we can assume that $h_{x,w,y} \neq 0$ for some $x \in W$. But then Lemma 2.5.8 shows that there exist some $z \in W$ and $d \in \tilde{\mathcal{D}}$ such that $z \sim_{\mathcal{LR}} d$, $h_{x,d,z} \neq 0$ and $\tilde{\gamma}_{z,w,y^{-1}} \neq 0$. In particular, by Lemma 1.6.5, $y$ and $w$ belong to the same left $\tilde{\mathbf{J}}$-cell. Thus, (a) is proved. The proof of (b) is analogous.

Now (a) shows that if $y \sim_{\mathcal{L}} w$, then $y, w$ belong to the same left $\tilde{\mathbf{J}}$-cell. Thus, the left Kazhdan–Lusztig cells coincide with the left $\tilde{\mathbf{J}}$-cells. Using (b), a similar statement holds for right cells. Now consider the two-sided cells. Let $y, w \in W$ be such that $y \sim_{\mathscr{LR}} w$. Then there is a sequence $y = y_0, y_1, \ldots, y_m = w$ in $W$ such that, for each $i \in \{1, \ldots, m\}$, we have $y_{i-1} \leqslant_{\mathscr{L}} y_i$ or $y_{i-1} \leqslant_{\mathscr{R}} y_i$. Since $y \sim_{\mathscr{LR}} w$, all elements $y_i$ belong to the same two-sided Kazhdan–Lusztig cell. Hence, by (a) and (b), all elements $y_i$ belong to the same two-sided $\tilde{\mathbf{J}}$-cell. In particular, $y, w$ belong to the same two-sided $\tilde{\mathbf{J}}$-cell. □

**Example 2.5.10.** Assume that (♠) holds. Let $\mathfrak{C}$ be a left Kazhdan–Lusztig cell of $W$. By Lemma 2.5.9, the set $\mathfrak{C}$ also is a left $\tilde{\mathbf{J}}$-cell. Then we claim that

$$\tilde{m}(\mathfrak{C}, \lambda) = m(\mathfrak{C}, \lambda) \qquad \text{for all } \lambda \in \Lambda,$$

where the left-hand side is defined in Theorem 1.8.1 and the right-hand side is defined in 2.2.2. Indeed, by the argument in the proof of Lemma 2.2.4, we have

$$\sum_{\mathfrak{s},\mathfrak{t} \in M(\lambda)} \sum_{w \in \mathfrak{C}} c_{w,\lambda}^{\mathfrak{st}}\, c_{w^{-1},\lambda}^{\mathfrak{ts}} = m(\mathfrak{C}, \lambda)\, d_\lambda\, f_\lambda.$$

By Theorem 1.8.1(b), the left-hand side also equals $d_\lambda \tilde{m}(\mathfrak{C}, \lambda) f_\lambda$, as required. We can now write the relations in Theorem 1.8.1 in the form

$$\sum_{w \in \mathfrak{C}} c_{w,\lambda}\, c_{w^{-1},\mu} = \begin{cases} m(\mathfrak{C}, \lambda) f_\lambda & \text{if } \lambda = \mu, \\ 0 & \text{otherwise.} \end{cases}$$

We close this section with some auxiliary results which will be useful at several places below. The proofs of some of these will only require the following weak version of (♣) which we already encountered at the beginning of Section 2.3 (p. 78):

$$(\clubsuit') \qquad E^\lambda \sim_L E^\mu \qquad \Rightarrow \qquad \mathbf{a}_\mu = \mathbf{a}_\lambda$$

**Lemma 2.5.11.** *Assume that (♣′) holds. Let $\mathfrak{T}$ be a two-sided Kazhdan–Lusztig cell and $a \in \Gamma_{\geqslant 0}$ be the common value of $\mathbf{a}_\lambda$, where $\lambda \in \Lambda$ is such that $\mathscr{F}_\lambda \subseteq \mathfrak{T}$. Then*

$$\varepsilon^a h_{x,y,z} \in \mathbb{Z}[\Gamma_{\geqslant 0}] \qquad \textit{and} \qquad \tilde{\gamma}_{x,y,z^{-1}} \equiv \varepsilon^a h_{x,y,z} \bmod \mathbb{Z}[\Gamma_{>0}]$$

*for all $x \in W$ and $y, z \in \mathfrak{T}$. In particular, $\tilde{\gamma}_{x,y,z^{-1}} \in \mathbb{Z}$.*

*Proof.* We have $h_{x,y,z} = \tau(C_x C_y D_{z^{-1}})$ and $\tau = \sum_{\lambda \in \Lambda} \mathbf{c}_\lambda^{-1} \chi^\lambda$. Furthermore, as in the proof of Lemma 2.3.10, $\mathbf{c}_\lambda^{-1} = f_\lambda^{-1} \varepsilon^{2\mathbf{a}_\lambda}/(1+g_\lambda)$, where $g_\lambda \in F[\Gamma_{>0}]$. This yields

$$\varepsilon^a h_{x,y,z} = \sum_{\lambda \in \Lambda} \sum_{\mathfrak{s},\mathfrak{t},\mathfrak{u} \in M(\lambda)} \frac{f_\lambda^{-1}}{1+g_\lambda} \big(\varepsilon^{\mathbf{a}_\lambda} \rho_{\mathfrak{st}}^\lambda(C_x)\big)\big(\varepsilon^{\mathbf{a}_\lambda} \rho_{\mathfrak{tu}}^\lambda(C_y)\big)\big(\varepsilon^a \rho_{\mathfrak{us}}^\lambda(D_{z^{-1}})\big).$$

Now assume that $\lambda \in \Lambda$ and $\mathfrak{s}, \mathfrak{t}, \mathfrak{u} \in M(\lambda)$ are such that all three terms

$$\varepsilon^{\mathbf{a}_\lambda} \rho_{\mathfrak{st}}^\lambda(C_x), \qquad \varepsilon^{\mathbf{a}_\lambda} \rho_{\mathfrak{tu}}^\lambda(C_y), \qquad \varepsilon^a \rho_{\mathfrak{us}}^\lambda(D_{z^{-1}})$$

in the above sum are non-zero. Let $\mathfrak{C}$ be a left Kazhdan–Lusztig cell such that $m(\mathfrak{C}, \lambda) > 0$. Then, by Lemma 2.3.9, there exist $y', z' \in \mathfrak{C}$ such that $y' \leqslant_{\mathscr{R}} y$ and

$z \leqslant_{\mathcal{R}} z'$. Since $y,z \in \mathfrak{T}$, we deduce that $\mathfrak{C} \subseteq \mathfrak{T}$. By Lemma 2.2.4, there exists some $w \in \mathfrak{C} \cap \mathscr{F}_\lambda$. Since $\mathfrak{C} \subseteq \mathfrak{T}$, this implies that $\mathscr{F}_\lambda \subseteq \mathfrak{T}$ and so $a = \mathbf{a}_\lambda$. Consequently, all of the above three terms lie in $\mathscr{O}_0$. Hence, the whole sum lies in $\mathscr{O}_0$ and its constant term can be computed term by term. Thus, we obtain

$$\varepsilon^a h_{x,y,z} \equiv \sum_{\lambda \in \Lambda} \sum_{\mathfrak{s},\mathfrak{t},\mathfrak{u} \in M(\lambda)} c_{x,\lambda}^{\mathfrak{st}} c_{y,\lambda}^{\mathfrak{tu}} c_{z^{-1},\lambda}^{\mathfrak{us}} \equiv \tilde{\gamma}_{x,y,z^{-1}} \bmod \mathfrak{m}.$$

Since $h_{x,y,z} \in \mathbb{Z}[\Gamma]$, we have $\tilde{\gamma}_{x,y,z^{-1}} \in \mathbb{Z}$ and the congruences are modulo $\mathbb{Z}[\Gamma_{>0}]$. □

**Proposition 2.5.12.** *Assume that we are in the equal-parameter case where $\Gamma = \mathbb{Z}$ and $L(s) = 1$ for all $s \in S$. Then property (♠) in 2.5.3 is a consequence of (♣′).*

*Proof.* By Remark 2.5.4, it is enough to prove (♠) assuming that $w = s \in S$. Thus, we must show that

(a) $$\sum_{z \in W} \tilde{\gamma}_{x,w',z^{-1}} h_{s,z,y} = \sum_{z \in W} h_{s,x,z} \tilde{\gamma}_{z,w',y^{-1}} \qquad \text{for all } s \in S,$$

where $w',x,y \in W$ are such that $x \sim_{\mathcal{LR}} y$ (in the Kazhdan–Lusztig pre-order). Let $\mathfrak{T}$ denote the two-sided Kazhdan–Lusztig cell such that $x,y \in \mathfrak{T}$. First we note that, by Lemmas 1.6.5 and 1.6.6, we can assume that $z \in \mathfrak{T}$ on both sides of the above identity; furthermore, we can also assume that $w' \in \mathfrak{T}$. Now we argue as follows.

If $L(s) = 0$, then $h_{s,z,y} = \delta_{zy}$; see Theorem 2.1.8. Hence, the left-hand side of (a) reduces to $\tilde{\gamma}_{x,w',y^{-1}}$. Similarly, since $h_{s,x,z} = \delta_{xz}$, the right-hand side reduces to $\tilde{\gamma}_{x,w',y^{-1}}$. Hence, the assertion is true in this case. We can assume from now on that $L(s) > 0$.

*Case 1*: $sx < x$. Then, by Theorem 2.1.8, we have $C_s C_x = -(v_s + v_s^{-1}) C_x$ and so the right-hand side of (a) reduces to $-(v_s + v_s^{-1}) \tilde{\gamma}_{x,w',y^{-1}}$. Now let $z \in W$ and assume that the corresponding term on the left-hand side of (a) is non-zero; that is, $\tilde{\gamma}_{x,w',z^{-1}} \neq 0$ and $h_{s,z,y} \neq 0$. By Lemma 1.6.5, this implies that $z \sim_{\mathcal{R}} x$ and so $sz < z$; see Remark 2.1.16. Hence, the left hand side also reduces to $-(v_s + v_s^{-1}) \tilde{\gamma}_{x,w',y^{-1}}$. Thus, the identity (a) holds in this case.

*Case 2*: $sx > x$. Let again $z \in W$ and assume that the corresponding term on the left-hand side of (a) is non-zero; that is, we have $\tilde{\gamma}_{x,w',z^{-1}} \neq 0$ and $h_{s,z,y} \neq 0$. Again, this implies that $z \sim_{\mathcal{R}} x$ and so $sz > z$. Hence, the sum on the left-hand side only needs to be extended over all $z \in W$ such that $sz > z$. But then, since we are in the equal-parameter case, we have $h_{s,z,y} \in \mathbb{Z}$; see Example 2.1.10(b). Now consider the usual associativity rule in $\mathbf{H}$: the identity $C_s(C_x C_{w'}) = (C_s C_x) C_{w'}$ yields

$$\sum_{z \in W} h_{x,w',z} h_{s,z,y} = \sum_{z \in W} h_{z,w',y} h_{s,x,z} \qquad \text{for all } y \in W.$$

Let $z \in W$ be such that the corresponding term on the left-hand side is non-zero. Then $h_{x,w',z} \neq 0$ and so $z \leqslant_{\mathcal{R}} x$; furthermore, $h_{s,z,y} \neq 0$ and so $y \leqslant_{\mathcal{L}} z$. Since $x \sim_{\mathcal{LR}} y$, we deduce that $z \sim_{\mathcal{LR}} x$. Thus, we can assume that $z \in \mathfrak{T}$ in the sum on the left-hand side. Now we use (♣′). Let $a \in \Gamma_{\geqslant 0}$ be the common value of $\mathbf{a}_\lambda$, where $\lambda \in \Lambda$ is such that $\mathscr{F}_\lambda \subseteq \mathfrak{T}$. By Lemma 2.5.11, we have

$$\varepsilon^a h_{x,w',z} \equiv \tilde{\gamma}_{x,w,z^{-1}} \mod \mathbb{Z}[\Gamma_{>0}].$$

Hence, since $h_{s,z,y} \in \mathbb{Z}$ for all $z$ such that $sz > z$, we have

$$\varepsilon^a \Big( \sum_{z \in W} h_{x,w',z} h_{s,z,y} \Big) = \sum_{z \in \mathfrak{T}} \big(\varepsilon^a h_{x,w',z}\big) h_{s,z,y} \in \mathbb{Z}[\Gamma_{>0}]$$

and the constant term of this expression equals the left hand side of (a). A similar argument applies to the right-hand side of the above associativity identity: the sum only needs to be extended over all $z \in \mathfrak{T}$. Furthermore, we have

$$\varepsilon^a h_{z,w',y} \equiv \tilde{\gamma}_{z,w',y^{-1}} \mod \mathbb{Z}[\Gamma_{>0}].$$

Hence, since $sx > x$ and $h_{s,x,z} \in \mathbb{Z}$, we have

$$\varepsilon^a \Big( \sum_{z \in W} h_{z,w',y} h_{s,x,z} \Big) = \sum_{z \in \mathfrak{T}} \big(\varepsilon^a h_{z,w',y}\big) h_{s,x,z} \in \mathbb{Z}[\Gamma_{>0}]$$

and the constant term equals the right hand side of (a). Thus, (a) is proved. □

**Lemma 2.5.13.** *Assume that (♣) holds. Then we have $\tilde{\gamma}_{x,y,z} \in \mathbb{Z}$ for all $x,y,z \in W$ and $\tilde{n}_w \in \mathbb{Z}$ for all $w \in W$.*

*Proof.* If $y, z^{-1}$ belong to the same two-sided Kazhdan–Lusztig cell, then we have $\tilde{\gamma}_{x,y,z} \in \mathbb{Z}$ by Lemma 2.5.11. Otherwise, we have $\tilde{\gamma}_{x,y,z} = 0$ by Proposition 2.1.20. It remains to consider $\tilde{n}_w$. Let $\lambda_0 \in \Lambda$ be such that $E^{\lambda_0} \leftrightsquigarrow_L w$. We have $\overline{P}^*_{1,w} = (-1)^{l(w)} \tau(C_w)$. Expressing $\tau$ as in the proof of Lemma 2.5.11, we obtain

$$(-1)^{l(w)} \varepsilon^{-\mathbf{a}_{\lambda_0}} \overline{P}^*_{1,w} = \sum_{\lambda \in \Lambda} \sum_{\mathfrak{s} \in M(\lambda)} \frac{f_\lambda^{-1}}{1+g_\lambda} \varepsilon^{\mathbf{a}_\lambda - \mathbf{a}_{\lambda_0}} \big(\varepsilon^{\mathbf{a}_\lambda} \rho^\lambda_{\mathfrak{s}\mathfrak{s}}(C_w)\big).$$

Let $\lambda \in \Lambda$ be such that $\rho^\lambda(C_w) \neq 0$. Then we claim that $\mathbf{a}_{\lambda_0} \leqslant \mathbf{a}_\lambda$. Indeed, let $\mathfrak{C}$ be a left Kazhdan–Lusztig cell such that $m(\mathfrak{C}, \lambda) > 0$. By Lemma 2.3.9(a), we have $y \leqslant_{\mathcal{R}} w$ for some $y \in \mathfrak{C}$. By Lemma 2.2.4, there also exists some $y' \in \mathfrak{C}$ such that $E^\lambda \leftrightsquigarrow_L y'$. In particular, we now have $y' \sim_{\mathcal{LR}} y \leqslant_{\mathcal{LR}} w$ and so $E^\lambda \preceq_L E^{\lambda_0}$. Since (♣) is assumed to hold, we can conclude that $\mathbf{a}_{\lambda_0} \leqslant \mathbf{a}_\lambda$, as required. This shows that the above sum lies in $\mathcal{O}_0$ and we have

$$(-1)^{l(w)} \varepsilon^{-\mathbf{a}_{\lambda_0}} \overline{P}^*_{1,w} \equiv \sum_{\substack{\lambda \in \Lambda \\ \mathbf{a}_\lambda = \mathbf{a}_{\lambda_0}}} \sum_{\mathfrak{s} \in M(\lambda)} f_\lambda^{-1} c^{\mathfrak{s}\mathfrak{s}}_{w,\lambda} \mod \mathfrak{m}.$$

But then the first sum can be extended over all $\lambda \in \Lambda$: just note that if $c^{\mathfrak{s}\mathfrak{s}}_{w,\lambda} \neq 0$, then $E^\lambda \leftrightsquigarrow_L w$ and so $E^\lambda \sim_L E^{\lambda_0}$; hence, $\mathbf{a}_\lambda = \mathbf{a}_{\lambda_0}$ in this case. So we conclude that

$$(-1)^{l(w)} \varepsilon^{-\mathbf{a}_{\lambda_0}} \overline{P}^*_{1,w} \equiv \tilde{n}_w \mod \mathfrak{m}.$$

Since the left-hand side lies in $\mathbb{Z}[\Gamma]$, we deduce that $\tilde{n}_w \in \mathbb{Z}$, as required. □

**Example 2.5.14.** Assume that (♠), (♣), (♦) are satisfied. Then Conjectures 1.5.12 and 1.6.18 hold. (Indeed, part (a) of Conjecture 1.5.12 holds by Lemma 2.5.13; using Lemma 2.5.9 and (♦), we see that Conjecture 1.6.18 holds; then part (b) of Conjecture 1.5.12 follows by the argument in Remark 1.6.19.) Now let $\mathfrak{C}$ be a left Kazhdan–Lusztig cell of $W$. Then, by Examples 1.8.6 and 2.5.10, we have

$$\sum_{\lambda\in\Lambda} f_\lambda^{-1} m(\mathfrak{C},\lambda) = 1.$$

This is a quite powerful statement. (It can also be easily deduced from [231, 21.4].) For example, it directly shows that if $f_\lambda = 1$ for all $\lambda \in \Lambda$, then $[\mathfrak{C}]_1 \in \mathrm{Irr}_{\mathbb{K}}(W)$.

*Remark 2.5.15.* Following Lusztig [223, 2.8], we can now also give a more direct proof of the fact that $\phi_K\colon \mathbf{H}_K \to \tilde{\mathbf{J}}_K$ is an isomorphism. Indeed, let us assume that both (♠) and (♣) hold. For each $w \in W$, choose some $\lambda \in \Lambda$ such that $E^\lambda \leftrightsquigarrow_L w$ and set $\mathbf{a}_w := \mathbf{a}_\lambda$. (This does not depend on the choice of $\lambda$, thanks to (♣).) Then

$$\varepsilon^{\mathbf{a}_w}\phi(C_w) = \sum_{\substack{z\in W,\, d\in\tilde{\mathscr{D}}\\ z\sim_{\mathscr{LR}} d}} \varepsilon^{\mathbf{a}_w}\,\tilde{n}_d\, h_{w,d,z} t_z = \sum_{z\in W}\Big(\sum_{d\in\tilde{\mathscr{D}}:z\sim_{\mathscr{LR}} d} \varepsilon^{\mathbf{a}_w}\,\tilde{n}_d\, h_{w,d,z}\Big) t_z.$$

Assume that $z$ and $d$ in the above sum are such that $h_{w,d,z} \neq 0$. Let $\mu \in \Lambda$ be such that $E^\mu \leftrightsquigarrow_L z$. Since $h_{w,d,z} \neq 0$ and $z \sim_{\mathscr{LR}} d$, we have $z \leqslant_{\mathscr{LR}} w$ and so $E^\mu \preceq_L E^\lambda$. Since (♣) holds, this implies that $\mathbf{a}_w \leqslant \mathbf{a}_z$, with equality only if $w \sim_{\mathscr{LR}} z$. Assume first that $w \sim_{\mathscr{LR}} z$. Then $\mathbf{a}_w = \mathbf{a}_z$ and Lemma 2.5.11 shows that $\varepsilon^{\mathbf{a}_w} h_{w,d,z}$ lies in $\mathbb{Z}[\Gamma_{\geqslant 0}]$ and has constant term $\tilde{\gamma}_{w,d,z^{-1}}$. Thus, we have

$$\sum_{d\in\tilde{\mathscr{D}}:z\sim_{\mathscr{LR}} d} \varepsilon^{\mathbf{a}_w}\,\tilde{n}_d\, h_{w,d,z} \equiv \sum_{d\in\tilde{\mathscr{D}}:z\sim_{\mathscr{LR}} d} \tilde{n}_d\,\tilde{\gamma}_{w,d,z^{-1}} \mod \mathbb{Z}[\Gamma_{>0}],$$

where $w \sim_{\mathscr{LR}} z$. Now, if $\tilde{\gamma}_{w,d,z^{-1}} \neq 0$, then $w,d,z$ belong to the same two-sided Kazhdan–Lusztig cell; see Proposition 2.1.20. So we can omit the condition $z \sim_{\mathscr{LR}} d$ in the above sum. Then Lemma 1.5.3 shows that

$$\sum_{d\in\tilde{\mathscr{D}}:z\sim_{\mathscr{LR}} d} \tilde{n}_d\,\tilde{\gamma}_{w,d,z^{-1}} = \sum_{d\in\tilde{\mathscr{D}}} \tilde{n}_d\,\tilde{\gamma}_{z^{-1},w,d} = \delta_{zw}.$$

Since $\tilde{n}_d \in \mathbb{Z}$ by Lemma 2.5.13, we finally obtain that

$$\begin{aligned}\varepsilon^{\mathbf{a}_w}\phi(C_w) = t_w &+ \mathbb{Z}[\Gamma_{>0}]\text{-combination of terms } t_z,\text{ where } z\sim_{\mathscr{LR}} w\\ &+ \mathbb{Z}[\Gamma]\text{-combination of terms } t_z,\text{ where } z\leqslant_{\mathscr{LR}} w,\ z\not\sim_{\mathscr{LR}} w.\end{aligned}$$

So, for a suitable ordering of the elements of $W$, the matrix of $\phi$ with respect to the basis $\{\varepsilon^{\mathbf{a}_w} C_w \mid w \in W\}$ of $\mathbf{H}$ and the basis $\{t_w \mid w \in W\}$ of $\tilde{\mathbf{J}}_A$ has a block triangular shape where the determinant of each diagonal block lies in $1+\mathbb{Z}[\Gamma_{>0}]$. Hence, the determinant of the whole matrix of $\phi$ lies in $1+\mathbb{Z}[\Gamma_{>0}]$. In particular, it is non-zero.

## 2.6 A Cellular Basis for H

We are now ready to define a new basis of **H** which will turn out to be a "cellular basis" in the sense of Graham and Lehrer [144]. We recall the basic definitions first.

**2.6.1.** Let $k$ be a commutative ring (with 1) and $\underline{H}$ be an associative $k$-algebra (with identity) which is finitely generated and free over $k$. Following Graham and Lehrer [144, Def. 1.1], a *cell datum* for $\underline{H}$ is a quadruple $(\underline{\Lambda}, \underline{M}, \underline{C}, *)$ satisfying the following conditions.

(C1) $\underline{\Lambda}$ is a partially ordered set, $\{\underline{M}(\lambda) \mid \lambda \in \underline{\Lambda}\}$ is a collection of finite sets and $\underline{C} = \{C^\lambda_{\mathfrak{s},\mathfrak{t}} \mid \lambda \in \underline{\Lambda},\, \mathfrak{s},\mathfrak{t} \in \underline{M}(\lambda)\}$ is a $k$-basis for $\underline{H}$.
(C2) There is a $k$-linear anti-involution, $h \mapsto h^*$, on $\underline{H}$ such that $(C^\lambda_{\mathfrak{s},\mathfrak{t}})^* = C^\lambda_{\mathfrak{t},\mathfrak{s}}$ for all $\lambda \in \underline{\Lambda}$ and all $\mathfrak{s},\mathfrak{t} \in \underline{M}(\lambda)$.
(C3) Denote by $\preceq$ the partial order on $\underline{\Lambda}$. If $\lambda \in \underline{\Lambda}$ and $\mathfrak{s},\mathfrak{t} \in \underline{M}(\lambda)$, then

$$hC^\lambda_{\mathfrak{s},\mathfrak{t}} \equiv \sum_{\mathfrak{s}' \in \underline{M}(\lambda)} r^\lambda_h(\mathfrak{s}',\mathfrak{s})\, C^\lambda_{\mathfrak{s}',\mathfrak{t}} \mod \underline{H}(\prec \lambda) \qquad \text{for all } h \in \underline{H},$$

where $r^\lambda_h(\mathfrak{s}',\mathfrak{s}) \in k$ is independent of $\mathfrak{t}$ and where $\underline{H}(\prec \lambda)$ is the $k$-submodule of $\underline{H}$ generated by $\{C^\mu_{\mathfrak{s}'',\mathfrak{t}''} \mid \mu \prec \lambda; \mathfrak{s}'',\mathfrak{t}'' \in \underline{M}(\mu)\}$.

If these conditions hold, we say that $\{C^\lambda_{\mathfrak{s},\mathfrak{t}}\}$ is a *cellular basis* of $\underline{H}$. Assume now that this is the case. Given $\lambda \in \underline{\Lambda}$, we can define a corresponding *cell representation* (or *cell module*) of $\underline{H}$ as follows. Let $W(\lambda)$ be a free $k$-module with basis $\{C_\mathfrak{s} \mid \mathfrak{s} \in \underline{M}(\lambda)\}$. Then, using (C3), $W(\lambda)$ is seen to be an $\underline{H}$-module with action given by

$$h.C_\mathfrak{s} = \sum_{\mathfrak{s}' \in M(\lambda)} r^\lambda_h(\mathfrak{s}',\mathfrak{s})\, C_{\mathfrak{s}'} \qquad \text{for } h \in \underline{H} \text{ and } \mathfrak{s} \in \underline{M}(\lambda).$$

This module is equipped with a canonical invariant bilinear form; see the following lemma.

**Lemma 2.6.2 (Graham and Lehrer [144, 2.4]).** *Let $\lambda \in \underline{\Lambda}$. Then there is a well-defined symmetric bilinear form $\langle\ ,\ \rangle_\lambda \colon W(\lambda) \times W(\lambda) \to k$ such that*

$$C^\lambda_{\mathfrak{u},\mathfrak{t}} C^\lambda_{\mathfrak{s},\mathfrak{v}} \equiv \langle C_\mathfrak{s}, C_\mathfrak{t}\rangle_\lambda\, C^\lambda_{\mathfrak{u},\mathfrak{v}} \mod \underline{H}(\prec \lambda) \qquad \textit{for all } \mathfrak{s},\mathfrak{t},\mathfrak{u},\mathfrak{v} \in \underline{M}(\lambda).$$

*Furthermore, we have $\langle h.C_\mathfrak{s}, C_\mathfrak{t}\rangle_\lambda = \langle C_\mathfrak{s}, h^*.C_\mathfrak{t}\rangle_\lambda$ for all $\mathfrak{s},\mathfrak{t} \in \underline{M}(\lambda)$ and $h \in \underline{H}$.*

*Proof.* This is a good exercise to see how the axioms are used. By (C3), we have

$$C^\lambda_{\mathfrak{u},\mathfrak{t}} C^\lambda_{\mathfrak{s},\mathfrak{v}} \equiv \sum_{\mathfrak{s}' \in \underline{M}(\lambda)} r^\lambda_{h_1}(\mathfrak{s}',\mathfrak{s})\, C^\lambda_{\mathfrak{s}',\mathfrak{v}} \mod \underline{H}(\prec \lambda), \qquad \text{where } h_1 = C^\lambda_{\mathfrak{u},\mathfrak{t}}.$$

On the other hand, by (C2), we have $C^\lambda_{\mathfrak{u},\mathfrak{t}} C^\lambda_{\mathfrak{s},\mathfrak{v}} = (C^\lambda_{\mathfrak{t},\mathfrak{u}})^* (C^\lambda_{\mathfrak{v},\mathfrak{s}})^* = (C^\lambda_{\mathfrak{v},\mathfrak{s}} C^\lambda_{\mathfrak{t},\mathfrak{u}})^*$. Applying (C3) to the product $C^\lambda_{\mathfrak{v},\mathfrak{s}} C^\lambda_{\mathfrak{t},\mathfrak{u}}$ and using (C2), we obtain that

$$C^{\lambda}_{\mathfrak{u},\mathfrak{t}} C^{\lambda}_{\mathfrak{s},\mathfrak{v}} \equiv \sum_{\mathfrak{t}' \in \underline{M}(\lambda)} r^{\lambda}_{h_2}(\mathfrak{t}',\mathfrak{t})\, C^{\lambda}_{\mathfrak{u},\mathfrak{t}'} \bmod \underline{H}(\prec \lambda), \qquad \text{where } h_2 = C^{\lambda}_{\mathfrak{v},\mathfrak{s}}.$$

Hence, we deduce that $C^{\lambda}_{\mathfrak{u},\mathfrak{t}} C^{\lambda}_{\mathfrak{s},\mathfrak{v}} \equiv \alpha C^{\lambda}_{\mathfrak{u},\mathfrak{v}} \bmod \underline{H}(\prec \lambda)$, where $\alpha := r^{\lambda}_{h_1}(\mathfrak{u},\mathfrak{s}) = r^{\lambda}_{h_2}(\mathfrak{v},\mathfrak{t})$. Note that $r^{\lambda}_{h_1}(\mathfrak{u},\mathfrak{s})$ does not depend on $\mathfrak{v}$, and $r^{\lambda}_{h_2}(\mathfrak{v},\mathfrak{t})$ does not depend on $\mathfrak{u}$. Consequently, $\alpha$ does not depend on $\mathfrak{u}$ and not on $\mathfrak{v}$. Now choose $\mathfrak{u} = \mathfrak{v}$. Then we also see that $\alpha$ is not affected if we exchange the roles of $\mathfrak{s}$ and $\mathfrak{t}$. Thus, we obtain a well-defined symmetric bilinear form $\langle\, ,\, \rangle_\lambda$, as required. It remains to show that this form has the desired invariance property. Let $h \in \underline{H}$ and $\mathfrak{s},\mathfrak{t} \in \underline{M}(\lambda)$. Then

$$\langle h.C_{\mathfrak{s}}, C_{\mathfrak{t}}\rangle_\lambda = \sum_{\mathfrak{s}' \in \underline{M}(\lambda)} r^{\lambda}_{h}(\mathfrak{s}',\mathfrak{s})\, \langle C_{\mathfrak{s}'}, C_{\mathfrak{t}}\rangle_\lambda .$$

Now let $\mathfrak{u},\mathfrak{v} \in \underline{M}(\lambda)$. Multiplying the above identity by $C^{\lambda}_{\mathfrak{u},\mathfrak{v}}$ and using the defining formula for $\langle\, ,\, \rangle_\lambda$, we obtain

$$\langle h.C_{\mathfrak{s}}, C_{\mathfrak{t}}\rangle_\lambda\, C^{\lambda}_{\mathfrak{u},\mathfrak{v}} \equiv \sum_{\mathfrak{s}' \in \underline{M}(\lambda)} r^{\lambda}_{h}(\mathfrak{s}',\mathfrak{s})\, C^{\lambda}_{\mathfrak{u},\mathfrak{s}'} C^{\lambda}_{\mathfrak{t},\mathfrak{v}} \quad \bmod \underline{H}(\prec \lambda).$$

On the other, by (C2) and (C3), we have

$$\sum_{\mathfrak{s}' \in \underline{M}(\lambda)} r^{\lambda}_{h}(\mathfrak{s}',\mathfrak{s})\, C^{\lambda}_{\mathfrak{u},\mathfrak{s}'} \equiv \left(h.C^{\lambda}_{\mathfrak{s},\mathfrak{u}}\right)^* \equiv C^{\lambda}_{\mathfrak{u},\mathfrak{s}} h^* \quad \bmod \underline{H}(\prec \lambda)$$

and so

$$\begin{aligned}\langle h.C_{\mathfrak{s}}, C_{\mathfrak{t}}\rangle_\lambda\, C^{\lambda}_{\mathfrak{u},\mathfrak{v}} &\equiv C^{\lambda}_{\mathfrak{u},\mathfrak{s}}\left(h^* C^{\lambda}_{\mathfrak{t},\mathfrak{v}}\right) \equiv \sum_{\mathfrak{t}' \in \underline{M}(\lambda)} r^{\lambda}_{h^*}(\mathfrak{t}',\mathfrak{t})\, C^{\lambda}_{\mathfrak{u},\mathfrak{s}} C^{\lambda}_{\mathfrak{t}',\mathfrak{v}} \\ &\equiv \sum_{\mathfrak{t}' \in \underline{M}(\lambda)} r^{\lambda}_{h^*}(\mathfrak{t}',\mathfrak{t})\, \langle C_{\mathfrak{s}}, C_{\mathfrak{t}'}\rangle_\lambda\, C^{\lambda}_{\mathfrak{u},\mathfrak{v}} \equiv \langle C_{\mathfrak{s}}, h^*.C_{\mathfrak{t}}\rangle_\lambda C^{\lambda}_{\mathfrak{u},\mathfrak{v}} \quad \bmod \underline{H}(\prec \lambda),\end{aligned}$$

as required. □

**Corollary 2.6.3.** *Let $\lambda \in \underline{\Lambda}$ and $\mathfrak{s},\mathfrak{s}',\mathfrak{t},\mathfrak{u} \in M(\lambda)$. Then*

$$r_h(\mathfrak{s}',\mathfrak{s}) = \delta_{\mathfrak{u}\mathfrak{s}'}\, \langle C_{\mathfrak{s}}, C_{\mathfrak{t}}\rangle_\lambda, \qquad \textit{where} \qquad h = C^{\lambda}_{\mathfrak{u},\mathfrak{t}}.$$

*Proof.* This is clear by Lemma 2.6.2 and the definiton of $r_h(\mathfrak{s}',\mathfrak{s})$ in (C3). □

**Definition 2.6.4.** Let $L^\lambda := W(\lambda)/\mathrm{rad}(\langle\, ,\, \rangle_\lambda)$ for any $\lambda \in \underline{\Lambda}$. Then $L^\lambda$ is a left $\underline{H}$-module since, by Lemma 2.6.2, the radical of $\langle\, ,\, \rangle_\lambda$ is an $\underline{H}$-submodule of $W(\lambda)$. Note that we may have $L^\lambda = \{0\}$; this happens if and only if $\langle\, ,\, \rangle_\lambda$ is identically zero.

Now we have the following two fundamental results of Graham and Lehrer [144] whose proof we will not give here. (See also Mathas [245, Chap. 2].)

**Theorem 2.6.5 (Graham and Lehrer [144, 3.4, 3.8]).** *Assume that k is a field. If $\langle\, ,\, \rangle_\lambda \neq 0$, then $L^\lambda$ is an absolutely irreducible $\underline{H}$-module; furthermore,*

$$\mathrm{Irr}(\underline{H}) = \{L^\mu \mid \mu \in \underline{\Lambda}^\circ\}, \quad \textit{where} \quad \underline{\Lambda}^\circ = \{\lambda \in \underline{\Lambda} \mid \langle\, ,\, \rangle_\lambda \neq 0\}.$$

*In particular, the algebra $\underline{H}$ is split. Finally, $\underline{H}$ is semisimple if and only if $\underline{\Lambda} = \underline{\Lambda}^\circ$ and $\langle\, ,\, \rangle_\lambda$ is non-degenerate for all $\lambda \in \underline{\Lambda}$.*

Recall that an algebra is called *split* if the endomorphism algebra of any irreducible representation consists just of the scalar multiples of the identity.

**Theorem 2.6.6 (Graham and Lehrer [144, 3.6]).** *Assume that $k$ is a field. For $\lambda \in \underline{\Lambda}$ and $\mu \in \underline{\Lambda}^\circ$, denote by $(W(\lambda) : L^\mu)$ the multiplicity of $L^\mu$ as a composition factor of $W(\lambda)$. Then*

$$(\Delta) \qquad (W(\mu) : L^\mu) = 1 \quad \textit{and} \quad (W(\lambda) : L^\mu) = 0 \quad \textit{unless} \quad \lambda \preceq \mu.$$

*Thus, the decomposition matrix $D = \big((W(\lambda) : L^\mu)\big)_{\lambda\in\underline{\Lambda},\mu\in\underline{\Lambda}^\circ}$ has a lower unitriangular shape, if the rows and columns are ordered according to the order relation $\preceq$.*

Let us now also assume that $\underline{H}$ is a *symmetric algebra*, with trace form $\tau\colon \underline{H} \to k$. Then, given a basis $\{C^\lambda_{\mathfrak{s},\mathfrak{t}} \mid \lambda \in \underline{\Lambda},\ \mathfrak{s},\mathfrak{t} \in \underline{M}(\lambda)\}$ as above, we have a corresponding dual basis $\underline{\hat{C}} := \{\hat{C}^\lambda_{\mathfrak{s},\mathfrak{t}} \mid \lambda \in \underline{\Lambda},\ \mathfrak{s},\mathfrak{t} \in \underline{M}(\lambda)\}$. We choose the notation such that

$$\tau\big(C^\lambda_{\mathfrak{s},\mathfrak{t}}\hat{C}^\mu_{\mathfrak{u},\mathfrak{v}}\big) = \begin{cases} 1 & \text{if } \lambda = \mu,\ \mathfrak{s} = \mathfrak{v},\ \mathfrak{t} = \mathfrak{u}, \\ 0 & \text{otherwise.} \end{cases}$$

To state the following result, note that if $V$ is a left $\underline{H}$-module, then $\mathrm{Hom}_k(V,k)$ also is a left $\underline{H}$-module where the action is given by $h.f(v) = f(h^*.v)$ for $h \in \underline{H}$, $f \in \mathrm{Hom}_k(V,k)$ and $v \in V$.

**Proposition 2.6.7 (Graham [143, 4.12]).** *Assume that $\underline{H}$ is symmetric with trace form $\tau\colon \underline{H} \to k$ such that $\tau(h^*) = \tau(h)$ for all $h \in \underline{H}$. Then, with the above notation, the following hold.*

*(a) The quadruple $(\underline{\Lambda}^{\mathrm{op}}, \underline{M}, \underline{\hat{C}}, *)$ also is a cell datum for $\underline{H}$, where $\underline{\Lambda}^{\mathrm{op}}$ is the set $\underline{\Lambda}$ endowed with the opposite partial order $\preceq_{\mathrm{op}}$ (that is, $\lambda \preceq_{\mathrm{op}} \mu \Leftrightarrow \mu \preceq \lambda$).*
*(b) Let $\lambda \in \underline{\Lambda}$ and $\hat{W}(\lambda)$ be the cell module with respect to the cell datum in (a). Then there is an isomorphism of left $\underline{H}$-modules $\hat{W}(\lambda) \cong \mathrm{Hom}_k(W(\lambda),k)$.*
*(c) If $k$ is a field and $\underline{H}$ is semisimple, then $W(\lambda) \cong \hat{W}(\lambda)$ for all $\lambda \in \Lambda$.*

We shall call $(\underline{\Lambda}^{\mathrm{op}}, \underline{M}, \underline{\hat{C}}, *)$ the *opposite cell datum* to $(\underline{\Lambda}, \underline{M}, \underline{C}, *)$.

*Proof.* Let us verify that (C1), (C2), (C3) hold for the quadruple in (a). First note that (C1) is clear and (C2) is easily seen to hold thanks to the assumption on $\tau$. To prove (C3), let $h \in \underline{H}$ and consider the product $h\hat{C}^\lambda_{\mathfrak{s},\mathfrak{t}}$. Let $\mu \in \underline{\Lambda}$ and $\mathfrak{u},\mathfrak{v} \in \underline{M}(\mu)$ be such that $\hat{C}^\mu_{\mathfrak{u},\mathfrak{v}}$ appears with a non-zero coefficient in the expansion of $h\hat{C}^\lambda_{\mathfrak{s},\mathfrak{t}}$ with respect to the basis $\underline{\hat{C}}$. Note that this coefficient is given by

$$\tau\big((h\hat{C}^\lambda_{\mathfrak{s},\mathfrak{t}})C^\mu_{\mathfrak{v},\mathfrak{u}}\big) = \tau\big(((h\hat{C}^\lambda_{\mathfrak{s},\mathfrak{t}})C^\mu_{\mathfrak{v},\mathfrak{u}})^*\big) = \tau\big((h^*C^\mu_{\mathfrak{u},\mathfrak{v}})\hat{C}^\lambda_{\mathfrak{t},\mathfrak{s}}\big).$$

Hence, by (C3) for the original cell datum, we must have $\lambda \preceq \mu$; furthermore, if $\lambda = \mu$, then the above expression evaluates to $\delta_{\mathfrak{v},\mathfrak{t}} r^{\lambda}_{h^*}(\mathfrak{s},\mathfrak{u})$. Thus, we have

$$h\hat{C}^{\lambda}_{\mathfrak{s},\mathfrak{t}} = \sum_{\mathfrak{u}\in\underline{M}(\lambda)} r^{\lambda}_{h^*}(\mathfrak{s},\mathfrak{u})\hat{C}^{\lambda}_{\mathfrak{u},\mathfrak{t}} \mod \underline{H}(\prec_{\mathrm{op}} \lambda).$$

This shows that (C3) holds. The above formula also proves (b). More precisely, if $\rho^{\lambda} \colon \underline{H} \to M_{d_\lambda}(k)$ is the matrix representation afforded by $W(\lambda)$ (with respect to its standard basis), then the matrix representation afforded by $\hat{W}(\lambda)$ (with respect to its standard basis) is given by $h \mapsto \rho(h^*)^{\mathrm{tr}}$ $(h \in \underline{H})$.

Finally, to prove (c), assume that $k$ is a field and $\underline{H}$ is semisimple. Let $\lambda \in \Lambda$ and $G^{\lambda}$ be the Gram matrix of the bilinear form $\langle\ ,\ \rangle_{\lambda}$ with respect to the standard basis of $W(\lambda)$. Then $G^{\lambda}$ is invertible by Theorem 2.6.5. On the other hand, the invariance condition in Lemma 2.6.2 implies that $G^{\lambda}\rho^{\lambda}(h) = \rho(h^*)^{\mathrm{tr}} G^{\lambda}$ for all $h \in \underline{H}$. Hence, the two representations are equivalent; that is, $W(\lambda) \cong \hat{W}(\lambda)$. □

We return to the situation where we consider the generic Iwahori–Hecke algebra $\mathbf{H} = \mathbf{H}_A(W,S,L)$ associated with a finite Coxeter group $W$ and a weight function $L\colon W \to \Gamma$. Recall that $\mathbf{H}$ is defined over $A = R[\Gamma]$, where $R \subseteq \mathbb{C}$ is a subring such that $\mathbb{Z}_W \subseteq R$; furthermore, we assume that there is a monomial order $\leqslant$ on $\Gamma$ such that $L(s) \geqslant 0$ for all $s \in S$. Let $\{C_w \mid w \in W\}$ be the associated Kazhdan–Lusztig basis of $\mathbf{H}$; see Section 2.1. Write

$$\mathrm{Irr}_{\mathbb{K}}(W) = \{E^{\lambda} \mid \lambda \in \Lambda\}, \qquad d_{\lambda} = \dim E^{\lambda},$$

and let $M(\lambda)$ be an indexing set for a basis of $E^{\lambda}$, as in Section 1.2; for each $E^{\lambda} \in \mathrm{Irr}_{\mathbb{K}}(W)$, we have a corresponding invariant $\mathbf{a}_{\lambda} \in \Gamma_{\geqslant 0}$. In Section 1.5, we used the leading matrix coefficients $c^{\mathfrak{st}}_{w,\lambda}$ to construct the ring $\tilde{\mathbf{J}}$.

**Definition 2.6.8 (Cf. [111, §3]).** Assume that $R \subseteq \mathbb{C}$ is $L$-good in the sense of Definition 1.5.9. Let $\underline{\Lambda} := \Lambda$ and $\underline{M}(\lambda) := M(\lambda)$ for all $\lambda \in \Lambda$. Let $\bar{\rho}^{\lambda}$ and $B^{\lambda}$ be as in Proposition 1.5.11. Let us write

$$\bar{\rho}^{\lambda}(t_w) = \big(c^{\mathfrak{st}}_{w,\lambda}\big)_{\mathfrak{s},\mathfrak{t}\in M(\lambda)} \qquad \text{and} \qquad B^{\lambda} = \big(\beta^{\lambda}_{\mathfrak{st}}\big)_{\mathfrak{s},\mathfrak{t}\in M(\lambda)}.$$

Then, for any $\lambda \in \Lambda$ and $\mathfrak{s},\mathfrak{t} \in M(\lambda)$, we define

$$\mathbf{C}^{\lambda}_{\mathfrak{s},\mathfrak{t}} := \sum_{w\in W} \sum_{\mathfrak{u}\in M(\lambda)} \beta^{\lambda}_{\mathfrak{tu}} c^{\mathfrak{us}}_{w^{-1},\lambda} C_w \quad \in \mathbf{H}.$$

We now show in several steps that (C1), (C2), (C3) hold for these data.

*Remark 2.6.9.* In the defining formula for $\mathbf{C}^{\lambda}_{\mathfrak{s},\mathfrak{t}}$, we can assume that the first sum runs over all $w \in \mathscr{F}_{\lambda}$ (where $\mathscr{F}_{\lambda}$ is defined in Proposition 1.6.11). Indeed, if $C_w$ appears with a non-zero coefficient in that sum, then $c^{\mathfrak{us}}_{w^{-1},\lambda} \neq 0$ for some $\mathfrak{u},\mathfrak{s} \in M(\lambda)$, and so $w^{-1} \in \mathscr{F}_{\lambda}$. But then Lemma 1.6.6 also shows that $w \in \mathscr{F}_{\lambda}$, as required.

**Lemma 2.6.10.** *The elements* $\{\mathbf{C}^{\lambda}_{\mathfrak{s},\mathfrak{t}} \mid \lambda \in \Lambda,\ \mathfrak{s},\mathfrak{t} \in M(\lambda)\}$ *form an A-basis of* **H**. *In fact, let* $y \in W$ *and* $\mathscr{F}$ *be the two-sided* $\tilde{\mathbf{J}}$*-cell containing* $y$. *Then* $C_y$ *is an R-linear combination of elements* $\mathbf{C}^{\lambda}_{\mathfrak{s},\mathfrak{t}}$, *where* $\lambda \in \Lambda$ *is such that* $E^{\lambda} \leftrightsquigarrow_L y$.

*Proof.* By the Artin–Wedderburn theorem, $|W| = \sum_{\lambda\in\Lambda} |M(\lambda)|^2$. Hence, the above set has the correct cardinality. It is now sufficient to show that the elements $\{\mathbf{C}^{\lambda}_{\mathfrak{s},\mathfrak{t}}\}$ span **H** as an $A$-module. Let us fix $y \in W$. We claim that

$$C_y = \sum_{\lambda\in\Lambda}\ \sum_{\mathfrak{s},\mathfrak{s}',\mathfrak{t}\in M(\lambda)} f_\lambda^{-1} c^{\mathfrak{s}\mathfrak{s}'}_{y,\lambda} \hat{\beta}^{\lambda}_{\mathfrak{s}'\mathfrak{t}} \mathbf{C}^{\lambda}_{\mathfrak{s},\mathfrak{t}}, \quad \text{where } (B^\lambda)^{-1} = (\hat{\beta}_{\mathfrak{s}\mathfrak{t}}).$$

Note that the coefficients in the above sum lie in $R$, since $f_\lambda$ and $\det(B^\lambda)$ are invertible in $R$ (since $R$ is $L$-good and by Proposition 1.5.11(b)). Furthermore, we have $E^{\lambda} \leftrightsquigarrow_L y$ if $\mathbf{C}^{\lambda}_{\mathfrak{s},\mathfrak{t}}$ occurs with non-zero coefficient in the above sum. Thus, it remains to prove the above identity. For this purpose, we insert the defining formula for $\mathbf{C}^{\lambda}_{\mathfrak{s},\mathfrak{t}}$ into the right-hand side; this yields

$$\begin{aligned}
&\sum_{w\in W}\sum_{\lambda\in\Lambda}\ \sum_{\mathfrak{s},\mathfrak{s}',\mathfrak{u}\in M(\lambda)} f_\lambda^{-1} c^{\mathfrak{s}\mathfrak{s}'}_{y,\lambda} \Big(\sum_{\mathfrak{t}\in M(\lambda)} \hat{\beta}^{\lambda}_{\mathfrak{s}'\mathfrak{t}} \beta^{\lambda}_{\mathfrak{t}\mathfrak{u}}\Big) c^{\mathfrak{u}\mathfrak{s}}_{w^{-1},\lambda} C_w \\
&\qquad = \sum_{w\in W}\Big(\sum_{\lambda\in\Lambda}\ \sum_{\mathfrak{s},\mathfrak{u}\in M(\lambda)} f_\lambda^{-1} c^{\mathfrak{s}\mathfrak{u}}_{y,\lambda} c^{\mathfrak{u}\mathfrak{s}}_{w^{-1},\lambda}\Big) C_w = C_y
\end{aligned}$$

as desired, where the last equality holds by Proposition 1.4.10(b). □

**Lemma 2.6.11.** *We have* $(\mathbf{C}^{\lambda}_{\mathfrak{s},\mathfrak{t}})^{\flat} = \mathbf{C}^{\lambda}_{\mathfrak{t},\mathfrak{s}}$ *for all* $\lambda \in \Lambda$ *and* $\mathfrak{s},\mathfrak{t} \in M(\lambda)$, *where* $\flat$ *is the anti-involution in 2.1.14.*

*Proof.* By 2.1.14, we have $C_w^{\flat} = C_{w^{-1}}$ for all $w \in W$. Thus, we obtain

$$(\mathbf{C}^{\lambda}_{\mathfrak{s},\mathfrak{t}})^{\flat} = \sum_{w\in W}\sum_{\mathfrak{u}\in M(\lambda)} \beta^{\lambda}_{\mathfrak{t}\mathfrak{u}} c^{\mathfrak{u}\mathfrak{s}}_{w^{-1},\lambda} C_{w^{-1}} = \sum_{w\in W} \big(B^\lambda.\bar{\rho}^\lambda(t_{w^{-1}})\big)_{\mathfrak{t},\mathfrak{s}} C_{w^{-1}}.$$

By Proposition 1.5.11, we have $B^\lambda.\bar{\rho}^\lambda(t_{w^{-1}}) = \bar{\rho}^\lambda(t_w)^{\mathrm{tr}}.B^\lambda$. This yields

$$(\mathbf{C}^{\lambda}_{\mathfrak{s},\mathfrak{t}})^{\flat} = \sum_{w\in W} \big(\bar{\rho}^\lambda(t_w)^{\mathrm{tr}}.B^\lambda\big)_{\mathfrak{t},\mathfrak{s}} C_{w^{-1}} = \sum_{w\in W}\sum_{\mathfrak{u}\in M(\lambda)} c^{\mathfrak{u}\mathfrak{t}}_{w,\lambda} \beta^{\lambda}_{\mathfrak{u}\mathfrak{s}} C_{w^{-1}} = \mathbf{C}^{\lambda}_{\mathfrak{t},\mathfrak{s}},$$

as required. (Recall that $B^\lambda$ is symmetric.) □

We can now state the main result of this chapter.

**Theorem 2.6.12 (Cf. [111, §3] [112, §5]).** *Assume that R is L-good and that* (♠) *in 2.5.3 holds. Then the elements* $\{\mathbf{C}^{\lambda}_{\mathfrak{s},\mathfrak{t}}\}$ *introduced in Definition 2.6.8 form a cellular basis of* **H** *with respect to the anti-involution* $T_w \mapsto T_w^{\flat} = T_{w^{-1}}$ *(see 2.1.14), and the partial order* $\trianglelefteq_L$ *on* $\Lambda$ *defined by*

$$\mu \trianglelefteq_L \lambda \quad \stackrel{\text{def}}{\Leftrightarrow} \quad \mu = \lambda \quad \text{or} \quad E^\mu \preceq_L E^\lambda,\ E^\mu \not\sim_L E^\lambda,$$

*where $\preceq_L$ and $\sim_L$ are as in Definition 2.2.1. If property (♣) in 2.5.3 also holds, then we have*

$$\mu \trianglelefteq_L \lambda \quad \Rightarrow \quad \lambda = \mu \quad \text{or} \quad \mathbf{a}_\mu > \mathbf{a}_\lambda.$$

*Proof.* Recall that (C1) holds by Lemma 2.6.10; (C2) holds by Lemma 2.6.11. In order to prove (C3), we need to consider a product $h\mathbf{C}_{\mathfrak{s},\mathfrak{t}}^\lambda$ where $h \in \mathbf{H}$ and $\lambda \in \Lambda$, $\mathfrak{s},\mathfrak{t} \in M(\lambda)$. It is sufficient to consider the case where $h = C_x$ for some $x \in W$. Now, by the definition of $\mathbf{C}_{\mathfrak{s},\mathfrak{t}}^\lambda$ and Remark 2.6.9, we have

$$h\mathbf{C}_{\mathfrak{s},\mathfrak{t}}^\lambda = \sum_{y \in W} r_y C_y \qquad \text{where} \qquad r_y = \sum_{w \in \mathscr{F}_\lambda} \sum_{\mathfrak{u} \in M(\lambda)} \beta_{\mathfrak{t}\mathfrak{u}}^\lambda\, c_{w^{-1},\lambda}^{\mathfrak{u}\mathfrak{s}}\, h_{x,w,y}.$$

Let $\mathfrak{T}_\lambda$ be the two-sided Kazhdan–Lusztig cell such that $\mathscr{F}_\lambda \subseteq \mathfrak{T}_\lambda$. Note that if $r_y \neq 0$, then there is some $w \in \mathscr{F}_\lambda \subseteq \mathfrak{T}_\lambda$ such that $h_{x,w,y} \neq 0$ and so $y \leqslant_{\mathscr{LR}} w$ (in the Kazhdan–Lusztig pre-order).

Assume first that $r_y \neq 0$ and $y \notin \mathfrak{T}_\lambda$. By Lemma 2.6.10, $C_y$ is a linear combination of elements $\mathbf{C}_{\mathfrak{u},\mathfrak{v}}^\mu$, where $E^\mu \leftrightsquigarrow_L y$. Hence, since $y \notin \mathfrak{T}_\lambda$, we conclude that $C_y \in \mathbf{H}(\lhd_L \lambda)$ and so we do not need to consider these terms in any more detail.

Thus, we can now assume that $y \in \mathfrak{T}_\lambda$. Then, by Lemma 2.5.8, we have

$$h_{x,w,y} = \sum_{\substack{z \in W, d \in \mathscr{D} \\ z \sim_{\mathscr{LR}} d}} \tilde{n}_d\, h_{x,d,z}\, \tilde{\gamma}_{z,w,y^{-1}} \qquad \text{for any } w \in \mathscr{F}_\lambda \subseteq \mathfrak{T}_\lambda.$$

We insert this formula for $h_{x,w,y}$ into the above expression for $r_y$; this yields

$$\begin{aligned} r_y &= \sum_{w \in \mathscr{F}_\lambda} \sum_{\substack{z \in W, d \in \mathscr{D} \\ z \sim_{\mathscr{LR}} d}} \sum_{\mathfrak{u} \in M(\lambda)} \beta_{\mathfrak{t}\mathfrak{u}}^\lambda\, c_{w^{-1},\lambda}^{\mathfrak{u}\mathfrak{s}}\, \tilde{n}_d\, h_{x,d,z}\, \tilde{\gamma}_{z,w,y^{-1}} \\ &= \sum_{\substack{z \in W, d \in \mathscr{D} \\ z \sim_{\mathscr{LR}} d}} \tilde{n}_d\, h_{x,d,z} \sum_{\mathfrak{u} \in M(\lambda)} \beta_{\mathfrak{t}\mathfrak{u}}^\lambda \Big( \sum_{w \in \mathscr{F}_\lambda} c_{w^{-1},\lambda}^{\mathfrak{u}\mathfrak{s}}\, \tilde{\gamma}_{z,w,y^{-1}} \Big). \end{aligned}$$

Now, by 1.6.10 and Proposition 1.6.11, the sum over $w \in \mathscr{F}_\lambda$ can be extended to a sum over all $w \in W$. Using the defining equation for $\tilde{\gamma}_{z,w,y^{-1}}$, we obtain

$$\begin{aligned} \sum_{w \in \mathscr{F}_\lambda} c_{w^{-1},\lambda}^{\mathfrak{u}\mathfrak{s}}\, \tilde{\gamma}_{z,w,y^{-1}} &= \sum_{w \in W} c_{w^{-1},\lambda}^{\mathfrak{u}\mathfrak{s}} \Big( \sum_{\mu \in \Lambda} \sum_{\mathfrak{s}',\mathfrak{v},\mathfrak{v}' \in M(\mu)} f_\mu^{-1}\, c_{z,\mu}^{\mathfrak{s}'\mathfrak{v}}\, c_{w,\mu}^{\mathfrak{v}\mathfrak{v}'}\, c_{y^{-1},\mu}^{\mathfrak{v}'\mathfrak{s}'} \Big) \\ &= \sum_{\mu \in \Lambda} \sum_{\mathfrak{s}',\mathfrak{v},\mathfrak{v}' \in M(\mu)} f_\mu^{-1}\, c_{z,\mu}^{\mathfrak{s}'\mathfrak{v}}\, c_{y^{-1},\mu}^{\mathfrak{v}'\mathfrak{s}'} \Big( \sum_{w \in W} c_{w^{-1},\lambda}^{\mathfrak{u}\mathfrak{s}}\, c_{w,\mu}^{\mathfrak{v}\mathfrak{v}'} \Big) = \sum_{\mathfrak{s}' \in M(\lambda)} c_{z,\lambda}^{\mathfrak{s}'\mathfrak{s}}\, c_{y^{-1},\lambda}^{\mathfrak{u}\mathfrak{s}'} \end{aligned}$$

where the last equality holds by Proposition 1.4.10(a). This yields

$$hC^{\lambda}_{\mathfrak{s},\mathfrak{t}} \equiv \sum_{y\in\mathfrak{T}_\lambda} r_y C_y \equiv \sum_{\substack{z\in W, d\in\tilde{\mathscr{D}}\\ z\sim_{\mathscr{LR}} d}} \tilde{n}_d h_{x,d,z} \sum_{\mathfrak{s}'\in M(\lambda)} c^{\mathfrak{s}'\mathfrak{s}}_{z,\lambda}\Big(\sum_{y\in W}\sum_{\mathfrak{u}\in M(\lambda)} \beta^{\lambda}_{\mathfrak{tu}} c^{\mathfrak{us}'}_{y^{-1},\lambda} C_y\Big)$$

$\bmod \mathbf{H}(\lhd_L \lambda)$. Since the parenthesised sum equals $\mathbf{C}^{\lambda}_{\mathfrak{s}',\mathfrak{t}}$, we see that

$$h\mathbf{C}^{\lambda}_{\mathfrak{s},\mathfrak{t}} \equiv \sum_{\mathfrak{s}'\in M(\lambda)} \Big( \sum_{\substack{z\in W, d\in\tilde{\mathscr{D}}\\ z\sim_{\mathscr{LR}} d}} \tilde{n}_d h_{x,d,z} c^{\mathfrak{s}'\mathfrak{s}}_{z,\lambda}\Big) \mathbf{C}^{\lambda}_{\mathfrak{s}',\mathfrak{t}} \mod \mathbf{H}(\lhd_L \lambda).$$

Thus, we have shown that, for $h = C_x$ ($x \in W$), we have

$$r^{\lambda}_h(\mathfrak{s}',\mathfrak{s}) = \sum_{\substack{z\in W, d\in\tilde{\mathscr{D}}\\ z\sim_{\mathscr{LR}} d}} \tilde{n}_d h_{x,d,z} c^{\mathfrak{s}'\mathfrak{s}}_{z,\lambda} \qquad \text{for all } \mathfrak{s},\mathfrak{s}' \in M(\lambda);$$

in particular, this expression does not depend on $\mathfrak{t}$, as required. □

The model for this theorem, namely the case where $W$ is the symmetric group $\mathfrak{S}_n$, will be considered in detail in Section 2.8.

*Remark 2.6.13.* Note that the ingredients for a cellular basis of **H** (that is, the elements $\{\mathbf{C}^{\lambda}_{\mathfrak{s},\mathfrak{t}}\}$ and the partial order $\unlhd_L$) are defined without reference to (♠); this property is only required for the proof.

*Remark 2.6.14.* Assume that we are in the equal-parameter case. Then we have seen in Proposition 2.5.12 that (♠) is a consequence of the following implication:

$$(♣') \qquad E^{\lambda} \sim_L E^{\mu} \quad \Rightarrow \quad \mathbf{a}_{\mu} = \mathbf{a}_{\lambda}.$$

Thus, in order to prove Theorem 2.6.12 in the equal-parameter case, we only need to assume that (♣′) holds. Recall that (♣′) does hold in type $I_2(m)$ (any $m \geqslant 2$), $H_3$, $H_4$ by Examples 2.2.8 and 2.2.9. Furthermore, (♣′) was already established by Lusztig [220, 5.27] (around 1985) for all finite Weyl groups.

*Remark 2.6.15.* Recall that, by Remark 2.2.11, we have the implication

$$E^{\lambda} \preceq_L E^{\mu} \quad \Rightarrow \quad E^{\mu^{\dagger}} \preceq_L E^{\lambda^{\dagger}}.$$

Now, by Examples 1.2.7 and 1.3.4, the following relation holds between $\mathbf{a}_{\lambda}$ and $\mathbf{a}_{\lambda^{\dagger}}$:

$$\mathbf{a}_{\lambda^{\dagger}} - \mathbf{a}_{\lambda} = N_{\lambda} = \sum_{s\in S'} \Big(\frac{N_s \operatorname{trace}(s,E^{\lambda})}{\dim E^{\lambda}}\Big) L(s).$$

It follows that if (♣) is satisfied, then we have the implication

$$\lambda \unlhd_L \mu \quad \Rightarrow \quad \lambda = \mu \quad \text{or} \quad N_{\lambda} < N_{\mu}.$$

Thus, Theorem 2.6.12 could be alternatively formulated using $N_{\lambda}$ instead of the invariants $\mathbf{a}_{\lambda}$. Note that $N_{\lambda}$ is much easier to define than $\mathbf{a}_{\lambda}$; also, $N_{\lambda}$ does not depend on the monomial order on $\Gamma$. The idea of using the invariants $N_\lambda$ appears, in a somewhat different context in Ginzburg et al. [137, §6]; see also Gordon [140].

The following result shows that, for any Iwahori–Hecke algebra associated with a finite Coxeter group, there does exist at least some cellular structure.

**Corollary 2.6.16.** *Let $k$ be a commutative ring (with 1) and $\{\xi_s \mid s \in S\} \subseteq k^\times$ a collection of elements such that $\xi_s = \xi_t$ whenever $s,t \in S$ are conjugate in $W$. Let $H_k = H_k(W,S,\{\xi_s\})$ be the corresponding Iwahori–Hecke algebra; see 1.1.2. Assume that $k$ is $L_0$-good (see Definition 1.5.9) for the "universal" weight function $L_0$ in Example 1.1.9. Then $H_k$ admits a cellular basis $\{C^\lambda_{\mathfrak{s},\mathfrak{t}} \mid \lambda \in \Lambda, \mathfrak{s},\mathfrak{t} \in M(\lambda)\}$ with respect to the anti-involution $T_w \mapsto T_w^\flat = T_{w^{-1}}$ and some partial order on $\Lambda$.*

*Proof.* Let $\Gamma_0$, $A_0$ and $\mathbf{H}_0$ be "universal", as in Example 1.1.9. Choose a monomial order on $\Gamma_0$ such that, on every irreducible component of type $B_n$, $F_4$ or $I_2(m)$ ($m$ even), we are in the "asymptotic case" in Example 1.1.11. Then, by Corollary 2.4.2, we know that **P1–P15** hold. Hence, as discussed in 2.5.3, the properties (♣) and (♠) also hold and so Theorem 2.6.12 applies. Thus, we obtain a cellular basis $\{\mathbf{C}^\lambda_{\mathfrak{s},\mathfrak{t}}\}$ for $\mathbf{H}_0$, where the partial order on $\Lambda$ is given by $\trianglelefteq_{L_0}$.

Since $k$ is $L_0$-good, there is a ring homomorphism $R \to k$. This extends to a ring homomorphism $\theta\colon A_0 \to k$ such that $\theta(v_s^\circ) = \xi_s$ for all $s \in S'$, where $\{v_s^\circ\}$ are the parameters of $\mathbf{H}_0$. Thus, $H_k = k \otimes_{A_0} \mathbf{H}_0$, where $k$ is regarded as an $A_0$-module via $\theta$. Since the elements $\{\mathbf{C}^\lambda_{\mathfrak{s},\mathfrak{t}} \mid \lambda \in \Lambda,\ \mathfrak{s},\mathfrak{t} \in M(\lambda)\}$ in $\mathbf{H}$ satisfy (C1), (C2), (C3), it is clear that the elements $\{C^\lambda_{\mathfrak{s},\mathfrak{t}} := 1 \otimes \mathbf{C}^\lambda_{\mathfrak{s},\mathfrak{t}} \mid \lambda \in \Lambda,\ \mathfrak{s},\mathfrak{t} \in M(\lambda))\}$ satisfy (C1), (C2), (C3) in $H_k$. Thus, we have constructed a cell datum for $H_k$. □

**Example 2.6.17.** Let $W$ be of type $I_2(4) = B_2$, where $S = \{s_1,s_2\}$ and $(s_1s_2)^4 = 1$. Assume that we are in the equal-parameter case, where $\Gamma = \mathbb{Z}$ and $L(s_1) = L(s_2) = 1$. Then $A = R[v,v^{-1}]$ is the ring of Laurent polynomials in one indeterminate $v = \varepsilon$. Now Theorem 2.6.12 applies where $R \subseteq \mathbb{C}$ can be any subring in which 2 is invertible. In order to determine a cellular basis, we need to work out the leading matrix coefficients of the irreducible representations of $\mathbf{H}_K$. The two-sided cells are given by $\{1_0\}$, $\{1_4\}$ and $W \setminus \{1_0, 1_4\}$, where we use the notation in Example 1.7.3. First consider the representation $\sigma_1$. By Example 1.3.7, we have $\mathbf{a}_{\sigma_1} = 1$ and so

$$v\sigma_1^\varepsilon(T_{1_1}) = \begin{pmatrix} -1 & 0 \\ 2v & v^2 \end{pmatrix} \equiv \begin{pmatrix} -1 & 0 \\ 0 & 0 \end{pmatrix} \bmod \mathfrak{m},$$

$$v\sigma_1^\varepsilon(T_{2_1}) = \begin{pmatrix} v^2 & v \\ 0 & -1 \end{pmatrix} \equiv \begin{pmatrix} 0 & 0 \\ 0 & -1 \end{pmatrix} \bmod \mathfrak{m},$$

$$v\sigma_1^\varepsilon(T_{1_2}) = \begin{pmatrix} -v & -1 \\ 2v^2 & v \end{pmatrix} \equiv \begin{pmatrix} 0 & -1 \\ 0 & 0 \end{pmatrix} \bmod \mathfrak{m},$$

$$v\sigma_1^\varepsilon(T_{2_2}) = \begin{pmatrix} v & v^2 \\ -2 & -v \end{pmatrix} \equiv \begin{pmatrix} 0 & 0 \\ -2 & 0 \end{pmatrix} \bmod \mathfrak{m},$$

$$v\sigma_1^\varepsilon(T_{1_3}) = \begin{pmatrix} -1 & -v \\ 0 & v^2 \end{pmatrix} \equiv \begin{pmatrix} -1 & 0 \\ 0 & 0 \end{pmatrix} \bmod \mathfrak{m},$$

$$v\sigma_1^\varepsilon(T_{2_3}) = \begin{pmatrix} v^2 & 0 \\ -2v & -1 \end{pmatrix} \equiv \begin{pmatrix} 0 & 0 \\ 0 & -1 \end{pmatrix} \bmod \mathfrak{m}.$$

A corresponding symmetric matrix is given by

$$B_{\sigma_1} = \begin{pmatrix} 2 & 0 \\ 0 & 1 \end{pmatrix}; \qquad \text{see Example 1.4.6.}$$

Performing similar (but much simpler) computations for the one-dimensional representations, we obtain the following expressions for $\mathbf{C}^{\lambda}_{\mathfrak{s},\mathfrak{t}}$ (see also [111, Exp. 4.3]):

$$\begin{aligned}
\mathbf{C}^{1_W}_{1,1} &= C_{1_0}, & \mathbf{C}^{\sigma_1}_{1,1} &= -2C_{1_1} - 2C_{1_3}, \\
\mathbf{C}^{\mathrm{sgn}}_{1,1} &= C_{1_4}, & \mathbf{C}^{\sigma_1}_{1,2} &= -2C_{1_2}, \\
\mathbf{C}^{\mathrm{sgn}_1}_{1,1} &= -C_{2_1} + C_{2_3}, & \mathbf{C}^{\sigma_1}_{2,1} &= -2C_{2_2}, \\
\mathbf{C}^{\mathrm{sgn}_2}_{1,1} &= -C_{1_1} + C_{1_3}, & \mathbf{C}^{\sigma_1}_{2,2} &= -C_{2_1} - C_{2_3}.
\end{aligned}$$

For the case of unequal parameters, see 2.8.19.

**Example 2.6.18.** Let $W$ be of type $I_2(6) = G_2$, where $S = \{s_1, s_2\}$ and $(s_1s_2)^6 = 1$. (a) Assume that we are in the equal-parameter case where $L(s_1) = L(s_2) > 0$. Then, again, Theorem 2.6.12 applies and so we have a cellular basis $\{\mathbf{C}^{\lambda}_{\mathfrak{s},\mathfrak{t}}\}$. In this case, we can take for $R$ any subring of $\mathbb{C}$ in which 2, 3 are invertible. Expressions for $\mathbf{C}^{\lambda}_{\mathfrak{s},\mathfrak{t}}$ have been worked out in [129, Exp. 2.7] (using computations similar to those in the previous example):

$$\begin{aligned}
\mathbf{C}^{1_W}_{1,1} &= C_{1_0}, & \mathbf{C}^{\mathrm{sgn}}_{1,1} &= C_{1_6}, \\
\mathbf{C}^{\mathrm{sgn}_1}_{1,1} &= C_{2_1} - C_{2_3} + C_{2_5}, & \mathbf{C}^{\mathrm{sgn}_2}_{1,1} &= C_{1_1} - C_{1_3} + C_{1_5}, \\
\mathbf{C}^{\sigma_1}_{1,1} &= 3C_{1_1} + 6C_{1_3} + 3C_{1_5}, & \mathbf{C}^{\sigma_2}_{1,1} &= C_{1_1} - C_{1_5}, \\
\mathbf{C}^{\sigma_1}_{1,2} &= -3C_{1_2} - 3C_{1_4}, & \mathbf{C}^{\sigma_2}_{1,2} &= -C_{1_2} + C_{1_4}, \\
\mathbf{C}^{\sigma_1}_{2,1} &= -3C_{2_1} - 3C_{2_4}, & \mathbf{C}^{\sigma_2}_{2,1} &= -C_{2_1} + C_{2_4}, \\
\mathbf{C}^{\sigma_1}_{2,2} &= C_{2_1} + 2C_{2_3} + C_{2_5}, & \mathbf{C}^{\sigma_2}_{2,2} &= C_{2_1} - C_{2_5}.
\end{aligned}$$

(b) Assume that the monomial order on $\Gamma$ is such that $L(s_1) > L(s_2) > 0$. By 2.4.1, **P1–P15** hold and, hence, by 2.5.3, the hypothesis of Theorem 2.6.12 is satisfied. In this case, we can take for $R$ any subring of $\mathbb{C}$ in which 2 is invertible. We find the following expressions for the cellular basis:

$$\begin{aligned}
\mathbf{C}^{1_W}_{1,1} &= C_{1_0}, & \mathbf{C}^{\sigma_1}_{1,1} &= C_{1_1} + C_{1_3}, & \mathbf{C}^{\sigma_2}_{1,1} &= C_{1_1} - C_{1_3}, \\
\mathbf{C}^{\mathrm{sgn}}_{1,1} &= C_{1_6}, & \mathbf{C}^{\sigma_1}_{1,2} &= -C_{1_2} - C_{1_4}, & \mathbf{C}^{\sigma_2}_{1,2} &= -C_{1_2} + C_{1_4}, \\
\mathbf{C}^{\mathrm{sgn}_1}_{1,1} &= C_{2_1}, & \mathbf{C}^{\sigma_1}_{2,1} &= -C_{2_2} - C_{2_4}, & \mathbf{C}^{\sigma_2}_{2,1} &= -C_{2_2} + C_{2_4}, \\
\mathbf{C}^{\mathrm{sgn}_2}_{1,1} &= C_{1_5}, & \mathbf{C}^{\sigma_1}_{2,2} &= C_{2_3} + C_{2_5}, & \mathbf{C}^{\sigma_2}_{2,2} &= C_{2_3} - C_{2_5}.
\end{aligned}$$

**Example 2.6.19.** Let $W$ be of type $I_2(5)$, where $S = \{s_1, s_2\}$ and $(s_1s_2)^5 = 1$. We are in the equal-parameter case; assume that $L(s_1) = L(s_2) > 0$. Then Theorem 2.6.12

applies and so we have a cellular basis $\{\mathbf{C}^\lambda_{\mathfrak{s},\mathfrak{t}}\}$. In this case, we can take for $R$ any subring of $\mathbb{C}$ in which 5 is invertible and $\alpha := \frac{1}{2}(-1+\sqrt{5}) \in R$. We find the following expressions for $\mathbf{C}^\lambda_{\mathfrak{s},\mathfrak{t}}$:

$$\begin{aligned}
\mathbf{C}^{1_W}_{1,1} &= C_{1_0}, & \mathbf{C}^{\mathrm{sgn}}_{1,1} &= C_{1_5},\\
\mathbf{C}^{\sigma_1}_{1,1} &= (2+\alpha)C_{1_1} + (3+2\alpha)C_{1_3}, & \mathbf{C}^{\sigma_2}_{1,1} &= (1-\alpha)C_{1_1} + (1-2\alpha)C_{1_3}\\
\mathbf{C}^{\sigma_1}_{1,2} &= -(2+\alpha)C_{1_2} - (1+\alpha)C_{1_4}, & \mathbf{C}^{\sigma_2}_{1,2} &= (\alpha-1)C_{1_2} + \alpha C_{1_4}\\
\mathbf{C}^{\sigma_1}_{2,1} &= -(2+\alpha)C_{2_2} - (1+\alpha)C_{2_4}, & \mathbf{C}^{\sigma_2}_{2,1} &= (\alpha-1)C_{2_2} + \alpha C_{2_4}\\
\mathbf{C}^{\sigma_1}_{2,2} &= C_{2_1} + (1+\alpha)C_{2_3}, & \mathbf{C}^{\sigma_2}_{2,2} &= C_{2_1} - \alpha C_{2_3}.
\end{aligned}$$

## 2.7 Further Properties of the Cellular Basis of H

Throughout this section (except for Corollary 2.7.14 at the very end), we assume that we are in the setting of Theorem 2.6.12, where properties (♣) and (♠) in 2.5.3 hold. Thus, we have a cellular basis $\{\mathbf{C}^\lambda_{\mathfrak{s},\mathfrak{t}}\}$ of $\mathbf{H}$, and the partial order on $\Lambda$ satisfies

$$\mu \trianglelefteq_L \lambda \qquad \Rightarrow \qquad \mu = \lambda \quad \text{or} \quad \mathbf{a}_\mu > \mathbf{a}_\lambda.$$

Let $\{W(\lambda) \mid \lambda \in \Lambda\}$ be the cell modules constructed from the cellular basis; see 2.6.1. By extension of scalars from $A$ to $K$, we obtain modules for $\mathbf{H}_K$ which we denote by $W_K(\lambda)$ ($\lambda \in \Lambda$). Since $\mathbf{H}_K$ is semisimple, Theorem 2.6.5 shows that

$$\mathrm{Irr}(\mathbf{H}_K) = \{W_K(\lambda) \mid \lambda \in \Lambda\}.$$

**Proposition 2.7.1.** *Let $\lambda \in \Lambda$. For any $h = C_w$ ($w \in W$) and $\mathfrak{s}, \mathfrak{s}' \in M(\lambda)$, we have*

$$\varepsilon^{\mathbf{a}_\lambda}\, r^\lambda_h(\mathfrak{s}', \mathfrak{s}) \in R[\Gamma_{\geqslant 0}], \quad \textit{with constant term equal to } c^{\mathfrak{s}'\mathfrak{s}}_{w,\lambda}.$$

*In particular, the representation of $\mathbf{H}_K$ afforded by $W_K(\lambda)$ (with respect to its standard basis) is balanced, and we have $W_K(\lambda) \cong E^\lambda_\varepsilon$.*

*Proof.* At the end of the proof of Theorem 2.6.12, we obtained the formula

$$r^\lambda_h(\mathfrak{s}', \mathfrak{s}) = \sum_{\substack{z \in W,\, d \in \tilde{\mathscr{D}} \\ z \sim_{\mathscr{LR}} d}} \tilde{n}_d\, h_{w,d,z}\, c^{\mathfrak{s}'\mathfrak{s}}_{z,\lambda}.$$

Now assume $z \in W$ and $d \in \tilde{\mathscr{D}}$ are such that $z \sim_{\mathscr{LR}} d$ and the corresponding terms in the above sum are non-zero; that is, $h_{w,d,z} \neq 0$ and $c^{\mathfrak{s}'\mathfrak{s}}_{z,\lambda} \neq 0$. Since $c^{\mathfrak{s}'\mathfrak{s}}_{z,\lambda} \neq 0$, we have $E^\lambda \leftrightsquigarrow_L z$. Hence, Lemma 2.5.11 shows that $\varepsilon^{\mathbf{a}_\lambda} h_{w,d,z}$ lies in $\mathbb{Z}[\Gamma_{\geqslant 0}]$ and has constant term $\tilde{\gamma}_{w,d,z^{-1}}$. Thus, we have $\varepsilon^{\mathbf{a}_\lambda}\, r^\lambda_h(\mathfrak{s}', \mathfrak{s}) \in R[\Gamma_{\geqslant 0}]$ and

$$\varepsilon^{\mathbf{a}_\lambda}\, r_h^\lambda(\mathfrak{s}',\mathfrak{s}) \equiv \sum_{z\in W}\Big(\sum_{d\in\tilde{\mathscr{D}}: z\sim_{\mathscr{LR}} d} \tilde{n}_d\, \tilde{\gamma}_{w,d,z^{-1}}\Big) c_{z,\lambda}^{\mathfrak{s}'\mathfrak{s}} \equiv c_{w,\lambda}^{\mathfrak{s}'} \mod R[\Gamma_{>0}],$$

where the last congruence follows as in Remark 2.5.15. Once this is established, it is clear that the representation afforded by $W_K(\lambda)$ is balanced. Furthermore, by Proposition 1.5.7, we also see that $W_K(\lambda) \cong E_\varepsilon^\lambda$. □

**Corollary 2.7.2.** *Let $\lambda \in \Lambda$ and denote by $G^\lambda = (g_{\mathfrak{st}}^\lambda)_{\mathfrak{s},\mathfrak{t}\in M(\lambda)}$ the Gram matrix of the bilinear form $\langle\ ,\ \rangle_\lambda : W(\lambda)\times W(\lambda) \to A$. Then*

$$\varepsilon^{\mathbf{a}_\lambda}\, g_{\mathfrak{st}}^\lambda \in R[\Gamma_{\geqslant 0}] \qquad \textit{and} \qquad \varepsilon^{\mathbf{a}_\lambda}\, g_{\mathfrak{st}}^\lambda \equiv f_\lambda \beta_{\mathfrak{st}}^\lambda \mod R[\Gamma_{>0}].$$

*Proof.* Recall that $g_{\mathfrak{st}}^\lambda = g_{\mathfrak{ts}}^\lambda = r_h^\lambda(\mathfrak{s},\mathfrak{s})$, where $h = \mathbf{C}_{\mathfrak{s},\mathfrak{t}}^\lambda$. Hence, using the defining formula for $\mathbf{C}_{\mathfrak{s},\mathfrak{t}}^\lambda$, we obtain

$$\varepsilon^{\mathbf{a}_\lambda}\, g_{\mathfrak{st}}^\lambda = \sum_{w\in W}\sum_{\mathfrak{u}\in M(\lambda)} \beta_{\mathfrak{tu}}^\lambda\, c_{w^{-1},\lambda}^{\mathfrak{us}} \big(\varepsilon^{\mathbf{a}_\lambda}\, r_{C_w}^\lambda(\mathfrak{s},\mathfrak{s})\big).$$

By Proposition 2.7.1, this expression lies in $R[\Gamma_{\geqslant 0}]$ and has constant term

$$\sum_{w\in W}\sum_{\mathfrak{u}\in M(\lambda)} \beta_{\mathfrak{tu}}^\lambda\, c_{w^{-1},\lambda}^{\mathfrak{us}}\, c_{w,\lambda}^{\mathfrak{ss}} = \sum_{\mathfrak{u}\in M(\lambda)} \beta_{\mathfrak{tu}}^\lambda \Big(\sum_{w\in W} c_{w^{-1},\lambda}^{\mathfrak{us}}\, c_{w,\lambda}^{\mathfrak{ss}}\Big) = \beta_{\mathfrak{ts}}^\lambda f_\lambda,$$

where we used the "Schur relations" in Proposition 1.4.10(a). □

**Example 2.7.3.** Let $\lambda \in \Lambda$ and consider the representation $\bar{\rho}^\lambda : \tilde{\mathbf{J}} \to M_{d_\lambda}(\mathbb{K})$, where $\bar{\rho}^\lambda(t_w) \in M_{d_\lambda}(R)$ for all $w \in W$. By first restricting $\bar{\rho}^\lambda$ to $\tilde{\mathbf{J}}_R$ and then extending scalars from $R$ to $A$, we can also regard $\bar{\rho}^\lambda$ as an $A$-algebra homomorphism

$$\bar{\rho}^\lambda : \tilde{\mathbf{J}}_A \to M_{d_\lambda}(A).$$

With this convention, the formula at the end of Theorem 2.6.12 means that

$$\big(r_h^\lambda(\mathfrak{s}',\mathfrak{s})\big)_{\mathfrak{s}',\mathfrak{s}\in M(\lambda)} = \bar{\rho}^\lambda(\phi(C_w)) \qquad (h = C_w).$$

By Proposition 2.7.1, we have $E_\varepsilon^\lambda \cong W_K(\lambda)$. Hence, the above formula shows that, for a suitable basis of $E_\varepsilon^\lambda$, the action of $C_w$ on $E_\varepsilon^\lambda$ is given by the matrix $\bar{\rho}^\lambda(\phi(C_w))$. We express this by saying that the action of **H** on $E_\varepsilon^\lambda$ factors through $\phi$.

**Example 2.7.4.** Let $\lambda \in \Lambda$ and assume that there exists a left Kazhdan–Lusztig cell $\mathfrak{C}$ such that $E_\varepsilon^\lambda \cong [\mathfrak{C}]_K$. (This is a very special situation, but we will see in Section 2.8 that it holds, for example, when $W \cong \mathfrak{S}_n$.) Let us write $\mathfrak{C} = \{x_\mathfrak{s} \mid \mathfrak{s} \in M(\lambda)\}$. Then we have a corresponding representation $\rho_\mathfrak{C} : \mathbf{H}_K \to M_{d_\lambda}(K)$ such that

$$\rho_\mathfrak{C}(C_w) = \big(h_{w,x_\mathfrak{t},x_\mathfrak{s}}\big)_{\mathfrak{s},\mathfrak{t}\in M(\lambda)} \qquad \text{for all } w \in W.$$

By Lemma 2.2.4, there exists some $w \in \mathfrak{C}$ such that $E^\lambda \leftrightsquigarrow_L w$. Since (♣) is assumed to hold, we can apply Lemma 2.5.11, which yields that $\rho_{\mathfrak{C}}$ is balanced and, for all $w \in W$ and $\mathfrak{s}, \mathfrak{t} \in M(\lambda)$, we have

$$\varepsilon^{\mathbf{a}_\lambda} h_{w,x_\mathfrak{t},x_\mathfrak{s}} \equiv \tilde{\gamma}_{w,x_\mathfrak{t},x_\mathfrak{s}^{-1}} \bmod \mathbb{Z}[\Gamma_{>0}].$$

Thus, we can assume that $\rho^\lambda = \rho_{\mathfrak{C}}$ and the leading matrix coefficients are given by

$$c_{w,\lambda}^{\mathfrak{s}\mathfrak{t}} = \tilde{\gamma}_{w,x_\mathfrak{t},x_\mathfrak{s}^{-1}} \qquad \text{for all } w \in W \text{ and } \mathfrak{s}, \mathfrak{t} \in M(\lambda).$$

Now, by Lemma 1.5.3, we have $\tilde{\gamma}_{w,x_\mathfrak{t},x_\mathfrak{s}^{-1}} = \tilde{\gamma}_{x_\mathfrak{t}^{-1},w^{-1},x_\mathfrak{s}} = \tilde{\gamma}_{w^{-1},x_\mathfrak{s},x_\mathfrak{t}^{-1}}$. This means that $\bar{\rho}^\lambda(t_w)^{\mathrm{tr}} = \bar{\rho}^\lambda(t_{w^{-1}})$ for all $w \in W$. Consequently, the conditions in Proposition 1.5.11 are satisfied where we take $B^\lambda$ to be the identity matrix. The formula for $r_h^\lambda(\mathfrak{s}', \mathfrak{s})$ in the proof of Proposition 2.7.1 now reads

$$r_h^\lambda(\mathfrak{s}', \mathfrak{s}) = \sum_{\substack{z \in W, d \in \tilde{\mathscr{D}} \\ z \sim_{\mathscr{LR}} d}} \tilde{n}_d \, h_{w,d,z} \, \tilde{\gamma}_{z,x_\mathfrak{s},x_{\mathfrak{s}'}^{-1}} = h_{w,x_\mathfrak{s},x_{\mathfrak{s}'}} \qquad (h = C_w),$$

where the last equality holds by Lemma 2.5.8. Thus, we have shown that $W(\lambda)$ is nothing but the left cell module $[\mathfrak{C}]_A$. Furthermore, the action of $\mathbf{H}$ on $[\mathfrak{C}]_A$ factors through $\phi$, as in Example 2.7.3.

**2.7.5.** One important feature of the definition of a "cell datum" is that it behaves well with respect to a *specialisation*; see [144, (1.8)]. Let $\theta \colon A \to k$ be a homomorphism into a field $k$. Let $\mathbf{H}_k = k \otimes_A \mathbf{H}$ be the corresponding specialised algebra over $k$. Assume that $\{\mathbf{C}_{\mathfrak{s},\mathfrak{t}}^\lambda\}$ satisfy (C1), (C2), (C3) in $\mathbf{H}$. Then the elements $\{1 \otimes \mathbf{C}_{\mathfrak{s},\mathfrak{t}}^\lambda\}$ will satisfy (C1), (C2), (C3) in $\mathbf{H}_k$. Hence, a cell datum for $\mathbf{H}$ automatically gives rise to a cell datum for $\mathbf{H}_k$. Note that then the cell representations of $\mathbf{H}_k$ are given by $W_k(\lambda) = k \otimes_A W(\lambda)$ $(\lambda \in \Lambda)$, and the bilinear form $\langle\ ,\ \rangle_\lambda$ on $W(\lambda)$ induces the corresponding form $\langle\ ,\ \rangle_{\lambda,k}$ on $W_k(\lambda)$. In particular, we have the following:

(a) Extending scalars from $A$ to $K$, we obtain a cell datum for $\mathbf{H}_K$. As already mentioned (see Proposition 2.7.1), since $\mathbf{H}_K$ is semisimple, we have

$$\mathrm{Irr}(\mathbf{H}_K) = \{W_K(\lambda) \mid \lambda \in \Lambda\}, \qquad \text{where} \qquad W_K(\lambda) \cong E_\varepsilon^\lambda \text{ for all } \lambda \in \Lambda.$$

(b) In general, given any map $\theta \colon A \to k$ as above, we set $L_k^\lambda = W_k(\lambda)/\mathrm{rad}(\langle\ ,\ \rangle_{\lambda,k})$ for $\lambda \in \Lambda$. Then Theorem 2.6.5 implies that

$$\mathrm{Irr}(\mathbf{H}_k) = \{L_k^\mu \mid \mu \in \Lambda_k^\circ\}, \qquad \text{where} \qquad \Lambda_k^\circ := \{\lambda \in \Lambda \mid \langle\ ,\ \rangle_{\lambda,k} \neq 0\}.$$

Furthermore, the composition multiplicities $(W_k(\lambda) : L_k^\mu)$ satisfy the conditions $(\Delta)$ in Theorem 2.6.6. Hence, since (♣) is assumed to hold, this means

$$(\Delta^a) \qquad \begin{cases} (W_k(\mu) : L_k^\mu) = 1 & \text{for all } \mu \in \Lambda_k^\circ, \\ (W_k(\lambda) : L_k^\mu) = 0 & \text{unless } \lambda = \mu \text{ or } \mathbf{a}_\lambda > \mathbf{a}_\mu. \end{cases}$$

Thus, our "fundamental problem" (p. 3) of determining $\mathrm{Irr}(\mathbf{H}_k)$ now takes the following more precise form (and this will be addressed in the subsequent chapters).

**Fundamental Problem (revised).** *Given a cell datum for* $\mathbf{H}_k$*, describe the subset* $\Lambda_k^\circ \subseteq \Lambda$ *and determine the dimension of* $L_k^\mu$ *for* $\mu \in \Lambda_k^\circ$.

Our next result provides an alternative characterisation of the subset $\Lambda_k^\circ \subseteq \Lambda$. In particular, this shows that $\Lambda_k^\circ$ does not depend on the choices involved in the definition of $\{\mathbf{C}_{\mathfrak{s},\mathfrak{t}}^\lambda\}$. (Recall that, for example, this definition relies on the balanced representations $\rho^\lambda$, and these are not uniquely determined.)

**Proposition 2.7.6.** *In the setting of 2.7.5, let* $\lambda \in \Lambda$. *Then the following three conditions are equivalent.*

*(a)* $\lambda \in \Lambda_k^\circ$.
*(b)* $\theta(\chi^\lambda(C_w)) \neq 0$ *for some* $w \in W$ *such that* $E^\lambda \leftrightsquigarrow_L w$.
*(c)* $C_w.W_k(\lambda) \neq \{0\}$ *for some* $w \in W$ *such that* $E^\lambda \leftrightsquigarrow_L w$.

*Proof.* "(a) $\Rightarrow$ (b)" If $\lambda \in \Lambda_k^\circ$, then $\langle\ ,\ \rangle_{\lambda,k} \neq 0$ and so $\langle C_\mathfrak{u}, C_\mathfrak{t}\rangle_{\lambda,k} \neq 0$ for some $\mathfrak{u},\mathfrak{t} \in M(\lambda)$. Now, by Proposition 2.7.1, we have $E_\varepsilon^\lambda \cong W_K(\lambda)$ and so

$$\chi^\lambda(h) = \mathrm{trace}(h, E_\varepsilon^\lambda) = \mathrm{trace}\big(h, W_K(\lambda)\big) = \sum_{\mathfrak{s}\in M(\lambda)} r_h^\lambda(\mathfrak{s},\mathfrak{s}) \qquad \text{for all } h \in \mathbf{H}.$$

Let $h = \mathbf{C}_{\mathfrak{u},\mathfrak{t}}^\lambda$ and apply Corollary 2.6.3. This yields $r_h^\lambda(\mathfrak{s},\mathfrak{s}) = \delta_{\mathfrak{u}\mathfrak{s}}\langle C_\mathfrak{s}, C_\mathfrak{t}\rangle_\lambda$ for all $\mathfrak{s} \in M(\lambda)$ and so $\chi^\lambda(h) = \langle C_\mathfrak{u}, C_\mathfrak{t}\rangle_\lambda$. Since $\langle C_\mathfrak{u}, C_\mathfrak{t}\rangle_{\lambda,k} \neq 0$, this shows that $\theta(\chi^\lambda(h)) \neq 0$. Finally, by Remark 2.6.9, $h$ is an $R$-linear combination of elements $C_w$, where $E^\lambda \leftrightsquigarrow_L w$. Hence, (b) follows.

"(b) $\Rightarrow$ (c)" As above, we have $\chi^\lambda(h) = \sum_{\mathfrak{s}\in M(\lambda)} r_h^\lambda(\mathfrak{s},\mathfrak{s})$ for all $h \in \mathbf{H}$. Since $W_k(\lambda) = k \otimes_A W(\lambda)$, this implies

$$\theta\big(\chi^\lambda(C_w)\big) = \mathrm{trace}\big(C_w, W_k(\lambda)\big) \qquad \text{for all } w \in W,$$

where $C_w$ is regarded as an element of $\mathbf{H}_k$ on the right-hand side. Hence, if (b) holds, then $C_w$ does not act as zero on $W_k(\lambda)$ and so (c) holds.

"(c) $\Rightarrow$ (a)" Let $w \in W$ be such that $E^\lambda \leftrightsquigarrow_L w$ and $C_w.W_k(\lambda) \neq \{0\}$. By Lemma 2.6.10, $C_w$ is an $R$-linear combination of elements $\mathbf{C}_{\mathfrak{u},\mathfrak{v}}^\mu$, where $\mu \in \Lambda$ is such that $E^\mu \leftrightsquigarrow_L w$. Hence, we also have $h.W_k(\lambda) \neq \{0\}$, where $h = \mathbf{C}_{\mathfrak{u},\mathfrak{v}}^\mu$ for some $\mu \in \Lambda$ and $\mathfrak{u},\mathfrak{v} \in M(\mu)$ such that $E^\lambda \sim_L E^\mu$. In particular, this implies that $h.W(\lambda) \neq \{0\}$. By the definition of the action of $\mathbf{H}$ on $W(\lambda)$, this means that there exist some $\mathfrak{s},\mathfrak{s}',\mathfrak{t} \in M(\lambda)$ such that $\mathbf{C}_{\mathfrak{s}',\mathfrak{t}}^\lambda$ appears with a non-zero coefficient in the decomposition of $h\mathbf{C}_{\mathfrak{s},\mathfrak{t}}^\lambda = \mathbf{C}_{\mathfrak{u},\mathfrak{v}}^\mu \mathbf{C}_{\mathfrak{s},\mathfrak{t}}^\lambda$. By (C2) and (C3), this implies that $\lambda \trianglelefteq_L \mu$. Since also $E^\lambda \sim_L E^\mu$, we conclude that $\lambda = \mu$.

Thus, we have $h.W_k(\lambda) \neq \{0\}$, where $h = \mathbf{C}^{\lambda}_{\mathfrak{u},\mathfrak{v}}$ for some $\mathfrak{u}, \mathfrak{v} \in M(\lambda)$. By the definition of the action of $\mathbf{H}_k$ on $W_k(\lambda)$ and Corollary 2.6.3, this implies that

$$\langle C_{\mathfrak{s}}, C_{\mathfrak{v}} \rangle_{\lambda,k} = \theta\big(r^{\lambda}_h(\mathfrak{u},\mathfrak{s})\big) \neq 0 \quad \text{for some } \mathfrak{s} \in M(\lambda).$$

Thus, $\langle \, , \, \rangle_{\lambda,k} \neq 0$ and so (a) holds. □

**Proposition 2.7.7.** *Let $\lambda \in \Lambda^{\circ}_k$. Then the following hold.*

*(a) We have $\theta(\chi^{\lambda}(C_w)) = \mathrm{trace}(C_w, L^{\lambda}_k)$ for all $w \in W$ such that $E^{\lambda} \leftrightsquigarrow_L w$.*
*(b) We have $C_w.L^{\lambda}_k \neq \{0\}$ for some $w \in W$ such that $E^{\lambda} \leftrightsquigarrow_L w$.*

*Proof.* (a) Let $w \in W$ be such that $E^{\lambda} \leftrightsquigarrow_L w$. As in the above proof, $\theta(\chi^{\lambda}(C_w)) = \mathrm{trace}(C_w, W_k(\lambda))$. Considering a composition series for $W_k(\lambda)$, we obtain

$$\theta(\chi^{\lambda}(C_w)) = \sum_{\mu \in \Lambda^{\circ}_k} (W_k(\lambda) : L^{\mu}_k)\,\mathrm{trace}(C_w, L^{\mu}_k).$$

Let $\mu \in \Lambda^{\circ}_k$ and assume that the corresponding terms in the sum are non-zero; that is, we have $(W_k(\lambda) : L^{\mu}_k) \neq 0$ and $\mathrm{trace}(C_w, L^{\mu}_k) \neq 0$. In particular, this means that $C_w.L^{\mu}_k \neq \{0\}$. We claim that this implies that $E^{\mu} \preceq_L E^{\lambda}$. Indeed, since $C_w.L^{\mu}_k \neq \{0\}$, we also have $C_w.W_k(\mu) \neq \{0\}$ and, hence, $C_w.W(\mu) \neq 0$. By Lemma 2.6.10, $C_w$ is an $R$-linear combination of elements $\mathbf{C}^{\nu}_{\mathfrak{u},\mathfrak{v}}$, where $\nu \in \Lambda$ is such that $E^{\nu} \leftrightsquigarrow_L w$. Hence, we also have $h.W(\mu) \neq \{0\}$, where $h = \mathbf{C}^{\nu}_{\mathfrak{u},\mathfrak{v}}$ for some $\nu \in \Lambda$ and $\mathfrak{u}, \mathfrak{v} \in M(\nu)$ such that $E^{\lambda} \sim_L E^{\nu}$. Arguing as in the above proof, this implies that $\mu \trianglelefteq_L \nu$. Since also $E^{\lambda} \sim_L E^{\nu}$, we conclude that $E^{\mu} \preceq_L E^{\lambda}$, as claimed. On the other hand, since $(W_k(\lambda) : L^{\mu}_k) \neq 0$, we have $\lambda \trianglelefteq_L \mu$; see Theorem 2.6.6. In combination with $E^{\mu} \preceq_L E^{\lambda}$, this implies that $\lambda = \mu$. Thus, since $(W_k(\lambda) : L^{\lambda}_k) = 1$, we have shown that $\theta(\chi^{\lambda}(C_w)) = \mathrm{trace}(C_w, L^{\lambda}_k)$.

(b) By Proposition 2.7.6, we have $\theta(\chi^{\lambda}(C_w)) \neq 0$ for some $w \in W$ such that $E^{\lambda} \leftrightsquigarrow_L w$. By (a), this implies that $\mathrm{trace}(C_w, L^{\lambda}_k) \neq 0$ and so $C_w.L^{\lambda}_k \neq \{0\}$. □

*Remark 2.7.8.* Once a cellular structure for $\mathbf{H}$ is available, it also natural to consider the "Jantzen filtration" on cell modules; for recent results and open problems in this direction, see James and Mathas [183], [184], Shan [277] (type $A$), Bonnafé and Jacon [27] and Policzew [266] (exceptional types).

Finally, we discuss the existence of $W$-graph representations, as already briefly mentioned at the end of Section 1.4. We begin with a preliminary result about the structure and the representations of $\tilde{\mathbf{J}}$.

**Lemma 2.7.9.** *In addition to (♣) and (♠), also assume that (♦) holds. Then:*

*(a) We have $\tilde{\gamma}_{x,y,z} \in \mathbb{Z}$ and $\tilde{n}_d = \pm 1$ for all $d \in \tilde{\mathscr{D}}$.*
*(b) The elements $\{\tilde{n}_d t_d \mid d \in \tilde{\mathscr{D}}\}$ are orthogonal idempotents.*

*Furthermore, for each $\lambda \in \Lambda$, the balanced representation $\rho^{\lambda}$ of $\mathbf{H}_K$ can be chosen such that the following holds for the corresponding representation $\bar{\rho}^{\lambda}$ of $\tilde{\mathbf{J}}$:*

*(c) The conditions in Proposition 1.5.11 hold where* $\bar{\rho}^\lambda(t_w) = (c^{\mathfrak{st}}_{w,\lambda}) \in M_{d_\lambda}(\mathbb{Z}_W)$ *for all* $w \in W$.

*(d) For any* $d \in \tilde{\mathscr{D}}$, *the matrix* $\bar{\rho}^\lambda(t_d)$ *is diagonal with* $0, \pm 1$ *on the diagonal.*

*Proof.* (a), (b) See Example 2.5.14. Once we know that $\tilde{\gamma}_{x,y,z}$ and $\tilde{n}_w$ are integers, the fact that $\tilde{n}_d = \pm 1$ follows from the formula in Remark 1.6.19(a).

(c), (d) We slightly refine the argument in Proposition 1.5.11. We can assume that $(W,S)$ is irreducible. First let $W$ be of type $I_2(m)$. In the proof of Proposition 1.5.11, we have seen that the representations in Example 1.3.7 satisfy (c). By Example 1.7.4, these representations also satisfy (d). Now assume that $W$ is not of type $I_2(m)$. Then $\mathbb{Z}_W$ is a principal ideal domain. As in the proof of Proposition 1.5.11, a general argument shows that $\rho^\lambda$ can be chosen such that (c) holds. Let $\mathbf{J}_{\mathbb{Z}_W} = \langle t_w \mid w \in W \rangle_{\mathbb{Z}_W}$ and let $\bar{E}^\lambda$ be a $\mathbf{J}_{\mathbb{Z}_W}$-module (finitely generated and free over $\mathbb{Z}_W$) which affords the representation $\bar{\rho}^\lambda$. Since the idempotents $\tilde{n}_d t_d$ ($d \in \tilde{\mathscr{D}}$) lie in $\mathbf{J}_{\mathbb{Z}_W}$ and since $\mathbb{Z}_W$ is a principal ideal domain, we have a direct sum decomposition $\bar{E}^\lambda = \bigoplus_{d \in \tilde{\mathscr{D}}} \bar{E}^\lambda_d$, where $\bar{E}^\lambda_d := \tilde{n}_d t_d . \bar{E}^\lambda$ is a $\mathbb{Z}_W$-submodule of $\bar{E}^\lambda$ which is finitely generated and free over $\mathbb{Z}_W$. Now choose a $\mathbb{Z}_W$-basis of $\bar{E}^\lambda$ which is adapted to this decomposition and perform a base change (over $\mathbb{Z}_W$) to this new basis. We replace $\rho^\lambda$ by the representation obtained via this base change (as in the proof of Proposition 1.5.11). This new representation is balanced, and it satisfies (c) and (d). $\square$

**2.7.10.** We keep the assumptions of Lemma 2.7.9. We shall consider the effect of Lusztig's homomorphism $\phi\colon \mathbf{H} \to \tilde{\mathbf{J}}_A$ (see Theorem 2.5.5) on $C_s$, where $s \in S$ is such that $L(s) > 0$. For any $d \in \tilde{\mathscr{D}}$, we have $C_s C_d = -(v_s + v_s^{-1}) C_d$ if $sd < d$; furthermore, if $sd > d$, then $h_{s,d,z} = 0$ unless $sz < z$, in which case we have $h_{s,d,z} = (-1)^{l(d)+l(z)+1} \mu^s_{z,d}$. Thus, the formula in Theorem 2.5.5 can be written as

$$\phi(C_s) = -(v_s + v_s^{-1}) \Big( \sum_{\substack{d \in \tilde{\mathscr{D}} \\ sd<d}} \tilde{n}_d t_d \Big) + \Big( \sum_{\substack{z \in W, d \in \tilde{\mathscr{D}} \\ sz<z<d<sd \text{ and } z \sim_{\mathscr{LR}} d}} (-1)^{l(z)+l(d)+1} \tilde{n}_d \mu^s_{z,d} t_z \Big).$$

Following Gyoja [150] (where this is discussed in the equal-parameter case, in a somewhat different setting), we define elements of $\tilde{\mathbf{J}}_A$ as follows:

$$\tilde{s}_0 := \sum_{\substack{d \in \tilde{\mathscr{D}} \\ sd<d}} \tilde{n}_d t_d \qquad \text{and} \qquad \tilde{s}_1 := \sum_{\substack{z \in W, d \in \tilde{\mathscr{D}} \\ sz<z<d<sd \text{ and } z \sim_{\mathscr{LR}} d}} (-1)^{l(z)+l(d)+1} \tilde{n}_d \mu^s_{z,d} t_z.$$

Thus, $\phi(C_s) = -(v_s + v_s^{-1})\tilde{s}_0 + \tilde{s}_1$. Now recall that $C_s = T_s - v_s T_1$. Then we have

$$\phi(T_s) = \phi(C_s) + v_s 1_{\tilde{\mathbf{J}}} = -v_s^{-1} \tilde{s}_0 + v_s (1_{\tilde{\mathbf{J}}} - \tilde{s}_0) + \tilde{s}_1 \qquad \text{for any } s \in S.$$

The collection of elements $\{\tilde{s}_0, \tilde{s}_1 \mid s \in S, L(s) > 0\}$ satisfies the following properties.

**Lemma 2.7.11 (Cf. Gyoja [150, 2.4]).** *In the above setting, we have*

$$\tilde{s}_0^2 = \tilde{s}_0, \qquad \tilde{s}_0 \tilde{t}_0 = \tilde{t}_0 \tilde{s}_0, \qquad \tilde{s}_0 \tilde{s}_1 = \tilde{s}_1, \qquad \tilde{s}_1 \tilde{s}_0 = 0$$

*for all* $s,t \in S$ *such that* $L(s) > 0$ *and* $L(t) > 0$.

*Proof.* By Lemma 2.7.9(b), the elements $\{\tilde{n}_d t_d \mid d \in \tilde{\mathscr{D}}\}$ are orthogonal idempotents in $\tilde{\mathbf{J}}$. This yields that $\tilde{s}_0^2 = \tilde{s}_0$ and $\tilde{s}_0\tilde{t}_0 = \tilde{t}_0\tilde{s}_0$ for all $s,t \in S$. Now consider

$$\tilde{s}_0\tilde{s}_1 = \sum_{\substack{z\in W, d'\in\tilde{\mathscr{D}} \\ sz<z<d'<sd' \text{ and } z\sim_{\mathscr{LR}} d'}} (-1)^{l(z)+l(d')+1}\tilde{n}_{d'}\,\mu_{z,d'}^s \Big(\sum_{\substack{d\in\tilde{\mathscr{D}} \\ sd<d}} \tilde{n}_d t_d t_z\Big).$$

In order to show that this equals $\tilde{s}_1$, it will be enough to show that $\sum_{d\in\tilde{\mathscr{D}}, sd<d} \tilde{n}_d t_d t_z = t_z$ for all $z \in W$ such that $sz < z$. Now, given any $d \in \tilde{\mathscr{D}}$, we have

$$\tilde{n}_d t_d t_z = \sum_{x\in W} \tilde{n}_d \tilde{\gamma}_{d,z,x^{-1}} t_x = \sum_{x\in W} \tilde{n}_d \tilde{\gamma}_{z,x^{-1},d} t_x = \begin{cases} t_z & \text{if } z^{-1} \sim_{\mathscr{L}} d, \\ 0 & \text{otherwise,} \end{cases}$$

where the second equality holds by Lemma 1.5.3(a) and the third equality holds by Remark 1.6.19(a). Hence, it remains to show that the unique $d \in \tilde{\mathscr{D}}$ such that $z^{-1} \sim_{\mathscr{L}} d$ satisfies $sd < d$. Now recall that $sz < z$. Using Lemma 2.1.16, we conclude that $s \in \mathscr{L}(z) = \mathscr{R}(z^{-1}) = \mathscr{R}(d)$. But, by Remark 1.6.19(a), we have $d^2 = 1$ and so $\mathscr{R}(d) = \mathscr{L}(d)$. Thus, $sd < d$, as required. Finally, consider

$$\tilde{s}_1\tilde{s}_0 = \sum_{\substack{d\in\tilde{\mathscr{D}} \\ sd<d}} \sum_{\substack{z\in W, d'\in\tilde{\mathscr{D}} \\ sz<z<d'<sd' \text{ and } z\sim_{\mathscr{LR}} d'}} (-1)^{l(z)+l(d')+1}\tilde{n}_{d'}\,\mu_{z,d'}^s \tilde{n}_d t_z t_d.$$

Assume, if possible, that this is non-zero. Then for some $z,d,d'$ in the above sum, we have $\mu_{z,d'}^s \neq 0$ and $t_z t_d \neq 0$. Arguing as above, the latter condition implies that $z \sim_{\mathscr{L}} d$. On the other hand, since $\mu_{z,d'}^s \neq 0$, we have $z \leqslant_{\mathscr{L}} d'$. Since we also have $z \sim_{\mathscr{LR}} d'$, we can conclude that $z \sim_{\mathscr{L}} d'$; see Lemma 2.5.9. Hence, (♦) yields that $d = d'$. But we have $sd < d$ and $sd' > d'$, which is a contradiction. Hence, the assumption was wrong and so we must have $\tilde{s}_1\tilde{s}_0 = 0$, as claimed. □

A version of the following result (for equal parameters, and without taking into account the cellular structure) was first shown by Gyoja [150, §2]; subsequently, Lusztig [224, 3.8] gave a slightly different argument based on his asymptotic algebra. Our proof is a combination of the techniques used in [150] and [224].

**Theorem 2.7.12.** *Recall our standing assumption that (♠) and (♣) hold; also assume that (♦) holds and that $L(s) > 0$ for all $s \in S$. Then the data in Definition 2.6.8 can be chosen such that the cell modules $\{W(\lambda) \mid \lambda \in \Lambda\}$ are afforded by $W$-graphs where the elements $\{m_{x,y}^s\} \subseteq A$ (see Definition 1.4.11) all lie in the subring $\mathbb{Z}_W[\Gamma] \subseteq A$.*

*Proof.* By Lemma 2.7.11, the elements $\{\tilde{s}_0 \mid s \in S\}$ pairwise commute with each other. Hence, we can define

$$F_I := \Big(\prod_{s\in I} \tilde{s}_0\Big)\Big(\prod_{s\in S\setminus I} (1_{\tilde{\mathbf{J}}} - \tilde{s}_0)\Big) \in \tilde{\mathbf{J}} \qquad \text{for any subset } I \subseteq S.$$

These elements have the following properties:

(a) $$1_{\tilde{\mathbf{J}}} = \sum_{I\subseteq S} F_I, \qquad F_I^2 = F_I \quad (I \subseteq S), \qquad F_I F_J = 0 \quad (I \neq J).$$

Indeed, by Lemma 2.7.11, the elements $\{\tilde{s}_0 \mid s \in S\}$ do not only commute with each other, but they are also idempotents. Hence, each $F_I$ is an idempotent (possibly zero). Furthermore, suppose that $I \neq J$. If $s \in I \setminus J$, then the factor $\tilde{s}_0$ occurs in $F_I$ and the factor $1_{\tilde{\mathbf{J}}} - \tilde{s}_0$ occurs in $F_J$. Hence, we have $F_I F_J = 0$. The argument is analogous if $s \in J \setminus I$. Finally, notice that $1_{\tilde{\mathbf{J}}} = 1_{\tilde{\mathbf{J}}}^{|S|} = \prod_{s\in S}(\tilde{s}_0 + (1_{\tilde{\mathbf{J}}} - \tilde{s}_0))$. Expanding the product yields that $1_{\tilde{\mathbf{J}}} = \sum_{I\subseteq S} F_I$, as required. Thus, (a) is proved.

Now assume that the balanced representations $\{\rho^\lambda \mid \lambda \in \Lambda\}$ are chosen such that they satisfy the properties in Lemma 2.7.9. In particular, for any $\lambda \in \Lambda$ and $d \in \tilde{\mathscr{D}}$, the matrix $\bar{\rho}^\lambda(t_d)$ is diagonal with $0, \pm 1$ on the diagonal. We conclude that $\bar{\rho}^\lambda(\tilde{s}_0)$ is a diagonal matrix for any $s \in S$ and, hence, $\bar{\rho}^\lambda(F_I)$ is a diagonal matrix for any $I \subseteq S$. Since $F_I$ is an idempotent, the diagonal coefficients of $\bar{\rho}^\lambda(F_I)$ will be $0, 1$. Since the elements $\{F_I \mid I \subseteq S\}$ are orthogonal idempotents whose sum is $1_{\tilde{\mathbf{J}}}$, we conclude that there is a well-defined partition

(b) $$M(\lambda) = \bigsqcup_{I\subseteq S} M_I(\lambda) \qquad \text{such that} \qquad F_I.e_{\mathfrak{t}} = e_{\mathfrak{t}} \Leftrightarrow \mathfrak{t} \in M_I(\lambda).$$

Let us now extend scalars from $R$ to $A$. Then $\bar{E}_A^\lambda := A \otimes_R \bar{E}^\lambda$ is a $\tilde{\mathbf{J}}_A$-module but it also becomes an **H**-module via Lusztig's homomorphism $\phi\colon \mathbf{H} \to \tilde{\mathbf{J}}_A$. The formula at the end of the proof of Theorem 2.6.12 shows that $h \in \mathbf{H}$ acts on $W(\lambda)$ in the same way as $\phi(h) \in \tilde{\mathbf{J}}_A$ acts on $\bar{E}_A^\lambda$. So let us identify $W(\lambda) = \bar{E}_A^\lambda$. Then, using the formula for $\phi(T_s)$ ($s \in S$) in 2.7.10, we see that the action of **H** on $W(\lambda)$ is given by

$$T_s.e_{\mathfrak{t}} = -v_s^{-1}\tilde{s}_0 e_{\mathfrak{t}} + v_s(1_{\tilde{\mathbf{J}}} - \tilde{s}_0)e_{\mathfrak{t}} + \tilde{s}_1 e_{\mathfrak{t}}, \qquad \text{where } \mathfrak{t} \in M(\lambda).$$

We will now show that this formula comes from a $W$-graph structure on $W(\lambda)$. First of all, the definition of a $W$-graph requires a map from $M(\lambda)$ to the set of all subsets of $S$. We define such a map as follows. Given $\mathfrak{t} \in M(\lambda)$, let $I(\mathfrak{t})$ be the unique subset $I \subseteq S$ such that $\mathfrak{t} \in M_I(\lambda)$; see (b). This definition implies that

(c) $$F_I.e_{\mathfrak{t}} = \begin{cases} e_{\mathfrak{t}} & \text{if } I = I(\mathfrak{t}), \\ 0 & \text{otherwise.} \end{cases}$$

Next we consider the action of $T_s$, where $s \in S$. If $s \in I(\mathfrak{t})$, then $\tilde{s}_1\tilde{s}_0 = 0$ and so $\tilde{s}_1 F_{I(\mathfrak{t})} = 0$; since $e_{\mathfrak{t}} = F_{I(\mathfrak{t})}.e_{\mathfrak{t}}$, we conclude that $\tilde{s}_1.e_{\mathfrak{t}} = 0$. Furthermore, $\tilde{s}_0.e_{\mathfrak{t}} = e_{\mathfrak{t}}$ (since $\tilde{s}_0^2 = \tilde{s}_0$ and so $\tilde{s}_0 F_{I(\mathfrak{t})} = F_{I(\mathfrak{t})}$). Hence, we obtain in this case

$$T_s.e_{\mathfrak{t}} = -v_s^{-1} e_{\mathfrak{t}},$$

as required in the definition of a $W$-graph. Now assume that $s \not\in I(\mathfrak{t})$. Then $\tilde{s}_0.e_{\mathfrak{t}} = 0$ (since $\tilde{s}_0 F_{I(\mathfrak{t})} = 0$) and so

$$T_s.e_{\mathfrak{t}} = v_s e_{\mathfrak{t}} + \tilde{s}_1.e_{\mathfrak{t}} = v_s e_{\mathfrak{t}} + \sum_{\mathfrak{u}\in M(\lambda)} m^s_{\mathfrak{u},\mathfrak{t}} e_{\mathfrak{u}},$$

where the terms $m^s_{\mathfrak{u},\mathfrak{t}} \in A$ are such that $\bar{m}^s_{\mathfrak{u},\mathfrak{t}} = m^s_{\mathfrak{u},\mathfrak{t}}$ and $v_s m^s_{\mathfrak{u},\mathfrak{t}} \in \mathbb{Z}_W[\Gamma_{>0}]$ (by the defining formula for $\tilde{s}_1$, Example 2.1.10 and Lemma 2.7.9(a) and (c)). So it remains to show that $m^s_{\mathfrak{u},\mathfrak{t}} = 0$ unless $s \in I(\mathfrak{u})$. But, for any $I \subseteq S$ such that $s \notin I$, we have $(1_{\mathbf{J}} - \tilde{s}_0)\tilde{s}_1 = 0$ by Lemma 2.7.11 and so $F_I\tilde{s}_1 = 0$. Hence, we have

$$\tilde{s}_1.e_{\mathfrak{t}} = \Big(\sum_{I\subseteq S} F_I\Big)\tilde{s}_1.e_{\mathfrak{t}} = \sum_{I\subseteq S, s\in I} F_I\tilde{s}_1.e_{\mathfrak{t}} \subseteq \langle F_I.e_{\mathfrak{u}} \mid \mathfrak{u}\in M(\lambda), s\in I\subseteq S\rangle_A.$$

By (c), the latter submodule is contained in $\langle e_{\mathfrak{u}} \mid s \in I(\mathfrak{u})\rangle_A$, as required. So the above formulae show that $W(\lambda)$ is afforded by a $W$-graph. □

The above result shows that the cell modules $\{W(\lambda) \mid \lambda \in \Lambda\}$ arising from our construction of a cellular basis of $\mathbf{H}$ are afforded by $W$-graphs. The following conjecture is a kind of converse to this statement.

**Conjecture 2.7.13 (Geck and Müller [129, 4.5]).** *Assume that $L(s) > 0$ for all $s \in S$ and that, for every $\lambda \in \Lambda$, we are given a $W$-graph affording an $\mathbf{H}$-module $V^\lambda$ such that $K \otimes_A V^\lambda \cong E^\lambda_\varepsilon$. Then the data in Definition 2.6.8 can be chosen such that $\{V^\lambda \mid \lambda \in \Lambda\}$ are the corresponding cell modules.*

In order to prove this conjecture, it would be sufficient to show that every $W$-graph representation of $\mathbf{H}$ factors through Lusztig's homomorphism $\phi\colon \mathbf{H} \to \tilde{\mathbf{J}}_A$. Somewhat related open problems are formulated by Gyoja [150, Remark 2.18].

The final result in this section holds without any assumptions on $W, L$.

**Corollary 2.7.14.** *Let $\mathbf{H} = \mathbf{H}_A(W,S,L)$ be any generic Iwahori–Hecke algebra, where $W$ is finite and the general assumptions in 1.2.1 hold; that is, $\Gamma$ admits a monomial ordering and $A = R[\Gamma]$, where $\mathbb{Z}_W \subseteq R \subseteq \mathbb{C}$. Then every irreducible representation of $\mathbf{H}_K$ can be realised over $\mathbb{Z}_W[\Gamma]$.*

*Proof.* Let $\mathbb{K}$ be the field of fractions of $R$ and set $k := \mathbb{K}[\Gamma]$. Then $k$ certainly is $L_0$-good and so, as in the proof of Corollary 2.6.16, there exists some cell datum for $\mathbf{H}_k$ which is obtained by extension of scalars from a cell datum in a "universal" algebra $\mathbf{H}_0$ over $A_0 = \mathbb{K}[\Gamma_0]$, where **P1–P15** are known to hold. Let $\{W_0(\lambda) \mid \lambda \in \Lambda\}$ be the corresponding cell modules of $\mathbf{H}_0$. We have a unique $\mathbb{K}$-linear ring homomorphism $\theta\colon A_0 \to k$ such that $\theta(v_s^\circ) = v_s$ for all $s \in S$, where $\{v_s^\circ\}$ are the parameters of $\mathbf{H}_0$ and $\{v_s \mid s \in S\}$ are the parameters of $\mathbf{H}$. Thus, $\mathbf{H}_k = k \otimes_{A_0} \mathbf{H}_0$, where $k$ is regarded as an $A_0$-module via $\theta$. Now $K$ (the field of fractions of $A$) equals the field of fractions of $k$. Since $\mathbf{H}_K = K \otimes_k \mathbf{H}_k$ is semisimple, we have

$$\operatorname{Irr}(\mathbf{H}_K) = \{W_K(\lambda) := K \otimes_{A_0} W_0(\lambda) \mid \lambda \in \Lambda\};$$

see Theorem 2.6.5 and 2.7.5. Since each $W_0(\lambda)$ is realised over $\mathbb{Z}_W[\Gamma_0]$ by Theorem 2.7.12, we conclude that $W_K(\lambda)$ is realised over the image of $\mathbb{Z}_W[\Gamma_0]$ under $\theta_0$; that is, over $\mathbb{Z}_W[\Gamma]$, as required. □

## 2.8 The Case of the Symmetric Group

The aim of this section is to give an elementary proof of the properties (♣), (♠) and (♦) in 2.5.3 when $W \cong \mathfrak{S}_n$. We will then see that the Kazhdan–Lusztig basis $\{C_w\}$ itself is a cellular basis in this case. Note that even if we were willing to admit from the beginning that **P1–P15** hold for $W$, then there would still be a substantial piece of work to do in order to determine the partial order $\trianglelefteq_L$ in Theorem 2.6.12.

We begin with a few general (and more or less well-known) remarks related to longest elements in parabolic subgroups. In 2.8.1–2.8.7, $W$ may be any finite Coxeter group $W$ and $L\colon W \to \Gamma$ a weight function such that $L(s) > 0$ for all $s \in S$. (Here, we explicitly exclude the possibility that $L(s) = 0$ for some $s \in S$.)

**2.8.1.** Let $I \subseteq S$ and consider the parabolic subgroup $W_I \subseteq W$. Let $X_I$ be the set of distinguished left coset representatives of $W_I$ in $W$; we have

$$X_I = \{w \in W \mid w \text{ has minimal length in } wW_I\}.$$

The map $X_I \times W_I \to W$, $(x,u) \mapsto xu$, is a bijection and we have $l(xu) = l(x) + l(u)$ for $u \in W_I$ and $x \in X_I$; see [132, §2.1]. Let $\mathbf{H}_I = \langle T_w \mid w \in W_I\rangle_A \subseteq \mathbf{H}$ be the corresponding parabolic subalgebra of $\mathbf{H}$. For any $w \in W_I$, we have $C_w \in \mathbf{H}_I$ and $C'_w \in \mathbf{H}_I$; hence, $\{C_w \mid w \in W_I\}$ and $\{C'_w \mid w \in W_I\}$ are the Kazhdan–Lusztig bases of $\mathbf{H}_I$.

**Lemma 2.8.2.** *Let $w_I \in W_I$ be the unique element of maximal length. We have*

$$C_{w_I} = (-1)^{l(w_I)}\varepsilon^{L(w_I)} \sum_{w \in W_I} (-1)^{l(w)}\varepsilon^{-L(w)} T_w.$$

*Furthermore, the following hold.*

*(a) For any $w \in W_I$, we have $T_w C_{w_I} = (-1)^{l(w)}\varepsilon^{-L(w)} C_{w_I}$.*
*(b) We have $C_{w_I}^2 = (-1)^{l(w_I)}\varepsilon^{-L(w_I)} P_I C_{w_I}$, where $P_I = \sum_{w \in W_I} \varepsilon^{2L(w)}$; cf. 1.2.11(c).*
*(c) The set $X_I w_I$ is a union of left cells in $W$; we have $X_I w_I = \{w \in W \mid w \leqslant_{\mathscr{L}} w_I\}$.*

*Proof.* The formula for $C_{w_I}$ already appears in Example 2.1.17. Next, we prove (a), by induction on $l(w)$. First assume that $w = s \in I$. Then we have $sw_I < w_I$ and so the multiplication rule in Theorem 2.1.8 shows that $C_s C_{w_I} = -(v_s + v_s^{-1})C_{w_I}$. Since $C_s = T_s - v_s T_1$, this yields $T_s C_{w_I} = -v_s^{-1} C_{w_I}$. If $l(w) > 1$, then write $w = w's$, where $s \in I$, $w' \in W_I$ and $l(w) = l(w') + 1$. We have $T_w = T_{w'} T_s$, and so the desired formula follows by induction. Once (a) is established, we compute

$$\begin{aligned} C_{w_I}^2 &= (-1)^{l(w_I)}\varepsilon^{L(w_I)} \sum_{w \in W_I} (-1)^{l(w)}\varepsilon^{-L(w)} T_w C_{w_I} \\ &= (-1)^{l(w_I)}\varepsilon^{L(w_I)} \sum_{w \in W_I} \varepsilon^{-2L(w)} C_{w_I} = (-1)^{l(w_I)}\varepsilon^{-L(w_I)} P_I C_{w_I}. \end{aligned}$$

To obtain the last equality, we used the formula $l(ww_I) = l(w_I) - l(w)$ for all $w \in W_I$. Thus, (b) is proved. Finally, consider (c). Let $w \in W$ be such that $w \leqslant_{\mathscr{L}} w_I$. Then $\mathscr{R}(w_I) \subseteq \mathscr{R}(w)$; see Lemma 2.1.16. Hence, since $\mathscr{R}(w_I) = I$, we can write $w = xw_I$, where $x \in X_I$. Conversely, if $x \in X_I$, then $l(xw_I) = l(x) + l(w_I)$ and so $xw_I \leqslant_{\mathscr{L}} w_I$. (This follows since $sw \leftarrow_{\mathscr{L}} w$ if $s \in S$, $w \in W$ are such that $sw > w$.) Thus, we obtain $X_I w_I = \{w \in W \mid w \leqslant_{\mathscr{L}} w_I\}$. This also shows that $X_I w_I$ is a union of left cells. □

**Lemma 2.8.3.** *Let $I \subseteq S$ and $\mathfrak{I}^{\mathscr{L}}_{w_I} \subseteq \mathbf{H}$ be the left ideal defined by the general procedure in 1.6.2, with respect to the basis $\{C_w \mid w \in W\}$ of $\mathbf{H}$. Then we have*

$$\mathfrak{I}^{\mathscr{L}}_{w_I} = \langle\, C_{xw_I} \mid x \in X_I \,\rangle_A = \langle\, T_x C_{w_I} \mid x \in X_I \,\rangle_A.$$

*Proof.* By definition, we have $\mathfrak{I}^{\mathscr{L}}_{w_I} = \langle\, C_w \mid w \leqslant_{\mathscr{L}} w_I \rangle_A$; this equals $\langle C_{xw_I} \mid x \in X_I \rangle_A$ by Lemma 2.8.2(c). Now set $\mathscr{M}_I := \langle T_x C_{w_I} \mid x \in X_I \rangle_A$. Since $\mathfrak{I}^{\mathscr{L}}_{w_I}$ is a left ideal, it is clear that $\mathscr{M}_I \subseteq \mathfrak{I}^{\mathscr{L}}_{w_I}$. Both $\mathscr{M}_I$ and $\mathfrak{I}^{\mathscr{L}}_{w_I}$ are free of the same rank over $A$; this already implies that $K \otimes_A \mathscr{M}_I = K \otimes_A \mathfrak{I}^{\mathscr{L}}_{w_I}$. But we also have that $\mathbf{H}/\mathfrak{I}^{\mathscr{L}}_{w_I}$ is free over $A$; furthermore, $\mathbf{H}$ is free as an $\mathbf{H}_I$-module and so $\mathbf{H}/\mathscr{M}_I$ is free as an $A$-module. Hence, we must have $\mathbf{H}_I = \mathfrak{I}^{\mathscr{L}}_{w_I}$. □

**Lemma 2.8.4.** *Let $I \subseteq S$ and $\mu \in \Lambda$ be such that $E^\mu$ is a constituent of the induced representation $\mathrm{Ind}_{W_I}^W(\mathrm{sgn}_I)$. Then $E^\mu \leftrightsquigarrow_L xw_I$ for some $x \in X_I$.*

*Proof.* Let $\mathrm{sgn}_I^\varepsilon$ denote the sign representation of $\mathbf{H}_I$. By Example 1.3.3, we have $\mathrm{sgn}_I^\varepsilon(T_w) = (-1)^{l(w)}\varepsilon^{-L(w)}$ for $w \in W_I$. So Lemma 2.8.2(a) shows that $\langle C_{w_I}\rangle_A \subseteq \mathbf{H}_I$ affords $\mathrm{sgn}_I^\varepsilon$. Now, the induction of representations is also defined on the level of $\mathbf{H}$; see [132, §9.1]. Hence, by Lemma 2.8.3, we have an isomorphism of left $\mathbf{H}$-modules

$$\mathfrak{I}^{\mathscr{L}}_{w_I} \xrightarrow{\sim} \mathrm{Ind}_{\mathbf{H}_I}^{\mathbf{H}}(\langle C_{w_I}\rangle_A), \qquad T_x C_{w_I} \mapsto T_x \otimes C_{w_I} \quad (x \in X_I).$$

By a specialisation argument (see Example 1.2.4), our assumption implies that $E^\mu_\varepsilon$ is a constituent of $\mathfrak{I}^{\mathscr{L}}_{w_I,K} := K \otimes_A \mathfrak{I}^{\mathscr{L}}_{w_I}$. Now, for any $w \in W$, we have $\mathrm{trace}(C_w, \mathfrak{I}^{\mathscr{L}}_{w_I}) = \sum_{x \in X_I} h_{w,xw_I,xw_I}$. Furthermore, by Lemma 2.8.2(a), we can write $X_I w_I = \mathfrak{C}_1 \cup \ldots \cup \mathfrak{C}_m$, where $\mathfrak{C}_1, \ldots, \mathfrak{C}_m$ are (pairwise distinct) left cells of $W$. Consequently, we have

$$\mathrm{trace}(C_w, \mathfrak{I}^{\mathscr{L}}_{w_I}) = \sum_{1 \leqslant i \leqslant m} \sum_{z \in \mathfrak{C}_i} h_{w,z,z} = \sum_{1 \leqslant i \leqslant m} \mathrm{trace}(C_w, [\mathfrak{C}_i]_A) \quad \text{for all } w \in W.$$

Since $\mathbf{H}_K$ is split semisimple, this implies that $\mathfrak{I}^{\mathscr{L}}_{w_I,K} \cong [\mathfrak{C}_1]_K \oplus \ldots \oplus [\mathfrak{C}_m]_K$. It follows that $E^\mu_\varepsilon$ is a constituent of $[\mathfrak{C}_i]_K$ for some $i$ and so $m(\mathfrak{C}_i, \mu) > 0$; see 2.2.2(b). Hence, by Lemma 2.2.4, there exists some $w \in X_I w_I$ such that $E^\mu \leftrightsquigarrow_L w$. □

**Lemma 2.8.5.** *Let $I \subseteq S$ and $\mu \in \Lambda$ be such that $\rho^\mu(C_{w_I}) \neq 0$. Then $E^\mu$ is a constituent of $\mathrm{Ind}_{W_I}^W(\mathrm{sgn}_I)$.*

*Proof.* As in the above proof, let $\mathrm{sgn}_I^\varepsilon$ denote the sign representation of $\mathbf{H}_I$. Using the formula for $C_{w_I}$ in Lemma 2.8.2, we obtain

$$\chi^\mu(C_{w_I}) = (-1)^{l(w_I)}\varepsilon^{L(w_I)} \sum_{w \in W_I} \mathrm{sgn}_I^\varepsilon(T_{w^{-1}})\, \chi^\mu(T_w),$$

where we also used the fact that $\mathrm{sgn}_I^\varepsilon(T_{w^{-1}}) = \mathrm{sgn}_I^\varepsilon(T_w)$ for all $w \in W_I$. Now, writing the restriction of $\chi^\mu$ to $\mathbf{H}_{I,K}$ as a sum of irreducible characters of $\mathbf{H}_{I,K}$ and using the Schur relations in Proposition 1.2.12, we conclude that

$$\chi^\mu(C_{w_I}) = (-1)^{l(w_I)} \varepsilon^{L(w_I)} \mathbf{c}_{\mathrm{sgn}_I}\, m(I,\mu),$$

where $m(I,\mu)$ denotes the multiplicity of $\mathrm{sgn}_I^\varepsilon$ in the restriction of $\chi^\mu$ to $\mathbf{H}_{I,K}$. By a specialisation argument (see Example 1.2.4), $m(I,\mu)$ also equals the multiplicity of $\mathrm{sgn}_I$ in the restriction of $E^\mu$ from $W$ to $W_I$. And by Frobenius reciprocity, this is the same as the multiplicity of $E^\mu$ as a constituent of $\mathrm{Ind}_I^W(\mathrm{sgn}_I)$.

Thus, it remains to show that $\chi^\mu(C_{w_I}) \neq 0$. Now, by Lemma 2.8.2(b), $C_{w_I}$ is a non-zero scalar multiple of an idempotent. Hence, $\rho^\mu(C_{w_I})$ will be conjugate to a non-zero scalar multiple of a diagonal matrix with 0 and 1 on the diagonal. Since $\rho^\mu(C_{w_I}) \neq 0$, we conclude that $\chi^\mu(C_{w_I}) = \mathrm{trace}(\rho^\mu(C_{w_I})) \neq 0$, as required. □

**Corollary 2.8.6.** *Let $I \subseteq S$ and $\lambda \in \Lambda$ be such that $\mathbf{a}_\lambda = L(w_I)$ and $E^\lambda$ is a constituent of $\mathrm{Ind}_{W_I}^W(\mathrm{sgn}_I)$. Then $w_I \in \mathscr{F}_\lambda$ (where $\mathscr{F}_\lambda$ is defined in Proposition 1.6.11).*

*Proof.* As in the above proof, $\chi^\lambda(C_{w_I}) = \pm\varepsilon^{L(w_I)} \mathbf{c}_{\mathrm{sgn}_I} m(I,\lambda)$. Now, we have

$$\mathbf{c}_{\mathrm{sgn}_I} = \sum_{w\in W_I} \mathrm{sgn}_I^\varepsilon(T_w)\mathrm{sgn}_I^\varepsilon(T_{w^{-1}}) = \sum_{w\in W_I} \varepsilon^{-2L(w)}.$$

Since $\mathbf{a}_\lambda = L(w_I)$ and $L(w) < L(w_I)$ for $w \in W$, $w \neq w_I$, we obtain that

$$\varepsilon^{\mathbf{a}_\lambda}\chi^\lambda(C_{w_I}) \equiv \pm m(I,\lambda) \sum_{w\in W_I} \varepsilon^{2(L(w_I)-L(w))} \equiv \pm m(I,\lambda) \mod \mathfrak{m}.$$

Since we also have $\varepsilon^{\mathbf{a}_\lambda}\chi^\lambda(T_{w_I}) \equiv \varepsilon^{\mathbf{a}_\lambda}\chi^\lambda(C_{w_I}) \mod \mathfrak{m}$ by 2.1.19, we conclude that $c_{w_I,\lambda} = \pm m(I,\lambda) \neq 0$ and so $w_I \in \mathscr{F}_\lambda$. □

**Corollary 2.8.7 (Cf. [107, 4.7, 4.8]).** *Let $I \subseteq S$ and define*

$$Z_{x,y}^I := P_I^{-1}\varepsilon^{L(w_I)} C_{xw_I} C_{w_I y^{-1}} \in \mathbf{H}_K \qquad \textit{for any } x,y \in X_I.$$

*Then the following hold:*

*(a) We have $Z_{x,y}^I = \overline{Z}_{x,y}^I \in \mathbf{H}$. Furthermore, $Z_{x,y}^I \in \mathfrak{I}_{w_I}^{\mathscr{LR}}$, where $\mathfrak{I}_{w_I}^{\mathscr{LR}}$ is defined by the general procedure in 1.6.2, with respect to the basis $\{C_w \mid w \in W\}$ of $\mathbf{H}$.*
*(b) If $\mu \in \Lambda$ is such that $\rho^\mu(Z_{x,y}^I) \neq 0$, then $E^\mu$ is a constituent of $\mathrm{Ind}_{W_I}^W(\mathrm{sgn}_I)$.*

*Proof.* (a) By Lemma 2.8.3, $C_{xw_I}$ is an $A$-linear combination of terms $T_{x_1}C_{w_I}$, where $x_1 \in X_I$. By 2.1.14, we have $C_{w_I y^{-1}} = C_{yw_I}^\flat$ and so $C_{w_I y^{-1}}$ is an $A$-linear combination of terms $C_{w_I}T_{y_1^{-1}}$, where $y_1 \in X_I$. Consequently, by Lemma 2.8.2(b), $C_{xw_I}C_{w_I y^{-1}}$ is an $A$-linear combination of terms $P_I T_{x_1} C_{w_I} T_{y_1^{-1}}$, where $x_1, y_1 \in X_I$. Hence, we have $Z_{x,y}^I \in \mathbf{H}$ and $Z_{x,y}^I \in \mathfrak{I}_{w_I}^{\mathscr{LR}}$. Since $\overline{P}_I = \varepsilon^{-2L(w_I)}P_I$, we also see that $Z_{x,y}^I = \overline{Z}_{x,y}^I$.

(b) Assume that $\rho^\mu(Z^I_{x,y}) \neq 0$. In the proof of (a), we have seen that $Z^I_{x,y}$ is an $A$-linear combination of terms $P_I T_{x_1} C_{w_I} T_{y_1^{-1}}$, where $x_1, y_1 \in X_I$. Hence, we must have $\rho^\mu(C_{w_I}) \neq 0$ and so the assertion follows from Lemma 2.8.5. $\square$

**2.8.8.** From now on, let $W = \mathfrak{S}_n$ be the symmetric group where $S = \{s_1, \ldots, s_{n-1}\}$ and $s_i = (i, i+1)$ for $1 \leqslant i \leqslant n-1$. The set $\Lambda$ consists of all partitions of $n$; we write $\lambda \vdash n$ to denote that $\lambda$ is a partition of $n$. Furthermore, we assume that $\Gamma = \mathbb{Z}$ and $L(s_i) = 1$ for $1 \leqslant i \leqslant n-1$. By 1.7.6, we have $f_\lambda = 1$ for all $\lambda \vdash n$. So $R = \mathbb{Z}$ is an $L$-good subring of $\mathbb{C}$; in particular, $A = \mathbb{Z}[v, v^{-1}]$, where $v = \varepsilon$ is an indeterminate. By Corollary 1.7.9, the balanced representations $\{\rho^\lambda \mid \lambda \vdash n\}$ can be chosen such that the corresponding leading matrix coefficients satisfy the following condition:

$$\text{(a)} \qquad c^{\mathfrak{st}}_{w,\lambda} = c^{\mathfrak{ts}}_{w^{-1},\lambda} \in \{0, \pm 1\} \qquad \text{for all } w \in W \text{ and } \mathfrak{s}, \mathfrak{t} \in M(\lambda).$$

Consequently, we have a bijection

$$\text{(b)} \qquad \bigsqcup_{\lambda \in \Lambda} (M(\lambda) \times M(\lambda)) \stackrel{1-1}{\longrightarrow} W, \qquad (\mathfrak{s}, \mathfrak{t}) \mapsto w_\lambda(\mathfrak{s}, \mathfrak{t}),$$

satisfying the properties in Theorem 1.7.10; in particular, we have

$$\text{(c)} \qquad \mathscr{F}_\lambda = \{w_\lambda(\mathfrak{s}, \mathfrak{t}) \mid \mathfrak{s}, \mathfrak{t} \in M(\lambda)\} \qquad \text{for all } \lambda \vdash n.$$

By Proposition 2.1.20, we already know that $\mathscr{F}_\lambda$ is contained in a two-sided Kazhdan–Lusztig cell. One of our aims is to show that the converse also holds.

**2.8.9.** For any subset $I \subseteq \{1, \ldots, n-1\}$, denote by $W_I \subseteq W$ the parabolic subgroup generated by $\{s_i \mid i \in I\}$. We now define, for any partition $\lambda \vdash n$, a particular parabolic subgroup of $W$. For this purpose, we set

$$I(\lambda) := \{1, \ldots, n\} \setminus \{\lambda_1^*, \lambda_1^* + \lambda_2^*, \lambda_1^* + \lambda_2^* + \lambda_3^*, \ldots\},$$

where $\lambda_1^* \geqslant \lambda_2^* \geqslant \lambda_3^* \geqslant \ldots \geqslant 0$ are the parts of $\lambda^*$, the conjugate partition of $\lambda$. (Thus, $W_{I(\lambda)} \subseteq W$ is the Young subgroup $\mathfrak{S}_{\lambda^*}$.) For example, if $\lambda = (1^n)$, then $\lambda^* = (n)$ and so $W_I = W$. Then *Young's rule* shows that, for any $\mu \vdash n$, we have

$$\text{(a)} \quad E^\mu \text{ is a constituent of } \operatorname{Ind}_{W_{I(\lambda)}}^W(\operatorname{sgn}_{I(\lambda)}) \quad \Leftrightarrow \quad \kappa_{\mu^*\lambda^*} \neq 0 \quad \Leftrightarrow \quad \mu \trianglelefteq \lambda,$$

where $\kappa_{\mu^*\lambda^*}$ is a *Kostka number* and $\trianglelefteq$ denotes the dominance order on partitions, as defined in Example 2.2.13. (For the first equivalence in (a), see Macdonald [236, p. 115]; the second equivalence is a combination of [236, I.6.5 and I.7.9]. Note also that $\lambda \trianglelefteq \mu$ if and only if $\mu^* \trianglelefteq \lambda^*$; see [236, I.1.11].) Consequently, we have

$$\text{(b)} \qquad \mathbf{a}_\lambda = l(w_{I(\lambda)}) \qquad \text{and} \qquad w_{I(\lambda)} \in \mathscr{F}_\lambda \qquad \text{for all } \lambda \vdash n,$$

where we use Corollary 2.8.6 and [132, 5.4.1, 5.4.3] to relate $\mathbf{a}_\lambda$ and $l(w_{I(\lambda)})$. Now we define a two-sided ideal of $\mathbf{H}_K$ by

(c) $\mathcal{N}_\lambda := \{h \in \mathbf{H}_K \mid \rho^\mu(h) = 0 \text{ for all } \mu \vdash n \text{ such that } \mu \ntrianglelefteq \lambda\}.$

We also set $\hat{\mathcal{N}}_\lambda := \{h \in \mathcal{N}_\lambda \mid \rho^\lambda(h) = 0\}$. Note that $\mathcal{N}_\lambda$ is the sum of all Wedderburn components of the split semisimple algebra $\mathbf{H}_K$ which correspond to the irreducible representations $E_\varepsilon^\mu$ where $\mu \trianglelefteq \lambda$.

**2.8.10.** We have just seen that $w_{I(\lambda)} \in \mathscr{F}_\lambda$ for any $\lambda \vdash n$. In particular, $w_{I(\lambda)} = w_\lambda(\mathfrak{t}_0, \mathfrak{t}_0)$ for a unique $\mathfrak{t}_0 \in M(\lambda)$. By Theorem 1.7.10(b) and Proposition 2.1.20, the set $\mathfrak{C}_0 := \{w_\lambda(\mathfrak{s}, \mathfrak{t}_0) \mid \mathfrak{s} \in M(\lambda)\}$ is contained in a left Kazhdan–Lusztig cell. Hence, by Lemma 2.8.2(c), we have $\mathfrak{C}_0 \subseteq X_{I(\lambda)} w_{I(\lambda)}$. Consequently, there is a subset $\{x_\mathfrak{s} \mid \mathfrak{s} \in M(\lambda)\} \subseteq X_{I(\lambda)}$ such that $w_\lambda(\mathfrak{s}, \mathfrak{t}_0) = x_\mathfrak{s} w_{I(\lambda)}$ for all $\mathfrak{s} \in M(\lambda)$. We now set

(a) $Z_w := Z_{x_\mathfrak{s}, x_\mathfrak{t}}^{I(\lambda)}, \qquad \text{where} \qquad w = w_\lambda(\mathfrak{s}, \mathfrak{t}) \text{ for } \mathfrak{s}, \mathfrak{t} \in M(\lambda).$

By Corollary 2.8.7(a), we have $Z_w = \overline{Z}_w \in \mathbf{H}$. We claim that

(b) $Z_w \in \mathcal{N}_\lambda \qquad \text{for all } w \in \mathscr{F}_\lambda.$

This is seen as follows. Let $\mu \vdash n$ and assume that $\rho^\mu(Z_w) \neq 0$. We must show that $\mu \trianglelefteq \lambda$. Now, by Corollary 2.8.7(b), $E^\mu$ is a constituent of $\mathrm{Ind}_{W_{I(\lambda)}}^W(\mathrm{sgn}_{I(\lambda)})$. By 2.8.9(a), this implies that $\mu \trianglelefteq \lambda$, as claimed.

**Lemma 2.8.11.** *Let $\lambda \in \Lambda$ and $\mathfrak{u}, \mathfrak{v} \in M(\lambda)$. Then, for any $w \in \mathscr{F}_\lambda$, we have*

$$v^{\mathbf{a}_\lambda} \rho_{\mathfrak{u}\mathfrak{v}}^\lambda(Z_w) \in \mathscr{O}_0 \qquad \text{and} \qquad v^{\mathbf{a}_\lambda} \rho_{\mathfrak{u}\mathfrak{v}}^\lambda(Z_w) \equiv \pm\delta_{\mathfrak{s}\mathfrak{u}}\delta_{\mathfrak{t}\mathfrak{v}} \mod \mathfrak{m},$$

*where $\mathfrak{s}, \mathfrak{t} \in M(\lambda)$ are such that $w = w_\lambda(\mathfrak{s}, \mathfrak{t})$.*

*Proof.* We have $\mathbf{a}_\lambda = l(w_{I(\lambda)})$ by 2.8.9(b). Hence we obtain

$$\begin{aligned} v^{\mathbf{a}_\lambda} \rho_{\mathfrak{u}\mathfrak{v}}^\lambda(Z_w) &= P_{I(\lambda)}^{-1} v^{2\mathbf{a}_\lambda} \rho_{\mathfrak{u}\mathfrak{v}}^\lambda\big(C_{x_\mathfrak{s} w_{I(\lambda)}} C_{w_{I(\lambda)} x_\mathfrak{t}^{-1}}\big) \\ &= P_{I(\lambda)}^{-1} v^{2\mathbf{a}_\lambda} \rho_{\mathfrak{u}\mathfrak{v}}^\lambda\big(C_{w_\lambda(\mathfrak{s},\mathfrak{t}_0)} C_{w_\lambda(\mathfrak{t}_0,\mathfrak{t})}\big) \\ &= P_{I(\lambda)}^{-1} \sum_{\mathfrak{r} \in M(\lambda)} \big(v^{\mathbf{a}_\lambda} \rho_{\mathfrak{u}\mathfrak{r}}^\lambda(C_{w_\lambda(\mathfrak{s},\mathfrak{t}_0)})\big)\big(v^{\mathbf{a}_\lambda} \rho_{\mathfrak{r}\mathfrak{v}}^\lambda(C_{w_\lambda(\mathfrak{t}_0,\mathfrak{t})})\big). \end{aligned}$$

First of all, this shows that the above expression lies in $\mathscr{O}_0$; note that $P_{I(\lambda)} \in 1 + \mathfrak{m}$. Furthermore, its constant term can be expressed by the leading matrix coefficients of $w_\lambda(\mathfrak{s}, \mathfrak{t}_0)$ and $w_\lambda(\mathfrak{t}_0, \mathfrak{t})$. Indeed, by 2.1.19 and Theorem 1.7.10, we have

$$\begin{aligned} v^{\mathbf{a}_\lambda} \rho_{\mathfrak{u}\mathfrak{r}}^\lambda(C_{w_\lambda(\mathfrak{s},\mathfrak{t}_0)}) &\equiv c_{w_\lambda(\mathfrak{s},\mathfrak{t}_0),\lambda}^{\mathfrak{u}\mathfrak{r}} \equiv \pm\delta_{\mathfrak{s}\mathfrak{u}}\delta_{\mathfrak{r}\mathfrak{t}_0} \mod \mathfrak{m}, \\ v^{\mathbf{a}_\lambda} \rho_{\mathfrak{r}\mathfrak{v}}^\lambda(C_{w_\lambda(\mathfrak{t}_0,\mathfrak{t})}) &\equiv c_{w_\lambda(\mathfrak{t}_0,\mathfrak{t}),\lambda}^{\mathfrak{r}\mathfrak{v}} \equiv \pm\delta_{\mathfrak{t}\mathfrak{v}}\delta_{\mathfrak{r}\mathfrak{t}_0} \mod \mathfrak{m}. \end{aligned}$$

Since $P_{I(\lambda)} \in 1 + \mathfrak{m}$, we obtain $v^{\mathbf{a}_\lambda} \rho_{\mathfrak{u}\mathfrak{v}}^\lambda(Z_w) \equiv \pm\delta_{\mathfrak{s}\mathfrak{u}}\delta_{\mathfrak{t}\mathfrak{v}} \mod \mathfrak{m}$, as desired. □

**Theorem 2.8.12 (Cf. [107, 4.10]).** *Recall that we are in the setting of 2.8.8, where $W = \mathfrak{S}_n$. Then the following hold for any partition $\lambda \vdash n$.*

(a) $\pm C_w \in Z_w + \hat{\mathcal{N}}_\lambda \subseteq \mathcal{N}_\lambda$ *for all* $w \in \mathcal{F}_\lambda$,

(b) $\mathcal{N}_\lambda = \langle C_w \mid w \in \mathcal{F}_\mu$ *for some* $\mu \vdash n$ *such that* $\mu \trianglelefteq \lambda\rangle_K$,

(c) $\hat{\mathcal{N}}_\lambda = \langle C_w \mid w \in \mathcal{F}_\mu$ *for some* $\mu \vdash n$ *such that* $\mu \triangleleft \lambda\rangle_K$,

*where* $\hat{\mathcal{N}}_\lambda \subseteq \mathcal{N}_\lambda$ *are the two-sided ideals of* $\mathbf{H}_K$ *defined in 2.8.9.*

*Proof.* We prove (a) by induction on the dominance order on partitions. The unique minimal element in this order is the partition $(1^n)$. We have $\mathcal{F}_{(1^n)} = \{w_0\}$ (where $w_0$ is the longest element of $W$), $I(1^n) = \{1,\ldots,n-1\}$, $X_{I(1^n)} = \{1\}$ and

$$Z_{w_0} = P_{I(1^n)}^{-1}\varepsilon^{l(w_0)}C_{w_0}^2 = (-1)^{l(w_0)}C_{w_0}; \qquad \text{see Lemma 2.8.2(b).}$$

Hence, (a) holds in this case. Now assume that $\lambda \neq (1^n)$ and that (a) holds for all partitions $\mu \vdash n$, where $\mu \triangleleft \lambda$. Let $w \in \mathcal{F}_\lambda$. By 2.8.10(b), we already know that $Z_w \in \mathcal{N}_\lambda$. Since $Z_w = \overline{Z}_w \in \mathbf{H}$ (see Corollary 2.8.7(a)), we can write

$$Z_w = \sum_{y\in W} \eta_y C_y, \qquad \text{where } \eta_y = \overline{\eta}_y \in \mathbb{Z}[v,v^{-1}] \text{ for all } y \in W.$$

Let $y \in W$ be such that $\eta_y \neq 0$; we have $y \in \mathcal{F}_\mu$ for a unique $\mu \vdash n$. If $\mu \triangleleft \lambda$, then $\pm C_y \in Z_y + \hat{\mathcal{N}}_\mu$ by induction. Furthermore, by 2.8.10(b), we have $Z_y \in \mathcal{N}_\mu$. By definition, it is also clear that $\mathcal{N}_\mu \subseteq \hat{\mathcal{N}}_\lambda$. Hence, we conclude that $C_y \in \mathcal{N}_\mu \subseteq \hat{\mathcal{N}}_\lambda$. So it remains to consider those $y$ where $y \in \mathcal{F}_\mu$, $\mu \not\triangleleft \lambda$. Let us write

$$\mathcal{C} := \{y \in W \mid \eta_y \neq 0 \text{ and } y \in \mathcal{F}_\mu \text{ where } \mu \not\triangleleft \lambda\}$$

and set $m := \max\{\deg(\eta_y) \mid y \in \mathcal{C}\}$. We claim that

$$(*) \qquad \{y \in \mathcal{C} \mid \deg(\eta_y) = m\} \subseteq \mathcal{F}_\lambda \qquad \text{and} \qquad m = 0.$$

This is seen as follows. Let $y_0 \in \mathcal{C}$ be such that $\deg(\eta_{y_0}) = m$; then $y_0 = w_\mu(\mathfrak{u},\mathfrak{v})$, where $\mu \vdash n$, $\mathfrak{u},\mathfrak{v} \in M(\mu)$ and $\mu \not\triangleleft \lambda$. By 2.1.19 and Theorem 1.7.10, we have

$$v^{\mathbf{a}_\mu}\rho_{\mathfrak{uv}}^\mu(C_y) \equiv c_{y,\mu}^{\mathfrak{uv}} \equiv \pm\delta_{yy_0} \bmod \mathfrak{m} \qquad \text{for any } y \in W.$$

It follows that

$$v^{\mathbf{a}_\mu+m}\rho_{\mathfrak{uv}}^\mu(Z_w) \equiv \sum_{y\in W}(v^m\eta_y)\left(v^{\mathbf{a}_\mu}\rho_{\mathfrak{uv}}^\mu(C_y)\right) \equiv (v^m\eta_{y_0})c_{y_0,\mu}^{\mathfrak{uv}} \not\equiv 0 \bmod \mathfrak{m}.$$

In particular, this yields that $\rho^\mu(Z_w) \neq 0$ and so $\mu \trianglelefteq \lambda$, since $Z_w \in \mathcal{N}_\lambda$ by 2.8.10(b). Combined with $\mu \not\triangleleft \lambda$, we conclude that $\mu = \lambda$ and so $y_0 \in \mathcal{F}_\lambda$, which proves the first part of $(*)$. Now the above congruence reads

$$v^m\left(v^{\mathbf{a}_\lambda}\rho_{\mathfrak{uv}}^\lambda(Z_w)\right) \not\equiv 0 \bmod \mathfrak{m}.$$

But then Lemma 2.8.11 implies that $m = 0$, as required. Thus, $(*)$ is proved. Consequently, we can now write

$$Z_w \equiv \sum_{y \in \mathscr{F}_\lambda} \eta_y C_y \mod \hat{\mathscr{N}_\lambda}, \qquad \text{where } \eta_y \in \mathbb{Z} \text{ for all } y \in \mathscr{F}_\lambda.$$

Let $\mathfrak{s}, \mathfrak{t} \in M(\lambda)$ be such that $w = w_\lambda(\mathfrak{s}, \mathfrak{t})$. Since $\rho^\lambda(h) = 0$ for all $h \in \hat{\mathscr{N}_\lambda}$, we have

$$\rho^\lambda_{\mathfrak{u}\mathfrak{v}}(Z_w) = \sum_{y \in \mathscr{F}_\lambda} \eta_y \rho^\lambda_{\mathfrak{u}\mathfrak{v}}(C_y) \qquad \text{for any } \mathfrak{u}, \mathfrak{v} \in M(\lambda).$$

We multiply this identity by $v^{\mathbf{a}_\lambda}$ and take constant terms. Using Theorem 1.7.10, Lemma 2.8.11 and 2.1.19, we deduce that

$$\pm \delta_{\mathfrak{s}\mathfrak{u}} \delta_{\mathfrak{t}\mathfrak{v}} = \sum_{y \in \mathscr{F}_\lambda} \eta_y c^{\mathfrak{u}\mathfrak{v}}_{y,\lambda} = \eta_{y_0} c^{\mathfrak{u}\mathfrak{v}}_{y_0,\lambda}, \qquad \text{where} \qquad y_0 = w_\lambda(\mathfrak{u}, \mathfrak{v}).$$

It follows that $\eta_w = \pm c^{\mathfrak{s}\mathfrak{t}}_{w,\lambda}$ and $\eta_y = 0$ for all $y \in \mathscr{F}_\lambda \setminus \{w\}$, as required. Thus, (a) is proved. Now let $\mathscr{M}_\lambda$ be the $K$-subspace of $\mathbf{H}_K$ defined by the right-hand side of the desired identity in (b). We claim that $\mathscr{M}_\lambda \subseteq \mathscr{N}_\lambda$. Indeed, let $w \in \mathscr{F}_\mu$, where $\mu \trianglelefteq \lambda$. By 2.8.10(b), we have $Z_w \in \mathscr{N}_\mu$. Hence, using (a), we see that $C_w \in \mathscr{N}_\mu$. Furthermore, by definition, it is clear that $\mathscr{N}_\mu \subseteq \mathscr{N}_\lambda$. Hence, we have $C_w \in \mathscr{N}_\lambda$, as claimed. Now notice that $\dim \mathscr{M}_\lambda = \sum_{\mu \trianglelefteq \lambda} |\mathscr{F}_\mu|$; furthermore, by 2.8.8(c), we have $|\mathscr{F}_\mu| = |M(\mu)|^2$. On the other hand, as already noted in 2.8.9, the ideal $\mathscr{N}_\lambda$ is the sum of all Wedderburn components of $\mathbf{H}_K$ which correspond to the irreducible representations $E^\mu_\varepsilon$ where $\mu \trianglelefteq \lambda$. Hence, we also have $\dim \mathscr{N}_\lambda = \sum_{\mu \trianglelefteq \lambda} |M(\mu)|^2$ and, consequently, $\mathscr{N}_\lambda = \mathscr{M}_\lambda$. Thus, (b) is proved. This also implies (c) since, by definition, $\hat{\mathscr{N}_\lambda} = \sum_\mu \mathscr{N}_\mu$, where the sum runs over all $\mu \vdash n$ such that $\mu \triangleleft \lambda$. $\square$

*Remark 2.8.13.* The above proof essentially follows [107, Theorem 4.10]. However, in [107], we referred to the results of Murphy [256], [257] in order to define the ideals $\mathscr{N}_\lambda$ and $\hat{\mathscr{N}_\lambda}$. The discussion here avoids that reference and, thus, is considerably more self-contained than that in [107].

**Corollary 2.8.14.** *Let $\lambda, \mu \vdash n$. Then we have $E^\mu \preceq_L E^\lambda$ if and only if $\mu \trianglelefteq \lambda$. In particular, the equivalence classes of* $\mathrm{Irr}_{\mathbb{K}}(\mathfrak{S}_n)$ *under $\sim_L$ are singleton sets. Consequently, the properties (♣) and (♠) (see 2.5.3) hold.*

*Proof.* Assume first that $\mu \trianglelefteq \lambda$. By 2.8.9(a) (Young's rule) and Lemma 2.8.4, this implies that $xw_{I(\lambda)} \in \mathscr{F}_\mu$ for some $x \in X_{I(\lambda)}$. But then we have $xw_{I(\lambda)} \leqslant_{\mathscr{L}} w_{I(\lambda)}$ and so $E^\mu \preceq_L E^\lambda$; recall that $w_{I(\lambda)} \in \mathscr{F}_\lambda$ by 2.8.9(b). Conversely, assume that $E^\mu \preceq_L E^\lambda$. This means that $y \leqslant_{\mathscr{LR}} w$, where $y \in \mathscr{F}_\mu$ and $w \in \mathscr{F}_\lambda$. By definition, we can find a sequence $y = y_0, y_1, \ldots, y_m = w$ such that, for each $i \in \{1, \ldots, m\}$, there exist some $x_i \in W$ such that $h_{x_i,y_i,y_{i-1}} \neq 0$ or $h_{y_i,x_i,y_{i-1}} \neq 0$. Now, by Theorem 2.8.12(b), we have $C_{y_m} = C_w \in \mathscr{N}_\lambda$. Since $\mathscr{N}_\lambda$ is a two-sided ideal, we have $C_{x_m}C_{y_m} \in \mathscr{N}_\lambda$ and $C_{y_m}C_{x_m} \in \mathscr{N}_\lambda$. Hence, we have $y_{m-1} \in \mathscr{F}_{\mu_{m-1}}$, where $\mu_{m-1} \trianglelefteq \lambda$; see Theorem 2.8.12(b). We repeat the argument with $x_{m-1}$, $y_{m-1}$ and find that

$y_{m-2} \in \mathscr{F}_{\mu_{m-2}}$, where $\mu_{m-2} \trianglelefteq \mu_{m-1} \trianglelefteq \lambda$. Continuing in this way, we eventually obtain that $\mu = \mu_0 \trianglelefteq \lambda$, as required. It is known that this implies $\mathbf{a}_\lambda \leqslant \mathbf{a}_\mu$, with equality only if $\lambda = \mu$; see 2.2.13. Consequently, (♣) holds. But then the weaker property (♣′) holds and so (♠) also holds; see Proposition 2.5.12. □

**Theorem 2.8.15.** *The algebra* $\mathbf{H}$ *admits a cellular basis as in Theorem 2.6.12. The data in Definition 2.6.8 can be chosen such that, for* $\lambda \vdash n$ *and* $\mathfrak{s},\mathfrak{t} \in M(\lambda)$, *we have*

$$\mathbf{C}^\lambda_{\mathfrak{s},\mathfrak{t}} = c^{\mathfrak{st}}_{w,\lambda}\, C_w = \pm C_w, \qquad \textit{where} \qquad w = w_\lambda(\mathfrak{s},\mathfrak{t}).$$

*Furthermore, the partial order* $\trianglelefteq_L$ *is given by the dominance order on partitions; we have* $\mathscr{N}_\lambda = \mathbf{H}_K(\trianglelefteq_L)$ *and* $\hat{\mathscr{N}_\lambda} = \mathbf{H}_K(\triangleleft_L)$.

*Proof.* Since (♣) and (♠) hold, we can apply Theorem 2.6.12 and so $\mathbf{H}$ admits a cellular basis where, by Definition 2.6.8, we have

$$\mathbf{C}^\lambda_{\mathfrak{s},\mathfrak{t}} = \sum_{w\in W} \sum_{\mathfrak{u}\in M(\lambda)} \beta^\lambda_{\mathfrak{tu}}\, c^{\mathfrak{us}}_{w^{-1},\lambda}\, C_w.$$

Now, since $c^{\mathfrak{st}}_{w,\lambda} = c^{\mathfrak{ts}}_{w^{-1},\lambda}$ for all $\mathfrak{s},\mathfrak{t} \in M(\lambda)$, we have $\bar{\rho}^\lambda(t_{w^{-1}}) = \bar{\rho}^\lambda(t_w)^{\mathrm{tr}}$ for all $w \in W$. Hence, we can take for $B^\lambda = (\beta^\lambda_{\mathfrak{st}})$ the identity matrix. Then the above sum reduces to $\mathbf{C}^\lambda_{\mathfrak{s},\mathfrak{t}} = c^{\mathfrak{st}}_{w,\lambda}\, C_w$, as required. Finally, by Corollary 2.8.14, the partial order $\trianglelefteq_L$ in Theorem 2.6.12 coincides with the dominance order. The identities $\mathscr{N}_\lambda = \mathbf{H}_K(\trianglelefteq_L)$, $\hat{\mathscr{N}_\lambda} = \mathbf{H}_K(\triangleleft_L)$ now follow from Theorem 2.8.12(b) and (c). □

*Remark 2.8.16.* This result was first stated by Graham and Lehrer [144, Example 1.2], but the argument is very sketchy, especially concerning the ordering $\trianglelefteq_L$. Some more details are contained in Graham's thesis [143, Example 4.3] and Williamson's Honours essay [295]. As far as we are aware, the first elementary proof of the characterisation of $\trianglelefteq_L$ in terms of the dominance order appeared in [107]. In Remark 2.8.18, we show how the signs in Theorem 2.8.15 can be fixed. A completely different construction of a cellular basis (with respect to the above ordering on $\Lambda$) is due to Murphy [256], [257]; the equivalence of the two constructions is shown in [107].

**2.8.17.** By Corollary 2.8.14 and Lemma 2.5.9, the Kazhdan–Lusztig cells of $W = \mathfrak{S}_n$ are given by Theorem 1.7.10. So, for any $\lambda \vdash n$, the following hold.

(a1) The set $\mathscr{F}_\lambda = \{w_\lambda(\mathfrak{s},\mathfrak{t}) \mid \mathfrak{s},\mathfrak{t} \in M(\lambda)\}$ is a two-sided Kazhdan–Lusztig cell.
(a2) For $\mathfrak{t} \in M(\lambda)$, the set $\{w_\lambda(\mathfrak{s},\mathfrak{t}) \mid \mathfrak{s} \in M(\lambda)\}$ is a left Kazhdan–Lusztig cell.
(a3) For $\mathfrak{s} \in M(\lambda)$, the set $\{w_\lambda(\mathfrak{s},\mathfrak{t}) \mid \mathfrak{t} \in M(\lambda)\}$ is a right Kazhdan–Lusztig cell.

In particular, we see that (♦) holds. We also obtain the following result originally due to Kazhdan and Lusztig [195, Theorem 1.4]: for any left Kazhdan–Lusztig cell $\mathfrak{C}$ of $W$, we have

(b) $$[\mathfrak{C}]_1 \in \mathrm{Irr}_{\mathbb{K}}(W) \qquad \text{and} \qquad [\mathfrak{C}]_1 \cong E^\lambda \Leftrightarrow \mathfrak{C} \subseteq \mathscr{F}_\lambda.$$

Indeed, let $\mathfrak{C}$ be a left cell and $\lambda \in \Lambda$ be such that $m(\mathfrak{C},\lambda) > 0$. Then $\mathfrak{C} \subseteq \mathscr{F}_\lambda$ by Lemma 2.2.4. Hence, $\mathfrak{C}$ is equal to a set as in (a2). But then $|\mathfrak{C}| = |M(\lambda)| = d_\lambda = \dim E^\lambda$ and so we must have $E^\lambda \cong [\mathfrak{C}]_1$. Finally, assume that $\mathfrak{C}$ is a left cell such that $\mathfrak{C} \subseteq \mathscr{F}_\lambda$. Then the same argument shows that $[\mathfrak{C}]_1 \cong E^\lambda$. (If we had $[\mathfrak{C}]_1 \cong E^\mu$, where $\mu \neq \lambda$, then $\mathfrak{C} \subseteq \mathscr{F}_\mu$, which is a contradiction.) Thus, (b) is proved. We can now also apply the discussion in Example 2.7.4, which shows that, for each $\lambda \vdash n$, the balanced representation $\rho^\lambda$ can be chosen such that

(c) $W(\lambda) = [\mathfrak{C}_\lambda]_A$, where $\mathfrak{C}_\lambda \subseteq \mathscr{F}_\lambda$ is a fixed left Kazhdan–Lusztig cell.

See also McDonough and Pallikaros [251], where the above cell modules are identified with the original "Specht modules" of Dipper and James [62].

*Remark 2.8.18.* Once Theorem 2.8.12, Corollary 2.8.14 and 2.8.17 are established, it is actually not too difficult to show that **P1–P15** hold for $W = \mathfrak{S}_n$; see [107, §5], [121, §4]. Furthermore, one can even show a tiny piece of "positivity" by elementary methods; namely, the fact that $\gamma_{x,y,z} \geqslant 0$ for all $x,y,z \in W$; see [107, Theorem 5.10]. (Recall that $\gamma_{x,y,z} = (-1)^{l(x)+l(y)+l(z)} c_{x,y,z}$; see Remark 2.3.6.) The argument relies on basic properties of the "Knuth–Robinson–Schensted correspondence" and the Kazhdan–Lusztig "star operations"; see Kazhdan and Lusztig [195, §5], Knuth [205, §5.1.4] and Ariki [8]. We will not go into any more detail here, as these results are not needed for the further discussions in this book.

Let us just explain how the signs in Theorem 2.8.15 can be fixed, *assuming* that **P1** and **P4** hold and that $\gamma_{x,y,z} \geqslant 0$ for all $x,y,z \in W$ (which is also known to be the case by 2.4.1(a)). This is done as follows. Choosing the balanced representation $\rho^\lambda$ as in Proposition 1.8.9, each coefficient $c_{w,\lambda}^{\mathfrak{s}\mathfrak{t}}$ is equal to a structure constant $\tilde{\gamma}_{w,x,y}$ for suitable $x,y \in \mathscr{F}_\lambda$. Now, by Proposition 2.3.16, Remark 2.3.5 and Proposition 2.3.14, we have

$$\tilde{\gamma}_{w,x,y} = (-1)^{l(w)+l(x)+l(y)} \gamma_{w,x,y} = (-1)^{\mathbf{a}(y)} \gamma_{w,x,y} \qquad \text{and} \qquad \mathbf{a}(y) = \mathbf{a}_\lambda.$$

Since $\gamma_{w,x,y} \geqslant 0$ and $c_{w,\lambda}^{\mathfrak{s}\mathfrak{t}} \in \{0, \pm 1\}$, we deduce that

$$c_{w,\lambda}^{\mathfrak{s}\mathfrak{t}} = (-1)^{\mathbf{a}_\lambda}, \qquad \text{where} \qquad w = w_\lambda(\mathfrak{s},\mathfrak{t}).$$

Arguing as in the proof of Theorem 2.8.15, we now let $B^\lambda = (\beta_{\mathfrak{s}\mathfrak{t}}^\lambda)$ be equal to $(-1)^{\mathbf{a}_\lambda}$ times the identity matrix. Then we obtain $\mathbf{C}_{\mathfrak{s},\mathfrak{t}}^\lambda = C_w$, as required.

For a further discussion of the combinatorics involved in the above constructions (Knuth–Robinson–Schensted correspondence, etc.), we refer the reader to the references cited in Remark 2.8.18. In a somewhat different context, we will have more to say about the combinatorics of Young tableaux in Section 3.5.

**2.8.19.** Having dealt with $W = \mathfrak{S}_n$, it is natural to ask what happens with the other cases in 1.7.6. So, let $W$ be of type $B_n$ and $L \colon W \to \Gamma$ a weight function given by

$B_n$

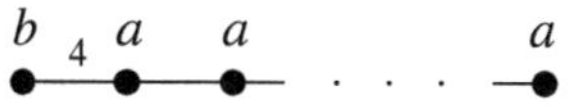

where $a, b > 0$ and $b \notin \{a, 2a, \ldots, (n-1)a\}$. Recall that $\Lambda$ is the set of all pairs of partitions $(\lambda, \mu)$ such that $|\lambda| + |\mu| = n$; see Example 1.3.9. As in the proof of Theorem 2.8.15, one sees that, for any $(\lambda, \mu) \in \Lambda$ and $\mathfrak{s}, \mathfrak{t} \in M(\lambda, \mu)$, we have

$$\mathbf{C}_{\mathfrak{s},\mathfrak{t}}^{(\lambda,\mu)} = \pm C_w, \qquad \text{where} \qquad w = w_{(\lambda,\mu)}(\mathfrak{s}, \mathfrak{t}).$$

However, property (♠) is not known in general, so we cannot conclude that the above elements form a cellular basis of $\mathbf{H}$.

Complete results are available in the *asymptotic case*, where $b > (n-1)a > 0$; denote by $L_{\text{asy}}$ the weight function in this case. Then **P1–P15** hold for $W, L_{\text{asy}}$ by the series of papers by Bonnafé, Geck, and Iancu [21], [26], [108], [114], [121]. Furthermore, as already mentioned in Example 2.2.17, we have

$$(\lambda, \mu) \trianglelefteq_{L_{\text{asy}}} (\lambda', \mu') \qquad \Leftrightarrow \qquad (\lambda, \mu) \trianglelefteq (\lambda', \mu').$$

Arguing as in 2.8.17, we obtain the following result originally due to Bonnafé and Iancu [26, Prop. 7.9]: for any left Kazhdan–Lusztig cell $\mathfrak{C}$ of $W$ (with respect to $L_{\text{asy}}$), we have

$$[\mathfrak{C}]_1 \in \operatorname{Irr}_{\mathbb{K}}(W) \qquad \text{and} \qquad [\mathfrak{C}]_1 \cong E^{\lambda} \Leftrightarrow \mathfrak{C} \subseteq \mathscr{F}_{(\lambda,\mu)}.$$

Furthermore, for $(\lambda, \mu) \in \Lambda$, the balanced representation $\rho^{(\lambda,\mu)}$ can be chosen such that $W(\lambda, \mu) = [\mathfrak{C}_{(\lambda,\mu)}]_A$, where $\mathfrak{C}_{(\lambda,\mu)} \subseteq \mathscr{F}_{(\lambda,\mu)}$ is a fixed left Kazhdan–Lusztig cell.

A completely different construction of a cellular basis is due to Dipper, James and Murphy [68]; but, by [124], the above cell modules in the asymptotic case are naturally isomorphic to the "Specht modules" of [68]. See also Chlouveraki, Gordon and Griffeth [49] for further realisations of these modules. The construction of [68] has been further generalised to Ariki–Koike algebras; see Dipper, James and Mathas [67] (and also Graham and Lehrer [144, §5] for a slightly different approach). We will describe these results on Ariki–Koike algebras in Section 5.3.

# Chapter 3
# Specialisations and Decomposition Maps

In Chapter 2 we constructed a "cell datum" for the Iwahori–Hecke algebra $H_k = H_k(W,S,\{\xi_s\})$ associated with a finite Coxeter group $W$. This yields a natural parametrisation of $\mathrm{Irr}(H_k)$ in terms of a certain subset $\Lambda_k^\circ \subseteq \Lambda$, where $\Lambda$ is an indexing set for $\mathrm{Irr}(W)$. The main theme of this chapter is that $\Lambda_k^\circ$ can be characterised in an entirely different way, using Brauer's classical theory of decomposition numbers. This alternative approach is important for (at least) three reasons:

- It applies in more general situations, e.g. not only for Iwahori–Hecke algebras associated with finite Coxeter groups, but also for Ariki–Koike algebras.
- It does not rely on the validity of the deep properties (♠) and (♣) in 2.5.3. For instance, we will be able to determine a subset $\Lambda_k^\circ$ as above for type $B_n$ with unequal parameters, where (♠) and (♣) are not yet known to hold in general.
- Brauer's decomposition numbers are independent of any "cellular structure", and efficient geometric and algorithmic methods are available for their computation.

After discussing some foundations in Section 3.1, we define in Section 3.2 a key concept for this book, the so-called "canonical basic sets". This may be seen as an axiomatic version of those features of a "cellular algebra" which do not explicitly refer to an underlying "cellular basis". We then have to deal with the various types of specialisations of our ground ring $A$ (which typically is a ring of Laurent polynomials with integer coefficients). A primary role will be played by those specialisations which arise from height 1 prime ideals. This leads us to consider so-called "principal specialisations" in Section 3.3. In this setting, one can show that canonical basic sets for so-called "blocks of defect 1" have a particularly simple description.

In Section 3.4, we develop a general strategy for determining canonical basic sets for algebras of classical type. A model case is given by $W \cong \mathfrak{S}_n$, and this will be treated in detail in Section 3.5. In Section 3.6 we prove a basic factorisation result which shows that any decomposition map always passes through a "principal specialisation". James's conjecture and its generalisations predict that "nothing happens" in the second step under certain precise conditions; see Section 3.7. Finally, Section 3.8 is devoted to the proof of a remarkable result which characterises the partition of $\mathrm{Irr}_{\mathbb{K}}(W)$ by Kazhdan–Lusztig cells in terms of certain "blocks".

M. Geck, N. Jacon, *Representations of Hecke Algebras at Roots of Unity*, Algebra and Applications 15, DOI 10.1007/978-0-85729-716-7_3,

## 3.1 Grothendieck Groups and Decomposition Maps

We begin with a brief discussion of decomposition maps in a general setting appropriate for our applications to Iwahori–Hecke algebras, following [132, Chap. 7].

**3.1.1.** Let $A$ be an integral domain and $\underline{H}$ be an associative $A$-algebra (with identity $1_{\underline{H}}$) which is finitely generated and free over $A$. Let $K$ be the field of fractions of $A$ and $\theta\colon A \to k$ be a ring homomorphism into a field $k$ such that $k$ is the field of fractions of $\theta(A)$. By extension of scalars, we obtain a $K$-algebra $\underline{H}_K := K \otimes_A \underline{H}$ (via the inclusion $A \subseteq K$) and a $k$-algebra $\underline{H}_k := k \otimes_A \underline{H}$ (where $k$ is regarded as an $A$-module via $\theta$). We wish to relate the representations of $\underline{H}_k$ with those of $\underline{H}_K$.

Denote by $R_0(\underline{H}_k)$ the Grothendieck group of finite-dimensional representations of $\underline{H}_k$. Recall that $R_0(\underline{H}_k)$ is the abelian group generated by expressions $[V]$, one for each finite-dimensional $\underline{H}_k$-module $V$ (up to isomorphism), with relations $[V] = [V'] + [V'']$ for each short exact sequence $\{0\} \to V' \to V \to V'' \to \{0\}$ of $\underline{H}_k$-modules. Two $\underline{H}_k$-modules $V, V'$ give rise to the same element in $R_0(\underline{H}_k)$ if and only if $V, V'$ have the same composition factors, counting multiplicities. It follows that $R_0(\underline{H}_k)$ is free abelian, with basis given by $\operatorname{Irr}(\underline{H}_k)$, the set of simple $\underline{H}_k$-modules (up to isomorphism). Let $R_0^+(\underline{H}_k)$ be the submonoid of $R_0(\underline{H}_k)$ consisting of elements $[V]$ where $V$ is an $\underline{H}_k$-module. Let $X$ be an indeterminate over $k$ and $\operatorname{Maps}(\underline{H}, k[X])$ be the $k$-algebra of maps from $\underline{H}$ to $k[X]$. Following [132, §7.4], [134], we define

$$\mathfrak{p}_k\colon R_0^+(\underline{H}_k) \to \operatorname{Maps}(\underline{H}, k[X])$$

by sending $[V] \in R_0^+(\underline{H}_k)$ to the map which assigns to each $h \in \underline{H}$ the characteristic polynomial of $1 \otimes h \in \underline{H}_k$ in its action on $V$. Considering $\operatorname{Maps}(\underline{H}, k[X])$ as a monoid with respect to multiplication, the map $\mathfrak{p}_k$ is a monoid homomorphism.

These definitions apply, in particular, to the inclusion $A \subseteq K$. Hence, we also have a monoid homomorphism $\mathfrak{p}_K\colon R_0^+(\underline{H}_K) \to \operatorname{Maps}(\underline{H}, K[X])$.

**Theorem 3.1.2 (Cf. [98, §2], [132, 7.4.3], [134, 2.11]).** *In the setting of 3.1.1, assume that $A$ is integrally closed in $K$ and that $\underline{H}_k$ is split. Then the following hold:*

*(a) The image of $\mathfrak{p}_K\colon R_0^+(\underline{H}_K) \to \operatorname{Maps}(\underline{H}, K[X])$ lies in $\operatorname{Maps}(\underline{H}, A[X])$. Thus, by restriction, we obtain a monoid homomorphism $\mathfrak{p}_A\colon R_0^+(\underline{H}_K) \to \operatorname{Maps}(\underline{H}, A[X])$.*

*(b) There exists a unique additive map $d_\theta^+\colon R_0^+(\underline{H}_K) \to R_0^+(\underline{H}_k)$ such that the following diagram is commutative:*

$$\begin{array}{ccc} R_0^+(\underline{H}_K) & \xrightarrow{\ \mathfrak{p}_A\ } & \operatorname{Maps}(\underline{H}, A[X]) \\ {\scriptstyle d_\theta^+}\downarrow & \circlearrowleft & \downarrow{\scriptstyle t_\theta} \\ R_0^+(\underline{H}_k) & \xrightarrow{\ \mathfrak{p}_k\ } & \operatorname{Maps}(\underline{H}, k[X]) \end{array}$$

*where $t_\theta\colon \operatorname{Maps}(\underline{H}, A[X]) \to \operatorname{Maps}(\underline{H}, k[X])$ is the natural map obtained by applying $\theta$ to the coefficients of the polynomials in $A[X]$.*

*(c) We have $d_\theta([K \otimes_A N]) = [k \otimes_A N]$ for any $\underline{H}$-module $N$ which is finitely generated and free over $A$.*

The assumption that $A$ is integrally closed in $K$ is used in the proof of Theorem 3.1.2(a). The assumption that $\underline{H}_k$ is split implies that $\mathfrak{r}_k$ is injective (a result originally due to Brauer and Nesbitt; see [132, 7.3.2, 7.3.3]). This is needed to establish the uniqueness part in Theorem 3.1.2(b). For further details, see [132, §7.4].

**Example 3.1.3.** Assume that $A$ is a *discrete valuation ring*; that is, $A$ is a noetherian local ring of positive dimension whose maximal ideal is principal; see [248, §11]. Let $k$ be the residue field and $\theta\colon A \to k$ be the canonical map. This is the usual setting in the representation theory of finite groups, where $\underline{H} = AG$ is the group algebra of a finite group $G$; e.g. see Curtis and Reiner [53, §16] and Feit [83, §I.17].

Note that a valuation ring is always integrally closed in its field of fractions. Furthermore, since $A$ is a principal ideal domain, a standard argument (see, for example, [132, 7.3.7]) shows that, for every finite-dimensional $\underline{H}_K$-module $V$, there exists an $\underline{H}$-module $N$, which is finitely generated and free over $A$, such that $V \cong K \otimes_A N$.

See Du et al. [73, §1] for a more general setting where $A$ is a regular local ring of Krull dimension at most 2.

**Definition 3.1.4.** Let $d_\theta$ be the canonical extension of $d_\theta^+$ to a group homomorphism $d_\theta\colon R_0(\underline{H}_K) \to R_0(\underline{H}_k)$. Then $d_\theta$ is called the *decomposition map* associated with the specialisation $\theta\colon A \to k$. It is determined by the equation

$$d_\theta([V]) = \sum_{M \in \mathrm{Irr}(\underline{H}_k)} d_{V,M}\,[M] \qquad \text{for any} \qquad V \in \mathrm{Irr}(\underline{H}_K),$$

where each $d_{V,M}$ is a non-negative integer. The matrix

$$D_\theta = \big(d_{V,M}\big)_{V \in \mathrm{Irr}(\underline{H}_K),\, M \in \mathrm{Irr}(\underline{H}_k)}$$

is called the *decomposition matrix* associated to $\theta$.

*Remark 3.1.5.* Assume that $\underline{H}_k$ is split and that $\hat{k} \supseteq k$ is a field extension. Then $\underline{H}_{\hat{k}}$ is split and we have a bijection (see [132, 7.3.4]):

(a) $$\mathrm{Irr}(\underline{H}_k) \to \mathrm{Irr}(\underline{H}_{\hat{k}}), \qquad M \mapsto \hat{k} \otimes_k M.$$

So there is no serious loss of generality involved in our standing assumption that $k$ should be the field of fractions of $\theta(A)$.

More generally, assume that $A$ is contained in a bigger integral domain $\hat{A}$ and $\theta$ extends to a homomorphism $\hat{\theta}\colon \hat{A} \to \hat{k}$ such that $\hat{k}$ is the field of fractions of $\hat{\theta}(\hat{A})$ and $\hat{\theta}|_A = \iota \circ \theta$ for a suitable embedding $\iota\colon k \hookrightarrow \hat{k}$. Let $\hat{K}$ be the field of fractions of $\hat{A}$. Assume that $\underline{H}_k$ is split and that either $K = \hat{K}$ or $\underline{H}_K$ is split. Also assume that $A$ is integrally closed in $K$ and $\hat{A}$ is integrally closed in $\hat{K}$. Then, formally, we have two decomposition maps $d_\theta\colon R_0(\underline{H}_K) \to R_0(\underline{H}_k)$ and $d_{\hat{\theta}}\colon R_0(\underline{H}_{\hat{K}}) \to R_0(\underline{H}_{\hat{k}})$. By (a), the extensions of scalars from $K$ to $\hat{K}$ and from $k$ to $\hat{k}$ (via $\iota$) give rise to bijections

$$\mathrm{Irr}(\underline{H}_K) \xrightarrow{1-1} \mathrm{Irr}(\underline{H}_{\hat{K}}), \qquad V \mapsto \hat{V} := \hat{K} \otimes_K V;$$
$$\mathrm{Irr}(\underline{H}_k) \xrightarrow{1-1} \mathrm{Irr}(\underline{H}_{\hat{k}}), \qquad M \mapsto \hat{M} := \hat{k} \otimes_k M.$$

By the characterisation of decomposition maps in Theorem 3.1.2(b), we have

(b) $\quad d_{V,M} = \hat{d}_{\hat{V},\hat{M}} \quad$ for all $V \in \mathrm{Irr}(\underline{H}_K)$ and $M \in \mathrm{Irr}(\underline{H}_k)$,

where $D_\theta = (d_{V,M})$ is the decomposition matrix associated with $\theta$ and $D_{\hat{\theta}} = (\hat{d}_{\hat{V},\hat{M}})$ is the decomposition matrix associated with $\hat{\theta}$. Thus, the decomposition numbers of $\underline{H}$ remain unchanged under "controlled" extensions of $\theta$.

**Example 3.1.6.** Let us keep the assumptions of Theorem 3.1.2. There exists a valuation ring $\hat{A} \subseteq K$ such that $A \subseteq \hat{A}$ and $A \cap \mathfrak{p} = \ker(\theta)$, where $\mathfrak{p}$ is the unique maximal ideal of $\hat{A}$. (See, for example, [138, §5]; note that, in general, $\hat{A}$ will not be a discrete valuation ring.) Let $\hat{k}$ be the residue field of $\hat{A}$. Then $\theta: A \to k$ extends to a homomorphism $\hat{\theta}: \hat{A} \to \hat{k}$ and there is an inclusion $\iota: k \hookrightarrow \hat{k}$ such that $\hat{\theta}|_A = \iota \circ \theta$. Thus, we are in the setting of Remark 3.1.5 where $\hat{K} = K$; note that any valuation ring is integrally closed in its field of fractions.

Now let $V \in \mathrm{Irr}(\underline{H}_K)$ and $M \in \mathrm{Irr}(\underline{H}_k)$. Since $\hat{A}$ is a valuation ring, we can find an $(\hat{A} \otimes_A \underline{H})$-module $N$, which is finitely generated and free over $\hat{A}$, such that $K \otimes_{\hat{A}} N \cong V$ (see [138, 5.4]). Then, by Theorem 3.1.2(c) and 3.1.5(b), we have

(a) $\quad d_{V,M} = \hat{d}_{\hat{V},\hat{M}} =$ multiplicity of $\hat{M}$ as a composition factor of $\hat{k} \otimes_{\hat{A}} N$.

If, for example, $A$ is a Dedekind domain, then the localisation of $A$ in $\ker(\theta)$ will be a discrete valuation ring, in which case we can take $\hat{A}$ to be that localisation. In our further study of Iwahori–Hecke algebras, this will not always be the case.

**3.1.7.** In a number of interesting situations, the decomposition matrix $D_\theta$ turns out to have a triangular shape with 1 on the diagonal. We formalise this as follows. In the setting of Theorem 3.1.2, assume that we are given a partial order $\sqsubseteq$ on $\mathrm{Irr}(\underline{H}_K)$ such that the following three conditions are satisfied:

- Given $M \in \mathrm{Irr}(\underline{H}_k)$, let $\mathscr{S}_\theta(M) := \{V \in \mathrm{Irr}(\underline{M}_K) \mid d_{V,M} \neq 0\}$. Then the set $\mathscr{S}_\theta(M)$ contains a unique minimal element (with respect to $\sqsubseteq$), which we denote by $V_M$.
- The map $\mathrm{Irr}(\underline{H}_k) \to \mathrm{Irr}(\underline{H}_K)$, $M \mapsto V_M$, is injective.
- We have $d_{V_M,M} = 1$ for all $M \in \mathrm{Irr}(\underline{H}_k)$.

If this holds, let $\underline{B}_\theta := \{V_M \mid M \in \mathrm{Irr}(\underline{H}_k)\}$. Thus, we obtain a bijection

$$\mathrm{Irr}(\underline{H}_k) \stackrel{1-1}{\longleftrightarrow} \underline{B}_\theta \subseteq \mathrm{Irr}(\underline{H}_K),$$

and $\underline{B}_\theta$ will be called a *canonical basic set* for $\underline{H}_k$ (with respect to $\sqsubseteq$). Note that, if a canonical basic set exists, then it is uniquely determined by the above conditions. In this situation, let us choose a labelling $\underline{B}_\theta = \{V_1, \ldots, V_r\}$ such that "$V_i \sqsubset V_j \Rightarrow i < j$". Then $\mathrm{Irr}(\underline{H}_k) = \{M_1, \ldots, M_r\}$, where $V_i = V_{M_i}$ for $1 \leqslant i \leqslant r$. Hence, the matrix

$$D_\theta^\circ := (d_{V_i,M_j})_{1 \leqslant i,j \leqslant r}$$

will have a square lower triangular shape with 1 on the diagonal.

**3.1.8.** In the setting of Theorem 3.1.2, the *Brauer graph* of $\underline{H}$ (with respect to $\theta$) is defined as follows: it has vertices labelled by $\mathrm{Irr}(\underline{H}_K)$ and there is an edge joining two vertices, labelled by $V \neq V'$ say, if there exists some $M \in \mathrm{Irr}(\underline{H}_k)$ such that $d_{V,M} \neq 0$ and $d_{V',M} \neq 0$. Considering the connected components of this graph, we obtain a partition of $\mathrm{Irr}(\underline{H}_K)$ into pieces called the *$\theta$-blocks* of $\mathrm{Irr}(\underline{H}_K)$. Let

$$\mathrm{Irr}(\underline{H}_K) = \mathrm{Irr}_1(\underline{H}_K) \sqcup \mathrm{Irr}_2(\underline{H}_K) \sqcup \cdots \sqcup \mathrm{Irr}_r(\underline{H}_K)$$

be the partition into $\theta$-blocks. For each $i$, let $\mathrm{Irr}_i(\underline{H}_k)$ be the set of all $M \in \mathrm{Irr}(\underline{H}_k)$ such that $d_{V,M} \neq 0$ for some $V \in \mathrm{Irr}_i(\underline{H}_K)$. Then we also have a partition

$$\mathrm{Irr}(\underline{H}_k) = \mathrm{Irr}_1(\underline{H}_k) \sqcup \mathrm{Irr}_2(\underline{H}_k) \sqcup \cdots \sqcup \mathrm{Irr}_r(\underline{H}_k).$$

Ordering the rows and columns of $D_\theta$ accordingly, we obtain a block shape

$$D_\theta = \begin{pmatrix} D_{\theta,1} & 0 & \dots & 0 \\ 0 & D_{\theta,2} & \ddots & \vdots \\ \vdots & \ddots & \ddots & 0 \\ 0 & \dots & 0 & D_{\theta,r} \end{pmatrix},$$

where $D_{\theta,i}$ has rows and columns labelled by the elements of $\mathrm{Irr}_i(\underline{H}_K)$ and $\mathrm{Irr}_i(\underline{H}_k)$ respectively. Thus, in order to determine $D_\theta$, we can proceed block by block. Of course, this remark will only be useful if there is more than one block and if there are alternative characterisations of blocks which do not rely on using $d_\theta$. In Lemma 3.1.10, we provide a basic criterion in term of *central characters*. (In Section 3.8, we shall discuss further characterisations in terms of *central idempotents*.)

**Definition 3.1.9.** Assume that $A$ is integrally closed in $K$ and $\underline{H}_K$ is split semisimple. Denote by $Z(\underline{H})$ the centre of $\underline{H}$. Let $V \in \mathrm{Irr}(\underline{H}_K)$ and $z \in Z(\underline{H})$. Then $z$ also lies in the centre of $\underline{H}_K$. Hence, by Schur's lemma, $z$ acts as multiplication by a scalar on $V$, which we denote by $\omega_V(z)$. By [132, 7.3.8], we have $\omega_V(z) \in A$. Thus, we obtain an $A$-algebra homomorphism $\omega_V : Z(\underline{H}) \to A$, called the *central character* of $V$.

**Lemma 3.1.10.** *In the setting of Theorem 3.1.2, assume that $\underline{H}_K$ is split semisimple and let $V, V' \in \mathrm{Irr}(\underline{H}_K)$. If $V, V'$ belong to the same $\theta$-block, then $\theta(\omega_V(z)) = \theta(\omega_{V'}(z))$ for all $z \in Z(\underline{H})$.*

(The converse holds under additional assumptions on $A$; see Example 3.8.5.)

*Proof.* It is enough to consider the case where $V, V'$ are directly linked in the Brauer graph of $\underline{H}$; that is, we have $d_{V,M} \neq 0$ and $d_{V',M} \neq 0$ for some $M \in \mathrm{Irr}(\underline{H}_k)$. Then, using the set-up and the notation in Example 3.1.6, we have

$$\begin{aligned} d_{V,M} &= \text{multiplicity of } \hat{M} \text{ as a composition factor of } \bar{N} := \hat{k} \otimes_{\hat{A}} N, \\ d_{V',M} &= \text{multiplicity of } \hat{M} \text{ as a composition factor of } \bar{N}' := \hat{k} \otimes_{\hat{A}} N', \end{aligned}$$

where $N, N'$ are $(\hat{A} \otimes_A \underline{H})$-modules (finitely generated and free over $\hat{A}$) such that $K \otimes_{\hat{A}} N \cong V$ and $K \otimes_{\hat{A}} N' \cong V'$. Now let $z \in Z(\underline{H})$. Then $z$ acts as $\omega_V(z)$ times the identity on $V$ and, hence, also as $\omega_V(z)$ times the identity on $N$; recall that $\omega_V(z) \in A$. Consequently, the central element $\bar{z} \in \underline{H}_{\hat{k}}$ (obtained by extending scalars from $A$ to $\hat{k}$ via $\hat{\theta}|_A$) acts as $\theta(\omega_V(z))$ times the identity on $\bar{N}$. Similarly, $\bar{z}$ acts as $\theta(\omega_{V'}(z))$ times the identity on $\bar{N}'$. The condition $d_{V,M} \neq 0$ implies that $\hat{M}$ is a composition factor of $\bar{N}$. So $\bar{z}$ must act as $\theta(\omega_V(z))$ times the identity on $\hat{M}$. Similarly, the condition $d_{V',M} \neq 0$ implies that $\bar{z}$ acts as $\theta(\omega_{V'}(z))$ times the identity on $\hat{M}$. Thus, we conclude that $\theta(\omega_V(z)) = \theta(\omega_{V'}(z))$, as required. □

**Example 3.1.11.** Let us keep the assumptions of Lemma 3.1.10. Then, by the Artin–Wedderburn theorem, the split semisimple algebra $\underline{H}_K$ is a direct sum of simple two-sided ideals where the direct factors correspond to the various irreducible representations of $\underline{H}_K$ (see [53, §3B]). Consequently, we can write uniquely

$$1_{\underline{H}_K} = \sum_{V \in \mathrm{Irr}(\underline{H}_K)} e_V,$$

where each $e_V$ acts as the identity on $V$ and as zero on any simple module not isomorphic to $V$; furthermore, $e_V$ is an idempotent (that is, we have $e_V^2 = e_V$) in the centre of $\underline{H}_K$. Assume now that $V \in \mathrm{Irr}(\underline{H}_K)$ satisfies the following condition:

$$(*) \qquad \alpha e_V \in \underline{H} \qquad \text{for some } \alpha \in A \text{ such that } \theta(\alpha) \neq 0.$$

Then we claim that $\{V\}$ is a $\theta$-block of $\mathrm{Irr}(\underline{H}_K)$. Indeed, assume that $V' \in \mathrm{Irr}(\underline{H}_K)$ belongs to the same $\theta$-block as $V$, where $V \not\cong V'$. Let $z := \alpha e_V \in Z(\underline{H})$. We have $\omega_V(z) = \alpha$ and $\omega_{V'}(z) = \alpha\omega_{V'}(e_V) = 0$. This implies that $\theta(\omega_V(z)) = \theta(\alpha) \neq 0$ and $\theta(\omega_{V'}(z)) = 0$, contradicting Lemma 3.1.10.

**3.1.12.** Recall the basic set-up from Section 1.2. Let $W$ be a finite Coxeter group with generating set $S$. Let $L \colon W \to \Gamma$ be a weight function where $\Gamma$ admits a monomial order $\leqslant$ such that $L(s) \geqslant 0$ for all $s \in S$. Let $R \subseteq \mathbb{C}$ be a subring such that $\mathbb{Z}_W \subseteq R$; let $\mathbb{K}$ be the field of fractions of $R$. We shall now also assume that

- the ring $R$ is noetherian and integrally closed in $\mathbb{K}$;
- the group $\Gamma$ is finitely generated.

As discussed in 1.1.10, these conditions imply that $A := R[\Gamma]$ is the ring of Laurent polynomials in finitely many variables over $R$; in particular, $A$ is noetherian.

Let $\mathbf{H} = \mathbf{H}_A(W, S, L)$ be the corresponding generic Iwahori–Hecke algebra with parameters $\{v_s \mid s \in S\}$, where $v_s = \varepsilon^{L(s)}$ for $s \in S$. Let $\mathbb{K}$ be the field of fractions of $R$ and $K$ be the field of fractions of $A$. Also recall from 1.2.1 that $\mathbf{H}_K = K \otimes_A \mathbf{H}$ is a split semisimple algebra which is abstractly isomorphic to the group algebra of $W$ over $K$. We have $\mathrm{Irr}(\mathbf{H}_K) = \{E_\varepsilon^\lambda \mid \lambda \in \Lambda\}$, where $\Lambda$ is an indexing set for $\mathrm{Irr}_{\mathbb{K}}(W)$.

Let $\theta \colon A \to k$ be a ring homomorphism, where $k$ is a field such that $k$ is the field of fractions of $\theta(A)$. We denote the corresponding specialised algebra by $\mathbf{H}_k$, or by $\mathbf{H}_{k,\theta}$ if it is necessary to specify $\theta$. We can now use our results on "cell data" from Chapter 2 to address the validity of the assumptions in Theorem 3.1.2.

**Table 3.1** Conditions for $R$ to be $L_0$-good (extracted from Table 1.4, p. 33)

| Type | Condition on $R$ |
|---|---|
| $A_{n-1}, B_n$ | no condition |
| $G_2, D_n$ | $2 \in R^\times$ |
| $F_4, E_6, E_7$ | $2,3 \in R^\times$ |
| $E_8$ | $2,3,5 \in R^\times$ |
| $I_2(m)$ $(m \geqslant 5, m \neq 6)$ | $m \in R^\times$ and $2\cos(2\pi/m) \in R$ |
| $H_3$ | $2,5 \in R^\times$ and $2\cos(2\pi/5) = \frac{1}{2}(-1+\sqrt{5}) \in R$ |
| $H_4$ | $2,3,5 \in R^\times$ and $2\cos(2\pi/5) = \frac{1}{2}(-1+\sqrt{5}) \in R$ |

**Lemma 3.1.13.** *In the above setting, assume that $R$ is $L_0$-good (see Table 3.1) where $L_0$ is the universal weight function in Example 1.1.9. Then the following hold.*

*(a) The algebra $\mathbf{H}_k$ is split.*
*(b) For every finite-dimensional $\mathbf{H}_K$-module $V$, there exists some $\mathbf{H}$-module $N$ which is free of finite rank over $A$ such that $V \cong K \otimes_A N$.*

*Proof.* Since $R$ is $L_0$-good, so is the ring $A$. Consequently, by Corollary 2.6.16, the algebra $\mathbf{H}$ admits a cellular basis $\{C^\lambda_{\mathfrak{s},\mathfrak{t}}\}$ with respect to some partial order on $\Lambda$. By 2.7.5, the algebra $\mathbf{H}_k$ also admits a cellular basis. So Theorem 2.6.5 shows that $\mathbf{H}_k$ is split. Thus, (a) is proved. Furthermore, since $\mathbf{H}_K$ is semisimple, it is enough to prove (b) assuming that $V$ is irreducible. Let $\{N^\lambda \mid \lambda \in \Lambda\}$ be the cell modules arising from the cell datum for $\mathbf{H}$. As remarked in 2.7.5(a), we have $\mathrm{Irr}(\mathbf{H}_K) = \{K \otimes_A N^\lambda \mid \lambda \in \Lambda\}$. Thus, we have $V \cong K \otimes_A N^\lambda$ for some $\lambda$. □

(It may be true that $\mathbf{H}_k$ is always split, even if $R$ is not $L_0$-good. However, at the moment, we do not see a general argument for proving this.)

**Theorem 3.1.14.** *Let $\mathbf{H} = \mathbf{H}_A(W,S,L)$ be a generic Iwahori–Hecke algebra as in 3.1.12; also assume that $R$ is $L_0$-good. Then, given any specialisation $\theta\colon A \to k$ (where $k$ is the field of fractions of $\theta(A)$), the assumptions of Theorem 3.1.2 are satisfied and so we have a well-defined decomposition map*

$$d_\theta\colon R_0(\mathbf{H}_K) \to R_0(\mathbf{H}_k) \quad \text{such that} \quad d_\theta([K \otimes_A N]) = [k \otimes_A N]$$

*for any $\mathbf{H}$-module $N$ which is free of finite rank over $A$. Furthermore, $d_\theta$ is surjective.*

We shall now write the equations determining $d_\theta$ in the form:

$$d_\theta([E^\lambda_\varepsilon]) = \sum_{M \in \mathrm{Irr}(\mathbf{H}_k)} d_{\lambda,M}\,[M], \qquad \text{where} \qquad d_{\lambda,M} \in \mathbb{Z}_{\geqslant 0}.$$

*Proof.* By Lemma 3.1.13(a), $\mathbf{H}_k$ is split. Furthermore, since $R$ is integrally closed in $\mathbb{K}$, the ring $A$ (which is a ring of Laurent polynomials in finitely many variables) will be integrally closed in $K$; see, for example, [132, Exercise 7.2]. Thus, we can apply Theorem 3.1.2, which yields the existence of $d_\theta$.

It remains to prove that $d_\theta$ is surjective. We use again the fact that $\mathbf{H}$ admits a cell datum; see the proof of Lemma 3.1.13. Let $\{N^\lambda \mid \lambda \in \Lambda\}$ be the corresponding cell modules. Since $\mathbf{H}_K$ is semisimple, we have $\mathrm{Irr}(\mathbf{H}_K) = \{N_K^\lambda := K \otimes_A N^\lambda \mid \lambda \in \Lambda\}$; see 2.7.5. Furthermore, we have an induced cell datum for $\mathbf{H}_k$ and so $\mathrm{Irr}(\mathbf{H}_k) = \{L_k^\mu \mid \mu \in \Lambda_k^\circ\}$, where $\Lambda_k^\circ$ is a certain subset of $\Lambda$. By the definition of $d_\theta$, we have $d_\theta([N_K^\lambda]) = [N_k^\lambda]$, where we write $N_k^\lambda := k \otimes_A N^\lambda$. Thus, the entries of the decomposition matrix $D_\theta$ are given by the multiplicities $(N_k^\lambda : L_k^\mu)$ defined in Theorem 2.6.6. The properties of these multiplicities then show that the square matrix with entries $(N_k^\lambda : L_k^\mu)$ (where $\lambda, \mu \in \Lambda_k^\circ$) has a triangular shape with 1 on the diagonal (for a suitable ordering of the rows and columns). Consequently, $d_\theta$ is surjective. □

**Example 3.1.15.** Without going into too much detail at this point, we mention three situations where decomposition maps and blocks of $\mathbf{H}$ have interesting applications. In all three cases, assume that $R$ is $L_0$-good and $\Gamma = \mathbb{Z}$. Thus, $A = R[v, v^{-1}]$ is the ring of Laurent polynomials in one indeterminate $v = \varepsilon$.

(a) Let $\theta\colon A \to k$ be such that $\mathrm{char}(k) = p > 0$ and $\theta(v) = \pm 1$. Then $\mathbf{H}_k$ is the group algebra of $W$ over $k$. We will see in Example 3.6.5 that $d_\theta$ essentially equals the usual $p$-modular decomposition map of $W$ in the theory of finite groups.
(b) Let $e \geqslant 1$ and $\zeta_{2e} := \exp(\pi\sqrt{-1}/e) \in \mathbb{C}$, the "canonical" primitive root of unity of order $2e$. Let $k = \mathbb{K}[\zeta_{2e}]$ and consider the specialisation $\theta_e\colon A \to k$ such that $\theta_e(v) = \zeta_{2e}$ and $\theta_e(r) = r$ for all $r \in R$. The corresponding decomposition map will be called the *$\Phi_e$-modular decomposition map*. The study of $d_{\theta_e}$ is one of the central themes of this book; see Chapter 4 for applications.
(c) Let $W$ be of type $B_n$; then $R := \mathbb{Z}_W = \mathbb{Z}$ is $L_0$-good. Assume that $L(s) = 1$ for all $s \in S$. (Note that $R$ is *not* $L$-good.) By reduction modulo 2, we obtain a specialisation $\theta\colon A \to k = \mathbb{F}_2(v)$. Here is the matrix $D_\theta$ for type $B_2 = I_2(4)$:

| $E^\lambda$ | $d_{\lambda,M}$ | | | |
|---|---|---|---|---|
| $1_W$ | 1 | 0 | 0 | 0 |
| $\mathrm{sgn}_1$ | 0 | 1 | 0 | 0 |
| $\sigma_1$ | 0 | 1 | 1 | 0 |
| $\mathrm{sgn}_2$ | 0 | 0 | 1 | 0 |
| $\mathrm{sgn}$ | 0 | 0 | 0 | 1 |

$\theta$-blocks:
$\{1_W\}$, $\{\mathrm{sgn}_1, \sigma_1, \mathrm{sgn}_2\}$, $\{\mathrm{sgn}\}$.

We notice that the $\theta$-blocks coincide with the sets $\mathrm{Irr}_{\mathbb{K}}(W \mid \mathscr{F})$ computed in Example 1.7.3! That this is true for any $n$ has been shown by Gyoja [151]. So it seems that there are close relations between decomposition maps and "cells". We will discuss this in more detail in Section 3.8.

**Example 3.1.16.** By 3.1.8, we have a partition of $\mathrm{Irr}(\mathbf{H}_K)$ into $\theta$-blocks. Since $\mathrm{Irr}(\mathbf{H}_K)$ is indexed by $\Lambda$, we can also interpret this as a partition of $\Lambda$ into $\theta$-blocks. Now recall that there is a unique $A$-algebra automorphism $\dagger\colon \mathbf{H} \to \mathbf{H}$ such that $T_s^\dagger = -T_s^{-1}$ for all $s \in S$; see 1.2.6. We now show that, in the setting of Theorem 3.1.14, $\dagger$ induces a symmetry of the Brauer graph of $\mathbf{H}$. This is seen as follows. By extension of scalars, $\dagger$ induces a $K$-algebra automorphism of $\mathbf{H}_K$ and a $k$-algebra automorphism of $\mathbf{H}_k$, which we denote by the same symbol. As in 1.2.6, given any finite-dimensional $\mathbf{H}_K$-module $V$, denote by $V^\dagger$ the $\mathbf{H}_K$-module with the same underlying vector space $V$, but where $h \in \mathbf{H}_K$ acts via $h^\dagger$. A similar operation is defined for finite-dimensional $\mathbf{H}_k$-modules. These operations induce maps

$$\dagger_K^*\colon R_0(\mathbf{H}_K) \to R_0(\mathbf{H}_K) \qquad \text{and} \qquad \dagger_k^*\colon R_0(\mathbf{H}_k) \to R_0(\mathbf{H}_k).$$

Then we have the following commutative diagram:

$$\begin{array}{ccc} R_0^+(\mathbf{H}_K) & \xrightarrow{\quad d_\theta \quad} & R_0^+(\mathbf{H}_k) \\ \dagger_K^* \big\downarrow & \circlearrowleft & \big\downarrow \dagger_k^* \\ R_0^+(\mathbf{H}_K) & \xrightarrow{\quad d_\theta \quad} & R_0^+(\mathbf{H}_k) \end{array}$$

(Indeed, note that $\mathfrak{x}_K$ is compatible with $\dagger_K^*$, in the sense that $\mathfrak{x}_K([V])(h^\dagger) = \mathfrak{x}_K([V^\dagger])(h)$ for any $[V] \in R_0^+(\mathbf{H}_K)$ and $h \in \mathbf{H}$. A similar remark applies to $\mathfrak{x}_k$. Then the characterisation of $d_\theta$ in Theorem 3.1.2(b) yields the commutativity of the above diagram.) Thus, $\dagger$ induces a symmetry of the Brauer graph of $\mathbf{H}$: two vertices labelled by $\lambda \neq \mu$ are connected by an edge if and only if the vertices labelled by $\lambda^\dagger, \mu^\dagger$ are connected (where $\lambda \mapsto \lambda^\dagger$ is defined in 1.2.6). In particular, the map $\lambda \mapsto \lambda^\dagger$ induces a permutation of the $\theta$-blocks of $\Lambda$.

In Chapter 1 we have seen that $\mathbf{H}$ carries a symmetrising trace $\tau\colon \mathbf{H} \to A$. In particular, we have a polynomial $\mathbf{c}_\lambda \in A$ for each $\lambda \in \Lambda$. The following result allows us to easily find "trivial" blocks. (The notion "defect 0" will become clear later.)

**Proposition 3.1.17 (Blocks of defect 0; see [132, 7.5.11]).** *In the setting of Theorem 3.1.14, let $\lambda \in \Lambda$ be such that $\theta(\mathbf{c}_\lambda) \neq 0$. Then $\{E_\varepsilon^\lambda\}$ is a $\theta$-block, and the corresponding decomposition matrix is the $(1 \times 1)$-matrix with entry* 1.

*Proof.* Since $\mathbf{H}_K$ is a symmetric algebra, we have the following explicit formula for the central idempotent $e_\lambda \in \mathbf{H}_K$ corresponding to $E_\varepsilon^\lambda$ (see [132, 7.2.7]):

$$e_\lambda = \mathbf{c}_\lambda^{-1} \sum_{w \in W} \operatorname{trace}(T_w, E_\varepsilon^\lambda)\, T_{w^{-1}}.$$

Now, since $\theta(\mathbf{c}_\lambda) \neq 0$, the condition $(*)$ in Example 3.1.11 is satisfied. Hence, $\{E_\varepsilon^\lambda\}$ is a $\theta$-block. The statement about the decomposition matrix is then clear, since $d_\theta$ is known to be surjective (see Theorem 3.1.14). □

**3.1.18.** In the setting of Theorem 3.1.14, a general semisimplicity criterion for symmetric algebras (see [132, 7.4.7]) shows that we have the following equivalence:

$$\mathbf{H}_k \text{ is semisimple} \qquad \Leftrightarrow \qquad \theta(\mathbf{c}_\lambda) \neq 0 \quad \text{for all } \lambda \in \Lambda.$$

Now recall from Remark 1.3.6 that

$$\tilde{P}_{W,L} := \Big(\prod_{\lambda \in \Lambda} \mathbf{c}_\lambda\Big) \in \mathbb{Z}[u_s^{\pm 1} \mid s \in S] \qquad (\text{where } u_s := v_s^2 \text{ for } s \in S).$$

Hence, $\mathbf{H}_k$ is semisimple if and only if $\theta(\tilde{P}_{W,L}) \neq 0$. Assume now that $\mathbf{H}_k$ is semisimple. Then Proposition 3.1.17 applies to all $\lambda \in \Lambda$ and so, up to reordering the columns, $D_\theta$ is the identity matrix. In particular, we obtain natural bijections

$$\operatorname{Irr}(\mathbf{H}_k) \overset{1-1}{\longleftrightarrow} \operatorname{Irr}(\mathbf{H}_K) \overset{1-1}{\longleftrightarrow} \operatorname{Irr}_{\mathbb{K}}(W),$$

where the first is induced by $d_\theta$ and the second is given by $E_\varepsilon^\lambda \leftrightarrow E^\lambda$. (This is one version of *Tits's deformation theorem* which we already saw in Section 1.2.)

**Table 3.2** Degrees of finite Coxeter groups

| Type | degrees $d_j$ |
|---|---|
| $A_{n-1}$ | $2,3,4,\ldots,n$ |
| $B_n$ | $2,4,6,\ldots,2n$ |
| $D_n$ | $2,4,6,\ldots,2(n-1),n$ |
| $I_2(m)$ | $2,m$ |
| $H_3$ | $2,6,10$ |

| Type | degrees $d_j$ |
|---|---|
| $H_4$ | $2,12,20,30$ |
| $F_4$ | $2,6,8,12$ |
| $E_6$ | $2,5,6,8,9,12$ |
| $E_7$ | $2,6,8,10,12,14,18$ |
| $E_8$ | $2,8,12,14,18,20,24,30$ |

**Example 3.1.19.** In the setting of Theorem 3.1.14, assume that $\Gamma = \mathbb{Z}$. Then $A = R[v, v^{-1}]$ is the ring of Laurent polynomials in one variable $v = \varepsilon$. By Remark 1.3.6, each $\mathbf{c}_\lambda \in A$ is a Laurent polynomial in $v$ all of whose (complex) roots are roots of unity. So, given $\theta \colon A \to k$, we have

$$\mathbf{H}_k \text{ not semisimple} \quad \Rightarrow \quad \theta(v) \in k \text{ is a root of unity or } R \text{ is not } L\text{-good.}$$

In particular, if $R$ is $L$-good, then $\mathbf{H}_k$ is semisimple unless $\theta(v)$ is a root of unity.

Now assume that we are in the equal parameter case where $L(s) = 1$ for all $s \in S$. By a classical result due to Solomon (see, for example, [44, §9.4]), we have

$$P_W := \sum_{w \in W} u^{l(w)} = \prod_{1 \leqslant j \leqslant |S|} (1 + u + u^2 + \cdots + u^{d_j - 1}) \qquad (u := v^2),$$

where $d_1, \ldots, d_{|S|}$ are the so-called *degrees* of $W$; we have $|W| = d_1 \cdots d_{|S|}$. The degrees for all types of $W$ are known (see [29]) and given in Table 3.2.

Note that $P_W = \mathbf{c}_{1_W}$ where $1_W$ denotes the unit representation of $W$. It is known that, for all $\lambda \in \Lambda$, we have $\mathbf{c}_\lambda \in R[u, u^{-1}]$ and $\mathbf{c}_\lambda$ divides $P_W$ in $\mathbb{K}[u, u^{-1}]$; see [132, 9.3.6] (if $W$ is of crystallographic type) and [132, 8.3.4 and App. E] (otherwise). Thus, assuming that $R$ is $L$-good, we conclude that

$$\mathbf{H}_k \text{ is semisimple} \quad \Leftrightarrow \quad \theta(P_W) \neq 0.$$

**Example 3.1.20.** Let $W$ be of type $B_n$ with generators $S = \{s_0, \ldots, s_{n-1}\}$ labelled as in Table 1.1 (p. 2). Let $L_0 \colon W \to \Gamma_0$ be the "universal" weight function, where

$\Gamma_0 = \mathbb{Z}^2$. Then $R = \mathbb{Z}_W = \mathbb{Z}$ is $L_0$-good and $A = \mathbb{Z}[V^{\pm 1}, v^{\pm 1}]$ is the ring of Laurent polynomials in two independent indeterminates $V := \varepsilon^{(1,0)}$ and $v := \varepsilon^{(0,1)}$; see Example 1.1.11(b). Thus, in $\mathbf{H}$, we have the quadratic relations

$$T_{s_0}^2 = T_1 + (V - V^{-1})T_{s_0} \quad \text{and} \quad T_{s_i}^2 = T_1 + (v - v^{-1})T_{s_i} \quad (i = 1, \ldots, n-1).$$

In this case, the analogue of the above polynomial $P_W$ is given by

$$P_{W,L} := \sum_{w \in W} \varepsilon^{L(w)} = \prod_{1 \leqslant j \leqslant n} (1 + u^{j-1}U) \prod_{1 \leqslant i \leqslant n} (1 + u + u^2 + \cdots + u^{i-1}),$$

where $U := V^2$ and $u := v^2$; see [132, 10.5.1]. Hoefsmit [157] obtained explicit formulae for all $\mathbf{c}_\lambda$. (See also Lusztig [212, 9.6] and Iancu [169] for alternative proofs.) We still have $\mathbf{c}_\lambda \in \mathbb{Z}[U^{\pm 1}, u^{\pm 1}]$, but it is no longer true that each $\mathbf{c}_\lambda$ divides $P_{W,L}$. In fact, Hoefsmit's formulae show that each $\mathbf{c}_\lambda$ is a product of terms of the form $U + u^{\pm r}$ (where $0 \leqslant r \leqslant n-1$) and $\Phi_d(u)$ (where $2 \leqslant d \leqslant n$); here, $\Phi_d(v) \in \mathbb{Z}[u]$ denotes the $d$th cyclotomic polynomial. Furthermore, each of these terms does occur in some $\mathbf{c}_\lambda$. Hence, we conclude that

$$\mathbf{H}_k \text{ is semisimple} \quad \Leftrightarrow \quad \theta(P_{W,L}) \neq 0 \quad \text{and} \quad \prod_{1 \leqslant j \leqslant n} \theta(U + u^{j-1}) \neq 0.$$

This semisimplicity criterion for type $B_n$ with general parameters was first established by Dipper and James [66, Theorem 5.5].

For similar descriptions in types $I_2(m)$ and $F_4$ with unequal parameters, see Sections 7.2 and 7.3. Hence, in all cases, we have an efficient criterion to determine whether a specialised algebra $\mathbf{H}_k$ is semisimple or not.

## 3.2 Canonical Basic Sets

We consider a generic Iwahori–Hecke algebra $\mathbf{H} = \mathbf{H}_A(W, S, L)$ and a specialisation $\theta\colon A \to k$ such that the conditions in Theorem 3.1.14 are satisfied and, hence, we have a corresponding decomposition map $d_\theta\colon R_0(\mathbf{H}_K) \to R_0(\mathbf{H}_k)$. The following definition is fundamental to the whole discussion in the subsequent chapters of this book. It provides the key technical notion which will allow us to obtain a new approach to the classification of $\mathrm{Irr}(\mathbf{H}_k)$, via $d_\theta\colon R_0(\mathbf{H}_K) \to R_0(\mathbf{H}_k)$. In Proposition 3.2.7 below, we will reconcile this approach with that in Chapter 2.

**Definition 3.2.1 (Geck and Rouquier [134], Geck [110, Def. 4.13]).** Consider the decompositon matrix $D_\theta = (d_{\lambda,M})$ associated with $\theta\colon A \to k$; see Theorem 3.1.14. Recall that $\leqslant$ is a *monomial order* on $\Gamma$ such that $L(s) \geqslant 0$ for all $s \in S$. By Section 1.3, we have a function $\mathrm{Irr}(W) \to \Gamma_{\geqslant 0}$, $E^\lambda \mapsto \mathbf{a}_\lambda$. Assume that the three conditions in 3.1.7 are satisfied for the partial order $\sqsubseteq$ on $\mathrm{Irr}(\mathbf{H}_K)$ defined by

$$E_\varepsilon^\lambda \sqsubseteq E_\varepsilon^\mu \quad \overset{\text{def}}{\Leftrightarrow} \quad \lambda = \mu \quad \text{or} \quad \mathbf{a}_\lambda < \mathbf{a}_\mu.$$

Explicitly, in the present context, this means:

- Given $M \in \operatorname{Irr}(\mathbf{H}_k)$, let $\mathscr{S}_\theta(M) := \{\lambda \in \Lambda \mid d_{\lambda,M} \neq 0\}$. Then the function $\mathscr{S}_\theta(M) \to \Gamma$, $\lambda \mapsto \mathbf{a}_\lambda$, reaches its minimum (with respect to $\leqslant$) at exactly one element of $\mathscr{S}_\theta(M)$, which we denote by $\lambda_M$.
- The map $\operatorname{Irr}(\mathbf{H}_k) \to \Lambda$, $M \mapsto \lambda_M$, is injective.
- We have $d_{\lambda_M,M} = 1$ for all $M \in \operatorname{Irr}(\mathbf{H}_k)$.

If this holds, let $\mathscr{B}_\theta := \{\lambda_M \mid M \in \operatorname{Irr}(\mathbf{H}_k)\}$. Thus, we obtain a bijection

$$\operatorname{Irr}(\mathbf{H}_k) \overset{1-1}{\longleftrightarrow} \mathscr{B}_\theta \subseteq \Lambda,$$

and $\underline{B}_\theta = \{E_\varepsilon^\mu \mid \mu \in \mathscr{B}_\theta\}$ is called a *canonical basic set* for $\mathbf{H}_k$; see 3.1.7. We shall also call $\mathscr{B}_\theta \subseteq \Lambda$ itself a *canonical basic set*. Note again that if a canonical basic set exists, then it is uniquely determined by the above conditions.

*Remark 3.2.2.* The above definition is an axiomatic version (without reference to any underlying "cell datum") of the properties in Theorem 2.6.6. Indeed, assume that we have a canonical basic set $\mathscr{B}_\theta$ as above. Then there is a unique labelling

$$\operatorname{Irr}(\mathbf{H}_k) = \{M^\mu \mid \mu \in \mathscr{B}_\theta\} \qquad \text{where} \qquad \lambda_{M^\mu} = \mu \quad (\mu \in \mathscr{B}_\theta).$$

Given $\lambda \in \Lambda$ and $\mu \in \mathscr{B}_\theta$, denote the decomposition number $d_{\lambda,M^\mu}$ by $d_{\lambda,\mu}$. Then the conditions in Definition 3.2.1 translate to:

$$(\Delta^a) \qquad \begin{aligned} d_{\mu,\mu} &= 1 && \text{for all } \mu \in \mathscr{B}_\theta, \\ d_{\lambda,\mu} &= 0 && \text{unless } \lambda = \mu \text{ or } \mathbf{a}_\lambda > \mathbf{a}_\mu. \end{aligned}$$

Thus, the *decomposition matrix* $D_\theta = \big(d_{\lambda,\mu}\big)_{\lambda \in \Lambda, \mu \in \mathscr{B}_\theta}$ has a lower unitriangular shape, if the rows and columns are ordered according to increasing values of $\mathbf{a}_\lambda$.

*Remark 3.2.3.* Suppose that, in some way, the decomposition matrix $D_\theta$ has been determined. (For example, by using the algorithmic methods described in Section 7.1.) The rows of $D_\theta$ are labelled by the set $\Lambda$. Let $r$ be the number of columns of $D_\theta$. Then $\operatorname{Irr}(\mathbf{H}_k)$ contains $r$ simple modules, which we denote by $M_1, \ldots, M_r$.

Assume that we are also given a monomial order $\leqslant$ on $\Gamma$ such that $L(s) \geqslant 0$ for all $s \in S$. Then a canonical basic set (if it exists) is found as follows. Consider the invariants $\mathbf{a}_\lambda$ and order the rows of $D_\theta$ according to increasing value of $\mathbf{a}_\lambda$. For each index $i \in \{1, \ldots, r\}$, find the first row in $D_\theta$ (from top to bottom) which has a non-zero entry in the $i$th column; let $\lambda_i \in \Lambda$ be the label of that row. Performing this procedure for all $i$, we obtain a subset $\mathscr{B}_\theta := \{\lambda_1, \ldots, \lambda_r\} \subseteq \Lambda$.

If there exists some $i \in \{1, \ldots, r\}$ and some $\lambda \in \Lambda$ such that $\lambda \neq \lambda_i$, $\mathbf{a}_\lambda = \mathbf{a}_{\lambda_i}$ and $d_{\lambda,M_i} \neq 0$, then the first condition in Definition 3.2.1 is not satisfied. Similarly, if

$\lambda_i = \lambda_j$ for some $i \neq j$, then the second condition is not satisfied. Finally, if $d_{\lambda_i,M_i} \neq 1$ for some $i$, then the third condition is not satisfied. If none of the above happens, then $\mathscr{B}_\theta$ is a canonical basic set for $\mathbf{H}_k$.

Thus, if $D_\theta$ is known, the question of showing the existence of a canonical basic set, and of explicitly determining it, is solved by a purely mechanical procedure.

**Example 3.2.4.** Consider the partitions of $\mathrm{Irr}(\mathbf{H}_K)$ and $\mathrm{Irr}(\mathbf{H}_k)$ into $\theta$-blocks as in 3.1.8. Correspondingly, we have a partition $\Lambda = \Lambda_1 \sqcup \cdots \sqcup \Lambda_r$, where $\Lambda_i$ labels the irreducible representations in the $\theta$-block $\mathrm{Irr}_i(\mathbf{H}_K)$. Then it is clear that, for a fixed $i$, the three conditions in Definition 3.2.1 can be formulated with respect to $\Lambda_i$ and $\mathrm{Irr}_i(\mathbf{H}_k)$. Just note that if $M \in \mathrm{Irr}_i(\mathbf{H}_k)$, then the set $\mathscr{S}_\theta(M)$ is automatically contained in $\Lambda_i$, by the definition of blocks.

Hence, a canonical basic set exists for $\mathbf{H}$ if and only if each $\theta$-block admits a canonical basic set. Furthermore, in this case, the canonical basic set for $\mathbf{H}$ is the union of the canonical basic sets for all $\theta$-blocks.

Assume, for example, that $\lambda \in \Lambda$ is such that $\theta(\mathbf{c}_\lambda) \neq 0$. By Proposition 3.1.17, the set $\{E_\varepsilon^\lambda\}$ is a $\theta$-block whose decomposition matrix is the $(1 \times 1)$-matrix with entry 1. Consequently, $\lambda$ belongs to the canonical basic set for $\mathbf{H}$ (if it exists).

**Example 3.2.5.** Let $W$ be of type $I_2(m)$ $(m \geqslant 3)$, with generators $s_1, s_2$ such that $(s_1s_2)^m = 1$. Let $\zeta \in \mathbb{C}$ be a root of unity of order $m$ such that $\zeta + \zeta^{-1} = 2\cos(2\pi/m)$. Let $A = R[\Gamma]$, where $m \in R^\times$ and $\zeta + \zeta^{-1} \in R \subseteq \mathbb{C}$. (Thus, $R$ is $L$-good.) The irreducible representations of $\mathbf{H}_K$ are described in Example 1.3.7. Let $\theta \colon A \to k$ be a specialisation where $k$ is the field of fractions of $\theta(A)$. In the following, we assume that $\xi := \theta(v_{s_1}) = \theta(v_{s_2})$. The corresponding decomposition matrices are determined by Müller [253, §3] (see also Section 7.2 in the appendix).

It turns out that non-trivial $\theta$-blocks (i.e., blocks which are not singleton sets), only exist if $\xi^2$ is a root of unity of order $e \geqslant 2$, where $e$ divides $m$. Furthermore:

(a) If $e = 2$, then $\{1_W, \mathrm{sgn}, \mathrm{sgn}_1, \mathrm{sgn}_2\}$ is a $\theta$-block (where $\mathrm{sgn}_1$ and $\mathrm{sgn}_2$ are omitted if $m$ is odd).
(b) If $e > 2$, then $\{1_W, \mathrm{sgn}, \sigma_{j(e)}\}$ is a $\theta$-block, where the index $j(e)$ is uniquely determined by the condition that $\xi^2 = \theta(\zeta)^{\pm j(e)}$.

The decomposition matrices of these blocks are, respectively, given by

| $E^\lambda$ | $d_{\lambda,M}$ |
|---|---|
| • $1_W$ | 1 |
| $(\mathrm{sgn}_1)$ | 1 |
| $(\mathrm{sgn}_2)$ | 1 |
| $\mathrm{sgn}$ | 1 |

| $E^\lambda$ | $d_{\lambda,M}$ | |
|---|---|---|
| • $1_W$ | 1 | 0 |
| • $\sigma_{j(e)}$ | 1 | 1 |
| $\mathrm{sgn}$ | 0 | 1 |

Now let $\leqslant$ be a monomial order on $\Gamma$ such that $L(s_1) > 0$ and $L(s_2) > 0$. Then each block admits a canonical basic set, which is marked by "•" in the above matrices. (Indeed, the table in Example 1.3.7 shows that, regardless of the exact values of $L$, we always have $\mathbf{a}_{1_W} < \mathbf{a}_{\mathrm{sgn}_1} \leqslant \mathbf{a}_{\sigma_j} \leqslant \mathbf{a}_{\mathrm{sgn}_2} < \mathbf{a}_{\mathrm{sgn}}$ if $L(s_1) \geqslant L(s_2) > 0$. A similar statement holds if $L(s_2) \geqslant L(s_1) > 0$, with the roles of $\mathrm{sgn}_1$ and $\mathrm{sgn}_2$ interchanged. It remains to apply the procedure described in Remark 3.2.3.)

**Example 3.2.6.** Let $W$ be of type $B_3$ with generators $s_0, s_1, s_2$ labelled as in Table 1.1. Assume that we are in the setting of Example 3.1.20, where $A = \mathbb{Z}[V^{\pm 1}, v^{\pm 1}]$ is the ring of Laurent polynomials in two independent indeterminates $V, v$ and, in $\mathbf{H}$, we have the quadratic relations

$$T_{s_0}^2 = T_1 + (V - V^{-1})T_{s_0} \quad \text{and} \quad T_{s_i}^2 = T_1 + (v - v^{-1})T_{s_i} \quad (i = 1, 2).$$

Consider a specialisation $\theta\colon A \to k$ such that $\theta(V) = -1$ and $\theta(v) = \sqrt{-1} \in k$, where $\operatorname{char}(k) \neq 2$. The matrix $D_\theta$ is given as follows (recall that, here, $\Lambda$ is the set of all pairs of partitions such that the total sum of their parts equals 3):

| $E^{(\lambda,\mu)}$ | $d_{(\lambda,\mu),M}$ | | | | $\mathbf{a}^{\text{lex}}_{(\lambda,\mu)}$ |
|---|---|---|---|---|---|
| $(3,\varnothing)$ | 1 | 0 | 0 | 0 | $(0,0)$ |
| $(21,\varnothing)$ | 0 | 1 | 0 | 0 | $(0,1)$ |
| $(111,\varnothing)$ | 1 | 0 | 0 | 0 | $(0,3)$ |
| $(2,1)$ | 0 | 1 | 1 | 0 | $(1,0)$ |
| $(11,1)$ | 0 | 1 | 1 | 0 | $(1,1)$ |
| $(1,2)$ | 1 | 0 | 0 | 1 | $(2,-1)$ |
| $(\varnothing,3)$ | 0 | 0 | 1 | 0 | $(2,2)$ |
| $(1,11)$ | 1 | 0 | 0 | 1 | $(3,-3)$ |
| $(\varnothing,21)$ | 0 | 0 | 0 | 1 | $(3,1)$ |
| $(\varnothing,111)$ | 0 | 0 | 1 | 0 | $(3,6)$ |

The column labelled $\mathbf{a}^{\text{lex}}_{(\lambda,\mu)}$ contains the invariants $\mathbf{a}$-invariants with respect to the monomial order $\leqslant$ on $\Gamma_0$ given by the lexicographic order (see Example 1.1.11(b)). Note that the rows of $D_\theta$ are already ordered according to increasing value of $\mathbf{a}^{\text{lex}}_{(\lambda,\mu)}$. Using Remark 3.2.3, we find that a canonical basic set exists and is given by

$$\mathscr{B}_\theta = \{(3,\varnothing),\ (21,\varnothing),\ (2,1),\ (1,2)\}.$$

Note also that, here, we are in the "asymptotic case" and so there is a cellular basis of $\mathbf{H}$ as in Theorem 2.6.12, with respect to the dominance order on $\Lambda$; see 2.8.19.

Now let us choose a different monomial order: consider the reverse lexicograhic order on $\Lambda$ (where $(0,1) > (r,0) > (0,0)$ for all $r \in \mathbb{Z}_{\geqslant 0}$). The corresponding $\mathbf{a}$-invariants are printed in the column labelled $\mathbf{a}^{\text{revlex}}_{(\lambda,\mu)}$ in the table below. Hence, reordering the rows of $D_\theta$ according to increasing value of $\mathbf{a}^{\text{revlex}}_{(\lambda,\mu)}$, we obtain

| $E^{(\lambda,\mu)}$ | $d_{(\lambda,\mu),M}$ | | | | $\mathbf{a}^{\text{revlex}}_{(\lambda,\mu)}$ |
|---|---|---|---|---|---|
| $(3,\varnothing)$ | 1 | 0 | 0 | 0 | $(0,0)$ |
| $(\varnothing,3)$ | 0 | 0 | 1 | 0 | $(1,0)$ |
| $(2,1)$ | 0 | 1 | 1 | 0 | $(0,1)$ |
| $(1,2)$ | 1 | 0 | 0 | 1 | $(1,1)$ |
| $(21,\varnothing)$ | 0 | 1 | 0 | 0 | $(-1,2)$ |
| $(\varnothing,21)$ | 0 | 0 | 0 | 1 | $(2,2)$ |
| $(11,1)$ | 0 | 1 | 1 | 0 | $(0,3)$ |
| $(1,11)$ | 1 | 0 | 0 | 1 | $(1,3)$ |
| $(111,\varnothing)$ | 1 | 0 | 0 | 0 | $(-2,6)$ |
| $(\varnothing,111)$ | 0 | 0 | 1 | 0 | $(3,6)$ |

We see that a corresponding canonical basic set exists and is given by

$$\mathscr{B}_\theta = \{(3,\varnothing),\ (\varnothing,3),\ (1,2),\ (2,1)\}.$$

This example shows that, in the general multiparameter case, $\mathscr{B}_\theta$ heavily depends on the chosen monomial order on $\Gamma$.

On the other hand, canonical basic sets do not always exist. Consider, for example, the decomposition matrix in Example 3.1.15(c). Since $\mathrm{sgn}_1$, $\mathrm{sgn}_2$ and $\sigma_1$ all have $\mathbf{a}_\lambda = 1$, the procedure in Remark 3.2.3 fails for the second and third columns of that matrix! Note that $R$ is not $L$-good in this case.

A general existence result for canonical basic sets is proved in [110, Theorem 6.6], following earlier work of Geck [97] and Geck and Rouquier [134]. Here, we obtain a new proof as an immediate consequence of the results in Chapter 2. This also shows how the condition that $R$ should be $L$-good naturally comes into play.

**Proposition 3.2.7.** *Assume that $R$ is $L$-good and that the conditions (♠) and (♣) in 2.5.3 hold. Thus, $\mathbf{H}$ admits a cellular basis $\{\mathbf{C}^\lambda_{\mathfrak{s},\mathfrak{t}}\}$ as in Theorem 2.6.12. Via $\theta: A \to k$, we obtain a cell datum for $\mathbf{H}_k$. So we have cell modules $\{W_k(\lambda) \mid \lambda \in \Lambda\}$ for $\mathbf{H}_k$ and there is a natural parametrisation $\mathrm{Irr}(\mathbf{H}_k) = \{L_k^\mu \mid \mu \in \Lambda_k^\circ\}$; see 2.7.5. Then the following hold.*

*(a) Let $M \in \mathrm{Irr}(\mathbf{H}_k)$ and $\mu \in \Lambda_k^\circ$ be such that $M \cong L_k^\mu$. Let $\lambda \in \Lambda$. Then we have*

$$d_{\lambda,M} = (W_k(\lambda) : L_k^\mu), \qquad \text{where } (W_k(\lambda) : L_k^\mu) \text{ is defined in Theorem 2.6.6.}$$

*(b) The subset $\mathscr{B}_\theta := \Lambda_k^\circ \subseteq \Lambda$ is a canonical basic set with respect to $\leqslant$. Thus, using the notation of Remark 3.2.2, we have*

$$\begin{aligned} d_{\mu,\mu} &= 1 && \text{for all } \mu \in \Lambda_k^\circ, \\ d_{\lambda,\mu} &= 0 && \text{unless } \lambda = \mu \text{ or } \mathbf{a}_\lambda > \mathbf{a}_\mu. \end{aligned}$$

*Furthermore, we have $d_{\lambda,\mu} = 0$ unless $\lambda = \mu$ or $N_\lambda < N_\mu$.*

*Proof.* By Proposition 2.7.1, for any $\lambda \in \Lambda$, the $\mathbf{H}_K$-modules $W_K(\lambda)$ and $E_\varepsilon^\lambda$ are isomorphic. Hence, we have $[E_\varepsilon^\lambda] = [W_K(\lambda)]$ in $R_0(\mathbf{H}_K)$. Since $W(\lambda)$ is an $\mathbf{H}$-module which is free of finite rank over $A$, we also have $d_\theta([W_K(\lambda)]) = [W_k(\lambda)]$ by Theorem 3.1.14. This yields the formula for $d_{\lambda,M}$ in (a). Now consider the conditions in Definition 3.2.1. Let $M \in \mathrm{Irr}(\mathbf{H}_k)$; then $M \cong L_k^\mu$ for a unique $\mu \in \Lambda_k^\circ$. The relations $(\Delta)$ in Theorem 2.6.6 show that $\mu \in \mathscr{S}_\theta(M)$, $d_{\mu,M} = 1$, and that $\lambda \trianglelefteq_L \mu$ for any $\lambda \in \mathscr{S}_\theta(M)$. Since, by Theorem 2.6.12, we also have the implication

$$\lambda \trianglelefteq_L \mu \qquad \Rightarrow \qquad \lambda = \mu \quad \text{or} \quad \mathbf{a}_\lambda > \mathbf{a}_\mu,$$

it follows that $\mu$ is the unique element of $\mathscr{S}_\theta(M)$ at which the function $\mathscr{S}_\theta(M) \to \Gamma$, $\lambda \mapsto \mathbf{a}_\lambda$, reaches its minimum. Thus, we have $\mu = \lambda_M$. Consequently, the three conditions in Definition 3.2.1 are satisfied, where $\mathscr{B}_\theta = \Lambda_k^\circ$. The statements in (b) concerning the invariants $N_\lambda$ then follow as in Remark 2.6.15. □

**Example 3.2.8.** Let us keep the assumptions of Proposition 3.2.7. The algebra $\mathbf{H}$ is symmetric (see 1.2.11) and, clearly, we have $\tau(T_w^\flat) = \tau(T_{w^{-1}}) = \tau(T_w)$ for all $w \in W$. Hence, $\tau$ satisfies the additional condition in Proposition 2.6.7. So we have an *oppositive cell datum* for $\mathbf{H}$; let $\{\hat{W}(\lambda) \mid \lambda \in \Lambda\}$ be the corresponding cell modules. Since $\mathbf{H}_K$ is semisimple, we have by Propositions 2.6.7 and 2.7.1:

(a) $$\hat{W}_K(\lambda) \cong W_K(\lambda) \cong E_\varepsilon^\lambda \qquad \text{for all } \lambda \in \Lambda.$$

Furthermore, by Theorem 2.6.5, we have a parametrisation $\operatorname{Irr}(\mathbf{H}_k) = \{\hat{L}_k^\mu \mid \mu \in \hat{\lambda}_k^\circ\}$, where $\hat{\Lambda}_k^\circ \subseteq \Lambda$ is defined with respect to the opposite cell datum. We claim that:

(b) $$\Lambda_k^\circ = \{\mu^\dagger \mid \mu \in \hat{\Lambda}_k^\circ\} \qquad \text{(with } \mu \mapsto \mu^\dagger \text{ as in Example 1.2.6).}$$

This is seen as follows. Since the two sets have the same cardinality, it is enough to prove one inclusion. Let $\mu \in \hat{\Lambda}_k^\circ$. By Proposition 3.2.7, it will be sufficient to show that $\mu^\dagger \in \mathscr{B}_\theta$. Let us denote $M := (\hat{L}_k^\mu)^\dagger \in \operatorname{Irr}(\mathbf{H}_k)$. The commutative diagram in Example 3.1.16 shows that

$$d_{\mu^\dagger,M} = d_{\mu,M^\dagger} = d_{\mu,\hat{L}_k^\mu} = 1 \quad \text{and so} \quad \mu^\dagger \in \mathscr{S}_\theta(M).$$

Now let $\lambda \in \mathscr{S}_\theta(M)$. Using Theorem 2.6.6, Proposition 3.2.7(a) and once more the commutative diagram in Example 3.1.16, we obtain

$$(\hat{W}_k(\lambda^\dagger) : \hat{L}_k^\mu) = d_{\lambda^\dagger,\hat{L}_k^\mu} = d_{\lambda^\dagger,M^\dagger} = d_{\lambda,M} \neq 0 \quad \text{and so} \quad \lambda^\dagger \trianglelefteq_{L,\mathrm{op}} \mu.$$

This yields $\mu \trianglelefteq_L \lambda^\dagger$ and so $\lambda \trianglelefteq_L \mu^\dagger$ (see Remark 2.2.11). Thus, we have $\mathbf{a}_\lambda \geqslant \mathbf{a}_{\mu^\dagger}$ with equality only if $\lambda = \mu^\dagger$. It follows that $\mu^\dagger \in \mathscr{B}_\theta$. Thus, (b) is proved.

**Example 3.2.9.** Let $(W,S)$ be of type $A_{n-1}$. Then $W \cong \mathfrak{S}_n$ and $S = \{s_1, \ldots, s_{n-1}\}$, where $s_i$ corresponds to the basic transposition $(i, i+1) \in \mathfrak{S}_n$ for all $i$. We are automatically in the equal parameter case; let $\Gamma = \mathbb{Z}$ and $L(s_i) = 1$ for all $i$. Now $A = R[v, v^{-1}]$ is the ring of Laurent polynomials in one indeterminate $v = \varepsilon$.

By Example 1.3.8, $\Lambda$ is the set of all partitions $\lambda \vdash n$. We have $f_\lambda = 1$ for all $\lambda \vdash n$, so any subring $R \subseteq \mathbb{C}$ is $L$-good. By Theorem 2.8.15, the Kazhdan–Lusztig basis $\{\pm C_w\}$ gives rise to a cell datum for $\mathbf{H}$. Hence, for any specialisation $\theta \colon A \to k$, we have a natural parametrization

$$\operatorname{Irr}(\mathbf{H}_k) = \{L_k^\mu \mid \mu \in \Lambda_k^\circ\} \qquad \text{with } \Lambda_k^\circ \text{ defined as in 2.7.5.}$$

By Proposition 3.2.7, the set $\Lambda_k^\circ$ is a canonical basic set. We have also seen in Theorem 2.8.15 that the order $\trianglelefteq_L$ on $\Lambda$ is given in terms of the *dominance order* on partitions. Thus, for $\lambda \vdash n$ and $\mu \in \Lambda_k^\circ$, we have

($\Delta$) $$d_{\mu,\mu} = 1 \quad \text{and} \quad d_{\lambda,\mu} = 0 \quad \text{unless} \quad \lambda \trianglelefteq \mu.$$

Similar statements first appeared (with rather different proofs) in the pioneering work of Dipper and James [62]. Note that the above discussion does not yet give us an explicit description of the set $\Lambda_k^\circ$; this problem will be solved in Theorem 3.5.14.

Let us just consider the example where $W \cong \mathfrak{S}_5$ is of type $A_4$ and $\Lambda$ consists of all partitions of 5. Let $\theta: A \to k$ be a specialisation such that $\theta(v^2) = -1 \neq 1$. Then, by James [181, p. 254], the matrix $D_\theta$ is given by

| $E^\lambda$ | $d_{\lambda,M}$ | | | $\mathbf{a}_\lambda$ |
|---|---|---|---|---|
| (5) | 1 | 0 | 0 | 0 |
| (41) | 0 | 1 | 0 | 1 |
| (32) | 0 | 0 | 1 | 2 |
| (311) | 1 | 0 | 1 | 3 |
| (221) | 0 | 0 | 1 | 4 |
| (2111) | 0 | 1 | 0 | 6 |
| (11111) | 1 | 0 | 0 | 10 |

The invariants $\mathbf{a}_\lambda$ are computed using the formula in Example 1.3.8. Applying Remark 3.2.3, we find that $\mathscr{B}_\theta = \{(5), (41), (32)\}$ is a canonical basic set.

**Example 3.2.10.** Let $(W,S)$ be of type $B_n$ and assume that we are in the "*asymptotic case*" as in 2.8.19, with weight function $L_{\text{asy}}: W \to \Gamma$ given by

$B_n$ 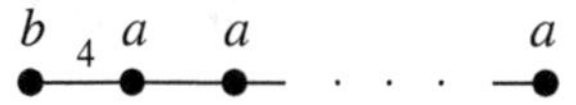

where $b > (n-1)a > 0$.

Recall that $\Lambda$ is the set of all pairs of partitions $(\lambda, \mu)$ such that $|\lambda| + |\mu| = n$. We have $f_{(\lambda,\mu)} = 1$ for all $(\lambda,\mu) \in \Lambda$, so any subring $R \subseteq \mathbb{C}$ is $L_{\text{asy}}$-good. As discussed in 2.8.19, the Kazhdan–Lusztig basis $\{\pm C_w\}$ gives rise to a cell datum for $\mathbf{H}$. Hence, for any specialisation $\theta: A \to k$, we have a natural parametrisation

$$\operatorname{Irr}(\mathbf{H}_k) = \{L_k^{(\lambda,\mu)} \mid (\lambda,\mu) \in \Lambda_k^\circ\} \qquad \text{with } \Lambda_k^\circ \text{ defined as in 2.7.5.}$$

By Proposition 3.2.7, the set $\Lambda_k^\circ$ is a canonical basic set. We have also seen in 2.8.19 that the order $\trianglelefteq_{L_{\text{asy}}}$ on $\Lambda$ is given in terms of the *dominance order* on pairs of partitions. Thus, for $(\lambda,\mu) \in \Lambda$ and $(\lambda',\mu') \in \Lambda_k^\circ$, we have

$$(\Delta) \qquad d_{(\lambda',\mu'),(\lambda',\mu')} = 1 \quad \text{and} \quad d_{(\lambda,\mu),(\lambda',\mu')} = 0 \quad \text{unless} \quad (\lambda,\mu) \trianglelefteq (\lambda',\mu').$$

Similar statements first appeared (with rather different proofs and in terms of the opposite order; see Example 3.2.8) in the work of Dipper, James and Murphy [68]. As in Example 3.2.9, this discussion does not yet give us an explicit description of the set $\Lambda_k^\circ$. This problem has been solved by Ariki [8], [9], [10]; see also Ariki and Mathas [15] and Ariki and Jacon [11]. We will obtain this result in Theorem 6.7.14. We shall also consider weight functions which do not correspond to the "asymptotic case"; this will be essential for applications to finite groups of Lie type in Section 4.4.

To state the following, quite useful technical result, recall the definition of the set $\mathscr{F}_\lambda = \{w \in W \mid E^\lambda \leftrightsquigarrow_L w\}$ for $\lambda \in \Lambda$; see Proposition 1.6.11. Also recall that if ($\spadesuit$) holds, then $\mathscr{F}_\lambda$ is a two-sided Kazhdan–Lusztig cell; see Lemma 2.5.9.

**Lemma 3.2.11.** *In the setting of Proposition 3.2.7, let $\lambda \in \Lambda$. Then $\lambda \in \Lambda_k^\circ$ if and only if $d_{\lambda,M} \neq 0$ and $C_w.M \neq \{0\}$ for some $M \in \mathrm{Irr}(\mathbf{H}_k)$ and some $w \in \mathscr{F}_\lambda$.*

*Proof.* Assume first that $\lambda \in \Lambda_k^\circ$. Then, setting $M := L_k^\lambda$, we have $d_{\lambda,M} \neq 0$ by 2.7.5 and $C_w.M \neq \{0\}$ for some $w \in \mathscr{F}_\lambda$ by Proposition 2.7.7. Conversely, assume that $d_{\lambda,M} \neq 0$ and $C_w.M \neq \{0\}$, where $M \in \mathrm{Irr}(\mathbf{H}_k)$ and $w \in \mathscr{F}_\lambda$. Let $\mu \in \Lambda_k^\circ$ be such that $M \cong L_k^\mu$. Then Proposition 3.2.7 shows that $(W_k(\lambda) : L_k^\mu) = d_{\lambda,M} \neq 0$. Hence, $M \cong L_k^\mu$ is a composition factor of $W_k(\lambda)$ and so $C_w.W_k(\lambda) \neq \{0\}$. Since $w \in \mathscr{F}_\lambda$, it follows that $\lambda \in \Lambda_k^\circ$; see Proposition 2.7.6. □

**3.2.12.** As in Section 2.4, let us now consider the case where $L(s) \geqslant 0$ for all $s \in S$ and the parabolic subgroup $\Omega = \langle t \in S \mid L(t) = 0 \rangle \subseteq W$ is non-trivial. Then, as in Proposition 2.4.5, we have a semidirect product decomposition $W = \Omega \ltimes W_1$ (where $W_1$ is a certain reflection subgroup of $W$) and two generic algebras $\mathbf{H}_1 \subseteq \mathbf{H}$ (where $\mathbf{H}_1$ is defined with respect to the restriction of $L$ to $W_1$). We shall write $\mathbf{H}_{K,1} = K \otimes_A \mathbf{H}_1$ and $\mathbf{H}_{k,1} = k \otimes_A \mathbf{H}_1$. Our aim is to show that, under appropriate assumptions, the canonical basic sets for the specialised algebras $\mathbf{H}_k$ and $\mathbf{H}_{k,1}$ determine each other; see Theorem 3.2.14 below. As in 2.4.11, we write

$$\mathrm{Irr}_{\mathbb{K}}(W_1) = \{E^{\lambda_1} \mid \lambda_1 \in \Lambda_1\} \qquad \text{and} \qquad \mathrm{Irr}_{\mathbb{K}}(W) = \{E^\lambda \mid \lambda \in \Lambda\}.$$

For any $\omega \in W$ and $w_1 \in W_1$, we have $T_\omega^{-1} T_{w_1} T_\omega = T_{\omega^{-1} w_1 \omega}$. Thus, conjugation by a fixed $T_\omega$ defines a $K$-algebra automorphism of $H_{K,1}$. Given any $\mathbf{H}_{K,1}$-module $V_1$, we can define a new $\mathbf{H}_{K,1}$-module structure on $V_1$ by composing the original action with the above automorphism. We denote that new $\mathbf{H}_{K,1}$-module by $V_1^{(\omega)}$. Thus, we have $V_1^{(\omega)} = V_1$ as $K$-vector spaces, but $h_1 \in \mathbf{H}_{K,1}$ acts on $V^{(\omega)}$ in the same way as $T_\omega^{-1} h_1 T_\omega$ acts on $V$. This notation is compatible with the action $\lambda_1 \mapsto \omega.\lambda_1$ defined in 2.4.11, that is, we have

(a) $$E_\varepsilon^{\omega.\lambda_1} \cong \big(E_\varepsilon^{\lambda_1}\big)^{(\omega)} \qquad \text{for all } \lambda_1 \in \Lambda_1 \text{ and } \omega \in \Omega.$$

In this set-up, we have the following compatibility property.

(b) Assume that (♠) holds. Then $\mathscr{F}_{\omega.\lambda_1} = \omega \mathscr{F}_{\lambda_1} \omega^{-1}$ for any $\lambda_1 \in \Lambda_1$ and $\omega \in \Omega$.

Indeed, let $\mathfrak{C}_1$ be a left Kazhdan–Lusztig cell of $W_1$ (with respect to $L|_{W_1}$) such that $m(\mathfrak{C}_1, \lambda_1) > 0$. As discussed in 2.4.11, the set $\omega \mathfrak{C}_1 \omega^{-1}$ also is a left Kazhdan–Lusztig cell and we have $m(\omega \mathfrak{C} \omega^{-1}, \omega.\lambda_1) = m(\mathfrak{C}_1, \lambda_1) > 0$. So, by Lemmas 2.2.4 and 2.5.9, we have $\mathfrak{C}_1 \subseteq \mathscr{F}_{\lambda_1}$ and $\omega \mathfrak{C}_1 \omega^{-1} \subseteq \mathscr{F}_{\omega.\lambda_1}$. The first inclusion implies that we also have $\omega \mathfrak{C}_1 \omega^{-1} \subseteq \omega \mathscr{F}_{\lambda_1} \omega^{-1}$. Hence, (b) holds.

**3.2.13.** In the above setting, let us now consider decomposition maps. We will require that $R$ is good for both $L$ and $L|_{W_1}$. Note that this implies that $|\Omega| \in R^\times$. (Indeed, let $\lambda \in \Lambda$ be such that $\dim E^\lambda = 1$; then the formula in Proposition 2.4.14 shows that $|\Omega|$ divides $f_\lambda$.) Now consider the decomposition maps

$$d_\theta \colon R_0(\mathbf{H}_K) \to R_0(\mathbf{H}_k) \qquad \text{and} \qquad d_\theta^1 \colon R_0(\mathbf{H}_{K,1}) \to R_0(\mathbf{H}_{k,1})$$

associated with $\theta\colon A \to k$; let $D_\theta = (d_{\lambda,M})$ and $D_\theta^1 = (d^1_{\lambda_1,M_1})$ be the corresponding decomposition matrices. Recall that conjugation by $T_\omega$ (where $\omega \in \Omega$) defines an $\mathbf{H}_{K,1}$-algebra automorphism, and this gives rise to an action of $\Omega$ on $\mathrm{Irr}(\mathbf{H}_{K,1})$. Similar remarks apply, of course, to $\mathbf{H}_{k,1}$-modules: given an $\mathbf{H}_{k,1}$-module $M_1$, we can define a new $\mathbf{H}_{k,1}$-module by composing the original action with the $k$-algebra automorphism $h_1 \mapsto T_\omega^{-1} h_1 T_\omega$ ($h_1 \in \mathbf{H}_{k,1}$). As above, this new $\mathbf{H}_{k,1}$-module will be denoted by $M_1^{(\omega)}$. The characterisation of decomposition maps in Theorem 3.1.2 immediately yields the following compatibility property:

(a) $$d^1_{\omega.\lambda_1,M_1^{(\omega)}} = d^1_{\lambda_1,M_1} \qquad \text{for all } \lambda_1 \in \Lambda_1 \text{ and } M_1 \in \mathbf{H}_{k,1}.$$

Now, the $\Omega$-graded Clifford system for $\mathbf{H}$ in Remark 2.4.6 certainly gives rise to an $\Omega$-graded Clifford for $\mathbf{H}_k$. Hence, there is also a version of *Clifford's theorem* for $\mathbf{H}_k$-modules (see [53, (11.16)]).

(b) Let $M \in \mathrm{Irr}(\mathbf{H}_k)$ and $M_1 \in \mathrm{Irr}(\mathbf{H}_{k,1})$ be a constituent of the restriction of $M$ to $\mathbf{H}_{k,1}$. Then that restriction is the direct sum of simple $\mathbf{H}_{k,1}$-modules which are all of the form $M_1^{(\omega)}$ for various $\omega \in \Omega$.

In particular, this shows that the restriction of $M \in \mathrm{Irr}(\mathbf{H}_k)$ to $\mathbf{H}_{k,1}$ is semisimple. More generally, the restriction of representations from $\mathbf{H}_K$ to $\mathbf{H}_{K,1}$ and from $\mathbf{H}_k$ to $\mathbf{H}_{k,1}$ certainly give rise to maps on the level of Grothendieck groups:

$$\mathrm{Res}_K\colon R_0(\mathbf{H}_K) \to R_0(\mathbf{H}_{K,1}) \qquad \text{and} \qquad \mathrm{Res}_k\colon R_0(\mathbf{H}_k) \to R_0(\mathbf{H}_{k,1}).$$

Then the characterisation in Theorem 3.1.2 shows that the following diagram is commutative:

$$\begin{array}{ccc} R_0^+(\mathbf{H}_K) & \xrightarrow{\quad d_\theta \quad} & R_0^+(\mathbf{H}_k) \\ {\scriptstyle \mathrm{Res}_K}\Big\downarrow & \circlearrowleft & \Big\downarrow{\scriptstyle \mathrm{Res}_k} \\ R_0^+(\mathbf{H}_{K,1}) & \xrightarrow{\quad d_\theta^1 \quad} & R_0^+(\mathbf{H}_{k,1}) \end{array}$$

With these preparations, we can now state the following theorem.

**Theorem 3.2.14 (Cf. [101, 5.5]).** *In the setting of 3.2.12 and 3.2.13, assume that* (♠) *and* (♣) *hold for* $W, L$ *and for* $W_1, L|_{W_1}$. *Using the corresponding cellular structures of* $\mathbf{H}$ *and of* $\mathbf{H}_1$, *we obtain subsets* $\Lambda_k^\circ \subseteq \Lambda$ *and* $\Lambda_{k,1}^\circ \subseteq \Lambda_1$ *such that*

$$\mathrm{Irr}(\mathbf{H}_k) = \{L_k^\mu \mid \mu \in \Lambda_k^\circ\} \qquad \textit{and} \qquad \mathrm{Irr}(\mathbf{H}_{k,1}) = \{L_k^{\mu_1} \mid \mu_1 \in \Lambda_{k,1}^\circ\};$$

*see 2.7.5. Then these two subsets determine each other, as follows:*

$$\begin{aligned} \Lambda_{k,1}^\circ &= \{\lambda_1 \in \Lambda_1 \mid E^{\lambda_1} \textit{ appears in the restriction of some } E^\lambda \textit{ where } \lambda \in \Lambda_k^\circ\}, \\ \Lambda_k^\circ &= \{\lambda \in \Lambda \mid \textit{the restriction of } E^\lambda \textit{ contains some } E^{\lambda_1} \textit{ where } \lambda_1 \in \Lambda_{k,1}^\circ\}. \end{aligned}$$

*In particular, if* $\lambda_1 \in \Lambda_{k,1}^\circ$, *then* $\omega.\lambda_1 \in \Lambda_{k,1}^\circ$ *for all* $\omega \in \Omega$.

*Proof.* First assume that $\lambda_1 \in \Lambda_{k,1}^\circ$ and that $\lambda \in \Lambda$ is such that $E^{\lambda_1}$ is a constituent of the restriction of $E^\lambda$ to $W_1$. We must show that $\lambda \in \Lambda_k^\circ$. This is seen as follows. By 2.4.11, $E_\varepsilon^{\lambda_1}$ is a constituent of the restriction of $E_\varepsilon^\lambda$ to $\mathbf{H}_{k,1}$. Since $\lambda_1 \in \Lambda_{k,1}^\circ$, let us apply Lemma 3.2.11: so there exists some $M_1 \in \mathrm{Irr}(\mathbf{H}_{k,1})$ and some $w_1 \in \mathscr{F}_{\lambda_1}$ such that $d_{\lambda_1,M_1} \neq 0$ and $C_{w_1}.M_1 \neq \{0\}$. Then the commutativity of the diagram in 3.2.13 shows that there exists some $M \in \mathrm{Irr}(\mathbf{H}_k)$ such that $d_{\lambda,M} \neq 0$ and $M_1$ is a constituent of the restriction of $M$ to $\mathbf{H}_{k,1}$. The latter condition implies that we also have $C_{w_1}.M \neq \{0\}$. Furthermore, $\mathscr{F}_{\lambda_1} \subseteq \mathscr{F}_\lambda$; see Corollary 2.4.12. So we can apply Lemma 3.2.11 again and conclude that $\lambda \in \Lambda_k^\circ$, as required.

Conversely, let $\lambda \in \Lambda_k^\circ$. As a first step, we show that there exists at least some $\lambda_1 \in \Lambda_{k,1}^\circ$ such that $E^{\lambda_1}$ is a constituent of the restriction of $E^\lambda$ to $W_1$. This is seen as follows. By Lemma 3.2.11, we have $d_{\lambda,M} \neq 0$ and $C_w.M \neq \{0\}$ for some $M \in \mathrm{Irr}(\mathbf{H}_k)$ and some $w \in \mathscr{F}_\lambda$. Writing $w = \omega w_1$, where $\omega \in \Omega$ and $w_1 \in W_1$, we have $C_w = T_\omega C_{w_1}$; see Proposition 2.4.5. Since $T_\omega$ is invertible, we also have $C_{w_1}.M \neq \{0\}$. Thus, $C_{w_1}$ does not act as zero on the restriction of $M$ to $\mathbf{H}_{k,1}$. By Clifford's theorem, that restriction is semisimple; see 3.2.13(b). Hence, there must exist some $M_1 \in \mathrm{Irr}(\mathbf{H}_{k,1})$ such that $C_{w_1}.M_1 \neq \{0\}$ and $M_1$ is a constituent of the restriction of $M$ to $\mathbf{H}_{k,1}$. By the commutativity of the diagram in 3.2.13, we will have $d^1_{\lambda_1,M_1} \neq 0$ for some $\lambda_1 \in \Lambda_1$ such that $E_\varepsilon^{\lambda_1}$ is a constituent of the restriction of $E_\varepsilon^\lambda$ to $\mathbf{H}_{K,1}$. Now, we have $E_\varepsilon^{\lambda_1} \cong W_K(\lambda_1)$ by Proposition 2.7.1. Hence, the fact that $d^1_{\lambda_1,M_1} \neq 0$ implies that $M_1$ is a composition factor of $W_k(\lambda_1)$; see Theorem 3.1.2(c). So, since $C_{w_1}.M_1 \neq \{0\}$, we conclude that $C_{w_1}.W_k(\lambda_1) \neq \{0\}$. Consequently, we also have $C_{w_1}.W(\lambda_1) \neq \{0\}$ and so $C_{w_1}.E_\varepsilon^{\lambda_1} \neq \{0\}$. But then Lemmas 2.3.9 and 2.2.4 show that $y_1 \leqslant_{\mathscr{LR}} w_1$ (with respect to $L|_{W_1}$), where $y_1 \in \mathscr{F}_{\lambda_1}$. On the other hand, by Corollary 2.4.12, we have $\mathscr{F}_{\lambda_1} \subseteq \mathscr{F}_\lambda$. So Remark 2.4.10 implies that $w_1 \in \mathscr{F}_{\lambda_1}$. Now we can apply Lemma 3.2.11 again and conclude that $\lambda_1 \in \Lambda_{k,1}^\circ$, as required.

By Clifford's theorem, as stated in 3.2.13(b), it now remains to show that, for any $\lambda_1 \in \Lambda_1$ and $\omega \in \Omega$, we have

$$\lambda_1 \in \Lambda_{k,1}^\circ \qquad \Rightarrow \qquad \omega.\lambda_1 \in \Lambda_{k,1}^\circ.$$

To prove this implication, we use again Lemma 3.2.11. This shows that if $\lambda_1 \in \Lambda_{k,1}^\circ$, then we have $d^1_{\lambda_1,M_1} \neq 0$ and $C_{w_1}.M_1 \neq \{0\}$ for some $M_1 \in \mathrm{Irr}(\mathbf{H}_{k,1})$ and some $w_1 \in \mathscr{F}_{\lambda_1}$. Now let $\omega \in \Omega$. Then, by 3.2.13(a) and the definition of $M_1^{(\omega)}$, we have

$$d^1_{\omega.\lambda_1,M_1^{(\omega)}} = d^1_{\lambda_1,M_1} \neq 0 \qquad \text{and} \qquad T_\omega C_{w_1} T_\omega^{-1}.M_1^{(\omega)} \neq \{0\}.$$

Now, we have $T_\omega C_{w_1} T_\omega^{-1} = C_{\omega w_1 \omega^{-1}}$; see 2.4.7. Furthermore, by 3.2.12(b), we have $\omega \mathscr{F}_{\lambda_1} \omega^{-1} = \mathscr{F}_{\omega.\lambda_1}$. Hence, another application of Lemma 3.2.11 shows that $\omega.\lambda_1 \in \Lambda_{k,1}^\circ$, as desired. □

**Example 3.2.15.** Let $(W,S)$ be of type $B_n$ with generators and weight function $L\colon W \to \mathbb{Z}$ given by

$B_n$ 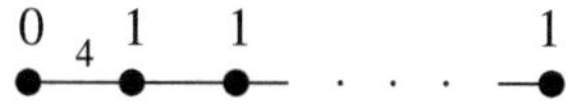

As in Example 2.4.4, we have $\Omega \cong \mathbb{Z}/2\mathbb{Z}$ and $(W_1, S_1)$ is of type $D_n$; furthermore, $L(\tilde{s}) = 1$ for all $\tilde{s} \in S_1$. By 2.4.1(a), Lusztig's conjectures **P1–P15** hold for $W_1, L|_{W_1}$. Hence, they will also hold for $W, L$; see 2.4.8. So, by 2.5.3, the properties (♠) and (♣) hold for **H** and for $\mathbf{H}_1$. By the table in Definition 1.5.9, every subring $R \subseteq \mathbb{C}$ in which 2 is invertible is good with respect to $L$ and with respect to $L|_{W_1}$.

Recall now that $\Lambda$ is the set of all pairs of partitions $(\lambda, \mu)$ such that $|\lambda|+|\mu| = n$; see Example 1.3.9. Given $(\lambda, \mu)$ and $(\lambda', \mu')$ in $\Lambda$, we write $(\lambda, \mu) \sim (\lambda', \mu')$ if $(\lambda, \mu) = (\lambda', \mu')$ or $(\lambda, \mu) = (\mu', \lambda')$. Let $[\lambda, \mu]$ denote the equivalence class of $(\lambda, \mu)$ and let $\bar{\Lambda}$ be the set of all equivalence classes. Then we have

$$\Lambda_1 = \{[\lambda, \mu] \in \bar{\Lambda} \mid \lambda \neq \mu\} \cup \{(\lambda, \pm) \mid [\lambda, \lambda] \in \bar{\Lambda}\};$$

see Example 1.3.10. Now assume that $\Lambda_k^\circ \subseteq \Lambda$ is known and let $\bar{\Lambda}_k^\circ$ be the set of all $[\lambda, \mu] \in \bar{\Lambda}$ such that $(\lambda, \mu) \in \Lambda_k^\circ$ or $(\mu, \lambda) \in \Lambda_k^\circ$. Then Theorem 3.2.14 shows that

$$\Lambda_{k,1}^\circ = \{[\lambda, \mu] \in \bar{\Lambda}_k^\circ \mid \lambda \neq \mu\} \cup \{(\lambda, \pm) \mid [\lambda, \lambda] \in \bar{\Lambda}_k^\circ\}.$$

(Note that $(\lambda, \mu) \in \Lambda_k^\circ$ if and only if $(\mu, \lambda) \in \Lambda_k^\circ$.) Thus, once we have found an explicit description for the subset $\Lambda_k^\circ \subseteq \Lambda$, then we immediately obtain an explicit description for the subset $\Lambda_{k,1}^\circ \subseteq \Lambda_1$. In this way, the problem of determining a canonical basic set in type $D_n$ is reduced to an analogous problem in type $B_n$ (with weight function $L$ as above). This result first appeared in [101, §6].

*Remark 3.2.16.* Assume that $(W, S)$ is irreducible. We have seen in 2.3.8 that (♠) and (♣) hold if the properties **P1**, **P4** and **P15** in Lusztig's Conjecture 2.3.2 hold. By the information summarized in 2.4.1, the only case where these are not yet known to hold is where $W$ is of type $B_n$ and we are neither in the equal parameter case nor in the "asymptotic case" nor in the case considered in Example 3.2.15. In Sections 5.8 and 6.7, we will nevertheless be able to construct explicitly canonical basic sets in type $B_n$ for *any* choice of $L$, under some assumptions on the target field of $\theta$.

## 3.3 Principal Specialisations and Blocks of Defect 1

We keep the basic setting of Section 3.1 where $\mathbf{H} = \mathbf{H}_A(W, S, L)$ is a generic Iwahori–Hecke algebra associated with a finite Coxeter group $W$; see 3.1.12. Now we discuss the important class of specialisations $\theta \colon A \to k$, where the localisation of $A$ in $\ker(\theta)$ is a discrete valuation ring. In this case, we are as close as possible to the classical setting of Brauer's modular representation theory (see, e.g., [53, §16]).

**3.3.1.** The product decomposition of the elements $\mathbf{c}_\lambda$ in Remark 1.3.6 leads us to consider the following set-up. Assume that there exists a subset $\mathscr{P} \subseteq A$ such that the following three conditions hold:

(P1) For each $\lambda \in \Lambda$, we have $\mathbf{c}_\lambda = f_\lambda \varepsilon^{\gamma_\lambda} \prod_{\Phi \in \mathscr{P}} \Phi^{n_{\lambda,\Phi}}$, where $\gamma_\lambda \in \Gamma$ and $n_{\lambda,\Phi} \geqslant 0$.

(P2) Each $\Phi \in \mathscr{P}$ is non-constant (i.e., $\Phi \notin R$) and generates a prime ideal in $A$ such that the quotient ring $A/(\Phi)$ is integrally closed in its field of fractions.
(P3) For $\Phi, \Phi' \in \mathscr{P}$, we have $(\Phi) \neq (\Phi')$ unless $\Phi = \Phi'$.

For each $\Phi \in \mathscr{P}$, the natural map $A \to A/(\Phi)$ gives rise to a specialisation $\theta_\Phi : A \to k_\Phi$, where $k_\Phi$ is the field of fractions of $A/(\Phi)$. One easily checks that the localisation $A_{(\Phi)}$ of $A$ in the prime ideal generated by $\Phi$ is a discrete valuation ring; see [132, Example 7.3.6]. The corresponding valuation $v_\Phi \colon K^\times \to \mathbb{Z}$ is defined by sending a non-zero $a \in A$ to the integer $n \geqslant 0$ such that $a \in (\Phi^n)$ but $a \notin (\Phi^{n+1})$.

**Example 3.3.2.** Let $W$ be a finite Weyl group and assume that we are in the equal parameter case where $\Gamma = \mathbb{Z}$ and $L(s) = 1$ for all $s \in S$. Then $A = R[v, v^{-1}]$ is the ring of Laurent polynomials in $v = \varepsilon$. Since $\mathbb{Z}_W = \mathbb{Z}$, we can assume that $R \subseteq \mathbb{Q} = \mathbb{K}$.

For any integer $d \geqslant 1$, denote by $\Phi_d(v) \in \mathbb{Z}[v]$ the $d$th cyclotomic polynomial. Thus, $\Phi_d(v)$ is irreducible in $\mathbb{Q}[v]$ and we have $v^d - 1 = \prod_{d'|d} \Phi_{d'}(v)$. Also note that:

$$\Phi_d(v^2) = \Phi_{2d}(v) \qquad \text{if } d \text{ is even;}$$
$$\Phi_d(v^2) = \Phi_d(v)\Phi_{2d}(v) \text{ and } \Phi_{2d}(v) = \pm\Phi_d(-v) \qquad \text{if } d \text{ is odd.}$$

Now let $u := v^2$. Then, as discussed in Example 3.1.19, each $\mathbf{c}_\lambda$ lies in $\mathbb{Z}[u, u^{-1}]$ and divides the polynomial $P_W$ in $\mathbb{Q}[u, u^{-1}]$. Using Solomon's formula for $P_W$ (see Example 3.1.19), we see that each $\mathbf{c}_\lambda$ can be written as

$$\mathbf{c}_\lambda = f_\lambda u^{-\mathbf{a}_\lambda} \prod_{d \in E(W)} \Phi_d(u)^{n_{\lambda,d}} \qquad (n_{\lambda,d} \geqslant 0),$$

where $E(W)$ is the set of all integers $d \geqslant 2$ such that $d$ divides some *degree* of $W$. Thus, the three conditions in 3.3.1 hold for

$$\mathscr{P} := \{\Phi_{2d}(v) \mid d \in E(W)\} \cup \{\Phi_d(v) \mid d \in E(W) \text{ odd}\}.$$

Note that, for $\Phi = \Phi_d(v) \in \mathscr{P}$, we have $A/(\Phi) \cong R[\zeta_d]$ where $\zeta_d \in \mathbb{C}$ is a root of unity of order $d$, and it is well-known (see, e.g., [53, (4.5)]) that this is integrally closed in its field of fractions.

**Example 3.3.3.** We generalise the setting in Example 3.3.2 as follows. Let $W$ be any finite Coxeter group, $\Gamma = \mathbb{Z}$ and $L\colon W \to \mathbb{Z}$ be any weight function such that $L(s) \geqslant 0$ for all $s \in S$. Recall that $R$ is assumed to be noetherian and integrally closed in $\mathbb{K}$. As before, $A = R[v, v^{-1}]$ is the ring of Laurent polynomials in one indeterminate $v = \varepsilon$. By Remark 1.3.6, each $\mathbf{c}_\lambda$ still is a Laurent polynomial in $v$ all of whose (complex) roots are roots of unity. Then the set

$$\mathscr{P} := \{\Phi \in A \mid \Phi \text{ is } \mathbb{K}\text{-cyclotomic and divides } \mathbf{c}_\lambda \text{ for some } \lambda \in \Lambda\}$$

satisfies the conditions in 3.3.1. Here, by a "$\mathbb{K}$-cyclotomic polynomial", we mean a non-constant monic polynomial in $R[v]$ which is irreducible in $\mathbb{K}[v]$ and all of whose complex roots are roots of unity. For example, if $\mathbb{K} = \mathbb{Q}[\sqrt{5}]$, then

$$\Phi_{5,a}(v) = v^2 + \tfrac{1}{2}(1+\sqrt{5})v + 1 \qquad \text{and} \qquad \Phi_{5,b} = v^2 + \tfrac{1}{2}(1-\sqrt{5})v + 1$$

are $\mathbb{K}$-cyclotomic polynomials such that $\Phi_{5,a}(v)\Phi_{5,b}(v) = \Phi_5(v) \in \mathbb{Q}[v]$. (These appear in $\mathscr{P}$ for $W$ of type $H_3$ or $H_4$.) We also set

$$E(W,L) := \{d \geqslant 1 \mid \Phi_d(v^2) \in A \text{ is divisible by some } \Phi \in \mathscr{P}\}.$$

Note that $d \geqslant 2$ for all $d \in E(W,L)$. (Indeed, if $1 \in E(W,L)$, then $\Phi = v \pm 1 \in \mathscr{P}$ and we could conclude that $\mathbf{H}_{k_\Phi} \cong \mathbb{K}W$ is not semisimple, which is a contradiction.) Also note that if $W, L$ and $R$ are as in Example 3.3.2, then $E(W,L) = E(W)$.

**Example 3.3.4.** Assume that $W$ is of type $B_n$ and we are in the setting of Example 3.1.20, where $R = \mathbb{Z}$ and $A$ is the ring of Laurent polynomials in two independent variables $V, v$. Then the three conditions in 3.3.1 hold for the set

$$\begin{aligned} \mathscr{P} = \{&\Phi_{2d}(v) \mid 2 \leqslant d \leqslant n\} \cup \{\Phi_d(v) \mid 2 \leqslant d \leqslant n, d \text{ odd}\} \\ &\cup \{V^2 + v^{\pm 2r} \mid 0 \leqslant r \leqslant n-1\}. \end{aligned}$$

Note that, if $\Phi = V^2 + v^{\pm 2r} \in \mathscr{P}$ then $A/(\Phi) \cong \mathbb{Z}[\sqrt{-1}, v^{\pm 1}]$, while if $\Phi = \Phi_d(v) \in \mathscr{P}$ then $A/(\Phi) \cong \mathbb{Z}[\zeta_d, V^{\pm 1}]$, where $\zeta_d \in \mathbb{C}$ is a root of unity of order $d$. In both cases, $A/(\Phi)$ is integrally closed in its field of fractions (see, e.g., [132, Exc. 7.2]).

We are now ready to introduce the following general notion.

**Definition 3.3.5.** Assume that $R$ is $L_0$-good. Let $\theta\colon A \to k$ be a specialisation where $k$ is the field of fractions of $\theta(A)$. Let

$$\mathscr{O} = \{a/b \in K \mid a, b \in A \text{ and } \theta(b) \neq 0\}$$

be the localization of $A$ in $\ker(\theta)$. We say that $\theta$ is a *principal specialisation* if $\mathscr{O}$ is a discrete valuation ring.

In this case, a general argument shows that for every finite-dimensional $\mathbf{H}_K$-module $V$ there exists some $\mathbf{H}_{\mathscr{O}}$-module which is free of finite rank over $\mathscr{O}$ such that $V \cong K \otimes_{\mathscr{O}} N$; see, for example, [132, 7.3.7]. We can also define a decomposition map $d_{\mathscr{O},\theta}\colon R_0(\mathbf{H}_K) \to R_0(\mathbf{H}_k)$ using $\mathscr{O}$ instead of $A$; note that $\theta$ has a natural extension to a map $\mathscr{O} \to k$. Then Remark 3.1.5 shows that $d_\theta = d_{\mathscr{O},\theta}$. We usually denote by $\Phi \in \mathscr{O}$ a generator for the unique maximal ideal of $\mathscr{O}$, and replace a subscript "$\theta$" by "$\Phi$". Thus, we write $d_\Phi$ instead of $d_\theta$, and $D_\Phi$ denotes the decomposition matrix. Furthermore, the $\theta$-blocks will be called $\Phi$-blocks.

Note that we can (and usually will) assume without loss of generality that $\Phi \in A$. We denote by $v_\Phi\colon K^\times \to \mathbb{Z}$ the valuation associated with $\mathscr{O}$. Thus, any $0 \neq x \in K$ can be written uniquely in the form $x = \Phi^n u$, where $n = v_\Phi(x)$ and $u \in \mathscr{O}^\times$. Furthermore, we have $\mathscr{O} = \{x \in K^\times \mid v_\Phi(x) \geqslant 0\} \cup \{0\}$.

We assume for the remainder of this section that $\theta$ is a principal specialisation, with corresponding valuation ring $\mathscr{O}$ and maximal ideal generated by $\Phi \in A$. For later reference, we shall now state the following result which hold thanks to this assumption; this is a very useful technical tool both for computational and for theoretical purposes. (See, for example, Proposition 3.3.12 and Theorem 3.7.16.)

**Proposition 3.3.6 (Geck and Rouquier; see [132, 7.5.3]).** *Given $M \in \mathrm{Irr}(\mathbf{H}_k)$, we define a function $P_M\colon Z(\mathbf{H}) \to K$ by the formula*

$$P_M(z) := \sum_{\lambda \in \Lambda} \mathbf{c}_\lambda^{-1} d_{\lambda,M}\, \omega_\lambda(z) \qquad (z \in Z(\mathbf{H})),$$

*where $\omega_\lambda\colon Z(\mathbf{H}) \to A$ denotes the central character of $E_\varepsilon^\lambda$. Then we have $P_M(z) \in \mathscr{O}$ for all $z \in Z(\mathbf{H})$. In particular, taking $z = 1$, we have $\sum_{\lambda\in\Lambda} \mathbf{c}_\lambda^{-1} d_{\lambda,M} \in \mathscr{O}$.*

(The proof relies on the fact that idempotents can be lifted from $\mathbf{H}_k$ to $\mathbf{H}_{\mathscr{O}}$.)

*Remark 3.3.7.* Recall the definition of the character table $X(\mathbf{H})$ and of the polynomials $\{f_{w,C}\}$ in Example 1.2.3. These polynomials can be used to construct a basis of $Z(\mathbf{H})$. Indeed, let us define

(a) $$z_C = \sum_{w\in W} \varepsilon^{-2L(w)} f_{w,C}\, \dot{T}_{w^{-1}} \qquad \text{for any } C \in \mathrm{Cl}(W).$$

Then, by a result due to Geck and Rouquier (see [132, 8.2.4]), the elements $\{z_C \mid C \in \mathrm{Cl}(W)\}$ form an $A$-basis of $Z(\mathbf{H})$. Using the above formula, the orthogonality relations in Proposition 1.2.12 can be rewritten as follows:

(b) $$\sum_{C\in\mathrm{Cl}(W)} \chi^\lambda(\dot{T}_{w_C})\, \omega_\mu(z_C) = \begin{cases} \mathbf{c}_\lambda & \text{if } \lambda = \mu, \\ 0 & \text{otherwise.} \end{cases}$$

(See also [132, Exc. 9.5].) Multiplying the defining formula for $P_M(z_C)$ by $\chi^\mu(\dot{T}_{w_C})$ and summing over all $C$, we obtain

(c) $$d_{\lambda,M} = \sum_{C\in\mathrm{Cl}(W)} \chi^\lambda(\dot{T}_{w_C})\, P_M(z_C) \qquad \text{for all } \lambda \in \Lambda \text{ and } M \in \mathrm{Irr}(\mathbf{H}_k).$$

**Example 3.3.8.** There is a recursive formula for computing the polynomials $f_{w,C}$; see [132, 8.2.7]. Hence, in concrete examples, one can work out the central elements $z_C$, but explicit closed formulae only seem to be known in some special cases. Here is one such special case. Let $C \in \mathrm{Cl}(W)$ be such that $C \cap S \neq \varnothing$; that is, $C$ consists of *reflections*. Then, by Iancu [170, §3], we have the explicit formula

$$z_C = \sum_{w\in C} \varepsilon^{-L(w_C)-L(w)} \dot{T}_w, \qquad \text{that is,} \qquad f_{w,C} = \begin{cases} \varepsilon^{L(w)-L(w_C)} & \text{if } w \in C, \\ 0 & \text{if } w \notin C. \end{cases}$$

The elements $\{z_C \mid C \in \mathrm{Cl}(W)$ such that $C \cap S \neq \varnothing\}$ appear to be suitable generalisations of the *Jucys–Murphy elements* in types $A$, $B$; see Example 3.4.8 and 3.5.1.

**Definition 3.3.9.** Let $\lambda \in \Lambda$ and recall that $0 \neq \mathbf{c}_\lambda \in A$. The integer $\mathbf{d}_\lambda := v_\Phi(\mathbf{c}_\lambda) \geqslant 0$ is called the *$\Phi$-defect* of $\lambda$. The *$\Phi$-defect* of a $\Phi$-block $\Lambda_1 \subseteq \Lambda$ is defined by

$$\mathbf{d}(\Lambda_1) := \max\{\mathbf{d}_\lambda \mid \lambda \in \Lambda_1\}.$$

As in the representation theory of finite groups (where these notions had their origin; see [53, §56]), the defect of a block is a rough measure of how complicated the representation theory (i.e., the decomposition matrix, etc.) of that block is.

Note that saying that $\lambda \in \Lambda$ has $\Phi$-defect 0 simply means that $\theta(\mathbf{c}_\lambda) \neq 0$. Thus, Proposition 3.1.17 describes the decomposition matrices of $\Phi$-blocks of defect 0.

There is also a sophisticated theory of blocks of defect 1, due to Brauer in the framework of the modular representation theory of finite groups; e.g. see [53, §62]. It was shown in [92] that this theory can be adapted to the framework of decomposition maps of Iwahori–Hecke algebras. Here, we shall present a simplified version of this adaptation, taking into account from an early point our results on cell data from Chapter 2. The discussion, closely following the original work of Brauer as presented in [138, §11], will rely on some subtle matrix-theoretic arguments. The following remarks are a preparation for this discussion.

**3.3.10.** Recall from Section 1.3 that $\mathrm{Irr}(\mathbf{H}_K) = \{E_\varepsilon^\lambda \mid \lambda \in \Lambda\}$. Choosing a basis of $E_\varepsilon^\lambda$, we obtain a matrix representation $\rho^\lambda : \mathbf{H}_K \to M_{d_\lambda}(K)$, where $d_\lambda := \dim E_\varepsilon^\lambda$. As before, let $M(\lambda)$ be an indexing set for that basis. Given $h \in \mathbf{H}$ and $\mathfrak{s}, \mathfrak{t} \in M(\lambda)$, the $(\mathfrak{s}, \mathfrak{t})$-entry of $\rho^\lambda(h)$ will be denoted by $\rho_{\mathfrak{st}}^\lambda(h)$. As already mentioned above, since $\mathscr{O}$ is a discrete valuation ring, we can choose the basis of $E_\varepsilon^\lambda$ such that

$$\rho_{\mathfrak{st}}^\lambda(T_w) \in \mathscr{O} \quad \text{for all } w \in W \text{ and } \mathfrak{s}, \mathfrak{t} \in M(\lambda).$$

Then denote by $\bar{\rho}^\lambda : \mathbf{H}_k \to M_{d_\lambda}(k)$ the representation of $\mathbf{H}_k$ obtained by extension of scalars via the map $\mathscr{O} \to k$. In general, $\bar{\rho}^\lambda$ will not be irreducible. But we can always choose the basis to be adapted to a composition series of $\bar{\rho}^\lambda$. More precisely, let $\mathrm{Irr}(\mathbf{H}_k) = \{M_1, \ldots, M_r\}$, where $\dim M_i = l_i$ for $1 \leqslant i \leqslant r$. Choosing a basis of $M_i$, we obtain a matrix representation $\sigma_i : \mathbf{H}_k \to M_{l_i}(k)$. Then, as in [138, (11.9)], the above basis of $E_\varepsilon^\lambda$ can be chosen such that the matrices $\bar{\rho}^\lambda(T_w)$ (where $T_w$ now also is a basis element of $\mathbf{H}_k$) will have an upper block triangular shape

$$\bar{\rho}^\lambda(T_w) = \begin{pmatrix} \sigma_{i_1}(T_w) & * & \cdots & * \\ 0 & \sigma_{i_2}(T_w) & \ddots & \vdots \\ \vdots & \ddots & \ddots & * \\ 0 & \cdots & 0 & \sigma_{i_t}(T_w) \end{pmatrix} \qquad \text{for all } w \in W,$$

where $i_1, \ldots, i_t \in \{1, \ldots, r\}$. Thus, the decomposition number $d_{\lambda, M_i}$ is given by

$$d_{\lambda, M_i} = \text{ number of indices in the list } (i_1, \ldots, i_t) \text{ which are equal to } i.$$

In what follows, we will assume that the above choices have been made.

**Lemma 3.3.11 (Brauer; cf. [138, 11.8], [92, 9.1]).** *Let $\lambda, \mu \in \Lambda$ be such that $\mathbf{d}_\lambda = \mathbf{d}_\mu = 1$. Assume there exist $\mathfrak{s}, \mathfrak{t} \in M(\lambda)$ and $\mathfrak{u}, \mathfrak{v} \in M(\mu)$ such that*

$$\rho_{\mathfrak{st}}^\lambda(h) \equiv \rho_{\mathfrak{uv}}^\mu(h) \bmod \Phi \qquad \textit{for all } h \in \mathbf{H}.$$

*Then we have* $\delta_{\mathfrak{st}}\,\mathbf{c}_\lambda - 2\delta_{\lambda\mu}\,\delta_{\mathfrak{s}\mathfrak{v}}\,\delta_{\mathfrak{t}\mathfrak{u}}\,\mathbf{c}_\lambda + \delta_{\mathfrak{u}\mathfrak{v}}\,\mathbf{c}_\mu \equiv 0 \bmod \Phi^2$.

*Proof.* Our assumption implies that we have $\rho^\lambda_{\mathfrak{st}}(T_w) - \rho^\mu_{\mathfrak{uv}}(T_w) \equiv 0 \bmod \Phi$ and $\rho^\lambda_{\mathfrak{st}}(T_{w^{-1}}) - \rho^\mu_{\mathfrak{uv}}(T_{w^{-1}}) \equiv 0 \bmod \Phi$ for all $w \in W$. For each fixed $w$, we multiply together these two congruences and then sum over all $w \in W$; the result is

$$\sum_{w\in W}\Big(\rho^\lambda_{\mathfrak{st}}(T_w) - \rho^\mu_{\mathfrak{uv}}(T_w)\Big)\Big(\rho^\lambda_{\mathfrak{st}}(T_{w^{-1}}) - \rho^\mu_{\mathfrak{uv}}(T_{w^{-1}})\Big) \equiv 0 \bmod \Phi^2.$$

We expand the terms in the sum on the left-hand side and consider the resulting four sums. Using the Schur relations in Proposition 1.2.12, we can evaluate these sums and obtain that $\delta_{\mathfrak{st}}\,\mathbf{c}_\lambda - 2\delta_{\lambda\mu}\,\delta_{\mathfrak{s}\mathfrak{v}}\,\delta_{\mathfrak{t}\mathfrak{u}}\,\mathbf{c}_\lambda + \delta_{\mathfrak{u}\mathfrak{v}}\,\mathbf{c}_\mu \equiv 0 \bmod \Phi^2$, as required. □

**Proposition 3.3.12 (Brauer; cf. [138, 11.9], [92, 9.3]).** *Let* $\Lambda_1 \subseteq \Lambda$ *be a* $\Phi$*-block of defect* 1 *and* $\mathrm{Irr}_1(\mathbf{H}_k)$ *the corresponding subset of* $\mathrm{Irr}(\mathbf{H}_k)$. *Assume that* $\mathrm{char}(k) \neq 2$. *Let* $M \in \mathrm{Irr}_1(\mathbf{H}_k)$. *Then the set* $\mathscr{S}_\Phi(M) := \{\lambda \in \Lambda_1 \mid d_{\lambda,M} \neq 0\}$ *contains precisely two elements, and we have* $d_{\lambda,M} \in \{0,1\}$ *for all* $\lambda \in \Lambda_1$.

*Proof.* First we show that $d_{\lambda,M} \leqslant 1$ for all $\lambda \in \Lambda_1$. Assume, if possible, that $d_{\lambda,M} \geqslant 2$. In the setting of 3.3.10, saying that $d_{\lambda,M_i} \geqslant 2$ (where $M \cong M_i$) means that two diagonal blocks in $\bar{\rho}^\lambda(T_w)$ are actually equal for all $w \in W$. Thus, there exist $\mathfrak{s} \neq \mathfrak{t}$ in $M(\lambda)$ such that $\bar{\rho}^\lambda_{\mathfrak{ss}}(T_w) = \bar{\rho}^\lambda_{\mathfrak{tt}}(T_w)$ or, in other words, $\rho^\lambda_{\mathfrak{ss}}(T_w) \equiv \rho^\lambda_{\mathfrak{tt}}(T_w) \bmod \Phi$ for all $w \in W$. Applying Lemma 3.3.11 with $\lambda = \mu$ yields that $\mathbf{c}_\lambda - 2\delta_{\mathfrak{st}}\delta_{\mathfrak{st}}\mathbf{c}_\lambda + \mathbf{c}_\lambda \equiv 2\mathbf{c}_\lambda \equiv 0 \bmod \Phi^2$. Since $\mathrm{char}(k) \neq 2$, this implies that $\Phi^2 \mid \mathbf{c}_\lambda$ and so $\mathbf{d}_\lambda \geqslant 2$, which is a contradiction. So all decomposition numbers with respect to $\Lambda_1$ are 0 or 1.

Next consider the set $\mathscr{S}_\Phi(M)$. Assume, if possible, that $\mathscr{S}_\Phi(M)$ contains only one element, $\lambda_0 \in \Lambda_1$ say. We have just proved that then $d_{\lambda_0,M} = 1$. Hence, the sum in Proposition 3.3.6 reduces to $\mathbf{c}_{\lambda_0}^{-1}$ and so we conclude that $\lambda_0$ has defect 0. So, by Proposition 3.1.17, $\{\lambda_0\}$ would form a $\Phi$-block by itself, which is a contradication. Consequently, $\mathscr{S}_\Phi(M)$ contains at least two elements, $\lambda \neq \mu$ say. Recall that $\rho^\lambda$ and $\rho^\mu$ are chosen as in 3.3.10. Then saying that $d_{\lambda,M_i} \neq 0$ and $d_{\mu,M_i} \neq 0$ (where $M \cong M_i$) means that $\sigma_i$ occurs as a diagonal block both in $\bar{\rho}^\lambda$ and in $\bar{\rho}^\mu$. In particular, there are $\mathfrak{s} \in M(\lambda)$ and $\mathfrak{u} \in M(\mu)$ such that $\bar{\rho}^\lambda_{\mathfrak{ss}}(T_w) = \bar{\rho}^\mu_{\mathfrak{uu}}(T_w)$ or, in other words, $\rho^\lambda_{\mathfrak{ss}}(T_w) \equiv \rho^\mu_{\mathfrak{uu}}(T_w) \bmod \Phi$ for all $w \in W$. Applying Lemma 3.3.11 yields that $\mathbf{c}_\lambda + \mathbf{c}_\mu \equiv 0 \bmod \Phi^2$ (since $\lambda \neq \mu$).

Finally assume, if possible, that $\mathscr{S}_\Phi(M)$ contains a third element, $\nu$ say. Then we can repeat the above argument with $\lambda,\nu$ and with $\mu,\nu$ instead of $\lambda,\mu$. So we conclude that $\mathbf{c}_\lambda \equiv -\mathbf{c}_\nu \bmod \Phi^2$ and $\mathbf{c}_\mu \equiv -\mathbf{c}_\nu \bmod \Phi^2$. These congruences would again imply that $2\mathbf{c}_\lambda \equiv 0 \bmod \Phi^2$, which is impossible since $\mathrm{char}(k) \neq 2$. Thus, we have shown that $\mathscr{S}_\Phi(M)$ contains precisely two elements. □

**Theorem 3.3.13 (Geck [92, §9], [97, §4], Geck and Rouquier [134, 4.4]).** *Let* $\leqslant$ *be a monomial order on* $\Gamma$ *such that* $L(s) \geqslant 0$ *for all* $s \in S$. *Assume that* $R$ *is* $L$*-good and* $\mathbf{H}$ *admits a cell datum as in Theorem 2.6.12. Also assume that* $\mathrm{char}(k) \neq 2$. *Let* $\Lambda_1$ *be a* $\Phi$*-block of defect* 1 *and* $\mathrm{Irr}_1(\mathbf{H}_k)$ *be the corresponding subset of* $\mathrm{Irr}(\mathbf{H}_k)$.

(a) *Suppose that* $|\Lambda_1| = n \geqslant 2$. *Then there is a unique labelling* $\Lambda_1 = \{\lambda_1,\ldots,\lambda_n\}$ *such that* $\lambda_n \lhd_L \lambda_{n-1} \lhd_L \cdots \lhd_L \lambda_1$ *(with* $\unlhd_L$ *defined as in Theorem 2.6.12).*

*(b) The Brauer graph of* $\Lambda_1$ *is a tree, as follows:*

*(c) The edges in the above tree are in bijection with* $\mathrm{Irr}_1(\mathbf{H}_k)$. *Thus,* $|\mathrm{Irr}_1(\mathbf{H}_k)| = n-1$ *and we can write* $\mathrm{Irr}_1(\mathbf{H}_k) = \{M_1, \ldots, M_{n-1}\}$, *where* $M_j$ *labels the edge joining* $\lambda_j$ *and* $\lambda_{j+1}$. *With this notation, we have*

$$d_{\lambda_i, M_j} = \begin{cases} 1 & \text{if } i = j \text{ or } i = j+1, \\ 0 & \text{otherwise.} \end{cases}$$

*Moreover, assuming that* (♣) *holds, then* $\mathbf{a}_{\lambda_1} < \mathbf{a}_{\lambda_2} < \cdots < \mathbf{a}_{\lambda_n}$ *and* $\mathscr{B}_{\Phi,1} = \{\lambda_1, \ldots, \lambda_{n-1}\}$ *is a canonical basic set for* $\Lambda_1$.

*Proof.* Our assumptions imply that $\mathrm{Irr}(\mathbf{H}_k) = \{L_k^\mu \mid \mu \in \Lambda_k^\circ\}$, with $\Lambda_k^\circ \subseteq \Lambda$ as in 2.7.5; furthermore, as in Proposition 3.2.7, for $\lambda \in \Lambda$ and $M \in \mathrm{Irr}(\mathbf{H}_k)$, we have $d_{\lambda,M} = (W_k(\lambda) : L_k^\mu)$, where $\mu \in \Lambda_k^\circ$ is such that $M \cong L_k^\mu$.

Let $\Lambda_{k,1}^\circ \subseteq \Lambda_k^\circ$ be the subset such that $\mathrm{Irr}_1(\mathbf{H}_k) = \{L_k^\mu \mid \mu \in \Lambda_{k,1}^\circ\}$. Let $M \in \mathrm{Irr}_1(\mathbf{H}_k)$. The following remark tells us how we can identify the element $\mu \in \Lambda_{k,1}^\circ$ such that $M \cong L_k^\mu$. By Proposition 3.3.12, the set $\mathscr{S}_\Phi(M) = \{\lambda \in \Lambda \mid d_{\lambda,M} \neq 0\}$ has precisely two elements. We claim that

$$(*) \qquad \mathscr{S}_\Phi(M) = \{\lambda_1, \lambda_2\}, \quad \text{where } \lambda_2 \lhd_L \lambda_1 \text{ and } M \cong L_k^{\lambda_1}.$$

Indeed, let $\lambda_1 \in \Lambda_{k,1}^\circ$ be such that $M \cong L_k^{\lambda_1}$. By Theorem 2.6.6, $(W_k(\lambda_1) : L_k^{\lambda_1}) = 1$ and so $\lambda_1 \in \mathscr{S}_\Phi(M)$. Let $\lambda_2$ be the second element of $\mathscr{S}_\Phi(M)$. Then $\lambda_2 \lhd_L \lambda_1$ since $(W_k(\lambda_2) : L_k^{\lambda_1}) \neq 0$; see again Theorem 2.6.6. Thus, $(*)$ is proved.

We use $(*)$ to eliminate certain configurations on the Brauer graph of $\Lambda_1$. First assume, if possible, that this Brauer graph contains a piece as follows:

(I)
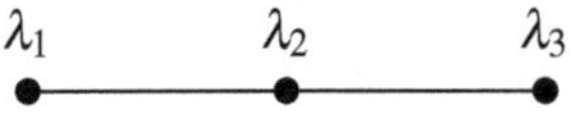

where $\lambda_1 \lhd_L \lambda_2$, $\lambda_3 \lhd_L \lambda_2$.

Let $M \in \mathrm{Irr}_1(\mathbf{H}_k)$ label the edge between $\lambda_1$ and $\lambda_2$. Then $\mathscr{S}_\Phi(M) = \{\lambda_1, \lambda_2\}$ and $(*)$ shows that $M \cong L_k^{\lambda_2}$. Similarly, let $M' \in \mathrm{Irr}_1(\mathbf{H}_k)$ label the edge between $\lambda_2$ and $\lambda_3$. Then $\mathscr{S}_\Phi(M') = \{\lambda_2, \lambda_3\}$ and $(*)$ shows, again, that $M' \cong L_k^{\lambda_2}$. Thus, $M \cong M'$ but $\mathscr{S}_\Phi(M) \neq \mathscr{S}_\Phi(M')$, which is a contradiction.

Next assume, if possible, that our Brauer graph contains a piece as follows:

(II)
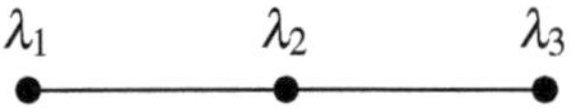

where $\lambda_2 \lhd_L \lambda_1$, $\lambda_2 \lhd_L \lambda_3$.

Then we use the automorphism † and its compatibility with blocks and decomposition numbers; see 3.1.16. Thus, $\Lambda_1^\dagger = \{\lambda^\dagger \mid \lambda \in \Lambda_1\}$ also is a $\Phi$-block and its Brauer graph contains a piece as follows:

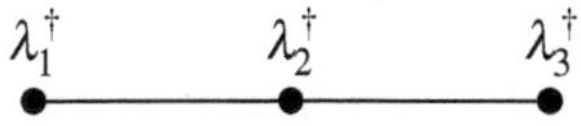

where $\lambda_2 \lhd_L \lambda_1$, $\lambda_2 \lhd_L \lambda_3$.

Note also that $\mathbf{c}_{\lambda^\dagger}$ equals $\mathbf{c}_\lambda$ (up to a monomial $\varepsilon^\gamma$ for some $\gamma \in \Gamma$); see Example 1.3.4. Hence, $\Lambda_1^\dagger$ is a block of $\Phi$-defect 1. By Remark 2.2.11, we have $\lambda_1^\dagger \trianglelefteq_L \lambda_2^\dagger$ and $\lambda_3^\dagger \trianglelefteq_L \lambda_2^\dagger$. Hence, we have found a piece in $\Lambda_1^\dagger$ that looks like (I), which is contradiction.

We now prove (a) and (b). Note that $(*)$ implies that, if $\lambda_1 \neq \lambda_2$ are connected by an edge in the Brauer graph of $\Lambda_1$, then $\mathscr{S}_\Phi(M) = \{\lambda_1, \lambda_2\}$ for some $M \in \mathrm{Irr}_1(\mathbf{H}_k)$ and so $\lambda_1 \triangleleft_L \lambda_2$ or $\lambda_2 \triangleleft_L \lambda_1$. Thus, the labels on any two vertices connected by an edge in the Brauer graph of $\Lambda_1$ are always related by $\trianglelefteq_L$. Then the impossibility of having any piece which looks like (I) or (II) means that, along every path in the Brauer graph of $\Lambda_1$, the labels of the vertices are either strictly increasing or strictly decreasing (with respect to $\trianglelefteq_L$). First of all, this shows that the Brauer graph of $\Lambda_1$ does not have any closed cyles; that is, it is a tree. Second, it shows that there is no bifurcation point. Indeed, if there was a bifurcation point, then there would exist four labels $\lambda_0, \lambda_a, \lambda_b, \lambda_c \in \Lambda_1$ with no edges among $\lambda_a, \lambda_b, \lambda_c$ but with each of these connected to $\lambda_0$. But then at least one path passing through $\lambda_0$ would contain a piece which looks like (I) or (II), which is a contradiction. This completes the proof of (a) and (b). Finally, consider (c). Let $1 \leqslant j \leqslant n-1$. Then $\mathscr{S}_\Phi(M) = \{\lambda_j, \lambda_{j+1}\}$ for some $M \in \mathrm{Irr}_1(\mathbf{H}_k)$, by $(*)$ and the definition of the Brauer graph. We claim that $M$ is uniquely determined (up to isomorphism). Indeed, if $\mathscr{S}_\Phi(M') = \{\lambda_j, \lambda_{j+1}\}$ and $M \not\cong M'$, then the two columns of the decomposition matrix of $\Lambda_1$ corresponding to $M$ and $M'$ would be equal, contradicting the fact that $d_\Phi$ is surjective (see Theorem 3.1.14). Thus, each edge determines a unique $M \in \mathrm{Irr}_1(\mathbf{H}_k)$. Every $M \in \mathrm{Irr}_1(\mathbf{H}_k)$ arises in this way, since $M$ must label some edge in the Brauer graph. Furthermore, each $M \in \mathrm{Irr}_1(\mathbf{H}_k)$ labels a unique edge, since $\mathscr{S}_\Phi(M)$ only has two elements. Hence, the edges in the tree in (b) correspond bijectively to $\mathrm{Irr}_1(\mathbf{H}_k)$. Thus, (c) is proved.

The final statement about the invariants $\mathbf{a}_\lambda$ and $\mathscr{B}_{\Phi,1}$ is now also clear. $\square$

*Remark 3.3.14.* The fact that the Brauer graph of a $\Phi$-block of defect 1 is an open polygon and that its decomposition matrix is given as in Theorem 3.3.13(c) can be shown without reference to (♠) or (♣); see [92, Theorem 9.6]. The references to (♠) and (♣) are needed to establish the ordering in Theorem 3.3.13(a).

**Example 3.3.15.** Assume that we are in the equal parameter case where $\Gamma = \mathbb{Z}$ and $L(s) = 1$ for all $s \in S$. In particular, $A = R[v, v^{-1}]$ is the ring of Laurent polynomials in $v = \varepsilon$. Let $e \geqslant 2$ and supppose that $\mathbb{K}$ contains all roots of unity of order $e$. We consider the *$\Phi_e$-modular specialisation*

$$\theta_e \colon A \to \mathbb{K}, \qquad v \mapsto \zeta_{2e} := \exp(\pi\sqrt{-1}/e) \in \mathbb{K};$$

see Example 3.1.15(b). Note that $\theta_e$ is a principal specialisation where $\Phi = v - \zeta_{2e}$. We shall denote extension of scalars from $A$ to $\mathbb{K}$ (via $\theta_e$) simply by a subscript "$(e)$". Thus, for example, $\mathbf{H}_{(e)} := \mathbb{K} \otimes_A \mathbf{H}$ is the corresponding specialised algebra. The assumptions of Proposition 3.2.7 are satisfied. Hence, we have a cell datum for $\mathbf{H}$ (see Theorem 2.6.12) and there is a natural parametrisation

$$\mathrm{Irr}(\mathbf{H}_{(e)}) = \{L_{(e)}^\mu \mid \mu \in \Lambda_{(e)}^\circ\}, \qquad \text{where } \Lambda_{(e)}^\circ \subseteq \Lambda \text{ is a canonical basic set.}$$

The corresponding decomposition matrix will be denoted by $D_{(e)}$; it has rows labelled by $\Lambda$ and columns labelled by $\Lambda^{\circ}_{(e)}$.

(a) Assume that $W$ is a finite Weyl group and let $e = h$ be the Coxeter number of $W$; see [132, §1.5] and note that $h$ is the largest *degree* of $W$. It is shown by Bleher, Geck and Kimmerle [20, §6] that there is precisely one $\Phi_h$-block of defect 1 and all other $\Phi_h$-blocks have defect 0. The Brauer tree of the block of defect 1 is given by

where $\lambda^{(i)}$ labels the $i$th exterior power of the reflection representation of $W$. (Thus, $\lambda^{(0)}$ corresponds to the unit representation and $\lambda^{(|S|)}$ to the sign representation.)

(b) Now assume that $(W,S)$ is of exceptional type $H_3$, $H_4$, $F_4$, $E_6$, $E_7$ or $E_8$. In each of these cases, the partition into $\Phi_e$-blocks has been determined in [132, Prop. 11.5.13]; the blocks of positive $\Phi_e$-defect are explicitly listed in [132, Appendix F]. For example, in type $H_3$, we find the following table:

| $e$ | $d$ | $\Phi_e$-blocks of defect $d$ |
|---|---|---|
| 2 | 3 | $\{1_r, \overline{3}'_s, 3'_s, 5_r, 5'_r, \overline{3}_s, 3_s, 1'_r\}$ |
| 3 | 1 | $\{1_r, 5_r, 4'_r\}$ $+$dual |
| 5 | 1 | $\{1_r, 4'_r, \overline{3}_s\}$ $+$dual |
| 6 | 1 | $\{1_r, 5_r, 5'_r, 1'_r\}$ |
| 10 | 1 | $\{1_r, 3'_s, 3_s, 1'_r\}$ |

(Here, "dual" indicates a block obtained by applying the map $\lambda \mapsto \lambda^{\dagger}$; see 3.1.16.) Thus, for $e \in \{3,5,6,10\}$, all blocks have defect 0 or 1. In the above table (and, more generally, in all tables in [132, App. F]), the representations belonging to a block of defect 1 are ordered according to increasing value of $\mathbf{a}_{\lambda}$. Hence, by Theorem 3.3.13, we can immediately write down the corresponding decomposition matrix. For example, the block for $e = 10$ (the Coxeter number) has a Brauer tree as described in (a). There is a further $\Phi_2$-block of defect 3. Its decomposition matrix has been determined by Müller [253, §4]; it is given by the first of the two matrices in Table 3.3. Taking into account the invariants $\mathbf{a}_{\lambda}$ (which Müller did not do), one can easily determine the corresponding canonical basic set; see Remark 3.2.3. In the second matrix, the canonical basic set is marked by "•".

*Remark 3.3.16.* Given $\lambda, \mu \in \Lambda$, it may be conjectured that $\mathbf{d}_{\lambda} = \mathbf{d}_{\mu}$ whenever $E^{\lambda}_{\varepsilon}$, $E^{\mu}_{\varepsilon}$ belong to the same $\Phi$-block. Under the assumption that $W$ is a finite Weyl group and we are in the equal parameter case, this is shown by a general argument in [92, Prop. 7.4], using an analogous result for blocks of finite groups of Lie type [91, 1.4]. For $W$ of exceptional type, this property can also be verified by inspection using the explicit knowledge of the $\Phi_e$-blocks; see [132, Appendix F]. It would be interesting to find a general argument purely in the framework of Iwahori–Hecke algebras.

In general, the determination of the decomposition matrix of a $\Phi$-block of defect $\geqslant 2$ is a hard problem. For $W$ of exceptional type, this problem can be attacked

**Table 3.3** A canonical basic set in type $H_4$

| $e=2$ | $d_{\lambda,M}$ | | |
|---|---|---|---|
| $1'_r$ | 1 | . | . |
| $1_r$ | 1 | . | . |
| $5'_r$ | 1 | 1 | 1 |
| $5_r$ | 1 | 1 | 1 |
| $3_s$ | 1 | 1 | . |
| $\overline{3}_s$ | 1 | . | 1 |
| $3'_s$ | 1 | 1 | . |
| $\overline{3}'_s$ | 1 | . | 1 |

$\rightsquigarrow$

| $e=2$ | $\mathbf{a}_\lambda$ | $d_{\lambda,M}$ | | |
|---|---|---|---|---|
| $\bullet\ 1_r$ | 0 | 1 | . | . |
| $\bullet\ 3'_s$ | 1 | 1 | 1 | . |
| $\bullet\ \overline{3}'_s$ | 1 | 1 | . | 1 |
| $5_r$ | 2 | 1 | 1 | 1 |
| $5'_r$ | 5 | 1 | 1 | 1 |
| $3_s$ | 6 | 1 | 1 | . |
| $\overline{3}_s$ | 6 | 1 | . | 1 |
| $1'_r$ | 15 | 1 | . | . |

(Here, and in all subsequent tables, the dot stands for 0)

by computational methods; see Section 7.1 for further details. A solution for $W$ of classical type $A_{n-1}$, $B_n$ and $D_n$ has been achieved through quite remarkable (and, perhaps, rather unexpected) connections with something completely different: the theory of canonical bases for quantized envelopping algebras. A first step in this direction will be taken in the following section.

## 3.4 Towards Canonical Basic Sets for Classical Types

In this section, we lay the theoretical foundations for studying canonical basic sets for Iwahori–Hecke algebras of classical type $X_n \in \{A_{n-1}, B_n, D_n\}$. The essential feature that we can exploit here is that we naturally have an infinite chain $W(X_1) \subseteq W(X_2) \subseteq W(X_3) \subseteq \cdots$ and so we may try to use inductive arguments. This will be formalised in the concept of an "abstract Fock datum"; in this form, it will also apply to the Ariki–Koike algebras in Chapter 5.

**3.4.1.** Let $\underline{H}$ be an $A$-algebra as in 3.1.1. Let $K$ be the field of fractions of $A$ and $\theta: A \to k$ be a ring homomorphism into a field $k$ such that $k$ is the field of fractions of $\theta(A)$. We assume that $A$ is integrally closed in $K$; furthermore, we assume that the $K$-algebra $\underline{H}_K$ is split semisimple and the $k$-algebra $\underline{H}_k$ is split. We have an associated decomposition map $d_\theta: R_0(\underline{H}_K) \to R_0(\underline{H}_k)$ as in Theorem 3.1.2. We wish to develop a method for constructing "good" approximations to the columns of the corresponding decomposition matrix $D_\theta$. By definition, an abstract *Fock datum* for $\underline{H}$ consists of a family $\{(\underline{H}_m, z_m) \mid 1 \leqslant m \leqslant n\}$ satisfying the following conditions.

(F1) $\underline{H}_1 := \langle 1_{\underline{H}} \rangle_A \subseteq \underline{H}_2 \subseteq \ldots \subseteq \underline{H}_n = \underline{H}$ and each $\underline{H}_i$ is an $A$-subalgebra of $\underline{H}$.
(F2) Each $\underline{H}_m$ is finitely generated and free over $A$; furthermore, $\underline{H}_m$ is free as a left $\underline{H}_{m-1}$-module.
(F3) For each $m$, we have $z_m \in \underline{H}_m$ and $z_m$ belongs to the centre of $\underline{H}_m$.
(F4) For each $m$, the $K$-algebra $\underline{H}_{K,m} := K \otimes_A \underline{H}_m$ is split semisimple and the $k$-algebra $\underline{H}_{k,m} := k \otimes_A \underline{H}_m$ is split.

For each $m$, we write $\operatorname{Irr}(\underline{H}_{K,m}) = \{V^\lambda \mid \lambda \in \underline{\Lambda}_m\}$, where $\underline{\Lambda}_m$ is a finite indexing set.

For any $\lambda \in \underline{\Lambda}_m$, denote by $\omega_\lambda$ the central character of $V^\lambda$; see Definition 3.1.9. Since $z_m \in Z(\underline{H}_m)$, we have $\omega_\lambda(z_m) \in A$. So we can define

$$\bar{\omega}_\lambda(z_m) := \theta(\omega_\lambda(z_m)) \qquad \text{for all } m \text{ and } \lambda \in \underline{\Lambda}_m.$$

For $m \geqslant 1$, denote by $\mathrm{Ind}_K^{(m)}$ the induction of representations from $\underline{H}_{K,m-1}$ to $\underline{H}_{K,m}$ (where $\underline{H}_0 = \{0\}$ by convention). Thus, for any $\underline{H}_{K,m-1}$-module $V$, we have

$$\mathrm{Ind}_K^{(m)}(V) := \underline{H}_{K,m} \otimes_{\underline{H}_{K,m-1}} V.$$

This operation can be refined as follows. For $\mu \in \underline{\Lambda}_{m-1}$ and $\lambda \in \underline{\Lambda}_m$, let

$$m(\lambda,\mu) := \text{ multiplicity of } V^\lambda \text{ as an irreducible constituent of } \mathrm{Ind}_K^{(m)}(V^\mu).$$

Then, for a fixed $\xi \in k$, we define

$$\xi\text{-}\mathrm{Ind}_K^{(m)}(V^\mu) := \bigoplus_\lambda (\underbrace{V^\lambda \oplus \ldots \oplus V^\lambda}_{m(\lambda,\mu) \text{ times}}),$$

where the sum runs over all $\lambda \in \underline{\Lambda}_m$ such that $\bar{\omega}_\lambda(z_m) - \bar{\omega}_\mu(z_{m-1}) = \xi$. By linearity, we can define $\xi\text{-}\mathrm{Ind}_K^{(m)}(V)$ for any $\underline{H}_{K,m}$-module $V$.

**Definition 3.4.2.** Given an abstract Fock datum for $\underline{H}$ as above, we define a family of representations $\mathscr{E}$ of $\underline{H}_K$ inductively as follows. For $m = 1$, let $\mathscr{E}_1 = \{\mathbf{1}\}$, where $\mathbf{1}$ denotes the trivial representation of $\underline{H}_{K,1} = \langle 1_{\underline{H}} \rangle_K$. Then set

$$\mathscr{E}_m = \{\xi\text{-}\mathrm{Ind}_K^{(m)}(V) \mid V \in \mathscr{E}_{m-1} \text{ and } \xi \in k\} \setminus \{0\} \quad \text{for } 2 \leqslant m \leqslant n.$$

Let us also set $\mathscr{E} := \mathscr{E}_n$. Thus, starting from $\mathbf{1}$, the representations in $\mathscr{E}$ are obtained by successive applications of the refined induction operators defined above.

*Remark 3.4.3.* Let $1 \leqslant m \leqslant n$ and $Y \in \mathscr{E}_m$. For $\lambda \in \underline{\Lambda}_m$, denote by $m(Y,\lambda)$ the multiplicity of $V^\lambda$ as an irreducible constituent of $Y$. Then the inductive definition of $\mathscr{E}_m$ immediately shows that, for any $\lambda, \lambda' \in \underline{\Lambda}_m$, we have the implication

$$m(Y,\lambda) > 0 \quad \text{and} \quad m(Y,\lambda') > 0 \qquad \Rightarrow \qquad \bar{\omega}_\lambda(z_m) = \bar{\omega}_{\lambda'}(z_m).$$

Thus, we can define $\xi(Y) \in k$ by the condition that $\bar{\omega}_\lambda(z_m) = \xi(Y)$ for all $\lambda \in \underline{\Lambda}_m$ such that $m(Y,\lambda) > 0$. We also define $d_Y := \big(m(Y,\lambda)\big)_{\lambda \in \underline{\Lambda}_m}$. The following key result shows how these vectors are related to the decomposition matrix $D_\theta$ of $\underline{H}$.

**Theorem 3.4.4.** *In the above setting, assume that $A$ is the ring of Laurent polynomials in finitely many indeterminates over a Dedekind domain. Then, for each $Y \in \mathscr{E}$, the vector $d_Y$ is a linear combination of the columns of $D_\theta$, where the coefficients are non-negative integers. In particular, we have*

$$|\,\mathrm{Irr}(\underline{H}_k)| \geqslant \mathrm{rank}\big(m(Y,\lambda)\big)_{Y \in \mathscr{E}, \lambda \in \underline{\Lambda}} \qquad (\underline{\Lambda} := \underline{\Lambda}_n).$$

*Proof.* The assumptions on $A$ imply that $A$ is a "regular" ring; see Matsumura [248, Theorem 19.5]. Consequently, one can apply the general construction in [248, Exc. 14.4]. This yields a discrete valuation ring $\hat{A} \subseteq K$ such that $A \subseteq \hat{A}$ and $\mathfrak{p} \cap A = \ker(\theta)$, where $\mathfrak{p}$ is the maximal ideal of $\hat{A}$. (See also [132, Exc. 8.7] for some further details.) Let $\hat{k}$ be the residue field of $\hat{\mathscr{O}}$. Since $\mathfrak{p} \cap A = \ker(\theta)$, there is an embedding $\iota \colon k \hookrightarrow \hat{k}$ and so $\theta$ extends to a map $\hat{\theta} \colon \hat{A} \to \hat{k}$ such that $\hat{\theta}|_A = \iota \circ \theta$. Then, by Remark 3.1.5, the decomposition matrices associated with $\theta$ and with $\hat{\theta}$ are the same, up to possibly permuting some columns. Thus, replacing $A$ by $\hat{A}$ and $k$ by $\hat{k}$, we can assume without loss of generality that $A$ is a discrete valuation ring; in particular, $\theta \colon A \to k$ is a principal specialisation in the sense of Section 3.3. Let

$$D_\theta = \big(d_{\lambda,M}\big)_{\lambda \in \underline{\Lambda}, M \in \mathrm{Irr}(\underline{H}_k)}$$

be the decomposition matrix of $\underline{H}$. We shall now also need some results about projective $\underline{H}$-modules. Since $\underline{H}_K$ is split semisimple, a result of Heller (see [53, 30.18]) shows that the Krull–Schmidt–Azuyama theorem holds for $\underline{H}$-modules which are finitely generated and free over $A$. (Such $\underline{H}$-modules will be called $\underline{H}$-lattices; if $M$ is an $\underline{H}$-lattice, we denote $M_K := K \otimes_A M$ and $M_k := k \otimes_A M$.) Furthermore, idempotents can be lifted from $\underline{H}_k$ to $\underline{H}$ (see [53, Exc. 6.16]). Thus, the general results from "Brauer's modular representation theory" are valid without passing to the completion of $A$. In particular, the following statement holds (see Feit [83, §I.17], [132, 7.5.2], and also 4.1.11 in Chapter 4). Let $P$ be a projective indecomposable $\underline{H}$-lattice. Then $P_k$ is a projective indecomposable $\underline{H}_k$-module and, hence, $M := P_k/\mathrm{rad}(P_k) \in \mathrm{Irr}(\underline{H}_k)$; furthermore, in $R_0(\underline{H}_K)$, we have

$$[P_K] = \sum_{\lambda \in \underline{\Lambda}} d_{\lambda,M}\,[V^\lambda] \qquad \text{"Brauer reciprocity".}$$

Thus, the vector containing the multiplicities of the various simple modules of $\underline{H}_K$ as irreducible constituents of $P_K$ is a column of the matrix $D_\theta$. Consequently, given any projective $\underline{H}$-lattice $P$ (not necessarily indecomposable), the corresponding vector of multiplicities will be a sum of some columns of $D_\theta$. Thus, we see that it is enough to show that, for each $Y \in \mathscr{E}$, there exists a projective $\underline{H}$-lattice $P$ such that $Y \cong P_K$.

We shall prove this by using the inductive definition of $\mathscr{E} = \mathscr{E}_n$; note that all of the above remarks hold not only for $\underline{H}$, but also for any subalgebra $\underline{H}_m$ where $m \geqslant 1$. First let $m = 1$. Then $\mathscr{E}_1 = \{\mathbf{1}\}$ and, clearly, $\mathbf{1}$ is obtained by scalar extension from a projective $\underline{H}_1$-lattice. Now assume that $m \geqslant 2$ and let $Y \in \mathscr{E}_m$. By definition, we have $Y = \xi\text{-}\mathrm{Ind}_K^{(m)}(Y')$ for some $Y' \in \mathscr{E}_{m-1}$ and some $\xi \in k$. So, arguing by induction, we already know that there exists a projective $\underline{H}_{m-1}$-lattice $P'$ such that $Y' \cong P'_K$. Now, the process of inducing representations also works over $A$, and (F2) implies that

$$\tilde{P} := \mathrm{Ind}_{\underline{H}_{m-1}}^{\underline{H}_m}(P') = \underline{H}_m \otimes_{\underline{H}_{m-1}} P' \quad \text{is a projective } \underline{H}_m\text{-lattice.}$$

We have $\mathrm{Ind}_K^{(m)}(Y') \cong \tilde{P}_K$; hence, we will try to find the required projective $\underline{H}_m$-lattice $P$ as a direct summand of $\tilde{P}$. For this purpose, consider the Brauer graph of

$\underline{H}_m$ and the corresponding $\theta$-blocks. Thus, as in 3.1.8, we have a partition

$$\mathrm{Irr}(\underline{H}_{K,m}) = \mathrm{Irr}_1(\underline{H}_{K,m}) \sqcup \cdots \sqcup \mathrm{Irr}_r(\underline{H}_{K,m}),$$

Now, by Brauer reciprocity, the irreducible constituents (over $K$) of a projective indecomposable $\underline{H}_m$-lattice all belong to the same $\theta$-block. Thus, we have a canonical decomposition

$$\tilde{P} = P_1 \oplus \cdots \oplus P_r,$$

where $P_i$ is the sum of all indecomposable direct summands of $\tilde{P}$ whose irreducible constituents (over $K$) belong to $\mathrm{Irr}_i(\underline{H}_{K,m})$. Consequently, by Lemma 3.1.10, for each $i \in \{1, \ldots, r\}$ there exists a well-defined $\xi_i \in k$ such that

$$\bar{\omega}_\lambda(z_m) = \xi_i \qquad \text{for all } \lambda \in \underline{\Lambda}_m \text{ such that } V^\lambda \text{ is a constituent of } P_{K,i}.$$

Thus, setting $P := \bigoplus_i P_i$, where the sum runs over all $i \in \{1, \ldots, r\}$ such that $\xi_i - \xi(Y') = \xi$, we have $Y \cong P_K$, as required. □

**Proposition 3.4.5.** *In the setting of Theorem 3.4.4, assume that $d_\theta \colon R_0(\underline{H}_K) \to R_0(\underline{H}_k)$ is surjective and that we are given a partial order $\sqsubseteq$ on $\mathrm{Irr}(\underline{H}_K)$. Furthermore, assume that there exists a subset $\underline{\Lambda}' \subseteq \underline{\Lambda}$ and a set of representations $\{P_\mu \mid \mu \in \underline{\Lambda}'\}$ of $\underline{H}_K$ such that the following conditions hold.*

*(a) We have $|\mathrm{Irr}(\underline{H}_k)| \leqslant |\underline{\Lambda}'|$.*
*(b) For each $\mu \in \underline{\Lambda}'$, some positive multiple of $P_\mu$ lies in $\mathscr{E}$.*
*(c) For each $\mu \in \underline{\Lambda}'$, we have the following relation in $R_0(\underline{H}_K)$:*

$$[P_\mu] = [V^\mu] + \sum_{\lambda \in \underline{\Lambda} : V^\lambda \sqsubset V^\mu} m_{\lambda\mu} [V^\lambda], \qquad \textit{where } m_{\lambda\mu} \in \mathbb{Z}_{\geqslant 0} \textit{ for all } \lambda.$$

*Then $|\mathrm{Irr}(\underline{H}_k)| = |\underline{\Lambda}'|$ and $\underline{B}_\theta := \{V^\mu \mid \mu \in \underline{\Lambda}'\}$ is a canonical basic set for $\underline{H}_k$ with respect to $\sqsubseteq$, in the sense of 3.1.7.*

*Proof.* For any $\mu \in \underline{\Lambda}'$, let $d_\mu$ denote the vector which contains the multiplicities of the various $V^\lambda$ ($\lambda \in \underline{\Lambda}$) as constituents of $P_\mu$. By (b) and Theorem 3.4.4, there exists a positive integer $p_\mu$ such that $p_\mu d_\mu$ is a sum of some columns of $D_\theta$. Now, the assumption that $d_\theta$ is surjective means that the elementary divisors of the matrix $D_\theta$ are all 1. This implies that $d_\mu$ must already be a sum of some columns of $D_\theta$. Using (c), we see that there exists a column of $D_\theta$ which has coefficient 1 in the row labelled by $\mu$; furthermore, the coefficient in a row labelled by any $\lambda \in \underline{\Lambda}$ is 0 unless $\lambda \sqsubseteq \mu$. Let us denote this column of $D_\theta$ by $d'_\mu$.

By (c), the vectors $\{d_\mu \mid \mu \in \underline{\Lambda}'\}$ are linearly independent. Hence, $D_\theta$ has at least $|\underline{\Lambda}'|$ columns. Using (a), we conclude that $D_\theta$ has exactly $|\underline{\Lambda}'|$ columns. It follows that $D_\theta$ consists precisely of the column vectors $\{d'_\mu \mid \mu \in \underline{\Lambda}'\}$. Consequently, we can write $\mathrm{Irr}(\underline{H}_k) = \{M^\mu \mid \mu \in \underline{\Lambda}'\}$ and $D_\theta = (d_{\lambda,\mu})_{\lambda \in \underline{\Lambda}, \mu \in \underline{\Lambda}'}$, where

$$(\Delta') \qquad \begin{aligned} d_{\mu,\mu} &= 1 && \text{for all } \mu \in \underline{\Lambda}', \\ d_{\lambda,\mu} &= 0 && \text{unless } \lambda = \mu \text{ or } V^\lambda \sqsubset V^\mu. \end{aligned}$$

Thus, the conditions in 3.1.7 are satisfied for the set $\underline{B}_\theta := \{V^\mu \mid \mu \in \underline{\Lambda}'\}$. □

*Remark 3.4.6.* In the above discussion of Fock data, we have worked with only one central element $z_m$ in each of the subalgebras $\underline{H}_m$, for $m = 1, \ldots, n$. Of course, one could also work with collections of central elements $\{z_{m,i} \mid i \in I_m\} \subseteq Z(\underline{H}_m)$ at each stage. The refined induction operator in 3.4.1 would then have to be defined as follows. For $\underline{\xi} := \{\xi_{ij} \mid i \in I_m, j \in I_{m-1}\} \subseteq k$, we let

$$\underline{\xi}\text{-}\mathrm{Ind}_K^{(m)}(V^\mu) := \bigoplus_\lambda (\underbrace{V^\lambda \oplus \ldots \oplus V^\lambda}_{m(\lambda,\mu) \text{ times}}),$$

where the sum runs over all $\lambda \in \underline{\Lambda}_m$ such that $\bar{\omega}_\lambda(z_{m,i}) - \bar{\omega}_\mu(z_{m-1,j}) = \xi_{ij}$ for all $i \in I_m$ and $j \in I_{m-1}$. With this modification, it is clear that Theorem 3.4.4 and Proposition 3.4.5 continue to hold without any further change.

**3.4.7.** We return to our basic set-up as in 3.1.12, where $\mathbf{H} = \mathbf{H}_A(W,S,L)$ is a generic Iwahori–Hecke algebra over $A = R[\Gamma]$ and $R$ is assumed to be $L_0$-good. Let $\theta \colon A \to k$ be a specialisation into a field $k$, where $k$ is the field of fractions of $\theta(A)$. By Theorem 3.1.14, we have a corresponding decomposition map $d_\theta \colon R_0(\mathbf{H}_K) \to R_0(\mathbf{H}_k)$; let $D_\theta$ be the decomposition matrix. In order to exhibit an abstract Fock datum for $\mathbf{H}$, let us consider a chain of parabolic subgroups $W_1 = \{1\} \subseteq W_2 \subseteq \cdots \subseteq W_n = W$ and the corresponding chain of parabolic subalgebras

$$\mathbf{H}_1 = \langle T_1 \rangle_A \subseteq \mathbf{H}_2 \subseteq \cdots \subseteq \mathbf{H}_n = \mathbf{H}.$$

Then (F1) and (F2) hold by general results on parabolic subalgebras; see [132, §9.1]. Since $R$ is $L_0$-good, (F4) holds by Lemma 3.1.13. So it remains to choose suitable central elements $z_m \in Z(\mathbf{H}_m)$ for $1 \leqslant m \leqslant n$. There are several candidates for $z_m$; perhaps the most natural idea is to use properties of the longest element in $W_m$.

**Example 3.4.8.** Let $n \geqslant 1$ and $(W,S)$ be of type $A_{n-1}$ or $B_n$ where $S = \{s_1, \ldots, s_{n-1}\}$ (in type $A_{n-1}$) or $S = \{s_0, s_1, \ldots, s_{n-1}\}$ (in type $B_n$), with the labelling in Table 1.1. Having fixed this labelling, there is a natural chain of parabolic subgroups. Indeed, as in 1.7.6, for any $m \in \{1, \ldots, n\}$, let $W_m \subseteq W$ be the parabolic subgroup generated by $s_0, s_1, \ldots, s_{m-1}$ (where we set $s_0 = 1$ in type $A_{n-1}$). Then $W_m$ is of type $A_{m-1}$ or $B_m$ respectively. In particular, we have $W_n = W$. We set

$$\mathscr{L}_1 := T_{s_0} \quad \text{and} \quad \mathscr{L}_m := T_{s_{m-1}} \cdots T_{s_2} T_{s_1} T_{s_0} T_{s_1} T_{s_2} \cdots T_{s_{m-1}} \qquad \text{for } 2 \leqslant m \leqslant n.$$

(Note that $T_{s_0}$ is the identity element in type $A_{n-1}$.) The elements $\mathscr{L}_m$ will be called the *Jucys–Murphy elements* of $\mathbf{H}_n$; see Hoefsmit [157], Mathas [245, §3.3] and Ram [268, (3.16), (4.20)]. We shall set

$$z_m := \mathscr{L}_1 + \mathscr{L}_2 + \cdots + \mathscr{L}_m \in \mathbf{H}_m \qquad \text{for } 1 \leqslant m \leqslant n.$$

The following result shows that $z_m \in Z(\mathbf{H}_m)$ for all $m$.

**Lemma 3.4.9.** *Let $w_0 \in W_n$ be the longest element. Then the following hold.*

*(a) If $(W,S)$ is of type $A_{n-1}$, then $T_{w_0}^2 = \mathscr{L}_1\mathscr{L}_2\cdots\mathscr{L}_n \in Z(\mathbf{H}_n)$.*
*(b) If $(W,S)$ is of type $B_n$, then $T_{w_0} = \mathscr{L}_1\mathscr{L}_2\cdots\mathscr{L}_n \in Z(\mathbf{H}_n)$.*
*(c) The elements $\mathscr{L}_1,\ldots,\mathscr{L}_n$ commute with each other.*
*(d) We have $z_n = \mathscr{L}_1 + \cdots + \mathscr{L}_n \in Z(\mathbf{H}_n)$.*

*Proof.* (a) We use induction on $n$. If $n=1$, then $W_1=\{1\}$ and $T_{w_0} = \mathscr{L}_1 = T_1$; so the assertion is clear. Now assume that $n \geqslant 2$. By the description of $w_0$ in [132, Exp. 1.5.4], we have $w_0 = w_0' s_{n-1}s_{n-2}\cdots s_1$, where $w_0'$ is the longest element in $W_{n-1} = \langle s_1,\ldots,s_{n-2}\rangle \cong \mathfrak{S}_{n-1}$. By induction, $T_{w_0'}^2 = \mathscr{L}_1\mathscr{L}_2\cdots\mathscr{L}_{n-1}$ and so

$$\mathscr{L}_1\mathscr{L}_2\cdots\mathscr{L}_n = T_{w_0'}\big(T_{w_0'}T_{s_{n-1}}\cdots T_{s_2}T_{s_1}\big)\big(T_{s_1}T_{s_2}\cdots T_{s_{n-1}}\big) = T_{w_0'}T_{w_0}T_{s_1}T_{s_2}\cdots T_{s_{n-1}}.$$

Now, for any $i \in \{1,\ldots,n-1\}$, we have $w_0 s_i = s_{n-i}w_0$ and so $T_{w_0}T_{s_i} = T_{s_{n-i}}T_{w_0}$. Applying this relation repeatedly, we obtain

$$T_{w_0}T_{s_1}T_{s_2}\cdots T_{s_{n-1}} = T_{s_{n-1}}T_{w_0}T_{s_2}\cdots T_{s_{n-1}} = \cdots = T_{s_{n-1}}\cdots T_{s_2}T_{s_1}T_{w_0}$$

and so $\mathscr{L}_1\mathscr{L}_2\cdots\mathscr{L}_n = T_{w_0'}T_{s_{n-1}}\cdots T_{s_2}T_{s_1}T_{w_0} = T_{w_0}^2$, as desired. Finally, we have already noted in Example 1.2.7 that $T_{w_0}^2 \in Z(\mathbf{H}_n)$.

(b) Again, we use induction on $n$. If $n=1$, then $W_1 = \langle T_{s_0}\rangle$ has order 2 and $T_{w_0} = \mathscr{L}_1 = T_{s_0}$; so the assertion is clear. Now assume that $n \geqslant 2$. By the description of $w_0$ in [132, Exp. 1.4.6], we have

$$w_0 = w_0' t_{n-1}, \qquad \text{where} \qquad t_{n-1} := s_{n-1}s_{n-2}\cdots s_1 s_0 s_1\cdots s_{n-2}s_{n-1}$$

and $w_0'$ is the longest element in $W_{n-1} = \langle s_0, s_1,\ldots,s_{n-2}\rangle$. Note that $\mathscr{L}_n = T_{t_{n-1}}$. By induction, $T_{w_0'} = \mathscr{L}_1\mathscr{L}_2\cdots\mathscr{L}_{n-1}$ and so $\mathscr{L}_1\mathscr{L}_2\cdots\mathscr{L}_n = T_{w_0'}T_{t_{n-1}} = T_{w_0}$, as required. Finally, by [132, Exp. 1.4.6], $w_0$ lies in the centre of $W_n$ and, hence, $T_{w_0} \in Z(\mathbf{H}_n)$.

(c) Let $h_m := \mathscr{L}_1\cdots\mathscr{L}_m \in \mathbf{H}_m$ for $1 \leqslant m \leqslant n$. By (a) and (b), we have $h_m \in Z(\mathbf{H}_m)$. It follows that the elements $h_1,\ldots,h_n$ commute with each other. Furthermore, each $h_m$ equals a power of a standard basis element of $\mathbf{H}_m$ and, hence, is invertible in $\mathbf{H}_m$. So, since $h_m = h_{m-1}\mathscr{L}_m$ for $m \geqslant 2$, the elements $\mathscr{L}_1,\ldots,\mathscr{L}_n$ also commute with each other. This argument also shows that $\mathscr{L}_m$ commutes with $T_{s_i}$ for $0 \leqslant i \leqslant m-2$.

(d) Again, we use induction on $n$. If $n=1$, then $z_1 = \mathscr{L}_1$ and the assertion is clear by (a) and (b). Now assume that $n \geqslant 2$. We have $z_n = z_{n-1} + \mathscr{L}_n$. By induction, $z_{n-1} \in Z(\mathbf{H}_{n-1})$ and so $z_{n-1}$ commutes with $T_{s_i}$ for $0 \leqslant i \leqslant n-2$. As already noted in the proof of (c), the element $\mathscr{L}_n$ also commutes with $T_{s_i}$ for $0 \leqslant i \leqslant n-2$. Hence, it remains to show that $z_n$ commutes with $T_{s_{n-1}}$. Now, it is clear that $T_{s_{n-1}}$ commutes with all elements in $\mathbf{H}_{n-2}$; in particular, $T_{s_{n-1}}$ will commute with $\mathscr{L}_1,\ldots,\mathscr{L}_{n-2}$. So, finally, it will be enough to show that $T_{s_{n-1}}$ commutes with $\mathscr{L}_{n-1} + \mathscr{L}_n$. Now, we have $\mathscr{L}_n = T_{s_{n-1}}\mathscr{L}_{n-1}T_{s_{n-1}}$ and so

$$\begin{aligned} T_{s_{n-1}}\mathscr{L}_n &= T_{s_{n-1}}^2\mathscr{L}_{n-1}T_{s_{n-1}} = \mathscr{L}_{n-1}T_{s_{n-1}} + (v-v^{-1})\mathscr{L}_n, \\ \mathscr{L}_nT_{s_{n-1}} &= T_{s_{n-1}}\mathscr{L}_{n-1}T_{s_{n-1}}^2 = T_{s_{n-1}}\mathscr{L}_{n-1} + (v-v^{-1})\mathscr{L}_n. \end{aligned}$$

This yields $T_{s_{n-1}}\mathscr{L}_n - \mathscr{L}_n T_{s_{n-1}} = \mathscr{L}_{n-1}T_{s_{n-1}} - T_{s_{n-1}}\mathscr{L}_{n-1}$ and so $\mathscr{L}_{n-1}+\mathscr{L}_n$ commutes with $T_{s_{n-1}}$, as required. □

For our applications to Iwahori–Hecke algebras, the following version of Proposition 3.4.5 will be useful.

**Proposition 3.4.10.** *In the setting of 3.4.7, let $z_m \in Z(\mathbf{H}_m)$ for $1 \leqslant m \leqslant n$ and consider the corresponding set of representations $\mathscr{E}$. Assume that it is known that a canonical basic set exists for $\mathbf{H}_k$; see Definition 3.2.1. Furthermore, assume that there exists a subset $\mathscr{E}' \subseteq \mathscr{E}$ such that the following conditions hold.*

*(a) We have $|\operatorname{Irr}(\mathbf{H}_k)| \leqslant |\mathscr{E}'|$.*

*(b) For each $Y \in \mathscr{E}'$, there exists some $\lambda_Y \in \Lambda$ such that, in $R_0(\mathbf{H}_K)$, we have*

$$[Y] = m(Y,\lambda_Y)[E_\varepsilon^{\lambda_Y}] + \sum_{\mu\in\Lambda:\mathbf{a}_\mu>\mathbf{a}_{\lambda_Y}} m(Y,\mu)\,[E_\varepsilon^{\mu}],$$

*where $m(Y,\mu) \in \mathbb{Z}_{\geqslant 0}$ for all $\mu$ and $m(Y,\lambda_Y) \neq 0$.*

*(c) We have $\lambda_Y \neq \lambda_{Y'}$ for $Y \neq Y'$ in $\mathscr{E}'$.*

*Then $|\operatorname{Irr}(\mathbf{H}_k)| = |\mathscr{E}'|$ and $\Lambda' := \{\lambda_Y \mid Y \in \mathscr{E}'\}$ is the canonical basic set for $\mathbf{H}_k$.*

*Proof.* Recall that $A = R[\Gamma]$ is the ring of Laurent polynomials in finitely many indeterminates. Furthermore, we can assume without loss of generality that either $\mathbb{K} = R$ or that $\mathbb{K}$ is a finite extension of $\mathbb{Q}$ and that $R$ is the ring of algebraic integers in $\mathbb{K}$. In both cases, the general assumption on $A$ in Theorem 3.4.4 is satisfied.

Now let $\Lambda' := \{\lambda_Y \mid Y \in \mathscr{E}'\} \subseteq \Lambda$. By (c), the map $\mathscr{E}' \to \Lambda'$, $Y \mapsto \lambda_Y$, is a bijection and so we can also write $\mathscr{E}' = \{Y_\mu \mid \mu \in \Lambda'\}$, where $\lambda_{Y_\mu} = \mu$ for all $\mu \in \Lambda'$.

For any $\mu \in \Lambda'$, let $d_\mu$ denote the vector which contains the multiplicities of the various $E_\varepsilon^\lambda$ ($\lambda \in \Lambda$) as constituents of $Y_\mu$. Using (b) we see that there exists a column of $D_\theta$ which has a non-zero coefficient in the row labelled by $\mu$; furthermore, the coefficient in a row labelled by any $\lambda \in \Lambda$ is 0 unless $\lambda = \mu$ or $\mathbf{a}_\lambda > \mathbf{a}_\mu$. Let us denote this column of $D_\theta$ by $d'_\mu$. As in the proof of Proposition 3.4.5, we can conclude that $D_\theta$ consists precisely of the column vectors $\{d'_\mu \mid \mu \in \Lambda'\}$. Furthermore, we can write $\operatorname{Irr}(\mathbf{H}_k) = \{M^\mu \mid \mu \in \Lambda'\}$ and $D_\theta = (d_{\lambda,\mu})_{\lambda\in\Lambda,\mu\in\Lambda'}$, where

$$(\Delta') \qquad \begin{array}{ll} d_{\mu,\mu} \neq 0 & \text{for all } \mu \in \Lambda', \\ d_{\lambda,\mu} = 0 & \text{unless } \lambda = \mu \text{ or } \mathbf{a}_\lambda > \mathbf{a}_\mu. \end{array}$$

Now, we are assuming that $\mathbf{H}_k$ admits a canonical basic set. Hence, comparing $(\Delta')$ with the relations $(\Delta^a)$ in Remark 3.2.2, we conclude that we must have $d_{\mu,\mu} = 1$ for all $\mu \in \Lambda'$, and $\Lambda'$ is the canonical basic set. □

We close this section by explaining a general method for computing $|\operatorname{Irr}(\mathbf{H}_k)|$.

**Proposition 3.4.11 ([132, 7.5.7]).** *Recall from Example 1.2.3 the definition of the character table $X(\mathbf{H})$. Then, in the setting of Theorem 3.1.14, we have*

$$|\operatorname{Irr}(\mathbf{H}_k)| = \text{rank of } X(\mathbf{H})_\theta,$$

*where $X(\mathbf{H})_\theta$ is the matrix obtained from $X(\mathbf{H})$ by applying $\theta$ to each entry.*

*Proof.* By Theorem 3.1.14, the columns of the corresponding decomposition matrix $D_\theta$ are linearly independent over $k$. Hence we can apply [132, Cor. 7.5.7], which yields that $|\operatorname{Irr}(\mathbf{H}_k)|$ can be expressed by the rank of the above matrix. □

This has been used in [132, Prop. 11.5.13] to determine the numbers $|\operatorname{Irr}(\mathbf{H}_k)|$ where $W$ is of exceptional type and $k = \mathbb{C}$. We shall now apply this to $W \cong \mathfrak{S}_n$.

**3.4.12.** Let $(W,S)$ be of type $A_{n-1}$ and keep the notation of Example 3.2.9. In particular, $A = R[v, v^{-1}]$ is the ring of Laurent polynomials in one indeterminate $v = \varepsilon$. Following Dipper and James [62], given a specialisation $\theta\colon A \to k$, we set

$$e := \min\{i \geqslant 2 \mid 1 + \theta(u) + \theta(u)^2 + \cdots + \theta(u)^{i-1} = 0\} \qquad (u := v^2).$$

(We let $e = \infty$ if $1 + \theta(u) + \theta(u)^2 + \cdots + \theta(u)^{i-1} \neq 0$ for all $i \geqslant 2$.) Thus, if $\theta(u) \neq 1$, then $e$ is the multiplicative order of $\theta(u) \in k^\times$. On the other hand, if $\theta(u) = 1$ and $\operatorname{char}(k) = \ell > 0$, then $e = \ell$; finally, if $\theta(u) = 1$ and $\operatorname{char}(k) = 0$, then $e = \infty$. By the discussion in Example 3.1.19, we have

$$\mathbf{H}_k \text{ is semisimple} \qquad \Leftrightarrow \qquad e > n.$$

(Recall that the degrees of $W \cong \mathfrak{S}_n$ are $2, 3, \ldots, n$.)

**Lemma 3.4.13 (Cf. [131, 4.3]).** *In the setting of 3.4.12, let $\Lambda_e'$ denote the set of all partitions of $n$ which have no part divisible by $e$. Then we have $|\operatorname{Irr}(\mathbf{H}_k)| \leqslant |\Lambda_e'|$, where equality holds if* $\operatorname{char}(k) = 0$.

*Proof.* First we deal with the case where $e = \infty$. Then $\mathbf{H}_k$ is semisimple by 3.4.12. So $|\operatorname{Irr}(\mathbf{H}_k)| = |\operatorname{Irr}_k(\mathfrak{S}_n)|$ is the number of all partitions of $n$; see 3.1.18. On the other hand, all partitions of $n$ are in $\Lambda_e'$ in this case. Thus, the statement is true if $e = \infty$. We can assume from now on that $e < \infty$.

Now, by Proposition 3.4.11, the problem is equivalent to computing the rank of the specialised matrix $X(\mathbf{H})_\theta$. We shall use *Starkey's rule* (see [99], [132, 9.2.11], [285]), which shows that the character table $X(\mathbf{H})$ is obtained by multiplying the ordinary character table of the group $W \cong \mathfrak{S}_n$ by the matrix $R := \big(r_{\lambda\mu}\big)_{\lambda,\mu \vdash n}$, where

$$r_{\lambda\mu} = \begin{cases} \dfrac{|C_\lambda \cap W_\mu|}{|W_\mu|}(u-1)^{-l(\mu)} \prod\limits_{i \geqslant 1}(u^{\lambda_i} - 1) & \text{if } C_\lambda \cap W_\mu \neq \varnothing, \\ 0 & \text{otherwise.} \end{cases}$$

Here, $l(\mu)$ denotes the number of non-zero parts of $\mu$; furthermore, $C_\lambda$ denotes the conjugacy class of elements of $W \cong \mathfrak{S}_n$ of cycle type $\lambda$ and $W_\mu \subseteq W$ denotes the parabolic subgroup corresponding to the Young subgroup $\mathfrak{S}_\mu \subseteq \mathfrak{S}_n$. In particular, all entries of $X(\mathbf{H})$ lie in the ring $B := \mathbb{Z}[u, u^{-1}]$.

Now assume first that $\operatorname{char}(k) = 0$ and let $q := \theta(u)$. Then $e \geqslant 2$ is the multiplicative order of $q \in k^\times$ and so we have, for any integer $m \geqslant 1$:

$$(*) \qquad \theta(u^m - 1) = 0 \qquad \Leftrightarrow \qquad e \text{ divides } m.$$

Now, we must show that $\operatorname{rank}(\bar{R}) = |\Lambda'_e|$, where $\bar{R} := \big(\theta(r_{\lambda\mu})\big)_{\lambda,\mu\vdash n}$.

This is seen as follows. Let $\lambda$ be a partition of $n$ which is not in $\Lambda'_e$; that is, some part $\lambda_i$ of $\lambda$ is divisible by $e$. Then $(*)$ shows that $\theta(u^{\lambda_i} - 1) = 0$ and so $\theta(r_{\lambda\mu}) = 0$ for all $\mu \vdash n$. Thus, we have $\operatorname{rank}(\bar{R}) \leqslant |\Lambda'_e|$. Next note that $R$ has a triangular shape for a suitable ordering of the rows and colums, since $r_{\lambda\mu} = 0$ unless $C_\lambda \cap \mathfrak{S}_\mu \neq \varnothing$; that is, $\lambda$ is a refinement of $\mu$. So it will be enough to show that $\theta_0(r_{\lambda\lambda}) \neq 0$ for all $\lambda \in \Lambda'_e$. But this is clear by $(*)$. Thus, we have $\operatorname{rank}(\bar{R}) = |\Lambda'_e|$, as required.

Finally, consider the case where $\operatorname{char}(k) = \ell > 0$. Let $\Phi_e \in \mathbb{Z}[u]$ be the $e$th cyclotomic polynomial and $B_e := \mathbb{Z}[u,u^{-1}]/(\Phi_e)$; let $\theta_e \colon B \to B_e$ be the canonical map. Then, by the definition of $e$, we have $\theta(\Phi_e(u)) = \Phi_e(q) = 0$ and so there is a factorisation $\theta|_B = \varphi \circ \theta_e$, where $\varphi \colon B_e \to k$ is a ring homomorphism. Clearly, the rank of $X(\mathbf{H})_\theta$ is less than or equal to the rank of $X(\mathbf{H})_{\theta_e}$ (over the field of fractions of $B_e$). Since the field of fractions of $B_e$ has characteristic 0, we already know that the rank of $X(\mathbf{H})_{\theta_e}$ is given by $|\Lambda'_e|$, as required. □

In the above proof, we employed in a special situation the technique of factorising a specialisation into a field of positive characteristic through a specialisation into a field of characteristic 0. This idea will be discussed in a wider context in Section 3.6.

**Example 3.4.14.** Let $n = 4$. The character table of $\mathbf{H}$ is given as follows:

$$X(\mathbf{H}) = \begin{bmatrix} 1 & -1 & 1 & 1 & -1 \\ 3 & u-2 & -2u+1 & -u+1 & u \\ 2 & u-1 & u^2+1 & -u & 0 \\ 3 & 2u-1 & u^2-2u & u^2-u & -u^2 \\ 1 & u & u^2 & u^2 & u^3 \end{bmatrix} \qquad \text{(see [132, §10.2])}.$$

Then, Starkey's rule (see the above proof) yields the product decomposition

$$X(\mathbf{H}) = \begin{bmatrix} 1 & -1 & 1 & 1 & -1 \\ 3 & -1 & -1 & 0 & 1 \\ 2 & 0 & 2 & -1 & 0 \\ 3 & 1 & -1 & 0 & -1 \\ 1 & 1 & 1 & 1 & 1 \end{bmatrix} \cdot \begin{bmatrix} 1 & \frac{u-1}{2} & \frac{(u-1)^2}{4} & \frac{(u-1)^2}{6} & \frac{(u-1)^3}{24} \\ 0 & \frac{u+1}{2} & \frac{u^2-1}{2} & \frac{u^2-1}{2} & \frac{(u^2-1)(u-1)}{4} \\ 0 & 0 & \frac{(u+1)^2}{4} & 0 & \frac{(u^2-1)(u+1)}{8} \\ 0 & 0 & 0 & \frac{u^2+u+1}{3} & \frac{u^3-1}{3} \\ 0 & 0 & 0 & 0 & \frac{u^3+u^2+u+1}{4} \end{bmatrix},$$

where the first factor is the character table of $W \cong \mathfrak{S}_4$ and the second factor is the matrix $R$. The partitions labelling the rows and columns of these matrices are ordered as follows: (1111), (211), (22), (31), (4). Thus, for example, the first row of $X(\mathbf{H})$ corresponds to the sign representation and the first column to the conjugacy class containing the identity element. Now consider a specialisation $\theta \colon A \to k$ where $\operatorname{char}(k) = 0$ and $e = 2$; that is, $\theta(u) = -1$. Then the second, third and fifth rows of $R$ specialise to 0 while the diagonal entries of the first and fourth rows, labelled by (1111) and (31), remain non-zero. Thus, we have $|\operatorname{Irr}(\mathbf{H}_k)| = \operatorname{rank}(X(\mathbf{H})_\theta) = 2$.

## 3.5 A Canonical Basic Set for the Symmetric Group

We are now ready to determine a canonical basic set for $W \cong \mathfrak{S}_n$. This was first achieved by Dipper and James [62]. Here, we will follow a different approach based on the ideas in the previous section. This will also provide a model for dealing with algebras of type $B_n$ and, more generally, Ariki–Koike algebras in Chapter 5.

**3.5.1.** Let $(W,S)$ be of type $A_{n-1}$ and keep the notation of Example 3.2.9. In particular, $W \cong \mathfrak{S}_n$ and $A = R[v, v^{-1}]$ is the ring of Laurent polynomials in one indeterminate $v = \varepsilon$. A Fock datum as in 3.4.1 is obtained as follows. For any $m \in \{1, \ldots, n\}$, consider the parabolic subgroup $W_m := \langle s_1, \ldots, s_{m-1} \rangle \cong \mathfrak{S}_m$ and denote the corresponding generic algebra by $\mathbf{H}_m$. Then we have a chain of parabolic subgroups and parabolic subalgebras

$$W_1 = \{1\} \subseteq W_2 \subseteq \cdots \subseteq W_n = W \quad \text{and} \quad \mathbf{H}_1 = \langle T_1 \rangle_A \subseteq \mathbf{H}_2 \subseteq \cdots \subseteq \mathbf{H}_n = \mathbf{H}.$$

For $1 \leqslant m \leqslant n$, let $\Lambda_m$ be the set of all partitions $\lambda \vdash m$. In Example 3.4.8, we already defined central elements $z_m \in \mathbf{H}_m$. Recall that $z_m = \mathscr{L}_1 + \cdots + \mathscr{L}_m$, where

$$\mathscr{L}_1 = T_1 \quad \text{and} \quad \mathscr{L}_m = T_{s_{m-1}} \cdots T_{s_2} T_{s_1} T_{s_1} T_{s_2} \cdots T_{s_{m-1}} \qquad \text{for } 2 \leqslant m \leqslant n.$$

Since the above definition involves $T_{s_1} T_{s_1}$, the formula can be rewritten as follows:

$$\begin{aligned} \mathscr{L}_m &= T_{s_{m-1}} \cdots T_{s_2} \big(T_1 + (v - v^{-1}) T_{s_1}\big) T_{s_2} \cdots T_{s_{m-1}} \\ &= T_{s_{m-1}} \cdots T_{s_3} T_{s_2} T_{s_2} T_{s_3} \cdots T_{s_{m-1}} + (v - v^{-1}) T_{s_{m-1} \cdots s_2 s_1 s_2 \cdots s_{m-1}}. \end{aligned}$$

We repeat the argument by writing $T_{s_2}^2 = T_1 + (v - v^{-1}) T_{s_2}$, and so on. Also note that $s_i \cdots s_{m-2} s_{m-1} s_{m-2} \cdots s_i = (i,m)$ for $1 \leqslant i < m$. Hence, eventually, we find that

$$\mathscr{L}'_m := \frac{\mathscr{L}_m - T_1}{v^2 - 1} = v^{-1}\big(T_{(1,m)} + T_{(2,m)} + \cdots + T_{(m-1,m)}\big) \in \mathbf{H}_m.$$

It will turn out that it is actually more convenient to work with $\mathscr{L}'_m$ instead of $\mathscr{L}_m$. (See also the comments in [245, Note 7, p. 54].) We shall set

$$z'_m := \mathscr{L}'_1 + \mathscr{L}'_2 + \cdots + \mathscr{L}'_m \in \mathbf{H}_m \qquad \text{for } 1 \leqslant m \leqslant n.$$

Since $z_m \in Z(\mathbf{H}_m)$ for all $m$, it is clear that we also have $z'_m \in Z(\mathbf{H}_m)$ for all $m$. Observe that $vz'_n = \sum_{w \in C} T_w$, where $C$ is the conjugacy class containing $S$; hence, $z_C = z'_n$ is the central basis element corresponding to $C$ as in Example 3.3.8. In what follows, we assume that the operators $\xi\text{-Ind}_K^{(m)}$ are defined with respect to the elements $\{z'_m\}$. Our next task is to find explicit formulae for $\xi\text{-Ind}_K^{(m)}$.

**3.5.2.** We introduce some useful combinatorial notions. Let $n \geqslant 0$ and $\lambda \vdash n$. Writing $\lambda = (\lambda_1 \geqslant \lambda_2 \geqslant \ldots \geqslant 0)$, the *Young diagram* of $\lambda$ is defined by

$$[\lambda] := \{(a,b) \mid a \geqslant 1 \text{ and } 1 \leqslant b \leqslant \lambda_a\} \subseteq \mathbb{Z}_{\geqslant 1} \times \mathbb{Z}_{\geqslant 1}.$$

The elements of $[\lambda]$ are called the *nodes* of $\lambda$. The *content* of a node $\gamma = (a,b) \in [\lambda]$ is defined by $c(\gamma) := b - a$. A Young diagram $[\lambda]$ will be visualised by an array of boxes in left-justified rows, with $\lambda_1$ boxes in the first row, $\lambda_2$ boxes in the second row and so on. The set of *border nodes* of $\lambda$ is defined by

$$\mathscr{N}_\lambda := \{(a, \lambda_a) \mid a \geqslant 1 \text{ and } \lambda_a > 0\} \subseteq [\lambda].$$

By definition, a $\lambda$-tableau is a filling of the boxes of $[\lambda]$ by the numbers $1, \ldots, n$. Thus, more formally, a $\lambda$-tableaux is a bijection

$$\mathfrak{t} \colon [\lambda] \to \{1, \ldots, n\}.$$

We say that $\mathfrak{t}$ is *row-standard* if the sequence $\mathfrak{t}(a,1), \mathfrak{t}(a,2), \ldots$ is strictly increasing for each $a$; we say that $\mathfrak{t}$ is *column-standard* if the sequence $\mathfrak{t}(1,b), \mathfrak{t}(2,b), \ldots$ is strictly increasing for each $b$. We say that $\mathfrak{t}$ is *standard* if $\mathfrak{t}$ is row-standard and column-standard. Here are some examples, where $n = 5$ and $\lambda = (3,2)$:

| | | |
|---|---|---|
| | | |
| | | |

($\lambda$-diagram)

| | | |
|---|---|---|
| 1 | 3 | 5 |
| 2 | 4 | |

(standard tableau)

| | | |
|---|---|---|
| 1 | 4 | 5 |
| 2 | 3 | |

(row-standard tableau)

Here, $\mathscr{N}_\lambda = \{(1,3),(2,2)\}$. Now let $n \geqslant 1$ and recall that, by 1.7.6, we have

$$\operatorname{Res}_{W_{n-1}}^{W_n}(E^\lambda) \cong \bigoplus_{\mu \in \Lambda'(\lambda)} E^\mu \qquad \text{for some subset } \Lambda'(\lambda) \subseteq \Lambda_{n-1}.$$

This subset can now be described more precisely as follows. We say that $\gamma \in [\lambda]$ is a *removable node* if $[\lambda] = [\mu] \cup \{\gamma\}$ for some partition $\mu \vdash n-1$; alternatively, we say that $\lambda$ can be obtained from $\mu$ by adding the node $\gamma$. Then the "branching rule" (see [132, 6.1.8]) can be stated as follows:

$$\Lambda'(\lambda) = \{\mu \vdash n-1 \mid [\lambda] = [\mu] \cup \{\gamma\} \text{ for some removable node } \gamma \in \mathscr{N}_\lambda\}.$$

Thus, $\Lambda'(\lambda)$ consists of all partitions of $n-1$ which can be obtained by removing a border node from $\Lambda$.

**Lemma 3.5.3.** *Let $n \geqslant 1$, $\lambda \vdash n$ and $\mu \in \Lambda'(\lambda)$. Then, writing $[\lambda] = [\mu] \cup \{\gamma\}$, where $\gamma = (a, \lambda_a) \in \mathscr{N}_\lambda$ is removable, we have*

$$\omega_\lambda(z_n') - \omega_\mu(z_{n-1}') = \frac{v^{2c(\gamma)} - 1}{v^2 - 1}, \qquad \textit{where} \qquad c(\gamma) = \lambda_a - a.$$

*Proof.* Let $\{\sigma^\lambda \mid \lambda \in \Lambda_m, 1 \leqslant m \leqslant n\}$ be a family of representations as in Proposition 1.7.7. Given $\lambda \in \Lambda_n$, let $\Lambda'(\lambda) = \{\mu_1, \ldots, \mu_r\} \subseteq \Lambda_{n-1}$; then $\sigma^\lambda(T_w)$ $(w \in W_{n-1})$ is a block diagonal matrix where the blocks are given by the matrices $\sigma^{\mu_l}(T_w)$ for $1 \leqslant l \leqslant r$. By Lemma 3.4.9(a), we have $T_{w_0}^2 = T_{w_0'}^2 \mathscr{L}_n$, where $w_0'$ is the longest element in $W_{n-1}$. Using Example 1.2.7, we obtain

$$v^{2N_\lambda} I_{d_\lambda} = \sigma^\lambda(T_{w_0}^2) = \begin{bmatrix} v^{2N_{\mu_1}} I_{d_1} & 0 & \dots & 0 \\ 0 & v^{2N_{\mu_2}} I_{d_2} & \ddots & \vdots \\ \vdots & \ddots & \ddots & 0 \\ 0 & \dots & 0 & v^{2N_{\mu_r}} I_{d_r} \end{bmatrix} \cdot \sigma^\lambda(\mathscr{L}_n),$$

where $d_l = \dim E^{\mu_l}$ for $l = 1,\dots,r$. This already shows that $\sigma^\lambda(\mathscr{L}_n)$ is a diagonal matrix with eigenvalues $v^{2(N_\lambda - N_{\mu_l})}$, where each of these occurs with multiplicity equal to $d_l$ for $1 \leqslant l \leqslant r$. Now, by Example 1.3.4, we have

$$N_\lambda = \mathbf{a}_{\lambda^\dagger} - \mathbf{a}_\lambda \qquad \text{and} \qquad N_{\mu_l} = \mathbf{a}_{\mu_l^\dagger} - \mathbf{a}_{\mu_l}.$$

Furthermore, write $[\lambda] = [\mu_l] \cup \{\gamma_l\}$, where $\gamma_l = (a_l, \lambda_{a_l}) \in \mathscr{N}_\lambda$ is removable. Then the formulae for the **a**-invariants in Example 1.3.8 directly imply that

$$\mathbf{a}_\lambda - \mathbf{a}_{\mu_l} = a_l - 1 \qquad \text{and} \qquad \mathbf{a}_{\lambda^\dagger} - \mathbf{a}_{\mu_l^\dagger} = \lambda_{a_l} - 1.$$

This yields $N_\lambda - N_{\mu_l} = (\lambda_{a_l} - 1) - (a_l - 1) = \lambda_{a_l} - a_l = c(\gamma_l)$. So $\sigma^\lambda(\mathscr{L}_n)$ has eigenvalues $v^{2c(\gamma_l)}$ where each of these occurs with multiplicity equal to $d_l$ for $1 \leqslant l \leqslant r$. Using that $\mathscr{L}_n = (v^2-1)\mathscr{L}_n' + T_1$, we see that $\sigma^\lambda(\mathscr{L}_n')$ is diagonal with eigenvalues $(v^{2c(\gamma_l)} - 1)/(v^2-1)$. On the other hand, since $z_n' = z_{n-1}' + \mathscr{L}_n'$, we also have

$$\omega_\lambda(z_n') I_{d_\lambda} = \begin{bmatrix} \omega_{\mu_1}(z_{n-1}') I_{d_1} & 0 & \dots & 0 \\ 0 & \omega_{\mu_2}(z_{n-1}') I_{d_2} & \ddots & \vdots \\ \vdots & \ddots & \ddots & 0 \\ 0 & \dots & 0 & \omega_{\mu_r}(z_{n-1}') I_{d_r} \end{bmatrix} + \sigma^\lambda(\mathscr{L}_n').$$

Hence, for each $l$, the difference $\omega_\lambda(z_n') - \omega_{\mu_l}(z_{n-1}')$ is an eigenvalue of $\sigma^\lambda(\mathscr{L}_n')$ which appears in the diagonal block corresponding to $\sigma^{\mu_l}$ in the above matrices; so this eigenvalue equals $(v^{2c(\gamma_l)} - 1)/(v^2-1)$, as desired. □

**3.5.4.** Let $\theta\colon A \to k$ be a specialisation such that $k$ is the field of fractions of $\theta(A)$. As in 3.4.12, we set

$$e := \min\{i \geqslant 2 \mid 1 + \theta(u) + \theta(u)^2 + \dots + \theta(u)^{i-1} = 0\} \qquad (u := v^2).$$

(We let $e = \infty$ if no such $i$ exists.) For any integer $b \in \mathbb{Z}$, we set

$$[b]_\theta := \begin{cases} 1 + \theta(u) + \theta(u)^2 + \dots + \theta(u)^{b-1} & \text{if } b > 0, \\ 1 & \text{if } b = 0, \\ -\theta(u)^b(1 + \theta(u) + \theta(u)^2 + \dots + \theta(u)^{-b-1}) & \text{if } b < 0. \end{cases}$$

Then we note that, for any $b, b' \in \mathbb{Z}$, we have

$$[b]_\theta = [b']_\theta \qquad \Leftrightarrow \qquad b \equiv b' \bmod e.$$

Furthermore, set $\xi_\iota := 1+\theta(u)+\cdots+\theta(u)^{\iota-1} \in k$ for $0 \leqslant \iota \leqslant e-1$. Then, for any $b \in \mathbb{Z}$, we have $[b]_\theta = \xi_\iota$, where $\iota \in \{0,1,\ldots,e-1\}$ is such that $b \equiv \iota \bmod e$. Now let $n \geqslant 1$ and $\mu \vdash n-1$. Using the formula in Lemma 3.5.3, we can write

$$\bar{\omega}_\lambda(z_n') - \bar{\omega}_\mu(z_{n-1}') = [c(\gamma)]_\theta, \qquad \text{where} \qquad [\lambda] = [\mu] \cup \{\gamma\}.$$

With these preparations, we can now state the following proposition.

**Proposition 3.5.5 (Cf. Lascoux, Leclerc and Thibon [208, 3.3]).** *Given any partition $\mu \vdash n-1$ (where $n \geqslant 1$) and any integer $\iota \in \{0,1,\ldots,e-1\}$, we have*

$$\xi_\iota\text{-}\mathrm{Ind}_K^{(n)}(E_v^\mu) \cong \bigoplus_{\lambda \vdash n \text{ such that } \mu \xrightarrow{\iota} \lambda} E_v^\lambda,$$

*where we write $\mu \xrightarrow{\iota} \lambda$ if $[\lambda] = [\mu] \cup \{\gamma\}$ and $\gamma \in \mathscr{N}_\lambda$ is a removable node such that $c(\gamma) \equiv \iota \bmod e$.*

*Proof.* By 3.5.2, Frobenius reciprocity and a general compatibility of induction with specialisations (see [132, §9.1]), the restriction formula in 3.5.2 translates to

$$\mathrm{Ind}_K^{(n)}(E_v^\mu) \cong \bigoplus_{\lambda \vdash n \text{ such that } \mu \in \Lambda'(\lambda)} E_v^\lambda.$$

Now let $\lambda \vdash n$ be such that $\mu \in \Lambda'(\lambda)$. Then $[\lambda] = [\mu] \cup \{\gamma\}$, where $\gamma \in [\lambda]$; see 3.5.2. Using Lemma 3.5.3, we obtain $\xi_\iota\text{-}\mathrm{Ind}_K^{(n)}(E_v^\mu) \cong \bigoplus_\lambda E_v^\lambda$, where $\lambda$ runs over all partitions of $n$ such that $[\lambda] = [\mu] \cup \{\gamma\}$ for some removable $\gamma \in \mathscr{N}_\lambda$ satisfying $[c(\gamma)]_\theta = \xi_\iota$. By 3.5.4, the latter condition is equivalent to $c(\gamma) \equiv \iota \bmod e$. □

The above result provides the basis for a purely combinatorial description of the sets $\mathscr{E}_1, \mathscr{E}_2, \mathscr{E}_3, \ldots$. Our aim will now be to understand how this works in detail.

**Example 3.5.6.** Given $e \geqslant 2$, we define a directed graph as follows. The vertices are given by the elements of the set $\Lambda_\infty := \bigcup_{n \geqslant 0} \Lambda_n$, and the edges are defined by the relation $\mu \xrightarrow{\iota} \lambda$ in Proposition 3.5.5. Then the sets of representations $\mathscr{E}_n$ are completely determined by this graph, which may be called the LLT-*graph*.

Let $\mathscr{I}_n^{(e)}$ be the set of all sequences $\eta = (\iota_1, \iota_2, \ldots, \iota_n)$ such that $\iota_1 = 1$ and $\iota_j \in \{0,1,\ldots,e-1\}$ for all $j \geqslant 2$. Given $\eta$, we obtain a sequence of representations:

$$Y_{(\iota_1)} := \mathbf{1}, \quad Y_{(\iota_1,\iota_2)} := \xi_{\iota_2}\text{-}\mathrm{Ind}_K^{(2)}(Y_{(\iota_1)}), \quad \ldots, \quad Y_\eta := \xi_{\iota_n}\text{-}\mathrm{Ind}_K^{(n)}(Y_{(\iota_1,\ldots,\iota_{n-1})}).$$

Hence, with this notation, we have $\mathscr{E}_n = \{Y_\eta \mid \eta \in \mathscr{I}_n^{(e)}\} \setminus \{0\}$.

For example, let $e = 2$. In Figure 3.1, we show the part of the LLT-graph corresponding to all partitions of $n \leqslant 5$. In the graph, each partition is represented by its Young diagram where the boxes are filled by $c(x) \bmod e$; furthermore, we just write $\lambda$ instead of $E_v^\lambda$ and $l \cdot \lambda$ instead of $E_v^\lambda \oplus \cdots \oplus E_v^\lambda$ ($l$ terms).

**Fig. 3.1** Part of the LLT-graph for $e = 2$

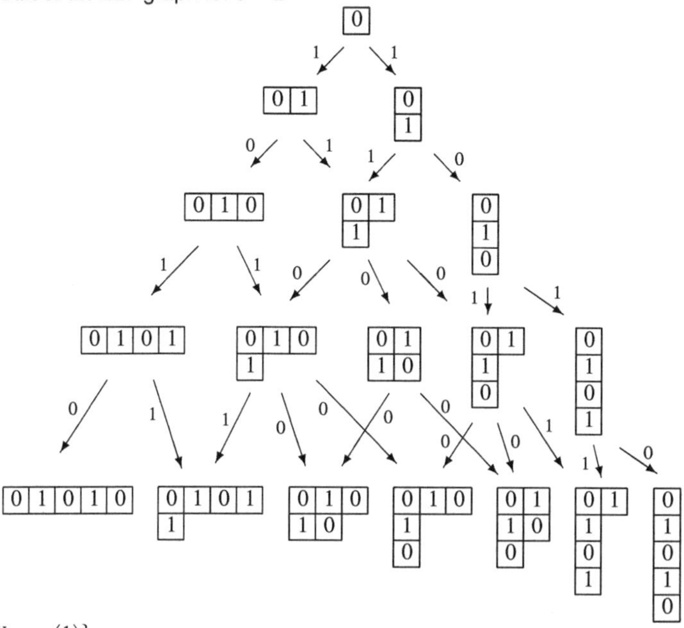

$$\begin{aligned}
\mathscr{E}_1 &= \{Y_{(0)} = (1)\},\\
\mathscr{E}_2 &= \{Y_{(0,1)} = (2) \oplus (11)\},\\
\mathscr{E}_3 &= \{Y_{(0,1,0)} = (3) \oplus (111),\ Y_{(0,1,1)} = 2\cdot(21)\},\\
\mathscr{E}_4 &= \{Y_{(0,1,0,1)} = (4) \oplus (31) \oplus (211) \oplus (1111),\ Y_{(0,1,1,0)} = 2\cdot(31) \oplus 2\cdot(22) \oplus 2\cdot(211)\},\\
\mathscr{E}_5 &= \{Y_{(0,1,0,1,0)} = (5) \oplus (32) \oplus 2\cdot(311) \oplus (221) \oplus (11111),\\
&\qquad Y_{(0,1,0,1,1)} = Y_{(0,1,1,0,1)} = 2\cdot(41) \oplus 2\cdot(2111),\ Y_{(0,1,1,0,0)} = 4\cdot(32) \oplus 4\cdot(311) \oplus 4\cdot(221)\}.
\end{aligned}$$

The LLT-graph is a remarkable combinatorial object; see Lascoux, Leclerc and Thibon [208], Kleshchev [202], [203] and Grojnowski [146] for further reading.

*Remark 3.5.7.* Let $\mu \vdash n$ and $\mathfrak{s}$ be a $\mu$-tableau. We define a corresponding sequence $\eta_e(\mathfrak{s}) \in \mathscr{I}_n^{(e)}$ as follows. For $j \in \{1,\ldots,n\}$, let $c_j$ be the content of the node which is filled by the number $j$ in $\mathfrak{s}$. Then $\eta_e(\mathfrak{s}) := (\iota_1,\ldots,\iota_n)$, where $\iota_j \equiv c_j \bmod e$ for $1 \leqslant j \leqslant n$. This sequence will be called the *e-residue sequence* of $\mathfrak{s}$.

Now let $\eta$ be an arbitrary sequence in $\mathscr{I}_n^{(e)}$ and consider the corresponding representation $Y_\eta \in \mathscr{E}_n$. Then the following two statements are equivalent:

(a) $E_v^\mu$ is an irreducible constituent of $Y_\eta$.
(b) There exists a standard $\mu$-tableau $\mathfrak{s}$ such that $\eta = \eta_e(\mathfrak{s})$.

To see this, note that a standard $\mu$-tableau $\mathfrak{s}$ is uniquely characterised by a sequence of partitions $\mu_{(1)},\ldots,\mu_{(n)} = \mu$ such that, for each $j \in \{1,\ldots,n\}$, we have

$$\mu_{(j)} \vdash j \qquad \text{and} \qquad [\mu_{(j)}] = [\mu_{(j-1)}] \cup \gamma_j, \quad \text{where} \quad \mathfrak{s}(\gamma_j) = j.$$

Thus, $\mu_{(n-1)}$ is obtained from $\mu = \mu_{(n)}$ by removing the unique node $\gamma_n$ which is filled by the number $n$; then $\mu_{(n-2)}$ is obtained from $\mu_{(n-1)}$ by removing the unique node $\gamma_{n-1}$ which is filled by the number $n-1$, and so on. The inductive definition of the sets $\mathscr{E}_n$ immediately yields the equivalence of (a) and (b).

Now the plan is to construct some distinguished representations in $\mathscr{E}_n$ and to apply Proposition 3.4.10. This construction involves the following type of partitions.

**Definition 3.5.8.** Let $e \geqslant 2$ and $\lambda \vdash n$. We write $\lambda = (1^{m_1}, 2^{m_2}, 3^{m_3}, \ldots)$, where the notation means that, for any $i \geqslant 1$, there are $m_i$ parts equal to $i$. Then we say that $\lambda$ is *e-regular* if $m_i < e$ for all $i$.

**3.5.9.** The construction of the desired distinguished representations involves a recursive procedure which associates with every $e$-regular partition $\lambda \vdash n$ a "canonical" standard $\lambda$-tableau $\mathfrak{t}_e(\lambda)$ and a corresponding sequence $\eta_e(\lambda) \in \mathscr{I}_n^{(e)}$. This procedure is a variation of (but not identical to) the *ladder method* used by Lascoux, Leclerc and Thibon [208, §6.2] (see also James and Kerber [182, 6.3.51, p. 283]).

First, we need some notation. Let $\lambda \vdash n$ be $e$-regular; let $\lambda_1 \geqslant \lambda_2 \geqslant \cdots \geqslant \lambda_l > 0$ be the non-zero parts of $\lambda$. Denote

$$\mathscr{N}_\lambda = \{\gamma_1, \ldots, \gamma_l\}, \qquad \text{where} \qquad \gamma_i = (i, \lambda_i) \in [\lambda] \text{ for } 1 \leqslant i \leqslant l.$$

For $b \in \mathbb{Z}$, let $\mathscr{N}_\lambda(b)$ be the set of all $\gamma_i \in \mathscr{N}_\lambda$ such that $c(\gamma_i) \equiv b \bmod e$.

Now let $r \geqslant 1$ be the index of the first node in $\mathscr{N}_\lambda$ which can be removed. Then $\lambda_1 = \cdots = \lambda_r > \lambda_{r+1}$; let $c_r \in \{0, 1, \ldots, e-1\}$ be such that $c(\gamma_r) \equiv c_r \bmod e$. Thus, $\gamma_r$ is a removable node in $\mathscr{N}_\lambda(c_r)$. Now consider the set $\mathscr{N}_\lambda(c_r - 1)$. If $\mathscr{N}(c_r - 1) \neq \varnothing$, let $i_0 \geqslant 1$ be the minimal element of $\mathscr{N}_\lambda(c_r - 1)$. If $\mathscr{N}(c_r - 1) = \varnothing$, set $i_0 = n+1$. Then note that $r < i_0$. Indeed, if $\mathscr{N}_\lambda(c_r - 1) = \varnothing$, this is clear since $i_0 = n+1$. On the other hand, if $\mathscr{N}_\lambda(c_r - 1) \neq \varnothing$, we have $c(\gamma_j) \equiv c_r + (r-j) \bmod e$ for $1 \leqslant j \leqslant r$. Since $\lambda$ is $e$-regular, we have $r < e$ and so $c(\gamma_j) \not\equiv c_r - 1 \bmod e$ for $1 \leqslant j \leqslant r$. This means that $r < i_0$, as claimed. Thus, we have shown that the set

$$\{i \in \{1, \ldots, l\} \mid i < i_0 \text{ and } \gamma_i \in \mathscr{N}_\lambda(c_r) \text{ is removable}\}$$

is non-empty. Let $i$ be the maximal element of this set. (We have $r \leqslant i < i_0$.) We start building the desired tableau $\mathfrak{t}_e(\lambda)$ by assigning $n$ to the node $\gamma_i$. We then remove the node $\gamma_i$ from the diagram of $\lambda$ and obtain a partition $\lambda' \vdash n-1$.

Next we claim that $\lambda'$ is also an $e$-regular partition. Assume, if possible, that this is not the case. Since $\lambda$ is $e$-regular, this could only happen if $\lambda_i = \lambda_{i+1} + 1$ and $\lambda_{i+1} = \lambda_{i+2} = \cdots = \lambda_{i+e-1} > \lambda_{i+e}$. Then $c(\gamma_i) \equiv c_r$, $c(\gamma_{i+1}) \equiv c_r - 2$, $c(\gamma_{i+2}) \equiv c_r - 3$, $\ldots$, $c(\gamma_{i+e-1}) \equiv c_r - e \equiv c_r$ where all congruences are modulo $e$. Hence, we have $\gamma_j \notin \mathscr{N}_\lambda(c_r - 1)$ for $i \leqslant j \leqslant i+e-1$. The definition of $i_0$ and the fact that $i < i_0$ now imply that $i + e \leqslant i_0$. But then $\gamma_{i+e-1} \in \mathscr{N}(c_r)$ is removable and $i + e - 1 < i_0$, contradicting the choice of $i$. Thus, the assumption was wrong and so $\lambda'$ is $e$-regular. We can now repeat the above procedure with $\lambda'$, find the appropriate node to be removed, fill that node by $n-1$, and continue. Eventually, after $n$ iterations, we have filled all the nodes in $[\lambda]$ and, thus, constructed a standard $\lambda$-tableau $\mathfrak{t}_e(\lambda)$. Then we define $\eta_e(\lambda)$ to be the $e$-residue sequence of $\mathfrak{t}_e(\lambda)$.

Here is an example where $e = 3$ and $\lambda = (44321) \vdash 14$. In the Young diagram of $\lambda$, each node $\gamma$ is filled by $c(\gamma)$ mod $e$.

$$\lambda: \begin{array}{|c|c|c|c|}\hline 0&1&2&0\\\hline 2&0&1&2\\\hline 1&2&0\\\cline{1-3} 0&1\\\cline{1-2} 2\\\cline{1-1}\end{array} \qquad \begin{array}{l} r=2,\ c_2=2,\ i_0=4\\ \mathcal{N}_\lambda(0)=\{1,3\}\\ \mathcal{N}_\lambda(1)=\{4\}\\ \mathcal{N}_\lambda(2)=\{2,5\}\end{array} \qquad \rightsquigarrow \qquad \lambda': \begin{array}{|c|c|c|c|}\hline 0&1&2&0\\\hline 2&0&1\\\cline{1-3} 1&2&0\\\cline{1-3} 0&1\\\cline{1-2} 2\\\cline{1-1}\end{array}$$

So, the node $\gamma_2 = (2,4) \in [\lambda]$ will be removed and, in $\mathfrak{t}_3(\lambda)$, the corresponding node will be filled by $n = 14$. We repeat the argument with $\lambda' = (43321) \vdash 13$: now, $r = 1$, $c_1 = 0$, $\mathcal{N}_{\lambda'}(0) = \{1,3\}$, $\mathcal{N}_{\lambda'}(1) = \{2,4\}$, $\mathcal{N}_{\lambda'}(2) = \{5\}$ and $i_0 = 5$. Hence, the nodes $\gamma_1 = (1,4)$ and $\gamma_3 = (3,3)$ belong to $\mathcal{N}_{\lambda'}(0)$ and are removable. Since $1 < 3 < i_0$, the node $\gamma_3$ will be removed and, in $\mathfrak{t}_3(\lambda)$, the corresponding node will be filled by $n - 1 = 13$. Continuing in this way, we eventually find that

$$\mathfrak{t}_3(\lambda) = \begin{array}{|c|c|c|c|}\hline 1&3&7&12\\\hline 2&5&10&14\\\hline 4&8&13\\\cline{1-3} 6&11\\\cline{1-2} 9\\\cline{1-1}\end{array} \qquad \eta_3(\lambda) = (0,2,1,1,0,0,2,2,2,1,1,0,0,2).$$

**3.5.10.** In order to state the crucial Lemma 3.5.11 below, it will be convenient to extend the notion of Young diagrams and tableaux to compositions. Recall that a *composition* of $n$ is a finite sequence $\mu = (\mu_1, \ldots, \mu_r)$ of non-negative integers such that $\sum_{1 \leqslant i \leqslant r} \mu_i = n$. The *height* of $\mu$ is defined as $\max\{i \mid \mu_i \neq 0\}$ if $\mu = \varnothing$, and as zero otherwise. As before, the set $[\mu] := \{(a,b) \mid 1 \leqslant a \leqslant r,\ 1 \leqslant b \leqslant \mu_a\}$ is called the *Young diagram* of $\mu$ and a bijection $\mathfrak{s}\colon [\mu] \to \{1, \ldots, n\}$ will be called a $\mu$-tableau. We also say that $\mu$ is the *shape* of $\mathfrak{s}$.

A $\mu$-tableau $\mathfrak{s}$ is called *row-standard* if the sequence $\mathfrak{s}(a,1), \mathfrak{s}(a,2), \ldots$ is strictly increasing for each $a$. (Note that, now, it does not really make sense to define the notion of a column-standard tableau.)

Now fix an integer $e \geqslant 2$. For each $\gamma = (a,b) \in [\mu]$, the number $c(\gamma) := b - a$ is called the *content* of $\gamma$. Furthermore, the unique $\imath \in \{0, 1, \ldots, e-1\}$ such that $c(\gamma) \equiv \imath$ mod $e$ is called the *e-residue* of $\gamma$ and denoted by $\mathrm{res}_e(\gamma)$. As before, the *e-residue sequence* of a $\mu$-tableau $\mathfrak{s}$ is defined by $\eta_e(\mathfrak{s}) = (\imath_1, \ldots, \imath_n)$, where $\imath_j$ is the $e$-residue of the node which is filled by the number $j$ in $\mathfrak{s}$.

Finally, we also extend the definition of the *dominance order*. Let $\mu, \mu'$ be compositions of $n$; then

$$\mu \trianglelefteq \mu' \quad \stackrel{\text{def}}{\Leftrightarrow} \quad \sum_{1 \leqslant i \leqslant d} \mu_i \leqslant \sum_{1 \leqslant i \leqslant d} \mu'_i \quad \text{for all } d \in \{1, \ldots, r\},$$

where $\mu = (\mu_1, \mu_2, \ldots, \mu_r)$ and $\mu' = (\mu'_1, \mu'_2, \ldots, \mu'_r)$. (We can assume that $\mu$, $\mu'$ have the same number of parts by adding zeroes at the end if necessary.)

We can now state the following result, which contains the combinatorial essence of our modified "ladder method" in 3.5.9. It is inspired by (but slightly stronger than) a similar result by Mathas [245, Lemma 3.40]; see also Murphy [255, §3].

**Lemma 3.5.11 (Jacon [174, §4]).** *Let $\lambda \vdash n$ be an e-regular partition and $\mu$ be a composition of n. Assume that there exists a row-standard $\mu$-tableau $\mathfrak{s}$ such that $\eta_e(\mathfrak{s}) = \eta_e(\lambda)$. Then $\mu \trianglelefteq \lambda$.*

*Proof.* Let us write $\mathfrak{t} := \mathfrak{t}_e(\lambda)$ and $\eta_e(\lambda) = (\iota_1, \ldots, \iota_n)$. We shall argue by downward induction on $d$, where $d$ is defined as follows. If $\mathfrak{t} = \mathfrak{s}$, then set $d := n+1$; otherwise, let $d \in \{1, \ldots, n\}$ be minimal such that $\mathfrak{t}^{-1}(d) \neq \mathfrak{s}^{-1}(d)$. Thus, if $d = n+1$, then there is nothing to prove. Now assume that $d \leqslant n$.

Let $a_\mathfrak{t}(d)$ be the row index of $\mathfrak{t}^{-1}(d)$ and $a_\mathfrak{s}(d)$ be the row index of $\mathfrak{s}^{-1}(d)$. Then $a_\mathfrak{t}(d) \neq a_\mathfrak{s}(d)$, by the minimality of $d$. We claim that

$$a_\mathfrak{t}(d) < a_\mathfrak{s}(d).$$

This is seen as follows. Let us successively remove the $n-d$ nodes which are filled by $n, n-1, \ldots, d+1$ from $\mathfrak{t} = \mathfrak{t}_e(\lambda)$. By the inductive construction in 3.5.9, the resulting tableau will be $\mathfrak{t}_e(\tilde{\lambda})$ for an $e$-regular partition $\tilde{\lambda} \vdash d$. Similarly, let us successively remove the $n-d$ nodes which are filled by $n, n-1, \ldots, d+1$ from $\mathfrak{s}$. The result will be a row-standard $\tilde{\mu}$-tableau $\tilde{\mathfrak{s}}$, for some composition $\tilde{\mu}$ of $d$. Furthermore, we have $\eta_e(\mathfrak{t}_e(\tilde{\lambda})) = \eta_e(\tilde{\mu})$. Let $\tilde{\lambda}' \vdash d-1$ be the partition obtained by removing the node filled by $d$ in $\mathfrak{t}_e(\tilde{\lambda})$ and let $\tilde{\mu}'$ be the composition of $d-1$ obtained by removing the node filled by $d$ in $\tilde{\mathfrak{s}}$. Since $d$ is minimal such that $\mathfrak{t}^{-1}(d) \neq \mathfrak{s}^{-1}(d)$, we have $\tilde{\lambda}' = \tilde{\mu}'$. Thus, both $\tilde{\lambda}$ and $\tilde{\mu}$ are obtained by adding a node with $e$-residue $\iota_d$ to the diagram of $\tilde{\lambda}'$. But, by the construction of $\mathfrak{t}_e(\tilde{\lambda})$ and $\tilde{\lambda}'$, all border nodes of $\tilde{\lambda}'$ with row index strictly less than $a_\mathfrak{t}(d)$ have $e$-residues which are not congruent to $\iota_d - 1$ modulo $e$. Hence, the only way to add a node of $e$-residue $\iota_d$ to $\tilde{\lambda}'$ is to add it to a row of index $a_\mathfrak{t}(d)$ or bigger. Thus, since $a_\mathfrak{t}(d) \neq a_\mathfrak{s}(d)$, we must have $a_\mathfrak{t}(d) < a_\mathfrak{s}(d)$, as claimed.

So, in the tableau $\mathfrak{s}$, we have the following configuration where "$*$" stands for a filling of the node by a number in $\{1, \ldots, d-1\}$ and where the node filled by $n_1$ has the same coordinates as the node filled by $d$ in $\mathfrak{t}$:

$$\begin{array}{lcccccl}
\text{row } a_\mathfrak{t}(d): & \boxed{*}\boxed{*} & \cdots & \boxed{*}\boxed{n_1}\boxed{n_2} & \cdots & \boxed{n_p} & d < n_1 < n_2 < \cdots < n_p \\
 & \vdots\;\vdots & & \vdots\;\vdots\;\vdots\;\vdots & & \vdots & \\
\text{row } a_\mathfrak{s}(d): & \boxed{*}\boxed{*} & \cdots & \boxed{*}\boxed{m_1}\boxed{m_2} & \cdots & \boxed{m_q} & d = m_1 < m_2 < \cdots < m_q
\end{array}$$

(The nodes filled by "$*$" also appear in the tableau $\mathfrak{t}$; note that the case $p = 0$ is allowed here, which just means that the row indexed by $a_\mathfrak{t}(d)$ only consists of the nodes filled by "$*$".) Since $\eta_e(\mathfrak{s}) = \eta_e(\mathfrak{t})$, the $e$-residues of the nodes filled by $m_1 = d$ and $n_1$ are equal. Consequently, for any $i$, the node filled by $n_i$ has the same $e$-residue as the node filled by $m_i$. We now perform some modifications on $\mathfrak{s}$ in order to obtain a new composition $\nu$ of $n$ and a corresponding row-standard $\nu$-tableau $\mathfrak{u}$. This is done as follows. We keep all nodes and their fillings in those rows of $\mathfrak{s}$ which are not indexed by $a_{\mathfrak{t}}(d)$ or $a_{\mathfrak{s}}(d)$. Thus, the modification will only take place in the two rows indexed by $a_{\mathfrak{t}}(d)$ and $a_{\mathfrak{s}}(d)$:

- in row $a_{\mathfrak{t}}(d)$, we replace $n_i$ by $\min\{n_i, m_i\}$ for $1 \leqslant i \leqslant \min\{p,q\}$;
- in row $a_{\mathfrak{s}}(d)$, we replace $m_i$ by $\max\{n_i, m_i\}$ for $1 \leqslant i \leqslant \min\{p,q\}$;
- if $q > p$, we move all nodes filled by $m_{p+1}, \ldots, m_q$ from row $a_{\mathfrak{s}}(d)$ to row $a_{\mathfrak{t}}(d)$.

Denote the resulting tableau by $\mathfrak{u}$ and let $\nu$ be the shape of $\mathfrak{u}$. One easily checks that $\mathfrak{u}$ is row-standard. (Note that this is just a statement about the reshuffling of two sequences of integers $n_1, \ldots, n_p$ and $m_1, \ldots, m_q$ according to the above rules; see Remark 3.5.12.) Taking into account the above remarks concerning $e$-residues, it is clear that $\eta_e(\mathfrak{u}) = \eta_e(\mathfrak{s}) = \eta_e(\lambda)$. Furthermore, the row of $\mathfrak{u}$ indexed by $a_{\mathfrak{t}}(d)$ has at least as many nodes as the row of $\mathfrak{s}$ indexed by $a_{\mathfrak{t}}(d)$; since $a_{\mathfrak{t}}(d) < a_{\mathfrak{s}}(d)$, the definition of the dominance order immediately shows that $\mu \trianglelefteq \nu$. Hence, if $\nu = \lambda$, then we are done. Otherwise note that, by construction, we have $\mathfrak{t}^{-1}(i) = \mathfrak{u}^{-1}(i)$ for $1 \leqslant i \leqslant d$. Hence, if we let $d' \in \{1, \ldots, n\}$ be minimal such that $\mathfrak{t}^{-1}(d') \neq \mathfrak{u}^{-1}(d')$, then $d' > d$. So, by induction, we have $\nu \triangleleft \lambda$ and, hence, $\mu \trianglelefteq \lambda$, as required. □

*Remark 3.5.12.* The above proof uses the following (very elementary) statement: Let $n_1 < n_2 < \cdots < n_p$ and $m_1 < m_2 < \cdots < m_q$ be two finite increasing sequences of integers. Let $p' := \max\{p,q\}$ and $q' := \min\{p,q\}$. For $1 \leqslant i \leqslant q'$, let $n_i' := \min\{n_i, m_i\}$ and $m_i' := \max\{n_i, m_i\}$. If $q > p$, let $n_i' := m_i$ for $p+1 \leqslant i \leqslant p'$. Then we have $n_1' < n_2' < \cdots < n_{p'}'$ and $m_1' < m_2' < \cdots < m_{q'}'$. (Proof left to the reader.)

**Example 3.5.13.** (a) To illustrate the above proof, let $\lambda = (422) \vdash 8$ and $e = 3$. Furthermore, let $\mu = (32111) \vdash 8$ and $\mathfrak{s}$ be the row-standard $\mu$-tableau given below:

$$\mathfrak{t}_e(\lambda) = \begin{array}{|c|c|c|c|} \hline 1 & 3 & 6 & 8 \\ \hline 2 & 5 \\ \cline{1-2} 4 & 7 \\ \cline{1-2} \end{array} \qquad \eta_3(\lambda) = (0,2,1,1,0,2,2,0) \qquad \mathfrak{s} = \begin{array}{|c|c|c|} \hline 1 & 3 & 7 \\ \hline 2 & 5 \\ \cline{1-2} 4 \\ \cline{1-1} 8 \\ \cline{1-1} 6 \\ \cline{1-1} \end{array}$$

We have $\eta_3(\mathfrak{s}) = \eta_3(\lambda)$. Now $d = 6$ is the smallest number which appears in different nodes of $\mathfrak{t} = \mathfrak{t}_3(\lambda)$ and of $\mathfrak{s}$; we have $a_{\mathfrak{t}}(6) = 1$ and $a_{\mathfrak{s}}(6) = 5$. With the notation of the above proof, $n_1 = 7$ and $m_1 = 6$ where $p = q = 1$. So we just swap 6 and 7 in the above tableau. Let $\mathfrak{u}$ be the resulting tableau; its shape is still $\nu = \mu$. Then $d' = 7$ is the smallest number which appears in different nodes of $\mathfrak{t}$ and of $\mathfrak{u}$; we have $a_{\mathfrak{t}}(7) = 3$ and $a_{\mathfrak{u}}(7) = 5$. With the notation of the above proof, $m_1 = 7$ where $p = 0$ and $q = 1$. So we just move 7 from the fifth row to the third row. The resulting composition is $\nu' = (3221)$; we have $\mu \triangleleft \nu' \triangleleft \lambda$ and we can go on by induction. (In more complicated examples, even if we start with two partitions, it may happen that genuine compositions appear in intermediate steps of the construction.)

(b) Let $n = 5$ and $\theta \colon A \to k$ be such that $\theta(v^2) = -1 \neq 1$, as in Example 3.2.9. Then $e = 2$. The 2-regular partitions of 5 are (5), (41), (32). We find that

$$\eta_2(5) = (0,1,0,1,0), \qquad \eta_2(41) = (0,1,0,1,1), \qquad \eta_2(32) = (0,1,1,0,0).$$

The corresponding representations $Y_{\eta_e(\lambda)}$ in $\mathscr{E}_5$ are computed in Example 3.5.6. Now, if $\lambda$ is 2-regular and $E^\mu$ is a constituent of $Y_{\eta_e(\lambda)}$, then there exists a standard $\mu$-tableau $\mathfrak{s}$ such that $\eta_2(\mathfrak{s}) = \eta_2(\lambda)$; see Remark 3.5.7. So Lemma 3.5.11 shows that $\mu \trianglelefteq \lambda$, which we can see explicitly in the formulae for $\mathscr{E}_5$ in Figure 3.1.

**Theorem 3.5.14 (Dipper and James [62]).** *Recall that we are in the setting of Example 3.2.9, where $W \cong \mathfrak{S}_n$. As in 3.5.4, let*

$$e := \min\{i \geqslant 2 \mid 1 + \theta(u) + \theta(u)^2 + \cdots + \theta(u)^{i-1} = 0\} \qquad (u := v^2).$$

*(We let $e = \infty$ if no such $i$ exists.) Then we have*

$$\Lambda_k^\circ = \{\lambda \vdash n \mid \lambda \text{ is } e\text{-regular}\} \qquad (\text{cf. Definition 3.5.8}).$$

*Proof.* First, it is known that the number of $e$-regular partitions of $n$ is equal to the number of partitions of $n$ with no part divisible by $e$. (This is one of the classical identities among partition counting functions; see Andrews [4, Cor. 1.3] and James and Kerber [182, Lemma 6.1.2]. For an alternative argument, see also James [181, 2.22, 2.27].) Hence, by Lemma 3.4.13, we already have that

$$|\operatorname{Irr}(\mathbf{H}_k)| \leqslant \text{number of } e\text{-regular partitions } \lambda \vdash n.$$

We set $\mathscr{E} = \mathscr{E}_n$. Let $\lambda \vdash n$ be $e$-regular and consider the tableau $\mathfrak{t}_e(\lambda)$ in 3.5.9. Let $\eta_e(\lambda)$ be its $e$-residue sequence and denote $Y_\lambda := Y_{\eta_e(\lambda)} \in \mathscr{E}$. We show that the set

$$\mathscr{E}' := \{Y_\lambda \mid \lambda \vdash n \text{ is } e\text{-regular}\} \subseteq \mathscr{E}$$

satisfies the conditions in Proposition 3.4.10. Recall from Example 3.2.9 that $\mathbf{H}$ admits a cell datum and, hence, a canonical basic set. Furthermore, the condition in Remark 3.5.7(b) is satisfied for $\lambda$ itself and so $E_v^\lambda$ is an irreducible constituent of $Y_\lambda$. Now let $\mu \vdash n$ be any partition such that $E_v^\mu$ is an irreducible constituent of $Y_\lambda$. Then Remark 3.5.7 shows that there exists a standard $\mu$-tableau $\mathfrak{s}$ such that $\eta_e(\mathfrak{s}) = \eta_e(\lambda)$ and so Lemma 3.5.11 implies that $\mu \trianglelefteq \lambda$. Thus, in $R_0(\mathbf{H}_K)$, we have

$$[Y_\lambda] = m(Y_\lambda, \lambda)[E_v^\lambda] + \sum_{\mu \vdash n :\, \mu \lhd \lambda} m(Y_\lambda, \mu)\,[E_v^\mu],$$

where $m(Y_\lambda, \mu) \in \mathbb{Z}_{\geqslant 0}$ for all $\mu$ and $m(Y_\lambda, \lambda) \neq 0$. It is known that if $\mu \trianglelefteq \lambda$, then $\mathbf{a}_\lambda \leqslant \mathbf{a}_\mu$, with equality only if $\lambda = \mu$; see 2.2.13. Hence, the three conditions in Proposition 3.4.10 are satisfied for $\mathscr{E}'$, as claimed. Thus, $\mathscr{E}'$ is a canonical basic set for $\mathbf{H}_k$. Finally, Proposition 3.2.7 shows that $\mathscr{E}' = \{Y_\lambda \mid \lambda \in \Lambda_k^\circ\}$. □

As mentioned earlier, the original proof of Dipper and James [62] is different from the one given here. For further variations, see Murphy [257], Mathas [245] and Xi [296]. Note that some authors (for example, Mathas [245, §3.4]) obtain a parametrisation of $\operatorname{Irr}(\mathbf{H}_k)$ in terms of "$e$-restricted" partitions. Since a partition is $e$-restricted if and only if the conjugate partition is $e$-regular, the two versions are seen to be equivalent by passing to the opposite cell datum and using Example 3.2.8.

## 3.6 Factorisation of Decomposition Maps

We keep the basic setting of the previous sections where $\mathbf{H} = \mathbf{H}_A(W,S,L)$ is a generic Iwahori–Hecke algebra associated with a finite Coxeter group $W$. Our aim now is to try to get a general overview of all possible specialisations and the associated decomposition maps. The basic tool for doing this is the factorisation result in Theorem 3.6.3 below. In the case where $\Gamma = \mathbb{Z}$, this will allow us to deal with any specialisation in a two-step procedure where, in a first step, we specialise to a field of characteristic 0 (via a principal specialisation) and then possibly specialise further into a field of positive characteristic. A first version of this result appeared in [92]; a systematic and more conceptual framework for establishing such factorisation results was developed by Geck and Rouquier [133].

We will then go on to discuss various questions related to specialisations of the parameters to roots of unity in a field of characteristic zero. We will see that these questions can be essentially reduced to the study of the $\Phi_e$-modular decomposition maps introduced in Example 3.1.15(b).

To give an idea of what kind of specialisations we will have to deal with in general, we mention the following example—which is, in any case, the most important as far as applications are concerned (see also, for example, Broué and Kim [33]).

**Example 3.6.1.** Assume that $\Gamma = \mathbb{Z}$ so that $A = R[v, v^{-1}]$ is the ring of Laurent polynomials in one indeterminate $v = \varepsilon$. Let $\theta\colon A \to k$ be a specialisation into a field $k$ such that $k$ is the field of fractions of $\theta(A)$.

Now, if $R$ is a field, then $A$ is a principal ideal domain; furthermore, $k$ has characteristic 0. More generally, assume that $R$ is the ring of algebraic integers in a finite field extension $\mathbb{K} \supseteq \mathbb{Q}$. Then the kernel of $\theta$ is a prime ideal in $A$, and one easily shows that the non-zero prime ideals in $A$ are given as follows (e.g. see [33, §2.B]):

- $fA$, where $f \in A$ is a non-constant, irreducible and non-invertible polynomial. In this case, the field of fractions $A/fA$ has characteristic 0.
- $\mathfrak{p}A$, where $\mathfrak{p}$ is a non-zero prime ideal in $R$; thus, $\mathfrak{p}A$ consists of all $\mathfrak{p}$-linear combinations of terms $v^n$ ($n \in \mathbb{Z}$). In this case we have $A/\mathfrak{p}A \cong \mathbb{F}[v, v^{-1}]$ where $\mathbb{F} = R/\mathfrak{p}$ is a finite field of characteristic $p > 0$ where $p \in \mathfrak{p}$.
- $\mathfrak{p}A + fA$, where $\mathfrak{p}$ is a prime ideal in $A$ and $f \in A$ is non-constant, irreducible and non-invertible modulo $\mathfrak{p}$. In this case, $\mathfrak{p}A + fA$ is maximal and the residue field has characteristic $p > 0$ where $p \in \mathfrak{p}$.

In particular, we see that $\dim A = 2$. The prime ideals in the first and second cases have height 1; those in the third case have height 2. Note that if $\ker(\theta)$ has height 1, then the localisation of $A$ in $\ker(\theta)$ will be a discrete valuation ring (see [248, §11]); in particular, $\theta$ is a principal specialisation. In the case where $\ker(\theta)$ has height 2, the situation is more complicated.

**3.6.2.** Recall our general assumption that $R$ is $L_0$-good. Let $\theta\colon A \to k$ and $\theta'\colon A \to k'$ be specialisations where $k$ is the field of fractions of $\theta(A)$ and $k'$ is the field of fractions of $\theta'(A)$. Let us assume that

(a) $$\ker(\theta) \subseteq \ker(\theta').$$

Then we can factorise $\theta' = \varphi \circ \theta$, where $\varphi$ is a ring homomorphism from $A' := \theta(A) \subseteq k$ to $k'$. (Explicitly, given $a' \in A'$, write $a' = \theta(a)$ for some $a \in A$ and set $\varphi(a') := \theta'(a)$; this does not depend on the choice of $a$ since $\ker(\theta) \subseteq \ker(\theta')$.) Clearly, the image of $\varphi$ equals the image of $\theta'$; hence, $k'$ also is the field of fractions of $\varphi(A')$. Theorem 3.6.3 below will show that the factorisation $\theta' = \varphi \circ \theta$ induces a factorisation of the corresponding decomposition maps. In order for this to work, an additional requirement on $\varphi$ is needed, which can be formulated as follows.

(b) $\varphi$ can be extended to a subring $B \subseteq k$ which is integrally closed in $k$.

We will say that $\theta'$ *lies above* $\theta$ if the above two conditions, (a) and (b), are satisfied.

The two standard situations in our applications are: (1) the case where $B = \theta(A)$ itself is integrally closed in $k$ and (2) the case where $\varphi$ can be extended to $k$.

**Theorem 3.6.3 (Geck [92, §4], [98, 2.6], Geck and Rouquier [133, 2.12]).** *Assume that $\theta'$ lies above $\theta$, in the sense defined in 3.6.2. Then there exists a unique additive map $d_{\theta'}^{\theta} \colon R_0^+(\mathbf{H}_k) \to R_0^+(\mathbf{H}_{k'})$ such that the following diagram is commutative:*

$$\begin{array}{ccc} R_0^+(\mathbf{H}_K) & \xrightarrow{\quad d_{\theta'}^+ \quad} & R_0^+(\mathbf{H}_{k'}) \\ & {\scriptstyle d_\theta^+} \searrow \quad \circlearrowleft \quad \nearrow {\scriptstyle d_{\theta'}^{\theta}} & \\ & R_0^+(\mathbf{H}_k) & \end{array}$$

*Thus, we also have a factorisation of the corresponding decomposition matrices as $D_{\theta'} = D_\theta . D_{\theta'}^{\theta}$, where $D_{\theta'}^{\theta}$ is a matrix all of whose entries are non-negative integers.*

Following James [181] (who considered Hecke and $q$-Schur algebras of type $A$), the matrix $D_{\theta'}^{\theta}$ will be called the *adjustment matrix* with respect to $\theta$ and $\theta'$.

*Proof.* Let $\tilde{\varphi} \colon B \to k'$ be an extension of $\varphi$ as in 3.6.2(b). Clearly, $k'$ also is the field of fractions of the image of $\tilde{\varphi}$. Since $B$ is integrally closed in $k$, we can apply Theorem 3.1.2. Thus, $\tilde{\varphi}$ induces a decomposition map $d_{\tilde{\varphi}}^+ \colon R_0^+(\mathbf{H}_k) \to R_0^+(\mathbf{H}_{k'})$ such that the following diagram is commutative:

$$\begin{array}{ccc} R_0^+(\mathbf{H}_k) & \xrightarrow{\quad \mathfrak{x}_B \quad} & \mathrm{Maps}(\mathbf{H}_B, B[X]) \\ {\scriptstyle d_{\tilde{\varphi}}^+} \downarrow & \circlearrowleft & \downarrow {\scriptstyle t_{B,\tilde{\varphi}}} \\ R_0^+(\mathbf{H}_{k'}) & \xrightarrow{\quad \mathfrak{x}_{B,k'} \quad} & \mathrm{Maps}(\mathbf{H}_B, k'[X]) \end{array}$$

where $t_{B,\tilde{\varphi}} \colon \mathrm{Maps}(\mathbf{H}_B, B[X]) \to \mathrm{Maps}(\mathbf{H}_B, k'[X])$ is the natural map obtained by applying $\tilde{\varphi}$ to the coefficients of the polynomials in $B[X]$.

We set $d_{\theta'}^{\theta} := d_{\tilde{\varphi}}^+$. Then it remains to show that $d_{\theta'}^+ = d_{\theta'}^{\theta} \circ d_\theta^+$. For this purpose, we first note that the natural map $\mathbf{H} \to \mathbf{H}_B$ induces monoid homomorphisms $\iota_B \colon \mathrm{Maps}(\mathbf{H}_B, B[X]) \to \mathrm{Maps}(\mathbf{H}, B[X])$ and $\iota_{k'} \colon \mathrm{Maps}(\mathbf{H}_B, k'[X]) \to \mathrm{Maps}(\mathbf{H}, k'[X])$. Hence, we have a commutative diagram:

$$\begin{array}{ccc}
\mathrm{Maps}(\mathbf{H}_B, B[X]) & \xrightarrow{\iota_B} & \mathrm{Maps}(\mathbf{H}, B[X]) \\
{\scriptstyle t_{B,\tilde{\varphi}}}\downarrow & \circlearrowleft & \downarrow{\scriptstyle t_{\tilde{\varphi}}} \\
\mathrm{Maps}(\mathbf{H}_B, k'[X]) & \xrightarrow{\iota_{k'}} & \mathrm{Maps}(\mathbf{H}, k'[X])
\end{array}$$

Noting that $\mathfrak{r}_k = \iota_B \circ \mathfrak{r}_B$ and $\mathfrak{r}_{k'} = \iota_{k'} \circ \mathfrak{r}_{B,k'}$, we obtain the following diagram in which each square is commutative:

$$\begin{array}{ccccc}
R_0^+(\mathbf{H}_K) & \xrightarrow{d_\theta^+} & R_0^+(\mathbf{H}_k) & \xrightarrow{d_{\tilde{\varphi}}^+} & R_0^+(\mathbf{H}_{k'}) \\
{\scriptstyle \mathfrak{r}_A}\downarrow & \circlearrowleft & \downarrow{\scriptstyle \mathfrak{r}_k} & \circlearrowleft & \downarrow{\scriptstyle \mathfrak{r}_{k'}} \\
\mathrm{Maps}(\mathbf{H}, A[X]) & \xrightarrow{t_\theta} & \mathrm{Maps}(\mathbf{H}, B[X]) & \xrightarrow{t_{\tilde{\varphi}}} & \mathrm{Maps}(\mathbf{H}, k'[X])
\end{array}$$

Now, the composition of the two maps of the bottom row is $t_{\theta'}\colon \mathrm{Maps}(\mathbf{H}, A[X]) \to \mathrm{Maps}(\mathbf{H}, k'[X])$. Then the characterisation in Theorem 3.1.2(b) shows that the composition of the two maps of the top row must be $d_{\theta'}^+$, as required. □

**Example 3.6.4.** Assume we are in the setting of 3.3.1, where we have factorisations of the elements $\mathbf{c}_\lambda$ in terms of a certain subset $\mathscr{P} \subseteq A$. Condition (P2) in 3.3.1 implies that whenever we have a specialisation $\theta\colon A \to k$ such that $\theta(\Phi) = 0$ for some $\Phi \in \mathscr{P}$, then $\theta$ *lies above* $\theta_\Phi$ in the sense of 3.6.2. So Theorem 3.6.3 yields a factorisation of decomposition matrices $D_\theta = D_\Phi . D_\theta^\Phi$, where $D_\theta^\Phi$ is the corresponding *adjustment matrix*.

**Example 3.6.5.** Assume that $\mathbb{K} \supseteq \mathbb{Q}$ is a finite extension and $R$ is a discrete valuation ring in $\mathbb{K}$ with residue field $\mathbb{F}$ of characteristic $p > 0$. By the classical modular representation theory of finite groups, we have a corresponding decomposition map

$$d_p\colon R_0(\mathbb{K}W) \to R_0(\mathbb{F}W),$$

where $\mathbb{K}W$ and $\mathbb{F}W$ are the group algebras of $W$ over $\mathbb{K}$ and $\mathbb{F}$ respectively; see Example 3.1.3. Let $D_p$ denote the decomposition matrix of $d_p$. We now use Theorem 3.6.3 to relate $d_p$ to decomposition maps of $\mathbf{H}$. For simplicity, let us assume that $\Gamma = \mathbb{Z}$ and $L(s) = 1$ for all $s \in S$ (equal parameter case). Also assume that $R$ is large enough so that our standard root of unity $\zeta_{2p} = \exp(\pi\sqrt{-1}/p)$ lies in $R$. Then consider the $\Phi_p$-modular decomposition map $d_{(p)}\colon R_0(\mathbf{H}_K) \to R_0(\mathbf{H}_{\mathbb{K}})$ associated with the specialisation $\theta_p\colon A \to \mathbb{K}$ such that $\theta_p(v) = \zeta_{2p}$, as in Example 3.3.15. Let $D_{(p)}$ denote the decomposition matrix of $d_{(p)}$. The rows of both $D_p$ and $D_{(p)}$ are labelled by the set $\Lambda$. We claim that

$$\boxed{D_p = D_{(p)} . D', \qquad \text{where } D' \text{ is a suitable "adjustment matrix";}}$$

that is, the entries of $D'$ are non-negative integers. This can be interpreted as saying that the $\Phi_p$-modular decomposition matrix $D_{(p)}$ of $\mathbf{H}$ is a first approximation to the usual $p$-modular decomposition matrix $D_p$ of $W$. This is proved as follows.

First, following [92, Prop. 4.6], consider the $\Phi_1$-modular specialisation $\theta_1 : A \to \mathbb{K}$ such that $\theta_1(v) = -1$. Thus, $\mathbf{H}_{\mathbb{K}} = \mathbb{K}W$; note also that $\theta_1(A) = R$ is integrally closed in $\mathbb{K}$. Furthermore, let $\theta_{\mathbb{F},1} : A \to \mathbb{F}$ be the composition of $\theta_1$ with the natural map $R \to \mathbb{F}$. Let $D_{\mathbb{F},1}$ denote the corresponding decomposition matrix. Clearly, we have $\ker(\theta_1) \subseteq \ker(\theta_{\mathbb{F},1})$, and so Theorem 3.6.3 applies. This yields a factorisation $D_{\mathbb{F},1} = D_1.D''$, where the adjustment matrix $D''$ is associated with the decomposition map $d'' : R_0(\mathbb{K}W) \to R_0(\mathbb{F}W)$ induced by the natural map $R \to \mathbb{F}$. Hence, we have $D'' = D_p$ and so $D_{\mathbb{F},1} = D_1.D_p$. Now, since $\mathbf{H}_{\mathbb{K}} = \mathbb{K}W$ is semisimple, $D_1$ is the identity matrix (for a suitable ordering of the rows and columns); see 3.1.18. Thus, we have $D_{\mathbb{F},1} = D_p$. Second, following [98, Exp. 3.2], consider the $\Phi_p$-modular specialisation $\theta_p : A \to \mathbb{K}$, $v \mapsto \zeta_{2p}$; again, we have $\theta_p(A) = R$. Now, since $\mathbb{F}$ has characteristic $p$, the image of $\zeta_{2p}$ in $\mathbb{F}$ is $-1$, and so $\ker(\theta_p) \subseteq \ker(\theta_{\mathbb{F},1})$. So we can apply Theorem 3.6.3 and this yields a factorisation $D_{\mathbb{F},1} = D_{(p)}.D'$, where $D'$ is a suitable adjustment matrix. Combining this with the identity $D_{\mathbb{F},1} = D_p$, we conclude that $D_p = D_{\mathbb{F},1} = D_{(p)}.D'$, as required.

Let us consider some concrete exanples. Let $W$ be of type $B_2 = I_2(4)$ and $p = 2$. The matrix $D_{(2)}$ is given by Example 3.2.5; since $W$ is a 2-group, it is well known that $D_2$ has only one column, with entries $\dim E^\lambda$ ($\lambda \in \Lambda$). Hence:

$D_{(2)}$ :

| $E^\lambda$ | $d_{\lambda,M}$ | |
|---|---|---|
| $1_W$ | 1 | . |
| $\mathrm{sgn}_1$ | 1 | . |
| $\mathrm{sgn}_2$ | 1 | . |
| $\mathrm{sgn}$ | 1 | . |
| $\sigma_1$ | . | 1 |

$D_2$ :

| $E^\lambda$ | $d_{\lambda,M}$ |
|---|---|
| $1_W$ | 1 |
| $\mathrm{sgn}_1$ | 1 |
| $\mathrm{sgn}_2$ | 1 |
| $\mathrm{sgn}$ | 1 |
| $\sigma_1$ | 2 |

$$D' = \begin{pmatrix} 1 \\ 2 \end{pmatrix}.$$

So $D_2$ and $D_{(2)}$ do not even have the same size! But even if $D_p$ and $D_{(p)}$ do have the same size, they will not necessarily be equal. For example, let $W \cong \mathfrak{S}_4$ be of type $A_3$ and $p = 2$. Then, by [181, p. 253], we have

$D_{(2)}$ :

| $E^\lambda$ | $d_{\lambda,M}$ | |
|---|---|---|
| (4) | 1 | . |
| (31) | 1 | 1 |
| (22) | . | 1 |
| (211) | 1 | 1 |
| (1111) | 1 | . |

$D_2$ :

| $E^\lambda$ | $d_{\lambda,M}$ | |
|---|---|---|
| (4) | 1 | . |
| (31) | 2 | 1 |
| (22) | 1 | 1 |
| (211) | 2 | 1 |
| (1111) | 1 | . |

$$D' = \begin{pmatrix} 1 & . \\ 1 & 1 \end{pmatrix}.$$

James [181] systematically computed $D_p$ and $D_{(p)}$ for $W \cong \mathfrak{S}_n$ and $1 \leqslant p \leqslant n \leqslant 10$. The results of these computations led him to formulate the following conjecture.

**Conjecture 3.6.6 (First version of James's conjecture [181]).** *In the setting of Example 3.6.5, assume that $W$ is of type $A_{n-1}$ so that $W \cong \mathfrak{S}_n$. If $p^2 > n$, then $D_p = D_{(p)}$; that is, the $\Phi_p$-modular decomposition matrix of $\mathbf{H}$ equals the usual $p$-modular decomposition matrix of $\mathfrak{S}_n$.*

Note that, by Theorem 3.5.14, both $D_p$ and $D_{(p)}$ have rows labelled by the set of all partitions of $n$ and columns labelled by the $p$-regular partitions of $n$. In general, the problem of determining $D_p$ for $W \cong \mathfrak{S}_n$ is completely open. On the other hand, as we will see in Section 6.2, the analogous problem for $D_{(p)}$ is solved,[1] thanks to the work of Ariki [7], [10]. We will discuss the relevance of the condition "$p^2 > n$", and generalisations of *James's conjecture*, in the next section.

We now consider two further applications of the factorisation of decomposition maps which will be useful later on. The following result shows that canonical basic sets behave well with respect to such factorisations.

**Lemma 3.6.7.** *Recall that $\leqslant$ is a monomial ordering on $\Gamma$ such that $L(s) \geqslant 0$ for all $s \in S$. In the setting of Theorem 3.6.3, assume that the following conditions hold:*

*(a) we have $|\operatorname{Irr}(\mathbf{H}_k)| = |\operatorname{Irr}(\mathbf{H}_{k'})|$ and*
*(b) there exists a canonical basic set $\mathscr{B}_{\theta'} \subseteq \Lambda$ for $\mathbf{H}_{k'}$ (see Definition 3.2.1).*

*Then $\mathscr{B}_{\theta'}$ also is a canonical basic set for $\mathbf{H}_k$.*

*Proof.* Let $D^\circ_{\theta'}$ be the submatrix of $D_{\theta'}$ with rows labelled by $\mathscr{B}_{\theta'}$; similarly, let $D^\circ_\theta$ be the submatrix of $D_\theta$ with rows labelled by $\mathscr{B}_{\theta'}$. By Theorem 3.6.3, we have

$$D_{\theta'} = D_\theta . D^\theta_{\theta'} \qquad \text{and} \qquad D^\circ_{\theta'} = D^\circ_\theta . D^\theta_{\theta'}.$$

As in Remark 3.2.2, we have a canonical labelling of $\operatorname{Irr}(\mathbf{H}_{k'})$ by the set $\mathscr{B}_{\theta'}$, and we can write

$$D_{\theta'} = \big(d'_{\lambda,\mu}\big)_{\lambda\in\Lambda,\,\mu\in\mathscr{B}_{\theta'}}.$$

Assume now that the elements of $\mathscr{B}_{\theta'}$ are ordered according to increasing value of $\mathbf{a}_\lambda$. Then, since $\mathscr{B}_{\theta'}$ is a canonical basic set for $\mathbf{H}_{k'}$, the matrix $D^\circ_{\theta'}$ is square and triangular with 1 along the diagonal. By condition (a), both $D^\circ_\theta$ and $D^\theta_{\theta'}$ are square matrices; furthermore, the entries of these matrices are non-negative integers. An elementary argument about matrices then shows that there is an ordering of the set $\operatorname{Irr}(\mathbf{H}_k)$ (that is, the indexing set of the columns of $D^\circ_\theta$ and of the rows of $D^\theta_{\theta'}$) such that both $D^\circ_\theta$ and $D^\theta_{\theta'}$ are triangular with 1 along the diagonal. This ordering defines a labelling $\operatorname{Irr}(\mathbf{H}_k) = \{M^\mu \mid \mu \in \mathscr{B}_{\theta'}\}$. We can now also write

$$D_\theta = \big(d_{\lambda,\mu}\big)_{\lambda\in\Lambda,\,\mu\in\mathscr{B}_{\theta'}} \qquad \text{and} \qquad D^\theta_{\theta'} = \big(a_{\mu,\nu}\big)_{\mu,\nu\in\mathscr{B}_{\theta'}},$$

where $d_{\nu,\nu} = a_{\nu,\nu} = 1$ for all $\mu \in \mathscr{B}_{\theta'}$. Now fix $\nu \in \mathscr{B}_{\theta'}$ and assume that $\lambda \in \Lambda$ is such that $d_{\lambda,\nu} \neq 0$. The above factorisation yields an equation

$$d'_{\lambda,\nu} = \sum_{\mu\in\mathscr{B}_{\theta'}} d_{\lambda,\mu} a_{\mu,\nu} = d_{\lambda,\nu} a_{\nu,\nu} + \sum_{\mu\in\mathscr{B}_{\theta'}\setminus\{\nu\}} d_{\lambda,\mu} a_{\mu,\nu}.$$

[1] This situation is reminiscent of that in the representation theory of simple algebraic groups over fields of positive characteristic; see Lusztig [227]; in this case, the formulation of the analogous problem in characteristic 0 involves the theory of quantised enveloping algebras.

Since all the terms in the above expression are non-negative integers, there are no cancellations and so $d'_{\lambda,\nu} \neq 0$. Since $\mathscr{B}_{\theta'}$ is a canonical basic set for $\mathbf{H}_{k'}$, it follows that $\lambda = \nu$ or $\mathbf{a}_\lambda > \mathbf{a}_\nu$. We have already seen that $d_{\nu,\nu} = 1$ for all $\nu \in \mathscr{B}_{\theta'}$. Thus, $\mathscr{B}_{\theta'}$ is a canonical basic set for $\mathbf{H}_k$. $\square$

The following example illustrates one of the most difficult cases in type $B_n$ that we will have to deal with.

**Example 3.6.8.** Let $W$ be of type $B_n$ with weight function $L\colon W \to \mathbb{Z}$ given by

$B_n$ 

where $c \geqslant 0$.

Now $A$ is the ring of Laurent polynomials in one indeterminate $v = \varepsilon$. Let $q \in \mathbb{Z}$ be a prime power and $\ell$ be a prime such that $\ell \nmid q$. Let $\theta\colon A \to k$ be a specialisation such that $\operatorname{char}(k) = \ell > 0$ and $\theta(u) = q1_k$, where $u := v^2$. Now recall that $\operatorname{Irr}_{\mathbb{K}}(W)$ is labelled by the set $\Lambda$ of all pairs of partitions $(\lambda, \mu)$ such that $|\lambda| + |\mu| = n$. By the end of this book, we will have determined a canonical basic set $\mathscr{B}_\theta \subseteq \Lambda$ for $\mathbf{H}_k$. This involves the following steps.

First of all, it is not yet known if the hypotheses of Proposition 3.2.7 (that is, the conjectural properties (♠) and (♣) in 2.5.3) are known to hold. We shall see in Example 4.5.15 (using an argument involving the representation theory of the finite unitary groups over $\mathbb{F}_q$) that there does exist a canonical basic set $\mathscr{B}_\theta \subseteq \Lambda$ for $\mathbf{H}_k$. However, this is only an existence result. The difficulty in obtaining an explicit description of $\mathscr{B}_\theta$ depends heavily on the parameter

$$e := \min\{i \geqslant 2 \mid 1 + q + q^2 + \cdots + q^{i-1} \equiv 0 \bmod \ell\}.$$

If $e = 2$ or $e$ is odd or $e$ is divisible by 4, then an explicit description of $\mathscr{B}_\theta$ will be obtained by elementary methods; see the discussion in Example 4.4.16. If $e = 2e'$, where $e' \geqslant 3$ is odd, then Proposition 3.6.7 will allow us to reduce the problem to characteristic 0, where completely new methods are available (involving among others the theory of canonical bases); see Example 4.5.15 for details.

Finally, we wish to show that the study of decomposition maps associated with specialisations where the parameters are sent to some roots of unity (in a field characteristic 0) can be reduced to the study of the $\Phi_e$-modular decomposition maps as defined in Example 3.1.15(b). For $W \cong \mathfrak{S}_n$, a statement of this kind explicitly appeared in Mathas [245, Cor. 6.24]. Owing to its nature, the discussion will be somewhat technical. We begin with some preparations.

**3.6.9.** As in Remark 1.1.8, let us set $\dot{T}_w := \varepsilon^{L(w)} T_w$ for all $w \in W$ and $u_s := v_s^2$ for all $s \in S$. Then the structure constants of $\mathbf{H}$ with respect to the basis $\{\dot{T}_w \mid w \in W\}$ lie in the subring $R[u_s \mid s \in S] \subseteq A$. Now let $\alpha\colon A \to A$ be a ring isomorphism such that $\alpha(R) = R$ and $\alpha(u_s) = u_s$ for all $s \in S$. Extending scalars via $\alpha$, we obtain a new algebra $\mathbf{H}^\alpha$. Since $\alpha(u_s) = u_s$ for all $s \in S$, this new algebra is canonically isomorphic to $\mathbf{H}$. So we will identify $\mathbf{H}^\alpha$ with $\mathbf{H}$. By Theorem 3.1.14, the composition of $\alpha$ with the inclusion $A \subseteq K$ gives rise to a group homomorphism $d_\alpha\colon R_0(\mathbf{H}_K) \to R_0(\mathbf{H}_K)$

which, by 3.1.18, is an isomorphism inducing a bijection of $\operatorname{Irr}(\mathbf{H}_K)$ onto itself. Thus, there exists a permutation $\lambda \mapsto \lambda^\alpha$ of $\Lambda$ such that

$$d_\alpha([E_\varepsilon^\lambda]) = E_\varepsilon^{\lambda^\alpha} \qquad \text{for all } \lambda \in \Lambda.$$

This permutation is uniquely determined by the condition that

$$\operatorname{trace}(T_w, E_\varepsilon^{\lambda^\alpha}) = \alpha\big(\operatorname{trace}(T_w, E_\varepsilon^\lambda)\big) \qquad \text{for all } w \in W.$$

Proposition 1.3.1 and the orthogonality relations in Proposition 1.2.12 imply that

$$\alpha(\mathbf{c}_\lambda) = \mathbf{c}_{\lambda^\alpha} \qquad \text{and} \qquad \mathbf{a}_\lambda = \mathbf{a}_{\lambda^\alpha} \quad \text{for all } \lambda \in \Lambda.$$

The permutation $\lambda \mapsto \lambda^\alpha$ is easily determined using the above condition on character values. Assume that $(W,S)$ is irreducible. Then $\lambda \mapsto \lambda^\alpha$ is given as follows.

(a) If $\alpha(r) = r$ for all $r \in \mathbb{Z}_W$ and $\alpha(v_s) = v_s$ ($s \in S$), then $\lambda \mapsto \lambda^\alpha$ is the identity. This follows from the fact that all values $\operatorname{trace}(T_w, E_\varepsilon^\lambda)$ lie in $\mathbb{Z}_W[\Gamma]$; see 1.2.1.
(b) If $\alpha(r) = r$ for all $r \in \mathbb{Z}_W$ and $\alpha(v_s) = -v_s$ ($s \in S$), then $\lambda = \lambda^\alpha$ unless $\lambda$ labels one of the following *exceptional representations* listed in [132, 9.2.3]:

$$\begin{array}{ll} H_3 & 4_r' \overset{\alpha}{\longleftrightarrow} 4_r \\ H_4 & 16_{rr} \overset{\alpha}{\longleftrightarrow} 16_r, \quad 16_r' \overset{\alpha}{\longleftrightarrow} 16_{rr}' \\ E_7 & 512_a' \overset{\alpha}{\longleftrightarrow} 512_a \\ E_8 & 4096_z \overset{\alpha}{\longleftrightarrow} 4096_x, \quad 4096_x' \overset{\alpha}{\longleftrightarrow} 4096_z'. \end{array}$$

(c) Let $W$ be of type $H_3$ or $H_4$, where $\mathbb{Z}_W = \mathbb{Z}[\frac{1}{2}(-1+\sqrt{5})]$. If $\alpha(v_s) = v_s$ ($s \in S$) and $\alpha(\sqrt{5}) = -\sqrt{5}$, then $\lambda \mapsto \lambda^\alpha$ is given by algebraic conjugation as specified in the tables in [132, Appendix C].
(d) Let $W$ be of type $I_2(m)$ ($m \geqslant 5$, $m \neq 6$). Then $\mathbb{Z}_W = \mathbb{Z}[2\cos(2\pi/m)] \not\subseteq \mathbb{Q}$ and so there are several maps $\alpha$ permuting the two-dimensional representations $\sigma_j$ in Example 1.3.7.

**3.6.10.** Let $\theta\colon A \to k$ be a specialisation where $k$ is the field of fractions of $\theta(A)$. Let $\alpha\colon A \to A$ be a ring isomorphism as in 3.6.9; we have just seen that this induces a permutation $\lambda \mapsto \lambda^\alpha$ of $\Lambda$. Now $\theta' = \theta \circ \alpha$ is another specialisation where $k$ is the field of fractions of $\theta(A)$. Formally, we have two specialised algebras $\mathbf{H}_{k,\theta}$ and $\mathbf{H}_{k,\theta'}$. However, since $\theta'(u_s) = \theta(u_s)$ for all $s \in S$, these two algebras are the same, so we can just denote them by $\mathbf{H}_k$. Using Theorem 3.6.3, we obtain a factorisation $d_{\theta'} = d_\theta \circ d_\alpha$. Hence, we have

$$d'_{\lambda,M} = d_{\lambda^\alpha,M} \qquad \text{for all } \lambda \in \Lambda \text{ and } M \in \operatorname{Irr}(\mathbf{H}_k),$$

where $D_\theta = \big(d_{\lambda,M}\big)$ and $D_{\theta'} = \big(d'_{\lambda,M}\big)$ are the decomposition matrices associated with $\theta$ and $\theta'$ respectively.

**3.6.11.** Let $\Gamma = \mathbb{Z}$ so that $A = R[v, v^{-1}]$ is the ring of Laurent polynomials in $v = \varepsilon$. Assume that $R$ is the ring of algebraic integers in some finite extension $\mathbb{K} \supseteq \mathbb{Q}$. Let

$\theta\colon A \to k$ be a specialisation such that

$$\mathbb{K} \subseteq k \quad \text{and} \quad k = \mathbb{K}[\theta(v)], \quad \text{where } \theta(v) \in k \text{ is } \textit{some} \text{ root of unity.}$$

We wish to relate the decomposition map $d_\theta$ to the $\Phi_e$-modular decomposition map $d_e$ (for a suitable $e$) in Example 3.1.15(b). First of all, by enlarging $\mathbb{K}$ if necessary, we can assume without loss of generality that $k = \mathbb{K}$ and $\theta(v) \in R$; see Remark 3.1.5.

Let $d \geqslant 1$ be the order of $\theta(v)$ and set $e := d/2$ (if $d$ is even) or $e := d$ (if $d$ is odd). Thus, $(-1)^d\theta(v)$ is a root of unity of order $2e$. Consequently, since we are in characteristic 0, there exists a field automorphism $\alpha_{\mathbb{K}} \in \operatorname{Aut}(\mathbb{K})$ such that $\alpha_{\mathbb{K}}(\zeta_{2e}) = (-1)^d\theta(v)$. Now, we have two specialised algebras: $\mathbf{H}_{\mathbb{K},\theta_e}$ (where $\mathbb{K}$ is regarded as an $A$-module via $\theta_e$) and $\mathbf{H}_{\mathbb{K},\theta}$ (where $\mathbb{K}$ is regarded as an $A$-module via $\theta$). Since $\alpha_{\mathbb{K}}(\zeta_{2e}) = \pm\theta(v)$, the algebra $\mathbf{H}_{\mathbb{K},\theta}$ can be identified with that obtained from $\mathbf{H}_{\mathbb{K},\theta_e}$ by scalar extension from $\mathbb{K}$ to $\mathbb{K}$ via $\alpha_{\mathbb{K}}$. Thus, $\alpha_{\mathbb{K}}$ induces a bijection

$$\operatorname{Irr}(\mathbf{H}_{\mathbb{K},\theta_e}) \to \operatorname{Irr}(\mathbf{H}_{\mathbb{K},\theta}), \qquad M \mapsto M',$$

which induces a group isomorphism $\alpha_{\mathbb{K}}^*\colon R_0(\mathbf{H}_{\mathbb{K},\theta_e}) \to R_0(\mathbf{H}_{\mathbb{K},\theta})$. Since $R$ is the ring of integers in $\mathbb{K}$, we have $\alpha_{\mathbb{K}}(R) = R$. Now note that $\theta(R) \subseteq \mathbb{K}$ and so $\theta$ induces a field automorphism $\theta_{\mathbb{K}} \in \operatorname{Aut}(\mathbb{K})$. Again, since $R$ is the ring of integers in $\mathbb{K}$, we have $\theta_{\mathbb{K}}(R) = R$. We can now define a ring homomorphism $\alpha\colon A \to A$ such that $\alpha(v) = (-1)^d v$ and $\alpha(r) = \theta_{\mathbb{K}}^{-1}(\alpha_{\mathbb{K}}(r))$ for all $r \in R$. Thus, $\alpha$ is a ring isomorphism as in 3.6.9, and we have $\alpha_{\mathbb{K}} \circ \theta_e = \theta \circ \alpha$. Then, using Remark 3.1.5 and 3.6.10, we have $\alpha_{\mathbb{K}}^* \circ d_{\theta_e} = d_{\alpha_{\mathbb{K}} \circ \theta_e} = d_{\theta \circ \alpha} = d_\theta \circ d_\alpha$, which means that the diagram

$$\begin{array}{ccc} R_0(\mathbf{H}_K) & \xrightarrow{\;d_{\theta_e}\;} & R_0(\mathbf{H}_{\mathbb{K},\theta_e}) \\ {\scriptstyle d_\alpha}\downarrow & \circlearrowleft & \downarrow{\scriptstyle \alpha_{\mathbb{K}}^*} \\ R_0(\mathbf{H}_K) & \xrightarrow{\;d_\theta\;} & R_0(\mathbf{H}_{\mathbb{K},\theta}) \end{array}$$

is commutative. Then let

$$D_{(e)} = \big(d_{\lambda,M}^{(e)}\big)_{\lambda \in \Lambda, M \in \operatorname{Irr}(\mathbf{H}_{\mathbb{K},\theta_e})} \qquad \text{and} \qquad D_\theta = \big(d_{\lambda,M'}\big)_{\lambda \in \Lambda, M' \in \operatorname{Irr}(\mathbf{H}_{\mathbb{K},\theta})}$$

be the decomposition matrices associated with $\theta_e$ and $\theta$ respectively. The commutativity of the above diagram means that

$$\boxed{d_{\lambda,M}^{(e)} = d_{\lambda^\alpha,M'} \quad \text{for all } \lambda \in \Lambda \text{ and } M \in \operatorname{Irr}(\mathbf{H}_{\mathbb{K},\theta_e}).}$$

Thus, we have found the desired relation between the decomposition matrices $D_\theta$ and $D_{(e)}$: up to a permutation of the rows, these two matrices are actually equal (where the columns are matched via the bijection $M \mapsto M'$).

**Corollary 3.6.12.** *In the setting of 3.6.11, let $\leqslant$ be a monomial order such that $L(s) \geqslant 0$ for all $s \in S$. Assume that there is a canonical basic set $\mathscr{B}_{\theta_e}$ with respect to $\theta_e$ and $\leqslant$. Thus, we can write*

$$\mathrm{Irr}(\mathbf{H}_{\mathbb{K},\theta_e}) = \{M^\mu \mid \mu \in \mathscr{B}_{\theta_e}\} \quad \textit{and} \quad d^{(e)}_{\lambda,\mu} := d^{(e)}_{\lambda,M^\mu} \quad (\lambda \in \Lambda, \mu \in \mathscr{B}_{\theta_e}).$$

*Define* $\mathscr{B}_\theta := \{\mu^\alpha \mid \mu \in \mathscr{B}_{\theta_e}\}$. *Then the following hold.*

*(a) The set* $\mathscr{B}_\theta$ *is a canonical basic set with respect to* $\theta$ *and* $\leqslant$. *Thus, we can also write* $\mathrm{Irr}(\mathbf{H}_{\mathbb{K},\theta}) = \{M^\nu \mid \nu \in \mathscr{B}_\theta\}$ *and* $d_{\lambda,\nu} := d_{\lambda,M^\nu}$ *for* $\lambda \in \Lambda$ *and* $\nu \in \mathscr{B}_\theta$.

*(b) With the notation in (a), the identity in 3.6.11 reads*

$$d_{\lambda^\alpha,\mu^\alpha} = d^{(e)}_{\lambda,\mu} \qquad \textit{for all } \lambda \in \Lambda \textit{ and } \mu \in \mathscr{B}_{\theta_e}.$$

*Proof.* (a) Let $M \in \mathrm{Irr}(\mathbf{H}_{\mathbb{K},\theta_e})$. Since $d^{(e)}_{\lambda,M} = d_{\lambda^\alpha,M'}$ for all $\lambda \in \Lambda$, we deduce that

$$\mathscr{S}_\theta(M') = \{\lambda^\alpha \mid \lambda \in \mathscr{S}_{\theta_e}(M)\}.$$

Now, we have already remarked in 3.6.9 that $\mathbf{a}_\lambda = \mathbf{a}_{\lambda^\alpha}$ for all $\lambda \in \Lambda$. Hence, if the function $\mathscr{S}_{\theta_e}(M) \to \Gamma$, $\lambda \mapsto \mathbf{a}_\lambda$, reaches its minimum at a uniquely determined $\lambda_M$, then the function $\mathscr{S}_\theta(M') \to \Gamma$, $\lambda' \mapsto \mathbf{a}_{\lambda'}$, reaches its minimum at a uniquely determined $\lambda_{M'}$, and we have $\lambda_{M'} = \lambda_M^\alpha$. It follows that there is a canonical basic set with respect to $\theta$, and we have $\mathscr{B}_\theta = \{\lambda^\alpha \mid \lambda \in \mathscr{B}_{\theta_e}\}$. The statement in (b) is then an immediate consequence of the identity in 3.6.11. □

**Example 3.6.13.** The essence of the above discussion is that, in most cases, we actually have $D_\theta = D_{(e)}$ and $\mathscr{B}_\theta = \mathscr{B}_{\theta_e}$. By 3.6.11, this applies whenever $W$ is a finite Weyl group with no component of type $E_7$ or $E_8$. Here are two examples where one can see the effect of a non-trivial permutation $\lambda \mapsto \lambda^\alpha$.

(a) Let $W$ be of type $I_2(m)$ ($m \geqslant 3$) and $L(s) = 1$ for $s \in S$ (equal parameter case). Consider an $R$-linear specialisation $\theta\colon A \to \mathbb{K}$ such that $\theta(v)$ is a root of unity of order $d \geqslant 2$. Define $e$ as above and assume that $e > 2$ and $e$ divides $m$. Then, by Example 3.2.5(b), there is a non-trivial $\theta$-block given by $\{1_W, \mathrm{sgn}, \sigma_{j(e)}\}$, where $j(e)$ is determined by the condition that $\theta(v)^2 = \zeta^{\pm j(e)}$.

Recall that $\zeta \in \mathbb{C}$ is a root of unity of order $m$ such that $\zeta + \zeta^{-1} = 2\cos(2\pi/m)$. Consequently, if $\theta = \theta_e$, then we find that $j(e) = m/e$. On the other hand, if $\theta \neq \theta_e$, then it may happen that $j(e) \neq m/e$. But then there will be an isomorphism $\alpha\colon A \to A$ as in 3.6.11 such that $j(e) \overset{\alpha}{\longleftrightarrow} m/e$. Thus, when speaking about the blocks and decomposition numbers for $W$ of type $I_2(m)$, it is important to say precisely which roots of unity are involved. Similar remarks apply to $W$ of type $H_3$ and $H_4$.

(b) Let $W$ be of type $E_7$ and $e = 5$. By Table F.5 in [132, Appendix F], there is a $\Phi_5$-block of defect 1 given by $\{7'_a, 378'_a, 512_a, 168'_a, 27'_a\}$. On the other hand, in [92, Theorem 12.6], we find that $\{7'_a, 378'_a, 512'_a, 168'_a, 27'_a\}$ is a $\Phi_5$-block of defect 1. This apparant contradiction is explained by the fact that, in [92], we used the $R$-linear specialisation $A \to \mathbb{K}$, $v \mapsto -1$, to establish a bijection $\mathrm{Irr}(\mathbf{H}_K) \leftrightarrow \mathrm{Irr}(W)$, whereas in [132] we used the specialisation $v \mapsto 1$. Thus, the two situations are transformed into each other using the ring isomorphism $\alpha\colon A \to A$ such that $\alpha(v) = -v$ and $\alpha(r) = r$ for all $r \in R$. As mentioned in 3.6.11(b), we have $512'_a = 512_a^\alpha$ in this case (and $\alpha$ fixes all other labels in $\Lambda$). A similar remark applies to several further blocks which appear in [92, §12] or [132, Appendix F].

*Remark 3.6.14.* Assume that $W$ is a Weyl group and $\Gamma = \mathbb{Z}$ so that $A = R[v, v^{-1}]$ is the ring of Laurent polynomials in $v = \varepsilon$. Let $\theta\colon A \to k$ and $\theta'\colon A \to k$ be two specialisations such that $k$ is the field of fractions of $\theta(A)$ and of $\theta'(A)$. Assume that $\theta(v)$ and $\theta'(v)$ are roots of unity of the same order in $k^\times$. Then one may conjecture that $D_\theta = D_{\theta'}$ (for a suitable matching of the columns). This was first stated by James [181, §4] for the case where $W \cong \mathfrak{S}_n$ (see also Mathas [245, 6.38]).

If $\mathrm{char}(k) = 0$, this easily follows (for any type of $W$) from the identity in 3.6.11. (Indeed, let $\lambda \mapsto \lambda^\alpha$ be the permutation associated with $\theta$ and $\lambda \mapsto \lambda^{\alpha'}$ be the one associated with $\theta'$. By the construction in 3.6.11, we have $\alpha(v) = \alpha'(v) = (-1)^d v$, where $d$ is the order of $\theta(v)$. Hence, since $W$ is a Weyl group, we have $\lambda^\alpha = \lambda^{\alpha'}$ for all $\lambda \in \Lambda$; see 3.6.9(a) and (b).) However, if $\mathrm{char}(k) = \ell > 0$, the above proofs will not work, as, in positive characteristic, two roots of unity of a given order are not necessarily conjugate under a field automorphism. See, however, the recent work of Brundan and Kleshchev [39, §6], where an entirely new approach to this problem is found (at least, as far as algebras of type $A_{n-1}$ and $B_n$ are concerned).

## 3.7 The General Version of James's Conjecture

In the previous section, we have seen a first version of James's conjecture for $W \cong \mathfrak{S}_n$. Our aim now is to try to understand how the hypothesis "$p^2 > n$" involved in that conjecture comes about, to formulate a general version for any type of $W$, and to establish some partial results supporting it. To begin with, let us state the full original version of *James's conjecture*.

**3.7.1.** Let $W \cong \mathfrak{S}_n$ be of type $A_{n-1}$. Let $\Gamma = \mathbb{Z}$ and $L(s) = 1$ for all $s \in S$. Then $A = R[v, v^{-1}]$ is the ring of Laurent polynomials in one indeterminate $v = \varepsilon$. Let $\theta\colon A \to k$ be a specialisation where $k$ is the field of fractions of $\theta(A)$. Let $D_\theta$ be the corresponding decomposition matrix. Then set

$$e := \min\{i \geqslant 2 \mid 1 + \theta(u) + \theta(u)^2 + \cdots + \theta(u)^{i-1} = 0\} \qquad (u := v^2).$$

(We let $e = \infty$ if $1 + \theta(u) + \theta(u)^2 + \cdots + \theta(u)^{i-1} \neq 0$ for all $i \geqslant 2$.) If $e > n$ or $e = \infty$, then $\mathbf{H}_k$ is semisimple; see 3.4.12. Hence, in this case, $D_\theta$ is the identity matrix, for a suitable ordering of the columns; see 3.1.18.

Now assume that $e < \infty$. Then consider the $\Phi_e$-modular specialisation such that $v \mapsto \zeta_{2e} \in \mathbb{C}$, as in Example 3.3.15. Let $D_{(e)}$ be the corresponding decomposition matrix. By Theorem 3.5.14, both $D_\theta$ and $D_{(e)}$ have rows labelled by the set $\Lambda$ consisting of all partitions of $n$ and columns labelled by the subset $\Lambda_k^\circ \subseteq \Lambda$ consisting of all $e$-regular partitions of $n$. Now we can state the following conjecture.

**Conjecture 3.7.2 (James [181, §4]).** *In the above setting, assume that* $\mathrm{char}(k) = \ell > 0$ *and* $e\ell > n$. *Then we have* $D_\theta = D_{(e)}$.

Note that the "first version" in Conjecture 3.6.6 is the special case where $\theta$ is such that $\theta(u) = 1$; we then have $e = \ell$. We also remark that James not only considers the algebra $\mathbf{H}$, but also the (slightly bigger) $q$-Schur algebra [64].

We shall see that an appropriate setting for understanding and generalising the above conjecture is given by the factorisation result in Theorem 3.6.3. In this framework, a general version for any type of $W$ first appeared in [92]. Taking into account our results on cell data from Chapter 2, we will give a more elegant formulation of that general version, which formally looks exactly like James's original one.

**3.7.3.** Throughout, we assume that $R$ is $L$-good and $L(s) \geqslant 0$ for all $s \in S$. Furthermore, we assume that there is a subset $\mathscr{P} \subseteq A$ such that the three conditions in 3.3.1 are satisfied. In particular, for $\Phi \in \mathscr{P}$, we have a corresponding principal specialisation $\theta_\Phi \colon A \to k_\Phi$, where $k_\Phi$ is the field of fractions of $A/(\Phi)$.

**3.7.4.** Assume that (♠) in 2.5.3 holds; then we have a cellular basis $\{\mathbf{C}^\lambda_{\mathfrak{s},\mathfrak{t}}\}$ as in Theorem 2.6.12. Let $\Phi \in \mathscr{P}$ and $\theta \colon A \to k$ be any specialisation such that $k$ is the field of fractions of $k$ and $\theta(\Phi) = 0$. (Thus, $\theta$ lies above $\theta_\Phi$; see Example 3.6.4).

Let $\{W(\lambda) \mid \lambda \in \Lambda\}$ be the cell representations defined in terms of the given cellular basis of $\mathbf{H}$. Let $G^\lambda = (g^\lambda_{\mathfrak{s}\mathfrak{t}})_{\mathfrak{s},\mathfrak{t}\in M(\lambda)}$ be the Gram matrix of the canonical bilinear form on $W(\lambda)$; see Lemma 2.6.2. By specialisation, we obtain matrices

$$G^\lambda_\Phi := \big(\theta_\Phi(g^\lambda_{\mathfrak{s}\mathfrak{t}})\big)_{\mathfrak{s},\mathfrak{t}\in M(\lambda)} \qquad \text{and} \qquad G^\lambda_k := \big(\theta(g^\lambda_{\mathfrak{s}\mathfrak{t}})\big)_{\mathfrak{s},\mathfrak{t}\in M(\lambda)}.$$

Then, setting $\Lambda^\circ_\Phi := \{\lambda \in \Lambda \mid G^\lambda_\Phi \neq 0\}$ and $\Lambda^\circ_k := \{\lambda \in \Lambda \mid G^\lambda_k \neq 0\}$, we have

$$\operatorname{Irr}(\mathbf{H}_{k_\Phi}) = \{L^\mu_\Phi \mid \mu \in \Lambda^\circ_\Phi\} \qquad \text{and} \qquad \operatorname{Irr}(\mathbf{H}_k) = \{L^\mu_k \mid \mu \in \Lambda^\circ_k\},$$

where $\dim L^\mu_\Phi = \operatorname{rank}(G^\mu_\Phi)$ and $\dim L^\mu_k = \operatorname{rank}(G^\mu_k)$; see Theorem 2.6.5 and 2.7.5. Here, we just write $L^\mu_\Phi$ instead of $L^\mu_{k_\Phi}$.

Now, since $\theta$ lies above $\theta_\Phi$, we can factorise $\theta = \varphi \circ \theta_\Phi$, where $\varphi \colon A/(\Phi) \to k$ is a ring homomorphism. Hence, $G^\lambda_k$ is obtained by applying $\varphi$ to the entries of $G^\lambda_\Phi$. First of all, this implies that if $G^\lambda_\Phi = 0$, then $G^\lambda_k = 0$. Furthermore, we certainly have $\operatorname{rank}(G^\lambda_k) \leqslant \operatorname{rank}(G^\lambda_\Phi)$. Thus, we conclude that

$$\boxed{\Lambda^\circ_k \subseteq \Lambda^\circ_\Phi \qquad \text{and} \qquad \dim L^\mu_k \leqslant \dim L^\mu_\Phi \quad \text{for all } \mu \in \Lambda^\circ_k.}$$

Recall from Example 3.6.4 that we also have a factorisation of decomposition maps $D_\theta = D_\Phi . D^\Phi_\theta$, where $D^\Phi_\theta$ is the corresponding *adjustment matrix*. Let us write

$$D^\Phi_\theta = \big(d'_{\mu\nu}\big)_{\mu\in\Lambda^\circ_\Phi,\, \nu\in\Lambda^\circ_k}.$$

The following observation shows that if the inclusion and the inequalities above are all equalities, then we can even say something more precise about the shape of $D^\Phi_\theta$. (A statement of this kind first appeared in [172, §3.1.B].)

**Proposition 3.7.5.** *In the setting of 3.7.4, assume that* $|\operatorname{Irr}(\mathbf{H}_{k_\Phi})| = |\operatorname{Irr}(\mathbf{H}_k)|$. *Then* $\Lambda_k^\circ = \Lambda_\Phi^\circ$ *and, for a suitable ordering of the elements of* $\Lambda_k^\circ = \Lambda_\Phi^\circ$, *the adjustment matrix* $D_\theta^\Phi$ *is lower triangular with* 1 *on the diagonal.*

*Furthermore,* $D_\theta^\Phi$ *is the identity matrix if* $\dim L_k^\mu = \dim L_\Phi^\mu$ *for all* $\mu \in \Lambda_k^\circ$.

*Proof.* Our assumption implies that $|\Lambda_\Phi^\circ| = |\Lambda_k^\circ|$. Since we already know that $\Lambda_k^\circ \subseteq \Lambda_\Phi^\circ$, we deduce that these two sets are actually equal. Let us set $\Lambda^\circ := \Lambda_\Phi^\circ = \Lambda_k^\circ$.

Now consider the factorisation of decomposition maps $D_\theta = D_\Phi . D_\theta^\Phi$. Let $D_\theta^\circ$ be the submatrix of $D_\theta$ obtained by deleting all rows corresponding to labels $\lambda \notin \Lambda^\circ$. Similarly, we define the submatrix $D_\Phi^\circ$ of $D_\phi$. Then $D_\theta^\circ$, $D_\Phi^\circ$ and $D_\theta^\Phi$ are square matrices with rows and columns labelled by $\Lambda^\circ$, and we have $D_\theta^\circ = D_\Phi^\circ . D_\theta^\Phi$.

The cell datum for $\mathbf{H}$ comes with a partial order $\trianglelefteq_L$ on $\Lambda$. If we choose a total order on $\Lambda^\circ$ which refines $\trianglelefteq_L$, then Theorem 2.6.6 shows that both $D_\theta^\circ$ and $D_\Phi^\circ$ are lower triangular with 1 along the diagonal. Hence, the same is true for $D_\theta^\Phi$ as well. It follows that, for any $\mu \in \Lambda^\circ$, we have equations

$$\dim L_\Phi^\mu = \dim L_k^\mu + \sum_{\nu \in \Lambda^\circ :\, \nu \triangleleft_L \mu} d'_{\mu\nu} \dim L_k^\nu, \quad \text{where} \quad D_\theta^\Phi = (d'_{\mu\nu})_{\mu,\nu \in \Lambda^\circ}.$$

Hence, if $\dim L_k^\mu = \dim L_\Phi^\mu$ for all $\mu \in \Lambda^\circ$, then $d'_{\mu\nu} = 0$ whenever $\nu \triangleleft_L \mu$. Since we already know that $D_\theta^\Phi$ has a lower triangular shape with 1 along the diagonal, we deduce that $D_\theta^\Phi$ is the identity matrix, as required. □

We now have the following general finiteness result. Note that it does not yield any explicit numerical bound. The general version of James's conjecture will be concerned precisely with specifying a numerical bound.

**Proposition 3.7.6 (Cf. [92, 5.5], [98, 2.7]).** *In the setting of 3.7.4, assume that A is noetherian and* $\dim A = 2$. *Then, for a fixed* $\Phi \in \mathscr{P}$, *the set*

$$\mathfrak{M}(\Phi) := \{\mathfrak{p} \in \operatorname{Spec}(A) \,|\, \mathfrak{p} = \ker(\theta) \textit{ where } \theta \textit{ lies above } \theta_\Phi \textit{ and } D_\theta^\Phi \neq \mathrm{Id}\}$$

*is finite (where* Id *denotes the identity matrix of the appropriate size).*

*Proof.* For $\mu \in \Lambda_\Phi^\circ$, let $r_\mu := \dim L_\Phi^\mu = \operatorname{rank}(G_\Phi^\mu)$. So there is an $(r_\mu \times r_\mu)$-submatrix $H^\mu$ of $G^\mu$ such that $\theta_\Phi(\det(H^\mu)) \neq 0$. Let $d := \prod_{\mu \in \Lambda_\Phi^\circ} \det(H^\mu) \in A$; note that $\theta_\Phi(d) \neq 0$. We claim that

$$\mathfrak{M}(\Phi) \subseteq \{\mathfrak{p} \in \operatorname{Spec}(A) \,|\, (\Phi) \subseteq \mathfrak{p} \text{ and } d \in \mathfrak{p}\}.$$

Indeed, let $\mathfrak{p} = \ker(\theta)$, where $\theta$ lies above $\Phi$ and $d \notin \mathfrak{p}$. We show that $\mathfrak{p} \notin \mathfrak{M}(\Phi)$. Now, since $\theta(d) \neq 0$, we also have $\theta(\det(H^\mu)) \neq 0$ for all $\mu \in \Lambda_\Phi^\circ$. In particular, this implies that $G_k^\mu \neq 0$ and so $\mu \in \Lambda_k^\circ$. Since we already know that $\Lambda_k^\circ \subseteq \Lambda_\Phi^\circ$, we conclude that $\Lambda_\Phi^\circ = \Lambda_k^\circ$. Furthermore, the fact that $\theta(\det(H^\mu)) \neq 0$ implies that $\det(H_k^\mu) \neq 0$ and so $\dim L_k^\mu = \operatorname{rank}(G_k^\mu) \geqslant r_\mu$. Since we already know that $\dim L_k^\mu \leqslant r_\mu$, we conclude that $\dim L_k^\mu = r_\mu$ for all $\mu \in \Lambda_\Phi^\circ$. So Proposition 3.7.5 implies that $D_\theta^\Phi$ is the identity matrix and, hence, $\mathfrak{p} \notin \mathfrak{M}(\Phi)$. Thus, the above claim is proved.

Now set $B := A/(\Phi)$. Then the prime ideals of $A$ containing $(\Phi)$ correspond bijectively to the prime ideals of $B$. Hence, we have

$$\mathfrak{M}(\Phi) \quad \overset{1-1}{\longleftrightarrow} \quad \{\mathfrak{m} \in \operatorname{Spec}(B) \mid 0 \neq \theta_\Phi(d) \in \mathfrak{m}\}.$$

Finally, we use our assumption that $A$ is noetherian and $\dim A = 2$. First note that $(\Phi)$ is a height 1 prime ideal in $A$. Hence, $B = A/(\Phi)$ is a one-dimensional noetherian ring. Consequently, every prime ideal of $B$ containing $\theta_\Phi(d) \neq 0$ is minimal with that property. But then it is well known that, in a noetherian ring, the set of prime ideals containing a given element has only finitely many elements; see, for example, [248, Exc. 4.12]. Hence, the set $\{\mathfrak{m} \in \operatorname{Spec}(B) \mid 0 \neq \theta_\Phi(d) \in \mathfrak{m}\}$ is finite. □

**Example 3.7.7.** In the setting of 3.7.4, let $\Gamma = \mathbb{Z}$; then $A = R[v, v^{-1}]$ is the ring of Laurent polynomials in one indeterminate $v = \varepsilon$. Assume that $\mathbb{K} \subseteq \mathbb{C}$ is a finite Galois extension of $\mathbb{Q}$ and that $R$ is the ring of algebraic integers in $\mathbb{K}$. Then $R$ is a Dedekind domain and so $A$ is noetherian and $\dim A = 2$. As in Example 3.3.3, the set $\mathscr{P}$ consists of $\mathbb{K}$-cyclotomic polynomials. Let us fix $\Phi \in \mathscr{P}$. Then $B = A/(\Phi)$ is the ring of algebraic integers in some finite extension of $\mathbb{Q}$. In the above proof, we have seen that all $\mathfrak{p} \in \mathfrak{M}(\Phi)$ correspond to maximal ideals in $B$. Hence, $A/\mathfrak{p}$ is a finite field for every $\mathfrak{p} \in \mathfrak{M}(\Phi)$. The finiteness statement in Proposition 3.7.6 then means that there exists a bound $N > 0$ (depending only on $W, L$) such that

$$\boxed{D_\theta^\Phi = \operatorname{Id} \text{ whenever } \theta\colon A \to k \text{ lies above } \theta_\Phi \text{ and } \operatorname{char}(k) = \ell > N.}$$

A statement of this kind first appeared in [92, Prop. 5.5]; see also [98, 2.7]. This shows that, in order to determine the matrices $D_\theta$ for all possible specialisations $\theta\colon A \to k$, it will be enough to solve the following three problems:

(1) Determine $D_\Phi$ for all $\Phi \in \mathscr{P}$.
(2) Find the bound $N$.
(3) Determine $D_\theta$ where $0 < \operatorname{char}(k) \leqslant N$.

Note that, although the proof of Proposition 3.7.6 does not give an explicit formula[2] for the bound $N$, it nevertheless leads to a straightforward algorithm for computing the set $\mathfrak{M}(\Phi)$ in concrete examples. This algorithm involves the following steps:

- Compute the Gram matrices $G^\mu$ ($\mu \in \Lambda_\Phi^\circ$) and determine $r_\mu = \operatorname{rank}(G_\Phi^\mu)$.
- For each $\mu \in \Lambda_\Phi^\circ$, find an $(r_\mu \times r_\mu)$-submatrix $H^\mu$ of $G^\mu$ such that $\theta_\Phi(H^\mu) \neq 0$.
- Let $d := \prod_{\mu \in \Lambda_\Phi^\circ} \det(H^\mu) \in A$ and denote by $\bar{d} \neq 0$ the image of $d$ in $B = A/(\Phi)$. Then find all maximal ideals $\mathfrak{m} \subsetneq B$ such that $\bar{d} \in \mathfrak{m}$.

This algorithm has been used by Geck and Müller [129] to determine $\mathfrak{M}(\Phi)$ for $W$ of exceptional type $G_2, F_4, E_6, E_7, E_8$ (in the equal parameter case). It turns out that, in these cases, $N$ can be taken to be the largest prime divisior of $W$.

---

[2] As far as the analogous problem for simple algebraic groups over fields of positive characteristic is concerned (see the footnote p. 185), Fiebig [84], [85] recently found an explicit bound.

We are now almost ready to state the general version of James's conjecture. We need one further piece of notation.

**Definition 3.7.8.** Let $\Phi, \Phi' \in \mathscr{P}$. We write $\Phi \sim \Phi'$ if $\Phi' = \alpha(\Phi)$, where $\alpha\colon A \to A$ is a ring isomorphism as in 3.6.9; that is, we have $\alpha(R) = R$ and $\alpha(v_s) = \pm v_s$ for all $s \in S$. In this case, the decomposition matrices $D_\Phi$ and $D_{\Phi'}$ are the same up to reordering the rows according to the permutation $\lambda \mapsto \lambda^\alpha$; see 3.6.10.

Now let $\theta\colon A \to k$ be a specialisation such that $k$ is the field of fractions of $\theta(A)$.

(a) We say that $\Phi \in \mathscr{P}$ is *$\theta$-isolated* if

$$\theta(\Phi) = 0 \qquad \text{but} \qquad \theta(\Phi') \neq 0 \text{ for all } \Phi' \in \mathscr{P} \text{ such that } \Phi' \not\sim \Phi.$$

(b) We say that $\Phi \in \mathscr{P}$ is *strongly $\theta$-isolated* if

$$\theta(\Phi) = 0 \qquad \text{but} \qquad \theta(\Phi') \neq 0 \text{ for all } \Phi \neq \Phi' \in \mathscr{P}.$$

(Slightly different versions of (a) and (b) appeared in Geck and Rouquier [133, 3.3].)

**Conjecture 3.7.9 (General version of James's conjecture; cf. Geck [92, 5.6], [98, 3.4], Geck and Rouquier [133, §5.2]).** *Recall our basic assumptions 3.7.3; also assume that (♠) holds; see 3.7.4. Let $\theta\colon A \to k$ and $\Phi \in \mathscr{P}$ be such that $\Phi$ is $\theta$-isolated; see Definition 3.7.8. Then*

$$\Lambda_\Phi^\circ = \Lambda_k^\circ \qquad \textit{and} \qquad \dim L_\Phi^\mu = \dim L_k^\mu \quad \textit{for all } \mu \in \Lambda_\Phi^\circ.$$

*Also, the adjustment matrix $D_\theta^\Phi$ is the identity matrix (see Proposition 3.7.5).*

We will see in Example 3.7.13 below that this contains James's original conjecture as a special case. First we need some preparations.

**3.7.10.** Assume we are in the setting of Example 3.3.3, where $A = R[v, v^{-1}]$ and $L\colon W \to \mathbb{Z}$ is a weight function such that $L(s) \geqslant 0$ for all $s \in S$. In order to avoid unnecessary technical complications, we shall assume that

$$(*) \qquad 1 \in \sum_{s \in S} \mathbb{Z} L(s).$$

(A consequence of $(*)$ is that, for any ring isomorphism $\alpha\colon A \to A$ as in Definition 3.7.8, we have $\alpha(v) = \pm v$ and, hence, $\alpha(\Phi_d(v^2)) = \Phi_d(v^2)$ for any $d \geqslant 1$.) Let $\theta\colon A \to k$ be a specialisation such that $\operatorname{char}(k) = \ell > 0$. We set

$$\mathscr{E}_\theta := \{d \geqslant 2 \mid \theta(\Phi) = 0 \text{ for some } \Phi \in \mathscr{P} \text{ such that } \Phi \mid \Phi_d(v^2)\}.$$

First we claim that $\mathscr{E}_\theta$ is non-empty unless $\mathbf{H}_k$ is semisimple. This is seen as follows. Assume that $\mathbf{H}_k$ is not semisimple. Then there is some $\lambda \in \Lambda$ such that $\theta(\mathbf{c}_\lambda) = 0$; see 3.1.18. Since $R$ is $L$-good, this implies that $\theta(\Phi) = 0$ for some $\Phi \in \mathscr{P}$. Since

$\Phi$ is $\mathbb{K}$-cyclotomic, we have $\Phi \mid \Phi_d(v^2)$ for some $d \geqslant 1$. In particular, $d \in E(W,L)$ and so $d \geqslant 2$; see Example 3.3.3. Thus, we have $\mathscr{E}_\theta \neq \varnothing$, as claimed.

Now, if $\mathbf{H}_k$ is semisimple, then we set $e = \infty$. Otherwise, we let $e$ be the minimal element of $\mathscr{E}_\theta$ and $\Phi \in \mathscr{P}$ be such that $\theta(\Phi) = 0$ and $\Phi \mid \Phi_e(v^2)$.

This definition of $e$ should be regarded as the appropriate generalisation of the definition of $e$ for $W$ of type $A_{n-1}$ in 3.7.1.

Next we need some standard facts about cyclotomic polynomials.

**3.7.11.** In the above setting, let $u := v^2$ and $q := \theta(u)$.

(a) Assume that $\Phi_d(q) = \Phi_{d'}(q) = 0$, where $d' \geqslant d$. Then $d' = d\ell^i$ for some $i \geqslant 0$.

To prove this, write $d = m\ell^a$, where $a \geqslant 0$ and $\ell \nmid m$. Now $\Phi_d(u)$ certainly divides $\Phi_m(u^{\ell^a})$ and so $\Phi_n(q^{\ell^a}) = 0$. Since the map $\alpha \mapsto \alpha^\ell$ is a ring homomorphism of $k$, we conclude that $\Phi_m(q) = 0$. A similar argument applies to $d'$: writing $d' = m'\ell^b$, where $b \geqslant 0$ and $\ell \nmid m'$, we conclude that $\Phi_{m'}(q) = 0$. Assume, if possible, that $m \neq m'$. Then $u^{mm'} - 1 = \Phi_m(u)\Phi_{m'}(u)p(u)$, where $p(u) \in \mathbb{Z}[u]$. Differentiating with respect to $u$ and then setting $u = q$ would yield $mm' = 0$ in $k$, which is a contradiction. Hence, we have $m = m'$ and so $d = m\ell^a$ and $d' = m'\ell^b$. This implies (a).

(b) Let $d \geqslant 1$. Then $\Phi_d(q) = 0$ if and only if $\Phi_{d\ell}(q) = 0$.

This is seen as follows. It is well known that

$$\Phi_d(u^\ell) = \begin{cases} \Phi_d(u)\,\Phi_{d\ell}(u) & \text{if } \ell \nmid d, \\ \Phi_{d\ell}(u) & \text{if } \ell \mid d. \end{cases}$$

Since the map $\alpha \mapsto \alpha^\ell$ is a ring homomorphism of $k$, we conclude that $\Phi_d(u^\ell) = \Phi_d(u)$ (in $k[u]$). Hence $\Phi_{d\ell}(u) = \Phi_d(u)^m$ (in $k[u]$), where $m = \ell$ or $m = \ell - 1$. Thus, we also have $\Phi_{d\ell}(q) = \Phi_d(q)^m$, as required.

**Lemma 3.7.12.** *In the above setting, assume that $\mathbf{H}_k$ is not semisimple and let $e \geqslant 2$ be as in 3.7.10. Let $\Phi \in \mathscr{P}$ be such that $\theta(\Phi) = 0$ and $\Phi \mid \Phi_e(v^2)$. Then the following hold.*

*(a) $\Phi$ is $\theta$-isolated if and only if $e\ell^i \notin E(W,L)$ for all $i \geqslant 1$.*
*(b) If $\Phi$ is $\theta$-isolated and $\ell \nmid 2e$, then $\Phi$ is strongly $\theta$-isolated.*
*(c) If $L(s) = 1$ for all $s \in S$ and $\ell \nmid |W|$, then $\Phi$ is strongly $\theta$-isolated.*

*Proof.* As above, we write $u := v^2$ and $q := \theta(u)$.

(a) First assume that $e\ell^i \notin E(W,L)$ for all $i \geqslant 1$. Let $d \in E(W,L)$ and assume that $\theta(\Phi') = 0$ for some $\Phi' \in \mathscr{P}$ such that $\Phi' \mid \Phi_d(v^2)$. Then $\Phi_d(q) = 0$ and so 3.7.11(a) shows that $d = e\ell^i$ for some $i \geqslant 0$. Our assumption implies that $i = 0$ and so $d = e$. Hence, both $\Phi$ and $\Phi'$ divide $\Phi_e(v^2)$.

If $e$ is even, then $\Phi_e(v^2) = \Phi_{2e}(v)$; thus, $\Phi$ and $\Phi'$ both divide $\Phi_{2e}(v)$. If $e$ is odd, then $\Phi_e(v^2) = \Phi_e(v)\Phi_{2e}(v)$, where $\Phi_{2e}(v) = \pm\Phi_e(-v)$. Thus, replacing $\Phi$ by $\Phi(\pm v)$ and $\Phi'$ by $\Phi'(\pm v)$ for appropriate signs, we can assume that both $\Phi$ and $\Phi'$ divide $\Phi_{2e}(v)$. (Note that $\Phi \sim \Phi(\pm v)$ and $\Phi' \sim \Phi'(\pm v)$.) But, since all roots of unity in $\mathbb{K}$ of a given order are algebraically conjugate, there will be a ring isomorphism $\alpha\colon A \to A$ such that $\alpha(R) = R$ and $\alpha(\Phi) = \Phi'$. Hence, $\Phi \sim \Phi'$. Thus, we have shown that $\Phi$ is $\theta$-isolated, as required.

To prove the converse assume, if possible, that $\Phi$ is $\theta$-isolated and that there is some $i \geqslant 1$ such that $e\ell^i \in E(W,L)$. Then 3.7.11(b) shows that $\Phi_{e\ell^i}(q) = 0$. Hence, there will be some $\Phi' \in \mathscr{P}$ such that $\theta(\Phi') = 0$ and $\Phi' \mid \Phi_{e\ell^i}(v^2)$. Since $\Phi$ is $\theta$-isolated, we must have $\Phi \sim \Phi'$. In particular, this implies that both $\Phi$ and $\Phi'$ divide $\Phi_e(v^2)$. (Here, assumption $(*)$ in 3.7.10 is used.) But, if $i \geqslant 1$, then $\Phi_{e\ell^i}(u)$ and $\Phi_e(u)$ are coprime in $\mathbb{Q}[u]$ and, hence, also coprime in $\mathbb{K}[v]$, which is a contradiction.

(b) Let $\Phi' \in \mathscr{P}$ and assume that $\theta(\Phi') = 0$. We must show that $\Phi' = \Phi$. Let $d \in E(W,L)$ be such that $\Phi' \mid \Phi_d(v^2)$. By 3.7.11(b), we have $d = e\ell^i$ for some $i \geqslant 0$. By (b), we must have $i = 0$. Thus, both $\Phi$ and $\Phi'$ divide $\Phi_e(v^2)$ and, hence, also $v^{2e} - 1$. Assume, if possible, that $\Phi \neq \Phi'$. Then we can write $v^{2e} - 1 = \Phi\Phi' f$ for some $f \in R[v]$. Differentiating with respect to $v$ and then applying $\theta$ would yield that $\theta(2e) = 0$; that is, $\ell \mid 2e$, which is contrary to our assumption.

(c) In this case, one easily checks that $E(W,L)$ consists of all integers $d \geqslant 2$ such that $d$ divides some degree of $W$; see Example 3.3.2 (for $W$ of crystallographic type) and [132, 8.3.4 and Appendix E] (otherwise). Since the product of all degrees is the order of $W$, the assertion follows from (a) and (b). □

**Example 3.7.13.** Let $W$ be a finite Weyl group and assume that we are in the equal parameter case where $L(s) = 1$ for all $s \in S$, as in Example 3.3.2. Since $\mathbb{Z}_W = \mathbb{Z}$, we can assume that $R \subseteq \mathbb{Q} = \mathbb{K}$. Assume that $\mathbf{H}_k$ is not semisimple. First, one easily checks that the definition of $e$ in 3.7.10 can be rewritten as follows:

$$e = \min\{i \in E(W) \mid 1 + \theta(u) + \theta(u)^2 + \cdots + \theta(u)^{i-1} = 0\} \qquad (u := v^2);$$

recall that $E(W)$ is the set of all integers $d \geqslant 2$ such $d$ divides some degree of $W$. Let $\Phi \in \mathscr{P}$ be such that $\theta(\Phi) = 0$ and $\Phi \mid \Phi_e(v^2)$; by the description of $\mathscr{P}$ in Example 3.3.2, we have $\Phi = \Phi_{2e}(v)$ (if $e$ is even) or $\Phi \in \{\Phi_e(v), \Phi_{2e}(v)\}$ (if $e$ if odd). By Lemma 3.7.12(a), we have

$$\boxed{\Phi \text{ is } \theta\text{-isolated if and only if } e\ell \text{ does not divide any degree of } W.}$$

A general condition of this kind first appeared in [133, §5]; see also [129, §3].

(a) Consider the special case where $W \cong \mathfrak{S}_n$ is of type $A_{n-1}$. By Example 3.1.19, the degrees of $W$ are $2, 3, \ldots, n$. Hence, we have $E(W) = \{2, 3, \ldots, n\}$ and so

$$\Phi \text{ is } \theta\text{-isolated if and only if } e\ell > n.$$

Thus, Conjecture 3.7.2 really is a special case of Conjecture 3.7.9.

(b) Now let $W$ be of type $F_4$ or $E_6$. In these cases, $R \subseteq \mathbb{Q}$ is $L$-good if $2, 3 \in R^\times$. So we are only considering specialisations $\theta\colon A \to k$ where $\operatorname{char}(k) = \ell \geqslant 5$. Comparing with the list of degrees of $W$ in Example 3.3.2, we see that $\Phi$ is automatically $\theta$-isolated. In these cases, Conjecture 3.7.9 was verified (assuming that $q = \theta(v)^2$ lies in the prime field of $k$) by Geck and Lux [126] (type $F_4$) and Geck [94] (type $E_6$). These results for groups of exceptional type provided a major motivation for formulating the general version of James's conjecture above.

(c) Using the algorithm already mentioned in Example 3.7.7, the general version of James's conjecture has been systematically verified by Geck and Müller [129] for $W$ of exceptional type $G_2$, $F_4$, $E_6$, $E_7$, $E_8$.

Here is an entirely different instance of James's conjecture which was described by Leclerc and Miyachi [210].

**Example 3.7.14.** Let $W$ be of type $B_n$ and assume that we are in the setting of Example 3.3.4, where $A = \mathbb{Z}[V^{\pm 1}, v^{\pm 1}]$ with two independent variables $V, v$ and

$$T_{s_0}^2 = T_1 + (V - V^{-1})T_{s_0} \quad \text{and} \quad T_{s_i}^2 = T_1 + (v - v^{-1})T_{s_i} \quad \text{for } 1 \leqslant i \leqslant n-1.$$

Now let $t$ be an indeterminate over $\mathbb{F}_2 = \mathbb{Z}/2\mathbb{Z}$ and $k = \mathbb{F}_2(t)$. Let $1 \leqslant r \leqslant n-1$. Then the natural map $\mathbb{Z} \to \mathbb{F}_2$ extends to a specialisation $\theta_r \colon A \to k$ such that $V \mapsto t^r$ and $v \mapsto t$. Furthermore, consider the principal specialisation defined by $\Psi_r := V^2 + v^{2r} \in \mathscr{P}$. We have $\theta_r(\Psi_r) = 0$, and one easily checks that $\Psi_r$ is strongly $\theta_r$-isolated. So, according to Conjecture 3.7.9, the adjustment matrix $D_{\theta_r}^{\Psi_r}$ should be the identity matrix. This has been shown to be the case by Leclerc and Miyachi [210, Prop. 14]. In their notation, $D_{\theta_r}$ is the matrix associated with the decomposition map $\mathbf{d}_{r,+}^2$ in [210, Prop. 14], and $D_{\theta_r}^{\Phi_r}$ corresponds to the map $\mathbf{d}_{r,-}^2$. Leclerc and Miyachi interpret $D_{\Psi_r}$ in terms of the *constructible representations* defined by Lusztig [231, Chap. 22]. This is related to the discussion in Example 3.8.10; see [106, §6] for further details.

Apart from partial results in special situations where much is known about decomposition numbers (see, for example, Fayers [78], [79], [81] and Geck and Müller [129], or the "defect 1" case treated in [92]), James's conjecture – in all its versions – is completely open. (See also [80], [258], [259], [260] and [290, Chap. VII] for further attempts at this conjecture.) We will now present a general argument for proving at least one piece of that conjecture; namely, the assertion that $\Lambda_\Phi^\circ = \Lambda_k^\circ$. For $W \cong \mathfrak{S}_n$, this has already been established in Theorem 3.5.14. In general, by Proposition 3.7.5, it will be sufficient to prove that $|\operatorname{Irr}(\mathbf{H}_{k_\Phi})| = |\operatorname{Irr}(\mathbf{H}_k)|$.

**Lemma 3.7.15.** *Let $\theta \colon A \to k$ and $\Phi \in \mathscr{P}$ be such that $\theta(\Phi) = 0$, so that $\theta$ lies above $\theta_\Phi$. Assume that, for any $M \in \operatorname{Irr}(\mathbf{H}_{k_\Phi})$, the function $P_M \colon Z(\mathbf{H}) \to K$ (see Proposition 3.3.6) takes its values in $A_\mathfrak{p}$, the localisation of $A$ in $\mathfrak{p} = \ker(\theta)$. Then we have $|\operatorname{Irr}(\mathbf{H}_k)| = |\operatorname{Irr}(\mathbf{H}_{k_\Phi})|$.*

*Proof.* By 3.7.4, we already know that $|\operatorname{Irr}(\mathbf{H}_k)| \leqslant |\operatorname{Irr}(\mathbf{H}_{k_\Phi})|$. So we only need to prove the reverse inequality. Write $D_\Phi = \big(d_{\lambda,M}\big)$. Then, by Remark 3.3.7(c), we have

$$d_{\lambda,M} = \sum_{C \in \operatorname{Cl}(W)} \chi^\lambda(\dot{T}_{w_C})\, P_M(z_C) \quad \text{for all } \lambda \in \Lambda \text{ and } M \in \operatorname{Irr}(\mathbf{H}_{k_\Phi}).$$

These identities can be written as a matrix equation $D_\Phi = X(\mathbf{H}).P^{\mathrm{tr}}$, where $P$ has entries $P_M(z_C)$ ($M \in \operatorname{Irr}(\mathbf{H}_{k_\Phi})$ and $C \in \operatorname{Cl}(W)$). Now, the entries of $X(\mathbf{H})$ lie in $A$ and those of $D_\Phi$ are rational integers. Hence, our assumption on the values of $P_M$

implies that all entries of the above matrices lie in $A_\mathfrak{p}$. So we can apply the canonical extension of $\theta$ to $A_\mathfrak{p}$ and obtain an identity among the specialised matrices over $k$. Since $D_\Phi$ and its specialisation have rank equal to $|\operatorname{Irr}(\mathbf{H}_{k_\Phi})|$ (see Theorem 3.1.14), we conclude that the rank of $X(\mathbf{H})_\theta$ is greater than or equal to $|\operatorname{Irr}(\mathbf{H}_{k_\Phi})|$. Hence, by Proposition 3.4.11, we also have $|\operatorname{Irr}(\mathbf{H}_k)| \geqslant |\operatorname{Irr}(\mathbf{H}_{k_\Phi})|$, as required. □

**Theorem 3.7.16 (Cf. Geck and Rouquier [100], [133, Theorem 3.3]).** *If $\Phi \in \mathscr{P}$ is strongly $\theta$-isolated, then $|\operatorname{Irr}(\mathbf{H}_{k_\Phi})| = |\operatorname{Irr}(\mathbf{H}_k)|$. Hence, in the setting of 3.7.4, we have $\Lambda_\Phi^\circ = \Lambda_k^\circ$.*

*Proof.* Assume that $\Phi$ is strongly $\theta$-isolated. We must show that, for any $M \in \operatorname{Irr}(\mathbf{H}_{k_\Phi})$, the condition on the values of $P_M$ in Lemma 3.7.15 holds. Now, we have

$$P_M(z_C) = \sum_{\lambda\in\Lambda} \mathbf{c}_\lambda^{-1}\, d_{\lambda,M}\, \omega_\lambda(z_C) \qquad (C \in \operatorname{Cl}(W)),$$

where the numbers $d_{\lambda,M}$ are the entries of $D_\Phi$ and $\omega_\lambda(z_C) \in A$. Recall from 3.3.1 that, for each $\lambda \in \Lambda$, we can write

$$\mathbf{c}_\lambda = \Phi^{n_{\lambda,\Phi}}\, \mathbf{c}'_\lambda, \qquad \text{where} \qquad \mathbf{c}'_\lambda := f_\lambda\, v^{\gamma_\lambda} \prod_{\Phi' \in \mathscr{P}\setminus\{\Phi\}} (\Phi')^{n_{\lambda,\Phi'}}.$$

Note that $\theta(\mathbf{c}'_\lambda) \neq 0$ for all $\lambda \in \Lambda$, since $\Phi$ is strongly $\theta$-isolated and $R$ is $L$-good.

Now we can argue as follows. Setting $f = \prod_\lambda \mathbf{c}'_\lambda$ and $n := \max\{n_{\lambda,\Phi} \mid \lambda \in \Lambda\}$, we have $f\Phi^n P_M(z_C) \in A$. So we can write $f\Phi^n P_M(z_C) = \Phi^m a$, where $m \geqslant 0$ and $a \in A$ is such that $\Phi \nmid a$. Since $\theta(\mathbf{c}'_\lambda) \neq 0$ for all $\lambda \in \Lambda$, we have $\Phi \nmid f$. We conclude that we must have $m-n = \nu_\Phi(P_M(z_C))$, where $\nu_\Phi$ is the discrete valuation associated with $\Phi$. But, by Proposition 3.3.6, we have $P_M(z_C) \in A_{(\Phi)}$ and, hence, $m-n \geqslant 0$. Finally, since $\theta(\mathbf{c}'_\lambda) \neq 0$ for all $\lambda \in \Lambda$, we also have $\theta(f) \neq 0$. So we conclude that $P_M(z_C) = f^{-1}\Phi^{m-n} a \in A_{\ker(\theta)}$, as required.

Finally, if we are in the setting of 3.7.4 and $|\operatorname{Irr}(\mathbf{H}_{k_\Phi})| = |\operatorname{Irr}(\mathbf{H}_k)|$, then Proposition 3.7.5 shows that we also have $\Lambda_\Phi^\circ = \Lambda_k^\circ$. □

We just mention that similar arguments can be used to show that if $\Phi \in \mathscr{P}$ is strongly $\theta$-isolated, then the partition of $\Lambda$ into $\Phi$-blocks coincides with the partition into $\theta$-blocks. (See [92, Prop. 7.6] for further details.)

## 3.8 Blocks and Bad Specialisations

The main result of this section will show that the relation $\sim_L$ on $\operatorname{Irr}_{\mathbb{K}}(W)$, which is defined in terms of the Kazhdan–Lusztig cells of $W$ (see Section 2.2), has a ring-theoretic interpretation in terms of certain "blocks" of $\mathbf{H}_K$ (in the general sense of the representation theory of associative algebras). These blocks are in some way related to specialisations of $A$ which arise from the "$L$-bad" prime ideals in $\mathbb{Z}_W$, in the sense of Definition 1.5.9. (So far, such specialisations have been excluded from our discussion as we generally assumed that $R$ should be an $L$-good subring of $\mathbb{C}$.)

We begin with some remarks on block idempotents. We have already encountered idempotents at various places above but, because they will play a crucial role in this section, let us collect here the basic facts and definitions in a general setting. (As before, the symbol $\underline{H}$ will eventually denote a generic Iwahori–Hecke algebra $\mathbf{H} = \mathbf{H}_A(W,S,L)$. Since we will have to deal with algebras over various different base rings, we shall use the symbol $\underline{\dot{H}}$ in the following general discussion.)

**3.8.1.** Let $\underline{\dot{H}}$ be an associative ring with identity. An element $e \in \underline{\dot{H}}$ is called an idempotent if $e^2 = e$. Two idempotents $e, f$ are called orthogonal if $ef = fe = 0$; an idempotent $e \neq 0$ is called primitive if $e$ cannot be expressed as the sum of two non-zero, orthogonal idempotents. Denote by $Z(\underline{\dot{H}})$ the centre of $\underline{\dot{H}}$. A *central idempotent* is an idempotent $e \in \underline{\dot{H}}$ which lies in $Z(\underline{\dot{H}})$. We say that $e$ is a *centrally primitive idempotent* if $e \neq 0$ is a central idempotent and if $e$ is primitive as an idempotent of $Z(\underline{\dot{H}})$. We set

$$\mathrm{Bl}(\underline{\dot{H}}) := \{ \text{ set of all centrally primitive idempotents in } \underline{\dot{H}} \ \};$$

the elements of $\mathrm{Bl}(\underline{\dot{H}})$ are also called *block idempotents*. Note that two block idempotents $b \neq b'$ are automatically orthogonal. Indeed, $b = bb' + (b - bb')$ is a decomposition of $b$ as a sum of two orthogonal central idempotents; since $b$ is centrally primitive, we have $bb' = 0$ or $bb' = b$. Similarly, $b' = bb' + (b' - bb')$ is a decomposition of $b'$ as a sum of two orthogonal central idempotents; since $b'$ is centrally primitive, we have $bb' = 0$ or $bb' = b'$. Hence, if we had $bb' \neq 0$, then $b = bb' = b'$, which is a contradiction. So we have $bb' = 0$, as claimed.

**3.8.2.** A useful method for finding the block idempotents in $\underline{\dot{H}}$ is as follows.

Assume that we are given a decomposition $1_{\underline{\dot{H}}} = \sum_{e \in P} e$, where $P$ is a finite set of pairwise orthogonal, primitive idempotents in $\underline{\dot{H}}$. (We do not assume that the idempotents in $P$ are central.) Given $e, e' \in P$, we write $e \leftrightarrow e'$ if there exists a sequence $e = e_0, e_1, \ldots, e_m = e'$ in $P$ such that, for each $i \in \{1, \ldots, m\}$, we have $e_{i-1}\underline{\dot{H}}e_i \neq \{0\}$ or $e_i\underline{\dot{H}}e_{i-1} \neq \{0\}$. Also note that

$$e\underline{\dot{H}}e' \cong \mathrm{Hom}_{\underline{\dot{H}}}(\underline{\dot{H}}e, \underline{\dot{H}}e') \qquad \text{for any idempotents } e, e' \in \underline{\dot{H}}.$$

Thus, $e\underline{\dot{H}}e' \neq \{0\}$ if and only if there exists a non-zero homomorphism $\underline{\dot{H}}e \to \underline{\dot{H}}e'$. Now let $P = P_1 \sqcup \cdots \sqcup P_r$ be the partition of $P$ into equivalences classes under $\leftrightarrow$. Then the following hold (see, for example, [48, 2.1.4]):

$$\mathrm{Bl}(\underline{\dot{H}}) = \{e_1, \ldots, e_r\} \qquad \text{where} \qquad e_i := \sum_{e \in P_i} e \qquad \text{for all } i.$$

Thus, the block idempotents of $\underline{\dot{H}}$ are obtained by summing up the idempotents in each equivalence class of $P$ under $\leftrightarrow$. Furthermore, for every central idempotent $b \in \underline{\dot{H}}$, there exists a subset $J \subseteq \{1, \ldots, r\}$ such that $b = \sum_{i \in J} e_i$.

**3.8.3.** Now assume that $\underline{\dot{H}}$ is finitely generated and free over an integral domain $A_0$. Let $K$ be a field which contains the field of fractions of $A_0$ and assume that the $K$-algebra $\underline{H}_K = K \otimes_{A_0} \underline{\dot{H}}$ is split semisimple. As in Example 3.1.11, we have a central idempotent $e_V \in \underline{H}_K$ corresponding to any $V \in \mathrm{Irr}(\underline{H}_K)$ where $e_V$ acts as the identity on $V$ and as zero on any simple module not isomorphic to $V$. It follows that

$$1_{\underline{H}_K} = \sum_{V \in \mathrm{Irr}(\underline{H}_K)} e_V \qquad \text{and} \qquad \mathrm{Bl}(\underline{H}_K) = \{e_V \mid V \in \mathrm{Irr}(\underline{H}_K)\}.$$

Now, we certainly have $Z(\underline{\dot{H}}) \subseteq Z(\underline{H}_K)$. Hence, if $b \in \mathrm{Bl}(\underline{\dot{H}})$, then $b \in Z(\underline{H}_K)$. Furthermore, $e_V b = e_V \Leftrightarrow e_V b \neq 0 \Leftrightarrow b.V \neq \{0\}$ for any $V \in \mathrm{Irr}(\underline{H}_K)$. It follows that

$$b = \sum_{V \in \mathrm{Irr}(\underline{H}_K \mid b)} e_V \qquad \text{where} \qquad \mathrm{Irr}(\underline{H}_K \mid b) := \{V \in \mathrm{Irr}(\underline{H}_K) \mid e_V b \neq 0\};$$

Conversely, if $\mathscr{S} \subseteq \mathrm{Irr}(\underline{H}_K)$ is any subset such that

$$\Big(\sum_{V \in \mathscr{S}} e_V\Big) \in \underline{\dot{H}} \qquad \text{and} \qquad \Big(\sum_{V \in \mathscr{S}'} e_V\Big) \notin \underline{\dot{H}} \quad \text{for any } \mathscr{S}' \subsetneqq \mathscr{S}, \mathscr{S}' \neq \varnothing,$$

then $\sum_{V \in \mathscr{S}} e_V$ is a block idempotent of $\underline{H}$ (see also [48, 2.1.6]). Thus, we obtain:

$$\mathrm{Irr}(\underline{H}_K) = \bigsqcup_{b \in \mathrm{Bl}(\underline{\dot{H}})} \mathrm{Irr}(\underline{H}_K \mid b) \qquad \text{and} \qquad 1_{\underline{\dot{H}}} = \sum_{b \in \mathrm{Bl}(\underline{\dot{H}})} b.$$

**Definition 3.8.4.** In the setting of 3.8.3, let $V, V' \in \mathrm{Irr}(\underline{H}_K)$. We say that $V, V'$ belong to the same *$\underline{\dot{H}}$-block* (or just to the same *$A_0$-block* if $\dot{H}$ is clear from the context), if there exists some $b \in \mathrm{Bl}(\underline{\dot{H}})$ such that $V, V' \in \mathrm{Irr}(\underline{H}_K \mid b)$.

**Example 3.8.5.** Let us go back to the setting in Section 3.1. Let $\underline{H}$, $K$ and $\theta\colon A \to k$ be as in 3.1.1. Assume that $A$ is integrally closed in $K$; furthermore, assume that $\underline{H}_k$ is split and $\underline{H}_K$ is split semisimple. Hence, by Theorem 3.1.2, we have a well-defined decomposition map $d_\theta\colon R_0(\underline{H}_K) \to R_0(\underline{H}_k)$. As discussed in 3.1.8, the corresponding Brauer graph gives rise to the partition of $\mathrm{Irr}(\underline{H}_K)$ into $\theta$-blocks.

Now let $A_0 \subseteq K$ be the localisation of $A$ in the prime ideal $\ker(\theta)$ and set $\underline{\dot{H}} := A_0 \otimes_A \underline{H}$. Then we are in the set-up of 3.8.3. Assume that $A_0$ is a *discrete valuation ring* and let $V, V' \in \mathrm{Irr}(\underline{H}_K)$. Then the following three conditions are equivalent:

(a) $V$ and $V'$ belong to the same $\theta$-block (see 3.1.8);
(b) $V$ and $V'$ belong to the same $A_0$-block (see Definition 3.8.4);
(c) $\theta(\omega_V(z)) = \theta(\omega_{V'}(z))$ for all $z \in Z(\underline{H})$ (see Lemma 3.1.10).

This is part of the classical representation theory of associative algebras; see, for example, [83, Cor. I.17.9], [132, 7.5.10]. (As already discussed in the proof of Theorem 3.4.4, it is not necessary to assume that $A_0$ is complete.)

In the following discussion of Iwahori–Hecke algebras, we shall apply Definition 3.8.4 to an algebra $\underline{\dot{H}}$ over a base ring $A_0$ which is not a local ring.

**3.8.6.** Now let again $W$ be a finite Coxeter group and $\mathbf{H} = \mathbf{H}_A(W, S, L)$ be the generic Iwahori–Hecke algebra associated with a weight function $L\colon W \to \Gamma$ such that

$L(s) \geqslant 0$ for all $s \in S$; see 3.1.12. Here, $A = R[\Gamma]$, where $R \subseteq \mathbb{C}$ is a subring such that $\mathbb{Z}_W \subseteq R$. Also recall our notation

$$\mathrm{Irr}_{\mathbb{K}}(W) = \{E^\lambda \mid \lambda \in \Lambda\} \qquad \text{and} \qquad \mathrm{Irr}(\mathbf{H}_K) = \{E_\varepsilon^\lambda \mid \lambda \in \Lambda\},$$

where $\mathbb{K}$ is the field of fractions of $R$ and $K$ is the field of fractions of $A$. As in Remark 1.7.8, consider the subring

$$\mathscr{R}(\Gamma) := \left\{ \frac{f}{g} \;\Big|\; f \in \mathbb{Z}[\Gamma],\ g \in 1 + \mathbb{Z}[\Gamma_{>0}] \right\} \subseteq K.$$

Since the structure constants of $\mathbf{H}$ lie in $\mathbb{Z}[\Gamma]$, we have an $\mathscr{R}(\Gamma)$-subalgebra

$$\mathbf{H}_{\mathscr{R}(\Gamma)} := \left\{ \sum_{w \in W} a_w T_w \mid a_w \in \mathscr{R}(\Gamma) \text{ for all } w \in W \right\} \subseteq \mathbf{H}_K.$$

Hence, setting $A_0 = \mathscr{R}(\Gamma)$ and $\underline{\dot{H}} := \mathbf{H}_{\mathscr{R}(\Gamma)}$, we are in the set-up of 3.8.3. Consequently, we can speak of the $\mathbf{H}_{\mathscr{R}(\Gamma)}$-blocks (or just the $\mathscr{R}(\Gamma)$-blocks) of $\mathrm{Irr}(\mathbf{H}_K)$.

The following remarkable result was first proved by Rouquier [272] in the equal-parameter case, following earlier work of Gyoja [151]. A version for the case of unequal parameters appeared in [106, §3].

**Theorem 3.8.7.** *In the setting of 3.8.6, assume that the properties (♣), (♠), (♦) in 2.5.3 hold. Let $\lambda, \mu \in \Lambda$. Then*

$$E^\lambda \sim_L E^\mu \qquad \Leftrightarrow \qquad E_\varepsilon^\lambda, E_\varepsilon^\mu \text{ belong to the same } \mathscr{R}(\Gamma)\text{-block},$$

*where the relation $\sim_L$ on $\mathrm{Irr}_{\mathbb{K}}(W)$ is defined in Definition 2.2.1.*

*Proof.* The basic idea is to reduce everything to a problem inside the asymptotic algebra $\tilde{\mathbf{J}}$. For this purpose, consider Lusztig's homomorphism $\phi \colon \mathbf{H} \to \tilde{\mathbf{J}}_A$, where

$$\phi(C_w) = \sum_{\substack{z \in W, d \in \tilde{\mathscr{D}} \\ z \sim_{\mathscr{L}\mathscr{R}} d}} h_{w,d,z} \tilde{n}_d t_z \qquad (w \in W);$$

see Theorem 2.5.5. Here, we already used that (♠) holds. Since (♣) also holds, we have $\tilde{\gamma}_{x,y,z} \in \mathbb{Z}$ for all $x, y, z \in W$ and $\tilde{n}_d \in \mathbb{Z}$ for all $d \in \tilde{\mathscr{D}}$; see Lemma 2.5.13. Hence, we can use the above formula to define an $\mathscr{R}(\Gamma)$-algebra homomorphism

$$\phi_{\mathscr{R}(\Gamma)} \colon \mathbf{H}_{\mathscr{R}(\Gamma)} \to \tilde{\mathbf{J}}_{\mathscr{R}(\Gamma)}$$

where $\tilde{\mathbf{J}}_{\mathscr{R}(\Gamma)} := \left\{ \sum_{w \in W} r_w t_w \mid r_w \in \mathscr{R}(\Gamma) \text{ for all } w \in W \right\}$. The first crucial observation is that, by Remark 2.5.15, this homomorphism is an isomorphism of $\mathscr{R}(\Gamma)$-algebras. Consequently, we have

$$\mathrm{Bl}(\tilde{\mathbf{J}}_{\mathscr{R}(\Gamma)}) = \{\phi_{\mathscr{R}(\Gamma)}(b) \mid b \in \mathrm{Bl}(\mathbf{H}_{\mathscr{R}(\Gamma)})\}.$$

Recall that the simple $\mathbf{H}_K$-modules afford balanced representations $\{\rho^\lambda \mid \lambda \in \Lambda\}$ and these give rise to the irreducible representations $\{\bar{\rho}^\lambda \mid \lambda \in \Lambda\}$ of $\tilde{\mathbf{J}}$. Extending scalars to $K$, we obtain irreducible representations $\bar{\rho}_K^\lambda \colon \tilde{\mathbf{J}}_K \to M_{d_\lambda}(K)$ such that $\bar{\rho}_K^\lambda(t_w) = \bar{\rho}^\lambda(t_w)$ for all $w \in W$. As in Example 2.7.3, we see that, for a suitable basis of $E_\varepsilon^\lambda$, the action of $C_w$ ($w \in W$) on $E_\varepsilon^\lambda$ is given by the matrix $\bar{\rho}_K^\lambda(\phi_{\mathscr{R}(\Gamma)}(C_w))$. Thus, for any block idempotent $b \in \mathrm{Bl}(\mathbf{H}_{\mathscr{R}(\Gamma)})$, we have

$$b.E_\varepsilon^\lambda \neq \{0\} \qquad \Leftrightarrow \qquad \bar{\rho}_K^\lambda(\phi_{\mathscr{R}(\Gamma)}(b)) \neq 0.$$

So, since $\phi_{\mathscr{R}(\Gamma)}(b) \in \mathrm{Bl}(\tilde{\mathbf{J}}_{\mathscr{R}(\Gamma)})$, we have

$$\begin{array}{c}E_\varepsilon^\lambda, E_\varepsilon^\mu \text{ belong to the} \\ \text{same } \mathscr{R}(\Gamma)\text{-block (of } \mathbf{H}_K)\end{array} \quad \Leftrightarrow \quad \begin{array}{c}\bar{\rho}_K^\lambda, \bar{\rho}_K^\mu \text{ belong to the} \\ \text{same } \mathscr{R}(\Gamma)\text{-block (of } \tilde{\mathbf{J}}_K).\end{array}$$

On the other hand, recall from Definition 2.2.1 that $E^\lambda \sim_L E^\mu$ if and only if the associated two-sided $\tilde{\mathbf{J}}$-cells are contained in the same two-sided Kazhdan–Lusztig cell. Now, since (♠) holds, the two-sided Kazhdan–Lusztig cells coincide with the two-sided $\tilde{\mathbf{J}}$-cells; see Lemma 2.5.9. Hence, we have $E^\lambda \sim_L E^\mu$ if and only if $\mathscr{F}_\lambda = \mathscr{F}_\mu$. Thus, it remains to show the following equivalence:

$$(*) \qquad \mathscr{F}_\lambda = \mathscr{F}_\mu \qquad \Leftrightarrow \qquad \bar{\rho}_K^\lambda, \bar{\rho}_K^\mu \text{ belong to the same } \mathscr{R}(\Gamma)\text{-block.}$$

To prove this, recall from Remark 1.6.13 that we have a central idempotent

$$1_{\mathscr{F}} = \sum_{d \in \tilde{\mathscr{D}} \cap \mathscr{F}} \tilde{n}_d t_d \qquad \text{for any two-sided } \tilde{\mathbf{J}}\text{-cell } \mathscr{F};$$

furthermore, we have $\mathscr{F}_\lambda = \mathscr{F}_\mu$ if and only if there exists some two-sided $\tilde{\mathbf{J}}$-cell $\mathscr{F}$ such that $\bar{\rho}^\lambda(1_{\mathscr{F}}) \neq 0$ and $\bar{\rho}^\mu(1_{\mathscr{F}}) \neq 0$. Consequently, in order to prove $(*)$, it is sufficient to prove the following two statements:

(a) For each $d \in \tilde{\mathscr{D}}$, the element $\tilde{n}_d t_d$ is a primitive idempotent in $\tilde{\mathbf{J}}_{\mathscr{R}(\Gamma)}$.
(b) The elements $\{1_{\mathscr{F}} \mid \mathscr{F} \text{ two-sided } \tilde{\mathbf{J}}\text{-cell}\}$ are the block idempotents of $\tilde{\mathbf{J}}_{\mathscr{R}(\Gamma)}$.

Note that, since (♦) holds, we already know that the elements in (a) are pairwise orthogonal idempotents and that $\tilde{n}_d = \pm 1$ for all $d \in \tilde{\mathscr{D}}$; see Lemma 2.7.9. Now, to prove (a), assume that $\tilde{n}_d t_d = e + e'$, where $e, e' \in \tilde{\mathbf{J}}_{\mathscr{R}(\Gamma)}$ are orthogonal idempotents and $e \neq 0$. Then we must show that $e' = 0$. For this purpose, we shall compute $\bar{\tau}(e)$ in two different ways, where $\bar{\tau}$ is the symmetrising trace in Proposition 1.5.6. First, writing $e = \sum_{w \in W} r_w t_w$, where $r_w \in \mathscr{R}(\Gamma)$, we find that

$$\bar{\tau}(e) = \sum_{w \in W} r_w \bar{\tau}(t_w) = \sum_{d \in \tilde{\mathscr{D}}} r_d \tilde{n}_d \in \mathscr{R}(\Gamma) = \sum_{d \in \tilde{\mathscr{D}}} (\pm r_d) \in \mathscr{R}(\Gamma).$$

On the other hand, by the defining formula for $\tilde{n}_w$, we have

$$\bar{\tau}(t_w) = \tilde{n}_w = \sum_{\lambda \in \Lambda} f_\lambda^{-1} \mathrm{trace}(\bar{\rho}^\lambda(t_w)) \qquad \text{for all } w \in W$$

and, hence, $\bar{\tau}(e) = \sum_{\lambda \in \Lambda} f_\lambda^{-1} \text{trace}(\bar{\rho}_K^\lambda(e))$. Now, since $e$ is an idempotent, the trace of $\bar{\rho}_K^\lambda(e)$ is a non-negative integer for all $\lambda \in \Lambda$, and it is non-zero for at least some $\lambda$. Since the elements $f_\lambda$ are strictly positive real numbers, we conclude that $\bar{\tau}(e)$ is a strictly positive real number. Thus, we have shown that $\bar{\tau}(e) \in \mathscr{R}(\Gamma) \cap \mathbb{R}_{>0}$. But, by the definition of $\mathscr{R}(\Gamma)$, we have $\mathscr{R}(\Gamma) \cap \mathbb{R} = \mathbb{Z}$. Hence, we deduce that $\bar{\tau}(e)$ is a strictly positive integer; in particular, $\bar{\tau}(e) \geqslant 1$. If $e' \neq 0$, then the same argument would apply to $e'$ and so $\bar{\tau}(e') \geqslant 1$. This would yield the contradiction $1 = \bar{\tau}(\tilde{n}_d t_d) = \bar{\tau}(e) + \bar{\tau}(e') \geqslant 2$. Thus, we must have $e' = 0$, as required.

To prove (b), we use the method in 3.8.2. We must show that two given elements $d, d' \in \tilde{\mathscr{D}}$ lie in the same two-sided $\tilde{\mathbf{J}}$-cell if and only if $d \leftrightarrow d'$. To prove the "only if" part, it is sufficient to consider an elementary step of $\leftrightarrow$; that is, we have $t_d \, \tilde{\mathbf{J}}_{\mathscr{R}(\Gamma)} \, t_{d'} \neq \{0\}$. Then $t_d t_w t_{d'} \neq 0$ for some $w \in W$ and so

$$0 \neq t_d t_w t_{d'} = \sum_{x \in W} \tilde{\gamma}_{d,w,x^{-1}} t_x t_{d'} = \sum_{x,y \in W} \tilde{\gamma}_{d,w,x^{-1}} \, \tilde{\gamma}_{x,d',y^{-1}} \, t_y.$$

Hence, there exist some $x, y \in W$ such that $\tilde{\gamma}_{d,w,x^{-1}} \neq 0$ and $\tilde{\gamma}_{x,d',y^{-1}} \neq 0$. By Corollary 1.6.7, this implies that $d, w, x, d', y$ all lie in the same two-sided $\tilde{\mathbf{J}}$-cell, as required. Conversely, assume that $d, d' \in \tilde{\mathscr{D}}$ belong to the same two-sided $\tilde{\mathbf{J}}$-cell. Let $\mathfrak{C}_0$ and $\mathfrak{C}'$ be the left $\tilde{\mathbf{J}}$-cells such that $d \in \mathfrak{C}_0$ and $d' \in \mathfrak{C}'$. Let $\lambda, \mu \in \Lambda$ be such that $m(\mathfrak{C}_0, \lambda) > 0$ and $m(\mathfrak{C}', \mu) > 0$. Then we have $\mathscr{F}_\lambda = \mathscr{F}_\mu$ by Example 1.8.2. So Lemma 1.8.4 shows that there exist a sequence $\lambda = \lambda_0, \lambda_1, \ldots, \lambda_m = \mu$ in $\Lambda$ and a sequence of left $\tilde{\mathbf{J}}$-cells $\mathfrak{C}_1, \ldots, \mathfrak{C}_m$ such that, for each $i \in \{1, \ldots, m\}$, we have $m(\mathfrak{C}_i, \lambda_{i-1}) > 0$ and $m(\mathfrak{C}_i, \lambda_i) > 0$. Consequently, $m(\mathfrak{C}_i, \lambda_{i-1}) > 0$ and $m(\mathfrak{C}_{i-1}, \lambda_{i-1}) > 0$ for $1 \leqslant i \leqslant m$. Let us also set $\mathfrak{C}_{m+1} := \mathfrak{C}'$ and recall that $\mu = \lambda_m$. Then $[\mathfrak{C}_{i-1}]_{\mathbb{K}}$ and $[\mathfrak{C}_i]_{\mathbb{K}}$ have an irreducible constituent in common and so

$$\operatorname{Hom}_{\mathbf{J}}([\mathfrak{C}_{i-1}]_{\mathbb{K}}, [\mathfrak{C}_i]_{\mathbb{K}}) \neq \{0\} \qquad \text{for } 1 \leqslant i \leqslant m+1.$$

Now, for each $i$, let $d_i \in \tilde{\mathscr{D}}$ be the unique element contained in $\mathfrak{C}_i$. Then we have $\tilde{\mathbf{J}} t_{d_i} \cong [\mathfrak{C}_i]_{\mathbb{K}}$; see Example 1.8.6. It follows that

$$t_{d_{i-1}} \tilde{\mathbf{J}} t_{d_i} \cong \operatorname{Hom}_{\tilde{\mathbf{J}}}(\tilde{\mathbf{J}} t_{d_{i-1}}, \tilde{\mathbf{J}} t_{d_i}) \neq \{0\} \qquad \text{for } 1 \leqslant i \leqslant m+1.$$

Note that then we also have $t_{d_{i-1}} \tilde{\mathbf{J}}_{\mathscr{R}(\Gamma)} t_{d_i} \neq \{0\}$ for $1 \leqslant i \leqslant m+1$ and so $d = d_0 \leftrightarrow d_{m+1} = d'$, as required. Thus, (a) and (b) are proved. □

*Remark 3.8.8.* By going once more through the above proof, we notice that the same argument works when we replace $\mathscr{R}(\Gamma)$ by any noetherian ring $\Theta$, where $\mathbb{Z}[\Gamma] \subseteq \Theta \subseteq K$, such that the following conditions hold:

$$1 + \mathbb{Z}[\Gamma_{>0}] \subseteq \Theta^\times \quad \text{and} \quad \Theta \cap \Big\{ \sum_{\lambda \in \Lambda} f_\lambda^{-1} m_\lambda \mid m_\lambda \in \mathbb{Z} \text{ for all } m_\lambda \in \mathbb{Z} \Big\} \subseteq \mathbb{Z}.$$

(This observation appeared in [106, §3].) Indeed, the first condition makes sure that $\phi_\Theta \colon \mathbf{H}_\Theta \to \tilde{\mathbf{J}}_\Theta$ is an isomorphism of $\Theta$-algebras. The second condition is required in the proof of the fact that $\tilde{n}_d t_d$ is a primitive idempotent for each $d \in \tilde{\mathscr{D}}$. (See (a) in the above proof.) Recall that we needed to evaluate $\bar{\tau}(e)$, where $e \in \tilde{\mathbf{J}}_\Theta$ is a non-zero idempotent. First, writing $e = \sum_{w \in W} r_w t_w$, where $r_w \in \Theta$, we see that $\bar{\tau}(e) \in \Theta$. Then, using $\bar{\tau}(e) = \sum_{\lambda \in \Lambda} f_\lambda^{-1} \operatorname{trace}(\bar{\rho}_K^\lambda(e))$, we see that $\bar{\tau}(e) = \sum_{\lambda \in \Lambda} f_\lambda^{-1} m_\lambda$ is a positive real number, where $m_\lambda \in \mathbb{Z}$ for all $\lambda \in \Lambda$. Hence, the second condition implies that $\bar{\tau}(e) \in \mathbb{Z}$. Now we can continue to argue as above. Thus, we obtain that

$$E^\lambda \sim_L E^\mu \qquad \Leftrightarrow \qquad E_\varepsilon^\lambda, E_\varepsilon^\mu \text{ belong to the same } \Theta\text{-block.}$$

We will illustrate this in Example 3.8.10 below.

**Table 3.4** Character table and Schur elements for $\mathbf{H}(G_2)$

| | $T_1$ | $T_{s_1}$ | $T_{s_2}$ | $T_{s_1s_2}$ | $T_{s_1s_2s_1s_2}$ | $T_{s_1s_2s_1s_2s_1}$ | $\mathbf{c}_\lambda$ |
|---|---|---|---|---|---|---|---|
| $1_W^\nu$ | $1$ | $u$ | $u$ | $u^2$ | $u^4$ | $u^6$ | $(u+1)^2(u^4+u^2+1)$ |
| $\operatorname{sgn}_1^\nu$ | $1$ | $u$ | $-1$ | $-u$ | $u^2$ | $-u^3$ | $3u^{-1}(u+1)^2$ |
| $\operatorname{sgn}_2^\nu$ | $1$ | $-1$ | $u$ | $-u$ | $u^2$ | $-u^3$ | $3u^{-1}(u+1)^2$ |
| $\operatorname{sgn}^\nu$ | $1$ | $-1$ | $-1$ | $1$ | $1$ | $1$ | $u^{-6}(u+1)^2(u^2+u^2+1)$ |
| $\sigma_1^\nu$ | $2$ | $u-1$ | $u-1$ | $u$ | $-u^2$ | $-2u^3$ | $6u^{-1}(u^2-u+1)$ |
| $\sigma_2^\nu$ | $2$ | $u-1$ | $u-1$ | $-u$ | $-u^2$ | $2u^3$ | $2u^{-1}(u^2+u+1)$ |

**Example 3.8.9.** Since $\mathbf{H}$ is a symmetric algebra, there is an explicit formula for the block idempotents of $\mathbf{H}_K$ in terms of the values of the irreducible characters. (This was already used in the proof of Proposition 3.1.17.) Hence, using the notation and the formulae in Remark 3.3.7, we find that

$$\begin{aligned} e_\lambda &= \mathbf{c}_\lambda^{-1} \sum_{w \in W} \varepsilon^{-2L(w)} \chi^\lambda(\dot{T}_w)\, \dot{T}_{w^{-1}} \\ &= \mathbf{c}_\lambda^{-1} \sum_{w \in W} \sum_{C \in \mathrm{Cl}(W)} \varepsilon^{-2L(w)} f_{w,C}\, \chi^\lambda(\dot{T}_{w_C})\, \dot{T}_{w^{-1}} \\ &= \mathbf{c}_\lambda^{-1} \sum_{C \in \mathrm{Cl}(W)} \chi^\lambda(\dot{T}_{w_C})\, z_C. \end{aligned}$$

Now let $\Theta \subseteq K$ be a subring as in Remark 3.8.8 and consider the partition of $\operatorname{Irr}(\mathbf{H}_K)$ into $\Theta$-blocks. By the discussion in 3.8.3, this partition is found by determining the non-empty subsets $\mathscr{S} \subseteq \Lambda$ which are minimal with respect to the property that

$$(*) \qquad \sum_{\lambda \in \mathscr{S}} \mathbf{c}_\lambda^{-1} \chi^\lambda(\dot{T}_{w_C}) \in \Theta \qquad \text{for all } C \in \mathrm{Cl}(W).$$

The following simple rules are now obvious where we assume that $\mathbb{Z}_W = \mathbb{Z}$:

(a) If $f_\lambda = 1$, then $\{E_\varepsilon^\lambda\}$ is a $\Theta$-block. (Indeed, in this case, we have $\mathbf{c}_\lambda \in 1 + \mathbb{Z}[\Gamma_{>0}]$ and so $e_\lambda \in \mathbf{H}_\Theta$.)

(b) Let $b \in \mathrm{Bl}(\mathbf{H}_\Theta)$ and $a := \max\{\mathbf{a}_\lambda \mid E_\varepsilon^\lambda \in \mathrm{Irr}(\mathbf{H}_K \mid b)\}$. Then $\sum_\lambda f_\lambda^{-1} \dim E^\lambda \in \mathbb{Z}$, where the sum runs over all $\lambda \in \Lambda$ such that $\mathbf{a}_\lambda = a$ and $E_\varepsilon^\lambda \in \mathrm{Irr}(\mathbf{H}_K \mid b)$. (Indeed, just consider the coefficient of $z_C$ in $(*)$ where $C = \{1\}$.)

In any case, since the polynomials $\{\mathbf{c}_\lambda\}$ and the character table $X(\mathbf{H})$ are known, we now have a practical algorithm for determining the $\Theta$-blocks.

For example, let $W$ be of type $G_2 = I_2(6)$ with generators $s_1, s_2$ such that $(s_1 s_2)^6 = 1$. Assume that we are in the equal parameter case where $\Gamma = \mathbb{Z}$ and $L(s_1) = L(s_2) = 1$. Then $A$ is the ring of Laurent polynomials in one indeterminate $v = \varepsilon$; let $u := v^2$. The character table of $\mathbf{H}$ and the polynomials $\{\mathbf{c}_\lambda\}$ are given in Table 3.4. Let $\Theta = \mathscr{R}$ (the original version of Rouquier's ring). By (a), we have the $\mathscr{R}$-blocks $\{1_W^v\}$ and $\{\mathrm{sgn}^v\}$. Let $\mathscr{S} := \mathrm{Irr}(\mathbf{H}_K) \setminus \{1_W^v, \mathrm{sgn}^v\}$. Then all irreducible representations in $\mathscr{S}$ have $\mathbf{a}_\lambda = 1$. Furthermore,

$$\sum_{E^\lambda \in \mathscr{S}} f_\lambda^{-1} \dim E^\lambda = \frac{1}{3} + \frac{1}{3} + \frac{2}{6} + \frac{2}{2}.$$

We notice that $\sigma_2^v$ does not form an $\mathscr{R}$-block by itself. Hence, there is no proper subset of $\mathscr{S}$ such that the corresponding sum of terms $f_\lambda^{-1} \dim E^\lambda$ would still lie in $\mathbb{Z}$. Hence, we conclude that $\mathscr{S}$ is an $\mathscr{R}$-block.

**Example 3.8.10.** Let $\Gamma = \mathbb{Z}$ so that $A = R[v, v^{-1}]$ is the ring of Laurent polynomials in one indeterminate $v = \varepsilon$. Let $W$ be of type $B_n$ and consider a weight function $L \colon W \to \mathbb{Z}$ such that we are not in one of the cases in 1.7.6; that is, we have

$B_n$ 

where $r \in \{0, 1, \ldots, n-1\}$.

Assume that (♣), (♠), (♦) hold for $W, L$. This is the case, for example, for $r = 1$ (the equal parameter case) and for $r = 0$ (the case relevant for dealing with type $D_n$). Since $\mathbb{Z}_W = \mathbb{Z}$, we can assume that $R \subseteq \mathbb{K} = \mathbb{Q}$. By Example 1.3.9, we know that

$$f_\lambda \in \{1, 2, 4, 8, 16, \ldots\} \qquad \text{for all } \lambda \in \Lambda.$$

So the subring $R = \mathbb{Z}$ is not $L$-good but it is $L_0$-good; see Table 3.1 (p. 139). Now consider the specialisation $\theta_2 \colon A \to k = \mathbb{F}_2(v)$ given by reducing the coefficients of the polynomials in $A$ modulo 2. We obtain a well-defined decomposition map

$$d_2 \colon R_0(\mathbf{H}_K) \to R_0(\mathbf{H}_k); \qquad \text{see Theorem 3.1.14.}$$

Let $\Theta \subseteq \mathbb{K}$ be the localisation of $A = \mathbb{Z}[v, v^{-1}]$ in the prime ideal generated by 2. Then $\Theta$ satisfies the conditions in Remark 3.8.8 and so we conclude that $E^\lambda \sim_L E^\mu$ if and only if $E_v^\lambda$, $E_v^\mu$ belong to the same $\Theta$-block. Now, since $\Theta$ is a discrete valuation ring, the $\Theta$-blocks coincide with the $\theta_2$-blocks defined in terms of the Brauer graph associated with $d_2$; see Example 3.8.5. Thus, we have shown the following result.

**Theorem 3.8.11 (Gyoja [151]).** *Let $W$ be of type $B_n$ and $L \colon W \to \mathbb{Z}$ be a weight function as in Example 3.8.10, where (♣), (♠) and (♦) are assumed to hold. Let $\lambda, \mu \in \Lambda$. Then we have*

$$E_\varepsilon^\lambda \sim_L E_\varepsilon^\mu \qquad \Leftrightarrow \qquad E_\varepsilon^\lambda, E_\varepsilon^\mu \textit{ belong to the same } \theta_2\textit{-block.}$$

In other words: the relation $\sim_L$ on $\mathrm{Irr}_{\mathbb{K}}(W)$ is determined by the Brauer graph associated with the "bad" specialisation $\theta_2 \colon A \to k = \mathbb{F}_2(v)$, which brings us back to a statement announced in Example 3.1.15(c). One can even go further and also interpret the decomposition matrix of $d_2$ in terms of left cells; see [151], [106, §6].

The special feature of the above example is that all $f_\lambda$ are powers of one fixed prime number. Thus, Rouquier's ring $\mathscr{R}(\Gamma)$ really is the optimal choice for dealing with the general situation where more than one prime number will be involved.

*Remark 3.8.12.* One can associate a "generic" algebra $\mathbf{H}$ to any finite complex reflection group $G$. Although many results from the theory of finite Coxeter groups extend to this more general setting, it is not yet known how to define "cells" à la Kazhdan–Lusztig. But it does make sense to consider blocks of $\mathrm{Irr}(\mathbf{H}_K)$ with respect to a suitable version of Rouquier's ring $\mathscr{R}$. And it turns out that these blocks have many properties that one would expect to hold if there was a suitable theory of Kazhdan–Lusztig cells for complex reflection groups.

We refer the reader to Broué and Kim [33], Broué, Malle and Michel [35], Chlouveraki [48], Malle [238], Malle and Rouquier [241] for further details, background and references on this area.

# Chapter 4
# Hecke Algebras and Finite Groups of Lie Type

In this chapter we show how the theory of Iwahori–Hecke algebras is applied to the representation theory of a finite group of Lie group $G$. Examples of such groups are $\mathrm{GL}_n(\mathbb{F}_q)$ (the finite general linear group) or $\mathrm{Sp}_{2n}(\mathbb{F}_q)$ (the finite symplectic group). Any $G$ as above is naturally defined over a finite field of characteristic $p > 0$. We shall work with representations of $G$ over a sufficiently large field $k$ of characteristic $\ell \geqslant 0$, where either $\ell = 0$ or $\ell > 0$ but $\ell \neq p$. The remaining case where $\ell = p$ is treated by different methods and will not be considered here. (For further details and references on this case, see Curtis and Reiner [53, §72], Cabanes and Enguehard [43, Chap. 6], Donkin [71] and Lusztig [227].) So, from now on, $\ell \neq p$.

If $\ell = 0$, then it is well-known that the irreducible representations of $G$ can be partitioned into *Harish-Chandra series*. A particular case is given by the "unipotent principal series representations" of $G$, which are precisely those irreducible representations of $G$ which admit non-zero vectors fixed by a Borel subgroup of $G$. By classical results due to Bourbaki, Iwahori, Tits, etc., these representations are in natural bijection with the irreducible representations of the Weyl group $W$ of $G$.

It is remarkable that, with suitable and quite natural modifications, there are analogous versions of these statements for "modular" representations; that is, for the case where $\ell > 0$. In Section 4.1 we give a self-contained account of the basic theory of Schur functors and Dipper's Hom functors. This provides the theoretical foundation for the discussion of "modular" Harish-Chandra series of $G$; a survey of this theory (mostly without proofs) will be given in Section 4.2. We then discuss the unipotent principal series representations in more detail. In Section 4.3 we show how generic Iwahori–Hecke algebras enter the picture. Here, we rely to some extent on the theory of algebraic groups (which underlies the definition of $G$). We then come to a highlight of this chapter: the "modular" analogue of the classification of the unipotent principal series representations; see Theorem 4.4.1. This is one of the main applications of our results on canonical basic sets from Chapter 3.

Finally, in Sections 4.5 and 4.6, we discuss a number of examples and open problems. In particular, we explain how Theorem 4.4.1 fits into a (conjectural) classification of all irreducible representations of $G$ in characteristic $\ell > 0$.

M. Geck, N. Jacon, *Representations of Hecke Algebras at Roots of Unity*,
Algebra and Applications 15, DOI 10.1007/978-0-85729-716-7_4,

## 4.1 The Schur Functor and its Variations

This section prepares the ground for the study of Harish-Chandra series. We assume that the reader has a basic familiarity with the representation theory of finite-dimensional associative algebras (Artin–Wedderburn theorem, projective modules and their relation with simple modules, etc.; e.g. see the relevant chapters of Curtis and Reiner [53] or Feit [83].) Throughout, let $\underline{A}$ be a finite-dimensional associative algebra over a field $k$. (In our applications, $\underline{A}$ will be the group algebra of a finite group.) We assume that $\underline{A}$ has an identity element and that all $\underline{A}$-modules are unital; furthermore, all $\underline{A}$-modules will be tacitly assumed to be finite-dimensional.

Let $M$ be a (finite-dimensional) $\underline{A}$-module and $\underline{H} := \mathrm{End}_{\underline{A}}(M)^\circ$ be the opposite algebra of its endomorphism algebra. In this set-up, we have the *Hom functor*

$$\mathfrak{F}_M \colon \underline{A}\text{-mod} \to \underline{H}\text{-mod}, \qquad V \mapsto \mathrm{Hom}_{\underline{A}}(M,V),$$

where $\underline{H}$ acts on $\mathfrak{F}_M(V)$ via $h.f = f \circ h$ (for $h \in \underline{H}$, $f \in \mathfrak{F}_M(V)$) and $\mathfrak{F}_M$ sends a map $\varphi\colon V \to V'$ to the map $\mathfrak{F}_M(\varphi) = \varphi_*\colon \mathfrak{F}_M(V) \to \mathfrak{F}_M(V')$, $f \mapsto \varphi \circ f$. Let

$$\begin{aligned} \mathrm{Irr}(\underline{A}) &= \text{ set of irreducible representations of } \underline{A} \text{ (up to isomorphism)},\\ \mathrm{Irr}(\underline{H}) &= \text{ set of irreducible representations of } \underline{H} \text{ (up to isomorphism)}. \end{aligned}$$

Our aim is to establish a relation, via $\mathfrak{F}_M$, between $\mathrm{Irr}(\underline{H})$ and the set

$$\mathrm{Irr}(\underline{A} \mid M) := \{V \in \mathrm{Irr}(\underline{A}) \mid V \text{ appears in } M/\mathrm{rad}(M)\}.$$

If $M$ is projective, then $\mathfrak{F}_M$ is an example of a *Schur functor*. In this case, $\mathfrak{F}_M$ induces a bijection between the two sets; see Proposition 4.1.3 and Remark 4.1.4. We will then show that the same conclusion holds under a weaker assumption on $M$. The original reference is Dipper [58], [59], but we have incorporated here some variations proposed by Cline, Parshall and Scott [51] and Schubert [279].

Apart from the use of standard properties of projective modules, Hom spaces and tensor products, our exposition will be self-contained.

*Remark 4.1.1.* To obtain a one-sided inverse to $\mathfrak{F}_M$, consider the functor

$$\mathfrak{G}_M \colon \underline{H}\text{-mod} \to \underline{A}\text{-mod}, \qquad E \mapsto M \otimes_{\underline{H}} E,$$

where $M$ is considered as a right $\underline{H}$-module in a natural way. (We have $m.h = h(m)$ for $m \in M$ and $h \in \underline{H}$.) A straightforward application of the "Change of Rings Lemma" (see [53, 2.19]) yields that $E \cong \mathfrak{F}_M\big(\mathfrak{G}_M(E)\big)$ for all $E \in \underline{H}$-mod. Thus, $\mathfrak{G}_M$ is a right inverse to $\mathfrak{F}_M$.

**Lemma 4.1.2.** *Let $M = M_1 \oplus \cdots \oplus M_n$ be a direct sum decomposition of $M$. Then, for each $i$, the following holds:*

$$\dim_k \mathrm{Hom}_{\underline{A}}(M_i, V) = \dim_k \mathrm{Hom}_{\underline{H}}(\mathfrak{F}_M(M_i), \mathfrak{F}_M(V)) \quad \textit{for any } V \in \underline{A}\text{-mod}.$$

*Proof.* We have a natural map

$$\Phi_i\colon \operatorname{Hom}_{\underline{A}}(M_i, V) \to \operatorname{Hom}_{\underline{H}}\big(\mathfrak{F}_M(M_i), \mathfrak{F}_M(V)\big), \quad \alpha \mapsto \alpha_* = (f \mapsto \alpha \circ f).$$

First we note that this map is injective. Indeed, if $\Phi_i(\alpha) = \alpha_* = 0$, then $\alpha \circ f = 0$ for all $f \in \mathfrak{F}_M(M_i)$. Now let $e_i\colon M \to M_i$ be the canonical projection. Then $e_i \in \mathfrak{F}_M(M_i)$ and so $\alpha(e_i(m)) = 0$ for all $m \in M$. Thus, we have $\alpha = 0$, as claimed. Now, since $\operatorname{Hom}_{\underline{A}}$ and $\mathfrak{F}_M$ preserve direct sums, we have

$$\begin{aligned}\dim_k \mathfrak{F}_M(V) &= \dim_k \operatorname{Hom}_{\underline{A}}(M,V) = \textstyle\sum_i \dim_k \operatorname{Hom}_{\underline{A}}(M_i, V),\\ &\leqslant \textstyle\sum_i \dim_k \operatorname{Hom}_{\underline{H}}\big(\mathfrak{F}_M(M_i), \mathfrak{F}_M(V)\big) \quad \text{(by the injectivity of } \Phi_i\text{)}\\ &= \dim_k \operatorname{Hom}_{\underline{H}}\big(\mathfrak{F}_M(M), \mathfrak{F}_M(V)\big).\end{aligned}$$

However, the Hom space in the last line certainly is isomorphic to $\mathfrak{F}_M(V)$. Hence, all maps $\Phi_i$ must be surjective. □

**Proposition 4.1.3.** *Assume that $M$ is projective. Then $\mathfrak{F}_M$ induces a bijection*

$$\mathfrak{F}_M\colon \operatorname{Irr}(\underline{A} \mid M) \xrightarrow{1-1} \operatorname{Irr}(\underline{H}).$$

*Proof.* Let $M = M_1 \oplus \cdots \oplus M_n$ be a decomposition of $M$ into indecomposable direct summands. Being projective, each $M_i$ has a unique simple factor module; we set $Y_i = M_i/\operatorname{rad}(M_i)$ for all $i$. It is well known that $Y_i \cong Y_j$ if and only if $M_i \cong M_j$; we write $i \sim j$ in this case. Then $\operatorname{Irr}(\underline{A} \mid M) = \{Y_i \mid i \in I\}$, where $I \subseteq \{1, \ldots, n\}$ is a set of representatives under the relation $\sim$.

Now let $Y_i \in \operatorname{Irr}(\underline{A} \mid M)$. We show that $\mathfrak{F}_M(Y_i)$ is a simple $\underline{H}$-module. Indeed, let $0 \neq f \in \mathfrak{F}_M(Y_i)$. Since $Y_i$ is simple, the homomorphism $f\colon M \to Y_i$ must be surjective. Hence, since $M$ is projective, the induced map $f_*\colon \mathfrak{F}_M(M) \to \mathfrak{F}_M(Y_i)$ (given by composition with $f$) also is surjective. This means that, as an $\underline{H}$-module, we have $\mathfrak{F}_M(Y_i) = \underline{H}.f$ and so $\mathfrak{F}_M(Y_i)$ is simple, as claimed.

Thus, $\mathfrak{F}_M$ induces a map from $\operatorname{Irr}(\underline{A} \mid M)$ to $\operatorname{Irr}(\underline{H})$.

We show that this map is injective. So assume that $\mathfrak{F}_M(Y_i) \cong \mathfrak{F}_M(Y_j)$. Since there is a non-zero map $M_i \to Y_i$, we also have a non-zero map $\mathfrak{F}_M(M_i) \to \mathfrak{F}_M(Y_i)$ by Lemma 4.1.2. Composing this with the given isomorphism $\mathfrak{F}_M(Y_i) \cong \mathfrak{F}_M(Y_j)$ yields that $\operatorname{Hom}_{\underline{H}}\big(\mathfrak{F}_M(M_i), \mathfrak{F}_M(Y_j)\big) \neq \{0\}$. But then, again by Lemma 4.1.2, we have $\operatorname{Hom}_{\underline{A}}(M_i, Y_j) \neq \{0\}$ and so $Y_i \cong Y_j$, as required.

Finally, we show that each $E \in \operatorname{Irr}(\underline{H})$ is isomorphic to $\mathfrak{F}_M(Y_i)$ for some $i$. Let $V := \mathfrak{G}_M(E) \in \underline{A}$-mod. By Remark 4.1.1, we have $E \cong \mathfrak{F}_M(V)$. So, since $E \neq \{0\}$, there is a non-zero homomorphism $M \to V$. Hence, if $V$ is simple, then $V \in \operatorname{Irr}(\underline{A} \mid M)$ and we are done. Otherwise, let $U \subseteq V$ be a non-trivial submodule. Since $M$ is projective, the functor $\mathfrak{F}_M$ sends an exact sequence of $\underline{A}$-modules to an exact sequence of $\underline{H}$-modules. Consequently, $E$ will be isomorphic to $\mathfrak{F}_M(U)$ or to $\mathfrak{F}_M(V/U)$. By an inductive argument, we will eventually find some simple subquotient $V'$ of $V$ such that $\mathfrak{F}_M(V') \cong E$. As before, this implies that $V' \in \operatorname{Irr}(\underline{A} \mid M)$. □

*Remark 4.1.4.* Assume that $M$ is projective and let $\underline{A}$-mod$_M$ be the full subcategory of $\underline{A}$-mod consisting of all $V \in \underline{A}$-mod such that

- every non-zero submodule of $V$ has a composition factor in $\mathrm{Irr}(\underline{A} \mid M)$ and
- every non-zero quotient module of $V$ has a composition factor in $\mathrm{Irr}(\underline{A} \mid M)$.

Then one can show that $\mathfrak{F}_M$ induces an equivalence of abelian categories between $\underline{A}$-mod$_M$ and $\underline{H}$-mod, with inverse being given by $\mathfrak{G}_M$. For further details and references, see Green [145, §6.2] and Brundan, Dipper and Kleshchev [38, §3.1].

The above results relied on the assumption that $M$ is projective. The following result will allow us to obtain similar conclusions under a weaker assumption on $M$.

**Lemma 4.1.5 (Schubert [279]; see also Cline, Parshall and Scott [51]).** *Assume that there is a surjective homomorphism of $\underline{A}$-modules $\beta\colon P \to M$, where $P$ is projective, such that the following holds:*

(S) $\qquad \ker(\beta) \subseteq \ker(\varphi) \qquad$ *for every* $\varphi \in \mathrm{Hom}_{\underline{A}}(P, M)$.

*Then $M$ is a projective $\underline{A}/I$-module, where $I$ denotes the annihilator of $M$ in $\underline{A}$.*

*Proof.* Denote (as usual) by $IP$ the $\underline{A}$-submodule of $P$ spanned by all terms $a.p$, where $a \in I$ and $p \in P$. Then $IP \subseteq \ker(\beta)$ and so $\beta$ induces a surjective homomorphism of $\underline{A}/I$-modules $\bar{\beta}\colon P/IP \to M$. Also note that $P/IP$ is a projective $\underline{A}/I$-module. Hence, it will be enough to show that $\bar{\beta}$ is injective or, in other words, that $\ker(\beta) \subseteq IP$. To see this, we use the following well-known characterisation of $P$ as a finitely generated projective module: there exist elements $p_1, \ldots, p_n \in P$ and $f_1, \ldots, f_n \in \mathrm{Hom}_{\underline{A}}(P, \underline{A})$ such that $p = \sum_{i=1}^n f_i(p).p_i$ for all $p \in P$. (This readily follows from the fact that $P$ is a direct summand of a direct sum of copies of $\underline{A}$.)

Now we argue as follows. Let $z \in \ker(\beta)$. We want to show that $z \in IP$. For this purpose, let $i \in \{1, \ldots, n\}$ and $m \in M$ be fixed. Then the map $P \to M$ defined by sending $p \in P$ to $f_i(p).m$ is an $\underline{A}$-module homomorphism. By condition (S), we have $f_i(z).m = 0$. Since this holds for all $m \in M$, we conclude that $f_i(z) \in I$. Thus, we see that $z = \sum_{i=1}^n f_i(z).p_i \in IP$, as required. □

*Remark 4.1.6.* Schubert [279] shows that the converse of Lemma 4.1.5 is also true (but we do not need this result here). One easily checks that (S) is equivalent to

(D) $\quad \ker(\beta)$ is invariant under any endomorphism in $\mathrm{End}_{\underline{A}}(P)$.

This condition is the one that appeared in Dipper's original approach [58].

The following observation provides an efficient tool for checking that the condition in Lemma 4.1.5 is satisfied. Together with its application to Harish-Chandra series (see Theorem 4.2.9), the "if" part first appeared in Geck, Hiss and Malle [120, §2]; see also [102, §2]. (The "only if" part is due to Dipper [58, (2.8)].)

**Lemma 4.1.7.** *Condition* (S) *in Lemma 4.1.5 holds if*

$$\dim_k \mathrm{Hom}_{\underline{A}}(P, M) = \dim_k \mathrm{Hom}_{\underline{A}}(M, M).$$

*Conversely, if* (S) *holds, then* $\dim_k \mathrm{Hom}_{\underline{A}}(P,V) = \dim_k \mathrm{Hom}_{\underline{A}}(M,V)$ *for any* $V \in \underline{A}$-mod *which is isomorphic to a factor module of* $M$.

*Proof.* Let $V \in \underline{A}$-mod be isomorphic to a factor module of $M$. Consider the $k$-linear map $\beta^*\colon \mathrm{Hom}_{\underline{A}}(M,V) \to \mathrm{Hom}_{\underline{A}}(P,V)$, $\alpha \mapsto \alpha \circ \beta$. Since $\beta$ is surjective, this map is injective. Now let $V = M$ and assume that the two Hom spaces have the same dimension. Then $\beta^*$ will be surjective. So, if $\varphi \in \mathrm{Hom}_{\underline{A}}(P,M)$, then $\varphi = \alpha \circ \beta$ for some $\alpha \in \mathrm{Hom}_{\underline{A}}(M,M)$ and, hence, $\ker(\beta) \subseteq \ker(\varphi)$; that is, condition (S) holds.

Conversely, assume that (S) holds. We must show that $\beta^*$ is surjective. Let $\varphi \in \mathrm{Hom}_{\underline{A}}(P,V)$. In the proof of Lemma 4.1.5, we have seen that $\ker(\beta) = IP$. Now our assumption on $V$ implies that $I$ annihilates $V$ and so $IP \subseteq \ker(\varphi)$. Hence, $\varphi$ induces a map $\bar{\varphi} \in \mathrm{Hom}_{\underline{A}}(M,V)$ such that $\varphi = \bar{\varphi} \circ \beta$. Thus, $\beta^*$ is surjective, as claimed. □

**Proposition 4.1.8 (Dipper [58, 2.28]).** *Let* $\beta\colon P \to M$ *be a surjective homomorphism of* $\underline{A}$*-modules where* $P$ *is projective. Assume that condition* (S) *in Lemma 4.1.5 is satisfied for* $\beta$. *Then the Hom functor* $\mathfrak{F}_M\colon \underline{A}$-mod $\to \underline{H}$-mod *induces a bijection*

$$\mathfrak{F}_M\colon \mathrm{Irr}(\underline{A} \mid M) \xrightarrow{1-1} \mathrm{Irr}(\underline{H}).$$

*Proof.* Let $\underline{A}' := \underline{A}/I$, where, as before, $I$ denotes the annihilator of $M$ in $A$. To have a separate notation, let us write $M'$ instead of $M$ when we consider $M$ as an $\underline{A}'$-module. By Lemma 4.1.5, $M'$ is a projective $\underline{A}'$-module. So we can apply Proposition 4.1.3 and there is an induced bijection

$$\mathfrak{F}_{M'}\colon \mathrm{Irr}(\underline{A}' \mid M') \xrightarrow{1-1} \mathrm{Irr}(\underline{H}'), \quad \text{where} \quad \underline{H}' := \mathrm{End}_{\underline{A}'}(M')^\circ.$$

Now note that a $k$-linear map from $M$ into itself is an $A$-module homomorphism if and only if it is an $\underline{A}'$-homomorphism. Hence, we have an equality $\underline{H} = \underline{H}'$.

Furthermore, let $V$ be a simple $\underline{A}$-module. If $V$ is a composition factor of $M$, then $I$ is contained in the annihilator of $V$ and so $V$ can also be regarded as an $\underline{A}'$-module; let us denote this $\underline{A}'$-module by $V'$. Then we have an equality

$$\mathfrak{F}_M(V) = \mathrm{Hom}_{\underline{A}}(M,V) = \mathrm{Hom}_{\underline{A}'}(M',V') = \mathfrak{F}_{M'}(V').$$

It also follows that we can naturally identify $\mathrm{Irr}(\underline{A} \mid M)$ and $\mathrm{Irr}(\underline{A}' \mid M')$. Thus, we see that $\mathfrak{F}_M$ induces a bijection $\mathrm{Irr}(\underline{A} \mid M) \to \mathrm{Irr}(\underline{H})$, as required. □

*Remark 4.1.9.* There are alternative ways to make sure that $\mathfrak{F}_M$ induces a bijection as in Proposition 4.1.8. For example, instead of the above condition (S) in Lemma 4.1.5, it would be enough to assume that $\underline{H}$ is a symmetric algebra and that the set of composition factors of $M/\mathrm{rad}(M)$ equals the set of composition factors of $\mathrm{soc}(M)$. For further details see [117, §2] and [102, §2]. Here, we chose to rely on the condition (S) in Lemma 4.1.5 because it is rather straightforward to verify, thanks to Lemma 4.1.7. Note, however, that the alternative conditions discussed in [102] yield further structural properties of $M$.

**Corollary 4.1.10 (Dipper [58, 2.32]).** *Let* $M = M_1 \oplus \cdots \oplus M_n$ *be a decomposition of* $M$ *into indecomposable direct summands. Assume that we are in the setting of*

*Proposition 4.1.8, so that we have a bijection*

$$\mathfrak{F}_M \colon \operatorname{Irr}(\underline{A} \mid M) \xrightarrow{1-1} \operatorname{Irr}(\underline{H}).$$

*Also assume that* $\operatorname{End}_{\underline{A}}(Y) = k\,\mathrm{id}_Y$ *for all* $Y \in \operatorname{Irr}(\underline{A} \mid M)$.

(a) *Each* $M_i$ *has a unique maximal submodule; let* $Y_i := M_i/\operatorname{rad}(M_i)$. *We have* $Y_i \cong Y_j$ *if and only if* $M_i \cong M_j$; *we write* $i \sim j$ *in this case. Then* $\operatorname{Irr}(\underline{A} \mid M) = \{Y_i \mid i \in I\}$, *where* $I \subseteq \{1,\ldots,n\}$ *is a set of representatives under the relation* $\sim$.
(b) *Let* $V \in \underline{A}$-mod *be isomorphic to a factor module of* $M$. *Then the dimension of* $\operatorname{Hom}_{\underline{A}}(M_i, V)$ *equals the multiplicity of* $Y_i$ *as a composition factor of* $V$.
(c) *The* $\underline{H}$*-module* $\mathfrak{F}_M(M_i)$ *is projective indecomposable, and its unique simple factor module is isomorphic to* $\mathfrak{F}_M(Y_i)$.
(d) *The Cartan matrix of* $\underline{H}$ *records the multiplicities of the various simple modules in* $\operatorname{Irr}(\underline{A} \mid M)$ *as composition factors of* $M$.

*Proof.* We use the notation in the proof of Proposition 4.1.8 where $I$ is the annihilator of $M$ and $A' = \underline{A}/I$. In particular, we can identify $\operatorname{Irr}(\underline{A} \mid M) = \operatorname{Irr}(\underline{A}' \mid M')$ and $\underline{H} = \underline{H}'$ where $M'$ is $M$ regarded as an $\underline{A}'$-module.

(a) Since $M_i$ is an indecomposable $\underline{A}$-module, the same is true when we regard $M_i$ as an $\underline{A}'$-module. But, by Lemma 4.1.5, the latter module is projective. Hence, all the statements follow from general properties of projective indecomposable modules.

(b) Since $V$ is isomorphic to a factor module of $M$, the annihilator of $V$ contains $I$. Hence, we can regard $V$ as an $\underline{A}'$-module. It remains to note that $\operatorname{Hom}_{\underline{A}}(M_i, V) = \operatorname{Hom}_{\underline{A}'}(M_i', V)$ and $M_i'$ is a projective indecomposable $\underline{A}'$-module.

(c) The decomposition $M = M_1 \oplus \cdots \oplus M_n$ gives rise to a decomposition $\underline{H} = \mathfrak{F}_M(M) = \mathfrak{F}_M(M_1) \oplus \cdots \oplus \mathfrak{F}_M(M_n)$. Hence, each $\mathfrak{F}_M(M_i)$ is a projective $\underline{H}$-module. Since $Y_i$ is a factor module of $M_i$ and since $\mathfrak{F}_M(Y_i)$ is simple, we also have a surjective homomorphism $\mathfrak{F}_M(M_i) \to \mathfrak{F}_M(Y_i)$ by Lemma 4.1.2. Thus, it remains to show that $\mathfrak{F}_M(M_i)$ is indecomposable. To see this, one could of course refer to the "Fitting correspondence" (e.g. see [53, §6]). Alternatively, one can use Lemma 4.1.2, which provides an isomorphism of vector spaces

$$\operatorname{End}_{\underline{A}}(M_i) = \operatorname{Hom}_{\underline{A}}(M_i, M_i) \cong \operatorname{Hom}_{\underline{H}}(\mathfrak{F}_M(M_i), \mathfrak{F}_M(M_i)) = \operatorname{End}_{\underline{H}}(\mathfrak{F}_M(M_i)).$$

We have seen that this isomorphism is given by $\Phi_i \colon \alpha \mapsto \alpha_* = (f \mapsto \alpha \circ f)$. Hence, in the present context, $\Phi_i$ even is a ring homomorphism and so the above two End-spaces are isomorphic as $k$-algebras. It remains to use the fact that the endomorphism ring of a module is a local ring if and only if the module is indecomposable.

(d) This follows from (b) and (c), where we take $V = Y_j$. □

For our applications to finite groups of Lie type, it will be essential to relate the above results to a situation where $k$ is the residue field of a discrete valuation ring in a field of characteristic 0. This will be done within the following general setting.

**4.1.11.** Let us assume that $k$ is the residue field of a discrete valuation ring $\mathscr{O}$ with field of fractions $\mathbb{K}$; furthermore, assume that $\underline{A}$ can be identified with $\mathscr{A}_k := k \otimes_{\mathscr{O}}$

$\mathscr{A}$, where $\mathscr{A}$ is an $\mathscr{O}$-algebra which is finitely generated and free over $\mathscr{O}$. In order to avoid any technical complications, we assume that

$$\mathscr{A}_k \text{ is split and } \mathscr{A}_{\mathbb{K}} := \mathbb{K} \otimes_{\mathscr{O}} \mathscr{A} \text{ is split semisimple.}^1$$

An $\mathscr{A}$-module which is finitely generated and free over $\mathscr{O}$ will be called an $\mathscr{A}$-*lattice*. For any $\mathscr{A}$-lattice $\tilde{V}$, we shall write $\tilde{V}_{\mathbb{K}} := \mathbb{K} \otimes_{\mathscr{O}} \tilde{V}$ and $\tilde{V}_k := k \otimes_{\mathscr{O}} \tilde{V}$. We are now in the classical setting of "modular representation theory". Thus, we have a decomposition map

$$d_{\mathscr{A}} \colon R_0(\mathscr{A}_{\mathbb{K}}) \to R_0(\mathscr{A}_k),$$

where $R_0(\cdot)$ denotes the appropriate Grothendieck group; see also Example 3.1.3. Recall that $d_{\mathscr{A}}$ is defined as follows. Given $\rho \in \mathrm{Irr}(\mathscr{A}_{\mathbb{K}})$, let $\tilde{V}$ be an $\mathscr{A}$-lattice such that $\rho \cong \tilde{V}_{\mathbb{K}}$. Then $d_{\mathscr{A}}([\rho]) := [\tilde{V}_k] \in R_0(\mathscr{A}_k)$. In order to separate the notation here from that used for Iwahori–Hecke algebras in Chapter 3, we shall write

$$d_{\mathscr{A}}([\rho]) = \sum_{Y \in \mathrm{Irr}(\mathscr{A}_k)} \langle \rho : Y \rangle_{\mathscr{A}} [Y] \qquad \text{for any } \rho \in \mathrm{Irr}(\mathscr{A}_{\mathbb{K}});$$

the numbers $\langle \rho : Y \rangle_{\mathscr{A}}$ are called the *decomposition numbers* of $\mathscr{A}$ and

$$\mathrm{Dec}(\mathscr{A}) = \big(\langle \rho : Y \rangle_{\mathscr{A}}\big)_{\rho \in \mathrm{Irr}(\mathscr{A}_{\mathbb{K}}), Y \in \mathrm{Irr}(\mathscr{A}_k)}$$

is called the *decomposition matrix* of $\mathscr{A}$. This matrix also has an interpretation in terms of projective modules, as follows. Let $\tilde{P}$ be a projective indecomposable $\mathscr{A}$-lattice. Then $\tilde{P}_k$ is a projective indecomposable $\mathscr{A}_k$-module and $Y := \tilde{P}_k/\mathrm{rad}(\tilde{P}_k) \in \mathrm{Irr}(\mathscr{A}_k)$; furthermore, in $R_0(\mathscr{A}_{\mathbb{K}})$, we have

$$[\tilde{P}_{\mathbb{K}}] = \sum_{\rho \in \mathrm{Irr}(\mathscr{A}_{\mathbb{K}})} \langle \rho : Y \rangle_{\mathscr{A}} [\rho] \qquad \text{“Brauer reciprocity”.}$$

(This reciprocity relation already played an essential role in our discussion of "abstract Fock data"; see Theorem 3.4.4.)

**4.1.12.** We shall need some compatibility relations between Hom spaces and scalar extension. Let $\tilde{V}$, $\tilde{V}'$ be $\mathscr{A}$-lattices. Since $\mathbb{K}$ is the field of fractions of $\mathscr{O}$, we have

$$\dim_{\mathbb{K}} \mathrm{Hom}_{\mathscr{A}_{\mathbb{K}}}(\tilde{V}_{\mathbb{K}}, \tilde{V}'_{\mathbb{K}}) = \dim_{\mathbb{K}} \big(\mathbb{K} \otimes_{\mathscr{O}} \mathrm{Hom}_{\mathscr{A}}(\tilde{V}, \tilde{V}')\big).$$

On the other hand, we have a natural injection $k \otimes_{\mathscr{O}} \mathrm{Hom}_{\mathscr{A}}(\tilde{V}, \tilde{V}') \hookrightarrow \mathrm{Hom}_{\mathscr{A}_k}(\tilde{V}_k, \tilde{V}'_k)$, which is an isomorphism if $\tilde{V}$ is projective. Thus, we obtain

$$\dim_{\mathbb{K}} \mathrm{Hom}_{\mathscr{A}_{\mathbb{K}}}(\tilde{V}_{\mathbb{K}}, \tilde{V}'_{\mathbb{K}}) \leqslant \dim_k \mathrm{Hom}_{\mathscr{A}_k}(\tilde{V}_k, \tilde{V}'_k),$$

where equality holds if $\tilde{V}$ is projective. Note also that $\tilde{V}$ is projective if and only if $\tilde{V}_k$ is projective. (For all this, see [53, §30].)

[1] Usually, it is also assumed that $\mathscr{O}$ is complete, but this assumption is unnecessary thanks to our assumption that $\mathscr{A}_{\mathbb{K}}$ is split semisimple. Indeed, idempotents can be lifted from $\underline{A} = \mathscr{A}_k$ to $\mathscr{A}$ by [53, Ex. 6.16] and the Krull–Schmidt–Azuyama theorem holds for $\mathscr{A}$-lattices by [53, (30.18)].

**Lemma 4.1.13 (Dipper [58, 4.7]).** *Let $\tilde{M}$ be an $\mathscr{A}$-lattice and assume that there exists a surjective homomorphism of $\mathscr{A}$-modules $\tilde{\beta}\colon \tilde{P} \to \tilde{M}$, where $\tilde{P}$ is finitely generated and projective, such that*

(D) $\quad \ker(\tilde{\beta})_{\mathbb{K}}$ *and* $\tilde{M}_{\mathbb{K}}$ *have no irreducible constituents in common.*

*Then condition* (S) *in Lemma 4.1.5 holds for the induced map $\tilde{\beta}_k\colon \tilde{P}_k \to \tilde{M}_k$. Furthermore, let $\mathscr{H} := \mathrm{End}_{\mathscr{A}}(\tilde{M})^\circ$; note that $\mathscr{H}$ is an $\mathscr{O}$-algebra which is finitely generated and free over $\mathscr{O}$. The natural injective map $\mathscr{H}_k \hookrightarrow \mathrm{End}_{\mathscr{A}_k}(\tilde{M}_k)^\circ$ is an isomorphism.*

*Proof.* First note that $\tilde{P}_k$ is projective and the induced map $\tilde{\beta}_k$ is surjective. Now consider the map $\tilde{\beta}_k^*\colon \mathrm{Hom}_{\mathscr{A}_k}(\tilde{M}_k,\tilde{M}_k) \to \mathrm{Hom}_{\mathscr{A}_k}(\tilde{P}_k,\tilde{M}_k)$, $\alpha \mapsto \alpha \circ \tilde{\beta}_k$. Since $\tilde{\beta}_k$ is surjective, this map is injective. Hence, we conclude

$$(*) \qquad \dim_k \mathrm{Hom}_{\mathscr{A}_k}(\tilde{M}_k,\tilde{M}_k) \leqslant \dim_k \mathrm{Hom}_{\mathscr{A}_k}(\tilde{P}_k,\tilde{M}_k).$$

By 4.1.12, the right-hand side equals $\dim_{\mathbb{K}} \mathrm{Hom}_{\mathscr{A}_{\mathbb{K}}}(\tilde{P}_{\mathbb{K}},\tilde{M}_{\mathbb{K}})$ and the left-hand side is greater than or equal to $\dim_{\mathbb{K}} \mathrm{Hom}_{\mathscr{A}_{\mathbb{K}}}(\tilde{M}_{\mathbb{K}},\tilde{M}_{\mathbb{K}})$. Thus, we deduce that equality holds in $(*)$ if and only if $\dim_{\mathbb{K}} \mathrm{Hom}_{\mathscr{A}_{\mathbb{K}}}(\tilde{M}_{\mathbb{K}},\tilde{M}_{\mathbb{K}}) = \dim_{\mathbb{K}} \mathrm{Hom}_{\mathscr{A}_{\mathbb{K}}}(\tilde{P}_{\mathbb{K}},\tilde{M}_{\mathbb{K}})$. Since $\tilde{P}_K$ is semisimple, the latter equality holds thanks to condition (D). Hence, we also have equality in $(*)$ and so (S) holds for $\tilde{\beta}_k$ by Lemma 4.1.7.

Finally, consider the natural injective map $\mathscr{H}_k := k \otimes_{\mathscr{O}} \mathscr{H} \hookrightarrow \mathrm{End}_{\mathscr{A}_k}(\tilde{M}_k)^\circ$. Going once more through the above argument, we see that the two spaces have the same dimension and so the map is also surjective. □

Now let us fix an $\mathscr{A}$-lattice $\tilde{M}$ and set $\mathscr{H} := \mathrm{End}_{\mathscr{A}}(\tilde{M})^\circ$. Since $\mathscr{H}$ is finitely generated and free over $\mathscr{O}$, we can apply the discussion in 4.1.11 to $\mathscr{H}$ and so there is a decomposition map

$$d_{\mathscr{H}}\colon R_0(\mathscr{H}_{\mathbb{K}}) \to R_0(\mathscr{H}_k).$$

Note also that, by 4.1.12 and Lemma 4.1.13, we have

$$\mathscr{H}_{\mathbb{K}} \cong \mathrm{End}_{\mathscr{A}_{\mathbb{K}}}(\tilde{M}_{\mathbb{K}})^\circ \qquad \text{and} \qquad \mathscr{H}_k \cong \mathrm{End}_{\mathscr{A}_k}(\tilde{M}_k)^\circ.$$

**Theorem 4.1.14 (Dipper [58, 4.10]).** *In the above setting, assume that there exists a surjective homomorphism of $\mathscr{A}$-modules $\tilde{\beta}\colon \tilde{P} \to \tilde{M}$, where $\tilde{P}$ is finitely generated and projective, such that condition* (D) *in Lemma 4.1.13 holds. Let $\mathscr{H} = \mathrm{End}_{\mathscr{A}}(\tilde{M})^\circ$ as above. Then the functors $\mathfrak{F}_{\mathbb{K}} := \mathfrak{F}_{\tilde{M}_{\mathbb{K}}}$ and $\mathfrak{F}_k := \mathfrak{F}_{\tilde{M}_k}$ induce bijections*

$$\mathfrak{F}_{\mathbb{K}}\colon \mathrm{Irr}(\mathscr{A}_{\mathbb{K}} \mid \tilde{M}_{\mathbb{K}}) \xrightarrow{1-1} \mathrm{Irr}(\mathscr{H}_{\mathbb{K}}), \qquad \mathfrak{F}_k\colon \mathrm{Irr}(\mathscr{A}_k \mid \tilde{M}_k) \xrightarrow{1-1} \mathrm{Irr}(\mathscr{H}_k),$$

*and the decomposition matrix of $\mathscr{H}$ is contained in the decomposition matrix of $\mathscr{A}$. More precisely, for any $\rho \in \mathrm{Irr}(\mathscr{A}_{\mathbb{K}} \mid \tilde{M}_{\mathbb{K}})$, we have*

$$d_{\mathscr{H}}([\mathfrak{F}_{\mathbb{K}}(\rho)]) = \sum_{Y \in \mathrm{Irr}_k(\mathscr{A}_k \mid \tilde{M}_k)} \langle \rho : Y \rangle_{\mathscr{A}} [\mathfrak{F}_k(Y)].$$

*Proof.* By condition (D), we have $\dim_{\mathbb{K}} \mathrm{Hom}_{\mathscr{A}_{\mathbb{K}}}(\tilde{M}_{\mathbb{K}}, \tilde{M}_{\mathbb{K}}) = \dim_{\mathbb{K}} \mathrm{Hom}_{\mathscr{A}_{\mathbb{K}}}(\tilde{P}_{\mathbb{K}}, \tilde{M}_{\mathbb{K}})$ and so condition (S) holds for the induced map $\tilde{\beta}_{\mathbb{K}}$ by Lemma 4.1.7. Hence, $\mathfrak{F}_{\mathbb{K}}$ induces a bijection as above. An analogous statement holds for $\mathfrak{F}_k$ by Lemma 4.1.13. It remains to prove the statement about the decomposition numbers.

Let $\tilde{M}_k = M_1 \oplus \cdots \oplus M_n$ be a decomposition into indecomposable direct summands and write $\mathrm{Irr}(\mathscr{A}_k \mid \tilde{M}_k) = \{Y_i \mid i \in I\}$, where $Y_i := M_i/\mathrm{rad}(M_i)$; see Corollary 4.1.10(a). Let us fix $\rho \in \mathrm{Irr}_{\mathbb{K}}(\mathscr{A}_{\mathbb{K}} \mid \tilde{M}_{\mathbb{K}})$; then $\rho$ is a direct summand of $\tilde{M}_{\mathbb{K}}$. We begin by establishing a reinterpretation of the decomposition numbers $\langle \rho : Y_i \rangle_{\mathscr{A}}$.

By [53, (23.7)], there is a pure $\mathscr{A}$-submodule $U \subseteq \tilde{M}$ such that, if we set $\tilde{V} := \tilde{M}/U$, then $\tilde{V}$ is an $\mathscr{A}$-lattice such that $\tilde{V}_{\mathbb{K}} \cong \rho$; furthermore, $\tilde{V}_k$ will be isomorphic to a factor module of $\tilde{M}_k$. Having chosen $\tilde{V}$ in this particular way, we can now apply Corollary 4.1.10(b) and this yields that $\dim_k \mathrm{Hom}_{\mathscr{A}_k}(M_i, \tilde{V}_k)$ equals the multiplicity of $Y_i$ as a composition factor of $\tilde{V}_k$. By the definition of $d_{\mathscr{A}}$, this multiplicity is $\langle \rho : Y_i \rangle_{\mathscr{A}}$. Using also Lemma 4.1.2, we conclude that

$$\langle \rho : Y_i \rangle_{\mathscr{A}} = \dim_k \mathrm{Hom}_{\mathscr{H}_k}\big(\mathfrak{F}_k(M_i), \mathfrak{F}_k(\tilde{V}_k)\big).$$

By Corollary 4.1.10(c), $\mathfrak{F}_k(M_i)$ is a projective indecomposable $\mathscr{H}_k$-module, and its unique simple factor module is isomorphic to $\mathfrak{F}_k(Y_i)$. Hence, we obtain that

$$\langle \rho : Y_i \rangle_{\mathscr{A}} = \text{ multiplicity of } \mathfrak{F}_k(Y_i) \text{ as a composition factor of } \mathfrak{F}_k(\tilde{V}_k).$$

Thus, by the definition of $d_{\mathscr{H}}$, it remains to show that

$$(*) \qquad d_{\mathscr{H}}\big([\mathfrak{F}_{\mathbb{K}}(\rho)]\big) = [\mathfrak{F}_k(\tilde{V}_k)] \in R_0(\mathscr{H}_k).$$

Now, we have $\mathfrak{F}_{\mathbb{K}}(\rho) \cong \mathrm{Hom}_{\mathscr{A}_{\mathbb{K}}}(\tilde{M}_{\mathbb{K}}, \tilde{V}_{\mathbb{K}}) \cong \tilde{X}_{\mathbb{K}}$, where $\tilde{X} = \mathrm{Hom}_{\mathscr{A}}(\tilde{M}, \tilde{V})$. Note that $\tilde{X}$ is an $\mathscr{H}$-lattice (where, as before, $\varphi \in \mathscr{H}$ acts via $f \mapsto f \circ \varphi$ for $f \in \tilde{X}$). Hence, $(*)$ will follow if we can show that $\mathfrak{F}_k(\tilde{V}_k) \cong \tilde{X}_k$. Since we certainly have an injection $\tilde{X}_k \hookrightarrow \mathrm{Hom}_{\mathscr{A}_k}(\tilde{M}_k, \tilde{V}_k)$, it will be enough to show that

$$(*') \qquad \dim_{\mathbb{K}} \mathrm{Hom}_{\mathscr{A}_{\mathbb{K}}}(\tilde{M}_{\mathbb{K}}, \tilde{V}_{\mathbb{K}}) = \dim_k \mathrm{Hom}_{\mathscr{A}_k}(\tilde{M}_k, \tilde{V}_k).$$

But, since $\tilde{P}$ is projective, we have $\dim_{\mathbb{K}} \mathrm{Hom}_{\mathscr{A}_{\mathbb{K}}}(\tilde{P}_{\mathbb{K}}, \tilde{V}_{\mathbb{K}}) = \dim_k \mathrm{Hom}_{\mathscr{A}_k}(\tilde{P}_k, \tilde{V}_k)$; see 4.1.12. This implies $(*')$ by Lemma 4.1.6, since (S) holds for $\tilde{\beta}_{\mathbb{K}}$ and $\tilde{\beta}_k$. □

The above proof is somewhat different from the original one given by Dipper [58]; yet another proof (due to Plesken) is contained in [153, 5.1.2].

## 4.2 Hom Functors and Harish-Chandra Series

We shall now apply the general theory of the previous section to the representation theory of finite groups of Lie type. This essentially relies on the existence of a suitable *BN*-pair (or Tits system) in any such group. So let $G$ be a finite group containing two subgroups $B,N$ which form a *split BN-pair* of characteristic $p$ (where $p$ is a prime number). Hence, the following properties hold:

- $G$ is generated by $B$ and $N$.
- $H := B \cap N$ is normal in $N$ and the quotient $W := N/H$ is a finite group generated by a set $S$ of elements of order 2. The group $W$ is called the *Weyl group* of $G$.
- $n_s B n_s \neq B$ if $s \in S$ and $n_s$ is a representative of $s$ in $N$.
- $n_s B n \subseteq B n_s n B \cup B n B$ for any $s \in S$ and $n \in N$.
- $\bigcap_{n \in N} nBn^{-1} = H$.
- We have $B = UH$, where $U$ is a normal $p$-subgroup of $B$ and $H$ is abelian of order prime to $p$.
- Chevalley's commutator relations hold; see [53, §69] and [43, Chap. 2].

If $G$ is a finite group of Lie type, then the above conditions hold; see Steinberg's lectures [287], or the remarks in [45, §1.18, §2.6]. Let $k$ be a field which is a splitting field for $G$ and all its subgroups. Let $\mathrm{Irr}_k(G)$ denote the set of simple $kG$-modules (up to isomorphism). All $kG$-modules will be tacitly assumed to be finite-dimensional.

Let $\ell \geqslant 0$ be the characteristic of $k$. Recall that we assume throughout that either $\ell = 0$ or $\ell > 0$ is a prime different from $p$. In this section, we describe (mostly without proofs) the classification of $\mathrm{Irr}_k(G)$ in terms of *Harish-Chandra series*; this provides the background and the motivation for the subsequent study of the *unipotent principal series representations*. For $\ell = 0$, all this is a classical story; see, for example, [45, Chap. 9], [53, §70] and [55, Chap. 6]. Here, we present a streamlined version of the expositions in [61], [120] and [154].

**4.2.1.** We recall some basic facts about Harish-Chandra induction and restriction.[2] Let $W$ be the Weyl group of $G$ and $S$ be the set of simple reflections determined by $B$. For each subset $J \subseteq S$, we have a corresponding (standard) parabolic subgroup $P_J$, with standard Levi decomposition $P_J = U_J \rtimes L_J$. The subgroup $U_J$ can be characterised as the largest normal $p$-subgroup of $P_J$. Now let

$$\mathscr{P}_G := \{{}^nP_J \mid J \subseteq S,\ n \in N\} \qquad \text{and} \qquad \mathscr{L}_G := \{{}^nL_J \mid J \subseteq S,\ n \in N\}.$$

(Here, for any subgroup $A \subseteq G$, we write ${}^nA = nAn^{-1}$.) Now let $L \in \mathscr{L}_G$ and $P \in \mathscr{P}_G$. We say that $L$ is a *Levi complement* in $P$ if $L \subseteq P$ and $P = U_P \rtimes L$, where $U_P$ denotes the largest normal $p$-subgroup of $P$. Note that $L$ itself is a finite group with a split $BN$-pair of characteristic $p$. Now we have functors

$$R^G_{L\subseteq P}\colon kL\text{-mod} \to kG\text{-mod} \qquad \text{and} \qquad {}^*R^G_{L\subseteq P}\colon kG\text{-mod} \to kL\text{-mod},$$

which are called *Harish-Chandra induction* and *Harish-Chandra restriction* respectively. Given a $kL$-module $X$ and a $kG$-module $Y$, these are defined by

$$R^G_{L\subseteq P}(X) := \mathrm{Ind}^G_P(\tilde{X}) \qquad \text{and} \qquad {}^*R^G_{L\subseteq P}(Y) := e_{U_P}.Y,$$

[2] These are special cases of the more general concept of induction and restriction with respect to subquotients in finite groups; see Cabanes and Enguehard [43, Part 1] for an up-to-date exposition.

where $\tilde{X}$ denotes the "inflation" of $X$ from $L$ to $P$ (via the natural map $P \to L$ with kernel $U_P$) and $\mathrm{Ind}_P^G$ denotes the usual induction of representations; furthermore, $e_{U_P} \in kU_P$ is the idempotent given by $e_{U_P} := |U_P|^{-1} \sum_{u \in U_P} u$. (Since $U_P$ is normalised by $L$, we have $le_{U_P} = e_{U_P}l$ for all $l \in L$ and so there is a natural $kL$-module structure on ${}^*R_{L\subseteq P}^G(Y)$.) These two functors are left and right adjoint to each other:

(a)
$$\begin{cases} \mathrm{Hom}_{kG}\big(R_{L\subseteq P}^G(X), Y\big) \cong \mathrm{Hom}_{kL}\big(X, {}^*R_{L\subseteq P}^G(Y)\big), \\ \mathrm{Hom}_{kG}\big(Y, R_{L\subseteq P}^G(X)\big) \cong \mathrm{Hom}_{kL}\big({}^*R_{L\subseteq P}^G(Y), X\big). \end{cases}$$

Now let $(M, Q)$ be another pair where $M \in \mathscr{L}_G$ is a Levi complement in $Q \in \mathscr{P}_G$. Assume that $P \subseteq Q$ and $L \subseteq M$. Then $L \in \mathscr{L}_M$ is a Levi complement in $P \cap M \in \mathscr{P}_M$. In this situation, we have the following transitivity:

(b) $$R_{L\subseteq P}^G(X) \cong R_{M\subseteq Q}^G\big(R_{L\subseteq P\cap M}^M(X)\big) \quad \text{and} \quad {}^*R_{L\subseteq P}^G(Y) \cong {}^*R_{L\subseteq P\cap M}^M\big({}^*R_{M\subseteq Q}^G(Y)\big).$$

The *Mackey formula* gives a direct sum decomposition of ${}^*R_{M\subseteq Q}^G\big(R_{L\subseteq P}^G(X)\big)$, where, again, $M \in \mathscr{L}_G$ is a Levi complement in $Q \in \mathscr{P}_G$. For each $n \in N$, we have ${}^nL \cap M \in \mathscr{L}_M$ and this is a Levi complement in ${}^nP \cap M \in \mathscr{P}_M$; similarly, ${}^nL \cap M \in \mathscr{L}_{{}^nL}$ is a Levi complement in ${}^nL \cap Q \in \mathscr{P}_{{}^nL}$. By "transport of structure", we can define an ${}^nL$-action on $X$ by $nln^{-1}\colon x \mapsto l.x$ ($l \in L, x \in X$). This $k^nL$-module will be denoted by ${}^nX$. Then the Mackey formula says that

(c) $$ {}^*R_{M\subseteq Q}^G\big(R_{L\subseteq P}^G(X)\big) \cong \bigoplus_{n \in D(Q,P)} R_{{}^nL\cap M\subseteq {}^nP\cap M}^M\big({}^*R_{{}^nL\cap M\subseteq {}^nL\cap Q}^{{}^nL}({}^nX)\big),$$

where $D(Q,P) \subseteq N$ is a set of $(Q,P)$-double coset representatives in $G$. The above three properties are rather straightforward generalisations of similar properties of ordinary induction and restriction; this was worked out in detail by Dipper and Fleischmann [61]. Finally, we have the following independence property, which was first established by Howlett and Lehrer [163] and Dipper and Du [60]:

(d) $$R_{L\subseteq P}^G(X) \cong R_{M\subseteq Q}^G(X') \qquad \text{if } M = {}^nL \text{ and } X' \cong {}^nX \text{ for some } n \in N.$$

In particular, if $L \in \mathscr{L}_G$ is a Levi complement in both $P \in \mathscr{P}_G$ and $P' \in \mathscr{P}_G$, then $R_{L\subseteq P}^G(X) \cong R_{L\subseteq P'}^G(X)$. Therefore, we will now omit $P$ from the notation and simply write $R_L^G(X)$ and ${}^*R_L^G(Y)$.

**Definition 4.2.2.** A $kG$-module $Y$ is said to be *cuspidal* if ${}^*R_L^G(Y) = \{0\}$ for all $L \in \mathscr{L}_G$ such that $L \subsetneq G$.

We shall need the following two technical results.

**Lemma 4.2.3.** *Let $Y \in \mathrm{Irr}_k(G)$. Let $L \in \mathscr{L}_G$ be minimal such that ${}^*R_L^G(Y) \neq \{0\}$. Then all composition factors of ${}^*R_L^G(Y)$ are cuspidal.*

*Proof.* Let $X \in \mathrm{Irr}_k(L)$ be any composition factor of ${}^*R_L^G(Y)$. To show that $X$ is cuspidal, let $M \in \mathscr{L}_G$ be such that $M \subseteq L$ and assume that ${}^*R_M^L(X) \neq \{0\}$. Then we also have ${}^*R^L_M(Y) \neq \{0\}$ and so ${}^*R^G_M(Y) \neq \{0\}$, by the transitivity of Harish-Chandra restriction. By the minimality of $L$, this implies $M = L$, as required. □

**Lemma 4.2.4.** *Let $Y \in \mathrm{Irr}_k(G)$. Let $L, M \in \mathscr{L}_G$, $X \in \mathrm{Irr}_k(L)$, $Z \in \mathrm{Irr}_k(M)$ and assume that the following conditions are satisfied:*

- *$L$ is minimal such that ${}^*R^G_L(Y) \neq \{0\}$ and $X$ is a composition factor ${}^*R^G_L(Y)$.*
- *$Z$ is cuspidal and $Y$ is isomorphic to a submodule or to a factor module of $R^G_M(Z)$.*

*Then there is some $n \in N$ such that $L = {}^nM$ and $X \cong {}^nZ$.*

*Proof.* Let $P_X$ be a projective cover of $X$. Since $X$ is a composition factor of ${}^*R^G_L(Y)$, we have $\mathrm{Hom}_{kL}(P_X, {}^*R^G_L(Y)) \neq \{0\}$. So, by the adjointness of Harish-Chandra induction and restriction, $Y$ is isomorphic to a factor module of $R^G_L(P_X)$. Let $P_Y$ be a projective cover of $Y$. Now one easily shows that $R^G_L(P_X)$ is projective. (Indeed, since $P_X$ is a direct summand of $kL$ and since $R^G_L$ respects direct sums, we have that $R^G_L(P_X)$ is a direct summand of $R^G_L(kL)$. Now assume that $L$ is a Levi complement in $P \in \mathscr{P}_G$; then the definition shows that $R^G_L(kL) = \mathrm{Ind}^G_{U_P}(k)$, which is projective since $|U_P|$ is invertible in $k$.) But then we conclude that $P_Y$ is a direct summand of $R^G_L(P_X)$. On the other hand, since $Y$ is a composition factor of $R^G_M(Z)$, there is a non-zero homomorphism $P_Y \to R^G_M(Z)$. Thus, we obtain

$$\mathrm{Hom}_{kG}\big(R^G_L(P_X), R^G_M(Z)\big) \neq \{0\}.$$

Using adjointness and the Mackey formula, we deduce that

$$\mathrm{Hom}_{kL}\big(P_X, R^L_{{}^nM\cap L}\big({}^*R^{{}^nM}_{{}^nM\cap L}({}^nZ)\big)\big) \neq \{0\} \qquad \text{for some } n \in N.$$

Now note that ${}^nZ$ also is a cuspidal $k^nM$-module. Hence, we must have ${}^nM \cap L = {}^nM$; that is, ${}^nM \subseteq L$. But, since $Y$ is a submodule or a factor module of $R^G_M(Z)$, we have $\mathrm{Hom}_{kG}(Y, R^G_M(Z)) \neq \{0\}$ or $\mathrm{Hom}_{kG}(R^G_M(Z), Y) \neq \{0\}$. Using the independence in 4.2.1, the same is true with $(M, Z)$ replaced by $({}^nM, {}^nZ)$. So, by adjointness, we have ${}^*R^G_{{}^nM}(Y) \neq \{0\}$ and then the minimality of $L$ shows that $L = {}^nM$. Returning to the above Hom space, we now also see that $\mathrm{Hom}_{kL}(P_X, {}^nZ) \neq \{0\}$. Since $P_X$ has a unique simple quotient (namely, $X$), we must have $X \cong {}^nZ$. □

**Definition 4.2.5.** Let $\mathscr{L}^\circ_G$ be the set of all pairs $(L, X)$, where $L \in \mathscr{L}_G$ and $X \in \mathrm{Irr}_k(L)$ is cuspidal. For $(L, X) \in \mathscr{L}^\circ_G$, the set

$$\mathrm{Irr}_k(G \mid (L,X)) := \left\{ Y \in \mathrm{Irr}_k(G) \;\middle|\; \begin{array}{l} L \in \mathscr{L}_G \text{ is minimal such that } {}^*R^G_L(Y) \neq \{0\} \\ \text{and } X \text{ is a composition factor of } {}^*R^G_L(Y) \end{array} \right\}$$

is called the *Harish-Chandra series* associated with $(L, X)$. (By 4.2.1(d) and adjointness, this does not depend on the choice of $P \in \mathscr{P}_G$ such that $P = U_P \rtimes L$.) Given $(L, X)$ and $(L, X')$ in $\mathscr{L}^\circ_G$, we write $(L, X) \sim_N (L', X')$ if there exists some $n \in N$ such that $L' = {}^nL$ and $X' \cong {}^nX$. This defines an equivalence relation on $\mathscr{L}^\circ_G$.

**Theorem 4.2.6 (Hiss [154]).** *We have a disjoint union*

$$\mathrm{Irr}_k(G) = \bigsqcup_{(L,X)\in\mathscr{L}_G^\circ/\sim_N} \mathrm{Irr}_k(G \mid (L,X)).$$

*Furthermore, given* $Y \in \mathrm{Irr}_k(G)$ *and a pair* $(L,X) \in \mathscr{L}_G^\circ$*, we have*

$$\begin{aligned} Y \in \mathrm{Irr}_k(G \mid (L,X)) \quad &\Leftrightarrow \quad Y \text{ is isomorphic to a factor module of } R_L^G(X) \\ &\Leftrightarrow \quad Y \text{ is isomorphic to a submodule of } R_L^G(X). \end{aligned}$$

*Proof.* First we show the equivalence of the above three statements. Assume that $Y \in \mathrm{Irr}_k(G \mid (L,X))$. Let $X' \in \mathrm{Irr}_k(L)$ be a submodule of ${}^*R_L^G(Y)$. By Lemma 4.2.3, $X'$ is cuspidal and, by adjointness, $Y$ is isomorphic to a factor module of $R_L^G(X')$. We now apply Lemma 4.2.4 with $(M,Z) = (L,X')$. So there is some $n \in N$ such that $L = {}^nL$ and $X \cong {}^nX'$. Hence, by the independence in 4.2.1, $Y$ is isomorphic to a factor module of $R_L^G(X) \cong R_L^G(X')$. A similar argument shows that, if $X' \in \mathrm{Irr}_k(L)$ is a factor module of ${}^*R_L^G(Y)$, then $Y$ is isomorphic to a submodule of $R_L^G(X) \cong R_L^G(X')$.

Conversely, assume that $Y$ is isomorphic to a submodule or to a factor module of $R_L^G(X)$. By adjointness, $X$ is a submodule or a factor module (in particular, a composition factor) of ${}^*R_L^G(Y)$. It remains to show that $L$ is minimal such that ${}^*R_L^G(Y) \neq \{0\}$. Choose $M \in \mathscr{L}_G$ to be minimal such that $M \subseteq L$ and ${}^*R_M^G(Y) \neq \{0\}$. Let $Z \in \mathrm{Irr}_k(M)$ be a composition factor of ${}^*R_M^G(Y)$; by Lemma 4.2.3, $Z$ is cuspidal. Then Lemma 4.2.4 (with the roles of $L,X$ and $M,Z$ reversed) shows that $M = {}^nL$. Since $M \subseteq L$, we conclude that $M = L$, as required.

By Lemma 4.2.3, it is clear that every $Y \in \mathrm{Irr}_k(G)$ belongs to some Harish-Chandra series. Finally, assume that $(L,X)$ and $(L',X')$ are two pairs in $\mathscr{L}_G^\circ$ such that $Y$ belongs to the corresponding Harish-Chandra series. By the above equivalences, $Y$ is isomorphic to a factor module of $R_{L'}^G(X')$. Applying Lemma 4.2.4 with $(M,Z) = (L',X')$ we obtain $(L,X) \sim_N (L',X')$, as desired. □

**Definition 4.2.7.** Let $(L,X) \in \mathscr{L}_G^\circ$. Let $P \in \mathscr{P}_G$ be such that $P = U_P \rtimes L$ and assume that $R_L^G$ is defined with respect to $P$. Then

$$\mathscr{H}(L,X) := \mathrm{End}_{kG}\big(R_L^G(X)\big)^\circ$$

is called the *Hecke algebra* associated with $(L,X)$. (Here, the superscript $^\circ$ denotes the opposite algebra.) Note that, by the independence in 4.2.1(d), we obtain an isomorphic algebra if we choose another subgroup $P' \in \mathscr{P}_G$ such that $P' = U_{P'} \rtimes L$.

**Example 4.2.8.** Consider a pair $(H,\eta) \in \mathscr{L}_G^\circ$, where $H = N \cap B$ and $\eta$ is a one-dimensional representation of $H$. (Every irreducible representation of $H$ is automatically cuspidal.) By Theorem 4.2.6, $Y \in \mathrm{Irr}_k(G)$ belongs to the Harish-Chandra series associated with $(H,\eta)$ if and only if $Y$ is isomorphic to a factor module of $R_H^G(\eta)$. By the definition of $R_H^G$, we have $R_H^G(\eta) \cong \mathrm{Ind}_B^G(\tilde{\eta})$, where $\tilde{\eta}$ denotes the "inflation" of $\eta$ from $H$ to $B$. Hence, by Frobenius reciprocity, we have

$$Y \in \mathrm{Irr}_k(G \mid (H,\eta)) \qquad \Leftrightarrow \qquad Y_\eta \neq \{0\},$$

where $Y_\eta := \{y \in Y \mid h.y = \eta(h)y \text{ for all } h \in H\}$. The representations of $G$ belonging to $\mathrm{Irr}_k(G \mid (H,\eta))$ for some $\eta \in \mathrm{Irr}_k(H))$ are called the *principal series representations* of $G$. If $\eta = k_H$ (the trivial representation of $H$), then $R_H^G(k_H) \cong k[G/B]$, the permutation representation of $G$ on the cosets of $B$. In this case, we denote $\mathrm{Irr}_k(G \mid (H,k_H))$ simply by $\mathrm{Irr}_k(G \mid B)$ and call it the set of *unipotent principal series representations* of $G$. The corresponding Hecke algebra is given by

$$\mathscr{H}_k := \mathscr{H}(H,k_H) = \mathrm{End}_{kG}(k[G/B])^\circ.$$

Returning to the general case, our next aim is to establish a relation between $\mathrm{Irr}_k(G \mid (L,X))$ and $\mathscr{H}(L,X)$, for a fixed $(L,X) \in \mathscr{L}_G^\circ$. We place ourselves in the setting of Section 4.1, where $\underline{A} = kG$ and $M = R_L^G(X)$. Then we have the *Hom functor*

$$\mathfrak{F}_{R_L^G(X)} \colon kG\text{-mod} \to \mathscr{H}(L,X)\text{-mod}, \qquad Y \mapsto \mathrm{Hom}_{kG}\big(R_L^G(X),Y\big).$$

In the following, we will simply write $\mathfrak{F}_k$ instead of $\mathfrak{F}_{R_L^G(X)}$.

**Theorem 4.2.9 (Geck, Hiss and Malle [120]).** *Let us fix $(L,X) \in \mathscr{L}_G^\circ$ (and a subgroup $P \in \mathscr{P}_G$ such $P = U_P \rtimes L$). Then the Hom functor $\mathfrak{F}_k$ induces a bijection*

$$\mathfrak{F}_k \colon \mathrm{Irr}_k(G \mid (L,X)) \xrightarrow{1-1} \mathrm{Irr}(\mathscr{H}(L,X)).$$

*Furthermore, the statements in Corollary 4.1.10 hold for $M = R_L^G(X)$.*

*Proof.* Let $P_X$ be a projective cover of $X$. Then $R_L^G(P_X)$ is a projective $kG$-module and the natural map $P_X \to X$ induces a surjective homomorphism of $kG$-modules $\beta \colon R_L^G(P_X) \to R_L^G(X)$. By the results in Section 4.1, it will be enough to verify the condition on Hom spaces in Lemma 4.1.7; that is, we have to prove

$$(*) \qquad \dim_k \mathrm{Hom}_{kG}\big(R_L^G(X),R_L^G(X)\big) = \dim_k \mathrm{Hom}_{kG}\big(R_L^G(P_X),R_L^G(X)\big).$$

Now, by adjointness and the Mackey formula, we have

$$\begin{aligned}\mathrm{Hom}_{kG}\big(R_L^G(X),R_L^G(X)\big) &\cong \mathrm{Hom}_{kG}\big(X,{}^*R_L^G\big(R_L^G(X)\big)\big)\\ &\cong \bigoplus_{n \in D(P,P)} \mathrm{Hom}_{kL}\big(X,{}^*R^L_{{}^nL\cap L}\big({}^*R^{{}^nL}_{{}^nL\cap L}({}^nX)\big)\big).\end{aligned}$$

Note again that ${}^nX$ is a cuspidal $k^nL$-module for any $n \in N$. Hence, we only need to consider terms where ${}^nL \cap L = L$; that is, $n \in D(P,P) \cap N_G(L)$. Thus, we obtain

$$\mathrm{Hom}_{kG}\big(R_L^G(X),R_L^G(X)\big) \cong \bigoplus_{n \in D(P,P)\cap N_G(L)} \mathrm{Hom}_{kL}(X,{}^nX).$$

We can repeat this argument to determine $\mathrm{Hom}_{kG}(R_L^G(P_X),R_L^G(X))$; this yields

$$\mathrm{Hom}_{kG}\big(R_L^G(P_X), R_L^G(X)\big) \cong \bigoplus_{n\in D(P,P)\cap N_G(L)} \mathrm{Hom}_{kL}(P_X, {}^nX).$$

But then we note that $P_X$ has a unique simple quotient (namely, $X$) and so we have

$$\mathrm{Hom}_{kL}(P_X, {}^nX) \cong \mathrm{Hom}_{kL}(X, {}^nX) \qquad \text{for any } n \in N.$$

Consequently, the terms corresponding to a given $n$ in the above two direct sum decompositions have the same dimensions. Thus, $(*)$ is proved.

The desired bijection induced by $\mathfrak{F}_k$ now follows from Proposition 4.1.8. □

**Proposition 4.2.10.** *Let $(L,X) \in \mathscr{L}_G^\circ$. Then $R_L^G(X)$ is a semisimple $kG$-module if and only if $\mathscr{H}(L,X)$ is a semisimple algebra. In this case, we have*

$$R_L^G(X) \cong \bigoplus_{E\in\mathrm{Irr}(\mathscr{H}(L,X))} \underbrace{Y_E \oplus \cdots \oplus Y_E}_{\dim_k E \text{ times}},$$

*where $Y_E$ denotes the unique simple module in $\mathrm{Irr}_k(G \mid (L,X))$ such that $\mathfrak{F}_k(Y_E) \cong E$.*

*Proof.* The endomorphism algebra of a semisimple module certainly is a semisimple algebra. The problem is to prove the reverse implication. So assume that $\mathscr{H}(L,X)$ is semisimple and let $R_L^G(X) = M_1 \oplus \cdots \oplus M_n$ be a decomposition into indecomposable direct summands. We must show that each $M_i$ is simple.

By Theorem 4.2.9, the statements in Corollary 4.1.10 hold for $R_L^G(X)$. Hence, setting $Y_i := M_i/\mathrm{rad}(M_i) \in \mathrm{Irr}_k(G \mid (L,X))$ for $1 \leqslant i \leqslant n$, we have $M_i \cong M_j$ if and only if $Y_i \cong Y_j$. Furthermore, the Cartan matrix of $\mathscr{H}(L,X)$ records the multiplicities of the various $Y_j$ as composition factors of the modules $M_i$. Since $\mathscr{H}(L,X)$ is assumed to be semisimple, that Cartan matrix is the identity matrix. Consequently, none of the composition factors of $\mathrm{rad}(M_i)$ belongs to $\mathrm{Irr}_k(G \mid (L,X))$. Now assume, if possible, that there is some $i$ such that $\mathrm{rad}(M_i) \neq \{0\}$. Let $Z \subseteq M_i$ be a simple submodule. As we have just seen, $Z \notin \mathrm{Irr}_k(G \mid (L,X))$. On the other hand, $Z$ is a simple submodule of $R_L^G(X)$ and so, by Theorem 4.2.6, we have $Z \in \mathrm{Irr}_k(G \mid (L,X))$, which is a contradiction.

Finally, the statement about the decomposition of $R_L^G(X)$ is now clear, since $\mathfrak{F}_k(R_L^G(X))$ is isomorphic to the left regular representation of $\mathscr{H}(L,X)$. □

**Proposition 4.2.11.** *Let $(L,X) \in \mathscr{L}_G^\circ$. Then $\dim_k \mathscr{H}(W,L) = |\mathscr{W}(L,X)|$, where*

$$\mathscr{W}(L,X) := \{n \in (N_G(L) \cap N)L \mid {}^nX \cong X\}/L.$$

*The group $\mathscr{W}(L,X)$ is called the inertia group of $X$.*

*Proof.* Arguing as in the proof of Theorem 4.2.9, we obtain

$$\mathscr{H}(L,X) = \mathrm{Hom}_{kG}\big(R_L^G(X), R_L^G(X)\big) \cong \bigoplus_{n\in D(P,P)\cap N_G(L)} \mathrm{Hom}_{kL}(X, {}^nX),$$

and so $\dim_k \mathscr{H}(L,X) = |\{n \in D(P,P) \cap N_G(L) \mid {}^nX \cong X\}|$. It remains to note that, by general results on double coset representatives of parabolic subgroups (see the proof of [45, 9.2.4]), we have $(N_G(L) \cap N)L/L \cong D(P,P) \cap N_G(L)$. □

**Theorem 4.2.12 (Howlett and Lehrer [161] ($\ell = 0$); Geck, Hiss and Malle [120] ($\ell > 0$)).** *Let us fix $(L,X) \in \mathscr{L}_G^\circ$ and let $\mathscr{W}(L,X)$ be defined as in Proposition 4.2.11. Then there is a Coxeter system $(W_1,S_1)$ and a finite group $\Omega$ acting on $(W_1,S_1)$ such that $\mathscr{W}(L,X) = \Omega \ltimes W_1$; furthermore, $\mathscr{H}(W,L)$ has a basis $\{\dot{T}_w \mid w \in \mathscr{W}(L,X)\}$ where the multiplication is given by the following rules:*

*(a) For all $w \in W$ and $w' \in \Omega$, we have $\dot{T}_w\dot{T}_{w'} = \mu(w,w')\dot{T}_{ww'}$ and $\dot{T}_{w'}\dot{T}_w = \mu(w',w)\dot{T}_{w'w}$ where $\mu\colon \mathscr{W}(L,X) \times \mathscr{W}(L,X) \to k^\times$ is a 2-cocycle.*
*(b) There are elements $\{p_{\tilde{s}} \mid \tilde{s} \in S_1\} \subseteq k \setminus \{0,1\}$ such that*

$$\dot{T}_{\tilde{s}}\dot{T}_w = \begin{cases} \dot{T}_{\tilde{s}w} & \text{if } l_1(\tilde{s}w) > l_1(w), \\ p_{\tilde{s}}\dot{T}_{\tilde{s}w} + (p_{\tilde{s}} - 1)\dot{T}_w & \text{if } l_1(\tilde{s}w) < l_1(w), \end{cases}$$

*for all $\tilde{s} \in S_1$ and $w \in W_1$; here, $l_1$ denotes the length function on $W_1$.*

*The above relations form a presentation of $\mathscr{H}(L,X)$ as a k-algebra.*

(The formulation of this result in [120] is slightly different from the original formulation in [161]; the two formulations are reconciled in [1].) We will not attempt to give a proof here, which would be rather long and would require a number of further preparations. For $\ell = 0$, this is discussed in detail by Carter [45, Chap. 10]; for $\ell > 0$, see Cabanes and Enguehard [43, Chap. 3] for an up-to-date exposition. Once the above result is established, one can also show the following proposition, whose proof is similar to the one given in [45, Prop. 10.9.1] (in the case where $\ell = 0$).

**Proposition 4.2.13 (Howlett and Lehrer [162] ($\ell = 0$); Geck and Hiss [117] ($\ell > 0$)).** *Define a linear map $\tau\colon \mathscr{H}(L,X) \to k$ by $\tau(\dot{T}_1) = 1$ and $\tau(\dot{T}_w) = 0$ for $w \neq 1$. Then $\tau$ is a symmetrizing trace on $\mathscr{H}(L,X)$.*

*Remark 4.2.14.* One may conjecture that the cocycle $\mu$ in Theorem 4.2.12(a) is always trivial. (For $\ell = 0$, this conjecture already appears in [161].) By Howlett and Kilmoyer [160], this is the case when $L = H$ (where $H = B \cap N$) and $X = \eta$ is one-dimensional. It is also known to be true when $\ell = 0$ and $G$ arises as the fixed point set of a connected reductive algebraic group under a Frobenius map (see Lusztig [220, Chap. 8] and [95].) The question is open for $\ell > 0$ in general.

*Remark 4.2.15.* Assume that $\ell > 0$ (and $l \neq p$). Then the classification of the cuspidal irreducible representations of $G$ is an open problem. (The analogous problem in characteristic 0 is solved by Lusztig [220].) For $G = \mathrm{GL}_n(\mathbb{F}_q)$, it is true that every cuspidal irreducible representation in characteristic $\ell$ can be lifted to a representation in characteristic 0; see Dipper [59, Cor. 5.23]. Similar results hold for finite classical groups and special characteristics; see Gruber and Hiss [149]. But, otherwise, there may exist cuspidal representations in characteristic $\ell$ which cannot be lifted to characteristic 0. (By Example 4.5.3 below, this already happens for the Suzuki groups.) See [117], [119], [120], [148], [153] for further examples and references.

## 4.3 Unipotent Principal Series Representations, I

We keep the general setting of the previous section where $k$ is a field of characteristic $\ell \geqslant 0$ ($\ell \neq p$), but now exclusively consider the Harish-Chandra series

$$\mathrm{Irr}_k(G \mid B) = \{Y \in \mathrm{Irr}_k(G) \mid Y \text{ admits non-zero vectors fixed by } B\};$$

see Example 4.2.8. Recall that the corresponding Hecke algebra is given by

$$\mathscr{H}_k := \mathscr{H}(H, k_H) = \mathrm{End}_{kG}(k[G/B])^\circ.$$

The aim of this section is to show how generic Iwahori–Hecke algebras enter the picture. This will allow us to apply the machinery developed in Chapter 3.

**4.3.1.** The structure of $\mathscr{H}_k$ was determined by Iwahori [171]. (In this case, the proof of Theorem 4.2.12 simplifies drastically; see [132, §8.4].) We have $\mathscr{W}(H, k_H) = W$, the Weyl group of $G$. Then $\mathscr{H}_k$ has a basis $\{\dot{T}_w \mid w \in W\}$ such that

$$\dot{T}_s \dot{T}_w = \begin{cases} \dot{T}_{sw} & \text{if } l(sw) > l(w), \\ \bar{q}_s \dot{T}_{sw} + (\bar{q}_s - 1)\dot{T}_w & \text{if } l(sw) < l(w), \end{cases}$$

for all $s \in S$ and $w \in W$; here, the parameters $\{\bar{q}_s\}$ are given by $\bar{q}_s = q_s 1_k$, where $q_s = |B\dot{s}B/B|$ for $s \in S$; see [132, 8.4.6]. As an endomorphism of $k[G/B] \cong R_H^G(k_H)$, the map $\dot{T}_w \colon k[G/B] \to k[G/B]$ is explicitly given as follows (see [132, 8.4.1]):

$$\dot{T}_w(xB) \;=\; \text{sum of all cosets } yB \in k[G/B] \text{ such that } x^{-1}y \in B\dot{w}B$$

(where, as usual, $\dot{w}$ denotes a representative of $w \in W$ in $N$). The parameters $\{\bar{q}_s\}$ satisfy the condition that $\bar{q}_s = \bar{q}_t$ whenever $s, t \in S$ are conjugate in $W$ (see [132, Exc. 4.7]). Since $k$ is large enough, we can find square roots $\bar{q}_s^{1/2} \in k$ ($s \in S$) such that $\bar{q}_s^{1/2} = \bar{q}_t^{1/2}$ whenever $s, t \in S$ are conjugate in $W$. Setting $T_s := \bar{q}_s^{-1/2}\dot{T}_s$ ($s \in S$) and $T_w := T_{s_1} \cdots T_{s_n}$ whenever $w = s_1 \cdots s_n$ ($s_i \in S$) is a reduced expression, we obtain

$$T_s T_w = \begin{cases} T_{sw} & \text{if } l(sw) > l(w), \\ T_{sw} + (\bar{q}_s^{1/2} - \bar{q}_s^{-1/2})T_w & \text{if } l(sw) < l(w), \end{cases}$$

for all $s \in S$ and $w \in W$. Thus, we see that $\mathscr{H}_k$ is obtained by specialisation from a suitable generic Iwahori–Hecke algebra associated with $W$ (see Example 1.1.9). This statement will be made more precise in 4.3.9.

Note that $\bar{q}_s \neq 0$ for all $s \in S$, but it may happen that $\bar{q}_s = 1$ for some $s \in S$. But then (in accordance with Theorem 4.2.12) we can further break down the structure of $\mathscr{H}_k$ as follows. Let $\Omega \subseteq W$ be the subgroup generated by all $s \in S$ such that $\bar{q}_s = 1$. Let $S_1 := \{\omega s \omega^{-1} \mid \omega \in \Omega, s \in S \text{ such that } \bar{q}_s \neq 1\}$ and $W_1 \subseteq W$ be the subgroup generated by $S_1$. Then, by 2.4.3, we have $W = \Omega \ltimes W_1$ and $(W_1, S_1)$ is a Coxeter system. Furthermore, the subspace $\mathscr{H}_{k,1} = \langle \dot{T}_w \mid w \in W_1 \rangle_k$ is a subalgebra of $\mathscr{H}_k$ and we have the following multiplication rules (see Proposition 2.4.5):

(a) We have $\dot{T}_w\dot{T}_{w'} = \dot{T}_{ww'}$ and $\dot{T}_{w'}\dot{T}_w = \dot{T}_{w'w}$ for all $w \in W$ and $w' \in \Omega$. (Thus, the cocycle $\mu$ in Theorem 4.2.12(a) is trivial in this case.)
(b) For $\tilde{s} \in S_1$ and $w \in W_1$, we have

$$\dot{T}_{\tilde{s}}T_w = \begin{cases} \dot{T}_{\tilde{s}w} & \text{if } l_1(\tilde{s}w) > l_1(w), \\ p_{\tilde{s}}\dot{T}_{\tilde{s}w} + (p_{\tilde{s}} - 1)\dot{T}_w & \text{if } l_1(\tilde{s}w) < l_1(w); \end{cases}$$

here, $l_1$ is the length function on $W_1$ and, for $\tilde{s} \in S_1$, the parameter $p_{\tilde{s}}$ is given by $p_{\tilde{s}} = \bar{q}_s$ if $\tilde{s} = \omega s \omega^{-1}$, where $\omega \in \Omega$ and $s \in S$ is such that $\bar{q}_s \neq 1$.

The following result provides a useful criterion for checking if $\mathcal{H}_k$ is semisimple. (It is implicitly contained in Hiss [153, Kap. 6] and Puig [267].)

**Lemma 4.3.2.** *We have the following equivalences:*

$$\mathcal{H}_k \text{ is semisimple} \Leftrightarrow k[G/B] \text{ is a semisimple } kG\text{-module} \Leftrightarrow [G:B]1_k \neq 0.$$

*Proof.* The first equivalence is a special case of Proposition 4.2.10. The second equivalence is clear if $\ell = 0$. So now asume that $\ell > 0$. If $\ell$ divides $[G:B]$, then the trivial module $k_G$ is not a direct summand of $k[G/B]$. (This is a well-known fact which holds for the permutation representation of any finite group on the cosets of a subgroup.) Hence, $k[G/B]$ is not semisimple. Conversely, assume that $\ell \nmid [G:B]$. Let $\psi\colon \mathcal{H}_k \to k$ be the linear map obtained by taking the trace of an element of $\mathcal{H}_k$ in its action on $k[G/B]$. By the description of $\dot{T}_w$ in 4.3.1, we see that $\psi(\dot{T}_1) = [G:B]1_k \neq 0$ and $\psi(\dot{T}_w) = 0$ for $w \neq 1$. Hence, $\psi$ is a non-zero scalar multiple of the standard symmetrizing trace on $\mathcal{H}_k$. Consequently, the associated bilinear form $\mathcal{H}_k \times \mathcal{H}_k \to k$, $(\varphi, \varphi') \mapsto \psi(\varphi.\varphi')$, is non-degenerate. Now it is a general fact that if a finite-dimensional associative algebra carries a non-degenerate bilinear form which is associated with the trace function of a representation, then that algebra must be semisimple. Hence, $k[G/B]$ is semisimple by the first equivalence. □

**Example 4.3.3.** In the setting of 4.3.1, we see that there are unique one-dimensional representations $\text{ind}\colon \mathcal{H}_k \to k$ and $\text{sgn}\colon \mathcal{H}_k \to k$ such that

$$\text{ind}(\dot{T}_s) = \bar{q}_s \qquad \text{and} \qquad \text{sgn}(\dot{T}_s) = -1 \qquad \text{for all } s \in S.$$

We have ind = sgn if $\bar{q}_s = -1$ for all $s \in S$. We claim that $\mathfrak{F}_k(k_G) \cong \text{ind}$.

Indeed, identifying $R_H^G(k_H) = k[G/B]$, we have $\mathfrak{F}_k(k_G) = \text{Hom}_{kG}(k[G/B], k_G)$. This space has dimension 1; it is spanned by the function $f_1\colon k[G/B] \to k$ which takes constant value 1 on all cosets. Now let $s \in S$. Then $\dot{T}_s.f_1 = f_1 \circ \dot{T}_s$ and so $(\dot{T}_s.f_1)(xB) = (f_1 \circ \dot{T}_s)(xB) = \sum_{yB} f_1(yB)$, where the sum runs over all cosets $yB \in G/B$ such that $x^{-1}y \in B\dot{s}B$. Thus, $(\dot{T}_s.f_1)(xB) = |B\dot{s}B/B|1_k = \bar{q}_s$ and we have shown that $\dot{T}_s.f_1 = \bar{q}_s f_1 = \text{ind}(\dot{T}_s)f_1$, as required.

It is much more difficult to describe a simple $kG$-module $Y$ such that $\mathfrak{F}_k(Y) \cong \text{sgn}$. Note that if $\mathcal{H}_k$ is semisimple, then $\mathfrak{F}_k(\text{St}_{G,k}) \cong \text{sgn}$, where

$$\text{St}_{G,k} = \{m \in k[G/B] \mid \dot{T}_w(m) = (-1)^{l(w)}m \text{ for all } w \in W\}$$

is the *Steinberg representation* [286].

**Example 4.3.4.** Assume that the Weyl group $W$ of $G$ has order 2. Examples are:

$$\begin{array}{ll} G = \mathrm{GL}_2(\mathbb{F}_q) & \text{(general linear group of type } A_1,\ q \text{ any prime power),} \\ G = \mathrm{GU}_3(\mathbb{F}_q) & \text{(general unitary group of type } {}^2A_2,\ q \text{ any prime power),} \\ G = \mathrm{Suz}(q^2) & \text{(Suzuki group of type } {}^2B_2,\ q = \sqrt{2}^{2f+1} \text{ for some } f \geqslant 0\text{),} \\ G = \mathrm{Ree}(q^2) & \text{(Ree group of type } {}^2G_2,\ q = \sqrt{3}^{2f+1} \text{ for some } f \geqslant 0\text{).} \end{array}$$

(See Carter [44] for the construction of these groups; see also Example 4.3.6 below.) We have $\mathscr{L}_G = \{G, H\}$, where, as above, $H = B \cap N$. Hence, all irreducible representations of $G$ are either cuspidal or lie in the principal series. Let us consider the series $\mathrm{Irr}_k(G \mid B) = \mathrm{Irr}_k(G \mid (H, k_H))$, where $k_H$ is the trivial representation of $H$.

Let $W = \{1, s\}$. By 4.3.1, we have $\mathscr{W}(H, k_H) = W$ and so $\mathscr{H}_k = \mathscr{H}(H, k_H) = \langle \dot{T}_1, \dot{T}_s \rangle$, where $\dot{T}_s^2 = \bar{q}_s \dot{T}_1 + (\bar{q}_s - 1)\dot{T}_s$; note that $[G : B] = 1 + q_s$ in this case. For the groups in the above list, we have $q_s = q, q^3, q^4, q^6$ respectively.

As we have seen in Example 4.3.3, there are two one-dimensional representations ind and sgn of $\mathscr{H}$. If $\bar{q}_s \neq -1$, then ind $\neq$ sgn and $\mathrm{Irr}(\mathscr{H}) = \{\mathrm{ind}, \mathrm{sgn}\}$; if $\bar{q}_s = -1$, then ind $=$ sgn and $\mathrm{Irr}(\mathscr{H}) = \{\mathrm{ind}\}$. Thus, we have two essentially different cases:

- If $\bar{q}_s \neq -1$, then $\mathscr{H}_k$ is semisimple and $R_H^G(k_H) \cong k_G \oplus \mathrm{St}_G$, where $\mathrm{St}_{G,k}$ is the Steinberg module which has dimension $q_s$ in this case; see [286]. Thus, we have $\mathrm{Irr}_k(G \mid B) = \{k_G, \mathrm{St}_{G,k}\}$; furthermore, $\mathfrak{F}_k(k_G) \cong \mathrm{ind}$ and $\mathfrak{F}_k(\mathrm{St}_{G,k}) \cong \mathrm{sgn}$.
- If $\bar{q}_s = -1$, then $\mathscr{H}_k$ is not semisimple and $R_H^G(k_H)$ is indecomposable, with a unique simple submodule and a unique simple factor module, both isomorphic to $k_G$. Thus, we have $\mathrm{Irr}_k(G \mid B) = \{k_G\}$ and $\mathfrak{F}_k(k_G) \cong \mathrm{ind}$.

In particular, we see that the series $\mathrm{Irr}_k(G \mid B)$ only depends on whether $\bar{q}_s \in k^\times$ has order 2 or not. We will establish a general statement of this kind for any $G$ in Theorem 4.4.8 below.

In order to proceed, it will be useful to assume that our group $G$ is realised as the fixed point set of an algebraic group under a Frobenius map.

**Definition 4.3.5.** Let $\overline{\mathbb{F}}_p$ be an algebraic closure of $\mathbb{F}_p = \mathbb{Z}/p\mathbb{Z}$. Let $\mathbf{G}$ be a connected reductive algebraic group over $\overline{\mathbb{F}}_p$ and $F\colon \mathbf{G} \to \mathbf{G}$ be a homomorphism of algebraic groups such that some power of $F$ is the Frobenius map relative to a rational structure on $\mathbf{G}$ over some finite subfield of $\overline{\mathbb{F}}_p$. Then the fixed point set

$$\mathbf{G}^F := \{g \in \mathbf{G} \mid F(g) = g\}$$

is a finite group which is called a *finite group of Lie type*. A split $BN$-pair in $G$ is obtained by taking $B = \mathbf{B}^F$ (where $\mathbf{B} \subseteq \mathbf{G}$ is an $F$-stable Borel subgroup) and $N = N_{\mathbf{G}}(\mathbf{T}_0)^F$ (where $\mathbf{T}_0 \subseteq \mathbf{B}$ is an $F$-stable maximal torus).

It is beyond the scope of this text to discuss the theory of algebraic groups in more detail. (See [104] for an introduction.) But we shall give some typical examples which the reader may keep in mind throughout this chapter.

**Example 4.3.6.** (a) Let $\mathbf{G} = \mathrm{GL}_n(\overline{\mathbb{F}}_p)$ be the general linear group of $n \times n$-matrices. Given a power $q$ of $p$, we have the "standard" Frobenius map $F_q\colon \mathbf{G} \to \mathbf{G}$, $(a_{ij}) \mapsto (a_{ij}^q)$. Let $\mathbb{F}_q \subseteq \overline{\mathbb{F}}_p$ be the unique subfield with $q$ elements. Then $\mathbf{G}^{F_q} = \mathrm{GL}_n(\mathbb{F}_q)$, the general linear group over $\mathbb{F}_q$. The subgroup $\mathbf{B} \subseteq \mathbf{G}$ consisting of all upper triangular matrices is an $F_q$-stable Borel subgroup, where the diagonal matrices form an $F_q$-stable maximal torus $\mathbf{T}_0 \subseteq \mathbf{B}$. We have

$$|\mathrm{GL}_n(\mathbb{F}_q)| = q^{n(n-1)/2}(q-1)(q^2-1)(q^3-1)\cdots(q^n-1).$$

(b) Let again $\mathbf{G} = \mathrm{GL}_n(\overline{\mathbb{F}}_p)$ and consider the $n \times n$ permutation matrix $Q_n$ with entry 1 at the positions $(1,n)$, $(2,n-1)$, $\ldots$, $(n,1)$; note that $Q_n = Q_n^{-1}$. Define an automorphism of algebraic groups $\alpha\colon \mathbf{G} \to \mathbf{G}$ by $\alpha(A) := Q_n \cdot (A^{\mathrm{tr}})^{-1} \cdot Q_n$. Then $\alpha$ commutes with $F_q$ and $\alpha^2$ is the identity. Hence, $F := F_q \circ \alpha$ also is a Frobenius map with respect to some $\mathbb{F}_q$-rational on $\mathbf{G}$ (see [104, Exc. 4.4]). Since $F^2 = F_{q^2}$, we have $\mathbf{G}^F \subseteq \mathrm{GL}_n(\mathbb{F}_{q^2})$. Now the restriction of $F_q$ to $\mathrm{GL}_n(\mathbb{F}_{q^2})$ is an automorphism of order 2, which we denote by $A \mapsto \bar{A}$. Then we obtain

$$\mathbf{G}^F = \mathrm{GU}_n(\mathbb{F}_q) := \{A \in \mathrm{GL}_n(\mathbb{F}_{q^2}) \mid \bar{A}^{\mathrm{tr}} \cdot Q_n \cdot A = Q_n\},$$

the general unitary group with respect to the hermitian form defined by $Q_n$. (The reason for choosing $Q_n$ as above is that then the Borel subgroup $\mathbf{B} \subseteq \mathbf{G}$ and the torus $\mathbf{T}_0$ in (a) are both $F_q$-stable and $F$-stable; see [104, 4.2.6].) We have

$$|\mathrm{GU}_n(\mathbb{F}_q)| = q^{n(n-1)/2}(q+1)(q^2-1)(q^3+1)\cdots(q^n-(-1)^n).$$

(c) Let $\mathbf{G} = \mathrm{Sp}_4(\overline{\mathbb{F}}_2)$ be the four-dimensional symplectic subgroup, defined with respect to the symplectic form given by the matrix $Q_4$ in (b). By [104, 3.3.5], we have $\mathbf{G} = \langle \mathbf{U}, \mathbf{U}' \rangle$, where $\mathbf{U} \subseteq \mathbf{G}$ is the subgroup consisting of all upper unitriangular matrices and $\mathbf{U}' \subseteq \mathbf{G}$ is the subgroup consisting of all lower unitriangular matrices. Explicitly, we have $\mathbf{U} = \{u(t_1,t_2,t_3,t_4) \mid t_i \in \overline{\mathbb{F}}_2\}$, where

$$u(t_1,t_2,t_3,t_4) := \begin{bmatrix} 1 & t_1 & t_3+t_1t_2 & t_4+t_1t_3 \\ 0 & 1 & t_2 & t_3 \\ 0 & 0 & 1 & t_1 \\ 0 & 0 & 0 & 1 \end{bmatrix}.$$

Similarly, $\mathbf{U}'$ consists of the transposes of the above matrices. By [104, 3.3.6], there is a unique homomorphism of algebraic groups $\alpha\colon \mathbf{G} \to \mathbf{G}$ such that

$$u(t_1,t_2,t_3,t_4) \mapsto \begin{bmatrix} 1 & t_2 & t_4 & t_3^2+t_2t_4 \\ 0 & 1 & t_1^2 & t_4+t_1^2t_2 \\ 0 & 0 & 1 & t_2 \\ 0 & 0 & 0 & 1 \end{bmatrix}$$

and $u(t_1,t_2,t_3,t_4)^{\mathrm{tr}}$ is sent to the transpose of the above matrix. Now $\alpha$ is bijective and $\alpha^2$ is the standard Frobenius map $F_2$ on $\mathbf{G}$. Let $r = 2^f$ $(f \geqslant 0)$ and consider the standard Frobenius map $F_r\colon \mathbf{G} \to \mathbf{G}$ (where $F_1 = \mathrm{id}$ by convention). Then $\alpha$ commutes with $F_r$ and, setting $F := \alpha \circ F_r$, we obtain $F^2 = F_2 \circ F_{r^2} = F_{2r^2}$. Let $q > 0$ be such that $q^2 = 2r^2$. The finite group $\mathbf{G}^F$ is called the *Suzuki group* (of type ${}^2B_2$) and denoted by $\mathrm{Suz}(q^2)$; we have

$$|\mathrm{Suz}(q^2)| = q^4(q^2-1)(q^4+1), \qquad \text{where} \qquad q = \sqrt{2}^{2f+1} \ (f \geqslant 0).$$

See [104, 4.6] for further details and references.

The groups $\mathrm{GL}_n(\mathbb{F}_q)$, $\mathrm{GU}_n(\mathbb{F}_q)$ and $\mathrm{Suz}(q^2)$ provide excellent examples for testing the general theory that we are now going to develop. From now on, assume that $G = \mathbf{G}^F$, where $\mathbf{G}$ and $F$ are as in Definition 4.3.5.

**Table 4.1** Weight functions in the quasi-split case

| $\mathbf{W}$ | $\lvert\gamma\rvert$ | $W$ | $L$ |
|---|---|---|---|
| $A_{2m-1}$ | 2 | $B_m$ | $1 \overset{4}{—} 2 — 2 — \cdots — 2$ |
| $A_{2m}$ | 2 | $B_m$ | $3 \overset{4}{—} 2 — 2 — \cdots — 2$ |
| $D_n$ | 2 | $B_{n-1}$ | $2 \overset{4}{—} 1 — 1 — \cdots — 1$ |
| $D_4$ | 3 | $G_2$ | $3 \overset{6}{—} 1$ |
| $E_6$ | 2 | $F_4$ | $1 — 1 \overset{4}{—} 2 — 2$ |

| $\mathbf{W}$ | $\lvert\gamma\rvert$ | $W$ | $L$ |
|---|---|---|---|
| $B_2$ | 2 | $A_1$ | $4$ |
| $G_2$ | 2 | $A_1$ | $6$ |
| $F_4$ | 2 | $I_2(4)$ | $4 \overset{8}{—} 2$ |

$\lvert\gamma\rvert$ denotes the order of $\gamma$; if $\gamma$=id, then $L(s)$=1 ($s\in S$).

**4.3.7.** Let $\mathbf{B} \subseteq \mathbf{G}$ be an $F$-stable Borel subgroup and $\mathbf{T}_0$ be an $F$-stable maximal torus contained in $\mathbf{B}$. Then the groups $\mathbf{B}$ and $\mathbf{N} := N_{\mathbf{G}}(\mathbf{T}_0)$ form a *BN*-pair in $\mathbf{G}$ and the fixed point sets $B := \mathbf{B}^F$ and $N := \mathbf{N}^F$ form a *BN*-pair in $G$ satisfying the properties mentioned at the beginning of Section 4.2. Let $\mathbf{W} = N_{\mathbf{G}}(\mathbf{T}_0)/\mathbf{T}_0$ be the Weyl group of $\mathbf{G}$, with set of simple reflections $\mathbf{S}$. Then $F$ induces an automorphism $\gamma\colon \mathbf{W} \to \mathbf{W}$ such that $\gamma(\mathbf{S}) = \mathbf{S}$. Now the Weyl group $W$ of $G$ can be identified with $\mathbf{W}^\gamma = \{w \in \mathbf{W} \mid \gamma(w) = w\}$. The set $S$ of simple reflections of $W$ can be identified with the set $\{w_I \mid I \in \bar{\mathbf{S}}\}$; here, $\bar{\mathbf{S}}$ denotes the set of orbits of $\mathbf{S}$ under the action of $\gamma$ and $w_I$ denotes the longest element in the parabolic subgroup of $\mathbf{W}$ generated by $I$. (For all this, see [104, 1.7.1].)

Let $\delta \geqslant 1$ be minimal such that $F^\delta$ is the Frobenius map relative to a rational structure on $\mathbf{G}$ over a finite subfield $k_0 \subseteq \overline{\mathbb{F}}_p$. Define $q > 0$ to be the unique real number such that $|k_0| = q^\delta$. (If $\mathbf{G}$ is simple modulo its centre, then $\delta = 1$ and $q$ is a power of $q$, except when $G$ is a Suzuki or Ree group, in which case $\delta = 2$ and $q$ is an odd power of $\sqrt{2}$ or $\sqrt{3}$.) Then we have

(a) $$|B\dot{s}B/B| = q^{c_s}, \qquad \text{where} \qquad c_s := \mathbf{l}(w_I) \text{ for } s = w_I \in S;$$

here, $\mathbf{l}$ is the length function of $\mathbf{W}$ with respect to $\mathbf{S}$; see Steinberg [287, pp. 131, 190]. Thus, we obtain the well-defined weight function

(b) $L\colon W \to \mathbb{Z}$ such that $L(s) = c_s$ for $s \in S$.

We have in fact $[G:B] = \sum_{w\in W} q^{L(w)}$. The weight functions arising in this way are what Lusztig calls the *quasi-split case* in [231, Chapter 16]; see Table 4.1 for a list of the various possibilities when $\mathbf{W}$ is irreducible.

**4.3.8.** We now place ourselves in the setting of 4.1.11, where $\ell > 0$ and $k$ is the residue field of a discrete valuation ring $\mathscr{O}$ with field of fractions $\mathbb{K} \subseteq \mathbb{C}$; let $\mathscr{A} := \mathscr{O}G$. We shall assume that $\mathbb{K}$ is a sufficiently large algebraic extension of $\mathbb{Q}$ of finite degree; in particular, $\mathbb{K}$ will be a splitting field for $G$ and all its subgroups. Finally, we shall assume that $\ell$ is *good* for $\mathbf{G}$ in the sense of the theory of algebraic groups (see, for example, [45, p. 28]). Recall that this means that $\ell$ is good for each simple factor involved in $\mathbf{G}$; the conditions for the various simple types are as follows.

$$\begin{aligned} A_n:&\quad \text{no condition},\\ B_n, C_n, D_n:&\quad \ell \neq 2,\\ G_2, F_4, E_6, E_7:&\quad \ell \neq 2,3,\\ E_8:&\quad \ell \neq 2,3,5. \end{aligned}$$

Let $R \subseteq \mathbb{K}$ be the subring obtained by adjoining the inverses of all bad primes for $\mathbf{G}$ to the ring of algebraic integers in $\mathbb{K}$; note that $R \subseteq \mathscr{O}$. Comparing with the conditions in Table 1.4 (p. 33), we see that $R$ is $L$-good, where $L\colon W \to \mathbb{Z}_{\geqslant 0}$ is the weight function in 4.3.7. Let $\Gamma = \mathbb{Z}$ and $\mathbf{H} = \mathbf{H}_A(W,S,L)$ be the generic Iwahori–Hecke algebra over the ring of Laurent polynomials $A = R[v, v^{-1}]$ in an indeterminate $v$. Thus, the algebra $\mathbf{H}$ has an $A$-basis $\{T_w \mid w \in W\}$ and, for $s \in S$ and $w \in W$, we have

$$T_sT_w = \begin{cases} T_{sw} & \text{if } l(sw) > l(w),\\ T_{sw} + (v^{L(s)} - v^{-L(s)})T_w & \text{if } l(sw) < l(w), \end{cases}$$

where $|B\dot{s}B/B| = q^{L(s)}$ for all $s \in S$.

**4.3.9.** Since $\mathbb{K}$ is assumed to be sufficiently large, we can find a square root $q^{1/2} \in R$. Let $\theta\colon A \to \mathscr{O}$ be the canonical ring homomorphism induced by the inclusion $R \subseteq \mathscr{O}$ and sending $v$ to $q^{1/2}$. Then the corresponding specialised algebra $\mathbf{H}_{\mathscr{O},\theta}$ is naturally isomorphic to

$$\mathscr{H}_{\mathscr{O}} := \operatorname{End}_{\mathscr{O}G}(\mathscr{O}[G/B])^{\circ}.$$

(Note that Iwahori's description in 4.3.1 works over any commutative ground ring instead of $k$; see [132, §8.4].) We obtain induced maps $\theta_{\mathbb{K}}\colon A \to \mathbb{K}$ (by composing $\theta$ with the inclusion $\mathscr{O} \subseteq \mathbb{K}$) and $\theta_k\colon A \to k$ (by composing $\theta$ with the natural map $\mathscr{O} \to k$). Let $\mathbf{H}_{\mathbb{K}}$ and $\mathbf{H}_k$ denote the corresponding specialised algebras. We have

$$\mathbf{H}_{\mathbb{K}} \cong \operatorname{End}_{\mathbb{K}G}(\mathbb{K}[G/B])^{\circ} \qquad \text{and} \qquad \mathbf{H}_k \cong \operatorname{End}_{kG}(k[G/B])^{\circ}.$$

Thus, the endomorphism algebras that we need to consider can all be naturally identified with specialisations of the generic algebra $\mathbf{H}$. Let $K = \mathbb{K}(v)$ be the field of fractions of $A$. Since $R$ is $L$-good, we obtain the decomposition maps (see Section 3.1)

$$d_{\theta_{\mathbb{K}}}\colon R_0(\mathbf{H}_K) \to R_0(\mathbf{H}_{\mathbb{K}}) \qquad \text{and} \qquad d_{\theta_k}\colon R_0(\mathbf{H}_K) \to R_0(\mathbf{H}_k).$$

**4.3.10.** We can now express the classification of $\mathrm{Irr}_{\mathbb{K}}(G \mid B)$ in terms of $\mathbf{H}$. Consider the decomposition map $d_{\theta_{\mathbb{K}}}\colon R_0(\mathbf{H}_K) \to R_0(\mathbf{H}_{\mathbb{K}})$. Being isomorphic to the endomorphism algebra of a semisimple module, $\mathbf{H}_{\mathbb{K}}$ is semisimple. Hence, by 3.1.18, the map $d_{\theta_{\mathbb{K}}}$ induces a bijection between $\mathrm{Irr}(\mathbf{H}_K)$ and $\mathrm{Irr}(\mathbf{H}_{\mathbb{K}})$. Thus, writing $\mathrm{Irr}_{\mathbb{K}}(W) = \{E^\lambda \mid \lambda \in \Lambda\}$ and $\mathrm{Irr}(\mathbf{H}_K) = \{E_v^\lambda \mid \lambda \in \Lambda\}$ as in Chapter 1, we have

$$\mathrm{Irr}(\mathbf{H}_{\mathbb{K}}) = \{E_{\mathbb{K}}^\lambda \mid \lambda \in \Lambda\}, \qquad \text{where} \qquad d_{\theta_{\mathbb{K}}}([E_v^\lambda]) = [E_{\mathbb{K}}^\lambda] \in R_0(\mathbf{H}_{\mathbb{K}}).$$

With this notation, Proposition 4.2.10 shows that

$$\mathrm{Irr}_{\mathbb{K}}(G \mid B) = \{\rho^\lambda \mid \lambda \in \Lambda\} \quad \text{and} \quad \mathbb{K}[G/B] \cong \bigoplus_{\lambda \in \Lambda} \underbrace{\rho^\lambda \oplus \cdots \oplus \rho^\lambda}_{\dim_{\mathbb{K}} E^\lambda \text{ times}},$$

where $\rho^\lambda \in \mathrm{Irr}_{\mathbb{K}}(G \mid B)$ is determined by the condition that $\mathfrak{F}_{\mathbb{K}}(\rho^\lambda) \cong E_{\mathbb{K}}^\lambda$. Furthermore, using the Schur elements $\mathbf{c}_\lambda$ in 1.2.11, we have

$$\dim_{\mathbb{K}} \rho^\lambda = \frac{[G:B]}{\theta_{\mathbb{K}}(\mathbf{c}_\lambda)};$$

see [132, §8.4]. Thus, we have established the classical fact that there is a natural bijection $\mathrm{Irr}_{\mathbb{K}}(G \mid B) \leftrightarrow \mathrm{Irr}_{\mathbb{K}}(W)$. We note that exactly the same arguments apply whenever $\mathbf{H}_k$ is semisimple. In this case, we can write again

$$\mathrm{Irr}(\mathbf{H}_k) = \{E_k^\lambda \mid \lambda \in \Lambda\}, \qquad \text{where} \qquad d_{\theta_k}([E_v^\lambda]) = [E_k^\lambda] \in R_0(\mathbf{H}_k).$$

Furthermore, we have

$$\mathrm{Irr}_k(G \mid B) = \{Y^\lambda \mid \lambda \in \Lambda\} \quad \text{and} \quad k[G/B] \cong \bigoplus_{\lambda \in \Lambda} \underbrace{Y^\lambda \oplus \cdots \oplus Y^\lambda}_{\dim_{\mathbb{K}} E^\lambda \text{ times}},$$

where $Y^\lambda \in \mathrm{Irr}_{\mathbb{K}}(G \mid B)$ is determined by the condition that $\mathfrak{F}_k(Y^\lambda) \cong E_k^\lambda$.

For further reading on the semisimple case, see Exercises 22–27 of Bourbaki [29, Chap. IV, §2], Curtis, Iwahori and Kilmoyer [52], Curtis and Reiner [53, §68], Carter [45, Chap. 10], Geck and Pfeiffer [132, §8.4] and Howlett–Lehrer [162].

*Remark 4.3.11.* Taking $q$ and $\ell$ as fixed, recall from 4.3.8 and 4.3.9 that we *chose* a square root $q^{1/2} \in R$ and a discrete valuation ring $\mathscr{O} \subseteq \mathbb{K}$ (with residue field of characteristic $\ell$). These choices can be made such that if we set

(a) $$e := \min\{i \geqslant 2 \mid 1 + q + q^2 + \cdots + q^{i-1} \in \mathrm{rad}(\mathscr{O})\},$$

then

(b) $$q^{1/2} - \zeta_{2e} \in \mathrm{rad}(\mathscr{O}), \qquad \text{where} \qquad \zeta_{2e} = \exp(\pi\sqrt{-1}/e) \in R \subseteq \mathbb{K}$$

is the "standard" $2e$th root of unity. Indeed, if $\delta = 1$, then $q$ is an integer and so

$$e = \min\{i \geqslant 2 \mid 1+q+q^2+\cdots+q^{i-1} \equiv 0 \bmod \ell\},$$

independently of any choices made. Hence, if $\mathfrak{p}$ is any maximal ideal of $R$ containing $\ell$, then $\Phi_e(q) \in \mathfrak{p}$. It follows that $q-\zeta \in \mathfrak{p}$ for some primitive $e$th root of unity $\zeta$. Since all these roots of unity are algebraically conjugate, we can choose $\mathfrak{p}$ such that $q-\zeta_{2e}^2 \in \mathfrak{p}$. Then we can take as $\mathscr{O}$ the localisation of $R$ in $\mathfrak{p}$. Replacing $q^{1/2}$ by $-q^{1/2}$ if necessary, we can make sure that (b) holds.

If $\delta = 2$, then we first define $e' := \{i \geqslant 2 \mid 1+q^2+q^4+\cdots+q^{2(i-1)} \equiv 0 \bmod \ell\}$. Now note that $\Phi_{e'}(q^2) = \Phi_{2e'}(q)$ (if $e'$ is even) or $\Phi_{e'}(q^2) = \Phi_{2e'}(q)\Phi_{e'}(q)$ (if $e'$ if odd); see Example 3.3.2. Hence, as above, we see that there is a maximal ideal $\mathfrak{p}$ of $R$ such that $\ell \in \mathfrak{p}$ and $q-\zeta_{2e}^2 \in \mathfrak{p}$, where $e = 2e'$ (if $e'$ is even) or $e \in \{e', 2e'\}$ (if $e'$ is odd). Again, let $\mathscr{O}$ be the localisation of $R$ in $\mathfrak{p}$. Replacing $q^{1/2}$ by $-q^{1/2}$ if necessary, we see that (b) holds. Then one also checks that $e$ satisfies the condition in (a); that is, $e \geqslant 2$ is minimal such that $1+q+q^2+\cdots+q^{e-1} \in \operatorname{rad}(\mathscr{O})$, as required.

Having fixed the above choices, let us now consider the $\Phi_e$-modular specialisation $\theta_e \colon A \to \mathbb{K}$ such that $\theta_e(v) = \zeta_{2e}$ (and $\theta_e(r) = r$ for all $r \in R$); denote by $\mathbf{H}_{(e)}$ the corresponding specialised algebra. By (b), we have $\ker(\theta_e) = (v-\zeta_{2e}) \subseteq \ker(\theta_k)$. Further note that $\theta_e(A) = R$ is integrally closed in $\mathbb{K}$. Hence, $\theta_k$ *lies above* $\theta_e$ in the sense of 3.6.2. So Theorem 3.6.3 yields that

(c) $d_{\theta_k}$ factors through the decomposition map $d_{\theta_e} \colon R_0(\mathbf{H}_K) \to R_0(\mathbf{H}_{(e)})$.

This remark will play an essential role in the subsequent discussion; see, for example, the proof of Theorem 4.4.8 and Example 4.5.15 below.

In the following section, we shall need the following ingredients from Lusztig's theory of representations of finite groups of Lie type.

**4.3.12.** Let $\rho \in \operatorname{Irr}_{\mathbb{K}}(G \mid B)$. Via the bijection $\operatorname{Irr}_{\mathbb{K}}(G \mid B) \leftrightarrow \operatorname{Irr}_{\mathbb{K}}(W)$ in 4.3.10, $\rho$ corresponds to a unique $E^\lambda \in \operatorname{Irr}_{\mathbb{K}}(W)$. In Chapter 1, using the generic algebra $\mathbf{H}$, we have attached a numerical invariant $\mathbf{a}_\lambda \in \mathbb{Z}_{\geqslant 0}$ to $E_\lambda$; thus, we can define $\mathbf{a}_\rho := \mathbf{a}_\lambda$. In fact, one can attach a numerical invariant (generalising $\mathbf{a}_\lambda$) to *any* $\rho \in \operatorname{Irr}_{\mathbb{K}}(G)$.

This requires the following definitions. Let $\mathbf{O}$ be an $F$-stable conjugacy class of $\mathbf{G}$ consisting of unipotent elements. Then the fixed point set $\mathbf{O}^F$ splits into conjugacy classes under the action of the finite group $G$. Let $u_1 \ldots, u_r \in \mathbf{O}^F$ be representatives of the $G$-conjugacy classes contained in $\mathbf{O}^F$. For each $j$, let $A(u_j)$ be the group of connected components of the centraliser of $u_j$ in $\mathbf{G}$. Since $F(u_j) = u_j$, there is an induced action of $F$ on $A(u_j)$ which we denote by the same symbol. We define the *average value* of $\rho \in \operatorname{Irr}_{\mathbb{K}}(G)$ on $\mathbf{O}^F$ by

(a) $$\operatorname{AV}(\mathbf{O}, \rho) := \sum_{1 \leqslant j \leqslant r} [A(u_j) : A(u_j)^F] \operatorname{trace}(u_j, \rho).$$

(Note that $\operatorname{AV}(\mathbf{O}, \rho)$ does not depend on the choice of the representatives $u_j$.) Assuming that $p, q$ are large enough, Lusztig [229] has shown that, given $\rho \in \operatorname{Irr}_{\mathbb{K}}(G)$, there exists a *unique* $F$-stable unipotent class $\mathbf{O}_\rho$ such that

- $\operatorname{AV}(\mathbf{O}_\rho, \rho) \neq 0$ and

- given any $F$-stable unipotent class $\mathbf{O}$ such that $\mathrm{AV}(\mathbf{O},\rho) \neq 0$, then $\mathbf{O} = \mathbf{O}_\rho$ or $\dim \mathbf{O} < \dim \mathbf{O}_\rho$.

The class $\mathbf{O}_\rho$ is called the *unipotent support* of $\rho$. The assumptions on $p,q$ have subsequently been removed by Geck and Malle [128]. Thus, every $\rho \in \mathrm{Irr}_{\mathbb{K}}(G)$ has a well-defined *unipotent support* $\mathbf{O}_\rho$. Using this concept, we associate with every $\rho \in \mathrm{Irr}_{\mathbb{K}}(G)$ a numerical invariant $\mathbf{a}_\rho$ by setting $\mathbf{a}_\rho := \dim \mathscr{B}_u$, where $\mathscr{B}_u$ is the variety of Borel subgroups of $\mathbf{G}$ containing an element $u \in \mathbf{O}_\rho$. (This variety already appeared in 2.2.12; this is not a coincidence, but we refer to [220] and [229] for further details.)

Now let $E^\lambda \in \mathrm{Irr}_{\mathbb{K}}(W)$ and consider the corresponding unipotent principal series representation $\rho = \rho^\lambda \in \mathrm{Irr}_{\mathbb{K}}(G \mid B)$; see 4.3.10. Then it is known that

(b) $\mathbf{a}_\rho$ (defined as above using the unipotent support of $\rho$) equals $\mathbf{a}_\lambda$ (defined using the Schur element $\mathbf{c}_\lambda$ as in Corollary 1.3.1).

This statement is contained in [229, §10] (see also [128, Theorem 3.7]). It relies on the fact that the map $\rho \mapsto \mathbf{O}_\rho$ has an alternative description in terms of the Springer correspondence and induction of representations of Weyl groups.

**4.3.13.** By the fundamental work of Deligne and Lusztig [54], we have a virtual representation $R_{\mathbf{T},\theta}$ of $G$ for any pair $(\mathbf{T},\theta)$ where $\mathbf{T}$ is an $F$-stable maximal torus $\mathbf{T}$ of $\mathbf{G}$ and $\theta \in \mathrm{Irr}_{\mathbb{K}}(\mathbf{T}^F)$; thus, $[R_{\mathbf{T},\theta}] \in R_0(\mathbb{K}G)$. The construction involves the theory of $\ell$-adic cohomology of certain varieties on which the finite group $G$ acts; see also the exposition by Carter [45, Chap. 7]. (Here, one has to identify $\mathbb{K}$ with a subfield of $\overline{\mathbb{Q}}_\ell$.) It is known that any $\rho \in \mathrm{Irr}_{\mathbb{K}}(G)$ appears in $R_{\mathbf{T},\theta}$ for at least some pair $(\mathbf{T},\theta)$; see [45, 7.5.8]. The irreducible representations of $G$ in the set

(a) $$\mathrm{Unip}_{\mathbb{K}}(G) := \{\rho \in \mathrm{Irr}_{\mathbb{K}}(G) \mid \rho \text{ appears in } R_{\mathbf{T},1} \text{ for some } \mathbf{T}\}$$

are called *unipotent representations*; here, 1 stands for the trivial representation of $\mathbf{T}^F$. Since $R_{\mathbf{T}_0,1} \cong \mathbb{K}[G/B]$ (see [45, 7.2.4]), we have

(b) $$\mathrm{Irr}_{\mathbb{K}}(G \mid B) \subseteq \mathrm{Unip}_{\mathbb{K}}(G).$$

If $G = \mathrm{GL}_n(\mathbb{F}_q)$, then equality holds (see [55, 15.8]); but in general, the inclusion is strict. The importance of $\mathrm{Unip}_{\mathbb{K}}(G)$ lies in the fact that, in a well-defined sense, the classification of $\mathrm{Irr}_{\mathbb{K}}(G)$ can be reduced to the classification of unipotent representations for $G$ itself and smaller groups (see Lusztig [220, 4.23] and [226] for details). By [220, Chap. 13] and [128, §4], the set of unipotent classes

(c) $$\{\mathbf{O}_\rho \mid \rho \in \mathrm{Unip}_{\mathbb{K}}(G)\}$$

(where $\mathbf{O}_\rho$ denotes the unipotent support) turns out to be the set of *special unipotent classes* defined by Lusztig [215, §9]; see also 2.2.12. These special unipotent classes are in bijection with the two-sided Kazhdan–Lusztig cells (see Section 2.1) of the Weyl group $\mathbf{W}$ of $\mathbf{G}$ with respect to the equal-parameter weight function.

## 4.4 Unipotent Principal Series Representations, II

In the previous section we saw that $\mathrm{Irr}_k(G \mid B)$ is naturally in bijection with $\mathrm{Irr}_k(W)$ whenever the algebra $\mathbf{H}_k$ is semisimple. The aim of this section is to formulate and prove the "modular" analogue of this statement. For $G = \mathrm{GL}_n(\mathbb{F}_q)$, this is originally due to Dipper and James; see [58] and [62]. Here, we follow the general approach in [109] based on the "canonical basic sets" discussed in Chapter 3.

Recall our basic set-up: $G = \mathbf{G}^F$ is a finite group of Lie type; see Definition 4.3.5. Furthermore, as in 4.3.8, we have $p \neq \ell > 0$ and $k$ is the residue field of a discrete valuation ring $\mathcal{O}$ in $\mathbb{K} \subseteq \mathbb{C}$. Here, $\mathbb{K}$ is a sufficiently large algebraic extension of $\mathbb{Q}$; in particular, $\mathbb{K}$ and $k$ will be splitting fields for $G$ and all its subgroups.

We are now ready to state the main results about the series $\mathrm{Irr}_k(G \mid B)$ in the case where $k[G/B]$ is not semisimple. Note that the statement does not refer to any properties of the Hecke algebra $\mathcal{H}_\mathcal{O}$; these will only be used in the proof.

**Theorem 4.4.1 (Cf. [109]).** *Recall that $\ell$ is assumed to be good for* $\mathbf{G}$. *As in 4.1.11, denote by $\langle \rho : Y \rangle_{\mathcal{O}G}$ (where $\rho \in \mathrm{Irr}_\mathbb{K}(G)$ and $Y \in \mathrm{Irr}_k(G)$) the entries of the decomposition matrix* $\mathrm{Dec}(\mathcal{O}G)$.

- *Given $Y \in \mathrm{Irr}_k(G \mid B)$, let $\hat{\mathcal{S}}_k(Y) := \{\rho \in \mathrm{Unip}_\mathbb{K}(G) \mid \langle \rho : Y \rangle_{\mathcal{O}G} \neq 0\}$. Then the function $\hat{\mathcal{S}}_k(Y) \to \mathbb{Z}_{\geqslant 0}$, $\rho \mapsto \mathbf{a}_\rho$ (defined using the unipotent support), reaches its minimum at exactly one element of $\hat{\mathcal{S}}_k(Y)$, which we denote by $\rho_Y$.*
- *The map $\mathrm{Irr}_k(G \mid B) \to \mathrm{Unip}_\mathbb{K}(G)$, $Y \mapsto \rho_Y$, is injective and its image is contained in $\mathrm{Irr}_\mathbb{K}(G \mid B)$.*
- *We have $\langle \rho_Y : Y \rangle_{\mathcal{O}G} = 1$ for all $Y \in \mathrm{Irr}_k(G \mid B)$.*

*Thus, setting $\mathcal{B}_k(G) := \{\rho_Y \mid Y \in \mathrm{Irr}_k(G \mid B)\}$, we obtain a natural bijection*

$$\mathrm{Irr}_k(G \mid B) \stackrel{1-1}{\longleftrightarrow} \mathcal{B}_k(G) \subseteq \mathrm{Irr}_\mathbb{K}(G \mid B) \subseteq \mathrm{Unip}_\mathbb{K}(G).$$

*In analogy to Definition 3.2.1, $\mathcal{B}_k(G)$ is called a canonical basic set for* $\mathrm{Irr}_k(G \mid B)$.

*Proof.* The natural map $\mathcal{O} \to k$ does not only induce the decomposition map $d_{\mathcal{O}G}\colon R_0(\mathbb{K}G) \to R_0(kG)$, but also a decomposition map $d_{\mathbf{H}_\mathcal{O}}\colon R_0(\mathbf{H}_\mathbb{K}) \to R_0(\mathbf{H}_k)$. Now the idea is to translate the above statements from $G$ to $\mathbf{H}_\mathcal{O}$ and then to use the known results on *canonical basic sets* from Section 3.2.

We begin by showing that $d_{\mathbf{H}_\mathcal{O}}$ can be "lifted" to the generic algebra $\mathbf{H}$. This is done by an argument entirely analogous to that in Example 3.6.5. Indeed, since the specialisation $\theta_k\colon A \to k$ is the composition of $\theta$ with the natural map $\mathcal{O} \to k$, it is clear that $\theta_k$ *lies above* $\theta_\mathbb{K}$ in the sense of 3.6.2. Thus, by Theorem 3.6.3, there is a factorisation $d_{\theta_k} = d_{\mathbf{H}_\mathcal{O}} \circ d_{\theta_\mathbb{K}}$. We have already seen in 4.3.10 that $d_{\theta_\mathbb{K}}$ induces a bijection between $\mathrm{Irr}(\mathbf{H}_K)$ and $\mathrm{Irr}(\mathbf{H}_\mathbb{K})$. Identifying $R_0(\mathbf{H}_K)$ and $R_0(\mathbf{H}_\mathbb{K})$ via this bijection, we see that $d_{\theta_k}$ is identified with the decomposition map $d_{\mathbf{H}_\mathcal{O}}$. Thus, using the notation in 4.3.10, we have

(1) $\quad d_{\Theta_k}([E_v^\lambda]) = d_{\mathbf{H}_\mathcal{O}}([E_\mathbb{K}^\lambda]) = d_{\mathbf{H}_\mathcal{O}}\big([\mathfrak{F}_\mathbb{K}(\rho^\lambda)]\big) \in R_0(\mathbf{H}_k) \qquad$ for all $\lambda \in \Lambda$.

Next we show that we are in a situation where Dipper's Theorem 4.1.14 applies. Write $B = U \rtimes H$, where $U$ is the largest normal $p$-subgroup of $B$ and $H = \mathbf{T}_0^F$. Since the order of $U$ is prime to $\ell$, the permutation module $\mathcal{O}[G/U]$ is projective and, since $U \subseteq B$, we have a natural surjective map

$$\tilde{\beta} : \tilde{P} \to \tilde{M} \qquad \text{where} \qquad \tilde{P} := \mathcal{O}[G/U] \quad \text{and} \quad \tilde{M} := \mathcal{O}[G/B].$$

Extending scalars from $\mathcal{O}$ to $\mathbb{K}$, we find that

$$\mathbb{K}[G/U] \cong \mathbb{K}[G/B] \oplus \ker(\tilde{\beta})_\mathbb{K}.$$

Furthermore, $\mathbb{K}[G/B] \cong R_H^G(\mathbb{K}_H)$ and $\mathbb{K}[G/U] \cong R_H^G(\mathbb{K}H)$, where $\mathbb{K}H$ denotes the (left) regular representation of $H$. Since $\mathbb{K}H \cong \bigoplus_{\eta \in \mathrm{Irr}_\mathbb{K}(H)} \eta$, we deduce that

$$\ker(\tilde{\beta})_\mathbb{K} \cong \bigoplus_{\mathbb{K}_H \neq \eta \in \mathrm{Irr}_\mathbb{K}(H)} R_H^G(\eta),$$

and so an easy application of the Mackey formula yields

$$\mathrm{Hom}_{\mathbb{K}G}(\mathbb{K}[G/B], \ker(\tilde{\beta})_\mathbb{K}) \cong \bigoplus_{\mathbb{K}_H \neq \eta \in \mathrm{Irr}_\mathbb{K}(H)} \mathrm{Hom}_{\mathbb{K}G}\big(R_H^G(\mathbb{K}_H), R_H^G(\eta)\big) = \{0\}.$$

Thus, condition (D) in Lemma 4.1.13 is satisfied for the above map $\tilde{\beta}$. Hence, Dipper's Theorem 4.1.14 applies and so the functor $\mathfrak{F}_k := \mathfrak{F}_{k[G/B]}$ induces a bijection

$$\mathfrak{F}_k : \mathrm{Irr}_k(G \mid B) \xrightarrow{1-1} \mathrm{Irr}(\mathbf{H}_k);$$

furthermore, the matrix $\mathrm{Dec}(\mathbf{H}_\mathcal{O})$ (associated with $d_{\mathbf{H}_\mathcal{O}} : R_0(\mathbf{H}_\mathbb{K}) \to R_0(\mathbf{H}_k)$) is contained in $\mathrm{Dec}(\mathcal{O}G)$. Combining this with (1), we obtain

(2) $$d_{\Theta_k}([E_v^\lambda]) = \sum_{Y \in \mathrm{Irr}_k(G|B)} \langle \rho^\lambda : Y \rangle_{\mathcal{O}G} [\mathfrak{F}_k(Y)] \qquad \text{for all } \lambda \in \Lambda.$$

Next we show that, for any $Y \in \mathrm{Irr}_k(G \mid B)$, we have

(3) $$\hat{\mathscr{S}}_k(Y) \subseteq \mathrm{Irr}_\mathbb{K}(G \mid B).$$

This is seen as follows. Let $\rho \in \hat{\mathscr{S}}_k(Y)$ and $\tilde{V}$ be an $\mathcal{O}G$-lattice such that $\tilde{V}_k \cong \rho$. Let $P_Y$ be a projective cover of $Y$. Thus, since $\langle \rho : Y \rangle_{\mathcal{O}G} \neq 0$, we have $\mathrm{Hom}_{kG}(P_Y, \tilde{V}_k) \neq \{0\}$. Since $Y$ is isomorphic to a factor module of $\tilde{M}_k$ and, hence, of $\tilde{P}_k$, we conclude that $P_Y$ is a direct summand of $\tilde{P}_k$. So we also have $\mathrm{Hom}_{kG}(\tilde{P}_k, \tilde{V}_k) \neq \{0\}$. Using 4.1.12 and the fact that $\tilde{P}$ is projective, we deduce that $\mathrm{Hom}_{\mathbb{K}G}(\tilde{P}_\mathbb{K}, \tilde{V}_\mathbb{K}) \neq \{0\}$. Now recall from the above computation that $\ker(\tilde{\beta})_K$ is a direct sum of terms $R_H^G(\eta)$, where $\mathbb{K}_H \neq \eta \in \mathrm{Irr}_\mathbb{K}(H)$. By [45, 7.2.4], we have $R_H^G(\eta) \cong R_{\mathbf{T}_0,\eta}$. Consequently, all irreducible constituents of $\ker(\tilde{\beta})_{\mathbb{K}}$ are non-unipotent (see [45, 7.3.8]). It follows that $\rho$ must be a constituent of $\mathbb{K}[G/B]$, as required.

Now we can complete the proof as follows. Given $Y \in \mathrm{Irr}_k(G \mid B)$, let $E := \mathfrak{F}_k(Y) \in \mathrm{Irr}(\mathbf{H}_k)$. Then (2) and (3) mean that

$$\hat{\mathscr{S}}_k(Y) = \{\rho^\lambda \mid \lambda \in \mathscr{S}_{\theta_k}(E)\},$$

with $\mathscr{S}_{\theta_k}(E)$ as in Definition 3.2.1. Hence, taking into account 4.3.12(b), we see that all the required statements concerning $\mathscr{B}_k(G)$ will hold if and only if $\mathbf{H}_k$ admits a canonical basic set $\mathscr{B}_{\theta_k}$ as in Definition 3.2.1. We have seen in Proposition 3.2.7 that such a canonical basic set for $\mathbf{H}_k$ exists if the properties (♠) and (♣) in 2.5.3 are satisfied for $\mathbf{H} = \mathbf{H}_A(W,S,L)$. Also recall from 2.5.3 that (♠) and (♣) can be deduced from **P1, P4, P15** in Conjecture 2.3.2. Now, by 2.4.1, these properties do hold in the "equal-parameter case", where $\gamma = \mathrm{id}$ and $L$ is the length function on $\mathbf{W} = W$. In [231, Chapter 16], Lusztig also sketches a proof of **P1, P4, P15** for the general quasi-split case; hence, (♠) and (♣) will hold in this case as well.[3] □

*Remark 4.4.2.* The above proof relies on the assumption that the properties (♠) and (♣) hold for $W, L$. But note that these are only used in the very last steps of the proof, after the translation of the original problem to a problem concerning Iwahori–Hecke algebras is achieved. In particular, the statements (1), (2) and (3) in the above proof do not rely at all on (♠) or (♣).

*Remark 4.4.3.* One may define the set of "modular" unipotent representations by

$$\mathrm{Unip}_k(G) := \{Y \in \mathrm{Irr}_k(G) \mid \langle \rho : Y \rangle_{\mathscr{O}G} \neq 0 \text{ for some } \rho \in \mathrm{Unip}_{\mathbb{K}}(G)\}.$$

We shall see in Remark 4.5.6 that the statement of Theorem 4.4.1 is "natural" in the sense that it fits into a general (conjectural) bijection between $\mathrm{Unip}_{\mathbb{K}}(G)$ and $\mathrm{Unip}_k(G)$, which itself is part of a global conjecture concerning all of $\mathrm{Irr}_k(G)$.

**Corollary 4.4.4.** *Write* $\mathrm{Irr}(\mathbf{H}_k) = \{L_k^\mu \mid \mu \in \Lambda_k^\circ\}$ *as in Proposition 3.2.7, where* $\Lambda_k^\circ \subseteq \Lambda$ *is the canonical basic set associated with* $\theta_k \colon A \to k$*. Then we have*

$$\mathrm{Irr}_k(G \mid B) = \{Y^\mu \mid \mu \in \Lambda_k^\circ\},$$

*where* $Y^\mu$ *is determined by the condition that* $\mathfrak{F}_k(Y^\mu) \cong L_k^\mu$*. Furthermore, we have*

$$d_{\lambda,\mu} = \langle \rho^\lambda : Y^\mu \rangle_{\mathscr{O}G} \quad \text{for all } \lambda \in \Lambda \text{ and } \mu \in \Lambda_k^\circ,$$

*where* $D_{\theta_k} = \big(d_{\lambda,\mu}\big)_{\lambda\in\Lambda,\,\mu\in\Lambda_k^\circ}$ *is the decomposition matrix of* $\mathbf{H}$ *with respect to* $\theta_k$*.*

*Proof.* This is just a reformulation of what we established in the proof of Theorem 4.4.1: under the bijection $\mathrm{Irr}_{\mathbb{K}}(G \mid B) \leftrightarrow \Lambda$ in 4.3.10, the subset $\mathscr{B}_k(G)$ corresponds to a canonical basic set for $\mathrm{Irr}(\mathbf{H}_k)$, in the sense of Definition 3.2.1. The

[3] Since not all details of the proof of **P1, P4, P15** for the quasi-split case are worked out in [231, Chapter 16], the incredulous reader should add the assumption to the theorem that if $W$ is of type $B_n$ and $L(s) > 0$ $(s \in S)$, then $\gamma = \mathrm{id}$ and $L = l$; the other cases where $\gamma \neq \mathrm{id}$ are covered by 2.4.1.

equality of decomposition numbers is expressed in equation (2) of the above proof. (It is a consequence of Dipper's Theorem 4.1.14.) □

**Example 4.4.5.** Let $G = \mathrm{GL}_n(\mathbb{F}_q)$. In this case, $W = \mathbf{W} \cong \mathfrak{S}_n$ and $\Lambda$ consists of all partitions $\lambda \vdash n$. Then, by Theorem 3.5.14, we have

$$\mathrm{Irr}(\mathbf{H}_k) = \{L_k^\mu \mid \mu \in \Lambda_k^\circ\}, \qquad \text{where} \qquad \Lambda_k^\circ = \{\mu \vdash n \mid \mu \text{ is } e\text{-regular}\}$$

and $e = \min\{i \geqslant 2 \mid 1+q+q^2+\cdots+q^{i-1} \equiv 0 \bmod \ell\}$. Hence, by Corollary 4.4.4, we also have a parametrisation of $\mathrm{Irr}_k(G \mid B)$ by the $e$-regular partitions of $n$.

In fact, much more is known in this case: Dipper and James [64], [65] describe the whole decomposition matrix of $G$ in terms of $q$-Schur algebras, and Dipper [59, Theorem 5.21] gives a complete description of the partition of $\mathrm{Irr}_k(G)$ into Harish-Chandra series. For further reading on $\mathrm{GL}_n$, see [38], [56], [57], [58], [59], [62], [64], [72], [119, §7], [180].

**Example 4.4.6.** Assume that $\ell \mid q^{L(s)} - 1$ for all $s \in S$. Then $\mathbf{H}_k \cong kW$, the group algebra of $W$ over $k$. By Example 3.6.5, the map $d_{\theta_k} \colon R_0(\mathbf{H}_K) \to R_0(\mathbf{H}_k)$ can be naturally interpreted as the usual $\ell$-modular decomposition map for the finite group $W$. Hence, Corollary 4.4.4 shows that the decomposition matrix $\mathrm{Dec}(\mathscr{O}W)$ embeds into $\mathrm{Dec}(\mathscr{O}G)$. For $G = \mathrm{GL}_n(\mathbb{F}_q)$, this (and a much more general statement) was first proved by Dipper [56]. For $G$ in general, see [86, 4.4] and [92, 4.6].

Here is a simple crtiterion for identifying the subset $\mathrm{Irr}_k(G \mid B) \subseteq \mathrm{Irr}_k(G)$.

**Lemma 4.4.7.** *Let $Y \in \mathrm{Irr}_k(G)$ and assume that $Y \in \mathrm{Irr}_k(G \mid B)$. Then $\langle \rho : Y \rangle_{\mathscr{O}G} = 0$ for all $\rho \in \mathrm{Unip}_{\mathbb{K}}(G)$ such that $\rho \notin \mathrm{Irr}_{\mathbb{K}}(G \mid B)$. Furthermore, we have that*

$$\ell^a \quad \textit{divides} \quad \sum_{\lambda \in \Lambda} \langle \rho^\lambda : Y \rangle_{\mathscr{O}G} \dim_{\mathbb{K}} \rho^\lambda,$$

*where $\ell^a$ ($a \geqslant 0$) is the largest power of $\ell$ dividing $[G : B]$.*

*Proof.* Using the notation in Corollary 4.4.4, we have $Y \cong Y^\mu$ for some $\mu \in \Lambda_k^\circ$. Now let $\rho \in \mathrm{Unip}_{\mathbb{K}}(G)$. Then, by (2) and (3) in the proof of Theorem 4.4.1, we have

$$\langle \rho : Y \rangle_{\mathscr{O}G} = \begin{cases} d_{\lambda,\mu} & \text{if } \rho \cong \rho^\lambda \text{ where } \lambda \in \Lambda, \\ 0 & \text{otherwise.} \end{cases}$$

Thus, the first condition holds. Now consider the algebra $\mathbf{H}_{\mathscr{O}}$ and the decomposition map $d_{\mathbf{H}_{\mathscr{O}}} \colon R_0(\mathbf{H}_{\mathbb{K}}) \to R_0(\mathbf{H}_k)$. By (1) in the proof of Theorem 4.4.1, the matrix $(d_{\lambda,\mu})$ is the decomposition matrix associated with $d_{\mathbf{H}_{\mathscr{O}}}$. Further note that, since $\mathscr{O}$ is a discrete valuation ring and $\mathbf{H}_{\mathscr{O}}$ is a symmtric algebra, where $\mathbf{H}_{\mathbb{K}}$ is split semisimple, the statement of Proposition 3.3.6 also holds for $\mathbf{H}_{\mathscr{O}}$ and $d_{\mathbf{H}_{\mathscr{O}}}$. (See [132, 7.5.3].) Thus, since the Schur elements of $\mathbf{H}_{\mathbb{K}}$ are given by $\theta_{\mathbb{K}}(\mathbf{c}_\lambda)$ ($\lambda \in \Lambda$), we have

$$\sum_{\lambda \in \Lambda} \theta_{\mathbb{K}}(\mathbf{c}_\lambda)^{-1} d_{\lambda,\mu} \in \mathscr{O} \qquad \text{for any } \mu \in \Lambda_k^\circ.$$

Finally, by the dimension formula in 4.3.10, we have $\dim_{\mathbb{K}} \rho^\lambda = [G:B]/\theta_{\mathbb{K}}(\mathbf{c}_\lambda)$. It follows that

$$\frac{1}{[G:B]}\Big(\sum_{\lambda\in\Lambda} d_{\lambda,\mu} \dim_{\mathbb{K}} \rho^\lambda\Big) \in \mathscr{O},$$

which yields the required statement. □

The next result shows that, in a well-defined sense, the classification of $\mathrm{Irr}_k(G \mid B)$ in Theorem 4.4.1 only depends on "$e$", as defined in Remark 4.3.11.

**Theorem 4.4.8 (Cf. [109]).** *In the setting of Theorem 4.4.1, assume that* $\mathbf{W}$ *is irreducible and that* $\ell$ *does not divide* $|\mathbf{W}|\cdot|\gamma|$. *Recall that* $q^{1/2}-\zeta_{2e} \in \mathrm{rad}(\mathscr{O})$, *where* $e \geqslant 2$ *is minimal such that* $1+q+q^2+\cdots+q^{e-1} \in \mathrm{rad}(\mathscr{O})$; *see Remark 4.3.11. Consider the* $\Phi_e$*-modular specialisation* $\theta_e\colon A \to \mathbb{K}$, $v \mapsto \zeta_{2e}$, *and denote by* $\mathbf{H}_{(e)}$ *the corresponding specialised algebra. Then, under the bijection* $\mathrm{Irr}_{\mathbb{K}}(G \mid B) \leftrightarrow \Lambda$ *in 4.3.10, the set* $\mathscr{B}_k(G)$ *corresponds to the canonical basic set* $\Lambda^\circ_{(e)} \subseteq \Lambda$ *for* $\mathbf{H}_{(e)}$.

*In this sense,* $\mathscr{B}_k(G)$ *only depends on e.*

*Proof.* We already know that, under the bijection $\mathrm{Irr}_{\mathbb{K}}(G \mid B) \leftrightarrow \Lambda$, the set $\mathscr{B}_k(G)$ corresponds to the canonical basic set $\Lambda_k^\circ \subseteq \Lambda$; see Corollary 4.4.4. So we must show that $\Lambda_k^\circ = \Lambda^\circ_{(e)}$. Now, we have seen in Remark 4.3.11(c) that $d_{\theta_k}$ factors through the decomposition map $d_{\theta_e}\colon R_0(\mathbf{H}_K) \to R_0(\mathbf{H}_{(e)})$. Hence, by Proposition 3.7.5 (or Lemma 3.6.7), it will be enough to show that $|\mathrm{Irr}(\mathbf{H}_k)| = |\mathrm{Irr}(\mathbf{H}_{(e)})|$.

Assume first that $\ell \nmid [G:B]$. Then Lemma 4.3.2 applies and so $\Lambda_k^\circ = \Lambda$. Since $|\mathrm{Irr}(\mathbf{H}_k)| \leqslant |\mathrm{Irr}(\mathbf{H}_{(e)})|$, we deduce that $\Lambda_k^\circ = \Lambda_e^\circ = \Lambda$ in this case, as required.

Thus, we can now assume that $\ell \mid [G:B]$ and, hence, $\mathbf{H}_k$ is not semisimple. Note that this implies that $\ell > 2$ and $q \not\equiv 1 \bmod \mathrm{rad}(\mathscr{O})$. (For, otherwise, we would have $q \equiv 1 \bmod \mathrm{rad}(\mathscr{O})$ and so $\mathbf{H}_k \cong kW$ would be semisimple since $\ell \nmid |W|$.) To deal with this case, we shall use Theorem 3.7.16. This requires some preparations. As mentioned in 4.3.7, we have $[G:B] = P_{W,L}(q)$, where $P_{W,L} \in \mathbb{Z}[u]$ is the "weighted" Poincaré polynomial $P_{W,L} := \sum_{w\in W} u^{L(w)}$. (Here, we set $u = v^2$.) Now the results in Example 3.3.2 generalise to the quasi-split case. More precisely, by Steinberg [287, pp. 190–191] (or [45, p. 75]), we have a factorisation

$$P_{W,L} = \prod_{1\leqslant j\leqslant r} \frac{1-\zeta_j u^{d_j}}{1-\zeta_j u},$$

where $d_1,\ldots,d_r$ are the degrees of $\mathbf{W}$ and $\zeta_1,\ldots,\zeta_r$ are certain roots of unity such that $\zeta_j^{|\gamma|} = 1$ for all $j$. In particular, $P_{W,L} \in \mathbb{Z}[u]$ is a product of cyclotomic polynomials. Let $\tilde{E}(W,L)$ be the set of all natural numbers $d \geqslant 2$ such that $\Phi_d(u)$ divides $P_{W,L}$. From the above factorisation of $P_{W,L}$, we deduce that

$$(*) \qquad d \in \tilde{E}(W,L) \quad\Rightarrow\quad d \text{ divides } |\mathbf{W}|\cdot|\gamma| \quad (\text{and, hence, } \ell \nmid d).$$

Now recall from 4.3.10 that $\dim_{\mathbb{K}} \rho^\lambda = [G:B]/\theta_{\mathbb{K}}(\mathbf{c}_\lambda)$ for all $\lambda \in \Lambda$. Then a standard argument (see, for example, Step 1 in the proof of [132, Cor. 8.3.6]) shows that the Schur elements $\mathbf{c}_\lambda$ divide $P_{W,L}$ in $\mathbb{K}[v,v^{-1}]$. Hence, with the notation in 3.3.1 (see also Example 3.3.3), we conclude that

$$\begin{aligned}\mathscr{P} :=&\{\Phi \in A \mid \Phi \text{ is } \mathbb{K}\text{-cyclotomic and divides } \mathbf{c}_\lambda \text{ for some } \lambda \in \Lambda\}\\ =&\{\Phi \in A \mid \Phi \text{ is } \mathbb{K}\text{-cyclotomic and divides } P_{W,L})\}.\end{aligned}$$

Since $\mathbb{K}$ is assumed to be sufficiently large, every polynomial in $\mathscr{P}$ will be of the form $v-\zeta$, where $\zeta \in \mathbb{K}$ is a root of unity. Hence, we have

$$\mathscr{P} = \{v-\zeta \in \mathbb{K}[v] \mid \Phi_d(\zeta^2)=0 \text{ for some } d \in \tilde{E}(W,L)\}.$$

Now, since $\ell \mid [G:B] = P_{W,L}(q)$, we have $\ell \mid \Phi_d(q)$ for some $d \in \tilde{E}(W,L)$; by $(*)$, we have $\ell \nmid d$. On the other hand, by the definition of $e$, we have $\ell \mid \Phi_e(q)$. Hence, 3.7.11(a) shows that $e = d\ell^i$ for some $i \in \mathbb{Z}$. As noted earlier, $q \not\equiv 1 \bmod \operatorname{rad}(\mathscr{O})$ and so $e$ is the order of the image of $q$ in $k^\times$; in particular, $\ell \nmid e$. Thus, we must have $d=e$ and so $\Phi := v-\zeta_{2e} \in \mathscr{P}$. Using $(*)$, we can now conclude that $\Phi$ is *strongly $\theta_k$-isolated* in the sense of Definition 3.7.8. (Indeed, if $\gamma \neq \mathrm{id}$ and $\mathbf{W}$ is of type $B_2$, $G_2$ or $F_4$, this is verified directly using the formulae in Example 4.4.9 below; for all the other cases, see Lemma 3.7.12(b).) So we can apply Theorem 3.7.16, and this yields that $|\operatorname{Irr}(\mathbf{H}_k)| = |\operatorname{Irr}(\mathbf{H}_{(e)})|$, as desired. □

**Example 4.4.9.** Let $\mathbf{W}$ be of type $B_2$, $G_2$ or $F_4$, and $\gamma\colon \mathbf{W} \to \mathbf{W}$ be the non-trivial automorphism of order 2. This gives rise to a weight function $L$ on $W \cong \mathbf{W}^\gamma$ as in Table 4.1. Using the notation in the above proof, we have

$$\begin{aligned}
P_{W,L} &= \frac{(1-u^2)(1+u^4)}{(1-u)(1+u)} && (\text{type } {}^2B_2),\\
P_{W,L} &= \frac{(1-u^2)(1+u^6)}{(1-u)(1+u)} && (\text{type } {}^2G_2),\\
P_{W,L} &= \frac{(1-u^2)(1+u^6)(1-u^8)(1+u^{12})}{(1-u)(1-u)(1+u)(1+u)} && (\text{type } {}^2F_4).
\end{aligned}$$

Hence, we find that $\tilde{E}(W,L)=\{8\}$ (type ${}^2B_2$), $\tilde{E}(W,L)=\{4,12\}$ (type ${}^2G_2$) and $\tilde{E}(W,L)=\{4,8,12,24\}$ (type ${}^2F_4$).

In the first two cases, $G^F$ is a Suzuki or Ree group of type ${}^2B_2$ or ${}^2G_2$ respectively. Here, $W=\mathbf{W}^\gamma$ is of type $A_1$ and so $\Lambda = \{(2),(11)\}$ (where $(2)$ labels the unit and $(11)$ labels the sign representation). Hence, we have (see also Example 4.3.4)

$$\Lambda_k^\circ = \begin{cases} \Lambda & \text{if } q^{L(s)} \not\equiv 1 \bmod \ell,\\ \{(2)\} & \text{otherwise.}\end{cases}$$

If $G^F$ is a Ree group of type ${}^2F_4$, then $W=\mathbf{W}^\gamma$ is of type $I_2(8)$ and the set $\Lambda_k^\circ$ is described in Table 7.6 (p. 373).

*Remark 4.4.10.* In the proof of Theorem 4.4.8 we have seen that $\Phi = v-\zeta_{2e}$ is strongly $\theta_k$-isolated in the sense of Definition 3.7.8. Hence, if the general version of James' Conjecture 3.7.9 was known to hold, then we could not only conclude that $\text{Irr}(\mathbf{H}_{(e)})$ and $\text{Irr}(\mathbf{H}_k)$ have the same cardinality, but also that the decomposition matrices of $\mathbf{H}$ corresponding to $\theta_e$ and to $\theta_k$ are equal. By Example 3.7.7 we know at least that this stronger conclusion holds whenever $\ell$ is "large enough".

The statement of Theorem 4.4.8 holds under somewhat weaker assumptions on $\ell$, but then the proof requires some case-by-case considerations and the results discussed in Section 4.6 concerning type $B_n$. The stronger statement is as follows.

**Theorem 4.4.11 (Cf. [109, 3.4]).** *In the statement of Theorem 4.4.8, instead of the assumption that $\ell$ does not divide $|\mathbf{W}| \cdot |\gamma|$, it is sufficient to assume that $\ell$ is good for* $\mathbf{G}$ *and that $\ell$ does not divide the order of the following parabolic subgroup of $W$:*

$$\Omega_k := \begin{cases} \langle s \in S \mid L(s) > 1 \text{ and } \theta_k(q)^{L(s)} = 1 \rangle & \text{if } \theta_k(q) \neq 1, \\ \{1\} & \text{if } \theta_k(q) = 1. \end{cases}$$

*In particular, if $L(s) = 1$ for all $s \in S$, then it is enough to assume that $\ell$ is good for* $\mathbf{G}$.

*Proof.* As in the proof of Theorem 4.4.8, it is enough to show that $|\text{Irr}(\mathbf{H}_k)| = |\text{Irr}(\mathbf{H}_{(e)})|$. Assume first that $W = \mathbf{W}^\gamma$ is of exceptional type. If $\ell$ does not divide $|\mathbf{W}| \cdot |\gamma|$, then we already know that the desired equality holds. Hence, for each type of $\mathbf{W}$, there is only a finite number of additional cases to be considered. These additional cases can be handled by explicit computations with the character tables of $\mathbf{H}_K$, using Proposition 3.4.11. In the case where $\gamma = \text{id}$, details of such computations are given in [100, §3]. Similar methods apply to the cases where $\mathbf{W}$ is of type $D_4$ or $E_6$, and $W = \mathbf{W}^\gamma$ is of type $G_2$ or $F_4$ respectively.

The cases where $\gamma \neq \text{id}$ and $\mathbf{W}$ is of type $B_2$, $G_2$, $F_4$ are already covered by Theorem 4.4.8. So it remains to consider $\mathbf{W}$ of type $A_{n-1}$, $B_n$ or $D_n$.

If $\mathbf{W}$ is of type $A_{n-1}$, where $\gamma = \text{id}$, the desired equality holds by Theorem 3.5.14. If $\mathbf{W}$ is of type $B_n$ or $D_n$, then the results in Section 4.6 show that $|\text{Irr}(\mathbf{H}_k)|$ only depends on $e$ and $\gamma$ and, hence, the desired equality holds; see the explicit consideration of these cases in Examples 4.4.13–4.4.15 below.

Now assume that $\mathbf{W}$ is of type $A_{n-1}$ and $\gamma$ has order 2, so that $W = \mathbf{W}^\gamma$ is of type $B_m$ (for suitable $m$). If $e \neq 2$, then the discussion in Example 4.4.16 below shows again that the desired equality holds. Finally, if $e = 2$, then $\theta_k(q) = -1$ and so $\Omega_k \cong \mathfrak{S}_m$. Hence, the condition on $\ell$ implies that $\ell \nmid |W|$. Consequently, in the setting of Section 4.6, we have $\tilde{e} = \ell > m$. Then Proposition 4.6.10 shows that $|\text{Irr}(\mathbf{H}_k)|$ and $|\text{Irr}(\mathbf{H}_{(e)})|$ are both given by the number of partitions of $m$. □

**Example 4.4.12.** Let $\mathbf{G}$ be of exceptional type $G_2$, $F_4$, $E_6$, $E_7$ or $E_8$. Let $F$ be a split Frobenius map so that $\gamma = \text{id}$. Then the assumptions of Theorem 4.4.11 are satisfied whenever $\ell$ is good for $\mathbf{G}$. Hence, the information in Table 7.1 (pp. 367) and Table 7.3 (p. 373) yields explicit descriptions for the canonical basic sets $\Lambda_k^\circ$ in these cases. The twisted exceptional groups $G^F = {}^3D_4(\mathbb{F}_q)$ and ${}^2E_6(\mathbb{F}_q)$ are covered by the information in Table 7.4 (p. 373) and Table 7.8 (p. 376).

To close this section, we shall now consider the various cases in Table 4.1 (p. 227) where $W$ is of type $B_n$. Recall that then $\mathrm{Irr}_{\mathbb{K}}(W)$ is parametrised by the set $\Lambda$ consisting of all pairs of partitions $(\lambda, \mu)$ such that $|\lambda| + |\mu| = n$. Let

$$e := \min\{i \geqslant 2 \mid 1 + q + q^2 + \cdots + q^{i-1} \equiv 0 \bmod \ell\}.$$

The Examples 4.4.13–4.4.16 contain a summary of results whose proofs rely on the theories developed later in Section 4.6 and Chapters 5 and 6.

**Example 4.4.13.** Let $G = \mathrm{Sp}_{2n}(\mathbb{F}_q)$ (the finite symplectic group) or $G = \mathrm{SO}_{2n+1}(\mathbb{F}_q)$ (the odd-dimensional orthogonal group); see [104, 4.2.6] and Grove [147]. Then $W = \mathbf{W}$ is of type $B_n$ and the associated weight function $L\colon W \to \mathbb{Z}$ is given by

$B_n$

That is, we are in the equal-parameter case. Since 2 is $L$-bad in this case, we shall assume that $\mathrm{char}(k) = \ell > 2$. The set $\Lambda_k^\circ \subseteq \Lambda$ is determined as follows.

If $e$ is odd, then the assumptions of Example 4.6.4 below are satisfied and so $\Lambda_k^\circ$ consists of all $(\lambda, \mu) \in \Lambda$ such that both $\lambda$ and $\mu$ are $e$-regular.

If $e$ is even, then $\theta_k(q) \in k^\times$ is a root of unity of order $e$ and so $\Lambda_k^\circ$ is given by the "FLOTW bipartitions" of Theorem 5.8.5 in Chapter 5.

**Example 4.4.14.** Let $\mathbf{G} = \mathrm{SO}_{2n}^\circ(\overline{\mathbb{F}}_p)$ be the connected component of the identity of the even-dimensional orthogonal group; see [104, 1.3.16] and Grove [147]; thus, the Weyl group $\mathbf{W}$ of $\mathbf{G}$ is of type $D_n$. Assume that $F$ is a split Frobenius map with respect to $\mathbb{F}_q$. Then $G = \mathbf{G}^F$ is the split orthogonal group; see Case 1 in [104, 4.2.6]. In this case $W = \mathbf{W}$ and $\Lambda_k^\circ$ is determined as in Example 3.2.15, by using the canonical basic set for a group of type $B_n$ with weight function

$B_n$

Since 2 is $L$-bad in this case, we shall assume that $\mathrm{char}(k) = \ell > 2$.

If $e$ is odd, then we can again apply Example 4.6.4 below, and so this canonical basic set (for type $B_n$) consists of all $(\lambda, \mu) \in \Lambda$ such that both $\lambda$ and $\mu$ are $e$-regular.

If $e$ is even, then $\theta_k(q) \in k^\times$ is a root of unity of order $e$, and so the canonical basic set is given by the "FLOTW bipartitions" of Theorem 5.8.15 in Chapter 5.

As explained in Example 3.2.15, this then yields an explicit description of $\Lambda_k^\circ$ (for $W$ of type $D_n$); see Theorem 5.8.19 in Chapter 5.

**Example 4.4.15.** Now let $\mathbf{G} = \mathrm{SO}_{2(n+1)}^\circ(\overline{\mathbb{F}}_p)$ and assume that $F$ is a non-split Frobenius map with respect to $\mathbb{F}_q$. Then $G = \mathbf{G}^F$ is the non-split orthogonal group; see Case 3 in [104, 4.2.6]. In this case, $W = \mathbf{W}^\gamma$ is of type $B_n$ with weight function

$B_n$

Since 2 is $L$-bad in this case, we shall assume that $\mathrm{char}(k) = \ell > 2$.

Again, if $e$ is odd, we can apply Example 4.6.4 below and so $\Lambda_k^\circ$ consists of all $(\lambda, \mu) \in \Lambda$ such that both $\lambda$ and $\mu$ are $e$-regular.

If $e \geqslant 4$ is even, then $\theta_k(q) \in k^\times$ is a root of unity of order $e$, and so $\Lambda_k^\circ$ is given by the "FLOTW bipartitions" of Theorem 5.8.13.

Finally, assume that $e = 2$; that is, $\theta_k(q) = -1$. To obtain a description of $\Lambda_k^\circ$, we use the fact that $\theta_k$ lies above the $\Phi_e$-modular specialisation $\theta_e \colon A \to \mathbb{K}$; see Remark 4.3.11(c). Let $\mathbf{H}_{(e)}$ be the corresponding specialised algebra. We shall see in Remark 6.7.12 that a canonical basic set $\mathscr{B}_{(e)} \subseteq \Lambda$ for $\mathbf{H}_{(e)}$ exists and that there is an explicit combinatorial description of $\mathscr{B}_{(e)}$ (using the theory of canonical bases for quantised enveloping algebras). Furthermore, Theorem 4.6.13 will show that $|\,\mathrm{Irr}(\mathbf{H}_k)| = |\,\mathrm{Irr}(\mathbf{H}_{(e)})|$. So Lemma 3.6.8 implies that $\Lambda_k^\circ = \mathscr{B}_{(e)}$.

**Example 4.4.16.** Let $G = \mathrm{GU}_N(\mathbb{F}_q)$. Recall that $G = \mathbf{G}^F$, where $\mathbf{G} = \mathrm{GL}_N(\overline{\mathbb{F}}_q)$ and $F \colon \mathbf{G} \to \mathbf{G}$ is the Frobenius map described in Example 4.3.6(b). We are in the "quasi-split case" where $\mathbf{W} \cong \mathfrak{S}_N$ and $W \cong \mathbf{W}^\gamma$ is of type $B_n$; here, $N = 2n + c$, where $c \in \{0, 1\}$. The associated weight function $L \colon W \to \mathbb{Z}$ is given by

$B_n$ 

There are no $L$-bad primes, so there is no restriction on $\mathrm{char}(k) = \ell$. The following distinction of cases works for arbitrary values of $c \geqslant 0$ (and not just $c \in \{0, 1\}$).

If $e = 2$, then $\theta_k(q)^2 = 1$ and $\theta_k(q)^{2c+1} = -1$. In this case, $\Lambda_k^\circ$ is given by Proposition 4.6.10 below. Now assume that $e \neq 2$; in particular, this means that $\ell \neq 2$.

If $e = 2e'$, where $e' \geqslant 3$ is odd, then $\theta_k(q)^2 \in k^\times$ is a root of unity of order $e'$ and so $\Lambda_k^\circ$ is given by the "FLOTW bipartitions" of Theorem 5.8.9 ($c = 0$) and Theorem 5.8.11 ($c = 1$) in Chapter 5. (For $c \geqslant 2$, see Example 4.5.15 below.)

Finally, assume that $e$ is not twice an odd number. Then $\theta_k(q^{2c+1} + q^{2j}) \neq 0$ for all $j \in \mathbb{Z}$. (Indeed, if $\theta_k(q)^{2c+1} = -\theta_k(q)^{2j}$, then $\theta_k(q)^{2c-2j+1} = -1$ and so $\theta_k(q) \in k^\times$ would be a root of unity whose order is twice an odd number, contrary to our assumption.) Thus, setting $\tilde{e} = e$ (if $e$ is odd) or $\tilde{e} = e/2$ (if $e$ is even), we have that $\Lambda_k^\circ$ consists of all $(\lambda, \mu) \in \Lambda$ such that both $\lambda$ and $\mu$ are $\tilde{e}$-regular; see Proposition 4.6.3 below.

We remark that, with the single exception of the case $e = 2$ in Example 4.4.15, the results cited in the above examples actually imply the existence of a canonical basic set for $\mathbf{H}_k$ (type $B_n$), without reference to the validity of (♠) and (♣) in 2.5.3.

## 4.5 Examples and Conjectures

The results obtained in the previous section raise the question as to what extent the restriction to unipotent principal series representations is really necessary. The difficulty in answering this question is that very little is known on decomposition numbers in general. We shall now formulate some conjectures which hint at a more general picture and then illustrate these with several examples. Note, however, that it is not the place here to provide a comprehensive survey of the state of knowledge on modular representations of finite groups of Lie type; throughout this section, we shall give references for further reading.

The starting point is the following result on "basic sets" for $G$. Let $\mathbf{T}$ be an $F$-stable maximal torus of $\mathbf{G}$. Since $\mathbf{T}^F$ is abelian, $\mathrm{Irr}_{\mathbb{K}}(\mathbf{T}^F)$ is an abelian group under taking tensor products; so we can speak of the *order* of $\theta \in \mathrm{Irr}_{\mathbb{K}}(\mathbf{T}^F)$.

**Theorem 4.5.1 (Geck and Hiss [116], [93]).** *Assume that $\ell$ is good for $\mathbf{G}$ and $\ell$ does not divide the order of $(Z_{\mathbf{G}}/Z_{\mathbf{G}}^{\circ})_F$. (Here, $Z_{\mathbf{G}}$ denotes the centre of $G$ and $(Z_{\mathbf{G}}/Z_{\mathbf{G}}^{\circ})_F$ denotes the largest quotient of $Z_{\mathbf{G}}/Z_{\mathbf{G}}^{\circ}$ on which $F$ acts trivially). Let*

$$\mathscr{E}'_\ell(G) := \left\{ \rho \in \mathrm{Irr}_{\mathbb{K}}(G) \;\middle|\; \begin{array}{l} \rho \text{ appears in some } R_{\mathbf{T},\theta} \text{ where} \\ \theta \in \mathrm{Irr}_{\mathbb{K}}(\mathbf{T}^F) \text{ has order prime to } \ell \end{array} \right\}.$$

*Then $|\mathscr{E}'_\ell(G)| = |\mathrm{Irr}_k(G)|$ and $\{d_{\mathscr{O}G}([\rho]) \mid \rho \in \mathscr{E}'_\ell(G)\}$ is a $\mathbb{Z}$-basis of $R_0(kG)$.*

Note that we always have $\mathrm{Unip}_{\mathbb{K}}(G) \subseteq \mathscr{E}'_\ell(G)$. If $\ell$ is not good for $G$, we have $|\mathscr{E}'_\ell(G)| \neq |\mathrm{Irr}_k(G)|$ in general. For further information on the cardinalities of the sets $|\mathscr{E}'_\ell(G)|$ in this case, see [117, 6.6]. The following conjecture is somewhat more precise than its precursors in [90], [109], [117]. The present version appeared in the first author's talk at Michel Broué's 60th birthday conference (ENS Paris, October 2006). Note that, implicitly, it relies on a basic result of Broué and Michel [37] concerning $\mathscr{O}$-blocks of $G$ (see Theorem 4.5.4 for a special case).

**Conjecture 4.5.2 (Cf. Geck [90], [109]; Geck and Hiss [117]).** *Assume that $\ell$ is good for $\mathbf{G}$ and $\ell$ does not divide the order $(Z_{\mathbf{G}}/Z_{\mathbf{G}}^{\circ})_F$.*

*(a) Given $Y \in \mathrm{Irr}_k(G)$, let $\hat{\mathscr{S}}_k(Y) := \{\rho \in \mathscr{E}'_\ell(G) \mid \langle \rho : Y \rangle_{\mathscr{O}G} \neq 0\}$. Then the function $\hat{\mathscr{S}}_k(Y) \to \mathbb{Z}_{\geqslant 0}$, $\rho \mapsto \mathbf{a}_\rho$ (defined using the unipotent support), reaches its minimum at exactly one element of $\hat{\mathscr{S}}_k(Y)$, which we denote by $\rho_Y$.*
*(b) The map $\mathrm{Irr}_k(G) \to \mathscr{E}'_\ell(G)$, $Y \mapsto \rho_Y$, is a bijection.*
*(c) We have $\langle \rho_Y : Y \rangle_{\mathscr{O}G} = 1$ for all $Y \in \mathrm{Irr}_k(G)$.*

The model case where this conjecture is known to hold is $G = \mathrm{GL}_n(\mathbb{F}_q)$; see Dipper and James [64]. (To be more precise, [64] contains all the essential ingredients to conclude that the conjecture, as formulated above, holds in this case.) It is also true for finite unitary groups and other finite groups of Lie type of low rank (where explicit computations are possible); see [109], [117], [152] and the references therein. Note the formal analogy with Definition 3.2.1; in fact, it was the above conjecture which motivated our work on "canonical basic sets" for Iwahori–Hecke algebras.

**Example 4.5.3.** Let $G = \mathrm{Suz}(q^2)$, where $q = \sqrt{2}^{2f+1}$ for some $f \geqslant 0$; see Example 4.3.6(c). Let us verify Conjecture 4.5.2 in this case. Using the notation in [104,

**Table 4.2** Irreducible representations of $\mathrm{Suz}(q^2)$ in characteristic 0

| $\rho$ | $\dim\rho$ | $\mathbf{a}_\rho$ | |
|---|---|---|---|
| $1_G$ | $1$ | 0 | (unit representation) |
| $\mathscr{W}$ | $\frac{1}{\sqrt{2}}q(q^2-1)$ | 1 | (cuspidal unipotent) |
| $\overline{\mathscr{W}}$ | $\frac{1}{\sqrt{2}}q(q^2-1)$ | 1 | (cuspidal unipotent) |
| $\mathrm{St}_G$ | $q^4$ | 4 | (Steinberg representation) |
| $\{\mathscr{X}_l\}$ | $q^4+1$ | 0 | (non-unipotent, various $l$) |
| $\{\mathscr{Y}_m\}$ | $(q^2-1)(q^2-q\sqrt{2}+1)$ | 0 | (non-unipotent, various $m$) |
| $\{\mathscr{Z}_n\}$ | $(q^2-1)(q^2+q\sqrt{2}+1)$ | 0 | (non-unipotent, various $n$) |

§4.6], the irreducible representations of $G$ over $\mathbb{K}$ are given in Table 4.2. Now let $\ell$ be a prime, $\ell \neq 2$. The set $\mathscr{E}'_\ell(G)$ consists of the unipotent representations and some of the representations $\mathscr{X}_l$, $\mathscr{Y}_m$, $\mathscr{Z}_n$ (for certain values of $l,m,n$). The decomposition numbers of $G$ are easily determined. (Note that the Sylow $\ell$-subgroups are always cyclic; see Burkhardt [42] for further details). For $\rho \in \mathscr{E}'_\ell(G) \setminus \{\mathrm{St}_G\}$, we have that $d_{\mathscr{O}G}([\rho])$ is the class of an irreducible $kG$-module (which we denote by $\rho_k$). For the Steinberg representation, the following hold:

- If $\ell \mid q^2-1$, then $d_{\mathscr{O}G}([\mathrm{St}_G])$ is the class of an irreducible $kG$-module.
- If $\ell \mid q^2+q\sqrt{2}+1$, then $d_{\mathscr{O}G}([\mathrm{St}_G]) = [1_G]+[\mathscr{W}_k]+[\overline{\mathscr{W}}_k]+[\psi]$, where $\psi \in \mathrm{Irr}_k(G)$ is cuspidal (and can be lifted to $\mathscr{Y}_m$ for suitable values of $m$).
- If $\ell \mid q^2-q\sqrt{2}+1$, then $d_{\mathscr{O}G}([\mathrm{St}_G]) = [1_G]+[\psi]$, where $\psi \in \mathrm{Irr}_k(G)$ is cuspidal of dimension $q^4-1$ (and cannot be lifted to characteristic 0).

Thus, the decomposition numbers of the unipotent representations are given by

| $\rho$ | $\dim\rho$ | $\mathbf{a}_\rho$ | dec. numbers | | | |
|---|---|---|---|---|---|---|
| $1_G$ | $1$ | 0 | 1 | . | . | . |
| $\mathscr{W}$ | $\frac{1}{\sqrt{2}}q(q^2-1)$ | 1 | . | 1 | . | . |
| $\overline{\mathscr{W}}$ | $\frac{1}{\sqrt{2}}q(q^2-1)$ | 1 | . | . | 1 | . |
| $\mathrm{St}_G$ | $q^4$ | 4 | $a$ | $b$ | $b$ | 1 |

where $a,b \in \{0,1\}$ according to the above cases. In particular, we see that Conjecture 4.5.2 holds in this case.

We also see the appearance of "new" cuspidal representations over $k$ which cannot be lifted to characteristic 0 (that is, which cannot be obtained by extension of scalars via the natural map $\mathscr{O} \to k$ from an $\mathscr{O}G$-lattice). This is not an isolated phenomenon; see Examples 4.5.9 and 4.5.18. (One can produce many more such examples using the techniques in [119], [120]).

The above conjecture admits, of course, a block-wise version. In this context, it is particularly interesting to consider those $\mathscr{O}$-blocks which contain unipotent representations. Note that the work of Bonnafé and Rouquier [28] provides an almost complete reduction to this case, similar to Lusztig's [220] "Jordan decomposition of characters" in characteristic 0. The union of these blocks is described by the following basic result.

**Theorem 4.5.4 (Broué and Michel [37]).** *The set*

$$\mathscr{E}_{\ell,1} := \left\{ \rho \in \mathrm{Irr}_{\mathbb{K}}(G) \;\middle|\; \begin{array}{l} \rho \text{ appears in some } R_{\mathbf{T},\theta} \text{ where} \\ \text{the order of } \theta \text{ is a power of } \ell \end{array} \right\}$$

*is a union of $\mathscr{O}$-blocks for $G$.*

**4.5.5.** Assume that $\ell$ is good for $\mathbf{G}$ and $\ell$ does not divide the order of $(Z_{\mathbf{G}}/Z_{\mathbf{G}}^{\circ})_F$. Consider the union of $\mathscr{O}$-blocks $\mathscr{E}_{\ell,1}$ described in the above theorem. Let

$$\mathrm{Unip}_k(G) := \{Y \in \mathrm{Irr}_k(G) \mid \langle \rho : Y \rangle_{\mathscr{O}G} \neq 0 \text{ for some } \rho \in \mathrm{Unip}_{\mathbb{K}}(G)\}.$$

Then the restriction of Theorem 4.5.1 to $\mathscr{E}_{\ell,1}$ shows that $|\mathrm{Unip}_{\mathbb{K}}(G)| = |\mathrm{Unip}_k(G)|$ and $\{d_{\mathscr{O}G}([\rho]) \mid \rho \in \mathrm{Unip}_{\mathbb{K}}(G)\}$ is a $\mathbb{Z}$-basis for the subgroup of $R_0(kG)$ generated by the classes of the simple modules in $\mathrm{Unip}_k(G)$. Furthermore, $\mathscr{E}_{\ell,1} \cap \mathscr{E}_\ell' = \mathrm{Unip}_{\mathbb{K}}(G)$. Now Conjecture 4.5.2 reads:

(a) Given $Y \in \mathrm{Unip}_k(G)$, let $\hat{\mathscr{S}}_k(Y) := \{\rho \in \mathrm{Unip}_{\mathbb{K}}(G) \mid \langle \rho : Y \rangle_{\mathscr{O}G} \neq 0\}$. Then the function $\hat{\mathscr{S}}_k(Y) \to \mathbb{Z}_{\geqslant 0}$, $\rho \mapsto \mathbf{a}_\rho$ (defined using the unipotent support), should reach its minimum at exactly one element of $\hat{\mathscr{S}}_k(Y)$, which we denote by $\rho_Y$.
(b) The map $\mathrm{Unip}_k(G) \to \mathrm{Unip}_{\mathbb{K}}(G)$, $Y \mapsto \rho_Y$, should be a bijection.
(c) We should have $\langle \rho_Y : Y \rangle_{\mathscr{O}G} = 1$ for all $Y \in \mathrm{Unip}_k(G)$.

*Remark 4.5.6.* Theorem 4.4.1 shows that the statements (a) and (c) in 4.5.5 hold for all $Y \in \mathrm{Irr}_k(G \mid B) \subseteq \mathrm{Unip}_k(G)$. Hence, in combination with Theorem 4.5.4, the properties (a) and (c) in Conjecture 4.5.2 are seen to hold for all $Y \in \mathrm{Irr}_k(G \mid B)$. Furthermore, the map $Y \mapsto \rho_Y$ is injective; see Theorem 4.4.1. In this sense, Conjecture 4.5.2 holds "as far as $\mathrm{Irr}_k(G \mid B)$ is concerned".

*Remark 4.5.7.* We note that, conversely to Remark 4.5.6, the statements in Theorem 4.4.1 are formal consequences of the statements (a), (b) and (c) in 4.5.5. (This immediately follows from (3) in the proof of Theorem 4.4.1.) Hence, by the last part of the proof of Theorem 4.4.1, we also see that the existence of a canonical basic set for $\mathbf{H}_k$ is a consequence of (a), (b) and (c) in 4.5.5.

*Remark 4.5.8.* Assume that the statements (a), (b) and (c) in 4.5.5 hold. In particular, this shows that, for any $Y, Y' \in \mathrm{Unip}_k(G)$, we have the implication

$$\langle \rho_Y, Y' \rangle_{\mathscr{O}G} \neq 0 \qquad \Rightarrow \qquad Y = Y' \quad \text{or} \quad \mathbf{a}_{\rho_Y} > \mathbf{a}_{\rho_{Y'}}.$$

One may conjecture that, in fact, the following stronger implication holds:

$$(*) \qquad \langle \rho_Y, Y' \rangle_{\mathscr{O}G} \neq 0 \qquad \Rightarrow \qquad Y = Y' \quad \text{or} \quad \mathbf{O}_{\rho_Y} \subsetneqq \overline{\mathbf{O}}_{\rho_{Y'}}.$$

It follows from the results in [113] that $(*)$ holds for all $Y, Y' \in \mathrm{Irr}_k(G \mid B)$, assuming that $G$ is of "split" type (that is, $\gamma = \mathrm{id}$ and $\mathbf{W} = W$). By Example 4.5.10 below, $(*)$ also holds for all $Y, Y' \in \mathrm{Unip}_k(G)$, where $G = \mathrm{GL}_n(\mathbb{F}_q)$ or $\mathrm{GU}_n(\mathbb{F}_q)$.

**Table 4.3** Decomposition numbers of $\mathrm{Unip}_{\mathbb{K}}(\mathrm{Sp}_4(q))$ where $2 \neq \ell \mid q+1$

| $\rho$ | $\dim\rho$ | $\mathbf{a}_\rho$ | decomposition numbers | | | | | |
|---|---|---|---|---|---|---|---|---|
| $1_G$ | $1$ | 0 | 1 | . | . | . | . | . |
| $\theta_{10}$ | $\frac{1}{2}q(q-1)^2$ | 1 | . | 1 | . | . | . | . |
| $\mathrm{sgn}_1$ | $\frac{1}{2}q(q^2+1)$ | 1 | 1 | . | 1 | . | . | . |
| $\mathrm{sgn}_2$ | $\frac{1}{2}q(q^2+1)$ | 1 | 1 | . | . | 1 | . | . |
| $\sigma_1$ | $\frac{1}{2}q(q+1)^2$ | 1 | . | . | . | . | 1 | . |
| $\mathrm{St}_G$ | $q^4$ | 4 | 1 | $a$ | 1 | 1 | . | 1 |

(If $\ell$=3 and $9 \nmid q+1$, then $a$=1; otherwise, $a$=2)

**Example 4.5.9.** Let $G = \mathrm{Sp}_4(\mathbb{F}_q)$. We have $|G| = q^4(q-1)^2(q+1)^2(q^2+1)$. Let $\ell \neq 2$ be a prime dividing the order of $G$ (but not dividing $q$), and consider the union of $\mathscr{O}$-blocks $\mathscr{E}_{\ell,1}$. This set contains six unipotent representations, five of which lie in the principal series (and are labelled by the corresponding representations of $W$ of type $B_2$) and one of which (denoted $\theta_{10}$) is cuspidal; see Srinivasan [284].

If $\ell \mid q^2+1$, then the Sylow $\ell$-subgroups are cyclic and the corresponding decomposition numbers are known by Fong and Srinivasan [87]. One checks that Conjecture 4.5.2 holds in this case (for the union of blocks $\mathscr{E}_{\ell,1}$). If $\ell \mid q-1$, then $d_{\mathscr{O}G}([\rho])$ is the class of an irreducible $kG$-module for any $\rho \in \mathrm{Unip}_{\mathbb{K}}(G)$; see White [293], [294]. Hence, Conjecture 4.5.2 holds in this case as well.

Now assume that $\ell \mid q+1$. Then the decomposition numbers are given in Table 4.3; see White [293], [294] (who showed that $a \geqslant 1$) and Okuyama and Waki [261, Theorem 2.3] (who determined the precise value of $a$). Again, we see that Conjecture 4.5.2 holds. By Example 3.2.5, we have $\mathrm{Irr}_k(G \mid B) = \{k_G, \bar{\sigma}_1\}$, where $\bar{\sigma}_1 \in \mathrm{Irr}_k(G)$ labels the fifth column of the above matrix. (We have $d_{\mathscr{O}G}([\sigma_1]) = [\bar{\sigma}_1]$.) The dimension of the modular irreducible representation corresponding to the last column of the above matrix is given by

$$\begin{array}{ll} (q-1)^2(q^2+1) & \text{if } a=2, \\ (q-1)^2(q^2+\frac{1}{2}q+1) & \text{if } a=1. \end{array}$$

One can check that this modular irreducible representation is cuspidal.

**Example 4.5.10.** Let $\mathbf{G} = \mathrm{GL}_n(\overline{\mathbb{F}}_p)$ and $F\colon \mathbf{G} \to \mathbf{G}$ be such that $\mathbf{G}^F = \mathrm{GL}_n(\mathbb{F}_q)$ or $\mathrm{GU}_n(\mathbb{F}_q)$; see Example 4.3.6. In both cases, there is a natural labelling $\mathrm{Unip}_{\mathbb{K}}(G) = \{\rho^\lambda \mid \lambda \vdash n\}$; see [55, §15.4]. For example, $\rho^{(n)}$ is the unit and $\rho^{(1^n)}$ is the Steinberg representation. The unipotent support (see 4.3.12) of $\rho^\lambda$ is given as follows:

$$\mathbf{O}_{\rho^\lambda} = \mathbf{O}_\lambda \qquad \text{and} \qquad \dim \mathscr{B}_u = \mathbf{a}_\lambda \quad (u \in \mathbf{O}_\lambda),$$

where $\mathbf{O}_\lambda$ denotes the conjugacy class of $\mathbf{G}$ consisting of unipotent matrices of Jordan type $\lambda$ and $\mathbf{a}_\lambda$ is given by the formula in Example 2.2.13. (This easily follows from the description of the map $\rho \mapsto \mathbf{O}_\rho$ in terms of the Springer correspondence, already mentioned in the remarks following 4.3.12(b).)

Now let $\ell$ be a prime not dividing $q$, and let us verify that Conjecture 4.5.2 holds as far as $\mathcal{E}_{\ell,1}$ is concerned. This can be done in a uniform way, as follows. Kawanaka [193] has associated with every unipotent class $\mathbf{O}_\mu$ (where $\mu \vdash n$) a so-called *generalised Gelfand–Graev representation* (GGGR for short). These GGGRs are obtained by inducing certain representations from unipotent subgroups of $G$. Denote by $\Gamma^\mu$ the GGGR associated with $\mathbf{O}_\mu$. Then, by [194, Theorem 2.4.1], we have

$$\begin{aligned}\Gamma^{\mu^*} \cong \rho^\mu &\oplus \text{(direct sum of unipotent representations } \rho^\lambda \text{ where } \mathbf{O}_\lambda \subsetneq \overline{\mathbf{O}}_\mu)\\ &\oplus \text{(direct sum of non-unipotent irreducible representations);}\end{aligned}$$

here, $\mu^*$ denotes the transpose partition. Since a GGGR is obtained by induction from a unipotent subgroup, we have $\Gamma^{\mu^*} \cong \Psi^\mu_{\mathbb{K}}$, where $\Psi^\mu$ is a finitely generated projective $\mathcal{O}G$-module. Let $\Upsilon^\mu$ be the unique indecomposable direct summand of $\Psi^\mu$ such that $\rho^\mu$ appears with multiplicity 1 in $\Upsilon^\mu_{\mathbb{K}}$. Then $\Upsilon^\mu$ is a finitely generated projective indecomposable $\mathcal{O}G$-module; let $Y^\mu$ be the corresponding irreducible $kG$-module. By Brauer reciprocity (see 4.1.11), we have

$$\begin{aligned}\langle \rho^\lambda : Y^\lambda \rangle_{\mathcal{O}G} &= 1 && \text{for all } \lambda \vdash n,\\ \langle \rho^\lambda : Y^\mu \rangle_{\mathcal{O}G} &= 0 && \text{unless } \lambda = \mu \text{ or } \mathbf{O}_\lambda \subsetneq \overline{\mathbf{O}}_\mu.\end{aligned}$$

It follows that $\mathrm{Unip}_k(G) = \{Y^\mu \mid \mu \vdash n\}$ and that Conjecture 4.5.2 holds as far as $\mathcal{E}_{\ell,1}$ is concerned, as claimed; note also that $\mathbf{O}_\lambda \subsetneq \overline{\mathbf{O}}_\mu \Rightarrow \mathbf{a}_\mu < \mathbf{a}_\lambda$. (This whole argument first appeared in [90, §2.5].) Let us also denote by $\Upsilon^\mu_{\mathrm{uni}}$ the projection of $\Upsilon^\mu$ onto the space of unipotent representations; that is, we have

$$\Upsilon^\mu \cong \Upsilon^\mu_{\mathrm{uni}} \oplus \text{(direct sum of non-unipotent irreducible representations).}$$

While the above description works uniformly for both $\mathrm{GL}_n$ and $\mathrm{GU}_n$, the situation is different for Harish-Chandra series (already over $\mathbb{K}$). Indeed, if $\mathbf{G}^F = \mathrm{GL}_n(\mathbb{F}_q)$, then we have $\mathrm{Irr}_{\mathbb{K}}(G \mid B) = \mathrm{Unip}_{\mathbb{K}}(G)$ (see [55, Prop. 15.8]). On the other hand, if $\mathbf{G}^F = \mathrm{GU}_n(\mathbb{F}_q)$, then the subset $\mathrm{Irr}_{\mathbb{K}}(G \mid B) \subseteq \mathrm{Unip}_{\mathbb{K}}(G)$ corresponds to the set of all partitions $\lambda \vdash n$ such that, in the Young diagram of $\lambda$, the number of boxes which have an odd hook length equals the number of boxes which have an even hook length; see [212, 9.6]. The picture is much more involved as far as modular Harish-Chandra series and decomposition numbers are concerned; see the references in Example 4.4.5 (for $\mathrm{GL}_n$) and [117], [119], [120] (for $\mathrm{GU}_n$). The references [89], [156] and [262] illustrate some of the difficulties that occur already for $\mathrm{GU}_3$.

**Example 4.5.11.** Let $n = 3$ and assume that $\ell \mid q+1$; then $e = 2$.

(a) If $G = \mathrm{GL}_3(\mathbb{F}_q)$, then the decomposition numbers in the union of blocks $\mathcal{E}_{\ell,1}$ are given by Table 4.4; see James [181, p. 253]. By Example 4.4.5, we have $\mathrm{Irr}_k(G \mid B) = \{Y^{(3)}, Y^{(21)}\}$; note also that $\mathrm{Unip}_{\mathbb{K}}(G) = \mathrm{Irr}_{\mathbb{K}}(G \mid B)$ in this case. One can check that $Y^{(111)}$ is non-cuspidal.

(b) If $G = \mathrm{GU}_3(\mathbb{F}_q)$, then the decomposition numbers in the union of blocks $\mathcal{E}_{\ell,1}$ are given by Table 4.5; see Geck [89] (who showed that $a \geqslant 2$ for $\ell \neq 2$), Okuyama

**Table 4.4** Decomposition numbers of $\mathrm{Unip}_{\mathbb{K}}(\mathrm{GL}_3(q))$ where $\ell \mid q+1$

| $\lambda$ | $\dim \rho^\lambda$ | $\mathbf{a}_\lambda$ | dec. num. | | | $\dim Y^\lambda_{\mathrm{uni}}$ |
|---|---|---|---|---|---|---|
| (3) | 1 | 0 | 1 | . | . | $(q+1)(q^2-q+1)$ |
| (21) | $q(q+1)$ | 1 | . | 1 | . | $q(q+1)$ |
| (111) | $q^3$ | 3 | 1 | . | 1 | $q^3$ |

**Table 4.5** Decomposition numbers of $\mathrm{Unip}_{\mathbb{K}}(\mathrm{GU}_3(q))$ where $\ell \mid q+1$

| $\lambda$ | $\dim \rho$ | $\mathbf{a}_\rho$ | dec. num. | | | $\dim Y^\lambda_{\mathrm{uni}}$ |
|---|---|---|---|---|---|---|
| (3) | 1 | 0 | 1 | . | . | $(q+1)(q^2-q+1)$ |
| (21) | $q(q-1)$ | 1 | . | 1 | . | $aq^3+q^2-q$ |
| $(1^3)$ | $q^3$ | 3 | 1 | $a$ | 1 | $q^3$ |

If $\ell = 2$ and $4 \mid q-1$, then $a = 1$;
in all other cases, $a = 2$.

and Waki [262] (who determined the precise value of $a$ for $\ell \neq 2$) and Hiss [156] (who dealt with $\ell = 2$). We have $\mathrm{Irr}_{\mathbb{K}}(G \mid B) = \{\rho^{(3)}, \rho^{(111)}\}$ and $\mathrm{Irr}_k(G \mid B) = \{Y^{(3)}\}$ (see Example 4.3.4). Furthermore, both $Y^{(21)}$ and $Y^{(111)}$ are cuspidal.

Here is a further conjecture which would provide a general source for cuspidal representations in positive characteristic.

**Conjecture 4.5.12 (Geck [90, 2.6.11], [92, 6.6]).** *Assume that $\ell$ is good for $\mathbf{G}$ and $\ell$ does not divide the order of $(Z_{\mathbf{G}}/Z^\circ_{\mathbf{G}})_F$. Let $\rho \in \mathscr{E}'_\ell(G)$ be cuspidal. Then $d_{\mathscr{O}G}([\rho])$ is the class of an irreducible kG-module (which is automatically cuspidal).*

(This applies, in particular, to all cuspidal $\rho \in \mathrm{Unip}_{\mathbb{K}}(G)$.)

By Example 4.5.3, this conjecture holds for $G = \mathrm{Suz}(q^2)$. A more striking example is given by the cuspidal unipotent representation of $G = \mathrm{GU}_n(\mathbb{F}_q)$, where $n = r(r+1)/2$ for some $r \geqslant 0$; see [119, Theorem 6.10]. (But note that $G = \mathrm{GU}_n(\mathbb{F}_q)$ can have many more cuspidal representations in characteristic $\ell$; see, for example, [120, §4].) The conjecture also holds for classical groups, where $\ell = 2$ (see [127]) and other groups of Lie type of low rank where explicit computations are possible.

*Remark 4.5.13.* Let $Y \in \mathrm{Irr}_k(G)$. Following Hiss [155], we say that $Y$ is *supercuspidal* if $Y$ does not occur as a composition factor in $R^G_L(X)$ for any $X \in kL$-mod where $L \in \mathscr{L}_G$ is such that $L \neq G$. Thus, clearly, if $Y$ is supercuspidal, then $Y$ is cuspidal.

Hiss [155, Conj. 3.1] conjectures that a supercuspidal $Y \in \mathrm{Irr}_k(G)$ can be lifted to characteristic 0. He also shows (see [155, Prop. 3.3]) that the truth of this conjecture is a consequence of the truth of Conjecture 4.5.12.

**4.5.14.** Assume that $\ell$ is good for $\mathbf{G}$ and $\ell$ does not divide the order of $(Z_{\mathbf{G}}/Z_{\mathbf{G}}^{\circ})_F$. Fix $L \in \mathscr{L}_G$ and assume that there is some cuspidal $\psi \in \mathrm{Unip}_{\mathbb{K}}(L)$ for which Conjecture 4.5.12 holds; that is, there is an $\mathscr{O}L$-lattice $\tilde{X}$ such that $\tilde{X}_{\mathbb{K}} \cong \psi$ and

(∗) $\tilde{X}_k$ is an irreducible (and, automatically, cuspidal) $kL$-module.

Then, by Theorem 4.2.9, we have bijections

$$\begin{aligned} \mathrm{Irr}_{\mathbb{K}}(G \mid (L,\tilde{X}_{\mathbb{K}})) &\overset{1-1}{\longrightarrow} \mathrm{Irr}(\mathscr{H}(L,\tilde{X}_{\mathbb{K}})), \\ \mathrm{Irr}_k(G \mid (L,\tilde{X}_k) &\overset{1-1}{\longrightarrow} \mathrm{Irr}(\mathscr{H}(L,\tilde{X}_k)), \end{aligned}$$

where $\mathscr{H}(L,\tilde{X}_{\mathbb{K}}) := \mathrm{End}_{\mathbb{K}G}\big(R_{L\subseteq P}^G(\tilde{X}_{\mathbb{K}})\big)^{\circ}$ and $\mathscr{H}(L,\tilde{X}_k) := \mathrm{End}_{kG}\big(R_{L\subseteq P}^G(\tilde{X}_k)\big)^{\circ}$; here, $P \in \mathscr{P}_G$ is such that $P = U_P \rtimes L$. Now let

$$\mathscr{H}(L,\tilde{X}) := \mathrm{End}_{\mathscr{O}G}\big(R_{L\subseteq P}^G(\tilde{X})\big)^{\circ}.$$

Clearly, we have $\mathscr{H}(L,\tilde{X}_{\mathbb{K}}) \cong \mathbb{K} \otimes_{\mathscr{O}} \mathscr{H}(L,\tilde{X})$. Now, by the argument in [102, Lemma 5.22], we have $\mathscr{W}(L,\tilde{X}_{\mathbb{K}}) = \mathscr{W}(L,\tilde{X}_k)$. By [102, Prop. 5.21], this implies that we also have

$$\mathscr{H}(L,\tilde{X}_k) \cong k \otimes_{\mathscr{O}} \mathscr{H}(L,\tilde{X})$$

and that $\mathrm{Dec}(\mathscr{H}(L,\tilde{X}))$ embeds into $\mathrm{Dec}(\mathscr{O}G)$ (as in Dipper's Theorem 4.1.14).

As shown by Lusztig [211, §5] (see also [220, Chap. 8]), the statement of Theorem 4.2.12 can be considerably strengthened in this situation. Namely, we have $\Omega = \{1\}$ and $\mathscr{W}(L,\tilde{X}_{\mathbb{K}}) = W_1$ is itself a Coxeter group; furthermore, there is a weight function $L_1 \colon W_1 \to \mathbb{Z}$, where $L_1(\tilde{s}) > 0$ for all $\tilde{s} \in S_1$, such that

$$\mathscr{H}(L,\tilde{X}_{\mathbb{K}}) \cong \mathbb{K} \otimes_A \mathbf{H}_A(W_1,S_1,L_1) \qquad \text{(via } \theta_{\mathbb{K}} \colon A \to \mathbb{K}\text{, as defined in 4.3.9)}$$

where $\mathbf{H}_A(W_1,S_1,L_1)$ is the associated generic Iwahori–Hecke algebra over $A = R[v,v^{-1}]$. In fact, by the argument in [102, Cor. 5.26], we actually have

$$\mathscr{H}(L,\tilde{X}) \cong \mathscr{O} \otimes_A \mathbf{H}_A(W_1,S_1,L_1) \qquad \text{(via } \theta \colon A \to \mathscr{O}\text{, as defined in 4.3.9).}$$

The various possibilities for $\mathbf{W}, W_1, L_1$ have been determined by Lusztig and are listed explicitly in [213, Table II, p. 35]; see also Carter [45, p. 464].

Now assume that $\mathbf{H}_k$ admits a canonical basic set. (Recall that this is the case when (♠) and (♣) in 2.5.3 are satisfied for $\mathbf{H}_A(W_1,S_1,L_1)$.) Then we can argue as in the proof of Theorem 4.4.1 to obtain the following conclusions.

- Given $Y \in \mathrm{Irr}_k(G \mid (L,\tilde{X}_k))$, let $\mathscr{S}_k(Y) := \{\rho \in \mathrm{Irr}_{\mathbb{K}}(G \mid (L,\tilde{X}_{\mathbb{K}})) \mid \langle \rho : Y \rangle_{\mathscr{O}G} \neq 0\}$. Then the function $\mathscr{S}_k(Y) \to \mathbb{Z}_{\geqslant 0}$, $\rho \mapsto \mathbf{a}_\rho$ (defined using the unipotent support), reaches its minimum at exactly one element of $\mathscr{S}_k(Y)$, which we denote by $\rho_Y$.
- The map $\mathrm{Irr}_k(G \mid (L,\tilde{X}_k)) \to \mathrm{Irr}_{\mathbb{K}}(G \mid (L,\tilde{X}_{\mathbb{K}}))$, $Y \mapsto \rho_Y$, is injective.
- We have $\langle \rho_Y : Y \rangle_{\mathscr{O}G} = 1$ for all $Y \in \mathrm{Irr}_k(G \mid (L,\tilde{X}_k))$.

Thus, we obtain a natural injection $\mathrm{Irr}_k(G \mid (L,\tilde{X}_k)) \hookrightarrow \mathrm{Irr}_{\mathbb{K}}(G \mid (L,\tilde{X}_{\mathbb{K}}))$. Since we also have bijections

$$\mathrm{Irr}_{\mathbb{K}}(G \mid (L,\tilde{X}_{\mathbb{K}})) \quad \stackrel{1-1}{\longleftrightarrow} \quad \mathrm{Irr}(\mathscr{H}(L,\tilde{X}_{\mathbb{K}})) \quad \stackrel{1-1}{\longleftrightarrow} \quad \mathrm{Irr}_{\mathbb{K}}(W_1),$$

we obtain a natural injection $\mathrm{Irr}_k(G \mid (L,\tilde{X}_k)) \hookrightarrow \mathrm{Irr}_{\mathbb{K}}(W_1)$. By an argument similar to that in Theorem 4.4.8, the image of this injection will only depend on $e$, if $\ell$ is sufficiently large. We omit further details.

**Example 4.5.15.** Let $G = \mathrm{GU}_N(\mathbb{F}_q)$ and assume that $N = 2n + \frac{1}{2}c(c+1)$, where $c \geqslant 0$ and $n \geqslant 1$. Then we have a Levi subgroup $L \in \mathscr{L}_G$ such that

$$L \cong (\mathbb{F}_{q^2}^{\times})^n \times \mathrm{GU}_{\frac{1}{2}c(c+1)}(\mathbb{F}_q).$$

By Lusztig [212, §9], $\mathrm{Unip}_{\mathbb{K}}(L)$ contains a unique cuspidal representation, denoted by $\psi$. The inertia group $\mathscr{W}(L,\psi) = W_1$ is of type $B_n$, with weight function given by

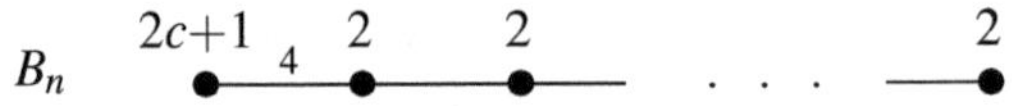


(see Lusztig [213, Table II, p. 35]); the situation in Example 4.4.16 corresponds to the special case where $c \in \{0,1\}$. It is known that Conjecture 4.5.12 holds for $\psi$; see [119, Theorem 6.10]. Hence, we are in the setting of 4.5.14 but note that it is not (yet) known if (♠) and (♣) hold for all values of $r$. However, by Example 4.5.10 we know that the properties (a), (b) and (c) in 4.5.5 hold. Hence, by Remark 4.5.7, we can conclude that a canonical basic set for $\mathbf{H}_k$ exists.

We can now distinguish cases as in Example 4.4.16. In order to obtain a canonical basic set in the "non-elementary" case where $e = 2e'$ and $e' \geqslant 3$ is odd, we argue as follow. We have seen in Remark 4.3.11 that $\theta_k$ lies above the $\Phi_e$-modular specialisation $\theta_e \colon A \to \mathbb{K}$. Let $\mathbf{H}_{(e)}$ be the corresponding specialised algebra. In Theorem 6.7.11 (Chapter 6) we shall establish the existence of a canonical basic set $\mathscr{B}_{(e)} \subseteq \Lambda$ for $\mathbf{H}_{(e)}$ and obtain an explicit combinatorial description of $\mathscr{B}_{(e)}$ (using the theory of canonical bases for quantised enveloping algebras). Furthermore, Theorem 4.6.13 will show that we have $|\mathrm{Irr}(\mathbf{H}_k)| = |\mathrm{Irr}(\mathbf{H}_{(e)})|$. So Lemma 3.6.8 implies that the subset $\mathscr{B}_{(e)} \subseteq \Lambda$ also is a canonical basic set for $\mathbf{H}_k$. Thus, we obtain an explicit description of the injection $\mathrm{Irr}_k(G \mid (L,\tilde{X}_k)) \hookrightarrow \mathrm{Irr}_{\mathbb{K}}(W_1)$.

**Example 4.5.16.** Let $G$ be a finite symplectic or orthogonal group over $\mathbb{F}_q$ where $\mathbf{W}$ is of type $B_N$ or $D_N$. By Lusztig [212, 8.11], for suitable values of $N$ and $n$ there exist an $L \in \mathscr{L}_G$ and a cuspidal $\psi \in \mathrm{Unip}_{\mathbb{K}}(L)$ such that the corresponding inertia group $\mathscr{W}(L,\psi) = W_1$ is of type $B_n$, with weight function given by

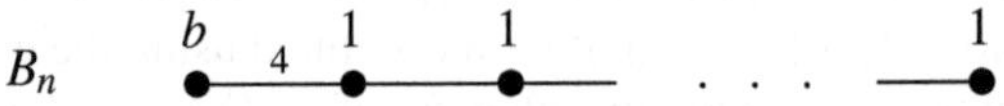


where $b$ is determined by $N$ and $n$. For example, if $\mathbf{W}$ is of type $B_N$, where $N = n + c^2 + c$ for some $c \geqslant 0$, then there is a pair $(L,\psi)$ as above such that $b = 2c+1$; see Lusztig [213, Table II, p. 35]. Similarly, if $\mathbf{W}$ is of type $D_N$, where $N = n + c^2$ for some $c \geqslant 1$, then there is a pair $(L,\psi)$ as above such that $b = 2c$.

However, in these cases, it is not known if Conjecture 4.5.12 holds for $\psi$; furthermore, it is not known if (♠) and (♣) hold for $W_1, L_1$. By Theorem 6.7.9, we know at least that a canonical basic set exists for the corresponding algebra $\mathbf{H}_{(e)}$ (see Remark 4.3.11). By Theorem 4.6.13, we also know that $|\operatorname{Irr}(\mathbf{H}_k)| = |\operatorname{Irr}(\mathbf{H}_{(e)})|$.

**Example 4.5.17.** Let $G = E_8(\mathbb{F}_q)$ and $L \in \mathscr{L}_G$ be of type $D_4$. By Lusztig [220, p. 359], $\operatorname{Unip}_{\mathbb{K}}(L)$ contains a unique cuspidal representation, denoted by $\psi$. The inertia group $\mathscr{W}(L,\psi) = W_1$ is of type $F_4$, with weight function given by

$F_4$ 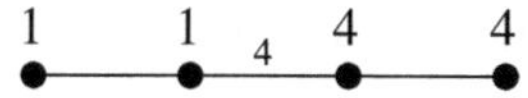


(see Lusztig [213, Table II, p. 35]). It is known that Conjecture 4.5.12 holds for $\psi$; see [130, Prop. 5.1] (for $q$ odd) and [96, Beisp. 4.5.4] (for $q$ a power of 2). Hence, we are in the setting of 4.5.14; the properties (♠) and (♣) hold by 2.4.1(c). Consequently, we obtain a natural injection $\operatorname{Irr}_k(G \mid (L, \tilde{X}_k)) \hookrightarrow \operatorname{Irr}_{\mathbb{K}}(W_1)$. The image of this injection can be extracted from the matrices in Table 7.9 (p. 377). Note that, by Bremke [30], these matrices coincide with the decomposition matrices of $\mathbf{H}_k$.

If $L \in \mathscr{L}_G$ has type $E_6$ or $E_7$, then $\operatorname{Unip}_{\mathbb{K}}(L)$ also contains cuspidal representations, where the inertia groups are of type $G_2$ (with weights $1, 9$) and $A_1$ (with weight 15) respectively. It is known that Conjecture 4.5.12 holds for the cuspidal unipotent representations in type $E_6$, at least if $q$ is not a power of 2 or 3; see [117, Theorem 7.4]. The corresponding decomposition matrices and canonical basic sets are contained in Table 7.5 (p. 373). The analogous question for type $E_7$ is open.

Our final example shows that some things can go wrong when $\ell$ is a bad prime.

**Table 4.6** Decomposition matrix of the union of blocks $\mathscr{E}_{2,1}$ of $\operatorname{Ree}(q^2)$

| $\rho$ | $\dim\rho$ | $\mathbf{a}_\rho$ | decomposition numbers | | | | | | |
|---|---|---|---|---|---|---|---|---|---|
| $1_G$ | $1$ | 0 | 1 | . | . | . | . | . | . |
| $\xi_6$ | $\frac{1}{2\sqrt{3}}q(q-1)(q^2-q\sqrt{3}+1)$ | 1 | . | 1 | . | . | . | . | . |
| $\xi_5$ | $\frac{1}{2\sqrt{3}}q(q-1)(q^2+q\sqrt{3}+1)$ | 1 | . | 1 | 1 | . | . | . | . |
| $\xi_7$ | $\frac{1}{2\sqrt{3}}q(q-1)(q^2+q\sqrt{3}+1)$ | 1 | . | . | 1 | 1 | . | . | . |
| $\xi_8$ | $\frac{1}{2\sqrt{3}}q(q-1)(q^2-q\sqrt{3}+1)$ | 1 | . | . | . | 1 | . | . | . |
| $\xi_9$ | $\frac{1}{\sqrt{3}}q(q^2-1)(q^2+1)$ | 1 | . | . | . | . | 1 | . | . |
| $\xi_{10}$ | $\frac{1}{\sqrt{3}}q(q^2-1)(q^2+1)$ | 1 | . | . | . | . | . | 1 | . |
| $\mathrm{St}_G$ | $q^6$ | 6 | 1 | 1 | 2 | 1 | . | . | 1 |
| $\rho_{\theta,1}$ | $q^4-q^2+1$ | 0 | 1 | . | 1 | . | . | . | . |
| $\rho_{\theta,2}$ | $q^2(q^4-q^2+1)$ | 2 | 1 | 1 | 1 | 1 | . | . | 1 |

**Example 4.5.18.** Let $G = \operatorname{Ree}(q^2)$ (type ${}^2G_2$), where $q = \sqrt{3}^{2f+1}$ for some $f \geqslant 0$; see Example 4.3.4. We have $|G| = q^6(q^2-1)(q^6+1)$. Let $\ell \neq 3$ be a prime which divides the order of $G$. If $\ell \neq 2$, then the Sylow $\ell$-subgroups of $G$ are cyclic and

the corresponding decomposition numbers are determined by Hiss [153, §D.2]. One checks that Conjectures 4.5.2 and 4.5.12 hold in this case. Now let $\ell = 2$ (which is a "bad" prime for **G**); in this case, the Sylow 2-subgroups are elementary abelian of order 8. The union of blocks $\mathscr{E}_{2,1}$ contains eight unipotent representations and two non-unipotent principal series representations which are constituents of $R_{\mathbf{T},\theta}$, where $\mathbf{T} \subseteq \mathbf{G}$ is an $F$-stable maximal torus contained in an $F$-stable Borel subgroup and $\theta \in \mathrm{Irr}(\mathbf{T}^F)$ is unique of order 2; the two constituents are denoted by $\rho_{\theta,1}$ and $\rho_{\theta,2}$. Following previous work of Fong, the decomposition matrix of $\mathscr{E}_{2,1}$ was completely determined by Landrock and Michler [207]; see Table 4.6. Note that the six unipotent representations $\xi_5, \xi_6, \ldots, \xi_{10}$ are all cuspidal! We notice that:

- Theorem 4.5.1 and, consequently, Conjecture 4.5.2 fail: the classes $\{d_{\mathscr{O}G}([\rho]) \mid \rho \in \mathrm{Unip}_{\mathbb{K}}(G)\}$ are not linearly independent. (A similar thing also happens, for example, for groups of type $D_4$ when $\ell = 2$; see [130, Prop. 5.1]. So the failure is not related to the fact that $\mathrm{Ree}(q^2)$ is a "very twisted" group.)
- The statement of Conjecture 4.5.12 fails for the cuspidal representations $\xi_5$ and $\xi_7$. (No other example for a failure of that statement is known.)

So, it seems that the assumption that $\ell$ should be a good prime is really necessary.

We also see from the decomposition matrix that the modular irreducible representation labelling the last column cannot be lifted to characteristic 0. Since this representation is not in the principal series, it must be cuspidal.

## 4.6 A First Approach to Type $B_n$

In the previous sections we have seen that Iwahori–Hecke algebras of type $B_n$ with various choices of unequal parameters occur in the theory of modular Harish-Chandra series of finite groups of Lie type. The study of the representation theory of these algebras has turned out to be an extremely difficult problem. However, certain cases can be dealt with by elementary methods, and these will be discussed in this section. This will also serve as a motivation for the following chapters.

Let $W$ be of type $B_n$ with weight function $L \colon W \to \Gamma$ given by

$B_n$  where $a, b \in \Gamma_{\geqslant 0}$.



We label the generators of $W$ by $s_0, s_1, \ldots, s_{n-1}$ as in Table 1.1 (p. 2). Recall that

$$\mathrm{Irr}_{\mathbb{K}}(W) = \{E^{(\lambda,\mu)} \mid (\lambda,\mu) \in \Lambda\},$$

where $\Lambda$ is the set of all pairs of partitions $(\lambda,\mu)$ such that $|\lambda| + |\mu| = n$. Let **H** be the generic Iwahori–Hecke algebra of $W$ over $A = R[\Gamma]$, where $R \subseteq \mathbb{C}$ is $L$-good. We have the quadratic relations

$$T_{s_0}^2 = T_1 + (\varepsilon^b - \varepsilon^{-b})T_{s_0} \quad \text{and} \quad T_{s_i}^2 = T_1 + (\varepsilon^a - \varepsilon^{-a})T_{s_i} \quad (i = 1, \ldots, n-1).$$

Let $\theta\colon A \to k$ be a specialisation into a field $k$ such that $k$ is the field of fractions of $\theta(A)$. The basic references for the study of $\mathbf{H}_k$ are Dipper and James [66] and Dipper, James and Murphy [68], [69]. We have nothing original to add to the excellent exposition in these articles. So we will not reproduce them here but merely state the results that we need and then explain in detail the implications on canonical basic sets. Note that these implications require some work since Dipper, James and Murphy did not take into account any monomial ordering[4] on $\Gamma$.

Already the article [66] shows that, independently of any monomial ordering on $\Gamma$, the representation theory of $\mathbf{H}_k$ essentially depends on two parameters. Let $U := \varepsilon^{2b}$ and $u := \varepsilon^{2a}$. Then, as in type $A$, the first parameter is given by

$$\tilde{e} := \min\{i \geqslant 2 \mid 1+\theta(u)+\theta(u)^2+\cdots+\theta(u)^{i-1} = 0\}.$$

(We let $\tilde{e} = \infty$ if $1+\theta(u)+\theta(u)^2+\cdots+\theta(u)^{i-1} \neq 0$ for all $i \geqslant 2$; we use the symbol $\tilde{e}$ because $e$ is already used as in Remark 4.3.11.) Furthermore, the difficulty in studying representations of $\mathbf{H}_k$ will depend very much on whether

$$f_n(\theta) := \prod_{-(n-1)\leqslant i\leqslant n-1} \theta(U+u^i)$$

equals zero or not. We shall see that the results of [66] actually yield canonically basic sets for any monomial ordering on $\Gamma$ provided that $\theta(u) = 1$ or $f_n(\theta) \neq 0$. (The cases where $\theta(u) \neq 1$ and $f_n(\theta) = 0$ require completely new methods; see Chapters 5 and 6.) In order to explain this, we first need to prepare some notation.

**4.6.1.** The elements $s_1, s_2, \ldots, s_{n-1}$ generate a parabolic subgroup of $W$ which we can (and will) identify with the symmetric group $\mathfrak{S}_n$. Let $r \in \{0,1,\ldots,n\}$ and consider the parabolic subgroup

(a) $$^rW = \langle s_1,\ldots,s_{r-1},s_{r+1},\ldots,s_{n-1}\rangle \cong \mathfrak{S}_r \times \mathfrak{S}_{n-r};$$

here, at the extreme cases, we have $^0W = {}^nW = \langle s_1,s_2,\ldots,s_{n-1}\rangle = \mathfrak{S}_n$. Let $^r\Lambda$ be the set of all pairs of partitions $(\lambda,\mu) \in \Lambda$ such that $|\lambda| = r$ (and, hence, $|\mu| = n-r$). Given $(\lambda,\mu) \in {}^r\Lambda$, we have corresponding irreducible representations $E^\lambda \in \mathrm{Irr}_{\mathbb{K}}(\mathfrak{S}_r)$ and $E^\mu \in \mathrm{Irr}_{\mathbb{K}}(\mathfrak{S}_{n-r})$. Taking the outer tensor product we obtain an irreducible representation of $\mathfrak{S}_r \times \mathfrak{S}_{n-r}$ which, via (a), defines an irreducible representation of $^rW$. Thus, we can write

(b) $$\mathrm{Irr}_{\mathbb{K}}({}^rW) = \{E^\lambda \boxtimes E^\mu \mid (\lambda,\mu) \in {}^r\Lambda\}.$$

Let $^r\mathbf{H} \subseteq \mathbf{H}$ be the parabolic subalgebra corresponding to $^rW$. Then $^r\mathbf{H}$ can be identified with the generic Iwahori–Hecke algebra associated with $\mathfrak{S}_r \times \mathfrak{S}_{n-r}$ and the weight function $^rL$ which assigns $a$ to each generator of $^rW$. Let $^rD_\theta$ be the decomposition matrix of $^r\mathbf{H}$ with respect to the specialisation $\theta\colon A \to k$. Since $^r\mathbf{H}$ is the tensor product of two algebras, a completely general argument shows that $D_\theta$

[4] In retrospective, one might say that their results (especially those in [68]) are most directly related to the asymptotic case where $b > (n-1)a > 0$; see also the remarks in 2.8.19.

is the Kronecker product of the decomposition matrices of the two algebras which are tensored together. First of all, by Theorem 3.5.14, this implies that $D_\theta$ has rows labelled by ${}^r\Lambda$ and columns labelled by the set ${}^r\Lambda'_\theta$ consisting of all $(\lambda,\mu) \in {}^r\Lambda$ such that both $\lambda$ and $\mu$ are $\tilde{e}$-regular. Let us write

$$ {}^rD_\theta = \big({}^rd_{(\lambda,\mu),(\lambda',\mu')}\big)_{(\lambda,\mu)\in {}^r\Lambda,\,(\lambda',\mu')\in {}^r\Lambda'_\theta}. $$

Then, furthermore, the relations in Example 3.2.9 imply that, given $(\lambda,\mu) \in {}^r\Lambda$ and $(\lambda',\mu') \in {}^r\Lambda'_\theta$, we have

$$ \text{(c)} \qquad \begin{cases} {}^rd_{(\lambda',\mu'),(\lambda',\mu')} = 1, & \\ {}^rd_{(\lambda,\mu),(\lambda',\mu')} = 0 & \text{unless } \lambda \trianglelefteq \lambda' \text{ and } \mu \trianglelefteq \mu'. \end{cases} $$

With this notation, we can now state the following result.

**Theorem 4.6.2 (Dipper and James [66, 5.8]).** *Assume that $f_n(\theta) \neq 0$ and let $D_\theta$ be the decomposition matrix of* **H** *with respect to $\theta\colon A \to k$. Then $D_\theta$ is a block diagonal matrix with diagonal blocks given by ${}^rD_\theta$ for $0 \leqslant r \leqslant n$.*

In particular, this shows that the irreducible representations of $\mathbf{H}_K$ which are labelled by all pairs of partitions in a set ${}^r\Lambda$ (where $r$ is fixed) form a union of $\theta$-blocks. Dipper and James [66, 4.7] actually show that $\mathbf{H}_k$ is Morita equivalent to the direct sum of algebras $\bigoplus_{0\leqslant r\leqslant n} {}^r\mathbf{H}_k$. Using this theorem, we can now establish the following result, without reference to the properties (♠) and (♣) in 2.5.3.

**Proposition 4.6.3 (Cf. [101, 6.8], [125, 3.1]).** *Assume that $f_n(\theta) \neq 0$. Then, independently of the monomial ordering on $\Gamma$, we have that*

$$ \mathscr{B}_\theta := \{(\lambda,\mu) \in \Lambda \mid \textit{both } \lambda \textit{ and } \mu \textit{ are } \tilde{e}\textit{-regular}\} $$

*is a canonical basic set for* $\mathbf{H}_k$.

*Proof.* Let us first deal with the case where $a=0$. Then $\theta(u)=1$ and so either $\tilde{e}=\infty$ or $\tilde{e}=\mathrm{char}(k)=\ell>0$. Since $R$ is $L$-good, we must have $\ell>n$ in the second case; see Table 1.4 (p. 33). Hence, in both cases, we have $\tilde{e}>n$. Since also $f_n(\theta)\neq 0$, it follows that $\mathbf{H}_k$ is semisimple; see Example 3.1.20. Hence, by 3.1.18, $D_\theta$ is the identity matrix and so $\mathscr{B}_\theta=\Lambda$ is a canonical basic set.

Let us now assume that $a>0$. By Theorem 4.6.2 and the discussion in 4.6.1, the matrix $D_\theta$ has columns labelled by $\mathscr{B}_\theta$ and, writing

$$ D_\theta = \big(d_{(\lambda,\mu),(\lambda',\mu')}\big)_{(\lambda,\mu)\in\Lambda,\,(\lambda',\mu')\in\mathscr{B}_\theta}, $$

we have for any $(\lambda,\mu)\in\Lambda$ and $(\lambda',\mu')\in\mathscr{B}_\theta$

$$ \begin{cases} d_{(\lambda',\mu'),(\lambda',\mu')} = 1, & \\ d_{(\lambda,\mu),(\lambda',\mu')} = 0 & \text{unless } |\lambda|=|\lambda'|,\ |\mu|=|\mu'|,\ \lambda\trianglelefteq\lambda',\ \mu\trianglelefteq\mu'. \end{cases} $$

Hence, in order to prove that $\mathscr{B}_\theta$ is a canonical basic set, it is enough to show that the following implication holds, for any $(\lambda,\mu) \in \Lambda$ and $(\lambda',\mu') \in \mathscr{B}_\theta$:

$$(\dagger) \quad \left\{ \begin{array}{ll} |\lambda| = |\lambda'|, & |\mu| = |\mu'|, \\ \lambda \trianglelefteq \lambda', & \mu \trianglelefteq \mu' \end{array} \right\} \quad \Rightarrow \quad (\lambda,\mu) = (\lambda',\mu') \quad \text{or} \quad \mathbf{a}_{(\lambda,\mu)} > \mathbf{a}_{(\lambda',\mu')}.$$

Thus, we are reduced to a purely combinatorial statement. We claim that, in fact, $(\dagger)$ holds for all pairs of partitions $(\lambda,\mu),(\lambda',\mu') \in \Lambda$. To prove this, we use the formulae for the **a**-invariants in Example 1.3.9.

Assume first that $b \geqslant ra$ for all $r \in \mathbb{Z}_{\geqslant 1}$. Then we have

$$\mathbf{a}_{(\lambda,\mu)} = \big(n(\lambda) + 2n(\mu) - n(\mu^*)\big)a + |\mu|b.$$

Now, if $\lambda \trianglelefteq \lambda'$ and $\mu \trianglelefteq \mu'$, then $n(\lambda) \geqslant n(\lambda')$ (with equality only if $\lambda = \lambda'$) and $n(\mu) \geqslant n(\mu')$ (with equality only if $\mu = \mu'$); see Example 2.2.13. Since we also have $\mu \trianglelefteq \mu'$ if and only if $\mu'^* \trianglelefteq \mu^*$ (see [236, (I.1.11)]), the above formula immediately shows that $\mathbf{a}_{(\lambda,\mu)} \geqslant \mathbf{a}_{(\lambda',\mu')}$, with equality only if $(\lambda,\mu) = (\lambda',\mu')$.

Now assume that there exists some $r \in \mathbb{Z}_{\geqslant 0}$ such that $ra \leqslant b < (r+1)a$, and set $b' = b - ra$. As in Case 3 of Example 1.3.9, we fix a large integer $N \geqslant 0$ and associate with $(\lambda,\mu)$ and $(\lambda',\mu')$ the multisets $Z_{a,b}^N(\lambda,\mu)$ and $Z_{a,b}^N(\lambda',\mu')$. Writing

$$\begin{aligned} Z_{a,b}^N(\lambda,\mu) &= \{z_1, z_2, \ldots, z_{2N+r}\}, \qquad \text{where} \qquad z_1 \geqslant z_2 \geqslant \ldots \geqslant z_{2N+r}, \\ Z_{a,b}^N(\lambda',\mu') &= \{z'_1, z'_2, \ldots, z'_{2N+r}\}, \qquad \text{where} \qquad z'_1 \geqslant z'_2 \geqslant \ldots \geqslant z'_{2N+r}, \end{aligned}$$

we have

$$\mathbf{a}_{(\lambda,\mu)} - \mathbf{a}_{(\lambda',\mu')} = \sum_{1 \leqslant i \leqslant 2N+r} (i-1)(z_i - z'_i).$$

Now note that, in order to prove $(\dagger)$, it is sufficient to do this in the case where $\lambda = \lambda'$ or $\mu = \mu'$. If $\lambda \triangleleft \lambda'$ and $\mu = \mu'$, then one can further reduce to the case where $\lambda'$ is obtained from $\lambda$ by increasing one part by 1 and by decreasing another part by 1; see [236, (I.1.16)]. A similar statement holds if $\lambda = \lambda'$ and $\mu \triangleleft \mu'$. Thus, it will be sufficient to consider the case where the multiset $Z_{a,b}^N(\lambda',\mu')$ is given by

$$Z_{a,b}^N(\lambda',\mu') = \{z_1, \ldots, z_{i-1}, z_i + a, z_{i+1}, \ldots, z_{j-1}, z_j - a, z_{j+1}, \ldots, z_{2N+r}\},$$

where $1 \leqslant i < j \leqslant 2N+r$; note, however, that this is just an equality as multisets. In order to be able to compare the **a**-invariants, we need to arrange the entries (written as above) in decreasing order. So we distinguish the following four cases:

(I) $z_{i-1} \geqslant z_i + a$ (if $i > 1$) and $z_j - a \geqslant z_{j+1}$ (if $j < 2N+r$).
(II) $i > 1$ and $z_{i-1} < z_i + a$ and $z_j - a \geqslant z_{j+1}$ (if $j < 2N+r$).
(III) $z_{i-1} \geqslant z_i + a$ (if $i > 1$) $j < 2N+r$ and $z_j - a < z_{j+1}$.
(IV) $i > 1$ and $j < 2N+r$ and $z_{i-1} < z_i + a$ and $z_j - a < z_{j+1}$.

In each of these cases, we must show that $\mathbf{a}_{(\lambda,\mu)} > \mathbf{a}_{(\lambda',\mu')}$. Now, if (I) holds, then the entries of $Z_{a,b}^N(\lambda',\mu')$ (written as above in terms of the entries of $Z_{a,b}^N(\lambda,\mu)$) already are arranged in decreasing order; that is, we have $z_i' = z_i + a$, $z_j' = z_j - a$ and $z_k' = z_k$ for $k \neq i, j$. This yields that

$$\mathbf{a}_{(\lambda,\mu)} - \mathbf{a}_{(\lambda',\mu')} = (i-1)(z_i - z_i') + (j-1)(z_j - z_j') = (j-i)a > 0,$$

as required. If (II) holds, then all entries are arranged in decreasing order, except for the two entries $z_{i-1}, z_i + a$. Now recall from Example 1.3.9 that the entries $\{z_k\}$ of $Z_{a,b}^N(\lambda,\mu)$ have the form of $z_k = n_k a + \delta_k b'$, where $n_k \in \mathbb{Z}_{\geqslant 1}$ and $\delta_k \in \{0,1\}$. Furthermore, if $b' > 0$, then all entries are distinct; if $b' = 0$, then it may happen that two entries are equal but no entry is repeated more than twice. These conditions imply that $z_{i-2} \geqslant z_i + a$ (if $i > 2$). Hence, all we need to do is to swap the entries $z_{i-1}$ and $z_i + a$; that is, we have $z_{i-1}' = z_i + a$, $z_i' = z_{i-1}$, $z_j' = z_j - a$ and $z_k' = z_k$ for all $k \neq i-1, i, j$. This yields that

$$\begin{aligned}\mathbf{a}_{(\lambda,\mu)} - \mathbf{a}_{(\lambda',\mu')} &= (i-2)(z_{i-1} - z_{i-1}') + (i-1)(z_i - z_i') + (j-1)(z_j - z_j')\\ &= (i-2)(z_{i-1} - z_i - a) + (i-1)(z_i - z_{i-1}) + (j-1)a\\ &= (z_i + a) - z_{i-1} + (j-i)a > 0,\end{aligned}$$

as required. Next, assume that (III) holds. Arguing as above, we see that all we need to do is to swap the entries $z_j - a$ and $z_{j+1}$; that is, we have $z_i' = z_i + a$, $z_j' = z_{j+1}$, $z_{j+1}' = z_j - a$ and $z_k' = z_k$ for $k \neq i, j, j+1$. This yields that

$$\begin{aligned}\mathbf{a}_{(\lambda,\mu)} - \mathbf{a}_{(\lambda',\mu')} &= (i-1)(z_i - z_i') + (j-1)(z_j - z_j') + j(z_{j+1} - z_{j+1}')\\ &= -(i-1)a + (j-1)(z_j - z_{j+1}) + j(z_{j+1} - z_j + a)\\ &= z_{j+1} - (z_j - a) + (j-i)a > 0,\end{aligned}$$

as required. Finally, assume that (IV) holds. In this case, we need to swap the entries $z_{i-1}, z_i + a$ and the entries $z_j - a, z_{j+1}$; that is, we have $z_{i-1}' = z_i + a$, $z_i' = z_{i-1}$, $z_j' = z_{j+1}$, $z_{j+1}' = z_j - a$ and $z_k' = z_k$ for $k \neq i-1, i, j, j+1$. This yields that

$$\begin{aligned}\mathbf{a}_{(\lambda,\mu)} - \mathbf{a}_{(\lambda',\mu')} &= (i-2)(z_{i-1} - z_i - a) + (i-1)(z_i - z_{i-1})\\ &\qquad + (j-1)(z_j - z_{j+1}) + j(z_{j+1} - z_j + a)\\ &= (z_i + a) - z_{i-1} + z_{j+1} - (z_j - a) + (j-i)a > 0,\end{aligned}$$

as required. Thus, (†) is proved and so $\mathscr{B}_\theta$ is a canonical basic set, as claimed. □

**Example 4.6.4.** Assume that the following three conditions hold:

(a) $\mathrm{char}(k) \neq 2$,
(b) $\tilde{e}$ is odd,
(c) $\theta(U) = \theta(u)^j$ for some $j \in \mathbb{Z}$.

We claim that $f_n(\theta) \neq 0$. Indeed, if $f_n(\theta) = 0$, then $\theta(U) = -\theta(u)^i$ for some $i$ and so $\theta(u)^{j-i} = -1$, using (c). By (a), this would imply that $\theta(u) \neq 1$ is a root of unity of even order and so $\tilde{e}$ would have to be even, contradicting (b). Thus, $f_n(\theta) \neq 0$ as claimed. Now Proposition 4.6.3 applies and so there exists a canonical basic set for $\mathbf{H}_k$, given by the set of all $(\lambda,\mu)\in\Lambda$ such that both $\lambda$ and $\mu$ are $\tilde{e}$-regular.

**4.6.5.** Assume that $\theta(u)=1$ and $f_n(\theta)=0$. Then we must have $\theta(U)=-1$ and so, in $\mathbf{H}_k$, we have the quadratic relations

$$(T_{s_0}+T_1)^2=0 \quad\text{and}\quad T_{s_i}^2=T_1 \quad (i=1,\ldots,n-1).$$

Let $\Omega\subseteq W$ be the parabolic subgroup generated by $s_1,\ldots,s_{n-1}$; we have $\Omega\cong\mathfrak{S}_n$. The corresponding parabolic subalgebra of $\mathbf{H}_k$ is nothing but the group algebra of $\Omega$ and will be denoted by $k\Omega\subseteq\mathbf{H}_k$.

Now consider an auxiliary weight function $L'\colon W\to\Gamma$ such that $L'(s_0)=b$ and $L'(s_i)=0$ for $1\leqslant i\leqslant n-1$. Let $\mathbf{H}'$ be the corresponding generic Iwahori–Hecke algebra over $A$. Thus, in $\mathbf{H}'$, we have the quadratic relations $T_{s_0}^2=T_1+(\varepsilon^b-\varepsilon^{-b})T_{s_0}$ and $T_{s_i}^2=T_1$ for $1\leqslant i\leqslant n-1$, as in Example 2.4.4(b). We define $t_0:=s_0$ and $t_i=s_it_{i-1}s_i$ for $1\leqslant i\leqslant n-1$. Then the submodule

$$\mathbf{H}'_1=\langle T_{t_0},T_{t_1},\ldots,T_{t_{n-1}}\rangle_A\subseteq\mathbf{H}'$$

is the generic Iwahori–Hecke algebra of a Coxeter group of type $A_1\times\cdots\times A_1$ ($n$ factors). Furthermore, as in Remark 2.4.6, the submodules

$$\{\mathbf{H}'_\omega:=\mathbf{H}'_1T_\omega\mid\omega\in\Omega\}\quad\text{form an }\Omega\text{-graded Clifford system in }\mathbf{H}'.$$

Now note that we can canonically identify the specialised algebras $\mathbf{H}'_k$ and $\mathbf{H}_k$; furthermore, the Clifford system in $\mathbf{H}'$ certainly gives rise to an $\Omega$-graded Clifford system in $\mathbf{H}'_k=\mathbf{H}_k$ with respect to the subalgebra $\mathbf{H}'_{k,1}=\langle T_{t_0},\ldots,T_{t_{n-1}}\rangle_k\subseteq\mathbf{H}_k$. Hence, as already noted in 3.2.13(b), the restriction of any irreducible representation of $\mathbf{H}_k$ to $\mathbf{H}'_{k,1}$ is a direct sum of irreducible representations. But $\mathbf{H}'_{k,1}$ is commutative, hence every irreducible representation of $\mathbf{H}'_{k,1}$ is one-dimensional. Since $(T_{s_0}+T_1)^2=0$, it follows that $T_{s_0}=T_{t_0}\in\mathbf{H}'_{k,1}$ acts as minus the identity in every irreducible representation of $\mathbf{H}_k$. We now obtain the following proposition.

**Proposition 4.6.6 (Dipper and James [66, 5.4]).** *In the setting of 4.6.5, every irreducible representation of $k\Omega\cong k\mathfrak{S}_n$ has a unique extension to an irreducible representation of $\mathbf{H}_k$; furthermore, all irreducible representations of $\mathbf{H}_k$ arise in this way. Hence, by Theorem 3.5.14,* $\mathrm{Irr}(\mathbf{H}_k)$ *is naturally parametrised by the $\tilde{e}$-regular partitions of $n$, where $\tilde{e}=\infty$ (if* $\mathrm{char}(k)=0$*) or $\tilde{e}=\ell$ (if* $\mathrm{char}(k)=\ell>0$*).*

*Proof.* By 4.6.5, $T_{s_0}$ acts as minus the identity in every irreducible representation of $\mathbf{H}_k$. Since $\mathbf{H}_k$ is generated by $T_{s_0}$ together with $T_{s_1},\ldots,T_{s_{n-1}}$, it is clear that the restriction of an irreducible representation of $\mathbf{H}_k$ to $k\Omega$ remains irreducible.

Conversely, if $\rho\colon k\Omega\to M_d(k)$ is an irreducible representation, then we obtain a representation $\hat{\rho}\colon\mathbf{H}_k\to M_d(k)$ by setting $\hat{\rho}(T_{s_0}):=-I_d$ (where $I_d$ denotes the identity matrix of size $d$) and $\hat{\rho}(T_{s_i}):=\rho(T_{s_i})$ for $1\leqslant i\leqslant n-1$. (One immediately checks that the defining relations for $\mathbf{H}_k$ are satisfied.) Clearly, $\hat{\rho}$ is irreducible.

Furthermore, this is the only possible way to extend $\rho$ from $k\Omega$ to $\mathbf{H}_k$, since $T_{s_0}$ has to act as minus the identity. $\square$

**Example 4.6.7.** In the setting of 4.6.5, assume that $a = 0$; that is, $L$ equals the auxiliary weight function $L'$ and so $\mathbf{H}' = \mathbf{H}$. As already noted, $\mathbf{H}'_1$ is the generic Iwahori–Hecke algebra of type $A_1 \times \cdots \times A_1$ ($n$ factors). Let $\mathrm{ind}_1 : \mathbf{H}'_1 \to A$ be the one-dimensional representation such that $\mathrm{ind}_1(T_{t_i}) = \varepsilon^b$ for $0 \leqslant i \leqslant n-1$. Specialising $\varepsilon^b$ to $-1 \in k$, we obtain the representation $\mathrm{ind}_{k,1} : \mathbf{H}'_{k,1} \to k$. By 4.6.5, we have $\mathrm{Irr}(\mathbf{H}'_{k,1}) = \{\mathrm{ind}_{k,1}\}$. The $\mathbf{a}$-invariant of $\mathrm{ind}_1$ is 0; hence, $\{\mathrm{ind}_1\}$ is the canonical basic set for $\mathbf{H}'_{k,1}$. Now the standard construction of $\mathrm{Irr}_{\mathbb{K}}(W)$ in [132, 5.5.4] (using Clifford theory) shows that, for any $(\lambda,\mu) \in \Lambda$, the restriction of $E_\varepsilon^{(\lambda,\mu)}$ to $\mathbf{H}'_{K,1}$ contains $\mathrm{ind}_{K,1}$ if and only if $\mu = \varnothing$. Hence, Theorem 3.2.14 implies that

$$\mathscr{B}_\theta = \{(\lambda,\varnothing) \mid \lambda \vdash n\} \subseteq \Lambda$$

is a canonical basic set for $\mathbf{H}_k$. Note that (♠) and (♣) hold for $\mathbf{H}$ and $\mathbf{H}'_1$ by 2.4.8 and 2.5.3.

**4.6.8.** In the setting of 4.6.5, assume now that $a > 0$. In order to describe a canonical basic set in this case as well, we need some further preparations. We consider the induction of representations from $\Omega \cong \mathfrak{S}_n$ to $W$. Recall that $\mathrm{Irr}_{\mathbb{K}}(\Omega) = \{E^\nu \mid \nu \vdash n\}$. By general results on Lusztig's $\mathbf{a}$-invariants, an induced representation $\mathrm{Ind}_\Omega^W(E^\nu)$ is a direct sum of various $E^{(\lambda,\mu)} \in \mathrm{Irr}_{\mathbb{K}}(W)$ where $\mathbf{a}_{(\lambda,\mu)} \geqslant \mathbf{a}_\nu$; furthermore, there will be equality for at least one pair $(\lambda,\mu)$. (These results can be found in [231, Chap. 20] or [103, §3].) Now, by Example 1.3.8, we have $\mathbf{a}_\nu = n(\nu)a$ and $f_\nu = 1$. Also note that $b \notin \mathbb{Z}a$. (Indeed, if we had $b = ra$ for some $r \in \mathbb{Z}$, then $= -1 = \theta(U) = \theta(u)^r = 1$ and so $\mathrm{char}(k) = 2$, contradicting the assumption that $R$ is $L$-good.) So we are in Case 3 of Example 1.3.9; in particular, $f_{(\lambda,\mu)} = 1$ for all $(\lambda,\mu) \in \Lambda$. In this sutuation, there is actually a unique $J(\nu) \in \Lambda$ such that $\mathbf{a}_{J(\nu)} = \mathbf{a}_\nu = n(\nu)a$ and

$$\mathrm{Ind}_\Omega^W(E^\nu) \cong E^{J(\nu)} \oplus \begin{cases} \text{direct sum of various } E^{(\lambda,\mu)} \in \mathrm{Irr}_{\mathbb{K}}(W) \\ \text{where } (\lambda,\mu) \in \Lambda \text{ and } \mathbf{a}_{(\lambda,\mu)} > \mathbf{a}_{J(\nu)} = \mathbf{a}_\nu; \end{cases}$$

see [231, 22.25] or [103, 3.8]. There is a purely combinatorial description of the map $\nu \mapsto J(\nu)$. To explain this, write $b = ar + b'$, where $r \geqslant 0$ and $0 < b' < a$. As in Example 1.3.9, choose a large integer $N$ and let $Z^0$ be the multiset whose entries are

$$0,\, a,\, 2a,\, \ldots,\, (N-1)a,\, b',\, a+b',\, 2a+b',\, \ldots,\, (N+r-1)a+b'.$$

Let $z_1^0, z_2^0, \ldots, z_{2N+r}^0$ be the entries of $Z^0$ arranged in decreasing order. Now write $\nu = (\nu_1 \geqslant \nu_2 \geqslant \ldots \geqslant \nu_t \geqslant 0)$ (where $t \leqslant n$) and consider the multiset

$$Z := \{z_1^0 + \nu_1 a,\, z_2^0 + \nu_2 a,\, \ldots,\, z_t^0 + \nu_t a,\, z_{t+1}^0,\, z_{t+2}^0,\, \ldots, z_{2N+r}^0\}.$$

Then, with the notation of Example 1.3.9 (Case 3), there is a unique pair of partitions $(\lambda,\mu) \in \Lambda$ such that $Z = Z_{a,b}^N(\lambda,\mu)$. We leave it as an exercise for the reader to check that, indeed, $J(\nu) = (\lambda,\mu)$. (See [231, 22.17] and [125, §3] for further details.)

**Example 4.6.9.** Assume that $b > (n-1)a > 0$. Then $r+1 \geqslant n$ and we see that the $n$ largest entries of $Z^0$ are

$$z_1^0 = (N+r-1)a+b', z_2^0 = (N+r-2)a+b', \ldots, z_n^0 = (N+r-n)a+b'.$$

Hence, the definition of $Z_{a,b}^N(\lambda,\mu)$ immediately shows that

$$J(\nu) = (\nu, \varnothing) \qquad \text{for all } \nu \vdash n.$$

So in this case, the statement of Proposition 4.6.10 below agrees with [68, 7.3]. Note that this is not necessarily true for other values of $a, b$. For example, assume that $a = 2$ and $b = 1$ (hence, $r = 0$ and $b' = 1$). Then for $n = 4$ we obtain

$$\begin{aligned} J(4) &= ((4), \varnothing), \\ J(31) &= ((3), (1)), \\ J(22) &= ((2), (2)), \\ J(211) &= ((21), (1)), \\ J(1111) &= ((11), (11)). \end{aligned}$$

**Proposition 4.6.10 (Cf. [125, 3.4]).** *In the setting of 4.6.5, assume that $a > 0$. Then*

$$\mathscr{B}_\theta := \{J(\nu) \mid \nu \vdash n \text{ is } \tilde{e}\text{-regular}\} \subseteq \Lambda$$

*is a canonical basic set for $\mathbf{H}_k$, where the map $\nu \mapsto J(\nu)$ is described in 4.6.8. Recall that, in this case, we have $\tilde{e} = \infty$ (if $\operatorname{char}(k) = 0$) or $\tilde{e} = \ell$ (if $\operatorname{char}(k) = \ell > 0$).*

*Proof.* Using the notation in 4.6.1, let ${}^n\mathbf{H} \subseteq \mathbf{H}$ be the parabolic subalgebra generated by $T_{s_1}, \ldots, T_{s_{n-1}}$. This is the generic Iwahori–Hecke algebra associated with $\mathfrak{S}_n$ and the weight function which assigns $a$ to each generator of $\mathfrak{S}_n$. We write

$$\operatorname{Irr}({}^n\mathbf{H}_K) = \{E_\varepsilon^\lambda \mid \lambda \vdash n\}.$$

As already noted, the specialised algebra ${}^n\mathbf{H}_k$ is isomorphic to $k\mathfrak{S}_n$. We have a corresponding decomposition matrix $(d_{\lambda,\nu})$, with rows labelled by all the partitions $\lambda \vdash n$ and columns labelled by the $\tilde{e}$-regular partitions $\nu \vdash n$; see Theorem 3.5.14. Given any $\tilde{e}$-regular partition $\nu \vdash n$, we define an ${}^n\mathbf{H}_K$-module $Q^\nu$ such that

$$[Q^\nu] = \sum_{\lambda \vdash n} d_{\lambda,\nu}[E_\varepsilon^\lambda] \qquad (\text{in } R_0({}^n\mathbf{H}_K)).$$

Now consider the induction of representations from ${}^n\mathbf{H}_K$ to $\mathbf{H}_K$. We simply denote this operation by Ind and define

$$P^\nu := \operatorname{Ind}(Q^\nu) \qquad \text{for any } \tilde{e}\text{-regular partition } \nu \vdash n.$$

Let $d_\nu$ denote the vector which contains the multiplicities of the various $E_\varepsilon^{(\lambda,\mu)} \in \operatorname{Irr}(\mathbf{H}_K)$ as constituents of $P^\nu$. We claim that

$$(*)\qquad \begin{cases} d_\nu \text{ is a linear combination of the columns of the decomposition} \\ \text{matrix } D_\theta \text{ of } \mathbf{H} \text{ where the coefficients are non-negative integers.} \end{cases}$$

Indeed, as in the proof of Proposition 3.4.10, we can assume without loss of generality that $A$ is the ring of Laurent polynomials in finitely many indeterminates over a Dedekind domain. Then $(*)$ follows by an argument completely analogous to that in the proof of Theorem 3.4.4, using Brauer reciprocity.

Let us now consider in more detail the vector $d_\nu$. By Example 3.2.9, we have $d_{\nu,\nu} = 1$ and $d_{\nu',\nu} = 0$ unless $\nu' = \nu$ or $\mathbf{a}_{\nu'} > \mathbf{a}_\nu$ (where $\nu' \vdash n$ is an arbitrary partition). This yields that

$$[Q^\nu] = [E_\varepsilon^\nu] + \sum_{\nu' \vdash n : \mathbf{a}_{\nu'} > \mathbf{a}_\nu} d_{\nu',\nu}\,[E_\varepsilon^{\nu'}].$$

Now apply Ind and take into account the results in 4.6.8; also recall that $\mathrm{Irr}(\mathbf{H}_K) = \{E_\varepsilon^{(\lambda,\mu)} \mid (\lambda,\mu) \in \Lambda\}$. Using the general compatibility of induction with specialisations (see [132, §9.1]), we deduce that

$$\begin{aligned}[P^\nu] = [\mathrm{Ind}(Q^\nu)] &= [\mathrm{Ind}(E_\varepsilon^\nu)] + \sum_{\nu' \vdash n : \mathbf{a}_{\nu'} > \mathbf{a}_\nu} d_{\nu',\nu}\,[\mathrm{Ind}(E_\varepsilon^{\nu'})] \\ &= [E_\varepsilon^{J(\nu)}] + \text{sum of various } [E_\varepsilon^{(\lambda,\mu)}] \text{ where } \mathbf{a}_{(\lambda,\mu)} > \mathbf{a}_{J(\nu)} = \mathbf{a}_\nu.\end{aligned}$$

Since $|\mathrm{Irr}(\mathbf{H}_k)|$ equals the number of $\tilde{e}$-regular partitions of $n$ (see Proposition 4.6.6), we can now argue as in the proof of Proposition 3.4.5 to conclude that $\mathscr{B}_\theta$ is a canonical basic set for $\mathbf{H}_k$. ☐

With no assumptions on $\theta(u)$ or $f_n(\theta)$, we have the following result which gives at least an upper bound on the number of irreducible representations of $\mathbf{H}_k$.

**Theorem 4.6.11 (Dipper, James and Murphy [68, 6.9]).** *We have*

$$|\mathrm{Irr}(\mathbf{H}_k)| \leqslant |\{(\lambda,\mu) \in \Lambda \mid \textit{both } \lambda \textit{ and } \mu \textit{ are } \tilde{e}\textit{-regular}\}|.$$

By Theorem 4.6.2, equality holds if $f_n(\theta) \neq 0$. In general, the inequality will be strict; see Example 3.2.6 where $n = 3$, $\tilde{e} = 2$ and $|\mathrm{Irr}(\mathbf{H}_k)| = 4$, but there are six pairs of partitions $(\lambda,\mu) \in \Lambda$ such that both $\lambda$ and $\mu$ are 2-regular. Thus, if $f_n(\theta) = 0$, then the inequality in the above theorem will be strict in general. As already mentioned, completely new methods will be required to study these questions, and this will be done in Chapters 5 and 6. Here is a brief outlook.

**4.6.12.** Assume that $\theta(u) \neq 1$ and $f_n(\theta) = 0$. Thus, there is some $\xi \in k^\times$ such that

$$\theta(u) = \xi \neq 1 \qquad \text{and} \qquad \theta(U) = -\xi^d, \qquad \text{where } d \in \mathbb{Z}.$$

Note that then $\tilde{e} \geqslant 2$ is the order of $\xi \in k^\times$. Let us set

$$\hat{T}_{s_0} := -\varepsilon^b T_{s_0} \qquad \text{and} \qquad \hat{T}_{s_i} = \varepsilon^a T_{s_i} \quad \text{for } 1 \leqslant i \leqslant n-1.$$

Then, in $\mathbf{H}_k$, the usual type $B_n$ braid relations hold for $\hat{T}_{s_0},\ldots,\hat{T}_{s_{n-1}}$, and we have the following quadratic relations (see also Lemma 1.1.12):

$$(\hat{T}_{s_0}-\xi^{c_1})(\hat{T}_{s_0}-\xi^{c_2})=0, \quad \text{where } c_1=d \text{ and } c_2=0,$$
$$(\hat{T}_{s_i}-\xi)(\hat{T}_{s_i}+1)=0 \quad \text{for } 1\leqslant i\leqslant n-1.$$

Thus, we see that $\mathbf{H}_k$ can be identified with an *Ariki–Koike algebra* as in Section 5.2 (where $l=2$ in the notation of 5.2.1). The following two chapters contain a detailed study of the representation theory of these algebras. For example, this will yield a purely combinatorial formula for the cardinality of the set $\mathrm{Irr}(\mathbf{H}_k)$; see Theorem 6.2.22. Let us just mention here the following consequence of that formula.

**Theorem 4.6.13 (Ariki and Mathas [15, Theorem A]).** *In the setting of 4.6.12, assume that $\tilde{e}<\infty$. Then $|\mathrm{Irr}(\mathbf{H}_k)|$ only depends on $\tilde{e}$ and the congruence class of $d$ modulo $\tilde{e}$.*

Let $\tilde{e}\geqslant 2$ and $d$ mod $\tilde{e}$ be fixed. Then Ariki and Mathas actually describe a generating function for the cardinalities $|\mathrm{Irr}(\mathbf{H}_k)|$ (as $n$ varies); see [15, Theorem D].

There are also versions of the above results for the case where $\tilde{e}=\infty$, but we will not pursue these here. (This is related to the situation considered in Example 3.7.14.)

# Chapter 5
# Representation Theory of Ariki–Koike Algebras

The aim of this chapter is to introduce a natural generalisation of Iwahori–Hecke algebras of type $A_{n-1}$ and $B_n$: the Ariki–Koike algebra or Hecke algebra associated with the complex reflection group of type $G(l,1,n)$. As we will see, many properties of Iwahori–Hecke algebras of finite Coxeter groups still hold for this class of algebras: they are finite dimensional, symmetric and even cellular. As a consequence, we also have a theory of cell modules which gives much information on the representation theory, both in the semisimple and in the modular case. In addition, the symmetric structure will allow us to define an **a**-invariant associated with each simple module of certain specialisations of Ariki–Koike algebras: the cyclotomic specialisations. One of the main goals of this chapter and the next one will be to study the representation theory of Ariki–Koike algebras with respect to this symmetric structure and to the **a**-invariants attached to the simple modules.

This chapter is organized as follows. We begin by reviewing the definition of the complex reflection group of type $G(l,1,n)$. The Ariki–Koike algebra is then defined in the second part as a deformation of this group algebra. In the next sections, we study a cellular structure and present a useful result by Dipper and Mathas showing that an Ariki–Koike algebra is always Morita equivalent to a direct sum of a tensor product of Ariki–Koike algebras of special kinds. In these first parts, several proofs will be omitted, as there are already good references on these subjects.

Section 5.5 is devoted to the study of certain specialisations of generic Ariki–Koike algebras: the cyclotomic Ariki–Koike algebras. These algebras will allow us to define an **a**-invariant for each simple module over $G(l,1,n)$. We can then formally introduce the notion of canonical basic sets in the context of Ariki–Koike algebras. Then, our main aim will be to prove the existence of canonical basic sets as for the symmetric group in Chapter 3. We define an operator of "$i$-induction" in the same spirit as in the case of Iwahori–Hecke algebras. Using this we construct an abstract "Fock datum" for our algebras which will eventually lead to the existence of canonical basic sets. At the end of this chapter we can state Theorem 5.8.2, which yields an explicit description of these basic sets for Ariki–Koike algebras and, in certain special cases, also for type $B_n$. However, the completion of the proof of this theorem will rely on the deep methods and results to be discussed in Chapter 6.

M. Geck, N. Jacon, *Representations of Hecke Algebras at Roots of Unity*,
Algebra and Applications 15, DOI 10.1007/978-0-85729-716-7_5,

## 5.1 The Complex Reflection Group of Type $G(l,1,n)$

We here define and study the complex reflection group of type $G(l,1,n)$ and its Hecke algebra in the same spirit as in Section 1.1. Further details and references can be found in [32], [48], [269].

**5.1.1.** In the classification of finite groups generated by complex reflections by Shephard and Todd [278], there are a single infinite family of groups $G(l,p,n)$ and exactly 34 further exceptional groups. In this chapter, we are interested in a special case of the "infinite series": the class of complex reflection groups of type $G(l,1,n)$. It may be seen as the wreath product $(\mathbb{Z}/l\mathbb{Z})\wr\mathfrak{S}_n$. There is a presentation of $G(l,1,n)$ by generators $S=\{s_i \mid i=0,1,\dots,n-1\}$ and relations

- $s_i^2=1$ for all $i=1,\dots,n-1$,
- $s_0^l=1$,

and the braid relations symbolised by the following braid diagram:

That is to say:

- $s_is_j=s_js_i$, if $|i-j|>1$,
- $s_{i+1}s_is_{i+1}=s_is_{i+1}s_i$ if $1\leqslant i\leqslant n-2$,
- $s_0s_1s_0s_1=s_1s_0s_1s_0$.

Note that the three last relations correspond to the type-$B$ braid relations. We also remark that if we set $l=1$ then we obtain the Weyl group of type $A_{n-1}$ and $G(2,1,n)$ is nothing but the Weyl group of type $B_n$. It is easy to see that the order of $G(l,1,n)$ is $l^n n!$. In addition, we may realize this group as the subgroup of $\mathrm{GL}_n(\mathbb{C})$ consisting of matrices such that

- the entries are either 0 or $l$th roots of unity;
- there is exactly one non-zero entry in each row and each column.

The identification is obtained as follows. Let $E_{i,j}$ be the $n\times n$ matrix with 1 in the $(i,j)$ position and all other entries 0. Then $s_0$ is identified with $\zeta_l E_{1,1}+\sum_{j\neq 1}E_{jj}$ and $s_i$ with $E_{i,i+1}+E_{i+1,i}+\sum_{j\neq i,i+1}E_{jj}$ for $i=1,\dots,n-1$, where $\zeta_l:=\exp(2\pi\sqrt{-1}/l)$. Finally, note that the subgroup generated by $\{s_i \mid i=0,1,\dots,n-2\}$ is isomorphic to $G(l,1,n-1)$ and that the elements of $\{s_i \mid i=1,\dots,n-1\}$ generate a natural subgroup $\mathfrak{S}_n$ isomorphic to a Weyl group of type $A_{n-1}$. Consequently, for each composition $\lambda$ of $n$, we have an associated Young subgroup $\mathfrak{S}_\lambda$ of $G(l,1,n)$. For $m\in\mathbb{Z}_{>0}$, we set $W_m:=G(l,1,m)$.

**5.1.2.** We introduce generalisations of some combinatorial objects we have a lready met in type $A_{n-1}$: the multicompositions and the multipartitions. An *$l$-composition* (or a *multicomposition*) of rank $n$ is an $l$-tuple of compositions $\lambda=(\lambda^1,\dots,\lambda^l)$ such that $|\lambda^1|+|\lambda^2|+\ldots+|\lambda^l|=n$. We denote $\lambda\models_l n$. If all the $\lambda^i$ are partitions, we say that $\lambda = (\lambda^1, \dots, \lambda^l)$ is an *l-partition* of $n$ (or a *multipartition*) and we denote $\lambda \vdash_l n$. For example, the set of 3-partitions of rank 3 is as follows:

$$\{(3,\varnothing),(\varnothing,3),\ (2.1,\varnothing),(\varnothing,2.1),\ (1.1.1,\varnothing),\\ (\varnothing,1.1.1),\ (2,1),\ (1,2),\ (1.1,1),\ (1,1.1)\}$$

**5.1.3.** In all our discussion, the parameter $l$ in $G(l,1,n)$ will generally be fixed. So we shall denote the group of type $G(l,1,n)$ simply by $W_n$ from now on. Let $R$ be a commutative ring with unity such that

$$\boxed{\mathbb{Z}[\zeta_l] \subseteq R \subset \mathbb{C} \text{ and let } \mathbb{K} \text{ denote the field of fractions of } R.}$$

Then $\mathbb{K}$ is a splitting field for $W_n$ (see [17] and [18] for a study of the splitting fields for all complex reflection groups). We shall now explain, following [239, Section 3] (for example), how the irreducible representations of $W_n$ may be obtained from the ones of the Weyl group of type $A_{n-1}$. The construction is similar to the case of the Weyl group of type $B_n$, as discussed in [132, (5.5.4)].

For $k = 1, \dots, l$, we consider the $l$ linear representations of $W_n$ which are uniquely defined as follows:

$$\begin{aligned} \sigma_k : S &\to \mathbb{K} \\ s_0 &\mapsto \zeta_l^{k-1} \\ s_j &\mapsto 1 \ (j = 1, \dots, n-1). \end{aligned}$$

Note that, for each $l$-composition $\lambda := (\lambda^1, \dots, \lambda^l)$ of $n$, we have an associated Young subgroup $\mathfrak{S}_\lambda = \mathfrak{S}_{\lambda^1} \times \dots \times \mathfrak{S}_{\lambda^l}$ of $W_n$.

Let $(n_1, \dots, n_l)$ be such that $\sum_{1 \leqslant i \leqslant l} n_i = n$ and for $1 \leqslant i \leqslant l$ let $\lambda^i$ be a partition of $n_i$ so that we have $\lambda = (\lambda^1, \dots, \lambda^l) \vdash_l n$. Let $W_n^\lambda$ be the subgroup $W_{n_1} \times \dots \times W_{n_l}$ of $W_n$ obtained from the natural embedding $\mathrm{GL}_{n_1}(\mathbb{C}) \times \dots \times \mathrm{GL}_{n_l}(\mathbb{C}) \subset \mathrm{GL}_n(\mathbb{C})$. Via the natural projection $W_{n_i} \to \mathfrak{S}_{n_i}$ induced by the semi-direct product, each representation $E^{\lambda^i}$ of $\mathfrak{S}_{n_i}$ may be seen as a representation for $W_{n_i}$. Consequently, we can define the following $\mathbb{K}W_n$-module:

$$E^\lambda := \mathrm{Ind}_{W_n^\lambda}^{W_n} (E^{\lambda^1} \boxtimes (E^{\lambda^2} \otimes \sigma_2) \boxtimes \dots \boxtimes (E^{\lambda^l} \otimes \sigma_l)),$$

where $\boxtimes$ denotes the outer tensor product. It turns out that this procedure gives us a complete set of non-isomorphic simple modules:

$$\mathrm{Irr}_{\mathbb{K}}(W_n) = \{E^\lambda \mid \lambda \vdash_l n\}.$$

(We refer to [236, Chapter 1, Appendix B] for further details.)

**Example 5.1.4.** There are exactly $2l$ irreducible representations of $W_n$ over $\mathbb{K}$ with dimension 1. We have already met the $\sigma_i$. One can define $l$ other representations uniquely as follows:

$$\begin{aligned} \mathrm{sgn}_i : S &\to \mathbb{K} \\ s_0 &\mapsto \zeta_l^{i-1} \\ s_j &\mapsto -1 \ (j = 1, \dots, n-1) \end{aligned}$$

for $i = 1, \dots, l$. The simple module $E^{\lambda}$ affords

- the representation $\sigma_i$ if $\lambda^j = \begin{cases} \varnothing & \text{if } j \neq i, \\ (n) & \text{if } j = i, \end{cases}$
- the representation $\mathrm{sgn}_i$ if $\lambda^j = \begin{cases} \varnothing & \text{if } j \neq i, \\ (1^n) & \text{if } j = i. \end{cases}$

Note that if we set $l = 1$ (and respectively $l = 2$), the above remarks are consistent with the fact that the simple modules of the Weyl group of type $A_{n-1}$ (and respectively of type $B_n$) are parametrised by the set of partitions (and respectively of bipartitions) of rank $n$.

**5.1.5.** We define some additional combinatorial notions. Let $\lambda = (\lambda^1, \dots, \lambda^l)$ be an $l$-composition of $n$. The *Young tableau* $[\lambda]$ of $\lambda$ is the subset

$$[\lambda] := \{(a,b,c) \mid a \geqslant 1, c \in \{1, \dots, l\} \text{ and } 1 \leqslant b \leqslant \lambda_a^c\} \subseteq \mathbb{Z}_{>0} \times \mathbb{Z}_{>0} \times \{1, \dots, l\}.$$

The elements of $[\lambda]$ are called the *nodes* of $\lambda$. The set of *border nodes* of $\lambda$ is defined by

$$\mathscr{N}_\lambda := \{(a, \lambda_a^c, c), \mid a \geqslant 1, c \in \{1, \dots, l\} \text{ and } \lambda_a^c > 0\} \subseteq [\lambda].$$

As in the case of partitions, it is convenient to visualize $[\lambda]$ as an $l$-tuple of arrays of boxes in the plane, as in the following example.

Assume now that $\lambda$ is an $l$-partition. We say that $\gamma \in [\lambda]$ is a *removable node* if $[\lambda] = [\mu] \cup \{\gamma\}$ for some $l$-partition $\mu$ of rank $n+1$. We denote $\gamma = [\lambda]/[\mu]$. In this case, $\gamma$ is called an *addable node* of $\mu$. The set of removable nodes of $\lambda$ is denoted by $\mathscr{R}(\lambda)$ and the set of addable nodes of $\lambda$ by $\mathscr{A}(\lambda)$.

**Example 5.1.6.** Let $\lambda = (2.1.1, 3.3, 1) \vdash_3 11$. The Young tableau of $\lambda$ is the tableau

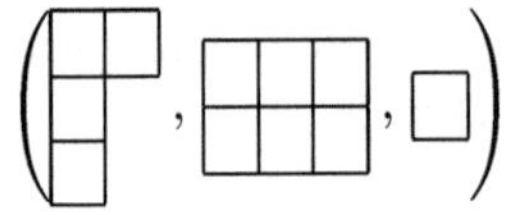

The set of border nodes is here given by

$$\mathscr{N}_\lambda = \{(1,2,1), (2,1,1), (3,1,1), (1,3,2), (2,3,2), (1,1,3)\}.$$

Furthermore, we obtain

$$\mathscr{R}(\lambda) = \{(1,2,1), (3,1,1), (2,3,2), (1,1,3)\},$$
$$\mathscr{A}(\lambda) = \{(1,3,1), (2,2,1), (4,1,1), (1,4,2), (3,1,2), (1,2,3), (2,1,3)\}.$$

**5.1.7.** As $W_{n-1}$ can be seen as a subgroup of $W_n$, we can define functors of induction $\mathrm{Ind}_{W_{n-1}}^{W_n}$ and restriction $\mathrm{Res}_{W_{n-1}}^{W_n}$ between the category of finitely generated $\mathbb{K}W_{n-1}$-modules and the category of finitely generated $\mathbb{K}W_n$-modules. The proposition below describes the branching rule for complex reflection groups of type $G(l,1,n)$. The proof is inspired from the paper [239, Lemma 3.3].

**Proposition 5.1.8.** *For all $\lambda \vdash_l n$, we have*

$$\mathrm{Ind}_{W_n}^{W_{n+1}}(E^{\lambda}) \simeq \bigoplus_{\mu} E^{\mu},$$

*where the sum is taken over all multipartitions $\mu$ such that $[\mu] = [\lambda] \cup \{\gamma\}$ for $\gamma \in \mathscr{A}(\lambda)$. In addition, we have*

$$\mathrm{Res}_{W_{n-1}}^{W_n}(E^{\lambda}) \simeq \bigoplus_{\mu} E^{\mu},$$

*where the sum is taken over all multipartitions $\mu$ such that $[\mu] = [\lambda] \setminus \{\gamma\}$ for $\gamma \in \mathscr{R}(\lambda)$.*

*Proof.* Let $\lambda \vdash_l n$ and denote by $V$ the $\mathbb{K}W_n$-module $E^{\lambda^1} \boxtimes (E^{\lambda^2} \otimes \sigma_2) \boxtimes \ldots \boxtimes (E^{\lambda^l} \otimes \sigma_l)$ so that we have

$$\mathrm{Ind}_{W_n}^{W_{n+1}}(E^{\lambda}) = \mathrm{Ind}_{W_n^{\lambda}}^{W_{n+1}}(V).$$

Now we have $\mathrm{Ind}_{W_n^{\lambda}}^{W_n^{\lambda} \times W_1}(V) = \bigoplus_{1 \leqslant j \leqslant l}(V \boxtimes \sigma_j)$ and, thus, we are reduced to describe $\mathrm{Ind}_{W_n^{\lambda} \times W_1}^{W_{n+1}}(\bigoplus_{1 \leqslant j \leqslant l} V \boxtimes \sigma_j)$. For $1 \leqslant j \leqslant l$, using the already known branching rule in type $A_{n-1}$, we conclude that the induction of each $V \boxtimes \sigma_j$ from $W_n^{\lambda} \times W_1$ to $W_{n_1} \times \ldots \times W_{n_j+1} \times \ldots \times W_{n_l}$ is the direct sum of all the modules of the form

$$E^{\lambda^1} \boxtimes (E^{\lambda^2} \otimes \sigma_2) \boxtimes \ldots \boxtimes (E^{\mu^j} \otimes \sigma_j) \boxtimes \ldots \boxtimes (E^{\lambda^l} \otimes \sigma_l),$$

where $\mu^j$ is obtained from $\lambda^j$ by adding an addable node (see 3.5.2). We can then conclude using the definition of the $E^{\lambda}$ and the transitivity of Ind. This gives the formula for the induction. The second part of the proposition follows using Frobenius reciprocity. □

**5.1.9.** More generally, one defines the group $G(l,p,n)$ as the group of $n \times n$ matrices such that

- the entries are either 0 or $l$th roots of unity,
- there is exactly one non-zero entry in each row and each column,
- the $(l/p)$-th power of the product of the non-zero entries is 1.

The Weyl group of type $D_n$ is a particular case of these groups (corresponding to the case $l = p = 2$). In fact, keeping the notation of the presentation of $W_n$ above, one can realize the group $G(l,p,n)$ as the subgroup of index $p$ in $W_n$ generated by $a = s_0^p$, $a_0 = s_0 s_1 s_0$ and $a_i = s_i$ for $1 \leqslant i \leqslant n-1$. Note that the reflections $a_i$ for $i = 1, \ldots, n-1$ again generate the symmetric group $\mathfrak{S}_n$. It is possible to obtain a complete set of irreducible representations for these groups using Clifford theory. We refer to [6], [135], [136], [239] for details.

## 5.2 Basic Properties of Ariki–Koike Algebras

In this section, we define a deformation of the group algebra of the complex reflection group $W_n$, which may be seen as an analogue of the Iwahori–Hecke algebra for $W_n$. This class of algebras has been independently introduced by Ariki and Koike [13] and Broué and Malle [34]. We study several basic properties of them here.

**5.2.1.** Let $l \in \mathbb{Z}_{>0}$ and $n \in \mathbb{Z}_{\geqslant 1}$. Let $k$ be any commutative ring with 1. Let $\xi$, $\xi_1$, $\xi_2$, $\ldots$, $\xi_l$ be elements in $k$ and assume that $\xi$ is invertible in $k$. The *Ariki–Koike algebra*

$$\mathbf{H}_{k,n} := \mathbf{H}_{k,n}(\xi;\xi_1,\ldots,\xi_l)$$

is the unital associative $k$-algebra with a presentation by generators $\{T_i \mid i = 0,\ldots,n-1\}$ and relations

- $(T_i - \xi)(T_i + 1) = 0$ for $i = 1,\ldots,n-1$,
- $(T_0 - \xi_1)(T_0 - \xi_2)\ldots(T_0 - \xi_l) = 0$,

and the braid relations symbolised by the following diagram:

That is to say:

- $T_iT_j = T_jT_i$ if $|i-j| \geqslant 1$,
- $T_iT_{i+1}T_i = T_{i+1}T_iT_{i+1}$ if $1 \leqslant i \leqslant n-2$,
- $T_0T_1T_0T_1 = T_1T_0T_1T_0$.

**Example 5.2.2.** (a) Assume that $\xi = 1$ and that $\xi_j = \zeta_l^{j-1}$ for $j = 1,\ldots,l$. Then the map $s_i \to T_i$ for $i = 0,1,\ldots,n-1$ defines an isomorphism of $k$-algebras from $kW_n$ onto $\mathbf{H}_{k,n}$ (where $\zeta_l := \exp(2\pi\sqrt{-1}/l)$.)

(b) Let $l = 1$ and assume that $k$ contains a square root $\xi^{1/2}$ of $\xi$; then, $\mathbf{H}_{k,n}$ is an Iwahori–Hecke algebra of type $A_{n-1}$ with parameter $\xi^{1/2}$. Indeed, if we put $\widetilde{T}_i := \xi^{-1/2}T_i$, we obtain the relation $(\widetilde{T}_i - \xi^{1/2})(\widetilde{T}_i + \xi^{1/2}) = 0$ for $1 \leqslant i \leqslant n-1$.

(c) Let $l = 2$. Assume that $k$ contains square roots $\xi_1^{1/2}$ and $\xi_2^{1/2}$ of $\xi_1$ and $\xi_2$ and a square root $\xi^{1/2}$ of $\xi$; then, $\mathbf{H}_{k,n}$ is an Iwahori–Hecke algebra of type $B_n$. Indeed, if we put $\widetilde{T}_i := \xi^{-1/2}T_i$ for $i \geqslant 1$ and $\widetilde{T}_0 = (\xi_2\xi_1)^{-1/2}T_0$, we obtain the relations

$$(\widetilde{T}_0 - \xi_1^{1/2}\xi_2^{-1/2})(\widetilde{T}_0 - \xi_1^{-1/2}\xi_2^{1/2}) \quad \text{and} \quad (\widetilde{T}_i - \xi^{1/2})(\widetilde{T}_i + \xi^{1/2}) = 0$$

for $i = 1,\ldots,n-1$. Thus, $\mathbf{H}_{k,n}$ is an Iwahori–Hecke algebra of type $B_n$ with parameters $\xi_1^{1/2}\xi_2^{-1/2}$ and $\xi^{1/2}$.

The above examples show that an Ariki–Koike algebra may be viewed as an analogue of an Iwahori–Hecke algebra for the complex reflection group $W_n$.

**Definition 5.2.3.** For $m = 1,2,\ldots,n$, we define the following elements:

$$L_m := \xi^{1-m} T_{m-1} \ldots T_1 T_0 T_1 \ldots T_{m-1}.$$

They are called the *Jucys–Murphy elements* for the Ariki–Koike algebra $\mathbf{H}_{k,n}$. (See also Example 3.4.8.)

*Remark 5.2.4.* One can easily check that, for all $1 \leqslant i \leqslant n-1$ and $1 \leqslant j \leqslant n$, we have the following relations involving the Jucys–Murphy elements (see [63, Sections 2.1, 2.2], [10, Lemma 13.2]):

1. $L_i$ and $L_j$ commute.
2. $T_i$ and $L_j$ commute if $i \neq j-1, j$.
3. $T_i$ commutes with $L_i L_{i+1}$ and with $L_i + L_{i+1}$.
4. For all $a \in k$, and $i \neq j$, $T_i$ commutes with $\prod_{1 \leqslant m \leqslant j}(L_m - a)$.

A basis for $\mathbf{H}_{k,n}$ has been given by Ariki and Koike.

**Theorem 5.2.5 (Ariki and Koike [13]).** *The algebra $\mathbf{H}_{k,n}$ is a free k-module with basis*

$$\left\{ L_1^{c_1} \ldots L_n^{c_n} T_w \mid w \in \mathfrak{S}_n \text{ and } 0 \leqslant c_m \leqslant l-1 \text{ for } m = 1,2,\ldots,n \right\}.$$

*In particular, we have* $\dim_k(\mathbf{H}_{k,n}) = l^n n!$.

We refer to [10, Cor. 13.9] for the proof. Several remarks must be added. First, we see that the subalgebra of $\mathbf{H}_{k,n}$ generated by $T_0, T_1, \ldots T_{n-2}$ is isomorphic to $\mathbf{H}_{k,n-1}$. In the general case where $l \in \mathbb{Z}_{>0}$, we see that the elements $T_1, \ldots, T_{n-1}$ generate an Iwahori–Hecke algebra of type $A_{n-1}$ which will be denoted by $\mathbf{H}_{k,n}(\mathfrak{S}_n)$. Consequently, if $w \in \mathfrak{S}_n$ and if $s_{i_1} s_{i_2} \ldots s_{i_r}$ is a reduced expression for $w$, we can (and we do) write $T_w := T_{i_1} T_{i_2} \ldots T_{i_r} \in \mathbf{H}_{k,n}(\mathfrak{S}_n) \subseteq \mathbf{H}_{k,n}$.

**5.2.6.** Now, it will be convient for us to work with a "generic Ariki–Koike algebra". Let $R \subset \mathbb{C}$ be a subring such that $\zeta_l := \exp(2\pi\sqrt{-1}/l) \in R$. Assume that $u, V_1, \ldots, V_l$ are independent indeterminates. Then let

$$\mathbf{H}_n := \mathbf{H}_{A,n}(u; V_1, \ldots, V_l), \qquad \text{where} \qquad A := R[u^{\pm 1}, V_1, \ldots, V_l].$$

We also denote $\varepsilon := (u, V_1, \ldots, V_l)$. Let $\mathbb{K}$ be the field of fractions of $R$ and let $K$ be the field of fractions of $A$. Under these hypotheses, by [10, Cor. 13.9], we have the following theorem.

**Theorem 5.2.7 (Ariki and Koike [13]).** *The algebra $\mathbf{H}_{K,n} := K \otimes_A \mathbf{H}_n$ is split semisimple.*

**5.2.8.** Let us consider the specialisation $\theta\colon A \to \mathbb{K}$ such that

$$\theta(u) = 1 \qquad \text{and} \qquad \theta(V_i) = \zeta_l^{i-1} \quad (i = 1, \ldots, l).$$

Then, the specialized algebra $\mathbf{H}_{\mathbb{K},n} := \mathbb{K} \otimes_A \mathbf{H}_n$ is isomorphic to the group algebra $\mathbb{K}W_n$. As $\mathbb{K}$ is a splitting field for $W_n$ (see 5.1.3) and as the characteristic of $\mathbb{K}$ is zero, the algebra $\mathbb{K}W_n$ is split semisimple. As a consequence, Tits's deformation theorem applies: the irreducible representations of $\mathbf{H}_{K,n}$ (up to isomorphism) are in bijection with $\mathrm{Irr}_{\mathbb{K}}(W_n)$; see [132, (8.1.7)]. By 5.1.3, we can write

$$\mathrm{Irr}(\mathbf{H}_{K,n}) = \{E_\varepsilon^\lambda \mid \lambda \vdash_l n\}.$$

**Example 5.2.9.** The $2l$ irreducible representations of $\mathbf{H}_{K,n}$ over $K$ of dimension 1 are defined uniquely as follows:

$$\begin{array}{lll} \sigma_i^\varepsilon : \mathbf{H}_{K,n} & \to K & \\ T_0 & \mapsto V_i & \\ T_j & \mapsto u \ (j = 1,\ldots,n-1), & \end{array} \qquad \begin{array}{lll} \mathrm{sgn}_i^\varepsilon : \mathbf{H}_{K,n} & \to K & \\ T_0 & \mapsto V_i & \\ T_j & \mapsto -1 \ (j = 1,\ldots,n-1). & \end{array}$$

By Example 5.1.4, the simple module $E_\varepsilon^\lambda$ affords

- the representation $\sigma_i^\varepsilon$ if $\lambda^m = \begin{cases} \varnothing & \text{if } m \neq i, \\ (n) & \text{if } m = i, \end{cases}$
- the representation $\mathrm{sgn}_i^\varepsilon$ if $\lambda^m = \begin{cases} \varnothing & \text{if } m \neq i, \\ (1^n) & \text{if } m = i. \end{cases}$

**5.2.10.** We can define a structure of symmetric algebras on Ariki–Koike algebras. The existence of such a structure is fundamental in our work. Let $\tau : \mathbf{H}_n \to A$ be the $A$-linear map determined by

$$\tau(L_1^{c_1} L_2^{c_2} \ldots L_n^{c_n} T_w) = \begin{cases} 1 & \text{if } c_1 = \ldots = c_n = 0 \text{ and } w = 1, \\ 0 & \text{otherwise.} \end{cases}$$

The proof of the fact that Ariki–Koike algebras are symmetric algebras was first established by Bremke and Malle in [31, Theorem 2.8], but for a different definition of the trace. The proof of the following result was then completed by Malle and Mathas in [240, Section 3].

**Theorem 5.2.11 (Bremke and Malle [31], Malle and Mathas [240]).** *The trace $\tau$ defined above is a trace function which makes $\mathbf{H}_n$ into a symmetric algebra.*

It is clear that this symmetrising form is a generalisation of the form constructed for Hecke algebras of type $A_{n-1}$. The fact that it also generalises the already-defined trace in type $B_n$ in 1.2.11 follows from a result by Malle and Mathas [240, Proposition 2.2].

**5.2.12.** Given the trace form $\tau$, we have corresponding Schur elements $\mathbf{c}_\lambda := \mathbf{c}_\lambda(u; V_1, \ldots, V_l) \in A$ indexed by the $l$-partitions of $n$ such that

$$\tau = \sum_{\lambda \vdash_l n} \mathbf{c}_\lambda^{-1} \chi_\lambda,$$

where $\chi_\lambda$ denotes the character associated with the simple $\mathbf{H}_{K,n}$-module $E_\varepsilon^\lambda$.

Explict expressions for the Schur elements $\mathbf{c}_\lambda$ have been conjectured by Malle in [237]. The proof was then obtained by Geck, Iancu and Malle [123, Theorem 1.3]. To give their expressions, we need to introduce several notations which will be also be useful in the next sections. Let $\lambda := (\lambda^1,\dots,\lambda^l)$ be an $l$-composition. For $j=1,\dots,l$, we denote by $h^j$ the height of the partition $\lambda^j$. The *height* of $\lambda$ is then the number

$$h_\lambda := \max\{h^j,\ j=1,\dots,l\}.$$

It is zero if and only if $\lambda$ is the empty $l$-composition. Let $h \geqslant h_\lambda$. The *ordinary symbol* $\mathbf{B}$ of size $s$ of $\lambda$ (first introduced in [237, Section 2]) is the collection of finite subsets $\mathbf{B} := (B^1, B^2, \dots, B^l)$ such that, for each $i=1,\dots,l$, we have

$$B^i = (B^i_s,\dots,B^i_1), \qquad \text{where} \qquad B^i_j = \lambda^i_j - j + s \quad \text{for all } j=1,\dots,s.$$

(We set $\lambda^i_j := 0$ if $j > h^i$.) This symbol is usually written as an $l$-row tableau, where for $i=1,\dots,l$, starting from the bottom, the $c$th row contains the sequence $B^c$ as in the following example.

**Example 5.2.13.** Let $\lambda := (2.1, 3.3.1, 4) \vdash_3 14$. Then we have $h_\lambda = 3$. Then the ordinary symbol of size 3 is $\mathbf{B} := (B^1, B^2, B^3)$ such that $B^1 := (4.2.0)$, $B^2 := (5.4.1)$, $B^3 := (6.1.0)$. We can write it as follows:

$$\mathbf{B} = \begin{pmatrix} 0\ 1\ 6 \\ 1\ 4\ 5 \\ 0\ 2\ 4 \end{pmatrix}.$$

We associate with a symbol as above the following elements:

$$\delta_{\mathbf{B}}(u, V_1,\dots,V_l) := \prod_{\substack{1\leqslant i\leqslant j\leqslant l \\ (\alpha,\beta)\in B^i\times B^j \\ \alpha>\beta \text{ if } i=j}} (u^\alpha V_i - u^\beta V_j),$$

$$\theta_{\mathbf{B}}(u, V_1,\dots,V_l) := \prod_{\substack{1\leqslant i,j\leqslant l \\ \alpha\in B^i \\ 1\leqslant k\leqslant\alpha}} (u^k V_i - V_j),$$

$$\nu_{\mathbf{B}}(u, V_1,\dots,V_l) := \prod_{1\leqslant i<j\leqslant l} (V_i - V_j)^h \theta_{\mathbf{B}}(u, V_1,\dots,V_l),$$

$$\sigma_{\mathbf{B}} := \binom{l}{2}\binom{h}{2} + n(l-1),$$

$$\tau_{\mathbf{B}} := \binom{l(h-1)+1}{2} + \binom{l(h-2)+1}{2} + \dots,$$

$$|\mathbf{B}| := \sum_{\substack{1\leqslant i\leqslant l \\ \alpha\in B^i}} \alpha.$$

The Schur elements are then given in the following proposition. This is a reformulation (given in [33, Proposition 3.7]) of the form of the Schur elements conjectured in [237].

**Proposition 5.2.14 (Geck, Iancu and Malle [123]).** *Let $\lambda$ be an $l$-partition of rank $n$ and let $\mathbf{B}$ be an ordinary symbol of size $h$ of this $l$-partition. Then*

$$\mathbf{c}_\lambda(u,V_1,\ldots,V_l) = \Big((u-1)\prod_{1\leqslant i\leqslant l} V_i\Big)^{-n}(-1)^{\sigma_{\mathbf{B}}}u^{\tau_{\mathbf{B}}-|\mathbf{B}|+n}\frac{\nu_{\mathbf{B}}(u,V_1,\ldots,V_l)}{\delta_{\mathbf{B}}(u,V_1,\ldots,V_l)}.$$

*This does not depend on the choice of $h \geqslant h_\lambda$.*

Note that Mathas has given another way to compute these Schur elements in [246, Theorem 5.18].

## 5.3 Ariki–Koike Algebras as Cellular Algebras

We have already seen that Ariki–Koike algebras may be viewed as natural generalisations of the Iwahori–Hecke algebras of type $A_{n-1}$ and $B_n$. It is then natural to ask for generalisations of the fundamental properties satisfied by the Iwahori–Hecke algebras of finite Coxeter groups. We are first interested on the cellularity. The aim of this part is thus to give a cellular basis for Ariki–Koike algebras and to study the consequences of the existence of such a structure for the representations of our algebras. We here work with the generic Ariki–Koike algebra $\mathbf{H}_n$ as in 5.2.6.

**5.3.1.** We first need to introduce several combinatorial notions. They essentially generalise the notation for partitions introduced in 3.5.2. Let $\lambda$ be an $l$-composition of $n$. A *$\lambda$-tableau* is a bijection $\mathfrak{t} : [\lambda] \to \{1,\ldots,n\}$. We say that $\mathfrak{t}$ is *column standard* if the sequence $\mathfrak{t}(1,b,c)$, $\mathfrak{t}(2,b,c)$, …is strictly increasing for each relevant $b$ and $c$. It is called *row-standard* if the sequence $\mathfrak{t}(a,1,c)$, $\mathfrak{t}(a,2,c)$, …is strictly increasing for each relevant $a$ and $c$. It is called *standard* if it is both row-standard and column-standard and if $\lambda$ is an $l$-partition.

The set of $\lambda$-tableaux will be denoted by $\mathrm{Tab}(\lambda)$. The set of row-standard $\lambda$-tableaux will be denoted by $\mathrm{Rstd}(\lambda)$. The set of standard $\lambda$-tableaux will be denoted by $\mathrm{Std}(\lambda)$. We also denote by $\mathrm{Std}(n)$ the set of all standard $\lambda$-tableaux such that $\lambda \vdash_l n$. One can visualize a $\lambda$-tableau as an array of boxes where each box is filled by the associated integer as in the following example.

**Example 5.3.2.** Let $\lambda = (2.1, 3.2, 1) \vdash_3 9$. The first of the following two $\lambda$-tableaux is standard, whereas the second is row-standard but not standard.

$$\left(\begin{array}{|c|c|}\hline 1&5\\\hline 4&\\\cline{1-1}\end{array},\ \begin{array}{|c|c|c|}\hline 2&7&8\\\hline 3&9&\\\cline{1-2}\end{array},\ \begin{array}{|c|}\hline 6\\\hline\end{array}\right),\qquad \left(\begin{array}{|c|c|}\hline 1&4\\\hline 6&\\\cline{1-1}\end{array},\ \begin{array}{|c|c|c|}\hline 3&5&8\\\hline 2&9&\\\cline{1-2}\end{array},\ \begin{array}{|c|}\hline 7\\\hline\end{array}\right)$$

**5.3.3.** Let $\lambda$ be an $l$-partition of rank $n$. Then note that a row-standard $\lambda$-tableau is equivalent to the datum of the sequence of $l$-compositions $\lambda^{\mathfrak{t}}[i] \models_l i$ such that $[\lambda^{\mathfrak{t}}[i]] = \mathfrak{t}^{-1}\{1,\ldots,i\}$ for all $i = 1,\ldots n$. With this notation, each $\lambda^{\mathfrak{t}}[i+1]$ is obtained from $\lambda^{\mathfrak{t}}[i]$ by adding the node $\mathfrak{t}^{-1}(i+1)$. In addition, $\mathfrak{t}$ is standard if and only if all the $l$-compositions $\lambda^{\mathfrak{t}}[i]$ are actually $l$-partitions.

We can define a partial order, called the *dominance order*, on the set of $l$-compositions of rank $n$. This generalises the dominance order on compositions defined in 3.5.10. Let $\lambda = (\lambda^1,\ldots,\lambda^l) \models_l n$ and $\mu = (\mu^1,\ldots,\mu^l) \models_l n$ be multicompositions of heights $h_\lambda$ and $h_\mu$ respectively. Then we set $\lambda \trianglerighteq \mu$ if and only if for all $c = 1,\ldots,l$ and $N = 1,\ldots,\max(h_\lambda, h_\mu)$ we have

$$\sum_{1\leqslant i\leqslant c}\sum_{1\leqslant j\leqslant N} \lambda^i_j \geqslant \sum_{1\leqslant i\leqslant c}\sum_{1\leqslant j\leqslant N} \mu^i_j.$$

(Here we set $\lambda^j_i := 0$ if $i$ is strictly greater than the height of $\lambda^j$.)

Let $\lambda \models_l n$. Then one can define a partial order on the set of row-standard $\lambda$-tableaux as follows. Let $\mathfrak{t}$ be a row-standard $\lambda$-tableau. For $(\mathfrak{t},\mathfrak{s}) \in \mathrm{Rstd}(\lambda)^2$, we set $\mathfrak{t} \triangleright \mathfrak{s}$ if for all $i = 1,\ldots,n$, we have $\lambda^{\mathfrak{t}}[i] \triangleright \lambda^{\mathfrak{s}}[i]$ for $i = 1,\ldots,n$. We write $\mathfrak{t} \trianglerighteq \mathfrak{s}$ if $\mathfrak{t} \triangleright \mathfrak{s}$ or $\mathfrak{t} = \mathfrak{s}$ (see Definition 2.2.13).

**Example 5.3.4.** Let $\lambda = (2, 3.1) \vdash_2 6$ and consider the row- standard $\lambda$-tableaux

$$\mathfrak{t} = \left(\begin{array}{|c|c|}\hline 1 & 2 \\ \hline\end{array}\,,\ \begin{array}{|c|c|c|}\hline 3 & 5 & 6 \\ \hline 4 \\ \cline{1-1}\end{array}\right) \qquad \text{and} \qquad \mathfrak{s} = \left(\begin{array}{|c|c|}\hline 1 & 5 \\ \hline\end{array}\,,\ \begin{array}{|c|c|c|}\hline 2 & 3 & 6 \\ \hline 4 \\ \cline{1-1}\end{array}\right).$$

Then, $\lambda^{\mathfrak{t}}[1] = (1,\varnothing)$, $\lambda^{\mathfrak{t}}[2] = (2,\varnothing)$, $\lambda^{\mathfrak{t}}[3] = (2,1)$, $\lambda^{\mathfrak{t}}[4] = (2,1.1)$, $\lambda^{\mathfrak{t}}[5] = (2,2.1)$, $\lambda^{\mathfrak{t}}[6] = (2,3.1)$ and $\lambda^{\mathfrak{s}}[1] = (1,\varnothing)$, $\lambda^{\mathfrak{s}}[2] = (1,1)$, $\lambda^{\mathfrak{s}}[3] = (1,2)$, $\lambda^{\mathfrak{s}}[4] = (1,2.1)$, $\lambda^{\mathfrak{s}}[5] = (2,2.1)$, $\lambda^{\mathfrak{s}}[6] = (2,3.1)$. Thus we have $\mathfrak{t} \triangleright \mathfrak{s}$.

For each $l$-composition $\lambda$ of rank $n$, there is a row-standard $\lambda$-tableau $\mathfrak{t}_\lambda$ of a special kind, the *canonical $\lambda$-tableau*: it is the $\lambda$-tableau such that

$$\mathfrak{t}_\lambda(a,b,c) = \sum_{j<c} |\lambda^j| + \sum_{i<a} \lambda^c_i + b \qquad \text{for each } (a,b,c) \in [\lambda].$$

It is easy to see that for all $\lambda$-tableaux $\mathfrak{t}$ we have $\mathfrak{t}_\lambda \trianglerighteq \mathfrak{t}$. Note that $\mathfrak{t}^\lambda$ is standard if and only if $\lambda$ is an $l$-partition.

**Example 5.3.5.** Let $\lambda = (3.2, 3.2, 1)$. We have

$$\mathfrak{t}_\lambda = \left(\begin{array}{|c|c|c|}\hline 1 & 2 & 3 \\ \hline 4 & 5 \\ \cline{1-2}\end{array}\,,\ \begin{array}{|c|c|c|}\hline 6 & 7 & 8 \\ \hline 9 & 10 \\ \cline{1-2}\end{array}\,,\ \begin{array}{|c|}\hline 11 \\ \hline\end{array}\right).$$

Note that $\mathfrak{S}_n$ acts naturally on the set of $\lambda$-tableaux by permuting the entries as follows:

$$\begin{array}{ccc} \mathfrak{S}_n \times \mathrm{Tab}(\lambda) & \to & \mathrm{Tab}(\lambda) \\ (\sigma, \mathfrak{t}) & \mapsto & \mathfrak{t}^\sigma \end{array}$$

such that $t^\sigma(a,b,c) = \sigma(t(a,b,c))$ for all $(a,b,c) \in [\lambda]$.

We now recall the construction of the cellular basis for Ariki–Koike algebras by Dipper, James and Mathas [67, Theorem 3.26]. This one is particularly well adapted to the theory of cyclotomic $q$-Schur algebras (quasi-hereditary covers of Ariki–Koike algebras; see [67], [274]). Another cellular basis is given in the original article of Graham and Lehrer [144, Section 5].

**Definition 5.3.6.** Let $\lambda = (\lambda^1, \ldots, \lambda^l)$ be an $l$-composition of $n$. We set

$$a_1 = 0 \qquad \text{and} \qquad a_m = \sum_{i=1}^{m-1} |\lambda^i| \quad \text{for } m = 2, \ldots, l.$$

Let $\mathbf{a} = (a_1, \ldots, a_l)$. Then we define

$$L^{\mathbf{a}} = \prod_{i=1}^{l} \prod_{m=1}^{a_i} (L_m - V_i).$$

Finally, we set

$$x_\lambda = \sum_{w \in \mathfrak{S}_\lambda} T_w \qquad \text{and} \qquad m_\lambda = x_\lambda L^{\mathbf{a}}.$$

**Example 5.3.7.** Assume that $l = 3$, $n = 6$ and $\lambda = (2.1, \varnothing, 3)$. Then we have $\mathbf{a} = (0,3,3)$ and we obtain that

$$L^{\mathbf{a}} = (L_1 - V_2)(L_2 - V_2)(L_3 - V_2)(L_1 - V_3)(L_2 - V_3)(L_3 - V_3).$$

**5.3.8.** Note that, by Remark 5.2.4, we have $m_\lambda = L^{\mathbf{a}} x_\lambda$. Let $\lambda$ be an $l$-composition of rank $n$. Under the action of $\mathfrak{S}_n$ on $\operatorname{Tab}(\lambda)$, the row stabiliser of $\mathfrak{t}_\lambda$ is the Young subgroup $\mathfrak{S}_\lambda$. Let $\mathfrak{s} \in \operatorname{Rstd}(\lambda)$. Then the element $d(\mathfrak{s}) \in \mathfrak{S}_n$ is uniquely defined by the equality $\mathfrak{s} = d(\mathfrak{s})\mathfrak{t}_\lambda$. The set

$$\mathscr{D}^\lambda := \{d(\mathfrak{s}) \in \mathfrak{S}_n \mid \mathfrak{s} \in \operatorname{Rstd}(\lambda)\}$$

is a complete set of left coset representatives for the action of $\mathfrak{S}_\lambda$ on $\mathfrak{S}_n$. In fact, they are the elements of minimal length in these cosets and thus, each $y \in \mathfrak{S}_n$ can be uniquely written as $d(\mathfrak{s})y_\lambda$ for $\mathfrak{s} \in \operatorname{Rstd}(\lambda)$ and $y_\lambda \in \mathfrak{S}_\lambda$ with $l(x) = l(d(\mathfrak{s})) + l(y_\lambda)$. The elements of $\mathscr{D}^\lambda$ are called the *distinguished left coset representatives* of $\mathfrak{S}_\lambda$ in $\mathfrak{S}_n$. Hence, we have a natural bijection between $\operatorname{Rstd}(\lambda)$ and the distinguished left coset representatives of $\mathfrak{S}_\lambda$ in $\mathfrak{S}_n$. We can now define the elements which will give the cellular structure for our algebra. To do this, we denote by $* : \mathbf{H}_n \to \mathbf{H}_n$ the $A$-linear anti-automorphism of $\mathbf{H}_n$ determined by

$$T_i^* = T_i \qquad \text{for all} \quad i \in \{0, 1, \ldots, n-1\}.$$

Thus, for all $w \in \mathfrak{S}_n$, we have $T_w^* = T_{w^{-1}}$. Moreover, each Jucys–Murphy element $L_j$ is self-dual with respect to this anti-automorphism.

**Definition 5.3.9.** Let $\lambda$ be an $l$-composition of $n$ and let $(\mathfrak{s},\mathfrak{t}) \in \mathrm{Rstd}(\lambda)^2$. Then, we define

$$m_{\mathfrak{s},\mathfrak{t}} = T_{d(\mathfrak{s})} m_\lambda T^*_{d(\mathfrak{t})}.$$

*Remark 5.3.10.* The elements defined above do not correspond to the ones defined in [67], [247]. This comes from the fact that we have chosen to deal with *left* modules in this book, contrary to [67], [247], where *right* modules are used throughout.

One of the main theorems of this section is the following one; we refer to [67, Theorem 3.26] for its proof.

**Theorem 5.3.11 (Dipper, James and Mathas [67]).** *We set*

$$\mathscr{M} = \left\{ m_{\mathfrak{s},\mathfrak{t}} \mid (\mathfrak{s},\mathfrak{t}) \in \mathrm{Std}(\lambda)^2,\ \lambda \vdash_l n \right\}.$$

*Then the quadruple $(\Pi^l(n), \mathrm{Std}(n), \mathscr{M}, *)$ is a cell datum for $\mathbf{H}_n$ in the sense of 2.6.1. Here, $\Pi^l(n)$ is the set of all $l$-partitions equipped with the partial order $\preceq$ defined as follows:*

$$\lambda \preceq \mu \qquad \Longleftrightarrow \qquad \lambda \trianglerighteq \mu.$$

As a consequence of the cellularity, note that if the algebra $\mathbf{H}_{k,n}$ is semisimple then it is split semisimple. Note also that, as a special case, this theorem implies the existence of a cellular basis for Hecke algebras of type $A_{n-1}$ and $B_n$; see the remarks in 2.8.16, 2.8.17 and 2.8.19.

*Remark 5.3.12.* With respect to the cellular basis of Dipper, James and Mathas (Theorem 5.3.11), we have a family of cell modules (see 2.6.1) which we denote as follows:

$$\{S^\lambda \mid \lambda \vdash_l n\}.$$

These modules are called the *Specht modules* of $\mathbf{H}_n$; they have been extensively used in the literature. (They are, for example, systematically used in the book of Ariki [10].) Recall that $K$ denotes the field of fractions of $A$. Then by Theorem 5.2.7, $\mathbf{H}_{K,n}$ is split semisimple so the set of $\mathbf{H}_{K,n}$-modules $S^\lambda_K := K \otimes_A S^\lambda$ with $\lambda \vdash_l n$ provides a complete set of non-isomorphic simple $\mathbf{H}_{K,n}$-modules. By [10, Theorem 13.21], for all $\lambda \vdash_l n$, we have $E^\lambda_\varepsilon \simeq S^\lambda_K$.

**5.3.13.** As in the case of Iwahori–Hecke algebras, it is possible to define an opposite cell datum for $\mathbf{H}_n$ (see Proposition 2.6.7). In the present situation, this requires a bit more work in order to establish the required properties of the trace form in Theorem 5.2.11.

**Lemma 5.3.14.** *We have $\tau(h^*) = \tau(h)$ for all $h \in \mathbf{H}_n$. Consequently, we have an opposite cell datum $(\Pi^l(n)^{\mathrm{op}}, \mathrm{Std}(n), \hat{\mathscr{M}}, *)$ as defined in Proposition 2.6.7.*

*Proof.* For all $i = 1, \ldots, n-1$ and $m = 1, \ldots, n$, one can check that we have the following relations:

$$T_i.L_m = \begin{cases} L_m T_i & \text{if } i \neq m-1, m, \\ (u-1)L_m + L_{m-1}T_{m-1} & \text{if } i = m-1, \\ L_{m+1}T_m - (u-1)L_{m+1} & \text{if } i = m. \end{cases}$$

Hence, if $w \in \mathfrak{S}_n$ and if $0 \leqslant c_m \leqslant l-1$ for $m = 1, 2, \ldots, n$, then a non-identity element $T_w L_n^{c_n} \ldots L_1^{c_1}$ is a linear combination of non-identity basis elements from

$$\{L_1^{c_1} \ldots L_n^{c_n} T_w \mid w \in \mathfrak{S}_n,\ 0 \leqslant c_m \leqslant l-1 \text{ for } m = 1, 2, \ldots, n\}.$$

This implies that we have

$$\tau(T_w L_n^{c_n} \ldots L_1^{c_1}) = \begin{cases} 1 & \text{if } c_1 = \ldots = c_n = 0 \text{ and } w = 1, \\ 0 & \text{otherwise.} \end{cases}$$

This concludes the proof, as $T_w^* = T_{w^{-1}}$ for all $w \in \mathfrak{S}_n$ and $L_j^* = L_j$ for all $j$. □

## 5.4 Decomposition Maps for Ariki–Koike Algebras

We now introduce the decomposition maps and the decomposition matrices for Ariki–Koike algebras. First, as in 3.1.12, we need to be more specific about the subring $R$ of $\mathbb{C}$. We thus shall now assume that $R$ is noetherian and integrally closed in $\mathbb{K}$. Recall that $K$ is the field of fractions of $A$ and that $\mathbf{H}_{K,n} := K \otimes_A \mathbf{H}_n$ is split semisimple.

**5.4.1.** Let $k$ be an arbitrary field and let $\theta : A \to k$ be a specialisation such that $k$ is the field of fractions of $\theta(A)$. Then the algebra $\mathbf{H}_{k,n} := k \otimes_A \mathbf{H}_n$ is cellular and thus it is split (see Theorem 2.6.5). By Theorem 3.1.2 one can consider the associated decomposition map

$$d_\theta : R_0(\mathbf{H}_{K,n}) \to R_0(\mathbf{H}_{k,n}).$$

We shall write the equations determining $d_\theta$ in the form

$$d_\theta(E_\varepsilon^\lambda) = \sum_{M \in \mathrm{Irr}(\mathbf{H}_{k,n})} d_{\lambda,M}[M], \qquad \text{where} \qquad d_{\lambda,M} \in \mathbb{Z}_{\geqslant 0}$$

We then have an associated decomposition matrix

$$D_\theta := (d_{\lambda,M})_{\lambda \vdash_l n,\, M \in \mathrm{Irr}(\mathbf{H}_{k,n})}.$$

As in 3.1.18, one sees that $D_\theta$ is the identity matrix (up to the ordering of the columns) if and only if $\mathbf{H}_{k,n}$ is semisimple. The following theorem proved in [5] gives a criterion of semisimplicity for the algebra $\mathbf{H}_{k,n}$. We refer to [10, Proposition 13.10] for its proof.

**Theorem 5.4.2 (Ariki [5]).** *Let $\theta : A \to k$ be a specialisation such that $k$ is the field of fractions of $\theta(A)$. Set $\xi := \theta(u)$ and $\xi_i := \theta(V_i)$ for $i = 1, \ldots, l$. Then the specialised algebra $\mathbf{H}_{k,n}$ is split semisimple if and only if*

$$\Big(\prod_{-n<d<n}\prod_{1\leqslant i<j\leqslant l}(\xi^d\xi_i-\xi_j)\Big)\Big(\prod_{1\leqslant i\leqslant n}(1+\xi+\xi^2+\ldots+\xi^{i-1})\Big)\neq 0.$$

The main problem of the rest of this chapter (and of the following one) will be to study the decomposition maps in the same spirit as in Chapter 3. To do this, a useful result by Dipper, James and Mathas will allow us to restrict ourselves to certain specialisations.

When $l=2$, the following theorem is due to Dipper and James [66, Theorem 4.17] and for general $l$, a special case has been proved by Du and Rui [74, Theorem 4.14]. In the general case, the theorem is proved in [70, Theorem 1.1].

**Theorem 5.4.3 (Dipper and Mathas [70]).** *Let $\theta: A\to k$ be a specialisation such that $k$ is the field of fractions of $\theta(A)$. Set $\xi:=\theta(u)$ and for $i=1,\ldots,l$, set $\xi_i:=\theta(V_i)$. We set $\mathcal{V}=\{\xi_1,\ldots,\xi_l\}$ and we assume that we have a partition*

$$\mathcal{V}:=\mathcal{V}_1\sqcup\mathcal{V}_2\sqcup\ldots\sqcup\mathcal{V}_s$$

*for subsets $\mathcal{V}_i$ of $\mathcal{V}$ ($i=1,\ldots,s$) such that*

$$f(\xi,\mathcal{V})=\prod_{1\leqslant\alpha<\beta\leqslant s}\ \prod_{(V_i,V_j)\in\mathcal{V}_\alpha\times\mathcal{V}_\beta}\ \prod_{-n<a<n}(\xi^a\xi_i-\xi_j)$$

*is non-zero in $k$. Then $\mathbf{H}_{k,n}$ is Morita equivalent to the algebra*

$$\bigoplus_{\substack{n_1+\ldots+n_s=n\\ n_1,\ldots,n_s\geqslant 0}}\mathbf{H}_{k,n_1}(\xi;\mathcal{V}_1)\otimes_k\mathbf{H}_{k,n_2}(\xi;\mathcal{V}_2)\otimes_k\ldots\otimes_k\mathbf{H}_{k,n_s}(\xi;\mathcal{V}_s).$$

**5.4.4.** Let us explain the consequences of the above theorem. Let $\theta: A\to k$ be a specialisation such that $\theta(V_i)=\xi_i\in k$ for all $i=1,\ldots,l$ and $\theta(u)=\xi\in k$. Let $D_\theta$ be the associated decomposition matrix. Then there exists a partition

$$\{V_1,\ldots,V_l\}=\mathfrak{V}_1\sqcup\mathfrak{V}_2\sqcup\ldots\sqcup\mathfrak{V}_s$$

such that, for all $i=1,\ldots,s$, the sets $\mathcal{V}_i:=\theta(\mathfrak{V}_i)$ satisfy the conditions of the above theorem. For $i=1,\ldots,s$, set $A_i:=R[u^{\pm1},\mathfrak{V}_i]$, $\mathbf{H}^i_{A_i,n_i}:=\mathbf{H}_{A_i,n_i}(u,\mathfrak{V}_i)$ and $\theta_i: A_i\to k_i$, where $k_i$ denotes the field of fractions of $\theta(A_i)$ and where $n_i\in\{1,\ldots,n\}$. Then we have associated decomposition maps $d_{\theta_i,n_i}$ and matrices $D_{\theta_i,n_i}$. The above theorem implies that, for a suitable ordering of the rows and columns, $D_\theta$ has the form of a block diagonal matrix where each block is given by a matrix $D_{\theta_1,n_1}\otimes\ldots\otimes D_{\theta_s,n_s}$ for $(n_1,\ldots,n_s)\in\mathbb{Z}^s_{\geqslant 0}$ such that $n_1+\ldots+n_s=n$. As a consequence, to compute the decomposition matrices for Ariki–Koike algebras, it will be sufficient to consider the case where each of the parameters $\xi_1,\ldots,\xi_l$ is a power of $\xi$. Hence, it will also be sufficient to know a labelling of the simple modules in this case to solve that problem in general. As in Chapter 3, we will try to solve the problem of labelling for simple modules using the notion of canonical basic sets.

*Remark 5.4.5.* It is also interesting to obtain a complete description of the blocks of Ariki–Koike algebras. This has been solved by Lyle and Mathas in [235].

## 5.5 Cyclotomic Ariki–Koike Algebras

We shall now introduce a certain one-parameter specialisation of $\mathbf{H}_n$. This will be used to define the **a**-invariants for the simple modules of $\mathbf{H}_{K,n}$. At the end of this section, we can then give a formal definition of canonical basic sets for Ariki–Koike algebras.

**5.5.1.** We set as before $\zeta_l = \exp(2\pi\sqrt{-1}/l)$. In addition, we fix a sequence of integers $(r^1,\ldots,r^l) \in \mathbb{Z}^l$. Let $R$ be as in 5.1.3. Let $y$ be an indeterminate and let $r \in \mathbb{Z}_{>0}$. We consider the Ariki–Koike algebra $\mathbf{H}_{R[y^{\pm 1}],n}$ defined over the ring $R[y^{\pm 1}]$ with the following choice of parameters:

$$\xi = y^r \qquad \text{and} \qquad \xi_j = y^{r^j}\zeta_l^{j-1} \quad \text{for } j = 1,\ldots,l.$$

We have

$$\Big(\prod_{-n<d<n}\prod_{1\leqslant i<j\leqslant l}(\xi^d\xi_i - \xi_j)\Big)\Big(\prod_{1\leqslant i\leqslant n}(1+\xi+\xi^2+\ldots+\xi^{i-1})\Big) \neq 0,$$

so the algebra $\mathbf{H}_{\mathbb{K}(y),n}$ is a (split) semisimple by Theorem 5.4.2. Following the terminology of [34], such an algebra is obtained from the generic Ariki–Koike algebra $\mathbf{H}_n$ by a "cyclotomic specialisation":

$$\begin{aligned}\theta_y : A &\to \mathbb{K}(y)\\ V_j &\mapsto y^{r^j}\zeta_l^{j-1} \text{ for } j = 1,\ldots,l,\\ u &\mapsto y^r.\end{aligned}$$

This justifies the name *cyclotomic Ariki–Koike algebras*. We define a sequence $\mathbf{m} = (m^1,\ldots,m^l)$ of rational numbers by setting $m^j := r^j/r$ for all $j = 1,\ldots,l$. This sequence will be referred to a *weight sequence*. To the datum of a weight sequence and a positive integer $r$ is uniquely associated a cyclotomic Ariki–Koike algebra. Let us now study in detail the representations of these algebras. Because of the semisimplicity of $\mathbf{H}_{\mathbb{K}(y),n}$, we have a set

$$\{E_y^\lambda \mid \lambda \vdash_l n\}$$

which gives a compete set of non-isomorphic simple $\mathbf{H}_{\mathbb{K}(y),n}$-modules. Following 5.2.6, these modules can be seen as reductions of the simple $\mathbf{H}_{K,n}$-modules $E_\varepsilon^\lambda$ via the specialisation $\theta_y$.

*Remark 5.5.2.* If we set $r = 1$ and $m^1 = 1$ and $m^j = 0$ for all $j = 2,\ldots,l$, then the associated cyclotomic Ariki–Koike algebra is known as the *spetsial Ariki–Koike algebra*. Such algebras play a central role in the program "Spets", whose goal is to give to complex reflection groups the role of Weyl groups of some as yet mysterious structures (see [35], [238]).

**5.5.3.** We can now define the **a**-*invariants* for the simple $\mathbf{H}_{K,n}$-modules by studying the Schur elements of the cyclotomic Ariki–Koike algebras. Let $\lambda$ be an $l$-partition of rank $n$. Fix $\mathbf{m} = (m^1,\dots,m^l) \in \mathbb{Z}^l$, $r \in \mathbb{Z}_{>0}$ and consider the associated cyclotomic Ariki–Koike algebra $\mathbf{H}_{\mathbb{K}(y),n}$ as in 5.5.1. Let $\mathbf{c}_\lambda$ be the Schur element of the simple $\mathbf{H}_{\mathbb{K}(y),n}$-module $E_y^\lambda$. This is a non-zero Laurent polynomial in $y$ which is the specialisation through $\theta_y$ of the Schur element of $E_\varepsilon^\lambda$ given in Proposition 5.2.14. Then, we define

$$\mathbf{a}^{(\mathbf{m},r)}(\lambda) := \min\{s \in \mathbb{Z} \mid y^s \mathbf{c}_\lambda \in R[y]\},$$

where, by convention, $\mathbf{a}^{(\mathbf{m},r)}(\varnothing) := 0$. Note that this **a**-invariant depends on the choice of the weight sequence **m** and on $r$. To give an explicit expression, it is convenient to introduce several additional combinatorial notions.

*Remark 5.5.4.* In [33], [46], [47], [48], the **a**-invariant is defined as the valuation of the associated Schur elements. This corresponds to the opposite of "our" **a**-invariant. Our definition is in agreement with [35], [132], [238].

**5.5.5.** Let $\mathbf{m} = (m^1,\dots,m^l) \in \mathbb{Q}^l$ be an $l$-tuple of positive rational numbers. Keeping the above notations, we will define the *shifted* **m**-*symbol* and the *ordinary* **m**-*symbol* of an $l$-partition $\lambda$ following [33]. Let $\beta = (\beta_1,\dots,\beta_k)$ be a sequence of integers and let $s$ be a non-negative rational number. We denote by $[s]$ the integer part of $s$; that is, the maximal integer $k$ such that $s - k \geqslant 0$. We set

$$\beta(s) := \begin{cases} (\beta_1,\dots,\beta_k) & \text{if } 0 \leqslant s < 1, \\ (s-[s], s-[s]+1,\dots,s-1,\beta_1+s,\dots,\beta_k+s) & \text{if } s > 1. \end{cases}$$

It is a sequence of rational numbers with exactly $k + [s]$ elements. Let now $\lambda$ be an $l$-composition and for $i = 1,\dots,l$, let $h^i$ be the heights of the compositions $\lambda^i$. For $i = 1,\dots,l$ we consider the sequence of $h^i$ non-negative integers

$$\beta^i = (\lambda^i_{h^i} - h^i + h^i, \dots, \lambda^i_j - j + h^i, \dots, \lambda^i_1 - 1 + h^i).$$

Note that this is a sequence of strictly increasing integers if and only if $\lambda$ is an $l$-partition. If $s$ is an integer greater than $h_\lambda$, note that the ordinary symbol of size $s \in \mathbb{Z}_{>0}$ which we have already defined in 5.5.3 is the family of sequences

$$\mathbf{B} = (B^1,\dots,B^l), \qquad \text{where} \qquad B^j = (\beta^j(s-h^j)) \quad \text{for } j = 1,\dots,l.$$

Now, for $i = 1,\dots,l$ we put

$$hc^i = h^i - m^i \qquad \text{and} \qquad hc^\lambda = \max(hc^1,\dots,hc^l).$$

Let $s$ be an integer such that $s \geqslant hc^\lambda$. The shifted **m**-symbol of $\lambda$ of size $s$ is the family of sequences

$$\mathfrak{B} = (\mathfrak{B}^1, \ldots, \mathfrak{B}^l) \qquad \text{such that} \qquad \mathfrak{B}^j = (\beta^j(s - hc^j)) \quad \text{for } j = 1, \ldots, l.$$

Each sequence $\mathfrak{B}^j$ contains exactly $s + [m^j]$ elements $(\mathfrak{B}^j_{s+[m^j]}, \ldots, \mathfrak{B}^j_1)$.

**Example 5.5.6.** Let $\mathbf{m} = (3,0,2,1)$ and let $\lambda = (2.2, 4.4, 2, 5.5.1) \vdash_4 25$. Then, we have

$$hc^1 = -1, \quad hc^2 = 2, \quad hc^3 = -1, \quad hc^4 = 2, \quad hc^\lambda = 2.$$

For $s = 3$, the shifted $\mathbf{m}$-symbol of $\lambda$ of size $s$ is then

$$\mathfrak{B} = \begin{pmatrix} 0\,2\,7\,8 \\ 0\,1\,2\,3\,6 \\ 0\,5\,6 \\ 0\,1\,2\,3\,6\,7 \end{pmatrix}.$$

Let $\mathbf{m} = (2, 1/2, 0)$ and let $\lambda = (2.2, 4.1, 5.1) \vdash_3 15$. Then, we have

$$hc^1 = 0, \quad hc^2 = \frac{3}{2}, \quad hc^3 = 2, \quad hc^\lambda = 2.$$

For $s = 3$, the shifted $\mathbf{m}$-symbol of $\lambda$ of size $s$ is then

$$\mathfrak{B} = \begin{pmatrix} 0 & 2 & 7 & \\ 1/2 & 5/2 & 13/2 & \\ 0 & 1 & 2 & 5\,6 \end{pmatrix}.$$

**5.5.7.** Continuing with the general discussion in 5.5.5, let now $s$ be an integer such that $s \geqslant hc^\lambda$. Let

$$\rho = [\max(m^1, \ldots, m^l)].$$

For $j = 1, \ldots, l$, we define a sequence $\widetilde{B}^j = (\widetilde{B}^j_{\rho+s}, \ldots, \widetilde{B}^j_1)$ such that, for $i = 1, \ldots, \rho + s$, we have

$$\widetilde{B}^j_i = \lambda^j_i - i + m^j + \rho + s.$$

The ordinary $\mathbf{m}$-symbol of $\lambda$ of size $s$ is then the family of sequences

$$\widetilde{\mathbf{B}} := (\widetilde{B}^1, \ldots, \widetilde{B}^l).$$

Note that the ordinary $\mathbf{m}$-symbol $\widetilde{\mathbf{B}}$ of size $s$ and the ordinary symbol $\mathbf{B}$ of size $s + \rho$ are connected as follows:

$$\widetilde{B}^j = (B^j_{\rho+s} + m^j, B^j_{\rho+s-1} + m^j, \ldots, B^j_1 + m^j).$$

The ordinary $\mathbf{m}$-symbol and the shifted $\mathbf{m}$-symbol (both of size $s$) of $\lambda$ are connected as follows:

$$\widetilde{B}^j = (m^j, \ldots, \rho + m^j - [m^j] - 1, \mathfrak{B}^j_{s+[m^j]} + \rho, \ldots, \mathfrak{B}^j_1 + \rho).$$

**Example 5.5.8.** Let $\mathbf{m} = (3,0,2,1)$ and let $\lambda = (2.2,4.4,2,5.5.1)$. Then the ordinary $\mathbf{m}$-symbol of size 3 is

$$\widetilde{\mathbf{B}} = \begin{pmatrix} 1 & 2 & 3 & 5 & 10 & 11 \\ 2 & 3 & 4 & 5 & 6 & 9 \\ 0 & 1 & 2 & 3 & 8 & 9 \\ 3 & 4 & 5 & 6 & 9 & 10 \end{pmatrix}.$$

Let $\mathbf{m} = (2,1/2,0)$ and let $\lambda = (2.2,4.1,5.1)$. Then the ordinary $\mathbf{m}$-symbol of size 3 is

$$\widetilde{\mathbf{B}} = \begin{pmatrix} 0 & 1 & 2 & 4 & 9 \\ 1/2 & 3/2 & 5/2 & 9/2 & 17/2 \\ 2 & 3 & 4 & 7 & 8 \end{pmatrix}.$$

The following result comes from [174, Proposition 3.2] and is largely inspired by the computations in [33, Proposition 3.18].

**Proposition 5.5.9.** *Let* $\mathbf{m} := (m^1,\dots,m^l) \in \mathbb{Q}^l$ *be a weight sequence and* $r \in \mathbb{Z}_{>0}$. *Set* $\rho = [\max(m^1,\dots,m^l)]$ *and let* $\mathbf{B}$ *be the ordinary symbol of size* $h := \rho + s$ *of* $\lambda$ *with* $s \in \mathbb{Z}_{>0}$. *Let* $\widetilde{\mathbf{B}} := (\widetilde{B}^1,\dots,\widetilde{B}^l)$ *be the ordinary* $\mathbf{m}$*-symbol of size* $s$ *associated with* $\lambda$. *Then, we have*

$$\mathbf{a}^{(\mathbf{m},r)}(\lambda) = r.\Big(f(n,h,\mathbf{m}) + \sum_{\substack{1\leqslant i\leqslant j\leqslant l \\ (a,b)\in \widetilde{B}^i\times\widetilde{B}^j \\ a>b \text{ if } i=j}} \min\{a,b\} - \sum_{\substack{1\leqslant i,j\leqslant l \\ a\in\widetilde{B}^i \\ 1\leqslant k\leqslant a}} \min\{k,m^j\}\Big),$$

*where*

$$f(n,h,\mathbf{m}) = n\Big(\sum_{j=1}^{l} m^j\Big) - \tau_{\mathbf{B}} + |\mathbf{B}| - n - h \sum_{1\leqslant i<j\leqslant l} \min\{m^i,m^j\} + \sum_{\substack{1\leqslant i,j\leqslant l \\ \alpha\in\widetilde{B}^i \\ 1\leqslant k\leqslant m^i}} \min\{k,m^j\}.$$

*Proof.* From the definition of the $\mathbf{a}$-invariants and the formulae for the Schur elements in Proposition 5.2.14, we get that

$$\mathbf{a}^{(\mathbf{m},r)}(\lambda) = r.\Big(n\sum_{j=1}^{l} m^j - \tau_{\mathbf{B}} + |\mathbf{B}| - n - h \sum_{1\leqslant i<j\leqslant l} \min\{m^i,m^j\} + \sum_{\substack{1\leqslant i\leqslant j\leqslant l \\ (\alpha,\beta)\in B^i\times B^j \\ \alpha>\beta \text{ if } i=j}} \min\{\alpha+m^i,\beta+m^j\} - \sum_{\substack{1\leqslant i,j\leqslant l \\ \alpha\in B^i \\ 1\leqslant k\leqslant \alpha}} \min\{k+m^i,m^j\}\Big).$$

We have

$$\sum_{\substack{1\leqslant i,j\leqslant l\\ \alpha\in B^i\\ 1\leqslant k\leqslant \alpha}} \min\{k+m^i,m^j\} = \sum_{\substack{1\leqslant i,j\leqslant l\\ \alpha\in B^i\\ 1\leqslant k\leqslant \alpha+m^i}} \min\{k,m^j\} - \sum_{\substack{1\leqslant i,j\leqslant l\\ \alpha\in B^i\\ 1\leqslant k\leqslant m^i}} \min\{k,m^j\}.$$

Using the definitions of the ordinary symbol and of the ordinary **m**-symbol, we get the formula. We can also check that this is in agreement with the convention that $\mathbf{a}^{(\mathbf{m},r)}(\varnothing)=0$. □

**5.5.10.** In this part we give a simple formula for $\mathbf{a}^{(\mathbf{m},r)}(\lambda)$ with $\lambda\vdash_l n$. Let $\mathfrak{B}$ be the shifted **m**-symbol of $\lambda$ of size $s$ and let

$$\kappa_1\geqslant\kappa_2\geqslant\ldots.\geqslant\kappa_t$$

be the elements of $\mathfrak{B}$ written in decreasing order (allowing repetition), where $t = ls+\sum_{1\leqslant i\leqslant l}[m^i]$. We set

$$\kappa_{\mathbf{m}}(\lambda)=(\kappa_1,\ldots,\kappa_t) \qquad \text{and} \qquad n^s_{\mathbf{m}}(\lambda)=\sum_{i=1}^{t}(i-1)\kappa_i.$$

Now we can state the following proposition.

**Proposition 5.5.11.** *Let $\lambda$ be an $l$-partition, $\mathbf{m}=(m^1,\ldots,m^l)$ be a weight sequence and $r\in\mathbb{Z}_{>0}$. For all $s\geqslant hc^\lambda$, we have*

$$\mathbf{a}^{(\mathbf{m},r)}(\lambda)=r.(n^s_{\mathbf{m}}(\lambda)-n^s_{\mathbf{m}}(\varnothing)).$$

*Proof.* The proof is by induction on $n$. When $n=0$, the result is obvious. Assume that $n>0$ and that $\lambda$ is an $l$-partition of rank $n$. Then there exists a removable node $R=(\alpha,\beta,\gamma)$ of $[\lambda]$. Let $\mu\vdash_l n$ be such that $[\mu]=[\lambda]\setminus\{R\}$. If $s\geqslant hc^\lambda$, we also have $s\geqslant hc^\mu$; so, by induction, we have

$$\mathbf{a}^{(\mathbf{m},r)}(\mu)=r.(n^s_{\mathbf{m}}(\mu)-n^s_{\mathbf{m}}(\varnothing)).$$

Note that

$$f(n+1,h,\mathbf{m})=\sum_{j=1}^{l}m^j+f(n,h,\mathbf{m}).$$

Let $\widetilde{B}(\lambda)$ (or $\widetilde{B}(\mu)$) be the ordinary **m**-symbol of size $s$ associated with $\lambda$ (or $\mu$). We denote by

$$\delta_1\geqslant\delta_2\geqslant\ldots.\geqslant\delta_{lh}$$

the elements of $\widetilde{B}(\lambda)$ written in decreasing order (with repetition), where $h:=s+\rho$. We have

$$\sum_{\substack{1\leqslant i\leqslant j\leqslant l\\ (a,b)\in\widetilde{B}(\lambda)^i\times\widetilde{B}(\lambda)^j\\ a>b \text{ if } i=j}} \min\{a,b\} = \sum_{1\leqslant i\leqslant lh}(i-1)\delta_i.$$

Similarly, we denote by

$$\delta'_1 \geqslant \delta'_2 \geqslant \dots \geqslant \delta'_{lh}$$

the elements of $\widetilde{B}(\mu)$ written in decreasing order (with repetition). We have

$$\sum_{\substack{1\leqslant i\leqslant j\leqslant l \\ (a,b)\in\widetilde{B}(\mu)^i\times\widetilde{B}(\mu)^j \\ a>b \text{ if } i=j}} \min\{a,b\} = \sum_{i=1}^{lh}(i-1)\delta'_i.$$

Let $k$ be the maximal integer such that

$$\delta'_k = \lambda^\gamma_\alpha - \alpha + m^\gamma + \rho + s.$$

Then we have

$$\sum_{i=1}^{lh}(i-1)\delta_i = \sum_{i=1}^{lh}(i-1)\delta'_i + (k-1).$$

In addition, we have that

$$\sum_{\substack{1\leqslant i,j\leqslant l \\ a\in\widetilde{B}(\lambda)^i \\ 1\leqslant k\leqslant a}} \min\{k,m^j\} - \sum_{\substack{1\leqslant i,j\leqslant l \\ a\in\widetilde{B}(\mu)^i \\ 1\leqslant k\leqslant a}} \min\{k,m^j\} = \sum_{1\leqslant j\leqslant l} \min\{\delta_k,m^j\} = \sum_{1\leqslant j\leqslant l} m^j$$

because, by definition, we have

$$\delta_k \geqslant 1 - h^\gamma + m^\gamma + m^j + h^\gamma - m^\gamma \geqslant m^j \qquad \text{for all } 1\leqslant j\leqslant l.$$

We use all these identities together to obtain that

$$\mathbf{a}^{(\mathbf{m},r)}(\lambda) = \mathbf{a}^{(\mathbf{m},r)}(\mu) + r.(k-1).$$

On the other hand, from the relation between the shifted **m**-symbols and the ordinary **m**-symbols, we also have that

$$n^s_{\mathbf{m}}(\lambda) = n^s_{\mathbf{m}}(\mu) + (k-1),$$

which completes the proof. □

**Example 5.5.12.** The above proposition shows that the preorder induced by $\mathbf{a}^{(\mathbf{m},r)}$ does not depend on $r \in \mathbb{Z}_{>0}$.

**Example 5.5.13.** Assuming that $l = 1$, the cyclotomic Ariki–Koike algebra $\mathbf{H}_{\mathbb{K}(y),n}$ associated with any weight sequence is nothing but the Hecke algebra of type $A_{n-1}$ with $a = r$ in the setting of Example 1.3.8. We recover that the **a**-invariant of $\lambda$ is here given by $a.n(\lambda)$.

**Example 5.5.14.** Assume that $l = 2$. Let $a \in \mathbb{Z}_{>0}$ and $b \in \mathbb{Z}_{\geqslant 0}$. We set $m^1 = b/a$, $m^2 = 0$ and $r = a$. Consider the cyclotomic Ariki–Koike algebra $\mathbf{H}_{\mathbb{K}(y),n}$ to the weight sequence $\mathbf{m} = (m^1, m^2)$ and $r$. Then, using the notations of 5.5.1, we have

$(r^1, r^2) = (b, 0)$. Thus, $\mathbf{H}_{\mathbb{K}(y),n}$ is nothing but the Hecke algebra of type $B_n$ with $L : W \to \mathbb{Z}$ given by

$B_n$

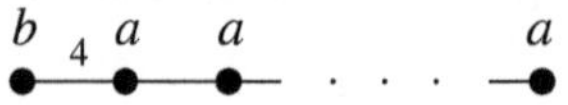


Let $(\lambda^1, \lambda^2)$ be a bipartition of $n$ and consider its shifted $\mathbf{m}$-symbol $\mathfrak{B} = (\mathfrak{B}^1, \mathfrak{B}^2)$ of size $N$. The elements appearing in $\mathfrak{B}$ correspond to the element of the multiset $Z^N_{a,b}(\lambda^1, \lambda^2)$ divided by $a$. Hence, we recover the formula for the $\mathbf{a}$-invariant in type $B_n$ in Example 1.3.9, Case 3.

**Example 5.5.15.** Let $s \in \{1, \ldots, l\}$ be such that $m^s = \max(m^1, \ldots, m^l)$. Let

$$\lambda = (\lambda^1, \ldots, \lambda^l), \qquad \text{where} \qquad \lambda^j = \begin{cases} \varnothing & \text{if } j \neq s, \\ (n) & \text{if } j = s. \end{cases}$$

Then $E^{\lambda}_{\varepsilon}$ affords the representation $\sigma^{\varepsilon}_{s}$ of Example 5.2.9. Proposition 5.5.11 gives us that $\mathbf{a}^{(\mathbf{m},r)}(\lambda) = 0$ for all $r \in \mathbb{Z}_{>0}$.

From this, we deduce the following useful result. (For this purpose, we extend the definition of the dominance order $\trianglelefteq$ to sequences of rational numbers.)

**Proposition 5.5.16.** *Assume that $\lambda$ and $\mu$ are two $l$-partitions of rank $n$ such that*

$$\kappa_{\mathbf{m}}(\mu) \rhd \kappa_{\mathbf{m}}(\lambda).$$

*Then we have*

$$\mathbf{a}^{(\mathbf{m},r)}(\lambda) > \mathbf{a}^{(\mathbf{m},r)}(\mu).$$

*Proof.* Let $s$ be an integer such that $s \geqslant hc^{\mu}$ and $s \geqslant hc^{\lambda}$ and set $t := ls + \sum_{1 \leqslant i \leqslant l} m^i$. Set $\kappa_{\mathbf{m}}(\mu) = (\kappa^{\mu}_1, \ldots, \kappa^{\mu}_t)$ and $\kappa_{\mathbf{m}}(\lambda) = (\kappa^{\lambda}_1, \ldots, \kappa^{\lambda}_t)$. Assume that $\kappa_{\mathbf{m}}(\mu) \rhd \kappa_{\mathbf{m}}(\lambda)$. By Proposition 5.5.11, we need to show that $n^s_{\mathbf{m}}(\lambda) > n^s_{\mathbf{m}}(\mu)$. We have

$$n^s_{\mathbf{m}}(\lambda) = \sum_{1 \leqslant i \leqslant t} (i-1)\kappa^{\lambda}_i = \sum_{2 \leqslant j \leqslant t} \sum_{j \leqslant i \leqslant t} \kappa^{\lambda}_i.$$

Now, because $\kappa_{\mathbf{m}}(\lambda)$ and $\kappa_{\mathbf{m}}(\mu)$ have the same rank and $\kappa_{\mathbf{m}}(\mu) \rhd \kappa_{\mathbf{m}}(\lambda)$, we have $\sum_{j \leqslant i \leqslant t} \kappa^{\lambda}_i \geqslant \sum_{j \leqslant i \leqslant t} \kappa^{\mu}_i$ for all $2 \leqslant j \leqslant t$. This proves that $n^s_{\mathbf{m}}(\lambda) \geqslant n^s_{\mathbf{m}}(\mu)$. As $\kappa_{\mathbf{m}}(\lambda) \neq \kappa_{\mathbf{m}}(\mu)$, we also have $n^s_{\mathbf{m}}(\lambda) \neq n^s_{\mathbf{m}}(\mu)$, which concludes the proof using Proposition 5.5.11. □

**Example 5.5.17.** Let $\mathbf{m} = (0, 2, 1)$, $r = 1$ and let $\lambda = (1, 4.1, 2)$. Let $s = 2$. The shifted $\mathbf{m}$-symbol of $\lambda$ of size $s$ and the shifted $\mathbf{m}$-symbol of $\varnothing$ of size $s$ are

$$\mathfrak{B} = \begin{pmatrix} 0\ 1\ 4 \\ 0\ 1\ 3\ 7 \\ 0\ 2 \end{pmatrix} \qquad \text{and} \qquad \begin{pmatrix} 0\ 1\ 2 \\ 0\ 1\ 2\ 3 \\ 0\ 1 \end{pmatrix}$$

respectively. Then we have

$$n_{\mathbf{m}}^s(\lambda) = 7\times 0+4\times 1+3\times 2+2\times 3+1\times 4+1\times 5 = 25,$$
$$n_{\mathbf{m}}^s(\varnothing) = 3\times 0+2\times 1+2\times 2+1\times 3+1\times 4+1\times 5 = 18.$$

Hence, we obtain $\mathbf{a}^{(\mathbf{m},r)}(\lambda) = 7$.

**Example 5.5.18.** Let $n \in \mathbb{Z}_{>1}$ and assume that $\mathbf{m} = (m^1,\ldots,m^l)$ is a weight sequence such that $m^i - m^j > n-1$ for all $i<j$. Let $\lambda \vdash_l n$ and write its shifted $\mathbf{m}$-symbol of size $s$:

$$\mathfrak{B} = (\mathfrak{B}^1,\ldots,\mathfrak{B}^l).$$

For all $c = 1,\ldots,l$, the elements of $\mathfrak{B}^c$ are the $\lambda_a^c - a + s + m^c$ for $a = 1,2,\ldots,s+[m^c]$. Let $c' > c$. Then an arbitrary element of $\mathfrak{B}^{c'}$ writes $\lambda_{a'}^{c'} - a' + s + m^{c'}$ for $a' = 1,2,\ldots,s+[m^{c'}]$. Let $a' \in \{1,2,\ldots,s+[m^{c'}]\}$ and $a \in \{1,2,\ldots,s+[m^c]\}$ be such that $\lambda_a^c \neq 0$ and $\lambda_{a'}^{c'} \neq 0$. Then, it is an easy combinatorial exercice to show that $(\lambda_a^c - a) - (\lambda_{a'}^{c'} - a') \geqslant 1-n$. As a consequence, we obtain that $\lambda_a^c - a + s + m^c > \lambda_{a'}^{c'} - a' + s + m^{c'}$. Now assume that $\lambda$ and $\mu$ are both $l$-partitions of $n$ such that $\mu \rhd \lambda$. Then using the discussion above, one can show that $\kappa_{\mathbf{m}}(\mu) \rhd \kappa_{\mathbf{m}}(\lambda)$. One way to see this is to use the same arguments as in [274, Prop. 6.4]. It is indeed sufficient to prove this statement for $l$-partitions such that there exist no $\nu \vdash_l n$ satisfying $\mu \rhd \nu \rhd \lambda$. The result then follows from the definition of $\kappa_{\mathbf{m}}$. We thus remark that for this choice of weight sequence we have

$$\mu \rhd \lambda \qquad \Rightarrow \qquad \mathbf{a}^{(\mathbf{m},r)}(\lambda) > \mathbf{a}^{(\mathbf{m},r)}(\mu).$$

But for general $\mathbf{m}$ this property is not true!

Having defined a partial order on the set of simple modules for $W_n$, we can now define the notion of basic set in the context of Ariki–Koike algebras.

**Definition 5.5.19.** Consider the decomposition matrix $D_\theta = (d_{\lambda,M})$ associated with $\theta : A \to k$. Let $(r^1,\ldots,r^l) \in \mathbb{Z}^l$ and $r \in \mathbb{Z}_{>0}$. For all $i = 1,\ldots,l$, we set $m^i := r^i/r$ and we define $\mathbf{m} := (m^1,\ldots,m^l)$. By 5.5.3, we have a function $\mathrm{Irr}_{\mathbb{K}}(W_n) \to \mathbb{Q}$, $E^\lambda \mapsto \mathbf{a}^{(\mathbf{m},r)}(\lambda)$. Assume that the three conditions in 3.1.7 are satisfied for the partial order

$$E_\varepsilon^\lambda \sqsubseteq E_\varepsilon^\mu \quad \overset{\text{def}}{\Leftrightarrow} \quad \lambda = \mu \quad \text{or} \quad \mathbf{a}^{(\mathbf{m},r)}(\lambda) < \mathbf{a}^{(\mathbf{m},r)}(\mu).$$

Explicitly, in the present context, this means:

- Given $M \in \mathrm{Irr}(\mathbf{H}_{k,n})$, let $\mathscr{S}_\theta(M) := \{\lambda \vdash_l n \mid d_{\lambda,M} \neq 0\}$. Then the function $\mathscr{S}_\theta(M) \to \mathbb{Q}$, $\lambda \mapsto \mathbf{a}^{(\mathbf{m},r)}(\lambda)$, reaches its minimum (with respect to $\leqslant$) at exactly one element of $\mathscr{S}_\theta(M)$, which we denote by $\lambda_M$.
- The map $\mathrm{Irr}(\mathbf{H}_{k,n}) \to \Pi^l(n)$, $M \mapsto \lambda_M$, is injective (where $\Pi^l(n) := \{\lambda \vdash_l n\}$).
- We have $d_{\lambda_M,M} = 1$ for all $M \in \mathrm{Irr}(\mathbf{H}_{k,n})$.

If this holds, let $\mathscr{B}_\theta := \{\lambda_M \mid M \in \mathrm{Irr}(\mathbf{H}_{k,n})\}$. Thus, we obtain a bijection

$$\mathrm{Irr}(\mathbf{H}_{k,n}) \stackrel{1-1}{\longleftrightarrow} \mathscr{B}_\theta \subseteq \Pi^l(n),$$

and $\underline{B}_\theta = \{E_\varepsilon^\mu \mid \mu \in \mathscr{B}_\theta\}$ is called a *canonical basic set* for $\mathbf{H}_{k,n}$; see 3.1.7. We shall also call $\mathscr{B}_\theta \subseteq \Pi^l(n)$ itself a *canonical basic set*. Note again that if a canonical basic set exists, then it is uniquely determined by the above conditions.

*Remark 5.5.20.* The above definition shows that the canonical basic set depends on the choice of the weight sequence $\mathbf{m}$. In the following, we will only deal with specific values of $\mathbf{m}$ which are sufficient for our applications on Iwahori–Hecke algebras; see 5.7.1. It should be interesting to study the existence of canonical basic sets for Ariki–Koike algebras in full generality.

*Remark 5.5.21.* According to 5.4.4, to study the decomposition matrices for Ariki–Koike algebras, it is sufficient to consider specialisations $\theta : A \to k$ such that $\theta(u)$ and all $\theta(V_i)$ are integral powers of one fixed element in $k$. In the following sections, we will thus only deal with specialisations $\theta$ which satisfy these assumptions.

## 5.6 A Fock Datum for Ariki–Koike Algebras

The goal of this section is to introduce a family of operators of "$i$-induction" similar the the ones constructed in Section 3.5 in type $A_{n-1}$. We then study the "branching" rule induced by these elements.

**5.6.1.** We consider the Ariki–Koike algebra $\mathbf{H}_n$ where $A := R[u^{\pm 1}, V_1, \ldots, V_l]$. Recall that $K$ denotes the field of fractions of $K$ and that

$$\mathrm{Irr}(\mathbf{H}_{K,n}) = \{E_\varepsilon^\lambda \mid \lambda \vdash_l n\}.$$

We now wish to define a Fock datum for $\mathbf{H}_n$ as introduced in 3.4.1. Let $\theta : A \to k$ be a specialisation such that

$$\theta(u) \neq 1 \qquad \text{and} \qquad \theta(V_j) = \theta(u)^{s_j} \quad \text{for some } s_j \in \mathbb{Z} \quad (1 \leqslant j \leqslant l);$$

see Remark 5.5.21. Let $e > 1$ be the order of $\theta(u) \in k^\times$ and write $\eta_e := \theta(u)$. Thus, if $e < \infty$, then $\eta_e$ is a primitive root of unity of order $e$.

We can define central elements in a similar way as in 3.5.1 by setting

$$z_m = L_1 + \ldots + L_m,$$

where the $L_i$ are the Jucys–Murphy elements given in 5.2.3. The fact that $z_m \in Z(\mathbf{H}_n)$ directly follows from Remark 5.2.4.

(We have excluded the case where $\theta(u) = 1$ because it is somewhat special in this context; it has been studied by Mathas in [244]. See Proposition 4.6.6, where the analogous problem in type $B_n$ is solved.)

**Proposition 5.6.2.** *The family* $\{(\mathbf{H}_m, z_m) \mid 1 \leqslant m \leqslant n\}$ *defines an abstract Fock datum in the sense of 3.4.1.*

*Proof.* First, we have $\mathbf{H}_1 \subseteq \mathbf{H}_2 \subseteq \ldots \subseteq \mathbf{H}_n$ so that (F1) is satisfied. By Theorem 5.2.5, the algebra $\mathbf{H}_n$ is finitely generated and free over $A$. In addition, we have

$$\mathbf{H}_n = \bigoplus_{0 \leqslant i \leqslant l-1} \bigoplus_{d \in \mathscr{D}^{n-1}} \mathbf{H}_{n-1} L_n^i T_d,$$

where $\mathscr{D}^{n-1}$ is the set of distinguished left coset representatives of $\mathfrak{S}_{n-1}$ in $\mathfrak{S}_n$. Hence, $\mathbf{H}_n$ is free as a left $\mathbf{H}_{n-1}$-module. So (F2) is satisfied. The property (F3) has already been established. Finally, the facts that $\mathbf{H}_{K,n}$ is split semisimple and that $\mathbf{H}_{k,n}$ is split follow from the cellularity of these algebras. Hence, (F4) holds. □

Given the above Fock datum, we can define operators $\zeta\text{-Ind}_K^{(n)}$ with respect to the central elements $\{z_m\}$ for all $\zeta \in k$. The aim of the following parts is to study these operators.

**5.6.3.** Let us denote by $\mathrm{Ind}_K^{(n)}$ the induction of representations from $\mathbf{H}_{K,n-1}$ to $\mathbf{H}_{K,n}$ (where $\mathbf{H}_{K,0} = \{0\}$ by convention). Thus, for any $\mathbf{H}_{K,n-1}$-module $V$, we have

$$\mathrm{Ind}_K^{(n)}(V) := \mathbf{H}_{K,n} \otimes_{\mathbf{H}_{K,n-1}} V.$$

Let $n \geqslant 1$ and $\lambda \vdash_l n$. Then we already know by Proposition 5.1.8 that the restriction of $E^\lambda$ to $W_{n-1}$ is the direct sum of all $E^\mu$, where $\mu$ runs over all the $l$-partitions of $n-1$ such that $[\mu] = [\lambda] \setminus \{\gamma\}$ with $\gamma \in \mathscr{R}(\lambda)$. By Frobenius reciprocity and a general criterion of compatibility of induction with specialisations, we obtain

$$\mathrm{Ind}_K^{(n)}(E_\varepsilon^\mu) \simeq \bigoplus_{\lambda \in \Lambda'(\mu)} E_\varepsilon^\lambda \qquad \text{for any } \mu \vdash_l n-1,$$

where $\Lambda'(\mu)$ is the set of all $\lambda \vdash_l n$ such that $[\lambda] = [\mu] \cup \{\gamma\}$ and $\gamma \in \mathscr{A}(\mu)$. By the discussion in 3.4.1, we can now define for all $\zeta \in k$ certain refined induction operators $\zeta\text{-Ind}_K^{(n)}$. We want to give explicit formulae for these operators. To do this, we need to study the elements $\omega_\lambda(z_m)$ for all $l$-partitions $\lambda$. Our strategy will be to use the cellular structure of Ariki–Koike algebras on $\mathbf{H}_n$ using the following result of [184, Proposition 3.7].

**Lemma 5.6.4 (James and Mathas [184]).** *Let* $\lambda \vdash_l n$. *Let* $\mathfrak{t}$ *be a standard* $\lambda$*-tableau and let* $i$ *be an integer such that* $1 \leqslant i \leqslant n$. *Assume that* $\mathfrak{t}^{-1}(i) = (a,b,c)$. *Then for each* $\mathfrak{s} \in \mathrm{Std}(\lambda)$ *there exists* $a_\mathfrak{s} \in R$ *such that*

$$L_i.m_{\mathfrak{t},\mathfrak{t}^\lambda} = V_c u^{b-a} m_{\mathfrak{t},\mathfrak{t}^\lambda} + \sum_{\mathfrak{s} \rhd \mathfrak{t}} a_\mathfrak{s} m_{\mathfrak{s},\mathfrak{t}^\lambda}.$$

*Proof.* Assume first that $\mathfrak{t} = \mathfrak{t}_\lambda$ so that $m_{\mathfrak{t},\mathfrak{t}^\lambda} = m_\lambda$. Let $j := \min\{\mathfrak{t}(\alpha,\beta,c) \mid \alpha \geqslant 1, 1 \leqslant \beta \leqslant \lambda_a^c\}$. Then we have $j \leqslant i$. We denote $T_{i,j} = T_{i-1}T_{i-2}\ldots T_j \in \mathfrak{S}_\lambda$ (with $T_{i,j} = 1$ if $j = i$) and $T_{j,i} = T_{i,j}^*$. As $T_{i,j}$ commutes with $L^{\mathbf{a}}$ (see Definition 5.3.9), we obtain that

$$\begin{aligned} L_i.m_\lambda &= L_i L^{\mathbf{a}} x_\lambda \\ &= u^{(j-i)} T_{i,j} L_j L^{\mathbf{a}} T_{j,i} x_\lambda \\ &= u^{(j-i)} T_{i,j}(L_j - V_c) L^{\mathbf{a}} T_{j,i} x_\lambda + V_c u^{(j-i)} L^{\mathbf{a}} T_{i,j} T_{j,i} x_\lambda. \end{aligned}$$

If we set $\mathbf{b} := (a_1,\ldots,a_{c-1},a_c+1,a_{c+1},\ldots,a_r)$, we obtain by definition $T_{i,j}(L_j - V_c)L^{\mathbf{a}}T_{j,i}x_\lambda = T_{i,j}L^{\mathbf{b}}T_{j,i}x_\lambda$. Note that we have $T_{i,j}L^{\mathbf{b}}T_{j,i}x_\lambda \in \mathbf{H}_n m_\lambda \mathbf{H}_n$.

By 2.6.1(C3), this implies that $T_{i,j}L^{\mathbf{b}}T_{j,i}x_\lambda \in \mathbf{H}_n(\trianglerighteq\lambda)$. We claim that we have in fact $T_{i,j}L^{\mathbf{b}}T_{j,i}x_\lambda \in \mathbf{H}_n(\triangleright\lambda)$. This can be seen as follows. If $c \neq 1$, consider the $l$-partition $\nu$ such that for all $i = 1,\ldots,l$ (observe that $\lambda^c \neq \varnothing$):

$$\nu^i = \begin{cases} (\underbrace{1.1.\ldots.1}_{|\lambda^i| \text{ times}}) & \text{if } i \neq c, c-1, \\ (\underbrace{1.1.\ldots.1}_{|\lambda^{c-1}|+1 \text{ times}}) & \text{if } i = c-1, \\ (\underbrace{1.1.\ldots.1}_{|\lambda^c|-1 \text{ times}}) & \text{if } i = c. \end{cases}$$

Then we have $T_{i,j}L^{\mathbf{b}}T_{j,i}x_\lambda \in \mathbf{H}_n m_\nu \mathbf{H}_n \subset \mathbf{H}_n(\trianglerighteq\nu)$. If $c = 1$, then one can set

$$\nu^i = \begin{cases} (\underbrace{1.1.\ldots.1}_{|\lambda^i| \text{ times}}) & \text{if } i \neq 1,2, \\ (\underbrace{1.1.\ldots.1}_{|\lambda^2|-1 \text{ times}}) & \text{if } i = 2, \\ (\underbrace{1.1.\ldots.1}_{|\lambda^1|+1 \text{ times}}) & \text{if } i = 1. \end{cases}$$

This leads to the same property: $T_{i,j}L^{\mathbf{b}}T_{j,i}x_\lambda \in \mathbf{H}_n(\trianglerighteq\nu)$ by the definition of cellular algebras. Note that in both cases $\lambda$ is not bigger than $\nu$ with respect to the dominance order. Hence, we obtain that $T_{i,j}L^{\mathbf{b}}T_{j,i}x_\lambda \in \mathbf{H}_n(\triangleright\lambda)$ and thus we can deduce that

$$L_i.m_\lambda \equiv V_c u^{(j-i)} T_{i,j} T_{j,i} L^{\mathbf{a}} x_\lambda \mod \mathbf{H}_n(\triangleright\lambda).$$

Now the element $u^{(j-i)}T_{i,j}T_{j,i}$ is a Jucys–Murphy element in type $A_{n-1}$. The action of such element on $x_\lambda$ can be deduced from [245, Theorem 3.32]. We obtain that

$$L_i.m_\lambda \equiv u^{b-a} V_c u^{(j-i)} m_\lambda \mod \mathbf{H}_n(\triangleright\lambda).$$

This completes the proof in the case $\mathfrak{t} = \mathfrak{t}_\lambda$. Assume now that $\mathfrak{t} \neq \mathfrak{t}_\lambda$. Then there exists an integer $k$ such that $\mathfrak{s} = s_k\mathfrak{t}$ and $\mathfrak{s} \rhd \mathfrak{t}$. In this case, one can show $m_{\mathfrak{t},\mathfrak{t}_\lambda} = T_k m_{\mathfrak{s},\mathfrak{t}_\lambda}$ (e.g. see [245, Corollary 3.4]). The lemma follows using exactly the same arguments as in [245, Lemma 3.29]. □

**Lemma 5.6.5.** *Let $n \geqslant 1$, $\mu \vdash_l n-1$ and $\lambda \in \Lambda'(\mu)$. Then writing $[\lambda] = [\mu] \cup \{\gamma\}$, where $\gamma = (a,b,c)$ is a removable node of $\lambda$ (or, equivalently, an addable node for $\mu$), we have*

$$\omega_\lambda(z_n) - \omega_\mu(z_{n-1}) = V_c u^{b-a}.$$

*Proof.* Let $\mathfrak{t}$ be a standard $\lambda$-tableau. By Lemma 5.6.4, for each $\mathfrak{s} \in \mathrm{Std}(\lambda)$ there exists $a_\mathfrak{s} \in R$ such that

$$z_n.m_{\mathfrak{t},\mathfrak{t}^\lambda} = \Big(\sum_{(a,b,c)\in[\lambda]} V_c u^{b-a}\Big) m_{\mathfrak{t},\mathfrak{t}^\lambda} + \sum_{\mathfrak{s}\rhd\mathfrak{t}} a_\mathfrak{s} m_{\mathfrak{s},\mathfrak{t}^\lambda}.$$

Now, by Remark 5.3.12, we have $E_\varepsilon^\lambda \simeq \widetilde{S}_K^\lambda$ and $\widetilde{S}_K^\lambda$ has a basis given by the elements $m_\mathfrak{t}$ with $\mathfrak{t} \in \mathrm{Std}(\lambda)$. The action of $z_n$ is given by (see 2.6.1)

$$z_n.m_\mathfrak{t} = \Big(\sum_{(a,b,c)\in[\lambda]} V_c u^{b-a}\Big) m_\mathfrak{t}.$$

We thus obtain that $\omega_\lambda(z_n) = \sum_{(a,b,c)\in[\lambda]} V_c u^{b-a}$, as required. □

We can now use the same strategy as in 3.5.4 to deduce the following result (see [10, Lemma 13.37]).

**Proposition 5.6.6 (Ariki [10]).** *Let $i \in \{0,1,\ldots,e-1\}$ and $n \geqslant 1$. Then we have*

$$\eta_e^i\text{-}\mathrm{Ind}_K^{(n)}(E_\varepsilon^\mu) \simeq \bigoplus_{\lambda\vdash_l n \text{ such that } \mu \xrightarrow{i} \lambda} E_\varepsilon^\lambda \qquad \text{for any } \mu \vdash_l n-1,$$

*where we write $\mu \xrightarrow{i} \lambda$ if $[\lambda] = [\mu] \cup \{x\}$ and $x = (a,b,c)$ is an addable node of $\mu$ such that $b-a+s_c \equiv i \bmod e$.*

*Proof.* Let $n \geqslant 1$, $\mu \vdash_l n-1$ and $\lambda \in \Lambda'(\mu)$. Using the specialisation $\theta$ of 5.6.1 and Lemma 5.6.5, we obtain $\overline{\omega}_\lambda(z_n) - \overline{\omega}_\mu(z_{n-1}) = \eta_e^{b-a+s_c}$. The proof is now entirely similar to that of Proposition 3.5.5, and follows from the definition of the operators $\eta_e^i\text{-}\mathrm{Ind}_K^{(n)}$. □

**5.6.7.** If $x = (a,b,c)$ is a node of an $l$-composition $\lambda$, we define $b-a+s_c \bmod e$ to be the *residue* of this node, we denote $i := \mathfrak{r}_e(x)$ and we say that $x$ is an $i$-node. In the following, we will often identify an element of $\mathbb{Z}/e\mathbb{Z}$ with its representant in $\{0,1,\ldots,e-1\}$. If, moreover, $\lambda$ is an $l$-partition, we denote by $\mathscr{A}_i(\lambda)$ the set of addable $i$-nodes of $\lambda$ and by $\mathscr{R}_i(\lambda)$ the set of removable $i$-nodes of $\lambda$. We have

$$\mathscr{A}(\lambda) = \bigsqcup_{i\in\mathbb{Z}/e\mathbb{Z}} \mathscr{A}_i(\lambda) \qquad \text{and} \qquad \mathscr{R}(\lambda) = \bigsqcup_{i\in\mathbb{Z}/e\mathbb{Z}} \mathscr{R}_i(\lambda).$$

Hence, in the above proposition, we have $\mu \xrightarrow{i} \lambda$ if and only if $[\lambda] = [\mu] \cup \{\gamma\}$ for $\gamma \in \mathscr{A}_i(\lambda)$. Similarly, we will often simply write $i$-$\mathrm{Ind}_K^{(n)}$ instead of $\eta_e^i$-$\mathrm{Ind}_K^{(n)}$.

**Fig. 5.1** Part of the LLT-graph for $l = 2$, $e = 2$, $s_1 = 0$ and $s_2 = 1$.

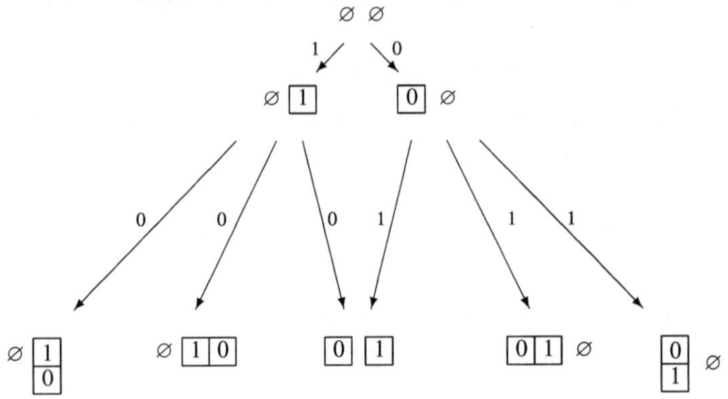

**Example 5.6.8.** In the same spirit as Example 3.5.6, one can define the LLT-graph as follows. The vertices are given by the all $l$-partitions and the edges are defined by the relation $\mu \xrightarrow{i} \lambda$ in Proposition 5.6.6.

**5.6.9.** Following 3.4.1, we can now define a family of representations $\mathscr{E}$ of $\mathbf{H}_{K,n}$ inductively as follows. For $m = 1$, let $\mathscr{E}_1 = \{\mathbf{1}\}$, where $\mathbf{1}$ denotes the trivial representation of $\mathbf{H}_{K,1} = \langle 1_{\mathbf{H}_{K,1}} \rangle_K$. Then set

$$\mathscr{E}_m = \{\xi\text{-}\mathrm{Ind}_K^{(m)}(V) \mid V \in \mathscr{E}_{m-1} \text{ and } \xi \in k\} \setminus \{0\} \quad \text{for } 2 \leqslant m \leqslant n.$$

As in type $A_{n-1}$, the LLT-graph allows us to understand the elements of $\mathscr{E}_1$, $\mathscr{E}_2$, ... for Ariki–Koike algebras. Let $\mathscr{I}_n^{(e)}$ be the set of all sequences $I = (i_1, i_2, \ldots, i_n)$ such that $i_j \in \{0, 1, \ldots, e-1\}$ for all $j \geqslant 1$. We obtain a sequence of representations:

$$Y_{(i_1)} := \mathbf{1}, \quad Y_{(i_1,i_2)} := i_2\text{-}\mathrm{Ind}_K^{(2)}\left(Y_{(i_1)}\right), \quad \ldots, \quad Y_I := i_n\text{-}\mathrm{Ind}_K^{(n)}\left(Y_{(i_1,\ldots,i_{n-1})}\right).$$

Hence, with this notation, we have $\mathscr{E}_n = \{Y_I \mid I \in \mathscr{I}_n^{(e)}\} \setminus \{0\}$.

**Example 5.6.10.** For example, let $e = 2$, $s_1 = 0$ and $s_2 = 1$. Figure 5.1 shows the part of the LLT-graph graph corresponding to all partitions of $n \leqslant 5$. We have

$$\mathscr{E}_1 = \{Y_{(0)} = (1, \varnothing), Y_{(1)} = (\varnothing, 1)\},$$
$$\mathscr{E}_2 = \{Y_{(1,0)} = (\varnothing, 1.1) \oplus (\varnothing, 2) \oplus (1,1); Y_{(0,1)} = (1.1, \varnothing) \oplus (2, \varnothing) \oplus (1,1)\},$$

where we just write $\lambda$ instead of $E_\varepsilon^\lambda$.

We end this section with a formula for the computation of the multiplicities of the simple modules in an $i$-induction process.

**Corollary 5.6.11.** *Let $\lambda \vdash_l n$ and $\mu \vdash_l n-1$, where $n \geqslant 1$. Let $i \in \{0,\dots,e-1\}$ and $s \in \mathbb{Z}_{>0}$. If $E_\varepsilon^\lambda$ appears with non-zero multiplicity as a direct summand in $(i\text{-Ind}_K^{(n)})^s(E_\varepsilon^\mu)$, then this multiplicity is $s!$.*

*Proof.* Assume that $E_\varepsilon^\lambda$ appears with non-zero multiplicity in $(i\text{-Ind}_K^{(n)})^s(E_\varepsilon^\mu)$. This means that there exist addable $i$-nodes $\gamma_1, \dots, \gamma_s$ for $\mu$ such that $[\lambda] = [\mu] \cup \{\gamma_1,\dots,\gamma_s\}$. Depending on the order we choose to add the $\gamma_i$, we see that we have exactly $s!$ possibilities to obtain $\lambda$ from $\mu$. Note, indeed, that as all the $\gamma_j$ are $i$-nodes, adding these nodes in *any* order to $\lambda$ still gives well-defined multipartitions. This concludes the proof. □

## 5.7 FLOTW Multipartitions

We continue with our programme to establish the existence of canonical basic sets for Ariki–Koike algebras. We have already introduced a Fock datum for these algebras. Recall that we assume throughout that $\theta(u) \neq 1$ and that there exists $(s_1,\dots,s_l) \in \mathbb{Z}^l$ such that $\theta(V_j) = \theta(u)^{s_j}$ for all $j = 1,\dots,n$. As before, we write $\eta_e = \theta(u)$, where $e > 1$ is the order of $\theta(u) \in k^\times$. Thus, if $e < \infty$, then $\eta_e$ is a primitive root of unity of order $e$. Note that each $s_j$ is only determined up to adding integer multiples of $e$ (if $e < \infty$). We shall now also assume that $\mathbf{s} := (s_1,\dots,s_l) \in \mathscr{S}_e^l$ where

$$\mathscr{S}_e^l := \{\mathbf{s} = (s_1,\dots,s_l) \in \mathbb{Z}^l \mid 0 \leqslant s_j - s_i < e \text{ for all } i < j\}.$$

Under these conditions, we wish to prove the existence of canonical basic sets in the sense of Definition 5.5.19. To do this, we use a similar strategy as for the symmetric group. This will require the following ingredients:

- We must fix a weight sequence $\mathbf{m}$ which induces the partial order $\ll_{\mathbf{m}}$ on the set of $l$-partitions.
- We must find analogues of the set of $e$-regular partitions in our setting.

**5.7.1.** In this part, we will make a special choice for the weight sequence. Let $\mathbf{v} = (v_1,\dots,v_l) \in \mathbb{Q}^l$ be such that

$$0 < v_j - v_i < e \qquad \text{for all } i < j.$$

Then we set

$$\mathbf{m} := (s_1 - v_1,\dots,s_l - v_l) \in \mathbb{Q}^l$$

and we consider a positive integer $r$ such that $r.\mathbf{m} \in \mathbb{Z}^l$. We then have an $\mathbf{a}$-function $\mathbf{a}^{(\mathbf{m},r)}$ which depends on the choices of $r$, $\mathbf{s}$ and $\mathbf{v}$. This choice for the weight sequence may look mysterious at the moment. Let us explain why we introduce such parameters. Recall that the **a**-invariants for finite Coxeter groups were introduced in Section 1.3 with respect to a choice of an arbitrary weight function $L : W \to \Gamma$. For Ariki–Koike algebras, these **a**-invariants were defined in a slightly different way, by studying the Schur elements of a cyclotomic Ariki–Koike algebra associated with the datum of **m**. The above choice of **m** (see also 6.6.5) is what we need to deduce in full generality the canonical basic sets for type $B_n$ from the properties that we will obtain at the level of Ariki–Koike algebras (see also Remark 5.7.4 below).

**Example 5.7.2.** Assume that $l = 1$, $e \in \mathbb{Z}_{>0}$ and consider the Hecke algebra of type $A_{n-1}$ with weight function $L$ such that $L(s) = a \in \mathbb{Z}_{>0}$ for all simple reflections $s$. Assume in addition that we have a specialisation $\theta$ such that $\theta(u) = \eta_e$. We can then set $s_1 = v_1 = 1$ and $r = a$. In this case, the generic Hecke algebra of type $A_{n-1}$ is nothing but the cyclotomic Ariki–Koike algebra associated with $\mathbf{m} = (0)$ and $r$. Thus, $\mathbf{a}^{(\mathbf{m},r)}$ is the usual **a**-invariant in type $A_{n-1}$.

**Example 5.7.3.** Let $W$ be the Weyl group of type $B_n$ with weight function $L : W \to \mathbb{Z}$ given by

$B_n$

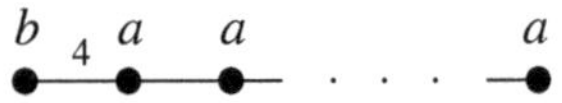


where $a > 0$ and $b \geqslant 0$.

Let $\mathbf{H}$ be the corresponding generic Iwahori–Hecke algebra over the ring $R[v, v^{-1}]$, where $v$ is an indeterminate; let $u = v^2$. We consider a specialisation $\theta : R[v, v^{-1}] \to k$ and set $\eta := \theta(u)$. In Section 4.6, we have already considered the cases where $\eta^a = 1$ or where $\eta^b \neq -\eta^{ad}$ for all $d \in \mathbb{Z}$. So now we can assume that $\eta^a \neq 1$ and that there exists an integer $d \in \mathbb{Z}$ such that $\eta^b = -\eta^{ad}$. Assume that $\eta^a$ is a primitive root of unity of order $e > 1$ and set $\eta_e := \eta^a$. Furthermore, assume that $d$ can be chosen such that the following condition holds:

$$(\star) \qquad d - e < b/a < d.$$

We then set $v_1 = d - b/a$ and $v_2 = e$. Then we have $0 < v_2 - v_1 < e$. We then also set $s_1 = d$, $s_2 = e$. Note that we do not have $(s_1, s_2) \in \mathscr{S}_e^2$ in general! Nevertheless, in all cases, the Ariki–Koike algebra $\mathbf{H}_{k,n}(\eta_e; \eta_e^{s_1}, \eta_e^{s_2})$ is isomorphic to the specialised algebra $\mathbf{H}_k$ of type $B_n$. In addition, if we set $r = a$ and $\mathbf{m} = (b/a, 0) = (s_1 - v_1, s_2 - v_2)$, then $\mathbf{a}^{(\mathbf{m},r)}$ is the usual **a**-function in type $B_n$; see Example 5.5.14.

*Remark 5.7.4.* The above choice of **m** comes from a generalisation of a very natural choice for the cyclotomic Ariki–Koike algebra (see [174, Section 4.1]). Indeed, set $r = l$ and for all $j = 1, \ldots, l$, $m^j = s_j - v_j$, where $v_j = je/l$. Then we have an associated cyclotomic Ariki–Koike algebra $\mathbf{H}_{R[y^{\pm 1}],n}$. We now consider the specialisation $\theta_y : R[y^{\pm 1}] \to k$ such that $\theta_y(y)$ is an $l$-root of $\eta_e$. Then the specialised algebra through this specialisation is nothing but $\mathbf{H}_{k,n} := \mathbf{H}_{k,n}(\eta_e; \eta_e^{s_1}, \ldots, \eta_l^{s_l})$. Thus, the weight sequence **m** defines a cyclotomic Ariki–Koike algebra which can be specialised to $\mathbf{H}_{k,n}$.

**5.7.5.** We now need to define a partial order on the set of $l$-compositions. Let $\mathbf{s} \in \mathbb{Z}^l$. Let $\lambda$ and $\mu$ be both $l$-compositions of rank $n$; we write

$$\lambda \ll_{\mathbf{m}} \mu \quad \stackrel{\text{def}}{\Longleftrightarrow} \quad \lambda = \mu \quad \text{or} \quad \kappa_{\mathbf{m}}(\lambda) \rhd \kappa_{\mathbf{m}}(\mu),$$

where we recall that $\mathbf{m} := (s_1 - v_1, \ldots, s_l - v_l)$ for $\mathbf{v} = (v_1, \ldots, v_l) \in \mathbb{Q}^l$ such that for all $i < j$ we have $0 < v_j - v_i < e$ (see 5.5.10) and $\mathbf{s} \in \mathbb{Z}^l$.

**Example 5.7.6.** Assume that $l = 1$ and that $\mathbf{m} \in \mathbb{Z}$; then, by the definition of $\ll_{\mathbf{m}}$, we have $\lambda \ll_{\mathbf{m}} \mu$ if and only if $\lambda \trianglerighteq \mu$.

**Proposition 5.7.7.** *Let* $\mathbf{s} \in \mathbb{Z}^l$. *If* $\lambda$ *and* $\mu$ *are two l-partitions such that* $\lambda \ll_{\mathbf{m}} \mu$ *and* $\lambda \neq \mu$, *then we have*

$$\mathbf{a}^{(\mathbf{m},r)}(\lambda) < \mathbf{a}^{(\mathbf{m},r)}(\mu).$$

*Proof.* This follows from Proposition 5.5.11. □

In this chapter, we will mainly use the partial order $\ll_{\mathbf{m}}$ in the case where $\mathbf{s} \in \mathscr{S}_e^l$.

We now consider the second problem of finding an appropriate generalisation of the notion of $e$-regular partitions. This is given by the set of FLOTW $l$-partitions which we now define.

**Definition 5.7.8 (Foda, Leclerc, Okado, Thibon and Welsh [88, 2.11]).** Let $l \in \mathbb{Z}_{>0}$ and $e \in \mathbb{Z}_{>1}$. Recall that we defined

$$\mathscr{S}_e^l := \{\mathbf{s} = (s_1, \ldots, s_l) \in \mathbb{Z}^l \mid 0 \leqslant s_j - s_i < e \text{ for all } i < j\}.$$

Assume that $\mathbf{s} = (s_1, \ldots, s_l) \in \mathscr{S}_e^l$. Then, the multipartition $\lambda = (\lambda^1, \ldots, \lambda^l)$ is called an FLOTW *l-partition* if and only if:

1. $\lambda$ is a *cylindrical multipartition*; that is, for every $k = 1, \ldots, l-1$ we have $\lambda_i^k \geqslant \lambda_{i+s_{k+1}-s_k}^{k+1}$ for all $i \geqslant 1$ (the partitions are taken with an infinite numbers of empty parts) and $\lambda_i^l \geqslant \lambda_{i+e+s_1-s_l}^1$ for all $i \geqslant 1$.
2. For all $r > 0$, among the residues appearing at the right ends of the length $r$ rows of $\lambda$, at least one element of $\{0, 1, \ldots, e-1\}$ does not appear.

We denote by $\Phi'_{(\mathbf{s},\mathbf{e})}$ the set of FLOTW $l$-partitions associated with $\mathbf{s} \in \mathscr{S}_e^l$ and by $\Phi'_{(\mathbf{s},\mathbf{e})}(n)$ the set of FLOTW $l$-partitions of rank $n$.

*Remark 5.7.9.* The second condition of Definition 5.7.8 can be alternatively stated as follows. For all $r > 0$, the map $\gamma := (a,b,r) \in \mathscr{N}(\lambda) \mapsto \mathfrak{r}_e(\gamma) \in \mathbb{Z}/e\mathbb{Z}$ is not surjective (see Definition 5.1.5 and 5.6.7).

**5.7.10.** Of course, this definition may not be very natural at the moment. However, we will really see in the next chapter why this is the kind of multipartition that we have to study in our framework. For instance, we can just note that if we take $l = 1$, the first condition above is empty and the second one is equivalent to the definition of "$e$-regularity" for partitions. Thus, the FLOTW $l$-partitions may been seen as generalisations of $e$-regular partitions.

**Example 5.7.11.** Let $l = 3$, $e = 4$, $s_1 = 0$ and $s_2 = s_3 = 1$; then we consider the 3-partition $(3.1.1, 3.1, 1)$ of 10. The Young diagram with residue is as follows:

$$\left( \begin{array}{|c|c|c|} \hline 0 & 1 & 2 \\ \hline 3 \\ \cline{1-1} 2 \\ \cline{1-1} \end{array}, \begin{array}{|c|c|c|} \hline 1 & 2 & 3 \\ \hline 0 \\ \cline{1-1} \end{array}, \begin{array}{|c|} \hline 1 \\ \hline \end{array} \right).$$

Then note that the set of residues appearing at the right ends of the part with length 1 is $\{0, 1, 2, 3\}$. Hence, the second condition of FLOTW 3-partitions is violated and $(3.1.1, 3.1, 1)$ is not an FLOTW 3-partition.

**Example 5.7.12.** Here is the list of FLOTW 2-partitions when $e = 4$, $s_1 = 0$ and $s_2 = 1$ when $n \leqslant 4$:

$$\begin{aligned}
&\Phi'_{((0,1),4)}(1) = \{(\varnothing, 1), (1, \varnothing)\}, \\
&\Phi'_{((0,1),4)}(2) = \{(\varnothing, 2), (1, 1), (1.1, \varnothing), (2, \varnothing)\}, \\
&\Phi'_{((0,1),4)}(3) = \{(\varnothing, 3), (1, 1.1), (1, 2), (1.1, 1), (1.1.1, \varnothing), (2, 1), (2.1, \varnothing), (3, \varnothing)\}, \\
&\Phi'_{((0,1),4)}(4) = \{(\varnothing, 4), (1, 2.1), (1, 3), (1.1, 1.1), (1.1, 2), (2, 1.1), (2, 2), (2.1, 1), \\
&\qquad\qquad (2.1.1, \varnothing), (2.2, \varnothing), (3, 1), (3.1, \varnothing), (4, \varnothing)\}.
\end{aligned}$$

In the rest of the chapter, we will show that these kinds of multipartitions enjoy similar properties as the ones given in Section 3.5 for the set of $e$-regular partitions. To do this, we need some preparatory results. Recall that $\mathscr{N}_\lambda$ denotes the set of border nodes of $\lambda$. We also denote by $\mathscr{N}_\lambda(k)$ the set of border nodes of $\lambda$ with residue $k \in \mathbb{Z}/e\mathbb{Z}$.

**Lemma 5.7.13.** *Let $e \in \mathbb{Z}_{>0}$. Assume that $\mathbf{s} \in \mathscr{S}_e^l$ and let $\lambda = (\lambda^1, \ldots, \lambda^l) \in \Phi'_{(\mathbf{s},\mathbf{e})}$ be a non-empty* FLOTW *$l$-partition. We set*

$$l_{max} := \max\{\lambda_1^1, \lambda_1^2, \ldots, \lambda_1^l\}.$$

*Then, there exists a removable $k$-node $\gamma_1$, for some $k$, on a part $\lambda_{j_1}^{i_1} = l_{max}$ which satisfies the following property. If $\gamma_2 \in \mathscr{N}_\lambda(k-1)$ is on a part $\lambda_{j_2}^{i_2}$, then*

$$\lambda_{j_1}^{i_1} > \lambda_{j_2}^{i_2}.$$

*Proof.* Let $\lambda^{l_1}, \ldots, \lambda^{l_r}$ be the partitions of $\lambda$ such that $\lambda_1^{l_1} = \ldots = \lambda_1^{l_r} = l_{\max}$ are the parts of maximal length. Let $k_1, \ldots, k_r$ be the residues of the removable nodes $\gamma_1, \ldots, \gamma_r$ on the parts of length $l_{\max}$. We want to show that there exists $1 \leqslant i \leqslant r$ such that there is no node with residue $k_i - 1$ on the border of a part with length $l_{\max}$.

Assume that, for each $1 \leqslant i \leqslant r$, there exists a node on the border of a part of length $l_{\max}$ with residue $k_i - 1$; see the first of the two diagrams below. Then, there exists a partition $\lambda^{l_{t_1}}$, for some $1 \leqslant t_1 \leqslant r$, with a $k_1 - 1$-node on the border of a part of length $l_{\max}$. Let $k_{t_1}$ be the residue of the removable node on the border of the part of $\lambda^{l_{t_1}}$ with length $l_{\max}$. We have $t_1 \neq 1$, otherwise the nodes on the border of the parts with length $l_{\max}$ on $\lambda^{l_{t_1}}$ would describe the set $\{0,\ldots,e-1\}$. This violates the second condition of FLOTW $l$-partition.

$$\lambda^{l_{t_1}} = \begin{array}{|c|c|}\hline \cdots & \cdots \\ \cdots & \cdots \\ \hline & k_1-1 \\ \hline \cdots & \cdots \\ \cdots & \cdots \\ \hline & k_{t_1} \\ \hline \cdots & \\ \cdots & \end{array} \qquad\qquad \lambda^{l_{t_2}} = \begin{array}{|c|c|}\hline \cdots & \cdots \\ \cdots & \cdots \\ \hline & k_{t_1}-1 \\ \hline \cdots & \cdots \\ \cdots & \cdots \\ \hline & k_{t_2} \\ \hline \cdots & \\ \cdots & \end{array}$$

We use the same strategy for the residue $k_{t_1}$. There exists $\lambda^{l_{t_2}}$, for some $1 \leqslant t_2 \leqslant r$, with a $k_{t_1}-1$-node on the border of a part of length $l_{\max}$; see the second of the above two diagrams. Let $k_{t_2}$ be the residue of the removable node on the border of the part of $\lambda^{l_{t_2}}$ with length $l_{\max}$. We have $t_2 \neq t_1$ (for the same reasons as above) and $t_2 \neq 1$, otherwise the nodes on the border of the parts with length $l_{\max}$ on $\lambda^{l_{t_2}}$ and on $\lambda^{l_{t_1}}$ would describe all the set $\{0,\ldots,e-1\}$. Continuing in this way, we finally obtain that there exists $1 \leqslant t_r \leqslant r$ such that

$$t_r \notin \{1,t_1,t_2,\ldots,t_{r-1}\}.$$

This is a contradiction, since $t_r \in \{1,\ldots,r\}$ and for all $i \neq j$, we have $t_i \neq t_j$.

So, there exists $1 \leqslant i \leqslant r$ such that there is no $k_i-1$-node on the border of the parts with maximal length. This concludes the proof. □

**5.7.14.** We now associate a canonical standard tableau and a corresponding $e$-residue sequence with each FLOTW $l$-partition. This generalises the work we have already done when $l=1$ in 3.5.9. First, we need some preparatory results which will help us to show the existence our sequence. In particular, we introduce a partial order on the set of $i$-nodes of a given $l$-composition $\lambda$. Let $\gamma$, $\gamma'$ be two $i$-nodes of an $l$-composition $\lambda$. We write

$$\gamma \prec_{(\mathbf{s},e)} \gamma' \qquad \overset{\text{def}}{\Longleftrightarrow} \qquad \begin{cases} \text{either} & b-a+s_c < b'-a'+s_{c'} \\ \text{or} & b-a+s_c = b'-a'+s_{c'} \text{ and } c > c'. \end{cases}$$

Note that this partial order becomes a total order if $\lambda$ is an $l$-partition. This order will play a crucial role in the next chapter, where it is used to construct an action of the quantum group of affine type $A$ on the vector space generated by all the $l$-partitions. It will also be helpful in our context. For the moment one can remark the following point. Assume that we have an $l$-tuple of integers $(v_1,\ldots,v_l)$ such that, for all $i<j$, we have $0 < v_j - v_i < e$. Set $\mathbf{m} := (s_1-v_1,\ldots,s_l-v_l)$. Then, if $\gamma$ and $\gamma'$ are two nodes of the same $l$-partition, we have

$$\gamma \prec_{(\mathbf{s},e)} \gamma' \qquad \Longleftrightarrow \qquad b-a+m^c < b'-a'+m^{c'}.$$

This comes from the fact that, in this case, there exists $m \in \mathbb{Z}$ such that $b - a + s_c = b' - a' + s_{c'} + me$. Then we have $\gamma \prec_{(\mathbf{s},e)} \gamma'$ if and only if $m \leqslant -1$ if $c < c'$ and $m \leqslant 0$ if $c \geqslant c'$. Note that if $b - a + s_c = b' - a' + s_{c'}$ and $c = c'$, then $\gamma$ and $\gamma'$ are not comparable (which does not happen if $\lambda$ is an $l$-partition).

The following proposition shows some sort of compatibility of this order with $\ll_{\mathbf{m}}$.

**Proposition 5.7.15.** *Let $\lambda$ be an $l$-composition. Let $k \in \mathbb{Z}_{>0}$ and let $i \in \mathbb{Z}/e\mathbb{Z}$. Assume that $\gamma_1, \ldots, \gamma_k, \gamma_1', \ldots, \gamma_k'$ are $i$-nodes satisfying the following conditions:*

- *$\mu$ and $\mu'$ given as follows are well-defined $l$-compositions:*

$$[\mu] = [\lambda] \cup \{\gamma_1, \ldots, \gamma_k\}, \qquad [\mu'] = [\lambda] \cup \{\gamma_1', \ldots, \gamma_k'\};$$

- *for all $j = 1, \ldots, k$, we have $\gamma_j \prec_{(\mathbf{s},e)} \gamma_j'$ or $\gamma_j = \gamma_j'$.*

*Then we have $\mu' \ll_{\mathbf{m}} \mu$.*

*Proof.* For all $j = 1, \ldots, k$, set $(a_j, b_j, c_j) := \gamma_j$ and $(a_j', b_j', c_j') := \gamma_j'$. Then, by assumption, we have

$$b_j - a_j + m^{c_j} \leqslant b_j' - a_j' + m^{c_j'} \qquad \text{for all } j,$$

because the nodes $\gamma_j$ and $\gamma_j'$ have all the same residues. We set $\kappa_{\mathbf{m}}(\lambda) = (\kappa_1, \ldots, \kappa_r)$, where $r$ is sufficiently large. Then we have $\kappa_{\mathbf{m}}(\mu) = (\kappa_1', \ldots, \kappa_r')$, where there exist $l_1, \ldots, l_k$ such that $\kappa_{l_i}' = \kappa_{l_i} + 1$ for $i = 1, \ldots, k$ and $\kappa_s = \kappa_s'$ otherwise. Similarly, we have $\kappa_{\mathbf{m}}(\mu') = (\kappa_1', \ldots, \kappa_r')$, where there exist $l_1', \ldots, l_k'$ such that $\kappa_{l_i'}' = \kappa_{l_i'} + 1$ for $i = 1, \ldots, k$ and $\kappa_s = \kappa_s'$ otherwise. Now by the definition of $\kappa_{\mathbf{m}}$, for all $i = 1, \ldots, k$, we have $\kappa_{l_i'} \geqslant \kappa_{l_i}$. This property implies that $\kappa_{\mathbf{m}}(\mu) \trianglelefteq \kappa_{\mathbf{m}}(\mu')$. Moreover, if $\mu \neq \mu'$, then there exists $j$ such that $\gamma_j \neq \gamma_j'$. We obtain that $b_j - a_j + m^{c_j} < b_j' - a_j' + m^{c_j'}$ and thus $\kappa_{\mathbf{m}}(\mu) \triangleleft \kappa_{\mathbf{m}}(\mu')$. Thus, we can conclude that $\mu' \ll_{\mathbf{m}} \mu$. □

The proof of the following proposition is entirely similar.

**Proposition 5.7.16.** *Let $\lambda$ be an $l$-composition. Let $k \in \mathbb{Z}_{>0}$ and let $i \in \mathbb{Z}/e\mathbb{Z}$. Assume that $\gamma_1, \ldots, \gamma_k, \gamma_1', \ldots, \gamma_k'$ are $i$-nodes satisfying the following conditions:*

- *$\mu$ and $\mu'$ given as follows are well-defined $l$-compositions:*

$$[\lambda] = [\mu] \cup \{\gamma_1, \ldots, \gamma_k\}, \; [\lambda] = [\mu'] \cup \{\gamma_1', \ldots, \gamma_k'\};$$

- *for all $j = 1, \ldots, k$, we have $\gamma_j \prec_{(\mathbf{s},e)} \gamma_j'$ or $\gamma_j = \gamma_j'$.*

*Then we have $\mu \ll_{\mathbf{m}} \mu'$.*

In the case where $\lambda$ is an FLOTW $l$-partition, the following result establishes a connection between the above order and the order $\prec_{(\mathbf{s},e)}$ on two removable or addable nodes with the same residue. For the proof, we will need the following basic property: if $\gamma = (a, b, c)$ is a node of an $l$-partition $\lambda$, we have

$$\lambda_a^c = \begin{cases} b & \text{if } \gamma \text{ is removable,} \\ b-1 & \text{if } \gamma \text{ is addable.} \end{cases}$$

**Lemma 5.7.17.** *Let $e \in \mathbb{Z}_{>0}$. Assume that $\mathbf{s} \in \mathscr{S}_e^l$ and let $\lambda = (\lambda^1, \ldots, \lambda^l) \in \Phi'_{(\mathbf{s},e)}$. Let $\gamma_1 = (a_1, b_1, c_1)$ be a removable or addable i-node of $\lambda$ and $\gamma_2 = (a_2, b_2, c_2)$ be a removable or addable i-node of $\lambda$. Assume that $\lambda_{a_1}^{c_1} < \lambda_{a_2}^{c_2}$ then $\gamma_1 \prec_{(\mathbf{s},e)} \gamma_2$.*

*Proof.* First, assume that $\gamma_2 \prec_{(\mathbf{s},e)} \gamma_1$. We want to show that $\lambda_{a_1}^{c_1} \geqslant \lambda_{a_2}^{c_2}$. First, we consider the case where $c_1 < c_2$. We then have

$$b_1 - a_1 + s_{c_1} \geqslant b_2 - a_2 + s_{c_2}.$$

Hence, we obtain $a_1 \leqslant b_1 - b_2 + a_2 + s_{c_1} - s_{c_2}$, which implies that

$$1 \leqslant b_1 - b_2 + a_2 + s_{c_1} - s_{c_2} \leqslant b_1 - b_2 + a_2$$

and that

$$\lambda_{a_1}^{c_1} \geqslant \lambda_{b_1-b_2+a_2+s_{c_1}-s_{c_2}}^{c_1}.$$

By the characterisation of the FLOTW $l$-partitions, we deduce

$$\lambda_{a_1}^{c_1} \geqslant \lambda_{b_1-b_2+a_2}^{c_2}.$$

If $b_1 - b_2 \leqslant 0$, then we obtain $\lambda_{a_1}^{c_1} \geqslant \lambda_{a_2}^{c_2}$, and the case $b_1 > b_2$ also gives the result because then $b_1 = \lambda_{a_1}^{c_1}$ or $b_1 = \lambda_{a_1}^{c_1} - 1$ and $b_2 = \lambda_{a_2}^{c_2}$ or $b_2 = \lambda_{a_2}^{c_2} - 1$.

The case $c_1 = c_2$ is trivial. We finally have to consider the case where $c_1 > c_2$. Because $\gamma_1$ and $\gamma_2$ have the same residue modulo $e$, we have

$$b_1 - a_1 + s_{c_1} \geqslant b_2 - a_2 + s_{c_2} + e.$$

Hence, we obtain $a_1 \leqslant b_1 - b_2 + a_2 + s_{c_1} - s_{c_2} + e$, which implies that

$$1 \leqslant b_1 - b_2 + a_2 + s_{c_1} - s_{c_2} + e \leqslant b_1 - b_2 + a_2$$

and that

$$\lambda_{a_1}^{c_1} \geqslant \lambda_{b_1-b_2+a_2+s_{c_1}-s_{c_2}+e}^{c_1}.$$

By the characterisation of the FLOTW $l$-partitions, we deduce

$$\lambda_{a_1}^{c_1} \geqslant \lambda_{b_1-b_2+a_2}^{c_2}.$$

Then the result follows exactly as above. □

**Lemma 5.7.18.** *Let $n \in \mathbb{Z}_{>0}$. Assume that $\mathbf{s} \in \mathscr{S}_e^l$ and let $\lambda = (\lambda^1, \ldots, \lambda^l) \in \Phi'_{(\mathbf{s},e)}(n)$ be a non-empty FLOTW $l$-partition. Let $\gamma_1$ be a removable k-node, for some $k \in \mathbb{Z}/e\mathbb{Z}$, which is chosen as in Lemma 5.7.13. Assume that*

$$\gamma_s := (a_s, b_s, c_s) \prec_{(\mathbf{s},e)} \cdots \prec_{(\mathbf{s},e)} \gamma_2 := (a_2, b_2, c_2) \prec_{(\mathbf{s},e)} \gamma_1 := (a_1, b_1, c_1)$$

*are the removable k-nodes of* $\lambda$.

*Let* $\gamma=(a,b,c)$ *be the maximal element of* $\mathcal{N}_\lambda(k-1)$ *with respect to* $\prec_{(e,\mathbf{s})}$. *Let* $u=max\{i\in\{1,\ldots,s\},\ b_i>b\}$. *Then, the multipartition* $\lambda'$ *such that* $[\lambda]=[\lambda']\sqcup\{\gamma_1,\ldots,\gamma_u\}$ *is in* $\Phi'_{(\mathbf{s},e)}(m)$ *for some* $m<n$.

*Proof.* The rank of $\lambda'$ is strictly smaller than the rank of $\lambda$ by Lemma 5.7.13. We now have to check the two conditions of the FLOTW $l$-partitions for $\lambda'$.

1. First, we have to check that if $\lambda_{a_j}^{c_j}=\lambda_{a_j+s_{c_j+1}-s_{c_j}}^{c_j+1}$ for some $j\leqslant u$, then there exists $t\leqslant u$ such that $\gamma_t=(a_j+s_{c_j+1}-s_{c_j},\lambda_{a_j+s_{c_j+1}-s_{c_j}}^{c_j+1},c_j+1)$ so that $\gamma_j$ and $\gamma_t$ must both be removed to obtain $\lambda'$. Observe that
$$\lambda_{a_j+s_{c_j+1}-s_{c_j}}^{c_j+1}-(a_j+s_{c_j+1}-s_{c_j})+s_{c_j+1}=\lambda_{a_j}^{c_j}-a_j+s_{c_j}.$$
Thus, the residue of the node $\eta:=(a_j+s_{c_j+1}-s_{c_j},\lambda_{a_j+s_{c_j+1}-s_{c_j}}^{c_j+1},c_j+1)$ is $k$. Moreover, this is a removable one, otherwise we would have
$$\lambda_{a_j+s_{i+1}-s_i+1}^{c_j+1}>\lambda_{a_j+1}^{c_j},$$
contradicting our assumption that $\lambda$ is an FLOTW $l$-partition. Since we have $\lambda_{a_j}^{c_j}=\lambda_{a_j+s_{c_j+1}-s_{c_j}}^{c_j+1}$ and $\lambda_{a_j+s_{c_j+1}-s_{c_j}}^{c_j+1}>b$, by construction, the node $\eta$ is a $\gamma_t$ for $t\leqslant u$. The result follows.
2. We now check the second condition of FLOTW $l$-partitions. The only problem may happen when there exists $t\in\{1,\ldots,u\}$ such that the set of residues of the nodes on the border of the parts of $\lambda$ with length $\lambda_{a_t}^{c_t}-1$ is equal to the following set:
$$\{0,\ldots,e-1\}\setminus\{k-1\}.$$
We want to show that, among the residues on the border of the parts of $\lambda'$ with length $\lambda_{a_t}^{c_t}-1$, $k$ does not occur.
There exists at least one $k$-node on the border of a part of $\lambda$ with length $\lambda_{a_t}^{c_t}-1$. Such a $k$-node must be a removable one. Indeed, otherwise, we would have a $k-1$-node on the border of a part of $\lambda$ with length $\lambda_{a_t}^{c_t}-1$, which contradicts our assumption.
We have $\lambda_{a_t}^{c_t}-1>b$ because there is no $k-1$-node on parts with length $\lambda_{a_t}^{c_t}-1$. So, all the $k$-nodes on the border of parts with length $\lambda_{a_t}^{c_t}-1$ must be removed. Then, the set of residues of the nodes on the border of parts of $\lambda'$ with length $\lambda_{a_t}^{c_t}-1$ is equal to
$$\{0,\ldots,e-1\}\setminus\{k\}.$$

Thus, $\lambda'$ is an FLOTW $l$-partition. □

**5.7.19.** Let us explain the consequences of the above combinatorial lemmas. First assume that we have a standard $\lambda$-tableau $\mathfrak{t}$. Then one can associate with this a sequence of elements in $\mathbb{Z}/e\mathbb{Z}$ called the *e-residue sequence* $\eta_{(\mathbf{s},e)}(\mathfrak{t})$ and which generalises the one we have already defined in Chapter 3 when $l = 1$. This is defined as follows:

$$\eta_{(\mathbf{s},e)}(\mathfrak{t}) = (\mathfrak{r}_e(t^{-1}(1)), \mathfrak{r}_e(t^{-1}(2)), \dots, \mathfrak{r}_e(t^{-1}(n))).$$

The result above gives us a way to define a remarkable $e$-residue sequence attached to each FLOTW $l$-partition. Assume that $s \in \mathscr{S}_e^l$ and let $\lambda = (\lambda^1, \dots, \lambda^l) \in \Phi'_{(\mathbf{s},e)}$ be a non-empty FLOTW $l$-partition. We will construct a canonical standard tableau $\mathfrak{t}_{(\mathbf{s},e)}(\lambda)$ and an associated $e$-residue sequence using the result described above. We first take a removable node $\gamma_1$ as in Lemma 5.7.13. If we have several choices for it, we take the maximal node with respect to $\prec_{(\mathbf{s},e)}$. Set $k := \mathfrak{r}_e(\gamma_1)$. Let $\gamma_1$, $\gamma_2, \dots$, $\gamma_s$ be removable $k$-nodes of $\lambda$ as in the previous lemma. For convenience, we now assume that these nodes are ordered such that

$$\gamma_s \succ_{(\mathbf{s},e)} \gamma_{s-1} \succ_{(\mathbf{s},e)} \cdots \succ_{(\mathbf{s},e)} \gamma_1.$$

We say that these nodes are the admissible nodes for $\lambda$ and that $k$ is the admissible residue. From this, we start building the tableau $\mathfrak{t}_{(\mathbf{s},e)}(\lambda)$ by assigning the number $n$ to $\gamma_1$ then $n-1$ to $\gamma_2, \dots, n-s+1$ to $\gamma_s$. Let $\lambda'$ be the $l$-partition such that

$$[\lambda] = [\lambda'] \cup \{\gamma_1, \gamma_2, \dots, \gamma_s\}.$$

Then $\lambda' \in \Phi'_{(\mathbf{s},e)}$ and the rank of $\lambda'$ is strictly smaller than the rank of $\lambda$ by Lemma 5.7.18. We can apply the same procedure as above for $\lambda'$. This allows us to define a standard $\lambda$-tableau $\mathfrak{t}_{(\mathbf{s},e)}(\lambda)$ by induction. Then we can define $\eta_{(\mathbf{s},e)}(\lambda)$ to be the $e$-residue sequence of $\mathfrak{t}_{(\mathbf{s},e)}(\lambda)$.

Here is an example. We assume that $l = 2$, $s_1 = 0$, $s_2 = 1$ and $e = 4$. The FLOTW 2-partitions are the 2-partitions $(\lambda^0, \lambda^1)$ which satisfy:

- $\lambda_i^1 \geqslant \lambda_{i+1}^2$ and $\lambda_i^2 \geqslant \lambda_{i+3}^1$ for all $i \in \mathbb{Z}_{>0}$;
- for all $k > 0$, among the residues appearing at the right ends of the length $k$ rows of $\lambda$, at least one element of $\{0, 1, 2, 3\}$ does not occur.

We consider the 2-partition $\lambda = (2.2, 2.2.1)$ with the following diagram with residues:

$$\left( \begin{array}{|c|c|} \hline 0 & 1 \\ \hline 3 & 0 \\ \hline \end{array}\, , \begin{array}{|c|c|} \hline 1 & 2 \\ \hline 0 & 1 \\ \hline 3 & \multicolumn{1}{c}{} \\ \cline{1-1} \end{array} \right).$$

Then $\lambda$ is an FLOTW 2-partition.

At the beginning the only candidates for $\gamma_1$ are the nodes $(2,2,1)$ and $(2,2,2)$. However, we have $\mathfrak{r}_e(2,2,2) = 1$, and there is a part of length 2 with a node on the border which has $k - 1 = 0 \pmod{e}$ as a residue. So, we have to take $\gamma_1 = (2,2,1)$ and $k = 0$. Note that we can remove this node, as 3-nodes on the border have maximal length 1. There is no other removable 0-node in $\lambda$. We can check that the 2-partition obtained by removing this node $(2.1, 2.2.1)$ is an FLOTW 2-partition. One has to assign the number 9 to the node $(2,2,1)$ in the standard $\lambda$-tableau $\mathfrak{t}(\lambda)$.

Now, the removable nodes on the part with maximal length have 1 as a residue and there is no 0-node, so we assign 7 and 8 to the nodes $(2,2,2)$ and $(1,2,1)$

respectively. Repeating the same procedure, the associated standard $\lambda$-tableau is

$$\mathfrak{t}_{((0,1),4)}(\lambda) = \left( \begin{array}{|c|c|} \hline 2 & 7 \\ \hline 4 & 9 \\ \hline \end{array} \, , \, \begin{array}{|c|c|} \hline 1 & 6 \\ \hline 3 & 8 \\ \hline 5 \\ \cline{1-1} \end{array} \right)$$

and $\eta_{((0,1),4)}(\lambda) = 1,0,0,3,3,2,1,1,0$. Note that, for *all* $m \in \{1,\ldots,9\}$, the 2-partitions $\lambda[m]$ such that $[\lambda[m]] = \mathfrak{t}_{((0,1),4)}(\lambda)^{-1}\{1,\ldots,m\}$ are FLOTW bipartitions:

$$\lambda[1] = (\varnothing,1), \lambda[2] = (1,1), \lambda[3] = (1,1.1), \lambda[4] = (1.1,1.1), \lambda[5] = (1.1,1.1.1),$$
$$\lambda[6] = (1.1,2.1.1), \lambda[7] = (2.1,2.1.1), \lambda[8] = (2.1,2.2.1), \lambda[9] = (2.2,2.2.1).$$

This is in fact a general property of our construction, as explained in the next result.

**Lemma 5.7.20.** *Let $\lambda$ be an* FLOTW *$l$-partition of $n$. For all $m \in \{1,\ldots,n\}$ define $\lambda[m]$ to be the $l$-partition such that*

$$[\lambda[m]] = \mathfrak{t}_{(\mathbf{s},e)}(\lambda)^{-1}\{1,\ldots,m\}.$$

*Then $\lambda[m]$ is a well-defined* FLOTW *$l$-partition. In addition, $\eta_{(\mathbf{s},e)}(\lambda[m])$ is the subsequence of the $m$ first elements of $\eta_{(\mathbf{s},e)}(\lambda)$ and $\mathfrak{t}_{(\mathbf{s},e)}(\lambda[m])$ is the restriction of $\mathfrak{t}_{(\mathbf{s},e)}(\lambda)$ to $[\lambda[m]] \subseteq [\lambda]$.*

*Proof.* First, note that, for all $i \in \{1,\ldots,n\}$, the $l$-composition $\lambda[i]$ is an $l$-partition because $\mathfrak{t}_{(\mathbf{s},e)}(\lambda)$ is standard (note that $\lambda[n] = \lambda$). Now, assume that $k$ is the admissible residue for $\lambda$ and that $\gamma_1, \ldots, \gamma_s$ are all the admissible $k$-nodes for this $l$-partition written in decreasing order with respect to $\prec_{(\mathbf{s},e)}$. By definition, the $l$-partition obtained from $\lambda$ by removing $\gamma_1, \ldots, \gamma_s$ is $\lambda[n-s]$ and it is an FLOTW $l$-partition by Lemma 5.7.18. Thus, by induction, it suffices to show that, for all $m \in \{n-s,\ldots,n\}$, the $l$-partition $\lambda[m]$ is an FLOTW $l$-partition. Assume to the contrary that $\lambda[m]$ is not an FLOTW $l$-partition. One can assume that $m$ is maximal for this property so that $\lambda[m+1]$ is an FLOTW $l$-partition. Write $\gamma_{n-m} = (a,b,c)$ so that $[\lambda[m]] \cup \{\gamma_{n-m}\} = [\lambda[m+1]]$. Then there are two cases to consider:

- $\lambda[m]$ violates the first condition to be an FLOTW $l$-partition. Assume that $c \neq l-1$. This means that $\lambda[m+1]_a^c = \lambda[m+1]_{a+s_{c+1}-s_c}^{c+1}$ and that we have to remove $\gamma_m$ from $\lambda[m+1]$ to obtain $\lambda[m]$. But in this case, the node $\gamma := (a+s_{c+1}-s_c, \lambda[m+1]_{a+s_{c+1}-s_c}^{c+1}, c+1)$ has the same residue as $\gamma_{n-m}$ does and by definition $\gamma_{n-m} \succ_{(\mathbf{s},e)} \gamma$. In addition, $\gamma$ is a removable node, otherwise we have $\lambda[m+1]_{a+s_{c+1}-s_c}^{c+1} = \lambda[m+1]_{a+s_{c+1}-s_c+1}^{c+1} = \lambda[m+1]_a^c > \lambda[m+1]_{a+1}^c$ which contradicts the fact that $\lambda[m+1]$ is an FLOTW $l$-partition. If $c = l-1$, then the result follows using the same argument.
- Assume that $\lambda[m]$ violates the second condition to be an FLOTW $l$-partition. The only possible way to have this is when the set of residues appearing at the right ends of the part with length $\lambda[m]_a^c$ of $\lambda[m]$ is equal to the set $\{0,1,\ldots,e-1\}$. In this case, as $\lambda$ is an FLOTW $l$-partition, the set of residues appearing at the right

ends of the parts with length $\lambda[m]_a^c$ of $\lambda$ is $\{0,1,\ldots,e-1\}\setminus\{k-1\}$. Hence, there exists a node $\gamma=(a',b',c')$ such that $\lambda[m]_{a'}^{c'}=\lambda[m]_a^c-1$ and $\mathfrak{r}_e(\gamma)=k$. This node is removable, otherwise $(a'+1,b',c')$ would be a $k-1$-node of $\lambda$ with $b'=\lambda[m]_a^c$. Hence, by Lemma 5.7.17, we obtain that $\gamma\prec_{(e,\mathbf{s})}\gamma_{n-m}$. One can easily see that this node is admissible and so $\gamma=\gamma_t$ for some $t<n-m$, which is a contradiction because it still belongs to $[\lambda[m]]$.

Now note that $k$ is the admissible residue for $\lambda[m]$, which completes the proof. □

The following result is the analogue of Lemma 3.5.11 for FLOTW $l$-partitions. We use the partial order $\ll_{\mathbf{m}}$ on $l$-compositions defined in 5.7.5.

**Lemma 5.7.21.** *Let $\lambda\vdash_l n$ be an FLOTW $l$-partition and let $\mu$ an $l$-composition of $n$. Assume that there exists a row-standard $\mu$-tableau $\mathfrak{s}$ such that $\eta_{(\mathbf{s},e)}(\mathfrak{s})=\eta_{(\mathbf{s},e)}(\lambda)$. Then $\lambda\ll_{\mathbf{m}}\mu$ if $\lambda\neq\mu$.*

*Proof.* Let us write $\mathfrak{t}:=\mathfrak{t}_{(\mathbf{s},e)}(\lambda)$ and assume that $\mu\neq\lambda$. We set $d_{\mathfrak{s}}:=n+1$ if $\mathfrak{t}=\mathfrak{s}$. Otherwise, $d_{\mathfrak{s}}\in\{1,\ldots,n\}$ is the smallest integer such that $\mathfrak{t}^{-1}(d_{\mathfrak{s}})\neq\mathfrak{s}^{-1}(d_{\mathfrak{s}})$. We now argue by downward induction on $d_{\mathfrak{s}}$. If $d_{\mathfrak{s}}=n+1$, there is nothing to prove. We now assume that if $\mathfrak{u}$ is a row-standard $\nu$-tableau such that $\eta_{(\mathbf{s},e)}(\mathfrak{u})=\eta_{(\mathbf{s},e)}(\lambda)$ and $d_{\mathfrak{u}}>d_{\mathfrak{s}}$, then $\lambda\ll_{\mathbf{m}}\nu$ if $\lambda\neq\nu$.

Set $d:=d_{\mathfrak{s}}$, $(a_1,b_1,c_1):=\mathfrak{t}^{-1}(d)$ and $(a_2,b_2,c_2):=\mathfrak{s}^{-1}(d)$. Then we claim that we have

$$b_1-a_1+m^{c_1}>b_2-a_2+m^{c_2}.$$

This is seen as follows. We remove successively the nodes filled by $n,\ldots,n-d+1$ from $\mathfrak{t}$. By Lemma 5.7.20, the resulting tableau will be $\mathfrak{t}_{(\mathbf{s},e)}(\widetilde{\lambda})$ for an FLOTW $l$-partition $\widetilde{\lambda}$. Similarly, if we remove the nodes filled by $n,\ldots,n-d+1$ from $\mathfrak{s}$ we obtain a row-standard $\widetilde{\mu}$ tableau $\widehat{\mathfrak{s}}$. We have $\eta_{(\mathbf{s},e)}(\widetilde{\lambda})=\eta_{(\mathbf{s},e)}(\widehat{\mathfrak{s}})$ by Lemma 5.7.20. If we remove the node filled by $d$ in $\mathfrak{t}_{(\mathbf{s},e)}(\widetilde{\lambda})$, we obtain a $\widetilde{\lambda}'$-tableau and if we remove the node filled by $d$ in $\widetilde{\mathfrak{s}}$, we obtain a $\widetilde{\mu}'$-tableau. Because of the minimality of $d$, we have in fact $\widetilde{\mu}'=\widetilde{\lambda}'$. Hence, both $\widetilde{\lambda}$ and $\widetilde{\mu}$ are obtained by adding a node $\gamma_1=(a_1,b_1,c_1)$ and $\gamma_2=(a_2,b_2,c_2)$ respectively, with the same residue on the same $l$-partition. By definition, we have $\gamma_2\prec_{(\mathbf{s},e)}\gamma_1$. As $\mathfrak{r}_e(\gamma_1)=\mathfrak{r}_e(\gamma_2)$ and by the definition of $\prec_{(\mathbf{s},e)}$, this implies that $b_1-a_1+m^{c_1}>b_2-a_2+m^{c_2}$, as claimed.

So, in the tableau $\mathfrak{s}$, we have the following configuration, where "$*$" stands for a filling of the node by a number in $\{1,\ldots,d-1\}$ and where the node filled by $n_1$ has the same coordinates as the node filled by $d$ in $\mathfrak{t}$:

row $a_1$ of $\mathfrak{s}^{c_1}$: $\boxed{*}\boxed{*}\cdots\boxed{*}\boxed{n_1}\boxed{n_2}\cdots\boxed{n_p}$ $\quad d<n_1<n_2<\ldots<n_p,$

row $a_2$ of $\mathfrak{s}^{c_2}$: $\boxed{*}\boxed{*}\cdots\boxed{*}\boxed{m_1}\boxed{m_2}\cdots\boxed{m_q}$ $\quad d=m_1<m_2<\ldots<m_q.$

Since $\eta_{(\mathbf{s},e)}(\mathfrak{s})=\eta_{(\mathbf{s},e)}(\lambda)$, the $e$-residues of the nodes filled by $m_1$ and $n_1$ are equal. Consequently, for any relevant $i$, the node filled by $n_i$ has the same $e$-residue as the node filled by $m_i$. We now perform some modifications on $\mathfrak{s}$ in order to obtain a new

$l$-composition $\nu$ of $n$ and a corresponding row-standard $\nu$-tableau $\mathfrak{u}$. This is done as follows. We keep all nodes and their fillings in the rows of $\mathfrak{s}$ except for the $a_1$th row of the $c_1$th component and the $a_2$th row of $c_2$th component. For these two rows we modify as follows:

- in the $a_1$th row of the $c_1$th component, we replace $n_i$ by $\min\{n_i, m_i\}$ for $1 \leqslant i \leqslant \min\{p,q\}$;
- in the $a_2$th row of the $c_2$th component, we replace $m_i$ by $\max\{n_i, m_i\}$ for $1 \leqslant i \leqslant \min\{p,q\}$;
- if $q > p$, we move all nodes filled by $m_{p+1}, \ldots, m_q$ from the $a_2$th row of the $c_2$th component to the $a_1$th row of the $c_1$th component.

We denote the resulting tableau by $\mathfrak{u}$. It is row-standard (see Remark 3.5.12) and we have $\eta_{(\mathbf{s},e)}(\mathfrak{u}) = \eta_{(\mathbf{s},e)}(\mathfrak{s}) = \eta_{(\mathbf{s},e)}(\lambda)$. Furthermore, the $a_1$th row of the $c_1$th component of $\mathfrak{u}$ has at least as many nodes as the $a_2$th row of the $c_2$th component of $\mathfrak{s}$. As we have $b_1 - a_1 + m^{c_1} > b_2 - a_2 + m^{c_2}$, if $\mu \neq \nu$, we obtain that $\nu \ll_{\mathbf{m}} \mu$. Note that, by construction, we have $\mathfrak{t}^{-1}(i) = \mathfrak{u}^{-1}(i)$ for $1 \leqslant i \leqslant d$. Hence, if $d_{\mathfrak{u}} \in \{1, \ldots, n\}$ is minimal such that $\mathfrak{t}^{-1}(d_{\mathfrak{u}}) \neq \mathfrak{u}^{-1}(d_{\mathfrak{u}})$, then $d_{\mathfrak{u}} > d$. So, we can conclude by induction. □

**Example 5.7.22.** To illustrate the above proof, take $s_1 = 0$, $s_2 = 1$, $e = 3$ and consider the FLOTW 2-partition $(3.2, 2) \vdash_2 7$. Let $(v_1, v_2)$ be rational numbers such that $0 < v_2 - v_1 < 3$. The Young diagram of $\lambda$ with residue is

$$\left( \begin{array}{|c|c|c|}\hline 0 & 1 & 2 \\ \hline 2 & 0 \\ \cline{1-2} \end{array}, \begin{array}{|c|c|}\hline 1 & 2 \\ \hline \end{array} \right).$$

We have

$$\mathfrak{t}_{((0,1),3)}(\lambda) = \left( \begin{array}{|c|c|c|}\hline 1 & 3 & 6 \\ \hline 2 & 5 \\ \cline{1-2} \end{array}, \begin{array}{|c|c|}\hline 4 & 7 \\ \hline \end{array} \right) \quad \text{and} \quad \eta_{((0,1),3)}(\lambda) = 0,2,1,1,0,2,2.$$

We consider the following row-standard tableau:

$$\mathfrak{s} = \left( \begin{array}{|c|}\hline 1 \\ \hline 2 \\ \hline 3 \\ \hline \end{array}, \begin{array}{|c|c|}\hline 4 & 6 \\ \hline 5 \\ \cline{1-1} 7 \\ \cline{1-1} \end{array} \right).$$

We have $\eta_{((0,1),3)}(\mathfrak{s}) = \eta_{((0,1),3)}(\lambda)$. The smallest number appearing in different nodes of $\mathfrak{t} := \mathfrak{t}_{((0,1),3)}(\lambda)$ and $\mathfrak{s}$ is 3. With the notation of the above proof, we have $m_1 = 3$, where $q = 1$ and $p = 0$. Hence, the tableau $\mathfrak{u}$ is just

$$\mathfrak{u} = \left( \begin{array}{|c|c|}\hline 1 & 3 \\ \hline 2 \\ \cline{1-1} \end{array}, \begin{array}{|c|c|}\hline 4 & 6 \\ \hline 5 \\ \cline{1-1} 7 \\ \cline{1-1} \end{array} \right).$$

We continue and take $d_{\mathfrak{u}} = 5$. With the above notation, we have $p = 0$ and $q = 1$ with $m_1 = 5$. So we have to move 5 from the second row of the second component to the second row of the first component, and we obtain the tableau

$$\mathfrak{u}_1 = \left( \begin{array}{|c|c|} \hline 1 & 3 \\ \hline 2 & 5 \\ \hline \end{array} \, , \begin{array}{l} \begin{array}{|c|c|} \hline 4 & 6 \\ \hline \end{array} \\ \\ \begin{array}{|c|} \hline 7 \\ \hline \end{array} \end{array} \right).$$

The shape of that tableau is $(2.1, 2.0.1)$. We now have $d_{\mathfrak{u}_1} = 6$, $p = 0$, $q = 1$ and $m_1 = 6$. So we move 6 from the first row of the second component to the first row of the first component. We obtain the tableau

$$\mathfrak{u}_2 = \left( \begin{array}{|c|c|c|} \hline 1 & 3 & 6 \\ \hline 2 & 5 \\ \cline{1-2} \end{array} \, , \begin{array}{l} \begin{array}{|c|} \hline 4 \\ \hline \end{array} \\ \\ \begin{array}{|c|} \hline 7 \\ \hline \end{array} \end{array} \right).$$

The shape of that tableau is $(3.2, 1.0.1)$. We have $d_{\mathfrak{u}_2} = 7$ and so we move 7 from the third row of the second component to the first row of the second component and we obtain $\mathfrak{t}$. As a conclusion, we have

$$(3.2, 2) \ll_{\mathbf{m}} (3.2, 1.0.1) \ll_{\mathbf{m}} (2.2, 2.0.1) \ll_{\mathbf{m}} (2.1, 2.1.1) \ll_{\mathbf{m}} (1.1.1, 2.1.1).$$

Using the above key result, we can now give the main results of this section.

**Proposition 5.7.23.** *Let* $\mathbf{s} = (s_1, \ldots, s_l) \in \mathscr{S}_e^l$ *and let* $(v_1, \ldots, v_l)$ *be such that* $0 < v_j - v_i < e$ *if* $i < j$. *Set* $\mathbf{m} = (s_1 - v_1, \ldots, s_l - v_l)$ *and let* $r \in \mathbb{Z}_{>0}$ *be such that* $r.\mathbf{m} \in \mathbb{Z}^l$. *Let* $\lambda \vdash_l n$ *be an* FLOTW $l$*-partition. Then there exists* $Y_\lambda \in \mathscr{E}$ *(see 5.6.9) such that*

$$[Y_\lambda] = m(Y_\lambda, \lambda)[E_\varepsilon^\lambda] + \sum_{\mu \vdash_l n : \mathbf{a}^{(\mathbf{m},r)}(\mu) > \mathbf{a}^{(\mathbf{m},r)}(\lambda)} m(Y_\lambda, \mu)[E_\varepsilon^\mu],$$

*where* $m(Y_\lambda, \mu) \in \mathbb{Z}_{\geqslant 0}$ *for all* $\mu \vdash_l n$ *and* $m(Y_\lambda, \lambda) \neq 0$.

*Proof.* Let $\lambda \vdash_l n$ be an FLOTW $l$-partition. Consider the $e$-residue sequence $\eta_{(\mathbf{s},e)}(\lambda)$ of 5.7.19. Then, by Lemma 5.7.21, the module $Y_\lambda := Y_{\eta_e(\lambda)}$ satisfies

$$[Y_\lambda] = m(Y_\lambda, \lambda)[E_\varepsilon^\lambda] + \sum_{\lambda \ll_{\mathbf{m}} \mu : \lambda \neq \mu} m(Y_\lambda, \mu)[E_\varepsilon^\mu]$$

and the results follow now from the compatibility of the **a**-invariant with the order $\ll_{\mathbf{m}}$ (see Proposition 5.7.7). □

**Proposition 5.7.24.** *In the setting of Proposition 5.7.23, for all* FLOTW $l$*-partitions* $\lambda \vdash_l n$ *and for all* $l$*-partitions* $\mu \vdash_l n$*, the number* $m(Y_\lambda, \lambda)$ *divides* $m(Y_\lambda, \mu)$.

*Proof.* Assume that we have

$$\eta_{(\mathbf{s},e)}(\lambda) = \underbrace{i_1, \ldots, i_1}_{a_1}, \underbrace{i_2, \ldots, i_2}_{a_2}, \ldots, \underbrace{i_s, \ldots, i_s}_{a_s},$$

where, for all $j = 1, \ldots, s-1$, we have $i_j \neq i_{j+1}$. We show by induction on $s$ that $m(Y_\lambda, \lambda) = \prod_{1 \leqslant i \leqslant s}(a_i!)$. Let $\lambda'$ be the $l$-partition obtained from $\lambda$ by removing the nodes filled by $n, n-1, \ldots, n-a_s+1$ in $\mathfrak{t}_{(\mathbf{s},e)}(\lambda)$. By Lemma 5.7.20, we have

$$\eta_{(\mathbf{s},e)}(\lambda') = \underbrace{i_1, \ldots, i_1}_{a_1}, \underbrace{i_2, \ldots, i_2}_{a_2}, \ldots, \underbrace{i_{s-1}, \ldots, i_{s-1}}_{a_{s-1}},$$

and by induction we may assume that we have $m(Y_\lambda, \lambda') = \prod_{1 \leqslant i \leqslant s-1} a_i!$. Now, by Corollary 5.6.11, $E_u^\lambda$ appears with multiplicity $a_s$ in $(i_s\text{-Ind}_K^{(n)})^{a_s}(E_u^{\lambda'})$. It thus suffices to show that the multiplicity of $E_u^\lambda$ in any of the $(i_s\text{-Ind}_K^{(n)})^{a_s}(E_u^{\mu'})$, where $m(Y_{\lambda'}, \mu') \neq 0$, is zero. Assume to the contrary that we can remove $a_s$ nodes with residue $i$ from $\lambda$ to obtain an $l$-partition $\mu'$ which satisfies $m(Y_{\lambda'}, \mu') \neq 0$. Using Proposition 5.7.16, we see that $\mu' \ll_{\mathbf{m}} \lambda'$, which is a contradiction to Proposition 5.7.23. Thus, the formula for $m(Y_\lambda, \lambda)$ is proved. Now, by Proposition 5.6.11, the integers $m(Y_\lambda, \mu)$ are multiples of $\prod_{1 \leqslant i \leqslant s} a_i!$. So the result follows. □

## 5.8 On Basic Sets for Ariki–Koike Algebras

We are now almost in position to prove the existence of canonical basic sets for our choice of weight sequence. We keep the setting of Section 5.7. Recall that we only consider the case where $\mathbf{m} = (s_1 - v_1, \ldots, s_l - v_l)$, $\mathbf{s} = (s_1, \ldots, s_l) \in \mathcal{S}_e^l$ and $\mathbf{v} = (v_1, \ldots, v_l)$ is such that $0 < v_j - v_i < e$ if $i < j$. At the end of the section, we will discuss the consequences for Iwahori–Hecke algebras of type $B_n$ and $D_n$. To show the existence of basic sets for $\mathbf{H}_{k,n}$, we need the following fundamental result.

**Theorem 5.8.1 (Ariki, Ariki–Mathas, Foda–Leclerc–Okado–Thibon–Welsh).** *In the above setting, the set* $\operatorname{Irr}(\mathbf{H}_{k,n})$ *has cardinality equal to the number of* FLOTW *$l$-partitions* $\Phi'_{(\mathbf{s},e)}(n)$.

The analogous result for the case of the symmetric group was established in Lemma 3.4.13. Unfortunately, contrary to that case, there is no elementary known proof of Theorem 5.8.1. This will be solved in the next chapter using Ariki's theorem and an interpretation of the FLOTW $l$-partitions in the context of crystal graph theory for quantum groups; see Theorem 6.3.2. In the case where $\operatorname{char}(k) = 0$, the following result appeared in [174, Theorem 4.1]. We can now prove this result for any field $k$. (This is a new result and appears here for the first time.)

**Theorem 5.8.2.** *Let $k$ be a field, $(s_1, \ldots, s_l) \in \mathcal{S}_e^l$ for $e \in \mathbb{Z}_{>1}$. Assume that we have a specialisation $\theta : A \to k$ such that $\theta(u)$ is a primitive root of unity of order $e$ and $\theta(V_j) = \theta(u)^{s_j}$ for all $j = 1, \ldots, l$. Set $r \in \mathbb{Z}_{>0}$ and* $\mathbf{m} = (s_1 - v_1, \ldots, s_l - v_l)$, *where*

*$\mathbf{v} = (v_1, \ldots, v_l)$ is such that $0 < v_j - v_i < e$ if $i < j$. With respect to the partial order induced by $\mathbf{a}^{(\mathbf{m},r)}$, the algebra $\mathbf{H}_{k,n} := \mathbf{H}_{k,n}(\eta_e; \eta_e^{s_1}, \ldots, \eta_e^{s_l})$ admits a canonical basic set $\Phi'_{(\mathbf{s},e)}(n)$ in the sense of Definition 5.5.19.*

*Proof.* Recall that $A$ is the ring of Laurent polynomials in finitely many indeterminates. Furthermore, we can assume without loss of generality that either $\mathbb{K} = R$ or that $\mathbb{K}$ is a finite extension of $\mathbb{Q}$ and that $R$ is the ring of algebraic integers in $\mathbb{K}$. In both cases, the general assumption on $A$ in Theorem 3.4.4 is satisfied.

We will use Proposition 3.4.5. First, recall that by our assumptions in Section 5.4, we have a well-defined decomposition map $d_\theta : R_0(\mathbf{H}_{K,n}) \to R_0(\mathbf{H}_{k,n})$; this map is surjective by the same argument as in Theorem 3.1.14. Now condition (a) of Proposition 3.4.5 is satisfied by Theorem 5.8.1. Next, recall the notation of Proposition 5.7.23. For each $\lambda \in \Phi'_{(\mathbf{s},e)}(n)$, define the following representation of $\mathbf{H}_{K,n}$:

$$P_\lambda = E_\varepsilon^\lambda \oplus \bigoplus_{\mathbf{a}^{(\mathbf{m},r)}(\mu) > \mathbf{a}^{(\mathbf{m},r)}(\lambda)} \frac{m(Y_\lambda, \mu)}{m(Y_\lambda, \lambda)} E_\varepsilon^\mu.$$

This is well defined by Proposition 5.7.24. Furthermore, by Proposition 5.7.23, these elements satisfy conditions (b) and (c) of Proposition 3.4.5. As a consequence, $\mathbf{H}_{k,n}$ admits a canonical basic set given by the set $\Phi'_{(\mathbf{s},e)}(n)$. □

*Remark 5.8.3.* In the case where $e = \infty$ and $\mathbf{s} = (s_1, \ldots, s_l)$ is such that $s_1 \leqslant s_2 \leqslant \ldots \leqslant s_l$, the algebra $\mathbf{H}_{k,n}$ still admits a canonical basic set with respect to $\mathbf{a}^{(\mathbf{m},r)}$. This canonical basic set is given by the set of FLOTW $l$-partitions $\Phi'_{(\mathbf{s},f)}(n)$ with $f$ sufficiently large; that is, the $l$-partitions $\lambda = (\lambda^1, \ldots, \lambda^l)$ are such that, for all $k = 1, \ldots, l-1$, we have $\lambda_i^k \geqslant \lambda_{i+s_{k+1}-s_k}^{k+1}$. The proof is easily adapted from the proof in the case where $e \in \mathbb{Z}_{>1}$.

We now examine the consequences of the above result for the Iwahori–Hecke algebras of type $B_n$ with $n \geqslant 2$. First, let us look at the "equal-parameter case".

**5.8.4.** Let $W$ be of type $B_n$ and consider the weight function $L : W \to \mathbb{Z}$ such that $L(t) = L(s_i) = 1$ for all $i = 1, \ldots, n-1$.

$B_n$ 

We will use the notation and results established in Example 5.7.3. Let $k$ be such that $\mathrm{char}(k) \neq 2$. We consider a specialisation $\theta : R[v, v^{-1}] \to k$ such that $\theta(u) = \eta_e$, a primitive root of unity of order $e$ where $e \in \mathbb{Z}_{>1}$. By Example 5.7.3, one can assume that there exists $d \in \mathbb{Z}$ such that $\eta_e = -\eta_e^d$. Since $\mathrm{char}(k) \neq 2$, this implies that $e$ must be even. Then we have to take $d = 1 + e/2$ so that the condition $(\star)$ in Example 5.7.3 is satisfied. We then set $l = 2$, $r = 1$, $\mathbf{s} = (1 + e/2, e)$, $\mathbf{v} = (e/2, e)$. Note that $\mathbf{s} \in \mathscr{S}_e^2$, so we can use Theorem 5.8.2 to compute the canonical basic set for $\mathbf{H}_{k,n}$.

The weight sequence is $\mathbf{m} = \mathbf{s} - \mathbf{v} = (1,0)$. Note that the cyclotomic Ariki–Koike algebra associated with $\mathbf{m} = (1,0)$ and $r = 1$ is the Hecke algebra of type $B_n$ associated with $L$ and thus that the **a**-invariant $\mathbf{a}^{(\mathbf{m},r)}$ is the usual **a**-invariant in type $B_n$ associated with $L$; see Example 5.5.14. In addition, the Ariki–Koike algebra $\mathbf{H}_{k,n}(\eta_e;\eta_e^{s_1},\eta_e^{s_2})$ is isomorphic to the specialised Iwahori–Hecke algebra of type $B_n$ over $k$. Theorem 5.8.2 thus implies the following result.

**Theorem 5.8.5.** *In the setting of 5.8.4, the algebra $\mathbf{H}_{k,n}$ admits a canonical basic set $\Phi'_{((1+e/2,e),e)}$.*

*Remark 5.8.6.* Note that we have $(\lambda^1,\lambda^2) \in \Phi'_{((1+e/2,e),e)}$ if and only if the two following conditions are satisfied:

1. We have $\lambda_i^1 \geqslant \lambda_{i+e/2-1}^2$ and $\lambda_i^2 \geqslant \lambda_{i+1+e/2}^1$.
2. We associate with each node $\gamma = (i,j,c)$ (where $i$ denotes the row, $j$ the column and $c \in \{1,2\}$ the component of $\gamma$) of the Young diagram of $(\lambda^1,\lambda^2)$ its residue

$$\mathfrak{r}_e(\gamma) := \begin{cases} j-i \bmod e & \text{if } c=2, \\ j-i+1+e/2 \bmod e & \text{if } c=1. \end{cases}$$

   Then, for all $r > 0$, the set of residues appearing at the right ends of the length $r$ rows of $(\lambda^1,\lambda^2)$ is a proper subset of $\mathbb{Z}/e\mathbb{Z}$.

**Example 5.8.7.** In the special case when $e = 2$, the associated basic set can be described in an easy way without the use of the notion of residues; that is, we have $(\lambda^1,\lambda^2) \in \Phi'_{((1+e/2,e),e)}$ if and only if the three following conditions are satisfied:

1. $\lambda_i^1 \geqslant \lambda_i^2$ and $\lambda_i^2 \geqslant \lambda_{i+2}^1$,
2. $\lambda^1$ and $\lambda^2$ are 2-regular,
3. for all $i \in \mathbb{Z}_{>0}$, we cannot have $\lambda_i^2 = \lambda_{i+1}^1 \neq 0$.

Indeed, if 1. is satisfied, one can easily check that 2. and 3. are equivalent to the second condition of FLOTW bipartitions.

**5.8.8.** We still assume that $W$ is a Weyl group of type $B_n$. Now we consider the weight function $L : W \to \mathbb{Z}$ such that $L(t) = 1$ and $L(s_i) = 2$ for all $i = 1,\dots,n-1$. (This case, and the following one in 5.8.10, occurs in the representation theory of the finite unitary groups; see Example 4.4.16.)

$B_n$

Let $k$ be a field of arbitrary characteristic. We consider a specialisation $\theta : R[v,v^{-1}] \to k$ such that $\theta(u) = \eta_f$, a root of unity of order $f > 1$. Assume that $f = 2e$, where $e \geqslant 3$ is odd and set $\eta_e := \eta_f^2$. Then $\eta_e$ is a primitive root of unity of order $e$. Let $d = (e+1)/2$. Then $\eta_f^{2d} = \eta_f^{e+1} = -\eta_f$ and the condition $(\star)$ in Example 5.7.3 is satisfied. We set $l = 2$, $r = 2$, $\mathbf{s} = ((e+1)/2, e)$ and $\mathbf{v} = (e/2, e)$. Note that we have $\mathbf{s} \in \mathscr{S}_e^2$, so we can use Theorem 5.8.2 to compute the canonical basic set for $\mathbf{H}_{k,n}$.

The weight sequence that we have to consider is thus $\mathbf{m} = \mathbf{s} - \mathbf{v} = (1/2, 0)$. Note that the cyclotomic Ariki–Koike algebra associated with $\mathbf{m} = (1/2, 0)$ and $r = 2$ is the Iwahori–Hecke algebra of type $B_n$ associated with $L$ and thus that the $\mathbf{a}$-invariant $\mathbf{a}^{(\mathbf{m},r)}$ is the usual $\mathbf{a}$-invariant in type $B_n$ associated with $L$; see Example 5.5.14. In addition, the Ariki–Koike algebra $\mathbf{H}_{k,n}(\eta_e; \eta_e^{s_1}, \eta_e^{s_2})$ is isomorphic to the specialised Iwahori–Hecke algebra of type $B_n$ over $k$, where $\eta_e = \eta^2$. Theorem 5.8.2 thus implies the following result.

**Theorem 5.8.9.** *In the setting of 5.8.8, the algebra* $\mathbf{H}_{k,n}$ *admits a canonical basic set* $\mathscr{B}_\theta = \Phi'_{(((e+1)/2,e),e)}$.

**5.8.10.** We still assume that $W$ is a Weyl group of type $B_n$. We now consider the weight function $L : W \to \mathbb{Z}$ such that $L(t) = 3$ and $L(s_i) = 2$ for all $i = 1, \ldots, n-1$.

$B_n$ 

Let $k$ be a field of arbitrary characteristic. We consider a specialisation $\theta : R[v, v^{-1}] \to k$ such that $\theta(u) = \eta_f$, a primitive root of unity of order $f > 1$. As in the previous example, assume that $f = 2e$, where $e \geqslant 3$ is odd, and set $\eta_e := \eta_f^2$. Then $\eta_e$ is a primitive root of unity of order $e$. Now let $d = (e+3)/2$. Then $\eta_f^{2d} = \eta_f^{e+3} = -\eta_f^3$ and the condition $(\star)$ in Example 5.7.3 is satisfied. We set $l = 2$, $r = 2$, $\mathbf{s} = ((e+3)/2, e)$ and $\mathbf{v} = (e/2, e)$. Note that we have $\mathbf{s} \in \mathscr{S}_e^2$, so we can use Theorem 5.8.2 to compute the canonical basic set for $\mathbf{H}_{k,n}$.

The weight sequence that we have to consider is thus $\mathbf{m} = \mathbf{s} - \mathbf{v} = (3/2, 0)$. Note that the cyclotomic Ariki–Koike algebra associated with $\mathbf{m} = (3/2, 0)$ and $r = 2$ is the Iwahori–Hecke algebra of type $B_n$ associated with $L$ and thus that the $\mathbf{a}$-invariant $\mathbf{a}^{(\mathbf{m},r)}$ is the usual $\mathbf{a}$-invariant in type $B_n$ associated with $L$; see Example 5.5.14. In addition, the Ariki–Koike algebra $\mathbf{H}_{k,n}(\eta_e; \eta_e^{s_1}, \eta_e^{s_2})$ is isomorphic to the specialised Iwahori–Hecke algebra of type $B_n$ over $k$, where $\eta_e = \eta^2$. Theorem 5.8.2 thus implies the following result.

**Theorem 5.8.11.** *In the setting of 5.8.10, the algebra* $\mathbf{H}_{k,n}$ *admits a canonical basic set* $\mathscr{B}_\theta = \Phi'_{(((e+3)/2,e),e)}$.

**5.8.12.** We still assume that $W$ is a Weyl group of type $B_n$. We now consider the weight function $L : W \to \mathbb{Z}$ such that $L(t) = 2$ and $L(s_i) = 1 \in \mathbb{Z}_{>0}$ for all $i = 1, \ldots, n-1$.

$B_n$ 

We assume that $k$ is a field such that $\mathrm{char}(k) \neq 2$. We consider a specialisation $\theta : R[v, v^{-1}] \to k$ such that $\theta(u) = \eta_e$, a primitive root of unity of order $e$, where $e \in \mathbb{Z}_{>1}$. By Example 5.7.3, one can assume that there exists $d \in \mathbb{Z}$ such that $\eta_e^2 = -\eta_e^d$. Hence, $e$ is even and we have to set $d = e/2 + 2$ so that the condition $(\star)$ in Example 5.7.3 is satisfied. We then set $l = 2$, $r = 1$, $\mathbf{s} = (2 + e/2, e)$, $\mathbf{v} = (e/2, e)$. Note that $\mathbf{s} \in \mathscr{S}_e^2$ only if $e \neq 2$. So we have to assume that $e \neq 2$ in order to use Theorem 5.8.2.

The weight sequence is $\mathbf{m} = \mathbf{s} - \mathbf{v} = (2,0)$. Note that the cyclotomic Ariki–Koike algebra associated with $\mathbf{m} = (2,0)$ and $r = 1$ is the Iwahori–Hecke algebra of type $B_n$ associated with $L$ and thus that the **a**-invariant $\mathbf{a}^{(\mathbf{m},r)}$ is the usual **a**-invariant in type $B_n$ associated with $L$; see Example 5.5.14. In addition, the Ariki–Koike algebra $\mathbf{H}_{k,n}(\eta_e; \eta_e^{s_1}, \eta_e^{s_2})$ is isomorphic to the specialised Iwahori–Hecke algebra of type $B_n$ over $k$. Theorem 5.8.2 thus implies the following result.

**Theorem 5.8.13.** *In the setting of 5.8.12, when $e \neq 2$, the algebra $\mathbf{H}_{k,n}$ admits a canonical basic set $\Phi'_{((2+e/2,e),e)}$.*

Note that the case where $e = 2$ is discussed in Example 4.4.15 and Remark 6.7.12.

**5.8.14.** We assume that $W$ is a Weyl group of type $B_n$. Finally, we consider the weight function $L : W \to \mathbb{Z}$ such that $L(t) = 0$ and $L(s_i) = 1$ for all $i = 1, \ldots, n-1$.

$B_n$ 

We assume that $k$ is a field such that $\operatorname{char}(k) \neq 2$. We consider a specialisation $\theta : R[v, v^{-1}] \to k$ such that $\theta(u) = \eta_e$, a primitive root of unity of order $e$, where $e \in \mathbb{Z}_{>1}$. By Example 5.7.3, one can assume that there exists $d \in \mathbb{Z}$ such that $\eta_e^0 = -\eta_e^d$. Hence, $e$ is even and we have to set $d = e/2$ so that the condition $(\star)$ in Example 5.7.3 is satisfied. We then set $l = 2$, $r = 1$, $\mathbf{s} = (e/2, e)$, $\mathbf{v} = (e/2, e)$. Note that we have $\mathbf{s} \in \mathscr{S}_e^2$. The weight sequence is $\mathbf{m} = \mathbf{s} - \mathbf{v} = (0,0)$. Note that the cyclotomic Ariki–Koike algebra associated with $\mathbf{m} = (0,0)$ and $r = 1$ is the Iwahori–Hecke algebra of type $B_n$ associated with $L$ and thus that the **a**-invariant $\mathbf{a}^{(\mathbf{m},r)}$ is the usual **a**-invariant in type $B_n$ associated with $L$. In addition, the Ariki–Koike algebra $\mathbf{H}_{k,n}(\eta_e; \eta_e^{s_1}, \eta_e^{s_2})$ is isomorphic to the specialised Iwahori–Hecke algebra of type $B_n$ over $k$. Theorem 5.8.2 thus implies the following result.

**Theorem 5.8.15.** *In the setting of 5.8.14, the algebra $\mathbf{H}_{k,n}$ admits a canonical basic set $\mathscr{B}_\theta = \Phi'_{((e/2,e),e)}$.*

*Remark 5.8.16.* Note that we have $(\lambda^1, \lambda^2) \in \Phi'_{((e/2,e),e)}$ if and only if the two following conditions are satisfied:

1. We have $\lambda_i^1 \geqslant \lambda_{i+e/2}^2$ and $\lambda_i^2 \geqslant \lambda_{i+e/2}^1$.
2. We associate with each node $\gamma = (i, j, c)$ (where $i$ denotes the row, $j$ the column and $c \in \{1,2\}$ the component of $\gamma$) of the Young diagram of $(\lambda^1, \lambda^2)$ its residue

$$\mathfrak{r}_e(\gamma) := \begin{cases} j - i \bmod e & \text{if } c = 2, \\ j - i + e/2 \bmod e & \text{if } c = 1. \end{cases}$$

   Then, for all $r > 0$, the set of residues appearing at the right ends of the length $r$ rows of $(\lambda^1, \lambda^2)$ is a proper subset of $\mathbb{Z}/e\mathbb{Z}$.

**Example 5.8.17.** In the special case when $e = 2$, the associated basic set can be described in an easy way without the use of the notion of residues. We have $(\lambda^1, \lambda^2) \in \mathscr{B}_\theta$ if and only if

1. $\lambda_i^1 \geqslant \lambda_{i+1}^2$ and $\lambda_i^2 \geqslant \lambda_{i+1}^1$;
2. $\lambda^1$ and $\lambda^2$ are 2-regular;
3. for all $i \in \mathbb{Z}_{>0}$, we cannot have $\lambda_i^2 = \lambda_i^1 \neq 0$.

Indeed, if 1. is satisfied, one can easily check that 2. and 3. are equivalent to the second condition of FLOTW bipartitions.

**5.8.18.** We now also obtain an explicit description of the canonical basic set for type $D_n$, first established in [173]. We place ourselves in the setting of Example 3.2.15. Let $W$ be a Weyl group of type $B_n$ with function $L$ as in 5.8.14; that is, $L(t) = 0$ and $L(s_i) = 1$ for all $i = 1, \dots, n-1$. Let $\tilde{W} \subseteq W$ be the reflection subgroup of type $D_n$ and $\tilde{L}\colon \tilde{W} \to \mathbb{Z}_{\geqslant 0}$ be the weight function such that $\tilde{W} = L|_{W_1}$; note that $\tilde{L}$ takes value 1 on each generator of $\tilde{W}$. Let $\tilde{\mathbf{H}}$ be the generic Iwahori–Hecke algebra associated with $\tilde{W}, \tilde{L}$ over the ring $R[v, v^{-1}]$. Let $u := v^2$.

Let $k$ be a field such that $\operatorname{char}(k) \neq 2$ and let $\theta : R[v, v^{-1}] \to k$ be a specialisation. As before, write $\eta_e = \theta(u)$, where $e \geqslant 1$ denotes the order of $\theta(u) \in k^\times$. If $e = 1$ or if $e \in \mathbb{Z}_{>1}$ is odd, then a canonical basic set for $\tilde{\mathbf{H}}_k$ is given by Example 3.2.15 and the results in Section 4.6.

**Theorem 5.8.19.** *In the setting of 5.8.18, and using the notation of Example 3.2.15, assume that $e \in \mathbb{Z}_{>1}$ is even. Then $\tilde{\mathbf{H}}_k$ admits a canonical basic set given by*

$$\begin{aligned}\mathscr{B}_\theta = &\{[\lambda^1, \lambda^2] \mid \lambda^1 \neq \lambda^2,\ (\lambda^1, \lambda^2) \in \Phi'_{((e/2,e),e)}(n)\}\\ &\cup \{(\lambda, \pm) \mid \lambda \textit{ is } (e/2)\textit{-regular of rank } n/2\} \quad (\textit{if } n \textit{ is even}).\end{aligned}$$

*Proof.* By Example 3.2.15 combined with Theorem 5.8.15, the canonical basic set in type $D_n$ is given by the set

$$\begin{aligned}\mathscr{B}_\theta = &\{[\lambda^1, \lambda^2] \mid \lambda^1 \neq \lambda^2,\ (\lambda^1, \lambda^2) \in \Phi'_{((e/2,e),e)}(n)\}\\ &\cup \{(\lambda, \pm) \mid (\lambda, \lambda) \in \Phi'_{((e/2,e),e)}(n)\} \quad (\text{if } n \text{ is even}).\end{aligned}$$

Now every bipartition of the form $(\lambda, \lambda)$ has the property 1. in Remark 5.8.16. Furthermore, one easily checks that property 2. is satisfied if and only if $\lambda$ is $(e/2)$-regular. □

*Remark 5.8.20.* By Remark 5.8.16, we have $(\lambda^1, \lambda^2) \in \Phi'_{((e/2,e),e)}$ if and only if $(\lambda^2, \lambda^1) \in \Phi'_{((e/2,e),e)}$.

Recall that the validity of the main results of this section depends on Theorem 5.8.1. In Chapter 6 we introduce the objects that will be needed to give a proof of this result.

# Chapter 6
# Canonical Bases in Affine Type $A$ and Ariki's Theorem

The main purpose of this chapter is to go further in the study of the representation theory of Ariki–Koike algebras. To do this, we will need to take an excursion to another area of representation theory: the theory of canonical bases and crystal bases for Kac–Moody algebras and quantum algebras. In fact, we will focus on a particular case: the quantum affine algebra $\mathcal{U}_q(\widehat{\mathfrak{sl}_e})$. A fundamental result in this theory is the existence, proved independently by Kashiwara and Lusztig, of a "canonical basis" for certain irreducible $\mathcal{U}_q(\widehat{\mathfrak{sl}_e})$-modules: the irreducible highest weight modules. We shall see that this theory has a deep interpretation in the modular representation theory of Iwahori–Hecke algebras. Indeed, the matrices associated with the canonical bases for these modules give the decomposition matrices for Ariki–Koike algebras in characteristic 0. This result has first been conjectured for Iwahori–Hecke algebras of type $A$ by Lascoux, Leclerc and Thibon [208] and proved for the wider class of Ariki–Koike algebras by Ariki in [7].

Of course, it is not the purpose of this book to give a complete survey on the theory of quantum algebras and canonical bases; we refer to [158], [185], [188], [190], [191] and [230] for details on this area. We will also not give the proof of Ariki's theorem, as there is already a book on this subject [10]. The main aim of this chapter will be to present an explicit way to compute the canonical bases for irreducible highest weight modules following [174]. As a consequence, using Ariki's theorem and some deep results by Uglov, we will be able to complete the description of the canonical basic sets in type $B_n$ and $D_n$.

The chapter is organized as follows. The first two parts give a brief introduction to the representation theory of the quantum algebra $\mathcal{U}_q(\widehat{\mathfrak{sl}_e})$ and the theory of crystal and canonical bases. In particular, we study the action(s) of $\mathcal{U}_q(\widehat{\mathfrak{sl}_e})$ on the $\mathbb{Q}(q)$-vector space generated by the set of multipartitions, namely the Fock space. This will allow us to give an explicit realisation of the irreducible highest weight $\mathcal{U}_q(\widehat{\mathfrak{sl}_e})$-modules. The end of the second part is devoted to Ariki's theorem and the connection of the theory of canonical bases with the modular representation theory of Iwahori–Hecke algebras. In the third section, we describe a fundamental object attached with each irreducible highest weight $\mathcal{U}_q(\widehat{\mathfrak{sl}_e})$-module: the crystal graph.

M. Geck, N. Jacon, *Representations of Hecke Algebras at Roots of Unity*,
Algebra and Applications 15, DOI 10.1007/978-0-85729-716-7_6,

We then study the "canonical bases" for irreducible highest weight modules and present an explicit combinatorial algorithm for computing them. These results will give a way to compute the decomposition matrices of Ariki–Koike algebras.

Ariki's theorem will allow us to complete the proof of a result which was necessary in Chapter 5 to obtain the canonical basic sets in certain cases for types $B_n$ and $D_n$. For the remaining cases, a much deeper study of the theory of Fock spaces will be necessary. This will be done in the remaining parts, where Uglov's theory of canonical bases for Fock spaces is studied. The chapter ends with the determination of the canonical basic sets in the remaining cases. We then close with some comments on recent developments and problems connecting Iwahori–Hecke algebras, the theory of canonical bases for quantum groups and some other class of algebras: the Cherednik algebras and the Khovanov–Lauda–Rouquier algebras.

## 6.1 The Quantum Affine Algebra $\mathscr{U}_q(\widehat{\mathfrak{sl}_e})$

We begin by recalling the definition of the algebra $\mathscr{U}_q(\widehat{\mathfrak{sl}_e})$ following [10, Chap. 3]. Then we study several aspects of its representation theory.

**6.1.1.** Let $e \in \mathbb{Z}_{>0}$ and let $\mathfrak{h}$ be the $\mathbb{Q}$-vector space with basis $\{h_0, \dots, h_{e-1}, D\}$. Let $\{\Lambda_0, \dots, \Lambda_{e-1}, \partial\}$ be a basis of $\mathfrak{h}^*$ defined with respect to the pairing

$$\langle \, , \rangle : \mathfrak{h}^* \times \mathfrak{h} \to \mathbb{Q},$$

such that, for $0 \leqslant i, j \leqslant e-1$, we have

$$\langle \Lambda_i, h_j \rangle = \delta_{i,j}, \quad \langle \Lambda_i, D \rangle = \langle \partial, h_i \rangle = 0, \quad \langle \partial, D \rangle = 1,$$

where we set $\delta_{i,j} = 0$ if $i \neq j$ and $\delta_{i,i} = 1$. In the following, if $i \in \mathbb{Z}$, we will put $\Lambda_i := \Lambda_{i \bmod e}$. In addition, for $l \in \mathbb{Z}_{>0}$ and $\mathbf{s} = (s_1, \dots, s_l) \in \mathbb{Z}^l$, we set $\Lambda_{\mathbf{s}} := \sum_{1 \leqslant i \leqslant l} \Lambda_{s_i}$. For any integer $i$ such that $0 \leqslant i \leqslant e-1$, we define the following elements in $\mathfrak{h}^*$:

$$\alpha_i := -\Lambda_{i-1} + 2\Lambda_i - \Lambda_{i+1} + \delta_{i,0}\partial.$$

**Definition 6.1.2.** The elements of $\{\Lambda_i \mid 0 \leqslant i \leqslant e-1\}$ are called the *fundamental weights* and the elements of $\{\alpha_i \mid 0 \leqslant i \leqslant e-1\}$ are called the *simple roots* for $\widehat{\mathfrak{sl}_e}$. We put

$$P := \bigoplus_{i=0}^{e-1} \mathbb{Z}\Lambda_i \oplus \mathbb{Z}\partial, \qquad P^+ := \bigoplus_{i=0}^{e-1} \mathbb{Z}_{\geqslant 0}\Lambda_i \oplus \mathbb{Z}\partial, \qquad Q := \bigoplus_{i=0}^{e-1} \mathbb{Z}\alpha_i.$$

Then $P$ is called the *weight lattice*, $P^+$ the set of *dominant weights* and $Q$ the *root lattice* of $\widehat{\mathfrak{sl}_e}$. The matrix $(a_{ij})_{0 \leqslant i,j \leqslant e-1}$ such that $a_{ij} = \langle \alpha_i, h_j \rangle$ for $0 \leqslant i, j \leqslant e-1$ is called the *Cartan matrix* of $\widehat{\mathfrak{sl}_e}$.

We will need the following standard notation. For $k \in \mathbb{Z}_{\geqslant 0}$, we denote

$$[k] = \frac{q^k - q^{-k}}{q - q^{-1}} \quad \text{and} \quad [k]! := [k][k-1]\dots[1].$$

For $0 \leqslant k \leqslant m$, we set

$$\begin{bmatrix} m \\ k \end{bmatrix} := \frac{[m]!}{[m-k]![k]!},$$

with $[0]! = 1$ by convention. Following [192], we obtain the following definition of the quantum affine algebra.

**Definition 6.1.3.** The *quantum affine algebra* $\mathcal{U}_q(\widehat{\mathfrak{sl}_e})$ is the unital associative $\mathbb{Q}(q)$-algebra with

- generators: $e_i, f_i, t_i, t_i^{-1}$ $(0 \leqslant i \leqslant e-1)$ and $\mathfrak{d}, \mathfrak{d}^{-1}$;
- relations:

$$\mathfrak{d}\mathfrak{d}^{-1} = \mathfrak{d}^{-1}\mathfrak{d} = 1, \quad t_i t_i^{-1} = t_i^{-1} t_i = 1, \quad t_i t_j = t_j t_i, \quad t_i e_j t_i^{-1} = q^{a_{ij}} e_j,$$

$$t_i f_j t_i^{-1} = q^{-a_{ij}} f_j, \quad e_i f_j - f_j e_i = \delta_{ij} \frac{t_i - t_i^{-1}}{q - q^{-1}} \quad (0 \leqslant i, j \leqslant e-1),$$

$$[\mathfrak{d}, e_i] = \delta_{i0} e_i, \quad [\mathfrak{d}, f_i] = -\delta_{i0} f_i, \quad [\mathfrak{d}, t_i] = 0;$$

and the following relations known as "Serre relations", where $i \neq j$:

$$\sum_{0 \leqslant k \leqslant 1-a_{ij}} (-1)^k \begin{bmatrix} 1-a_{ij} \\ k \end{bmatrix} e_i^{1-a_{ij}-k} e_j e_i^k = 0,$$

$$\sum_{0 \leqslant k \leqslant 1-a_{ij}} (-1)^k \begin{bmatrix} 1-a_{ij} \\ k \end{bmatrix} f_i^{1-a_{ij}-k} f_j f_i^k = 0.$$

Moreover, we denote by $\mathcal{U}_q'(\widehat{\mathfrak{sl}_e})$ the subalgebra of $\mathcal{U}_q(\widehat{\mathfrak{sl}_e})$ generated by $e_i, f_i, t_i^{\pm 1}$ with $0 \leqslant i \leqslant e-1$.

We have an analogue of the Poincaré–Birkhoff–Witt theorem for this algebra. Indeed, Let $\mathcal{U}_q(\mathfrak{n}^+)$ (or $\mathcal{U}_q(\mathfrak{n}^-)$, or $\mathcal{U}_q(\mathfrak{h})$) be the subalgebra of $\mathcal{U}_q'(\widehat{\mathfrak{sl}_e})$ generated by the $e_i$ (or $f_i$, or $t_i^{\pm 1}$) with $0 \leqslant i \leqslant e-1$. Then we have an isomorphism of $\mathbb{Q}(q)$-vector spaces (see [10, Theorem 6.2]):

$$\begin{array}{ccc} \mathcal{U}_q(\mathfrak{n}^+) \otimes \mathcal{U}_q(\mathfrak{h}) \otimes \mathcal{U}_q(\mathfrak{n}^-) & \to & \mathcal{U}_q'(\widehat{\mathfrak{sl}_e}) \\ n^+ \otimes h \otimes n^- & \to & n^+ h n^-. \end{array}$$

*Remark 6.1.4.* There is a coproduct $\Delta$ on $\mathcal{U}_q(\widehat{\mathfrak{sl}_e})$ which turns it into a Hopf algebra. This is given by the following formulae, where $0 \leqslant i \leqslant e-1$:

$$\begin{aligned}\Delta(e_i) &= e_i \otimes t_i^{-1} + 1 \otimes e_i,\\ \Delta(f_i) &= f_i \otimes 1 + t_i \otimes f_i,\\ \Delta(t_i) &= t_i \otimes t_i,\\ \Delta(t_i^{-1}) &= t_i^{-1} \otimes t_i^{-1},\\ \Delta(\mathfrak{d}) &= \mathfrak{d} \otimes 1 + 1 \otimes \mathfrak{d}.\end{aligned}$$

**6.1.5.** For $0 \leqslant i \leqslant e-1$ and $k \in \mathbb{Z}_{\geqslant 0}$, we define the *divided powers* of $e_i$ and $f_i$ as follows:

$$e_i^{(k)} = \frac{e_i^k}{[k]!} \quad \text{and} \quad f_i^{(k)} = \frac{f_i^k}{[k]!}.$$

Moreover, if $x \in \mathscr{U}_q(\widehat{\mathfrak{sl}_e})$ and $k \in \mathbb{Z}_{\geqslant 0}$, we also set

$$\{x\} = \frac{x - x^{-1}}{q - q^{-1}}, \quad \left\{ \begin{matrix} x \\ 0 \end{matrix} \right\} = 1, \quad \left\{ \begin{matrix} x \\ k \end{matrix} \right\} = \frac{\{x\}\{q^{-1}x\}\dots\{q^{-(k-1)}x\}}{[k]!}.$$

We now introduce a $\mathbb{Q}$-form for $\mathscr{U}_q(\widehat{\mathfrak{sl}_e})$. This is the $\mathbb{Q}[q,q^{-1}]$-subalgebra $\mathscr{U}_q(\widehat{\mathfrak{sl}_e})_{\mathbb{Q}}$ of $\mathscr{U}_q(\widehat{\mathfrak{sl}_e})$ generated by the divided powers $e_i^{(k)}$, $f_i^{(k)}$ and by $\left\{ \begin{matrix} t_i \\ k \end{matrix} \right\}$, $t_i$, where $k \in \mathbb{Z}_{\geqslant 0}$ and $0 \leqslant i \leqslant e-1$. If $M$ is an $\mathscr{U}_q(\widehat{\mathfrak{sl}_e})$-module, we denote by $M_{\mathbb{Q}}$ the associated $\mathscr{U}_q(\widehat{\mathfrak{sl}_e})_{\mathbb{Q}}$-module.

**6.1.6.** We briefly recall the theory of $\mathscr{U}_q(\widehat{\mathfrak{sl}_e})$-modules and then focus on a particular class of $\mathscr{U}_q(\widehat{\mathfrak{sl}_e})$-modules: the integrable modules. Let $M$ be a $\mathscr{U}_q(\widehat{\mathfrak{sl}_e})$-module and let $\Lambda = \sum_{0 \leqslant i \leqslant e-1} a_i \Lambda_i + d\partial \in P$ for $(a_0, \dots, a_{e-1}, d) \in \mathbb{Z}^{e+1}$. The $\mathbb{Q}(q)$-vector space generated by the set

$$\{m \in M \mid \mathfrak{d}.m = dm \text{ and } t_i m = q^{a_i} m \text{ for all } i \in \{0, 1, \dots, e-1\}\}$$

is called the $\Lambda$-*weight space* and it is denoted by $M_\Lambda$. If $M_\Lambda \neq 0$, we say that $\Lambda$ is a *weight* for $M$ and the non-zero elements in $M_\Lambda$ are called *weight vectors* of weight $\Lambda$. We say that a non-zero element $u \in M_\Lambda$ is a *highest weight vector* with weight $\Lambda$ if $e_i u = 0$ for all $i \in \{0, 1, \dots, e-1\}$. We say that a $\mathscr{U}_q(\widehat{\mathfrak{sl}_e})$-module $M$ is a *cyclic module* with highest weight $\Lambda$ if there exists a highest weight vector $m_\Lambda$ with weight $\Lambda$ such that $M = \mathscr{U}_q(\widehat{\mathfrak{sl}_e}).m_\Lambda$. Similarly, we say that a $\mathscr{U}'_q(\widehat{\mathfrak{sl}_e})$-module $M$ is cyclic with highest weight $\Lambda$ if there exists a highest weight vector $m_\Lambda$ with weight $\Lambda$ such that $M = \mathscr{U}'_q(\widehat{\mathfrak{sl}_e}).m_\Lambda$. In this case, by the Poincaré–Birkhoff–Witt theorem, the weight space $M_\Lambda$ has dimension 1. If

$$\Lambda = \sum_{0 \leqslant i \leqslant e-1} a_i \Lambda_i + d\partial \in P^+,$$

then the positive integer $l = \sum_{0 \leqslant i \leqslant e-1} a_i$ is called the *level* of $M$. For each $\Lambda \in P$, there exists a unique simple and cyclic $\mathscr{U}_q(\widehat{\mathfrak{sl}_e})$-module with highest weight $\Lambda$, which we denote by $V(\Lambda)$. The restriction of $V(\Lambda)$ to $\mathcal{U}'_q(\widehat{\mathfrak{sl}_e})$ is denoted by $V'(\Lambda)$. We have that (see [10, Lemma 6.9]):

- $V(\Lambda^1) \simeq V(\Lambda^2)$ as $\mathcal{U}_q(\widehat{\mathfrak{sl}_e})$-modules if and only if $\Lambda^1 = \Lambda^2$,
- $V'(\Lambda^1) \simeq V'(\Lambda^2)$ as $\mathcal{U}'_q(\widehat{\mathfrak{sl}_e})$-modules if and only if $\Lambda^1 - \Lambda^2 \in \mathbb{Z}\partial$.

**Definition 6.1.7.** We say that a $\mathcal{U}_q(\widehat{\mathfrak{sl}_e})$-module $M$ is an *integrable module* if and only if we have:

- $M = \bigoplus_{\Lambda \in P} M_\Lambda$,
- for all $\Lambda \in P$, $M_\Lambda$ is a finite-dimensional vector space,
- for all $i \in \{0,1,\dots,e-1\}$, $e_i$ and $f_i$ act locally nilpotently on $M$; that is, for any $m \in M$, there exists $N \in \mathbb{Z}_{>0}$ such that $e_i^N.m = f_i^N.m = 0$.

The category of integrable $\mathcal{U}_q(\widehat{\mathfrak{sl}_e})$-modules $M$ such that $\mathcal{U}_q(\mathfrak{n}^+)u$ is finite dimensional for all $u \in M$ is denoted by $\mathcal{O}_{\text{int}}$.

**6.1.8.** By the work of Lusztig (see [191, Section 3.3] or [10, Lemma 6.9]), we have the following properties:

- For each $\Lambda \in P^+$, the highest weight $\mathcal{U}_q(\widehat{\mathfrak{sl}_e})$-module $V(\Lambda)$ is irreducible and is in $\mathcal{O}_{\text{int}}$.
- Each irreducible highest weight module in $\mathcal{O}_{\text{int}}$ is isomorphic to a $V(\Lambda)$ with $\Lambda \in P^+$.
- The category $\mathcal{O}_{\text{int}}$ is semisimple: each integrable $\mathcal{U}_q(\widehat{\mathfrak{sl}_e})$-module $M$ in $\mathcal{O}_{\text{int}}$ is isomorphic to a direct sum of $V(\Lambda)$ with $\Lambda \in P^+$.

**6.1.9.** We give a brief introduction to the theory of crystal bases, crystal graphs and canonical bases following [10, Sections 4–9] (see also [189], [230].) Let $M$ be an integrable $\mathcal{U}_q(\widehat{\mathfrak{sl}_e})$-module and let $\Lambda \in P$. Let $x \in M_\Lambda$ and $i \in \{0,1,\dots,e-1\}$. Set $N := \langle \Lambda, h_i \rangle$. Then one can show that $x$ can be written uniquely in the form

$$x = \sum_{s \geqslant \max(0,-N)} f_i^{(s)} x_s,$$

where $e_i x_s = 0$, $t_i x_s = q^{\langle \Lambda, h_i \rangle + 2s} x_s$ and only finitely many $x_s$ are non-zero (see [10, Section 4.2].)

**Definition 6.1.10.** Keeping the above notations, we define

$$\widetilde{e}_i x = \sum_{s \geqslant \max(0,-N)} f_i^{(s-1)} x_s \quad \text{and} \quad \widetilde{f}_i x = \sum_{s \geqslant \max(0,-N)} f_i^{(s+1)} x_s$$

for $i \in \{0,1,\dots,e-1\}$. The operators $\widetilde{e}_i$ and $\widetilde{f}_i$ are called the *Kashiwara operators*.

To formulate the following definition, we set

$$\mathbb{A} := \left\{ \frac{a(q)}{b(q)} \in \mathbb{Q}(q) \;\middle|\; a(q), b(q) \in \mathbb{Q}[q], b(0) \neq 0 \right\}.$$

**Definition 6.1.11.** A pair $(L,B)$ is called a *crystal basis* for the integrable $\mathscr{U}_q(\widehat{\mathfrak{sl}_e})$-module $M$ (and for the $\mathscr{U}_q'(\widehat{\mathfrak{sl}_e})$-module $M'$) if it satisfies the following conditions:

1. $L$ is a free $\mathbb{A}$-module such that $\mathbb{Q}(q) \otimes_{\mathbb{A}} L \simeq M$.
2. $B$ is a basis of the $\mathbb{Q}$-vector space $L/qL$.
3. $L = \bigoplus_{\Lambda \in P} L_\Lambda$ and $B = \bigcup_{\Lambda \in P} B_\Lambda$, where $L_\Lambda = L \cap M_\Lambda$ and $B_\Lambda = B \cap (L_\Lambda / qL_\Lambda)$.
4. $\widetilde{e}_i L \subset L$, $\widetilde{f}_i L \subset L$ for all $0 \leqslant i \leqslant e-1$.
5. $\widetilde{e}_i B \subset B \cup \{0\}$, $\widetilde{f}_i B \subset B \cup \{0\}$.
6. For any $i \in \{0,1,\ldots,e-1\}$ and $u$, $v$ in $B$, we have $u = \widetilde{e}_i v$ if and only if $v = \widetilde{f}_i u$.

One can associate with any crystal basis a coloured oriented graph as follows.

**Definition 6.1.12.** Let $(L,B)$ be a crystal basis for an integrable $\mathscr{U}_q(\widehat{\mathfrak{sl}_e})$-module $M$. The *crystal graph* $\mathscr{G}$ (or the *crystal*) associated with $(L,B)$ is the coloured oriented graph with

- vertices: the elements in $B$;
- edges: for $u$ and $v$ in $B$ we have an arrow $u \xrightarrow{i} v$ coloured by $i \in \{0,1,\ldots,e-1\}$ if and only if $v = \widetilde{f}_i u$.

An example of such a graph is given in Example 6.2.18.

The following theorem shows the existence of a crystal basis for every irreducible highest weight $\mathscr{U}_q(\widehat{\mathfrak{sl}_e})$-modules. We refer to [190, Theorem 2] for its proof.

**Theorem 6.1.13 (Kashiwara, Lusztig).** *Let $\Lambda \in P^+$ and set*

$$L(\Lambda) = \sum \mathbb{A} \widetilde{f}_{i_1} \widetilde{f}_{i_2} \ldots \widetilde{f}_{i_k} m_\Lambda,$$
$$B(\Lambda) = \left\{ \widetilde{f}_{i_1} \widetilde{f}_{i_2} \ldots \widetilde{f}_{i_k} m_\Lambda \bmod qL(\Lambda)) \right\} \setminus \{0\}.$$

*Then $(L(\Lambda), B(\Lambda))$ is a crystal basis for $V(\Lambda)$. Moreover, any crystal basis of $V(\Lambda)$ coincides with $(L(\Lambda), B(\Lambda))$ up to a scalar multiple.*

**6.1.14.** If $M_1$ and $M_2$ are integrable $\mathscr{U}_q(\widehat{\mathfrak{sl}_e})$-modules and if $(L_1, B_1)$ and $(L_2, B_2)$ are crystal bases of $M_1$ and $M_2$ respectively, then we say that $(L_1, B_1)$ and $(L_2, B_2)$ are isomorphic if there exists an isomorphism of $\mathscr{U}_q(\widehat{\mathfrak{sl}_e})$-modules $\Psi : M_1 \to M_2$ such that the following conditions hold:

- $\Psi$ induces an isomorphism of $\mathbb{A}$-lattices $\Phi : L_1 \to L_2$.
- We have $\overline{\Phi}(B_1) = B_2$, where $\overline{\Phi}$ is the induced isomorphism of vector spaces $\overline{\Phi} : L_1/qL_1 \to L_2/qL_2$.

Note that if $(L_1, B_1)$ and $(L_2, B_2)$ are isomorphic, the associated bijection

$$\overline{\Phi} : B_1 \sqcup \{0\} \to B_2 \sqcup \{0\}$$

(with $\overline{\Phi}(0) = 0$) has the the following property. For all $i \in \{0,\ldots,e-1\}$ and $b_1 \in B_1$, we have $\overline{\Phi}(\widetilde{f}_i b) = \widetilde{f}_i \overline{\Phi}(b)$ and $\overline{\Phi}(\widetilde{e}_i b) = \widetilde{e}_i \overline{\Phi}(b)$. This implies that we have an arrow

$b \xrightarrow{i} b'$ in the crystal graph $\mathscr{G}_1$ associated with $(L_1, B_1)$ if and only if we have an arrow $\overline{\Phi}(b) \xrightarrow{i} \overline{\Phi}(b')$ in the crystal graph $\mathscr{G}_2$ associated with $(L_2, B_2)$. If a bijection between $B_1$ and $B_2$ satisfies such a property, then we say that there is an *isomorphism of crystal graphs* between $\mathscr{G}_1$ and $\mathscr{G}_2$.

The following result shows the compatibility of the crystal bases theory with the direct sum of $\mathscr{U}_q(\widehat{\mathfrak{sl}_e})$-modules in $\mathscr{O}_{\text{int}}$. The first assertion implies the existence of a crystal basis for the $\mathscr{U}_q(\widehat{\mathfrak{sl}_e})$-modules in $\mathscr{O}_{\text{int}}$ by Theorem 6.1.13 and the second one the uniqueness up to isomorphism. We refer to [190, Theorem 3] for its proof.

**Proposition 6.1.15 (Kashiwara).**

1. *Let $M_1$ and $M_2$ be integrable $\mathscr{U}_q(\widehat{\mathfrak{sl}_e})$ modules in $\mathscr{O}_{int}$ equipped with crystal bases $(L_1, B_1)$ and $(L_2, B_2)$. Then $(L_1, B_1) \oplus (L_2, B_2) := (L_1 \oplus L_2, B_1 \sqcup B_2)$ is a crystal basis for $M = M_1 \oplus M_2$.*
2. *Let $M$ be an integrable $\mathscr{U}_q(\widehat{\mathfrak{sl}_e})$-module in $\mathscr{O}_{int}$ equipped with a crystal basis $(L, B)$. Then there exists an isomorphism between $M$ and a direct sum $\bigoplus_\Lambda V(\Lambda)$, where the $V(\Lambda)$ are certain irreducible highest weight modules, which induces an isomorphism between $(L, B)$ and $\bigoplus_\Lambda (L(\Lambda), B(\Lambda))$.*

**6.1.16.** Now, the theorem below, independently proved by Kashiwara and Lusztig, shows that the basis $B(\Lambda)$ of $L(\Lambda)/qL(\Lambda)$ can be lifted to obtain a basis of $V(\Lambda)$. To state the theorem, we first need to introduce an involution on $\mathscr{U}_q(\widehat{\mathfrak{sl}_e})$. This is given as follows:

$$\overline{q} = q^{-1}, \quad \overline{e_i} = e_i, \quad \overline{f_i} = f_i, \quad \overline{t_i} = t_i^{-1}, \quad \overline{\partial} = \partial$$

for $0 \leqslant i < e$. This induces an involution on $V(\Lambda)_{\mathbb{Q}}$ with $\Lambda \in P^+$ by setting

$$\overline{u.m_\Lambda} = \overline{u}.m_\Lambda \qquad \text{for all } u \in \mathscr{U}_q(\widehat{\mathfrak{sl}_e})_{\mathbb{Q}}.$$

**Theorem 6.1.17 (Kashiwara [190], Lusztig [230]).** *Let $\Lambda \in P^+$, let $(L(\Lambda), B(\Lambda))$ be the crystal basis for $V(\Lambda)$. The module $V(\Lambda)_{\mathbb{Q}}$ has a unique $\mathbb{Q}[q, q^{-1}]$-basis*

$$\mathscr{B}(\Lambda) = \{G(b) \mid b \in B(\Lambda)\}$$

*such that for all $b \in B(\Lambda)$ we have*

$$G(b) = \overline{G(b)} \qquad \text{and} \qquad G(b) \equiv b \bmod qL(\Lambda).$$

*This basis is called the Kashiwara–Lusztig canonical basis (or the global crystal basis) for the $\mathscr{U}_q(\widehat{\mathfrak{sl}_e})$-module $V(\Lambda)$ and for the $\mathscr{U}'_q(\widehat{\mathfrak{sl}_e})$-module $V'(\Lambda)$.*

Note that the specialisation at $q = 1$ of the above canonical basis leads to a canonical basis for the irreducible highest weight modules of $\mathscr{U}(\widehat{\mathfrak{sl}_e})$.

## 6.2 The Fock Space and Ariki's Theorem

The aim of this section is to give a realisation of the irreducible highest weight $\mathscr{U}_q(\widehat{\mathfrak{sl}_e})$-modules. To do this, we construct an action of $\mathscr{U}_q(\widehat{\mathfrak{sl}_e})$ on a remarkable $\mathbb{Q}(q)$-vector space called the Fock space. We then deduce the structure of the desired modules and give the associated crystals. Finally, we present the connections with the representation theory of Ariki–Koike algebras given by Ariki's theorem.

**Definition 6.2.1.** Let $l \in \mathbb{Z}_{>0}$ and let $\mathbf{s} \in \mathbb{Z}^l$. The *Fock space* $\mathscr{F}_\mathbf{s}$ is the $\mathbb{Q}(q)$-vector space with a distinguished basis $|\lambda, \mathbf{s}\rangle$ with $\lambda \vdash_l n$. In other words, we denote

$$\mathscr{F}_\mathbf{s}^n = \bigoplus_{\lambda \vdash_l n} \mathbb{Q}(q)|\lambda, \mathbf{s}\rangle \qquad \text{for } n \in \mathbb{Z}_{\geqslant 0},$$

$$\mathscr{F}_\mathbf{s} = \bigoplus_{n \in \mathbb{Z}_{\geqslant 0}} \mathscr{F}_\mathbf{s}^n.$$

The positive integer $l$ is called the *level of the Fock space* $\mathscr{F}_\mathbf{s}$ and $\mathbf{s}$ is called the *$l$-charge* or the *multicharge* of $\mathscr{F}_\mathbf{s}$.

**6.2.2.** Let $\mathbf{s} \in \mathbb{Z}^l$ and let $i \in \{0, \ldots, e-1\}$. Recall the definition of addable and removable $i$-nodes for an $l$-partition in 5.1.5. Recall also from 5.7.14 the total order that we have already introduced on the set of addable and removable $i$-nodes of an $l$-partition. Let $\gamma$, $\gamma'$ be two removable or addable $i$-nodes of $\lambda$. We denote

$$\gamma \prec_{(\mathbf{s},e)} \gamma' \quad \overset{\text{def}}{\Longleftrightarrow} \quad \begin{cases} \text{either } b-a+s_c < b'-a'+s_{c'}, \\ \text{or} \quad\;\; b-a+s_c = b'-a'+s_{c'} \text{ and } c > c'. \end{cases}$$

*Remark 6.2.3.* The version of the order $\prec_{(\mathbf{s},e)}$ used in [291] is slightly different. This is given as follows: we have $\gamma <_{(\mathbf{s},e)} \gamma'$ if we have $b-a+s_c < b'-a'+s_{c'}$ or if $b-a+s_c = b'-a'+s_{c'}$ and $c < c'$. One can recover all the results in [291] by reversing the order of components of $\lambda$ on the basis vectors $|\lambda, \mathbf{s}\rangle$ of the Fock space. Our convention is in agreement with [88].

**Example 6.2.4.** Fix $n \in \mathbb{Z}_{>1}$ and assume that $\mathbf{s} \in \mathbb{Z}^l$ is such that $s_i - s_j > n-1-e$ for all $i < j$. Let $\gamma = (a,b,c)$, $\gamma' = (a',b',c')$ be two removable $i$-nodes of the same $l$-partition $\lambda \vdash_l n$. Assume that $c' < c$. Then, as already observed in Example 5.5.18, we have $b'-a'-(b-a) \geqslant 1-n$. We thus have $b'-a'+s_{c'}-(b-a+s_c) > -e$. As $\gamma$ and $\gamma'$ have the same residue, this implies that $b'-a'+s_{c'}-(b-a+s_c) \geqslant 0$. We conclude that we have $\gamma \prec_{(\mathbf{s},e)} \gamma'$. Hence, for this choice of $\mathbf{s}$, we have $\gamma \prec_{(\mathbf{s},e)} \gamma'$ if and only if $c' < c$ or if $c = c'$ and $a' < a$ (note, however, that this is only valid for $l$-partitions of rank $\leqslant n$). This order on $i$-nodes is connected with the one used in Ariki's book [10, Definition 10.7] (but now for general choice of $\lambda \vdash_l n$ and $n \in \mathbb{Z}_{>0}$!). It can be seen as an "asymptotic" version of our choice of order.

Let $\lambda$ and $\mu$ be two $l$-partitions of rank $n$ and $n+1$ such that there exists an $i$-node $\gamma$ such that $[\mu] = [\lambda] \cup \{\gamma\}$. We define the following numbers:

$$\begin{aligned}
N_i^{\succ}(\lambda,\mu) =& \sharp\{\text{addable } i-\text{nodes } \gamma' \text{ of } \lambda \text{ such that } \gamma' \succ_{(\mathbf{s},e)} \gamma\} \\
&- \sharp\{\text{removable } i-\text{nodes } \gamma' \text{ of } \mu \text{ such that } \gamma' \succ_{(\mathbf{s},e)} \gamma\}, \\
N_i^{\prec}(\lambda,\mu) =& \sharp\{\text{addable } i-\text{nodes } \gamma' \text{ of } \lambda \text{ such that } \gamma' \prec_{(\mathbf{s},e)} \gamma\} \\
&- \sharp\{\text{removable } i-\text{nodes } \gamma' \text{ of } \mu \text{ such that } \gamma' \prec_{(\mathbf{s},e)} \gamma\}, \\
N_i(\lambda) =& \sharp\{\text{addable } i-\text{nodes of } \lambda\} - \sharp\{\text{removable } i-\text{nodes of } \lambda\}, \\
N_{\mathfrak{d}}(\lambda) =& \sharp\{0-\text{nodes of } \lambda\}.
\end{aligned}$$

These numbers allow us to define an action of $\mathscr{U}_q(\widehat{\mathfrak{sl}_e})$ on the Fock space.

**Theorem 6.2.5 (Jimbo, Misra, Miwa and Okado [186, Prop. 3.5]).** *Let* $\mathbf{s} \in \mathbb{Z}^l$. *Then* $\mathscr{F}_{\mathbf{s}}$ *is an integrable* $\mathscr{U}_q(\widehat{\mathfrak{sl}_e})$*-module with action*

$$e_i|\lambda,\mathbf{s}\rangle = \sum_{\mathfrak{r}_e([\lambda]/[\mu])=i} q^{-N_i^{\prec}(\mu,\lambda)}|\mu,\mathbf{s}\rangle, \qquad f_i|\lambda,\mathbf{s}\rangle = \sum_{\mathfrak{r}_e([\mu]/[\lambda])=i} q^{N_i^{\succ}(\lambda,\mu)}|\mu,\mathbf{s}\rangle,$$

$$t_i|\lambda,\mathbf{s}\rangle = q^{N_i(\lambda)}|\lambda,\mathbf{s}\rangle, \qquad \mathfrak{d}|\lambda,\mathbf{s}\rangle = -(\Delta(\mathbf{s})+N_{\mathfrak{d}}(\lambda))|\lambda,\mathbf{s}\rangle,$$

*where* $0 \leqslant i \leqslant e-1$ *and* $\Delta(\mathbf{s}) \in \mathbb{Q}$ *is defined in [291, Section 2.1] (recall also that* $\mathfrak{r}_e(\gamma)$ *denotes the residue of the node* $\gamma$*).*

*Proof.* One checks that the above formulae give a well-defined $\mathscr{U}_q(\widehat{\mathfrak{sl}_e})$-module structure and that $\mathscr{F}_{\mathbf{s}}$ is in $\mathscr{O}_{\text{int}}$, directly from the definition of the action. □

*Remark 6.2.6.* The action of $\mathfrak{d}$ is not really important for our purpose, as we will mainly deal with the action of $\mathscr{U}_q'(\widehat{\mathfrak{sl}_e})$.

The following proposition generalises the definition of $N_i^{\succ}(\lambda,\mu)$ and $N_i^{\prec}(\lambda,\mu)$ and gives the action of the divided powers on the standard basis of the Fock space. It is largely inspired from [245, Proposition 6.16] and it can be seen as a quantisation of Proposition 5.6.11.

**Proposition 6.2.7.** *Let* $\lambda$ *be an* $l$*-partition,* $i \in \{0,\dots,e-1\}$ *and* $j \in \mathbb{Z}_{>0}$.

1. *We have*
$$f_i^{(j)}|\lambda,\mathbf{s}\rangle = \sum q^{N_i^{\succ}(\lambda,\mu)}|\mu,\mathbf{s}\rangle,$$
*where the sum is taken over all the* $|\mu,\mathbf{s}\rangle$ *such that* $\mu$ *is obtained from* $\lambda$ *by adding* $j$ *nodes with residue* $i$*. We have*
$$N_i^{\succ}(\lambda,\mu) := \sum_{\gamma\in[\mu]/[\lambda]} \left(\sharp\left\{\begin{array}{c}\textit{addable } i-\textit{nodes } \gamma' \\ \textit{of } \mu \textit{ such that } \gamma' \succ_{(\mathbf{s},e)} \gamma\end{array}\right\} - \sharp\left\{\begin{array}{c}\textit{removable } i-\textit{nodes } \gamma' \\ \textit{of } \lambda \textit{ such that } \gamma' \succ_{(\mathbf{s},e)} \gamma\end{array}\right\}\right).$$
2. *We have*
$$e_i^{(j)}|\mu,\mathbf{s}\rangle = \sum q^{-N_i^{\prec}(\lambda,\mu)}|\lambda,\mathbf{s}\rangle,$$

*where the sum is taken over all the* $|\lambda,\mathbf{s}\rangle$ *obtained from* $\mu$ *by removing* $j$ *nodes with residue* $i$. *We have*

$$N_i^{\prec}(\lambda,\mu) := \sum_{\gamma\in[\mu]/[\lambda]} \left(\sharp\left\{\begin{array}{c}\textit{addable } i-\textit{nodes } \gamma' \\ \textit{of } \mu \textit{ such that } \gamma' \prec_{(\mathbf{s},e)} \gamma\end{array}\right\} - \sharp\left\{\begin{array}{c}\textit{removable } i-\textit{nodes } \gamma' \\ \textit{of } \lambda \textit{ such that } \gamma' \prec_{(\mathbf{s},e)} \gamma\end{array}\right\}\right).$$

*Proof.* We only give the proof of the first assertion. The second one is completely analogous. We argue by induction on $j\in\mathbb{Z}_{>0}$. If $j=1$, this is the definition of the action in Theorem 6.2.5. So now assume that $j>1$.

Let $\mu$ be an $l$-partition such that $\mu=\lambda\cup\{\gamma_1,\dots,\gamma_j\}$ for $i$-nodes $\gamma_1,\dots,\gamma_j$. We can assume that

$$\gamma_1 \succ_{(\mathbf{s},e)} \gamma_2 \succ_{(\mathbf{s},e)} \cdots \succ_{(\mathbf{s},e)} \gamma_j.$$

Let $\mu_s$, $s=1,\dots,j$, be the $l$-partitions such that $[\mu]/[\mu_s]=\gamma_s$. By induction, the coefficient of $|\mu_s,\mathbf{s}\rangle$ in $f_i^{(j-1)}|\lambda,\mathbf{s}\rangle$ is $q^{N_i^{\succ}(\lambda,\mu_s)}$. Hence, the coefficient of $|\mu,\mathbf{s}\rangle$ in $f_if_i^{(j-1)}|\lambda,\mathbf{s}\rangle$ is $\sum_{1\leqslant s\leqslant j} q^{N_i^{\succ}(\lambda,\mu_s)+N_i^{\succ}(\mu_s,\mu)}$.

Now, we define the following numbers:

$$N_i^{\succ}(\lambda,\mu,\gamma_s) := \left(\sharp\left\{\begin{array}{c}\text{addable } i-\text{nodes } \eta \\ \text{of } \mu \text{ such that } \eta \succ_{(\mathbf{s},e)} \gamma_s\end{array}\right\} - \sharp\left\{\begin{array}{c}\text{removable } i-\text{nodes } \eta \\ \text{of } \lambda \text{ such that } \eta \succ_{(\mathbf{s},e)} \gamma_s\end{array}\right\}\right),$$

where $s=1,\dots,j$. We have

$$\begin{aligned} N_i^{\succ}(\lambda,\mu_s) &= \sum_{1\leqslant t\leqslant s-1} N_i^{\succ}(\lambda,\mu_s,\gamma_t) + \sum_{s+1\leqslant t\leqslant j} N_i^{\succ}(\lambda,\mu_s,\gamma_t) \\ &= \sum_{1\leqslant t\leqslant s-1} N_i^{\succ}(\lambda,\mu,\gamma_t) + \sum_{s+1\leqslant t\leqslant j} (N_i^{\succ}(\lambda,\mu,\gamma_t)+1) \\ &= N_i^{\succ}(\lambda,\mu) - N_i^{\succ}(\lambda,\mu_s,\gamma_s) + j - s, \\ N_i^{\succ}(\mu_s,\mu) &= N_i^{\succ}(\lambda,\mu_s,\gamma_s) - s + 1. \end{aligned}$$

Hence, the coefficient of $|\mu,\mathbf{s}\rangle$ in $f_if_i^{(j-1)}|\lambda,\mathbf{s}\rangle$ is $q^{N_i^{\succ}(\lambda,\mu)}\sum_{1\leqslant s\leqslant j} q^{j+1-2s}$. Now, we have $f_if_i^{(j-1)}|\lambda,\mathbf{s}\rangle=[j]_qf_i^{(j)}$. Thus, the coefficient of $|\mu,\mathbf{s}\rangle$ in $f_i^{(j)}|\lambda,\mathbf{s}\rangle$ is $q^{N_i^{\succ}(\lambda,\mu)}$. This concludes the proof. □

**6.2.8.** An interesting consequence of the above constructions is that we will have several ways to realise the same irreducible highest weight $\mathcal{U}_q(\widehat{\mathfrak{sl}_e})$-modules. To be more precise, we introduce the extended affine symmetric group. As this object will play a crucial role in the next parts, we give several useful comments on this group.

Let $r \in \mathbb{Z}_{>0}$. The *affine symmetric group* $\widetilde{\mathfrak{S}}_r$ of type $A_{r-1}^{(1)}$ is by definition the Coxeter group with presentation by:

- Generators: $\sigma_i$ with $i = 0, 1, \dots, r-1$.
- Relations:

$$\begin{aligned} \sigma_i \sigma_{i+1} \sigma_i &= \sigma_{i+1} \sigma_i \sigma_{i+1} && \text{for all } i = 0, \dots, r-1, \\ \sigma_i \sigma_j &= \sigma_j \sigma_i && \text{if } i - j \not\equiv 1 \bmod r, \\ \sigma_i^2 &= 1 && \text{for all } i = 0, \dots, r-1, \end{aligned}$$

where the subscripts are understood to be modulo $r$. The *extended affine symmetric group* $\widehat{\mathfrak{S}}_r$ is the semi-direct product $\widetilde{\mathfrak{S}}_r \rtimes \langle \tau \rangle$ such that $\langle \tau \rangle \simeq \mathbb{Z}$ and $\tau \sigma_i = \sigma_{i+1} \tau$ for all $i = 0, \dots, r-1$. Hence it has a presentation by:

- Generators: $\tau$, $\sigma_i$ with $i = 0, 1, \dots, r-1$.
- Relations:

$$\begin{aligned} \sigma_i \sigma_{i+1} \sigma_i &= \sigma_{i+1} \sigma_i \sigma_{i+1} && \text{for all } i = 0, \dots, r-1, \\ \sigma_i \sigma_j &= \sigma_j \sigma_i && \text{if } i - j \not\equiv 1 \bmod r, \\ \sigma_i^2 &= 1 && \text{for all } i = 0, \dots, r-1, \\ \tau \sigma_i &= \sigma_{i+1} \tau && \text{if } i = 0, \dots, r-1, \end{aligned}$$

where, again, the subscripts are understood to be modulo $r$. Note that the subgroup of $\widehat{\mathfrak{S}}_r$ generated by the elements $\sigma_i$, $i = 1, \dots, r-1$ can be identified with the symmetric group $\mathfrak{S}_r$ of rank $r$.

There is another way to define this group. Let $P_r := \mathbb{Z}^r$ and let $\{y_i \mid i = 1, \dots, r\}$ be the standard basis of $P_r$. The group $\widehat{\mathfrak{S}}_r$ can then be seen as the semi-direct product $P_r \rtimes \mathfrak{S}_r$, where the relations of the semidirect product are given by $\sigma_i y_j = y_j \sigma_i$ for $j \neq i, i+1$ and $\sigma_i y_i \sigma_i = y_{i+1}$. In other words, we have a presentation by:

- Generators: $\sigma_i$ with $i = 1, \dots, r-1$ and $y_i$ with $i = 1, \dots, r$.
- Relations:

$$\begin{aligned} \sigma_i \sigma_{i+1} \sigma_i &= \sigma_{i+1} \sigma_i \sigma_{i+1} && \text{for all } i = 1, \dots, r-1, \\ \sigma_i \sigma_j &= \sigma_j \sigma_i && \text{if } i - j \not\equiv 1 \bmod r, \\ \sigma_i^2 &= 1 && \text{for all } i = 1, \dots, r-1, \\ y_i y_j &= y_j y_i && \text{for all } i, j = 1, \dots, r, \\ \sigma_i y_j &= y_j \sigma_i && \text{for all } i = 1, \dots, r-1, j = 1, \dots, r \text{ and} \\ & && j \not\equiv i, i+1 \bmod r, \\ \sigma_i y_i \sigma_i &= y_{i+1} && \text{for all } i = 1, \dots, r-1. \end{aligned}$$

For $\mathbf{a} = (a_1, \dots, a_r) \in \mathbb{Z}^r$, we denote $y^{\mathbf{a}} = y_1^{a_1} \dots y_r^{a_r}$. One can go from one presentation to the other using the following formulae:

$$\tau = \sigma_1 \dots \sigma_{r-1} y_r, \quad \sigma_0 = \sigma_{r-1} \dots \sigma_2 \sigma_1 \sigma_2 \dots \sigma_{r-1} y_1^{-1} y_r.$$

As noted above, $\widetilde{\mathfrak{S}}_r$ is a Coxeter group and, therefore, it has an associated Bruhat–Chevalley order and a length function. But $\widehat{\mathfrak{S}}_r$ is *not* a Coxeter group. However, one can extend the Bruhat–Chevalley order and the length function from $\widetilde{\mathfrak{S}}_r$ to $\widehat{\mathfrak{S}}_r$: let $(w_1, w_2) \in \widehat{\mathfrak{S}}_r^2$; then there exist unique elements $(k_1, k_2) \in \mathbb{Z}^2$, $(x_1, x_2) \in \mathfrak{S}_r$ such that $w_1 = \tau^{k_1} x_1$ and $w_2 = \tau^{k_2} x_2$. We then define $w_1 < w_2$ if $k_1 = k_2$ and $x_1 < x_2$ and we set $l(w_1) := l(x_1)$. Hence the length of $w$ is 0 if and only if $w \in \langle \tau \rangle$.

**6.2.9.** We now return to the $\mathcal{U}_q(\widehat{\mathfrak{sl}_e})$-modue $\mathcal{F}_{\mathbf{s}}$ defined in Definition 6.2.1. Let $\mathbf{s} \in \mathbb{Z}^l$, then one easily check that the vector $|\varnothing, \mathbf{s}\rangle$ is a highest weight vector for the action of $\mathcal{U}_q(\widehat{\mathfrak{sl}_e})$ on $\mathcal{F}_{\mathbf{s}}$. Hence, the module

$$V(\Lambda_{\mathbf{s}})^{\mathbf{s}} := \mathcal{U}_q(\widehat{\mathfrak{sl}_e})|\varnothing, \mathbf{s}\rangle$$

is an irreducible highest weight $\mathcal{U}_q(\widehat{\mathfrak{sl}_e})$-module. The $\mathcal{U}_q'(\widehat{\mathfrak{sl}_e})$-module $V'(\Lambda_{\mathbf{s}})^{\mathbf{s}} := \mathcal{U}_q'(\widehat{\mathfrak{sl}_e})|\varnothing, \mathbf{s}\rangle$ is an irreducible highest weight $\mathcal{U}_q'(\widehat{\mathfrak{sl}_e})$-module with weight $\Lambda_{\mathbf{s}}$ (modulo $\partial$). Thus, it is isomorphic to $V'(\Lambda_{\mathbf{s}})$ (see 6.1.6). Let us give an interesting consequence of this result. Consider $e$ a fixed positive integer. Then we have an action of $\mathfrak{S}_l$ on $\mathbb{Z}^l$ by setting for any $\mathbf{s} = (s_1, \ldots, s_l) \in \mathbb{Z}^l$

$$\begin{aligned} \sigma_c.\mathbf{s} &= (s_1, \ldots, s_{c-1}, s_{c+1}, s_c, s_{c+2}, \ldots, s_l) \text{ for } c = 1, \ldots, r-1 \text{ and} \\ y_c.\mathbf{s} &= (s_1, \ldots, s_{c-1}, s_c + e, s_{c+1} \ldots, s_l) \quad \text{for } c = 1, \ldots, r. \end{aligned}$$

A fundamental domain for this action is given by

$$\{(s_1, \ldots, s_l) \in \mathbb{Z}^l \mid 1 \leqslant s_1 \leqslant \ldots \leqslant s_l \leqslant e\},$$

which is contained in the set $\mathcal{S}_e^l$ that we have already met in Section 5.7 (see in particular Definition 5.7.8.) In particular, note that if $\mathbf{s}^1 = (s_1^1, \ldots, s_l^1)$ and $\mathbf{s}^2 = (s_1^2, \ldots, s_l^2)$ are $l$-tuples of integers in the same orbit with respect to this action, then $V'(\Lambda_{\mathbf{s}^1})^{\mathbf{s}^1}$ and $V'(\Lambda_{\mathbf{s}^2})^{\mathbf{s}^2}$ are both irreducible highest weight $\mathcal{U}_q'(\widehat{\mathfrak{sl}_e})$-modules with the same weight (modulo $\partial$). As a consequence, we have an isomorphism

$$V'(\Lambda_{\mathbf{s}^1})^{\mathbf{s}^1} \to V'(\Lambda_{\mathbf{s}^2})^{\mathbf{s}^2},$$

such that the canonical basis of the first module is mapped to the canonical basis of the latter by this isomorphism (see 6.1.6).

**6.2.10.** Let $\mathbf{s} \in \mathbb{Z}^l$. As the Fock space $\mathcal{F}_{\mathbf{s}}$ is an integrable $\mathcal{U}_q(\widehat{\mathfrak{sl}_e})$-module, by Theorem 6.1.15, it is equipped with an associated crystal basis and a crystal graph. We begin this section with the description of this graph. We then deduce the crystal graph of an arbitrary irreducible highest weight $\mathcal{U}_q(\widehat{\mathfrak{sl}_e})$-module.

Let $\lambda$ be an $l$-partition. We can consider its set of addable and removable $i$-nodes. Let $w_i(\lambda)$ be the word obtained first by writing the addable and removable $i$-nodes of $\lambda$ in increasing order with respect to $\prec_{(\mathbf{s},e)}$ and then by encoding each addable $i$-node by the letter $A$ and each removable $i$-node by the letter $R$. Write $\widetilde{w}_i(\lambda) = A^p R^q$ for the word derived from $w_i$ by deleting as many subwords of type $RA$ as possible. $w_i(\lambda)$ is called the *i-word* of $\lambda$ and $\widetilde{w}_i(\lambda)$ the *reduced i-word* of $\lambda$. The addable $i$-nodes in $\widetilde{w}_i(\lambda)$ are called the *normal addable i-nodes*. The removable $i$-nodes in $\widetilde{w}_i(\lambda)$ are called the *normal removable i-nodes*. If $p > 0$, let $\gamma$ be the rightmost addable $i$-node in $\widetilde{w}_i$. The node $\gamma$ is called the *good addable i-node*. If $q > 0$, the leftmost removable $i$-node in $\widetilde{w_i}$ is called the *good removable i-node*. Note that this notion depends on the order $\prec_{e,\mathbf{s}}$ and, thus, on the choice of $\mathbf{s} \in \mathbb{Z}^l$. These definitions are due to Kleshchev [204] (when $l = 1$) who used them to describe the branching rules for symmetric group in positive characteristic.

**Example 6.2.11.** We consider the 2-partition $\lambda = (2.2, 2.1.1)$. Set $e = 4$ and $\mathbf{s} = (0,1)$; then $\lambda$ has the following diagram with residues:

$$\left( \begin{array}{|c|c|} \hline 0 & 1 \\ \hline 3 & 0 \\ \hline \end{array} \, , \, \begin{array}{|c|c|} \hline 1 & 2 \\ \hline 0 \\ \cline{1-1} 3 \\ \cline{1-1} \end{array} \right).$$

We have $w_2 = AARA$. Hence, we have $\widetilde{w_2} = AA$ and the 2-node $(3,1,1)$ is a good addable 2-node with respect to $\prec_{(\mathbf{s},e)}$. There is no normal removable 2-node (and thus no removable good $i$-node). If we now take $\mathbf{s} = (0,5)$, then $\lambda$ has the same diagram with residues. However, the word $w_2$ is not the same because the order $\prec_{(\mathbf{s},e)}$ has been modified: in this case, we have $w_2 = AAAR$. We then have $\widetilde{w_2} = AAAR$ and the 2-node $(4,1,2)$ is a good addable 2-node with respect to $\prec_{(\mathbf{s},e)}$. The node $(1,2,2)$ is a good removale 2-node.

Using the definition of the Kashiwara operators, we now get the following result.

**Theorem 6.2.12 (Jimbo, Misra, Miwa and Okado [186, §3]).** *Let $\lambda$ be an $l$-partition. Then, we have*

$$\widetilde{e}_i|\lambda,\mathbf{s}\rangle = \begin{cases} |\lambda \setminus \{\gamma\},\mathbf{s}\rangle & \text{if } \gamma \text{ is a good removable } i\text{-node}, \\ 0 & \text{if } \lambda \text{ has no good removable } i\text{-node}, \end{cases}$$

$$\widetilde{f}_i|\lambda,\mathbf{s}\rangle = \begin{cases} |\lambda \cup \{\gamma\},\mathbf{s}\rangle & \text{if } \gamma \text{ is a good addable } i\text{-node}, \\ 0 & \text{if } \lambda \text{ has no good addable } i\text{-node}, \end{cases}$$

*where the action is given modulo $qL(\Lambda_{\mathbf{s}})$. In other words, the crystal graph $\mathscr{G}_e(\Lambda_{\mathbf{s}})^{\mathbf{s}}$ of $\mathscr{F}_{\mathbf{s}}$ is the coloured oriented graph with*

- *vertices: the symbols $|\lambda,\mathbf{s}\rangle$ with $\lambda \vdash_l n$ and $n \in \mathbb{Z}_{\geqslant 0}$;*
- *arrows: $|\lambda,\mathbf{s}\rangle \xrightarrow{i} |\mu,\mathbf{s}\rangle$ if and only if $\mu$ is obtained by adding to $\lambda$ a good addable $i$-node.*

**Definition 6.2.13.** Let $\mathbf{s} \in \mathbb{Z}^l$. The set of *Uglov $l$-partitions $\Phi_{(\mathbf{s},e)}$* is defined recursively as follows.

- We have $\varnothing := (\varnothing, \varnothing, \ldots, \varnothing) \in \Phi_{(\mathbf{s},e)}$.
- If $\lambda \in \Phi_{(\mathbf{s},e)}$ with $\lambda \neq \varnothing$, there exist $i \in \{0, \ldots, e-1\}$ and a good removable $i$-node $\gamma$ such that, if we remove $\gamma$ from $\lambda$, the resulting $l$-partition is in $\Phi_{(\mathbf{s},e)}$.

We also set $\Phi_{(\mathbf{s},e)}(n)$ for the elements in $\Phi_{(\mathbf{s},e)}$ with rank $n$.

The following result is a direct consequence of Theorems 6.1.13 and 6.2.12.

**Proposition 6.2.14.** *The crystal graph of the irreducible highest weight $\mathcal{U}_q(\widehat{\mathfrak{sl}_e})$-module $V(\Lambda_{\mathbf{s}})^{\mathbf{s}}$ is the connected component $(\mathcal{G}_e(\Lambda_{\mathbf{s}})^{\mathbf{s}})^\circ$ of the graph $\mathcal{G}_e(\Lambda_{\mathbf{s}})^{\mathbf{s}}$ containing $|\varnothing,\mathbf{s}\rangle$. Hence, this is the graph with*

- *vertices: the symbols $|\lambda,\mathbf{s}\rangle$ with $\lambda \in \Phi_{(\mathbf{s},e)}$;*
- *arrows: $|\lambda,\mathbf{s}\rangle \xrightarrow{i} |\mu,\mathbf{s}\rangle$ if and only if $\mu$ is obtained by adding to $\lambda$ its good addable $i$-node.*

*In particular, $(L(\Lambda_{\mathbf{s}}), \Phi_{(\mathbf{s},e)})$ is the crystal basis of $V(\Lambda_{\mathbf{s}})^{\mathbf{s}}$.*

*Remark 6.2.15.* Note that the residue of a node does not change if we replace $\mathbf{s} = (s_1,\dots,s_l)$ by $\mathbf{s}^+ = (s_1+e,\dots,s_l+e)$, nor does the order between two arbitrary nodes with the same residue. As the notion of Uglov $l$-partitions only depends on these data, we deduce that $\Phi_{(\mathbf{s},e)} = \Phi_{(\mathbf{s}^+,e)}$.

**Example 6.2.16.** Fix $n \in \mathbb{Z}_{>1}$ and assume that $\mathbf{s} \in \mathbb{Z}^l$ is such that $s_i - s_j \geqslant n-1-e$ for all $i<j$. In this case, we say that $\mathbf{s}$ is a *very dominant multicharge*. Let $\lambda$ be an $l$-partition and let $\gamma$ be a good removable $i$-node with respect to $\prec_{(\mathbf{s},e)}$. Let $\gamma' = (a',b',c')$ be any removable or addable $i$-nodes of the same $l$-partition $\lambda \vdash_l n$. Then we have seen in Example 6.2.4 that $\gamma \prec_{(\mathbf{s},e)} \gamma'$ if and only if $c' < c$ or if $c = c'$ and $a' < a$. Thus, in the definition of good removable $i$-node, one can replace $\prec_{(\mathbf{s},e)}$ with this order (for the $l$-partitions of rank less than $n$). In this case, the Uglov $l$-partitions (of rank $n$) are known as the *Kleshchev $l$-partitions*. They coincide with the ones in [10, Section 11.2] up to transformation $(\lambda^1,\dots,\lambda^l) \mapsto (\lambda^{1*},\dots,\lambda^{l*})$. Note that this class of $l$-partitions only depends on the equivalence classes of the $s_j$ modulo $e$. We denote them by $\overline{\Phi}_{(\mathbf{s} \bmod e,e)}$. A non-recursive definition of them in the case $l=2$ is derived from results of Ariki, Kreiman and Tsuchioka [14].

**6.2.17.** By the discussion in 6.2.9, if $w \in \widehat{\mathfrak{S}}_l$, and $\mathbf{s} \in \mathbb{Z}^l$, we have an isomorphism between $V'(\Lambda_{\mathbf{s}})^{\mathbf{s}}$ and $V'(\Lambda_{w.\mathbf{s}})^{w.\mathbf{s}}$. One can check that this implies that the crystal graphs $(\mathcal{G}_e(\Lambda_{\mathbf{s}})^{\mathbf{s}})^\circ$ and $(\mathcal{G}_e(\Lambda_{w.\mathbf{s}})^{w.\mathbf{s}})^\circ$ of $V(\Lambda_{\mathbf{s}})^{\mathbf{s}}$ and $V(\Lambda_{w.\mathbf{s}})^{w.\mathbf{s}}$ respectively are isomorphic in the sense of 6.1.14. However, they are really different in general. Indeed, the $l$-partitions labelling these graphs, $\Phi_{(\mathbf{s},e)}$ and $\Phi_{(w.\mathbf{s},e)}$, are distinct (see Example 6.2.18), even if there is a bijection

$$\varphi_{\mathbf{s},w.\mathbf{s}} : \Phi_{(\mathbf{s},e)} \to \Phi_{(w.\mathbf{s},e)}$$

induced by the isomorphism of crystal graphs. This bijection may be described by induction as follows.

- We have $\varphi_{\mathbf{s},w.\mathbf{s}}|\varnothing,\mathbf{s}\rangle = |\varnothing,w.\mathbf{s}\rangle$.
- Let $\lambda \in \Phi_{(\mathbf{s},e)}(n)$ for $n \in \mathbb{Z}_{>0}$; then there exist $i \in \{0,1,\dots,e-1\}$, a good removable $i$-node $\gamma$ for $\lambda$ (with respect to the order $\prec_{(\mathbf{s},e)}$; see 6.2.10) and $\lambda' \in \Phi_{(\mathbf{s},e)}(n-1)$ such that $[\lambda] = [\lambda'] \cup \{\gamma\}$. Now $\mu' := \varphi_{\mathbf{s},w.\mathbf{s}}(\lambda')$ is known by induction. Then there exist a good addable $i$-node $\gamma'$ for $\mu'$ (with respect to the order $\prec_{(w.\mathbf{s},e)}$) and $\mu \in \Phi_{(\mathbf{s},e)}(n)$ such that $[\mu] = [\mu'] \cup \{\gamma'\}$. We then have $\varphi_{\mathbf{s},w.\mathbf{s}}(\lambda) = \mu$.

A non-recursive combinatorial description of $\varphi_{\mathbf{s},w.\mathbf{s}}$ has been given in [177, Theorem 4.6] for $l=2$ and [178, Theorem 5.4.2] in the general case. We illustrate the above phenomenon in the next example.

**Fig. 6.1** Part of the crystal graph $(\mathscr{G}_2(\Lambda_{(0,0,1)})^{(0,0,1)})^{\circ}$

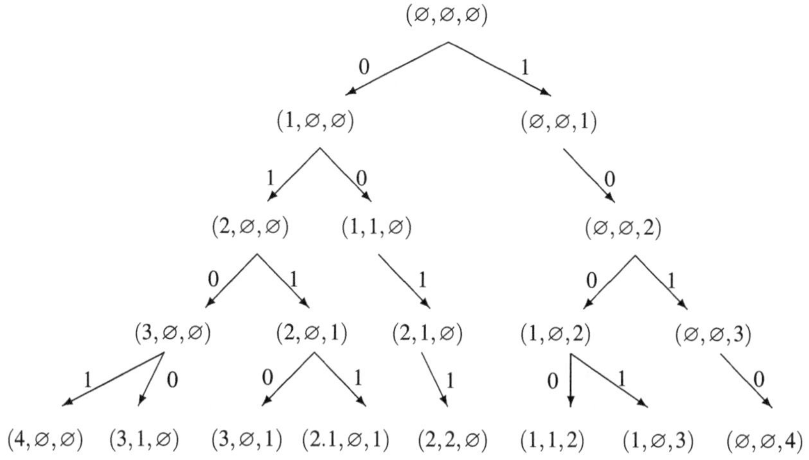


**Fig. 6.2** Part of the crystal graph $(\mathscr{G}_4(\Lambda_{(1,0,4)})^{(1,0,4)})^{\circ}$

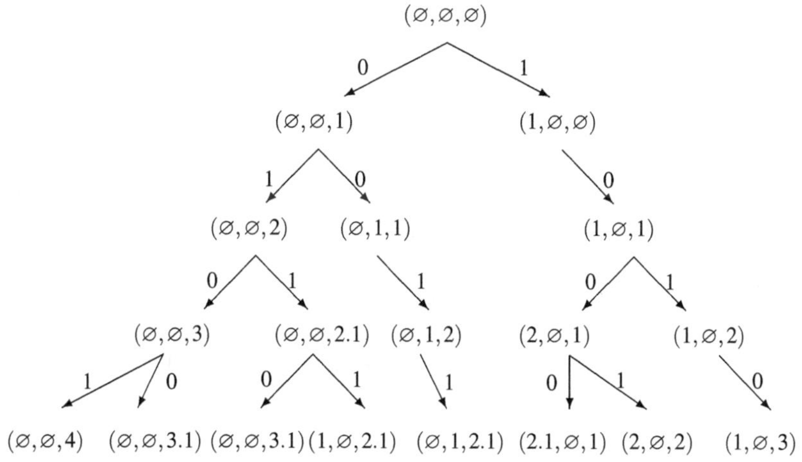


**Example 6.2.18.** In Figure 6.1 we show part of the crystal graph $(\mathscr{G}_2(\Lambda_{(0,0,1)})^{(0,0,1)})^{\circ}$ of the irreducible highest weight module $(V(2\Lambda_0+\Lambda_1)^{(0,0,1)})$ with $e=2$. To simplify the notation, we identify the symbols $|\boldsymbol{\lambda},\mathbf{s}\rangle$ with $\boldsymbol{\lambda}$. The 3-partitions appearing in this graph are the Uglov 3-partitions of $\Phi_{((0,0,1),2)}$ of rank less than or equal to 4.

Now consider the graph $(\mathscr{G}_4(\Lambda_{(1,0,4)})^{(1,0,4)})^\circ$ of the irreducible highest weight module $V(2\Lambda_0+\Lambda_1)^{(1,0,4)}$ with $e=2$. Here, $V'(2\Lambda_0+\Lambda_1)^{(1,0,4)}$ is isomorphic to $V'(2\Lambda_0+\Lambda_1)^{(0,0,1)}$ (see 6.1.6) and the associated graph for $l$-partitions of rank at most 4 is given in Figure 6.2.

We see that the sets of Uglov $l$-partitions are different in these cases.

**6.2.19.** Let $\mathbf{s}\in\mathbb{Z}^l$. By the above results, the module $V(\Lambda_{\mathbf{s}})^{\mathbf{s}}$ generated by the empty $l$-partition is an irreducible highest weight module. Hence, it admits a canonical basis labelled by the set of Uglov $l$-partitions:

$$\mathscr{B}_{\mathbf{s}}^q=\left\{G_q(\mu,\mathbf{s})\mid \mu\in\Phi_{(\mathbf{s},e)}\right\}.$$

For all $\mu\in\Phi_{(\mathbf{s},e)}(n)$, we have

$$G_q(\mu,\mathbf{s})=\sum_{\lambda\vdash_l n} d_{\lambda,\mu}(q)|\lambda,\mathbf{s}\rangle\in V(\Lambda_{\mathbf{s}})^{\mathbf{s}}_{\mathbb{Q}}.$$

Consider the $\mathscr{U}_q(\widehat{\mathfrak{sl}_e})_{\mathbb{Q}}$-module $(\mathscr{F}_{\mathbf{s}})_{\mathbb{Q}}$. The isomorphism between $\mathscr{U}(\widehat{\mathfrak{sl}_e})$ and $\mathbb{Q}\otimes_{\mathbb{Q}[q,q^{-1}]}\mathscr{U}_q(\widehat{\mathfrak{sl}_e})_{\mathbb{Q}}/\langle t_i-1\,;\,i=0,\ldots,e-1\rangle$ induces a structure of $\mathscr{U}(\widehat{\mathfrak{sl}_e})$-module on $(\mathscr{F}_{\mathbf{s}})_{\mathbb{Q}}$. The associated module is then denoted by $(\mathscr{F}_{\mathbf{s}})_{\mathbb{Q},1}$ and the submodule generated by the empty $l$-partition by $V(\Lambda_{\mathbf{s}})^{\mathbf{s}}_{\mathbb{Q},1}$. It is in fact an irreducible highest weight $\mathscr{U}(\widehat{\mathfrak{sl}_e})$-module. A basis of $V(\Lambda_{\mathbf{s}})^{\mathbf{s}}_{\mathbb{Q},1}$ is naturally obtained by specialising the canonical basis of $V(\Lambda_{\mathbf{s}})^{\mathbf{s}}$ at $q=1$. This basis is also called the canonical basis of $V(\Lambda_{\mathbf{s}})_{\mathbb{Q},1}$. It is the set

$$\mathscr{B}_{\mathbf{s}}=\left\{G_1(\mu,\mathbf{s})\mid \mu\in\Phi_{(\mathbf{s},e)}\right\}.$$

For all $\mu\in\Phi_{(\mathbf{s},e)}(n)$, we have

$$G_1(\mu,\mathbf{s})=\sum_{\lambda\vdash_l n} d_{\lambda,\mu}(1)|\lambda,\mathbf{s}\rangle.$$

**6.2.20.** We now come back to the representation theory of Ariki–Koike algebras. Recall one of the main problems of this chapter: the determination of the decomposition numbers for Ariki–Koike algebras. By Theorem 5.4.3 and the discussion in 5.4.4, we can restrict ourselves to specialisations $\theta: A\to k$ such that $\theta(u)=\eta_e$, a primitive root of order $e\in\mathbb{Z}_{>1}$ and $\theta(V_i)=\eta_e^{s_i}$ for $i\in\{1,2,\ldots,l\}$ with $\mathbf{s}=(s_1,\ldots,s_l)\in\mathbb{Z}^l$. The specialised algebra $\mathbf{H}_{k,n}$ is, in general, non-semisimple and we have an associated decomposition matrix:

$$D_\theta=(d_{\lambda,M})_{\lambda\vdash_l n, M\in\mathrm{Irr}(\mathbf{H}_{k,n})}.$$

For any simple $\mathbf{H}_{k,n}$ module $M$ with $n\in\mathbb{Z}_{\geqslant 0}$, we define the following elements of the $\mathscr{U}(\widehat{\mathfrak{sl}_e})$-module $(\mathscr{F}_{\mathbf{s}})_{\mathbb{Q},1}$:

$$B(M,\mathbf{s})=\sum_{\lambda\vdash_l n} d_{\lambda,M}|\lambda,\mathbf{s}\rangle.$$

Then we consider the subspace of $(\mathscr{F}_{\mathbf{s}})_{\mathbb{Q},1}$ generated by all these elements:

$$\mathscr{M}^{\mathbf{s}} := \langle B(M,\mathbf{s}) \mid M \in \mathrm{Irr}(\mathbf{H}_{k,n}),\ n \in \mathbb{Z}_{\geqslant 0} \rangle_{\mathbb{Q}},$$

where we denote $B(M,\mathbf{s}) = |\varnothing,\mathbf{s}\rangle$ if $n=0$. The following fundamental result asserts that these elements correspond to the canonical basis elements of the irreducible highest weight modules.

**Theorem 6.2.21 (Ariki [7], [10, Theorem 12.5]).** *In the above setting, assume that the characteristic of k is zero. Then, the $\mathscr{U}(\widehat{\mathfrak{sl}_e})$-module $\mathscr{M}^{\mathbf{s}}$ is equal to the $\mathscr{U}(\widehat{\mathfrak{sl}_e})$-module $V(\Lambda_{\mathbf{s}})^{\mathbf{s}}_{\mathbb{Q},1}$. In addition, we have*

$$\mathscr{B}_{\mathbf{s}} = \left\{ B(M,\mathbf{s}) \mid M \in \mathrm{Irr}(\mathbf{H}_{k,n}),\ n \in \mathbb{Z}_{\geqslant 0} \right\}.$$

As a consequence, the problem of computing the decomposition matrices for Ariki–Koike algebras in characteristic 0 is equivalent to the problem of determining the canonical basis for irreducible highest weight $\mathscr{U}_q(\widehat{\mathfrak{sl}_e})$-modules. In fact, in positive characteristic, the first part of this theorem is still true as shown by Ariki and Mathas. The precise statement is as follows.

**Theorem 6.2.22 (Ariki and Mathas [15, Theorems A, C and 2.3]).** *Without any condition on the characteristic of k, the $\mathscr{U}(\widehat{\mathfrak{sl}_e})$-module $\mathscr{M}^{\mathbf{s}}$ still is equal to the $\mathscr{U}(\widehat{\mathfrak{sl}_e})$-module $V(\Lambda_{\mathbf{s}})^{\mathbf{s}}_{\mathbb{Q},1}$. In particular, we have $|\mathrm{Irr}(\mathbf{H}_{k,n}))| = |\Phi_{(\mathbf{s},e)}(n)|$.*

*Thus, the cardinality of the set $\mathrm{Irr}(\mathbf{H}_{k,n})$ only depends on $\mathbf{s}$, e and n, but not on the particular field k.*

This theorem gives us a first result concerning the problem of computing the number of simple modules for $\mathbf{H}_{k,n}$.

*Remark 6.2.23.* Let $\mathbf{s} := (s_1,\dots,s_l) \in \mathbb{Z}^l$ and consider the Ariki–Koike algebra $\mathbf{H}_{k,n} := \mathbf{H}_{k,n}(\eta_e;\eta_e^{s_1},\dots,\eta_e^{s_l})$ as above. Let now $w \in \widehat{\mathfrak{S}}_l$ and define $\mathbf{s}' := (s'_1,\dots,s'_l)$ such that $\mathbf{s}' = w.\mathbf{s}$ for the action of $\widehat{\mathfrak{S}}_l$ on $\mathbb{Z}^l$ given in 6.2.9. Then it is easy to check that we have $\mathbf{H}_{k,n} = \mathbf{H}'_{k,n}$, where $\mathbf{H}'_{k,n} := \mathbf{H}_{k,n}(\eta_e;\eta_e^{s'_1},\dots,\eta_e^{s'_l})$. From this and Theorem 6.2.22, we deduce that the cardinality of the set $\mathrm{Irr}(\mathbf{H}_{k,n})$ actually only depends on $e$, $n$ and the class of $\mathbf{s}$ modulo the action of $\widehat{\mathfrak{S}}_l$ on $\mathbb{Z}^l$.

We will now be interested in the description of $\Phi_{(\mathbf{s},e)}(n)$.

## 6.3 Crystal and Canonical Bases for Highest Weight Modules

We will now be interested in the explicit description of the crystal basis and the canonical basis of an arbitrary irreducible highest weight $\mathscr{U}_q(\widehat{\mathfrak{sl}_e})$-module. The aim will be to give an explicit algorithm for computing this remarkable basis. To do this, we first need to study the $l$-partitions labelling the crystal basis: the Uglov $l$-partitions. This is done in this section.

**6.3.1.** Recall that we have an action of $\widehat{\mathfrak{S}}_l$ on $\mathbb{Z}^l$ and that a fundamental domain for this action is given by $\{(s_1,\dots,s_l)\in\mathbb{Z}^l \mid 1\leqslant s_1\leqslant\dots\leqslant s_l\leqslant e\}\subset\mathcal{S}_e^l$, where

$$\mathcal{S}_e^l := \{\mathbf{s}=(s_1,\dots,s_l)\in\mathbb{Z}^l \mid 0\leqslant s_j-s_i<e \text{ for all } i<j\}.$$

This means that if $\mathbf{t}\in\mathbb{Z}^l$ then there exists $\mathbf{s}\in\mathcal{S}_e^l$ with $V'(\Lambda_\mathbf{s})^\mathbf{s}\simeq V'(\Lambda_\mathbf{t})^\mathbf{t}$. Recall that there is an isomorphism of crystal graphs between $(\mathcal{G}_e(\Lambda_\mathbf{s})^\mathbf{s})^\circ$ and $(\mathcal{G}_e(\Lambda_\mathbf{t})^\mathbf{t})^\circ$. In particular, there is a bijection between the sets of $l$-partitions labelling these graphs: $\Phi_{(\mathbf{s},e)}$ and $\Phi_{(\mathbf{t},\mathbf{e})}$; see 6.2.17. We will first be interested in the explicit determination of $\Phi_{(\mathbf{s},e)}$ when $\mathbf{s}\in\mathcal{S}_e^l$. To do this, we will use some combinatorial notions and results that we have studied in Chapter 5. The goal of this section is to show that $\Phi_{(\mathbf{s},e)}$ is exactly the set of $l$-partitions we have already met at the end of Chapter 5: the FLOTW $l$-partitions (see Definition 5.7.8).

**Theorem 6.3.2 (Foda, Leclerc, Okado, Thibon and Welsh [88, 2.10]).** *Assume that $\mathbf{s}\in\mathcal{S}_e^l$. Then, the multipartition $\boldsymbol{\lambda}=(\lambda^1,\dots,\lambda^l)$ belongs to $\Phi_{(\mathbf{s},e)}$ if and only if $\boldsymbol{\lambda}$ is an FLOTW multipartition in the sense of Definition 5.7.8.*

To prove this result, we need to establish several combinatorial properties.

**Lemma 6.3.3.** *Assume that $\mathbf{s}\in\mathcal{S}_e^l$ and let $\boldsymbol{\lambda}=(\lambda^1,\dots,\lambda^l)\in\Phi_{(\mathbf{s},e)}$. Let $\gamma_1=(a_1,b_1,c_1)$ be a removable $i$-node of $\boldsymbol{\lambda}$ and let $\gamma_2=(a_2,b_2,c_2)$ be an addable $i$-node of $\boldsymbol{\lambda}$. Assume that $b_2=b_1+1$ (that is, $\lambda_{a_2}^{c_2}=\lambda_{a_1}^{c_1}$). Then we have $\gamma_1\prec_{(\mathbf{s},e)}\gamma_2$.*

*Proof.* Assume to the contrary that we have $\gamma_2\prec_{(\mathbf{s},e)}\gamma_1$. The case $c_1=c_2$ leads clearly to a contradiction. Assume first that $c_1<c_2$. If $a_2=1$, then

$$\lambda_{a_2}^{c_2}-a_2+s_{c_2}=b_2-1+s_{c_2}=b_1+s_{c_2}=\lambda_{a_1}^{c_1}+s_{c_2}>\lambda_{a_1}^{c_1}-a_1+s_{c_1}$$

which is a contradiction. Hence, we can assume that $a_2>1$, as in the proof of Lemma 5.7.17, we obtain:

$$\lambda_{a_1}^{c_1}\geqslant\lambda_{a_1+s_{c_2}-s_{c_1}}^{c_2}\geqslant\lambda_{b_1-b_2+a_2}^{c_2}.$$

Hence, we have $\lambda_{a_1}^{c_1}\geqslant\lambda_{a_2-1}^{c_2}$. Then we obtain $\lambda_{a_2}^{c_2}\geqslant\lambda_{a_2-1}^{c_2}$. Thus, we have $\lambda_{a_2}^{c_2}=\lambda_{a_2-1}^{c_2}$, which is a contradiction since $\gamma_2$ is an addable node. The case $c_1>c_2$ is similar. □

**Lemma 6.3.4.** *Assume that $\mathbf{s}\in\mathcal{S}_e^l$. Let $\boldsymbol{\lambda}=(\lambda^1,\dots,\lambda^l)\in\Phi_{(\mathbf{s},e)}$ be a non-empty FLOTW $l$-partition. Then there exist $i\in\{0,1,\dots,e-1\}$ and a good removable $i$-node $\gamma$ for $\boldsymbol{\lambda}$. Moreover, if we remove $\gamma$ from $\boldsymbol{\lambda}$, the resulting $l$-partition is an FLOTW $l$-partition.*

*Proof.* We take a removable $i$-node $\gamma=(a,b,c)$ of $\lambda$ for some $i\in\mathbb{Z}/e\mathbb{Z}$ as in Lemma 5.7.13. By definition, for all removable or addable $i$-nodes $\gamma'=(a',b',c')$ distinct from $\gamma$, we have that $\lambda_a^c>\lambda_{a'}^{c'}$. This implies by Lemma 5.7.17 that $\gamma'\prec_{(\mathbf{s},e)}\gamma$. As a consequence, $\gamma$ is a normal removable $i$-node for $\lambda$. Hence, there exists a good removable $i$-node for $\lambda$. Assume now that $\gamma$ is a good removable $i$-node for $\lambda$. Let $\mu$ be the $l$-partition such that $[\mu]=[\lambda]\setminus\{\gamma\}$. We must show that $\mu$ is an FLOTW $l$-partition. To do this, we have to check the two conditions to be an FLOTW $l$-partition. Assume to the contrary that $\mu$ is not an FLOTW $l$-partition.

1. If $\mu$ violates the first condition to be an FLOTW $l$-partition, this means that:
   (i) there exist $j>0$, $c\in\{1,\dots,l-1\}$ such that $\lambda_j^c=\lambda_{j+s_{c+1}-s_c}^{c+1}$ and that $\gamma=(j,\lambda_j^c,c)$ is a good removable $i$-node for $\lambda$;
   (ii) or there exist $j>0$ such that $\lambda_j^l=\lambda_{j+s_1-s_l+e}^1$ and that $\gamma=(j,\lambda_j^l,l)$ is a good removable $i$-node for $\lambda$.

   Assume that $\lambda$ is as in case (i). The node $\gamma'=(j+s_{c+1}-s_c,\lambda_{j+s_{c+1}-s_c}^{c+1},c+1)$ is then a removable node. Otherwise, as $\lambda_j^c>\lambda_{j+1}^c$ because $\gamma$ is removable, we have $\lambda_{j+1}^c<\lambda_{j+s_{c+1}-s_c+1}^{c+1}$ and this contradicts the first condition of FLOTW $l$-partition. Note also that $\lambda_{j+s_{c+1}-s_c}^{c+1}-(j+s_{c+1}-s_c)+s_{c+1}=\lambda_j^c-j+s_c$. This implies in particular that $\gamma'$ is a removable $i$-node. Moreover, we have $\gamma'\prec_{(\mathbf{s},e)}\gamma$ and there cannot be any removable or addable $i$-node $\eta$ such that $\gamma'\prec_{(\mathbf{s},e)}\eta\prec_{(\mathbf{s},e)}\gamma$. This implies that $\gamma$ is not a good removable $i$-node. This is a contradiction. Case (ii) is similar.
2. Assume that $\mu$ violates the second condition to be an FLOTW $l$-partition. This implies that there exists a good removable $i$-node $\gamma=(j,\lambda_j^c,c)$ for $\lambda$ and that the set of residues of the nodes on the border of the parts of $\lambda$ with length $\lambda_j^c-1>0$ is equal to the set $\{0,1,\dots,e-1\}\setminus\{i-1\}$. Let $\gamma'$ be (one of) the $i$-node $(j',\lambda_{j'}^{c'},c')$ such that $\lambda_{j'}^{c'}=\lambda_j^c-1$. This node must be a removable node otherwise there exists an $i-1$-node on the border of the parts of $\lambda$ with length $\lambda_j^c-1$. Assume that there exists an addable $i$-node $\eta=(j'',\lambda_{j''}^{c''}+1,c'')$ such that $\gamma'\prec_{(\mathbf{s},e)}\eta\prec_{(\mathbf{s},e)}\gamma$. By Lemma 5.7.17, we have $\lambda_j^c\geqslant\lambda_{j''}^{c''}\geqslant\lambda_{j'}^{c'}$. By the hypothesis on the set of residues of the parts of $\lambda$ with length $\lambda_j^c-1$, we must have $\lambda_j^c=\lambda_{j''}^{c''}$. Using Lemma 6.3.3, we obtain a contradiction. This implies that we have no addable $i$-node $\eta$ such that $\gamma'\prec_{(\mathbf{s},e)}\eta\prec_{(\mathbf{s},e)}\gamma$. By the definition of good $i$-nodes, $\gamma$ cannot be a good removable $i$-node, which is a contradiction.

□

**Lemma 6.3.5.** *Let $\mathbf{s}\in\mathscr{S}_e^l$ and let $\lambda=(\lambda^1,\dots,\lambda^l)\in\Phi_{(\mathbf{s},e)}(n)$ be an FLOTW $l$-partition. Assume that there exist $i\in\{0,1,\dots,e-1\}$ and a good addable $i$-node $\gamma$ for $\lambda$. Then the $l$-partition $\lambda\cup\{\gamma\}$ is an FLOTW $l$-partition.*

*Proof.* We must show that $\mu=\lambda\cup\{\gamma\}$ is an FLOTW $l$-partition. To do this, we have to check the two conditions to be an FLOTW $l$-partition. The proof is entirely similar to the proof of Lemma 6.3.4. □

We are now ready to prove the main result of this section.

**6.3.6.** *Proof of Theorem 6.3.2.* Let $n \in \mathbb{Z}_{\geqslant 0}$. We show that $\lambda$ belongs to $\Phi_{(\mathbf{s},e)}(n)$ if and only if $\lambda$ is an FLOTW $l$-partition. This is done by induction on $n$. If $n = 0$, the result is obvious. Let $n > 0$ and let $\lambda$ be an FLOTW $l$-partition. By Lemma 6.3.4, there exist $i \in \{0, 1, \ldots, e-1\}$ and a good $i$-node $\gamma$ such that $\mu := \lambda \setminus \{\gamma\}$ is an FLOTW $l$-partition. By induction, $\mu$ is in $\Phi_{(\mathbf{s},e)}(n-1)$. Hence, $\lambda \in \Phi_{(\mathbf{s},e)}(n)$. Conversely, let $\lambda \in \Phi_{(\mathbf{s},e)}(n)$; then there exist $i \in \{0, 1, \ldots, e-1\}$, a good removable $i$-node $\gamma$ and an $l$-partition $\mu \in \Phi_{(\mathbf{s},e)}(n-1)$ such that $\lambda = \mu \cup \{\gamma\}$. By induction, $\mu$ is an FLOTW $l$-partition. By Lemma 6.3.5, $\lambda$ is an FLOTW $l$-partition. □

Note that, from now, we can use the notation $\Phi_{(\mathbf{s},e)}$ instead of $\Phi'_{(\mathbf{s},e)}$ for the FLOTW $l$-partitions (see Definition 5.7.8).

**6.3.7.** As a corollary, we can now derive a proof for Theorem 5.8.1 of Chapter 5. This is indeed a direct consequence of Proposition 6.3.2 and Theorem 6.2.22. This was the missing piece to complete the proof of the main Theorem 5.8.2 and of Theorems 5.8.5, 5.8.9, 5.8.11, 5.8.13, 5.8.15 and 5.8.19.

## 6.4 Computing Decomposition Matrices of Ariki–Koike Algebras

We shall now discuss the problem of obtaining an explicit algorithm for the computation of the canonical basis of irreducible highest weight $\mathscr{U}_q(\widehat{\mathfrak{sl}_e})$-modules. As a consequence of Ariki's theorem, this will lead to an algorithm for the computation of decomposition matrices for Ariki–Koike algebras.

**6.4.1.** Now that we have an explicit description of the elements of $\Phi_{(\mathbf{s},e)}(n)$ in the case where $\mathbf{s} \in \mathscr{S}_e^l$, we can use the results in Chapter 5 to present an algorithm for computing the elements of the canonical basis. To do this, we will need the partial order $\ll_{\mathbf{m}}$ defined in 5.7.5. Here, we have $\mathbf{m} = (s_1 - v_1, \ldots, s_l - v_l)$, where $\mathbf{v} = (v_1, \ldots, v_l)$ is an arbitrary $l$-tuple of rational numbers chosen such that, for all $i < j$, we have $0 < v_j - v_i < e$.

Assume that $\lambda \in \Phi_{(\mathbf{s},e)}(n)$. Recall that we have associated with $\lambda$ a certain residue sequence $\eta_{(\mathbf{s},e)}(\lambda)$. Assume that

$$\eta_{(\mathbf{s},e)}(\lambda) = \underbrace{i_1, \ldots, i_1}_{a_1}, \underbrace{i_2, \ldots, i_2}_{a_2}, \ldots, \underbrace{i_s, \ldots, i_s}_{a_s},$$

where, for all $j = 1, \ldots, s-1$, we have $i_j \neq i_{j+1}$. We denote

$$A(\lambda, \mathbf{s}) := f_{i_s}^{(a_s)} f_{i_{s-1}}^{(a_{s-1})} \ldots f_{i_1}^{(a_1)} |\varnothing, \mathbf{s}\rangle.$$

From the definition of the action of the operators $f_i$ on the set of $l$-partitions, $|\mu, \mathbf{s}\rangle$ appears with non-zero coefficient in the expansion of $A(\lambda, \mathbf{s})$ if and only if there exists a row-standard $\mu$-tableau $\mathfrak{s}$ for an $l$-partition $\mu$ such that $\eta_e(\mathfrak{s}) = \eta_e(\lambda)$.

**Theorem 6.4.2 (Jacon [174, Prop. 4.6]).** *Let $\mathbf{s} \in \mathscr{S}_e^l$; recall that this means that $\mathbf{s} = (s_1, \ldots, s_l) \in \mathbb{Z}^l$, where $0 \leqslant s_j - s_i < e$ for all $i < j$. Let $\lambda$ be an* FLOTW *$l$-partition. Then we have*

$$A(\lambda, \mathbf{s}) = |\lambda, \mathbf{s}\rangle + \sum_{\lambda \ll_{\mathbf{m}} \mu, \mu \neq \lambda} c_{\lambda,\mu}(q)|\mu, \mathbf{s}\rangle,$$

*where $c_{\lambda,\mu}(q) \in \mathbb{Z}[q, q^{-1}]$.*

*Proof.* By the above discussion together with Lemma 5.7.21, we have

$$f_{i_s}^{(a_s)} f_{i_{s-1}}^{(a_{s-1})} \ldots f_{i_1}^{(a_1)} |\varnothing, \mathbf{s}\rangle = c_{\lambda,\lambda}(q)|\lambda, \mathbf{s}\rangle + \sum_{\lambda \ll_{\mathbf{m}} \mu, \lambda \neq \mu} c_{\lambda,\mu}(q)|\mu, \mathbf{s}\rangle,$$

where $c_{\lambda,\mu}(q) \in \mathbb{Z}[q, q^{-1}]$ is non-zero. So, we have to show that

$$c_{\lambda,\lambda}(q) = 1.$$

Let $\nu$ be the $l$-partition obtained by removing the $a_s$ greatest removable $i_s$-nodes $\gamma_j$ with respect to $\prec_{(\mathbf{s},e)}$ (for $j = 1, \ldots, a_s$.) By construction of the sequence $\eta_{(\mathbf{s},e)}(\lambda)$, $\nu$ is a FLOTW $l$-partition. Hence, we have by induction

$$f_{i_{s-1}}^{(a_{s-1})} \ldots f_{i_1}^{(a_1)} |\varnothing, \mathbf{s}\rangle = |\nu, \mathbf{s}\rangle + \sum_{\nu \ll_{\mathbf{m}} \alpha, \alpha \neq \nu} c_{\nu,\alpha}(q)|\alpha, \mathbf{s}\rangle.$$

Let $\alpha$ be an $l$-partition obtained by removing $a_s$ arbitrary removable $i_s$-nodes from $\lambda$ and assume that $\alpha \neq \nu$. We have that $\alpha \ll_{\mathbf{m}} \nu$ by Proposition 5.7.16. Hence, we have $c_{\nu,\alpha}(q) = 0$. This implies that if $c_{\nu,\alpha}(q) \neq 0$ then the element $|\lambda, \mathbf{s}\rangle$ appears with non-zero coefficient in the expansion of $f_{i_s}^{(a_s)} |\alpha, \mathbf{s}\rangle$ if and only if $\alpha = \nu$. By Proposition 6.2.7, it is now sufficient to prove that $N_{i_s}^{\succ}(\nu, \lambda) = 0$. There is no addable $i_s$-node of $\lambda$ greater than the $\gamma_l$ with respect to $\prec_{(\mathbf{s},e)}$ and there is no removable $i_s$-node of $\nu$ greater than the nodes $\gamma_l$ with respect to $\prec_{(\mathbf{s},e)}$. Hence, the result follows by the definition of $N_{i_s}^{\succ}(\nu, \lambda)$. □

**Corollary 6.4.3.** *Let $\mathbf{s} \in \mathscr{S}_e^l$. Then $\{A(\lambda, \mathbf{s}) \mid \lambda \in \Phi_{(\mathbf{s},e)}\}$ is a basis of $V(\Lambda_{\mathbf{s}})_{\mathbb{Q}}^{\mathbf{s}}$.*

*Proof.* By Theorem 6.4.2, the $A(\lambda, \mathbf{s})$ are linearly independent. Now we fix $n \in \mathbb{Z}_{\geqslant 0}$ and we consider the $\mathbb{Q}(q)$-vector space $(V(\Lambda_{\mathbf{s}})_{\mathbb{Q}}^{\mathbf{s}})^n := V(\Lambda_{\mathbf{s}})_{\mathbb{Q}}^{\mathbf{s}} \cap \mathscr{F}_{\mathbf{s}}^n$. We have $V(\Lambda_{\mathbf{s}})_{\mathbb{Q}}^{\mathbf{s}} = \oplus_{n \in \mathbb{Z}_{\geqslant 0}} (V(\Lambda_{\mathbf{s}})_{\mathbb{Q}}^{\mathbf{s}})^n$. Let $n \in \mathbb{Z}_{\geqslant 0}$. It is known that the set

$$\{G_q(\lambda, \mathbf{s}) \mid \lambda \in \Phi_{(\mathbf{s},e)}(n)\}$$

is a basis of $(V(\Lambda_{\mathbf{s}})_{\mathbb{Q}}^{\mathbf{s}})^n$. This comes from a general property of the canonical basis (see [10, Theorem 7.3 (1)]), but this can also be seen as a direct consequence of Theorem 6.2.21. Hence, we have

$$\dim(V(\Lambda_{\mathbf{s}})_{\mathbb{Q}}^{\mathbf{s}})^n = |\Phi_{(\mathbf{s},e)}(n)|,$$

and this implies that $\{A(\lambda,\mathbf{s}) \mid \lambda \in \Phi_{(\mathbf{s},e)}(n)\}$ is a basis of $(V(\Lambda_{\mathbf{s}})^{\mathbf{s}}_{\mathbb{Q}})^n$. □

The proof of the following result is inspired from the proof of [245, Theorem 6.21] and is a generalisation of a result by Lascoux, Leclerc and Thibon [208] (corresponding to the case $l = 1$).

**Theorem 6.4.4 (Lascoux–Leclerc–Thibon, Jacon).** *Let* $\mathbf{s} \in \mathscr{S}_e^l$. *Then, we have*

$$G_q(\mu,\mathbf{s}) = |\mu,\mathbf{s}\rangle + \sum_{\mu \ll_{\mathbf{m}} \lambda, \mu \neq \lambda} d_{\lambda,\mu}(q)|\lambda,\mathbf{s}\rangle \qquad \text{for all } \mu \in \Phi_{(\mathbf{s},\mathbf{e})},$$

*where* $d_{\lambda,\mu}(q) \in q\mathbb{Z}[q]$ *if* $\mu \ll_{\mathbf{m}} \lambda$ *and* $\mu \neq \lambda$.

*Proof.* By Corollary 6.4.3, for all $\mu \in \Phi_{(\mathbf{s},e)}(n)$, there exist Laurent polynomials $(\gamma_{\mu,\nu}(q))_{\nu \in \Phi_{(\mathbf{s},e)}(n)}$ in $\mathbb{Q}[q,q^{-1}]$ such that

$$G_q(\mu,\mathbf{s}) = \sum_{\nu \in \Phi_{(\mathbf{s},e)}(n)} \gamma_{\mu,\nu}(q)A(\nu,\mathbf{s}).$$

As both the canonical basis elements and the $A(\nu,\mathbf{s})$ are bar invariant, we have that $\gamma_{\mu,\nu}(q) = \gamma_{\mu,\nu}(q^{-1})$ for all $\nu \in \Phi_{(\mathbf{s},e)}(n)$ by Corollary 6.4.3. We consider an arbitrary total order $\leqslant_{\mathbf{m}}$ which is a refinement of our partial order $\ll_{\mathbf{m}}$. Let $\lambda$ be the minimal element with respect to $\leqslant_{\mathbf{m}}$ such that $\gamma_{\mu,\lambda}(q) \neq 0$. That is, we have $\gamma_{\mu,\nu}(q) = 0$ unless $\lambda \leqslant_{\mathbf{m}} \nu$. Then the coefficient of $|\lambda,\mathbf{s}\rangle$ in $G_q(\mu,\mathbf{s})$ is $\sum_{\nu \in \Phi_{(\mathbf{s},e)}(n)} \gamma_{\mu,\nu}(q)c_{\nu,\lambda}(q)$ (see Theorem 6.4.2 for the definition of the Laurent polynomials $c_{\nu,\lambda}(q)$.) By hypothesis and by Theorem 6.4.2, this coefficient is equal to $\gamma_{\mu,\lambda}(q)$. But by the definition of the canonical basis, it is in $q\mathbb{Z}[q]$ if $\lambda \neq \mu$. As it is also bar invariant we obtain that $\mu = \lambda$ and $\gamma_{\mu,\mu}(q) = 1$. In addition, we have $\gamma_{\mu,\nu}(q) = 0$ unless $\mu \leqslant_{\mathbf{m}} \nu$.

We now argue by induction on $\leqslant_{\mathbf{m}}$. If $\mu \in \Phi_{(\mathbf{s},e)}(n)$ is maximal with respect to $\leqslant_{\mathbf{m}}$ then we have $G_q(\mu,\mathbf{s}) = A(\mu,\mathbf{s})$ and we are done. We now assume that the theorem is true for all $\nu \in \Phi_{(\mathbf{s},e)}(n)$ such that $\mu \leqslant_{\mathbf{m}} \nu$ and $\nu \neq \mu$. By Theorem 6.4.2 and by the above discussion, there exist bar-invariant Laurent polynomials $\alpha_{\mu,\nu}(q)$ such that

$$G_q(\mu,\mathbf{s}) = A(\mu,\mathbf{s}) + \sum_{\nu \geqslant_{\mathbf{m}} \mu, \nu \neq \mu} \alpha_{\mu,\nu}(q)G_q(\nu,\mathbf{s}).$$

If $\nu \neq \mu$ and $\alpha_{\mu,\nu}(q) \neq 0$, we have $\mu \ll_{\mathbf{m}} \nu$. Indeed, assume to the contrary that $\nu \neq \mu$, $\alpha_{\mu,\nu}(q) \neq 0$ and that $\mu$ and $\nu$ are not comparable with respect to $\ll_{\mathbf{m}}$. We can assume that $\nu$ is minimal with respect to $\leqslant_{\mathbf{m}}$ with this property. We have $\nu \geqslant_{\mathbf{m}} \mu$. The coefficient of $|\nu,\mathbf{s}\rangle$ in $G_q(\mu,\mathbf{s})$ is $c_{\mu,\nu}(q) + \sum_{\nu' \geqslant_{\mathbf{m}} \mu, \nu' \neq \mu} \alpha_{\mu,\nu'}(q)d_{\nu,\nu'}(q)$. This last term is equal to $c_{\mu,\nu}(q) + \alpha_{\mu,\nu}(q)$ by induction hypothesis. But we also have $c_{\mu,\nu}(q) = 0$ by hypothesis. Thus, we have $\alpha_{\mu,\nu}(q) = 0$ because it is both bar-invariant and in $q\mathbb{Z}[q]$. This is a contradiction. Hence the theorem is proved. □

We conclude this section with a necessary and sufficient condition to be an FLOTW $l$-partition.

**Definition 6.4.5.** Assume that $\mathbf{s} \in \mathscr{S}_e^l$. We say that an $l$-partition $\lambda$ is $(\mathbf{s}, e)$-regular if there exists a standard $\lambda$-tableau $\mathfrak{t}(\lambda)$ satisfying the following property. For all $l$-partition $\mu$ and standard $\mu$-tableau $\mathfrak{s}$ such that $\eta(\mathfrak{s}) = \eta(\mathfrak{t}(\lambda))$, we have $\lambda \ll_{\mathbf{m}} \mu$.

**Corollary 6.4.6.** *Assume that* $\mathbf{s} \in \mathscr{S}_e^l$. *Then* $\lambda \in \Phi_{(\mathbf{s},e)}$ *if and only if it is* $(\mathbf{s}, e)$-*regular.*

*Proof.* If $\lambda \in \Phi_{(\mathbf{s},e)}$ then it is $(\mathbf{s}, e)$-regular by the discussion in 6.4.1 and Theorem 6.4.2. Conversely, assume that $\lambda$ is an $(\mathbf{s}, e)$-regular $l$-partition and denote by $\mathfrak{t}(\lambda)$ the $\lambda$-tableau as in Definition 6.4.5. Set

$$\eta(\mathfrak{t}) = \underbrace{i_1, \ldots, i_1}_{a_1}, \underbrace{i_2, \ldots, i_2}_{a_2}, \ldots, \underbrace{i_s, \ldots, i_s}_{a_s}.$$

There exist Laurent polynomials $b_{\lambda,\mu}(q) \in \mathbb{Z}[q, q^{-1}]$ such that

$$f_{i_s}^{(a_s)} f_{i_{s-1}}^{(a_{s-1})} \cdots f_{i_1}^{(a_1)} |\varnothing, \mathbf{s}\rangle = b_{\lambda,\lambda}(q)|\lambda, \mathbf{s}\rangle + \sum_{\lambda \ll_{\mathbf{m}} \mu, \mu \neq \lambda} b_{\lambda,\mu}(q)|\mu, \mathbf{s}\rangle,$$

with $b_{\lambda,\lambda}(q) \neq 0$. Now by Corollary 6.4.3, there exist Laurent polynomials $e_{\lambda,\mu}(q) \in \mathbb{Q}[q, q^{-1}]$ such that

$$f_{i_s}^{(a_s)} f_{i_{s-1}}^{(a_{s-1})} \cdots f_{i_1}^{(a_1)} |\varnothing, \mathbf{s}\rangle = \sum_{\mu \in \Phi_{(\mathbf{s},e)}} e_{\lambda,\mu}(q) A(\mu, \mathbf{s}).$$

Let $\nu \in \Phi_{(\mathbf{s},e)}$ be (one of) the minimal element (with respect to $\ll_{\mathbf{m}}$) satisfying $e_{\lambda,\nu}(q) \neq 0$. Then, by Proposition 6.4.2, we have $\nu = \lambda$. $\square$

*Remark 6.4.7.* Assume that $l = 1$ and that $\mathbf{s} \in \mathbb{Z}$. Then by 5.7.10 we have $\lambda \in \Phi_{(\mathbf{s},e)}$ if and only if $\lambda$ is $e$-regular. Then, by Example 5.7.6, the "only if" part of Corollary 6.4.6 had already been established in Lemma 3.5.11.

*Remark 6.4.8.* If $l = 2$, then a similar characterization of the Kleshchev 2-partitions conjectured by Dipper, James and Murphy has been obtained in [11, Corollary 4.3].

We can now derive an algorithm for the computation of the canonical basis.

**6.4.9.** We fix $n \in \mathbb{Z}_{\geqslant 0}$, $l \in \mathbb{Z}_{>0}$, $e \in \mathbb{Z}_{>0}$ and $\mathbf{s} \in \mathscr{S}_e^l$. The aim of the following algorithm is to compute the canonical basis

$$\{G_q(\mu, \mathbf{s}) \mid \mu \in \Phi_{(\mathbf{s},e)}(n)\}$$

of $V(\Lambda_{\mathbf{s}})_{\mathbb{Q}}^{\mathbf{s}}$. It was first described in [175, Section 3].

**Step 1**. For each $\mu \in \Phi_{(\mathbf{s},e)}(n)$, we denote

$$\eta_{(\mathbf{s},\mathbf{e})}(\mu) = \underbrace{i_1, \ldots, i_1}_{a_1}, \underbrace{i_2, \ldots, i_2}_{a_2}, \ldots, \underbrace{i_s, \ldots, i_s}_{a_s}.$$

Then, we compute the elements $A(\mu,\mathbf{s})$ of 6.4.1 using the action of the quantum group described in Proposition 6.2.7:

$$A(\mu,\mathbf{s}) := f_{i_s}^{(a_s)} f_{i_{s-1}}^{(a_{s-1})} \dots f_{i_1}^{(a_1)} |\varnothing,\mathbf{s}\rangle.$$

These elements have a triangular decomposition with respect to $\ll_{\mathbf{m}}$ by Theorem 6.4.2. We obtain a basis of $(V(\Lambda_{\mathbf{s}})_{\mathbb{Q}}^{\mathbf{s}})^n$

$$\{A(\mu,\mathbf{s}) \mid \lambda \in \Phi_{(\mathbf{s},e)}(n)\}.$$

Since $\overline{f_i} = f_i$ for all $i=0,\dots,e-1$, we have $\overline{A(\mu,\mathbf{s})} = A(\mu,\mathbf{s})$.

**Step 2**. Let $\nu \in \Phi_{(\mathbf{s},e)}(n)$ be a maximal element with respect to $\ll_{\mathbf{m}}$. Then, by the definition of the canonical basis, we have

$$G_q(\nu,\mathbf{s}) = A(\nu,\mathbf{s}).$$

**Step 3**. Let $\mu \in \Phi_{(\mathbf{s},e)}(n)$. The elements $G_q(\lambda,\mathbf{s})$ with $\mu \ll_{\mathbf{m}} \lambda$ are known by induction. There exist polynomials $\alpha_{\lambda,\mu}(q)$ such that

$$G_q(\mu,\mathbf{s}) = A(\mu,\mathbf{s}) - \sum_{\mu \ll_{\mathbf{m}} \lambda, \mu\neq\lambda} \alpha_{\mu,\lambda}(q) G_q(\lambda,\mathbf{s}).$$

We want to compute $\alpha_{\mu,\lambda}(q)$ for all $\lambda \in \Phi_{(\mathbf{s},e)}(n)$. By Theorem 6.4.2, we have

$$(\star) \qquad A(\mu,\mathbf{s}) = |\mu,\mathbf{s}\rangle + \sum_{\mu \ll_{\mathbf{m}} \lambda, \mu\neq\lambda} c_{\mu,\lambda}(q) |\lambda,\mathbf{s}\rangle.$$

Now, since $\overline{G_q(\lambda,\mathbf{s})} = G_q(\lambda,\mathbf{s})$ and $\overline{A(\lambda,\mathbf{s})} = A(\lambda,\mathbf{s})$ for all $\lambda \in \Phi_{(\mathbf{s},e)}(n)$, we must have $\alpha_{\mu,\lambda}(q) = \alpha_{\mu,\lambda}(q^{-1})$ for all $\lambda$ in $\Phi_{(\mathbf{s},e)}(n)$.

Let $\nu \vdash_l n$ be one of the minimal $l$-partitions with respect to $\ll_{\mathbf{m}}$ such that $c_{\mu,\nu}(q) \notin q\mathbb{Z}[q]$. If $\nu$ does not exist, then, by unicity, we have $G_q(\mu,\mathbf{s}) = A(\mu,\mathbf{s})$. Otherwise, by the existence of the canonical basis, we have $\nu \in \Phi_{(\mathbf{s},e)}(n)$. Assume now that we have

$$c_{\mu,\nu}(q) = a_i q^i + a_{i-1} q^{i-1} + \dots + a_0 + \dots + a_{-i} q^{-i},$$

where $(a_j)_{j\in\{-i,-i+1,\dots,i-1,i\}}$ is a sequence of elements in $\mathbb{Z}$ and $i$ is a positive integer. Then, we define

$$\alpha_{\mu,\nu}(q) = a_{-i} q^i + a_{-i+1} q^{i-1} + \dots + a_0 + \dots + a_{-i} q^{-i}.$$

We have $\alpha_{\mu,\nu}(q^{-1}) = \alpha_{\mu,\nu}(q)$. Then, in $(\star)$, we replace the element $A(\mu,\mathbf{s})$ by $A(\mu,\mathbf{s}) - \alpha_{\mu,\nu}(q) G_q(\nu,\mathbf{s})$, which is bar-invariant, and we repeat this step until we have $G_q(\mu,\mathbf{s}) = A(\mu,\mathbf{s})$. We finally obtain elements which verify Theorem 6.1.17; that is, the canonical basis elements. □

Note that we also obtain an algorithm for computing the canonical basis of the $\mathscr{U}(\widehat{\mathfrak{sl}_e})$-module with weight $\Lambda_{\mathbf{s}}$ by specializing the above elements at $q = 1$.

**6.4.10.** The algorithm in 6.4.9 is only concerned with the case where the multicharge $\mathbf{s}$ is in $\mathscr{S}_e^l$. We here explain how we can deduce an algorithm for the computation of the canonical basis associated with an arbitrary multicharge following [12, Section 4.2]. We have seen in 6.2.9 that there is an action of the (extended) affine symmetric group $\widehat{\mathfrak{S}}_l$ on $\mathbb{Z}^l$ such that $\mathscr{S}_e^l$ contains a fundamental domain for this action.

For any $\mathbf{t} := (t_1, \ldots, t_l) \in \mathbb{Z}^l$, there exist $\mathbf{s} := (s_1, \ldots, s_l) \in \mathscr{S}_e^l$ and $w \in \widehat{\mathfrak{S}}_r$ such that $\mathbf{t} = w.\mathbf{s}$. Since $\mathbf{t}$ and $\mathbf{s}$ yield the same dominant weight, we have an isomorphism $\phi_{\mathbf{s},\mathbf{t}}$ of $\mathscr{U}_q'(\widehat{\mathfrak{sl}_e})$-modules from $V'(\Lambda_{\mathbf{s}})^{\mathbf{s}}$ to $V'(\Lambda_{\mathbf{t}})^{\mathbf{t}}$. We can assume that $\phi_{\mathbf{s},\mathbf{t}}(|\varnothing,\mathbf{s}\rangle) = |\varnothing,\mathbf{t}\rangle$. For each $\lambda \in \Phi_{(\mathbf{s},e)}$, recall that we have already defined an element:

$$A(\lambda,\mathbf{s}) := f_{i_s}^{(a_s)} f_{i_{s-1}}^{(a_{s-1})} \cdots f_{i_1}^{(a_1)} |\varnothing,\mathbf{s}\rangle \in V'(\Lambda_{\mathbf{s}})^{\mathbf{s}}.$$

Moreover, there exist coefficients $\gamma_{\lambda,\mu}(q) \in \mathbb{Z}[q,q^{-1}]$ such that

$$(\star\star) \qquad G_q(\lambda,\mathbf{s}) = \sum_{\mu \in \Phi_{(\mathbf{s},e)}} \gamma_{\lambda,\mu}(q) A(\mu,\mathbf{s}),$$

which can be explictly obtained using our previous algorithm. Now, we fix $\lambda \in \Phi_{(\mathbf{s},e)}$ and set

$$A(\lambda,\mathbf{t}) = f_{i_s}^{(a_s)} \cdots f_{i_1}^{(a_1)} |\varnothing,\mathbf{t}\rangle \in V(\Lambda_{\mathbf{t}})^{\mathbf{t}}.$$

Then we have $\phi_{\mathbf{s},\mathbf{t}}(A(\lambda,\mathbf{s})) = A(\lambda,\mathbf{t})$. By the uniqueness of the crystal basis proved by Kashiwara, we also have $\phi_{\mathbf{s},\mathbf{t}}(G_q(\lambda,\mathbf{s})) = G_q(\varphi_{\mathbf{s},\mathbf{t}}(\lambda),\mathbf{t})$, where $\varphi_{\mathbf{s},\mathbf{t}}$ is the bijection $\Phi_{(\mathbf{s},e)}$ to $\Phi_{(\mathbf{t},e)}$ as in 6.2.17. By applying $\phi_{\mathbf{s},\mathbf{t}}$ to $(\star\star)$, we obtain

$$G_q(\nu,\mathbf{t}) = \sum_{\mu \in \Phi_{(\mathbf{s},e)}} \gamma_{\varphi_{\mathbf{s},\mathbf{t}}^{-1}(\nu),\mu}(q) A(\mu,\mathbf{t}),$$

for $\nu \in \Phi_{(\mathbf{t},e)}$, and it follows that

$$\Big\{ \sum_{\mu \in \Phi_{(\mathbf{s},e)}} \gamma_{\lambda,\mu}(q) A(\mu,\mathbf{t}) \mid \lambda \in \Phi_{(\mathbf{s},e)} \Big\}$$

is the canonical basis for the module $V(\Lambda_{\mathbf{t}})^{\mathbf{t}}$.

*Remark 6.4.11.* Another algorithm has been recently proposed by Fayers [82, Section 4] for computing the canonical bases of the highest weight $\mathscr{U}_q'(\widehat{\mathfrak{sl}_e})$-modules. In this case, this module is realiszed as the tensor product of level-one Fock spaces. This is connected with the case where the multicharge is very dominant (see Example 6.2.16).

**6.4.12.** Let us give the consequences of the previous results on the representation theory of Ariki–Koike algebras. We assume that the characteristic of $k$ is zero. Let $e \in \mathbb{Z}_{>0}$ and let $\mathbf{s} \in \mathbb{Z}^l$. We consider a specialisation

$$\theta : A \to k$$

such that $\theta(V_i) = \eta_e^{s_i}$ for all $i = 1, \ldots, l$ and $\theta(u) = \eta_e$ is a primitive root of unity of order $e \in \mathbb{Z}_{>1}$. The specialised algebra $\mathbf{H}_{k,n}(\eta_e; \eta_e^{s_1}, \ldots, \eta_e^{s_l})$ is non-semisimple and we have an associated decomposition matrix $D_\theta$. For studying this matrix, without loss of generality, one can assume that $\mathbf{s} \in \mathscr{S}_e^l$. By Ariki's theorem, $D_{(\mathbf{s},e)} := D_\theta$ is the matrix of the canonical basis of the $\mathscr{U}(\widehat{\mathfrak{sl}}_e)$-module with weight $\Lambda_{\mathbf{s}}$. Taking this into account, using 5.4.4 and 6.4.9, we deduce the following theorem.

**Theorem 6.4.13 (Lascoux–Leclerc–Thibon, Uglov, Jacon).** *There is an explicit and purely combinatorial algorithm for computing the decomposition matrices of Ariki–Koike algebras when the characteristic of k is zero.*

## 6.5 Uglov's Theory of Fock Spaces

In this section we survey Uglov's theory [291] concerning canonical bases for Fock spaces. The determination of canonical basic sets for all choices of parameters requires some new information that we are not able to obtain directly from the theory exposed in Section 6.2. This section is inspired by the thesis of Yvonne [298].

**6.5.1.** The first new object that we need to define is the *affine Hecke algebra of type* A. This can be seen as a deformation of the extended affine symmetric group which we introduced in the first section. We then study several actions of this algebra. In the next section, this will yield the definition of a new space: the semi-infinite $q$-wedge product which will play a crucial role in the following. Let $\widehat{H}_r$ be the *affine Hecke algebra of type* A over $\mathscr{K} = \mathbb{Q}(q)$. As a $\mathscr{K}$-algebra, $\widehat{H}_r$ is presented by:

- Generators: $T_i$ (with $0 \leqslant i \leqslant r-1$), $\tau$, $\tau^{-1}$.
- Relations: for all $0 \leqslant i, j \leqslant r-1$,

$$\begin{aligned}
&T_i T_{i+1} T_i = T_{i+1} T_i T_{i+1} \\
&T_i T_j = T_j T_i \quad \text{if } i - j \neq 1 \bmod r \\
&(T_i - q^{-1})(T_i + q) = 0 \\
&\tau T_i = T_{i+1} \tau,
\end{aligned}$$

where the indices are read modulo $r$. Each $w \in \widehat{\mathfrak{S}}_r$ can be written as a product $\tau^k \sigma_{i_1} \ldots \sigma_{i_t} \in \widehat{H}_r$ with $k \in \mathbb{Z}$ and $\sigma_{i_1} \ldots \sigma_{i_t} \in \mathfrak{S}_r$ is a decomposition of minimal length. We then set $T_w = \tau^k T_{i_1} \ldots T_{i_t}$. As a $\mathscr{K}$-vector space, a basis of $\widehat{H}_r$ is then given by the set

$$\left\{ T_\sigma \mid \sigma \in \widehat{\mathfrak{S}}_r \right\}$$

and the multiplication is characterized by the following property. Let $(\sigma,\sigma') \in \widehat{\mathfrak{S}}_r^2$; then

$$\begin{cases} T_\sigma T_{\sigma'} = T_{\sigma\sigma'} & \text{if } l(\sigma\sigma') = l(\sigma) + l(\sigma'), \\ (T_{\sigma_i} - q^{-1})(T_{\sigma_i} + q) = 0 & \text{if } 0 \leqslant i \leqslant r-1. \end{cases}$$

There is a canonical involution $x \mapsto \overline{x}$ of $\widehat{H}_r$ defined as the unique $\mathbb{Q}$-algebra automorphism such that

$$\overline{q} = q^{-1}, \qquad \overline{T_w} = T_{w^{-1}}^{-1} \qquad \text{for all } w \in \widehat{\mathfrak{S}}_r.$$

By the work of Bernstein, we can obtain another system of generators as follows. Let $\mathbf{p} \in P_r := \mathbb{Z}^r$. Recall the elements $y^{\mathbf{p}}$ defined in 6.2.8. We write $\mathbf{p} = \mathbf{p}^+ - \mathbf{p}^-$, where $\mathbf{p}^+$, $\mathbf{p}^-$ are in $P_r^+ := \{(p_1,\ldots,p_r) \in \mathbb{Z}^r \mid p_1 \geqslant \ldots \geqslant p_r\}$. Then one can show that the element $X^{\mathbf{p}} := T_{y^{\mathbf{p}^-}} T_{y^{\mathbf{p}^+}}^{-1}$ is well defined. Moreover, for all $\mathbf{p}$, $\mathbf{q}$ in $P_r$, we have that

$$X^{\mathbf{p}} X^{\mathbf{q}} = X^{\mathbf{p}+\mathbf{q}} = X^{\mathbf{q}} X^{\mathbf{p}}.$$

Then the elements $X^{\mathbf{p}}$ with $\mathbf{p} \in P_r$ and $T_1,\ldots,T_{r-1}$ generate $\widehat{H}_r$. We refer to [291, Lemma 3.13] and [200] for a proof of this assertion.

**6.5.2.** We now study two actions of the extended affine symmetric group on $P_r$. These actions will then be used to construct actions of the affine Hecke algebra. Let $l \in \mathbb{Z}_{>0}$. We can put a left action of $\widehat{\mathfrak{S}}_r$ on $P_r$ depending on $l$ and defined as follows. For all $(b_1,\ldots,b_r) \in P_r$, we have

$$\begin{cases} \sigma_i.(b_1,\ldots,b_r) = (b_1,\ldots,b_{i-1}b_{i+1},b_i,b_{i+2},\ldots,b_r) & \text{for all } i = 1,\ldots,r-1, \\ y_j.(b_1,\ldots,b_r) = (b_1,..,b_{j-1},b_j+l,b_{j+1},\ldots,b_r) & \text{for all } j = 1,\ldots,r. \end{cases}$$

A fundamental domain for this action is

$$B^l := \{(b_1,\ldots,b_r) \in P_r \mid l \geqslant b_1 \geqslant \ldots. \geqslant b_l \geqslant 1\}.$$

Let $\mathbf{b} \in B^l$. We denote by $(\widehat{\mathfrak{S}}_r)_{\mathbf{b}}$ the isotropy group of $\mathbf{b}$ for the action of $\widehat{\mathfrak{S}}_r$ on $P_r$. It is clear that we have $(\widehat{\mathfrak{S}}_r)_{\mathbf{b}} \subset \mathfrak{S}_r$. In fact, if we set $I_{\mathbf{b}} := \{1 \leqslant i \leqslant r-1 \mid b_i = b_{i+1}\}$, we have $(\widehat{\mathfrak{S}}_r)_{\mathbf{b}} = \langle \sigma_i \mid i \in I_{\mathbf{b}} \rangle$. Each element of $\widehat{\mathfrak{S}}_r/(\widehat{\mathfrak{S}}_r)_{\mathbf{b}}$ has a unique representative with minimal length. Let $(\widehat{\mathfrak{S}}_r)^{\mathbf{b}}$ be the set of these representatives.

- For all $w \in \widehat{\mathfrak{S}}_r$, there exist unique elements $y \in (\widehat{\mathfrak{S}}_r)^{\mathbf{b}}$ and $u \in (\widehat{\mathfrak{S}}_r)_{\mathbf{b}}$ such that $w = yu$ and we have $l(w) = l(y) + l(u)$.
- For all $\mathbf{p} \in P_r$, there exist unique elements $\mathbf{b} \in B^l$ and $y \in (\widehat{\mathfrak{S}}_r)^{\mathbf{b}}$ such that $\mathbf{p} = y.\mathbf{b}$.

Let $e \in \mathbb{Z}_{>0}$. By analogy, one can define a right action of $\widehat{\mathfrak{S}}_r$ on $P_r$ as follows. For all $(a_1,\ldots,a_r) \in P_r$, we have

$$\begin{cases} (a_1,\ldots,a_r).\sigma_i = (a_1,\ldots,a_{i+1},a_i,\ldots,a_r) & \text{for all } i = 1,\ldots,r-1, \\ (a_1,\ldots,a_r).y_j = (a_1,..,a_j+e,\ldots,a_r) & \text{for all } j = 1,\ldots,r. \end{cases}$$

A fundamental domain for this action is

$$A^e := \{(a_1,\dots,a_r) \in P_r \mid 1 \leqslant a_1 \leqslant \dots \leqslant a_r \leqslant e\}.$$

One can define groups $(\widehat{\mathfrak{S}}_r)_{\mathbf{a}}$ and ${}^{\mathbf{a}}(\widehat{\mathfrak{S}}_r)$ as above. For $\mathbf{a} \in A^e$, we denote by $(\widehat{\mathfrak{S}}_r)_{\mathbf{a}}$ the isotropy group of $\mathbf{a}$ for the right action of $\widehat{\mathfrak{S}}_r$ on $P_r$. It is clear that we have $(\widehat{\mathfrak{S}}_r)_{\mathbf{a}} \subset \mathfrak{S}_r$. Recall that $I_{\mathbf{a}} = \{1 \leqslant i \leqslant r-1 \mid a_i = a_{i+1}\}$; then we have $(\widehat{\mathfrak{S}}_r)_{\mathbf{a}} = \langle \sigma_i \mid i \in I_{\mathbf{a}} \rangle$. Each element of $(\widehat{\mathfrak{S}}_r)_{\mathbf{a}} \setminus \widehat{\mathfrak{S}}_r$ has a unique representative of minimal length. Let ${}^{\mathbf{a}}(\widehat{\mathfrak{S}}_r)$ be the set of these representatives.

- For all $w \in W$, there exist unique elements $y \in {}^{\mathbf{a}}(\widehat{\mathfrak{S}}_r)$ and $u \in (\widehat{\mathfrak{S}}_r)_{\mathbf{a}}$ such that $w = uy$ and we have $l(w) = l(u) + l(y)$.
- For all $\mathbf{p} \in P^r$, there exist unique elements $\mathbf{a} \in A^e$ and $y \in {}^{\mathbf{a}}(\widehat{\mathfrak{S}}_r)$ such that $\mathbf{p} = \mathbf{a}.y$.

**6.5.3.** We turn now to the affine Hecke algebra. We define a left and then a right action of $\widehat{H}_r$ on two $\mathscr{K}$-vector spaces $\mathscr{P}_r^{\mathscr{L}} := \bigoplus_{\mathbf{p}\in P_r} \mathscr{K}|\mathbf{p})$ generated by the symbols $|\mathbf{p})$ with $\mathbf{p} \in P_r$ and $\mathscr{P}_r^{\mathscr{R}} := \bigoplus_{\mathbf{p}\in P_r} \mathscr{K}(\mathbf{p}|$ generated by the symbols $(\mathbf{p}|$ with $\mathbf{p} \in P_r$.

- For $\mathbf{b} \in B^l$, we consider the parabolic subalgebra $H_{\mathbf{b}}$ of $\widehat{H}_r$ generated by the $T_w$ with $w \in (\widehat{\mathfrak{S}}_r)_{\mathbf{b}}$. We denote by $\mathscr{K}_{-q}$ the left $H_{\mathbf{b}}$-module with dimension 1 such that $T_i$ with $i \in I_{\mathbf{b}}$ acts by multiplication with $-q$. We then consider the left $\widehat{H}_r$-module $\widehat{H}_r \otimes_{H_{\mathbf{b}}} \mathscr{K}_{-q}$. As a $\mathscr{K}$-vector space, it has a basis given by the $\overline{T_y} \otimes 1$, where $y \in (\widehat{\mathfrak{S}}_r)^{\mathbf{b}}$. Now, there is an isomorphism of $\mathscr{K}$-vector spaces

$$(\bullet) \qquad \begin{array}{rcl} \widehat{H}_r \otimes_{H_{\mathbf{b}}} \mathscr{K}_{-q} & \to & \bigoplus_{y\in\widehat{\mathfrak{S}}_r^{\mathbf{b}}} \mathscr{K}|y.\mathbf{b}), \\ \overline{T_y} \otimes_{H_{\mathbf{b}}} 1 \quad (y \in (\widehat{\mathfrak{S}}_r)^{\mathbf{b}}) & \mapsto & |y.\mathbf{b}). \end{array}$$

From now, we can (and we do) identify $|y.\mathbf{b})$ with $\overline{T_y} \otimes_{H_{\mathbf{b}}} 1$ for $y \in (\widehat{\mathfrak{S}}_r)^{\mathbf{b}}$. We thus have an action of $\widehat{H}_r$ on the space $\bigoplus_{y\in\widehat{\mathfrak{S}}_r^{\mathbf{b}}} \mathscr{K}|y.\mathbf{b})$. By linearity and by the definition of $(\widehat{\mathfrak{S}}_r)^{\mathbf{b}}$, we obtain a left action of $\widehat{H}_r$ on the space $\mathscr{P}_r^{\mathscr{L}} := \bigoplus_{\mathbf{b}\in B^l} \bigoplus_{y\in(\widehat{\mathfrak{S}}_r)^{\mathbf{b}}} \mathscr{K}|y.\mathbf{b})$. One can check that this action is explicitly given as follows (see [291, Proposition 3.12] or [298, Proposition 1.29]). For all $\mathbf{p} \in P_r$, and $i = 0, 1, \dots, r-1$, we have

$$T_i.|\mathbf{p}) = \begin{cases} |\sigma_i.\mathbf{p}) & \text{if } p_i < p_{i+1} \\ -q|\mathbf{p}) & \text{if } p_i = p_{i+1} \\ |\sigma_i.\mathbf{p}) - (q-q^{-1})|\mathbf{p}) & \text{if } p_i > p_{i+1} \end{cases}$$

and

$$\tau.|\mathbf{p}) = |\tau.\mathbf{p}).$$

This action is thus obtained using the identification induced by $(\bullet)$.

- By analogy, let $H_{\mathbf{a}}$ be the parabolic subalgebra of $\widehat{H}_r$ generated by the $T_w$ with $w \in (\widehat{\mathfrak{S}}_r)_{\mathbf{a}}$. We denote by $\mathscr{K}_{q^{-1}}$ the right $H_{\mathbf{a}}$-module with dimension 1 such that $T_i$ with $i \in I_{\mathbf{a}}$ acts by multiplication with $q^{-1}$. A basis of $\mathscr{K}_{q^{-1}} \otimes_{H_{\mathbf{a}}} \widehat{H}_r$ is then given by the elements $1 \otimes_{H_{\mathbf{a}}} T_y$ with $y \in {}^{\mathbf{a}}\widehat{\mathfrak{S}}_r$. Using this, we obtain a right action of $\widehat{H}_r$ on $\mathscr{K}_{q^{-1}} \otimes_{H_{\mathbf{a}}} \widehat{H}_r$. Now consider the isomorphism of vector spaces:

$$\begin{array}{rcl} \mathscr{K}_{q^{-1}} \otimes_{H_{\mathbf{a}}} \widehat{H}_r & \to & \bigoplus_{y \in \widehat{\mathfrak{S}}_r^{\mathbf{b}}} \mathscr{K}(\mathbf{a}, y|, \\ 1 \otimes_{H_{\mathbf{a}}} T_y \quad (y \in {}^{\mathbf{a}}\widehat{\mathfrak{S}}_r) & \mapsto & (\mathbf{a}.y|. \end{array}$$

From now, we can (and we do) identify $(\mathbf{a}.y|$ with $1 \otimes_{H_{\mathbf{a}}} T_y$ for $y \in {}^{\mathbf{a}}\widehat{\mathfrak{S}}_r$. We thus obtain by linearity a right action of $\widehat{H}_r$ on $\mathscr{P}_r^{\mathscr{R}} = \bigoplus_{\mathbf{a} \in A^e} \bigoplus_{y \in {}^{\mathbf{a}}\widehat{\mathfrak{S}}_r} \mathscr{K}(\mathbf{a}.y|$. It is explicitly given as follows.

$$(\mathbf{p}|.T_i = \begin{cases} (\mathbf{p}.\sigma_i| & \text{if } p_i < p_{i+1}, \\ q^{-1}(\mathbf{p}| & \text{if } p_i = p_{i+1}, \\ (\mathbf{p}.\sigma_i| - (q - q^{-1})(\mathbf{p}| & \text{if } p_i > p_{i+1}, \end{cases}$$
$$(\mathbf{p}|.\tau = (\mathbf{p}.\tau|.$$

**6.5.4.** Using the above constructions, we will now be able to define a new space which will have strong connections with our main objects of interest. Let $e \in \mathbb{Z}_{>0}$, $l \in \mathbb{Z}_{>0}$, $\mathbf{a} \in A^e$ and $\mathbf{b} \in B^l$. We set

$$\Lambda(\mathbf{a}, \mathbf{b}) = \mathscr{K}_{q^{-1}} \otimes_{H_{\mathbf{a}}} \widehat{H}_r \otimes_{H_{\mathbf{b}}} \mathscr{K}_{-q},$$
$$\Lambda(\mathbf{b}) = \bigoplus_{\mathbf{a} \in A^e} \Lambda(\mathbf{a}, \mathbf{b}),$$
$$\Lambda_q^r V = \bigoplus_{\mathbf{b} \in B^l} \Lambda(\mathbf{b}).$$

This last $\mathscr{K}$-vector space is called the *q-wedge product*. The main aim of this section will be to give a basis of $\Lambda(\mathbf{b})$ for all $\mathbf{b} \in B^l$ and thus of $\Lambda_q^r V$. To do this, we need some preparatory results. We define a linear map

$$\Phi_{\mathbf{b}} : \mathscr{P}_r^{\mathscr{R}} \to \Lambda(\mathbf{b})$$

as follows. Let $\mathbf{p} \in P_r$. Then there exist unique elements $\mathbf{a} \in A^e$ and $x \in {}^{\mathbf{a}}\widehat{\mathfrak{S}}_r$ such that $\mathbf{p} = \mathbf{a}.x$. Define $[\mathbf{p}]_{\mathbf{b}} := (-q^{-1})^{l(w_{\mathbf{b}})} 1 \otimes_{H_{\mathbf{a}}} T_x \otimes_{H_{\mathbf{b}}} 1 = (-q^{-1})^{l(w_{\mathbf{b}})} (\mathbf{p}| \otimes_{H_{\mathbf{b}}} 1$, where $w_{\mathbf{b}}$ is the element in $\widehat{\mathfrak{S}}_r^{\mathbf{b}}$ of maximal length. We can set $\Phi_{\mathbf{b}}((\mathbf{p}|) = [\mathbf{p}]_{\mathbf{b}}$. One can easily show that $\mathrm{Ker}(\Phi_{\mathbf{b}}) = \sum_{j \in I_{\mathbf{b}}} \mathscr{P}_r^{\mathscr{R}}(T_j + q)$. We define

$$P_{r,\mathbf{b}}^{++} := \{(p_1, \ldots, p_r) \in P^r \mid i \in I_{\mathbf{b}} \Rightarrow p_i > p_{i+1}\}.$$

Then we have the following result.

**Proposition 6.5.5 (Uglov).** *The set* $\{[\mathbf{p}]_{\mathbf{b}} \mid \mathbf{p} \in P_{r,\mathbf{b}}^{++}\}$ *is a basis of* $\Lambda(\mathbf{b})$ *(as a* $\mathscr{K}$*-vector space).*

*Proof.* We first show that $\{[\mathbf{p}]_{\mathbf{b}} \mid \mathbf{p} \in P_{r,\mathbf{b}}^{++}\}$ is a generating set for $\Lambda(\mathbf{b})$. To do this, it suffices to show that an arbitrary element $1 \otimes_{H_{\mathbf{a}}} T_x \otimes_{H_{\mathbf{b}}} 1$ with $x \in \widehat{\mathfrak{S}}_r$ can be written as a linear combination of this set. Let $x \in \widehat{\mathfrak{S}}_r$ and let $\mathbf{a} \in A^e$. We set $\mathbf{p} := \mathbf{a}.x \in P_r$. Let $y \in {}^{\mathbf{a}}(\widehat{\mathfrak{S}}_r)$ be such that $\mathbf{p} = \mathbf{a}.y$. Then we have $u := x.y^{-1} \in (\widehat{\mathfrak{S}}_r)_{\mathbf{a}}$ and $l(x) = l(u) + l(y)$. It follows that

$$1 \otimes_{H_{\mathbf{a}}} T_x \otimes_{H_{\mathbf{b}}} 1 = 1 \otimes_{H_{\mathbf{a}}} T_u T_y \otimes_{H_{\mathbf{b}}} 1 = T_u \otimes_{H_{\mathbf{a}}} T_y \otimes_{H_{\mathbf{b}}} 1 = q^{-l(u)} 1 \otimes_{H_{\mathbf{a}}} T_y \otimes_{H_{\mathbf{b}}} 1.$$

Set $\mathbf{p} = (p_1, \ldots, p_r)$; then we have $1 \otimes_{H_{\mathbf{a}}} T_x \otimes_{H_{\mathbf{b}}} 1 = (-q)^{l(w_{\mathbf{b}})} [\mathbf{p}]_{\mathbf{b}}$. As we have $\mathrm{Ker}(\Phi_{\mathbf{b}}) = \sum_{j \in I_{\mathbf{b}}} \mathscr{P}_r^{\mathscr{R}} (T_j + q)$, we obtain that

$$[\mathbf{p}]_{\mathbf{b}} = \begin{cases} 0 & \text{if } p_i = p_{i+1}, \\ -q^{-1} [\mathbf{p}.\sigma_i]_{\mathbf{b}} & \text{if } p_i < p_{i+1}, \end{cases}$$

which leads to our claim.

We now show that the set $\left\{ [\mathbf{p}]_{\mathbf{b}} \mid \mathbf{p} \in P_{r,\mathbf{b}}^{++} \right\}$ is free over $\mathscr{K}$. Assume that we have $\sum_{\mathbf{p} \in P_{r,\mathbf{b}}^{++}} a_{\mathbf{p}} [\mathbf{p}]_{\mathbf{b}} = 0$. For $\mathbf{p} \in P_{r,\mathbf{b}}^{++}$, we have $a_{\mathbf{p}} \in \mathscr{K}$ and there is a finite subset of $P_{r,\mathbf{b}}^{++}$ such that $a_{\mathbf{p}} = 0$ unless $\mathbf{p}$ is in this subset. It follows that we have $\sum_{\mathbf{p} \in P_{r,\mathbf{b}}^{++}} a_{\mathbf{p}} (\mathbf{p}| \in \mathrm{Ker}(\Phi_{\mathbf{b}}) = \sum_{j \in I_{\mathbf{b}}} \mathscr{P}_r^{\mathscr{R}} (T_j + q)$. Now, we consider the element $C_{w_{\mathbf{b}}} := \sum_{y \in (\widehat{\mathfrak{S}}_r)_{\mathbf{b}}} (-q)^{l(y) - l(w_{\mathbf{b}})} T_y$. This is an element of the Kazhdan–Lusztig basis for $\mathfrak{S}_r$; see Example 2.1.17 (where $\varepsilon = q^{-1}$). Thus, it satisfies $C_{w_{\mathbf{b}}} = \overline{C}_{w_{\mathbf{b}}}$. It is an easy computation to show that, for all $i \in I_{\mathbf{b}}$, we have $(T_i + q) C_{w_{\mathbf{b}}} = 0$. Hence, we obtain that

$$\sum_{\mathbf{p} \in P_{r,\mathbf{b}}^{++}} a_{\mathbf{p}} (\mathbf{p}| . C_{w_{\mathbf{b}}} = 0 = \sum_{\mathbf{p} \in P_{r,\mathbf{b}}^{++}} a_{\mathbf{p}} (\mathbf{p}| . \overline{C_{w_{\mathbf{b}}}} = 0;$$

that is,

$$\sum_{\mathbf{p} \in P_{r,\mathbf{b}}^{++}, y \in (\widehat{\mathfrak{S}}_r)_{\mathbf{b}}} (-q)^{-l(y) + l(w_{\mathbf{b}})} a_{\mathbf{p}} (\mathbf{p}| . T_y^{-1} = 0.$$

By the definition of the right action of the affine Hecke algebra on $\mathscr{P}_r^{\mathscr{R}}$, we deduce that

$$\sum_{\mathbf{p} \in P_{r,\mathbf{b}}^{++}, y \in (\widehat{\mathfrak{S}}_r)_{\mathbf{b}}} (-q)^{-l(y) + l(w_{\mathbf{b}})} a_{\mathbf{p}} (\mathbf{p}.y| = 0,$$

which implies that all the $a_{\mathbf{p}}$ are zero as all the elements $(\mathbf{p}.y|$ with $\mathbf{p}$ in $P_{r,\mathbf{b}}^{++}$ and $y \in (\widehat{\mathfrak{S}}_r)_{\mathbf{b}}$ are distinct. □

**6.5.6.** We can now give a basis of $\Lambda_q^r V$. To do this, we need to introduce some more notation. Let $\mathbf{k} = (k_1, \ldots, k_r) \in P_r$. For $i = 1, \ldots, r$, there exist unique elements $c_i \in \{1, \ldots, e\}$, $d_i \in \{1, \ldots, l\}$ and $m_i \in \mathbb{Z}$ such that

$$k_i = c_i + e(d_i - 1) + elm_i.$$

We then associate with $\mathbf{k}$ the following elements:

$$\begin{aligned} \mathbf{c} &:= \mathbf{c}(\mathbf{k}) = (c_1, \ldots, c_r), \\ \mathbf{d} &:= \mathbf{d}(\mathbf{k}) = (d_1, \ldots, d_r), \\ \mathbf{m} &:= \mathbf{m}(\mathbf{k}) = (m_1, \ldots, m_r). \end{aligned}$$

In addition, there exist elements which are uniquely defined as follows: $\mathbf{a} := \mathbf{a}(\mathbf{k}) \in A^e$, $\mathbf{b} := \mathbf{b}(\mathbf{k}) \in B^l$, $u := u(\mathbf{k}) \in {}^{\mathbf{a}}(\widehat{\mathfrak{S}}_r)$ and $v := v(\mathbf{k}) \in (\widehat{\mathfrak{S}}_r)^{\mathbf{b}}$ such that $\mathbf{c} = \mathbf{a}.u$ and $\mathbf{d} = v.\mathbf{b}$. We also set $x(\mathbf{k}) := uy^{\mathbf{m}}v \in {}^{\mathbf{a}}(\widehat{\mathfrak{S}}_r)$ (where $y^{\mathbf{m}} \in \widehat{\mathfrak{S}}_r$) and $\mathbf{p}(\mathbf{k}) := \mathbf{a}.x \in P_r$. The element $w_{\mathbf{b}} \in \widehat{\mathfrak{S}}_r$ is the longest element in the parabolic subgroup $(\widehat{\mathfrak{S}}_r)_{\mathbf{b}}$.

**Example 6.5.7.** Assume that we have $e = 4$, $l = 3$, $r = 4$ and $\mathbf{k} = (1, -11, -17, 2)$. Then we obtain

$$\mathbf{c} = (1,1,3,2), \quad \mathbf{d} = (1,1,2,1), \quad \mathbf{m} = (0,-1,-2,0), \quad \mathbf{a} = (1,1,2,3),$$
$$\mathbf{b} = (2,1,1,1), \ u = \sigma_3, \ v = \sigma_2\sigma_1, \ x = \sigma_3 y^{(0,-1,-2,0)}\sigma_2\sigma_1, \ \mathbf{p}(\mathbf{k}) = (-5,1,-3,2).$$

Consider now $\mathbf{k} = (12,5,2,1)$. Then, with the same values for $e$, $l$ and $r$, we obtain in this case

$$\mathbf{c} = (0,1,2,1), \quad \mathbf{d} = (1,2,1,1), \quad \mathbf{m} = (1,0,0,0), \quad \mathbf{a} = (1,1,2,4),$$
$$\mathbf{b} = (2,1,1,1), \quad u = \sigma_3\sigma_2\sigma_1\sigma_3 y^{(-1,0,0,0)}, \quad v = \sigma_1.$$

We obtain $x = \sigma_3\sigma_2\sigma_3$ and $\mathbf{p}(\mathbf{k}) = (1,4,2,1)$. Note that, in this situation, we have $I_{\mathbf{b}} = \{2,3\}$ and if $i \in I_{\mathbf{b}}$, we have $p_i > p_{i+1}$ for $\mathbf{p}(\mathbf{k}) = (p_1, p_2, p_3, p_4)$. This phenomenon can be explained by looking at the proof of the next theorem.

**6.5.8.** For $\mathbf{k} \in P_r$, we now define the following element of $\Lambda(\mathbf{a},\mathbf{b}) \subset \Lambda_q^r V$:

$$(\bullet\bullet) \qquad u_{\mathbf{k}} = u_{k_1} \wedge \ldots \wedge u_{k_r} = (-q^{-1})^{l(w_{\mathbf{b}})} 1 \otimes_{H_{\mathbf{a}}} T_u X^{-\mathbf{m}} (T_{v^{-1}})^{-1} \otimes_{H_{\mathbf{b}}} 1;$$

this is called a *q-wedge*. If $\mathbf{k} \in P_r$ is in the set

$$P_r^{++} := \{\mathbf{k} = (k_1, \ldots, k_r) \mid k_i > k_{i+1} \text{ for all } i \in \{1, \ldots, r-1\}\},$$

then we say that $u_{\mathbf{k}}$ is an *ordered q-wedge*. The following result gives a basis of our space $\Lambda_q^r V$. As the proof is quite long, we only give a sketch of it following [298, Proposition 2.8]. A different proof can be found in [291, Proposition 3.21].

**Theorem 6.5.9 (Uglov).** *The set $\{u_{\mathbf{k}} \mid \mathbf{k} \in P_r^{++}\}$ is a basis of $\Lambda_q^r V$.*

*Proof.* We set $S := \{(\mathbf{b},\mathbf{p}) \in B^l \times P_r \mid \mathbf{p} \in P_{r,\mathbf{b}}^{++}\}$ and consider the map

$$\Psi \colon P_r^{++} \to S, \qquad \mathbf{k} \mapsto (\mathbf{b}(\mathbf{k}), \mathbf{p}(\mathbf{k})).$$

We show that this is well defined (that is, $\mathbf{p}(\mathbf{k}) \in P_{r,\mathbf{b}}^{++}$ if $\mathbf{k} \in P_r^{++}$) and that this is a bijection. This can be done by purely combinatorial arguments. We then show that for all $\mathbf{k} \in P_r^{++}$ we have $u_{\mathbf{k}} = [\mathbf{p}(\mathbf{k})]_{\mathbf{b}(\mathbf{k})}$. We can then conclude using Proposition 6.5.5 and the definition of $\Lambda_q^r V$. □

**6.5.10.** The theorem shows that the set of $q$-wedges is not linearly independent. However, each $q$-wedge can be expressed uniquely in terms of ordered $q$-wedges. The way to express them is given by the following rules. First we assume that $r = 2$.

Then any $q$-wedge of the form $u_{k_1} \wedge u_{k_2}$ can be expressed as a linear combination of ordered $q$-wedges using the following rules (R1), (R2), (R3) and (R4).

Let $k_1 \leqslant k_2$ and, for $i = 1, 2$, put $k_i = a_i + e(l - b_i) - elm_i$, where $a_i \in \{1, \dots, e\}$, $b_i \in \{1, \dots, l\}$ and $m_i \in \mathbb{Z}$. We define $\alpha := (a_2 - a_1) \bmod el$ and $\beta := e(b_1 - b_2) \bmod el$. Then the relations (R1), (R2), (R3) and (R4) are given as follows.

(R1) if $\alpha = 0$ and $\beta = 0$:

$$u_{k_1} \wedge u_{k_2} = -u_{k_2} \wedge u_{k_1};$$

(R2) if $\alpha \neq 0$ and $\beta = 0$:

$$\begin{aligned} u_{k_1} \wedge u_{k_2} = & -q^{-1} u_{k_2} \wedge u_{k_1} \\ & + (q^{-2} - 1) \sum_{m \geqslant 0} q^{-2m} u_{k_2 - \alpha - elm} \wedge u_{k_1 + \alpha + elm} \\ & - (q^{-2} - 1) \sum_{m \geqslant 1} q^{-2m+1} u_{k_2 - elm} \wedge u_{k_1 + elm}; \end{aligned}$$

(R3) if $\alpha = 0$ and $\beta \neq 0$:

$$\begin{aligned} u_{k_1} \wedge u_{k_2} = & q u_{k_2} \wedge u_{k_1} \\ & + (q^2 - 1) \sum_{m \geqslant 0} q^{2m} u_{k_2 - \beta - elm} \wedge u_{k_1 + \beta + elm} \\ & + (q^2 - 1) \sum_{m \geqslant 1} q^{2m-1} u_{k_2 - elm} \wedge u_{k_1 + elm}; \end{aligned}$$

(R4) if $\alpha \neq 0$ and $\beta \neq 0$:

$$\begin{aligned} u_{k_1} \wedge u_{k_2} = & u_{k_2} \wedge u_{k_1} \\ & + (q - q^{-1}) \sum_{m \geqslant 0} \frac{q^{2m+1} + q^{-2m-1}}{q + q^{-1}} u_{k_2 - \beta - elm} \wedge u_{k_1 + \beta + elm} \\ & + (q - q^{-1}) \sum_{m \geqslant 0} \frac{q^{2m+1} + q^{-2m-1}}{q + q^{-1}} u_{k_2 - \alpha - elm} \wedge u_{k_1 + \alpha + elm} \\ & + (q - q^{-1}) \sum_{m \geqslant 0} \frac{q^{2m+2} - q^{-2m-2}}{q + q^{-1}} u_{k_2 - \beta - \alpha - elm} \wedge u_{k_1 + \beta + \alpha + elm} \\ & + (q - q^{-1}) \sum_{m \geqslant 1} \frac{q^{2m} - q^{-2m}}{q + q^{-1}} u_{k_2 - elm} \wedge u_{k_1 + elm}, \end{aligned}$$

where the summations are taken over the set of ordered $q$-wedge. The proofs for these relations can be obtained by induction using ($\bullet\bullet$). We refer to [298, Section 2.5] for a complete proof of them. Thus, for $r \in \mathbb{Z}_{\geqslant 0}$, an arbitrary $q$-wedge can be expressed as a linear combinaison of ordered $q$-wedges using the relations (R1)–(R4) in every adjacent pair of the factors. We have the following remarkable result which will be useful in the following and which is a direct consequence of the above relations.

**Proposition 6.5.11 (Uglov).** *Let $(a,b) \in \mathbb{Z}^2$ be such that $a \geqslant b$. Then we have*

$$u_b \wedge u_a \wedge u_{a-1} \wedge \ldots \wedge u_b = 0 \quad \textit{and} \quad u_a \wedge u_{a-1} \wedge \ldots \wedge u_b \wedge u_a = 0.$$

**6.5.12.** Let $\mathbf{a} \in A^e$ and $\mathbf{b} \in B^l$. We denote

$$\mathscr{I}_{\mathbf{a}} := \sum_{i \in I_{\mathbf{a}}} (T_i - q^{-1}).\widehat{H}_r \qquad \text{and} \qquad \mathscr{I}_{\mathbf{b}} := \sum_{i \in I_{\mathbf{b}}} \widehat{H}_r.(T_i + q).$$

We have an isomorphism of vector spaces between $\Lambda(\mathbf{a},\mathbf{b})$ and $\mathscr{I}_{\mathbf{a}} \setminus \widehat{H}_r / \mathscr{I}_{\mathbf{b}}$. Using this, we can define an involution on the $q$-wedge product. To do this, we will use the canonical involution of $\widehat{H}_r$. As for all $i = 1, \ldots, r-1$, we have $T_i^{-1} = T_i + (q - q^{-1})$, the elements $T_i + q$ and $T_i - q^{-1}$ are invariant with respect to the involution of $\widehat{H}_r$. As a consequence, the submodules $\mathscr{I}_{\mathbf{a}}$ and $\mathscr{I}_{\mathbf{b}}$ of $\widehat{H}_r$ are invariant with respect to this involution. This implies that the natural involution of the affine Hecke algebra induces an involution on $\Lambda(\mathbf{a},\mathbf{b})$ and hence on $\Lambda_q^r V$. For $\mathbf{p} := (p_1, \ldots, p_r) \in P_r^{++}$ we set

$$\upsilon(\mathbf{p}) := \sharp\{(i,j) \in \mathbb{Z}_{>0} \mid 1 \leqslant i < j \leqslant r,\ p_i = p_j\}.$$

Let $\mathbf{k} \in P_r$ be such that $u_{\mathbf{k}} \in \Lambda(\mathbf{a},\mathbf{b})$. Then we can show that (see [291, Proposition 3.23, Remark 3.24])

$$(\circ) \qquad \overline{u_{\mathbf{k}}} = (-q)^{\upsilon(\mathbf{b})} (q^{-1})^{\upsilon(\mathbf{a})} u_{k_r} \wedge \ldots \wedge u_{k_1}.$$

We thus obtain an involution on the space $\Lambda_q^r V$. Using the relations (R1)–(R4), we can write

$$\overline{u_{\mathbf{k}}} = \sum_{\mathbf{l} \in P_r^{++}} a_{\mathbf{k},\mathbf{l}}(q) u_{\mathbf{l}}.$$

The involution can be encoded in a matrix:

$$A := (a_{\mathbf{k},\mathbf{l}}(q))_{\mathbf{k},\mathbf{l} \in P_r^{++}}.$$

*Remark 6.5.13.* If we define $\mathscr{L}_r^+ = \bigoplus_{\mathbf{k} \in P_r^{++}} \mathbb{Q}[q] u_{\mathbf{k}}$, then one can show the existence of a unique basis $\{G(\mathbf{k}) \mid \mathbf{k} \in P_r^{++}\}$ of $\Lambda_q^r V$, such that the following two conditions are satisfied:

- for all $\mathbf{k} \in P_r^{++}$, we have $\overline{G(\mathbf{k})} = G(\mathbf{k})$,
- for all $\mathbf{k} \in P_r^{++}$, we have $G(\mathbf{k}) = u_{\mathbf{k}} \bmod q\mathscr{L}_r^+$.

(The proof is analogous to the one of Theorem 6.6.11.) The elements of the basis $\{G(\mathbf{k}) \mid \mathbf{k} \in P_r^{++}\}$ are called the canonical basis elements of $\Lambda_q^r V$. The existence of such a basis comes from the unitriangularity of the matrix of the involution. There exists a matrix $\Delta := (d_{\mathbf{k},\mathbf{l}}(q))_{\mathbf{k},\mathbf{l} \in P_r^{++}}$ such that $G(\mathbf{k}) = \sum_{\mathbf{l} \in P_r^{++}} d_{\mathbf{k},\mathbf{l}}.(q) u_{\mathbf{l}}$ for all $\mathbf{k} \in P_r^{++}$. It is possible to show that the coefficients $d_{\mathbf{k},\mathbf{l}}(q)$ can be expressed in terms of Kazhdan–Lusztig polynomials for $\widehat{\mathfrak{S}}_r$ (see [291, Theorem 3.26] and [292]).

**6.5.14.** We now show that it is possible to define an action of $\mathscr{U}_q(\widehat{\mathfrak{sl}_e})$ on the space $\Lambda_q^r V$ which is "compatible" with the involution defined by ($\bullet\bullet$) in 6.5.12. For a statement $S$, we here denote $\delta_S = 1$ if $S$ is true and 0 if $S$ is false. First, we have an action of $\mathscr{U}_q(\widehat{\mathfrak{sl}_e})$ on $\Lambda_q^1 V$. To define it, we first define an action of $\mathscr{U}_q(\widehat{\mathfrak{sl}_e})$ on $V_{e,l} := \mathscr{K}[X, X^{-1}] \otimes \mathscr{K}^e \otimes \mathscr{K}^l$ (where we denote $\mathscr{K} := \mathbb{Q}(q)$). We fix a basis $(v_1, \ldots, v_e)$ of $\mathscr{K}^e$ and a basis $(w_1, \ldots, w_l)$ of $\mathscr{K}^l$ and define for $0 \leqslant i \leqslant e-1$, $1 \leqslant c \leqslant e$ and $1 \leqslant d \leqslant l$:

$$
\begin{aligned}
t_i(X^m \otimes v_c \otimes w_d) &= q^{\delta_{i \equiv c \bmod e} - \delta_{i+1 \equiv c \bmod e}} (X^m \otimes v_c \otimes w_d), \\
e_i(X^m \otimes v_c \otimes w_d) &= \delta_{i+1 \equiv c \bmod e} (X^{m - \delta_{i,0}} \otimes v_{c-1} \otimes w_d), \\
f_i(X^m \otimes v_c \otimes w_d) &= \delta_{i \equiv c \bmod e} (X^{m + \delta_{i,0}} \otimes v_{c+1} \otimes w_d), \\
\mathfrak{d}(X^m \otimes v_c \otimes w_d) &= -mX^m \otimes v_c \otimes w_d,
\end{aligned}
$$

(where we set $v_0 := v_e$ and $v_{e+1} := v_1$). This gives an action of $\mathscr{U}_q(\widehat{\mathfrak{sl}_e})$ on $\Lambda_q^1 V$ using the isomorphism of vector spaces

$$
V_{e,l} \to \Lambda_q^1 V, \qquad X^m \otimes v_c \otimes w_d \mapsto u_{c+e(d-1)+elm}.
$$

Now we use the coproduct $\Delta$ of $\mathscr{U}_q(\widehat{\mathfrak{sl}_e})$ to obtain an action of $\mathscr{U}_q(\widehat{\mathfrak{sl}_e})$ on $(\Lambda_q^1 V)^{\otimes r}$ (see Remark 6.1.4). Explicitly, this action is given as follows:

$$
\begin{aligned}
t_i(u_{k_1} \otimes \ldots \otimes u_{k_r}) &= t_i(u_{k_1}) \otimes t_i(u_{k_2}) \otimes \ldots \otimes t_i(u_{k_r}), \\
e_i(u_{k_1} \otimes \ldots \otimes u_{k_r}) &= \sum_{j=1}^{r} u_{k_1} \otimes \ldots \otimes u_{k_{j-1}} \otimes e_i(u_{k_j}) \otimes t_i^{-1}(u_{k_{j+1}}) \otimes \ldots \otimes t_i^{-1}(u_{k_r}), \\
f_i(u_{k_1} \otimes \ldots \otimes u_{k_r}) &= \sum_{j=1}^{r} t_i(u_{k_1}) \otimes \ldots \otimes t_i(u_{k_{j-1}}) \otimes f_i(u_{k_j}) \otimes u_{k_{j+1}} \otimes \ldots \otimes u_{k_r}, \\
\mathfrak{d}(u_{k_1} \otimes \ldots \otimes u_{k_r}) &= \sum_{j=1}^{r} u_{k_1} \otimes \ldots \otimes u_{k_{j-1}} \otimes \mathfrak{d}(u_{k_j}) \otimes u_{k_{j+1}} \otimes \ldots \otimes u_{k_r}.
\end{aligned}
$$

Now $\Lambda_q^r V$ can be seen as the quotient of $(\Lambda_q^1 V)^{\otimes r}$ by the relations (R1)–(R4). Using this we check that we obtain a well-defined action of $\mathscr{U}_q(\widehat{\mathfrak{sl}_e})$ on $\Lambda_q^r V$, which is explictly given as follows:

$$
\begin{aligned}
t_i(u_{k_1} \wedge \ldots \wedge u_{k_r}) &= t_i(u_{k_1}) \wedge t_i(u_{k_2}) \wedge \ldots \wedge t_i(u_{k_r}), \\
e_i(u_{k_1} \wedge \ldots \wedge u_{k_r}) &= \sum_{j=1}^{r} u_{k_1} \wedge \ldots \wedge u_{k_{j-1}} \wedge e_i(u_{k_j}) \wedge t_i^{-1}(u_{k_{j+1}}) \wedge \ldots \wedge t_i^{-1}(u_{k_r}), \\
f_i(u_{k_1} \wedge \ldots \wedge u_{k_r}) &= \sum_{j=1}^{r} t_i(u_{k_1}) \wedge \ldots \wedge t_i(u_{k_{j-1}}) \wedge f_i(u_{k_j}) \wedge u_{k_{j+1}} \wedge \ldots \wedge u_{k_r}, \\
\mathfrak{d}(u_{k_1} \wedge \ldots \wedge u_{k_r}) &= \sum_{j=1}^{r} u_{k_1} \wedge \ldots \wedge u_{k_{j-1}} \wedge \mathfrak{d}(u_{k_j}) \wedge u_{k_{j+1}} \wedge \ldots \wedge u_{k_r}.
\end{aligned}
$$

**6.5.15.** We now use the above results to construct a new space called the semi-infinite $q$-wedge product. We will then see later the connection with the Fock space. First, we fix $s \in \mathbb{Z}$. Let $t > r$ be two positive integers. We define a map

$$\begin{array}{rcl} \Lambda_q^r V & \to & \Lambda_q^t V, \\ v & \mapsto & v \wedge u_{s-r} \wedge u_{s-r-1} \wedge \ldots \wedge u_{s-t+1}. \end{array}$$

We now consider the inductive limit

$$\Lambda^s = \varinjlim(\Lambda_q^r V),$$

which is called the *semi-infinite $q$-wedge product* with charge $s$. For $v \in \Lambda_q^r V$, we denote by

$$v \wedge u_{s-r} \wedge u_{s-r-1} \wedge u_{s-r-2} \wedge \ldots$$

the image of the elements $v$ under the canonical map $\Lambda_q^r V \to \Lambda^s$. Such an element is called a *semi-infinite $q$-wedge*. We denote by $P(s)$ the sequence of integers $\mathbf{k} = (k_i)_{i \geqslant 1}$ such that $k_i = s - i + 1$ if $i$ is sufficiently large. From the definition, it is clear that the set of semi-infinite $q$-wedges

$$u_{\mathbf{k}} = u_{k_1} \wedge u_{k_2} \wedge \ldots$$

with $\mathbf{k} = (k_i)_{i \geqslant 1} \in P(s)$ is a generating set for $\Lambda^s$. Let $P^{++}(s)$ be the elements $\mathbf{k} = (k_i)_{i \geqslant 1}$ in $P(s)$ such that $k_i > k_{i+1}$ for all $i \geqslant 1$. If $\mathbf{k} \in P^{++}(s)$, the semi-infinite $q$-wedge $u_{\mathbf{k}} \in \Lambda^s$ is said to be ordered.

**Proposition 6.5.16 (Takemura and Uglov [288, 4.1]).** *The set $\{u_{\mathbf{k}} \mid \mathbf{k} \in P^{++}(s)\}$ is a basis of $\Lambda^s$.*

*Proof.* Let $u \in \Lambda^s$. Then, there exists $v \in \Lambda_q^r V$ such that

$$u = v \wedge u_{s-r} \wedge u_{s-r-1} \wedge u_{s-r-2} \wedge \ldots$$

Now, by Theorem 6.5.9, $v$ is a linear combination of ordered $q$-wedges. Hence, $u$ is a linear combination of elements of the form

$$u_{k_1} \wedge u_{k_2} \wedge \ldots \wedge u_{k_r} \wedge u_{s-r} \wedge u_{s-r-1} \wedge u_{s-r-2} \wedge \ldots$$

with $k_1 > \ldots > k_r$. Such an element is zero if $k_r \leqslant s - r$ by Proposition 6.5.11. Otherwise, it is a semi-infinite ordered $q$-wedge. This shows that the semi-infinite ordered $q$-wedges span $\Lambda^s$. The fact that the semi-infinite ordered $q$-wedges are linearly independent follows directly from the definition of the inductive limit together with Theorem 6.5.9. □

Hence, one can express an arbitrary semi-infinite $q$-wedge as a sum of semi-infinite ordered $q$-wedges using the formulae (R1)–(R4).

**6.5.17.** We give an alternative indexation for the elements of the basis of $\Lambda^s$ which will be useful to connect $\Lambda^s$ with the Fock space. This indexation uses the notion of abaci. Let us define this new object. A 1-runner abacus is a subset $\underline{A}$ of $\mathbb{Z}$ such that $-k \in \underline{A}$ and $k \notin \underline{A}$ for large enough $k \in \mathbb{Z}_{>0}$. This can be represented as follows: each $k \in \underline{A}$ corresponds to a the position of a bead on a horizontal abacus which is full of beads on the left and empty on the right (see Figure 6.3). An $l$-abacus is an $l$-tuple of abaci $(\underline{A}_1, \dots, \underline{A}_l)$. It is identified with the subset

$$\{(k,c) \mid k \in \underline{A}_c, 1 \leqslant c \leqslant l\} \subset \mathbb{Z} \times \{1, \dots, l\}$$

(see Figure 6.4).

Fig. 6.3 A 1-runner abacus

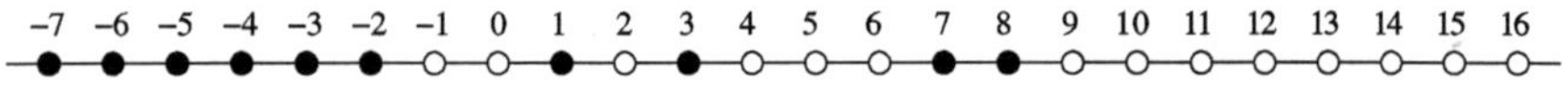


Fig. 6.4 A 3-runner abacus

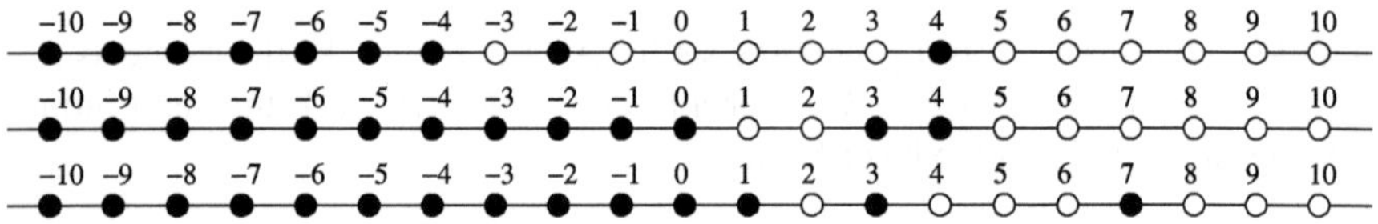


Now, let $\lambda = (\lambda^1, \dots, \lambda^l) \vdash_l n$ and $\mathbf{s} = (s_1, \dots, s_l) \in \mathbb{Z}^l$ be such that $\sum_{1 \leqslant i \leqslant l} s_i = s$. We associate with this datum an $l$-abacus:

$$\underline{A}_{\mathbf{s}}^{\lambda} := \{(\lambda_i^{(b)} - i + s_b + 1, b) \mid i \geqslant 1, 1 \leqslant b \leqslant l\}.$$

If we denote by $\mathscr{A}^l$ the set of $l$-abaci, we check that the map

$$\Pi^l(n) \times \mathbb{Z}^l \to \mathscr{A}^l, \qquad (\lambda, \mathbf{s}) \mapsto \underline{A}_{\mathbf{s}}^{\lambda}$$

is a bijection. Consequently, each semi-infinite ordered $q$-wedge can be labelled by a symbol $|\lambda, \mathbf{s}\rangle$, where $\lambda$ is an $l$-partition of rank $n$ and $\mathbf{s} = (s_1, \dots, s_l) \in \mathbb{Z}^l$ is such that $\sum_{1 \leqslant i \leqslant l} s_i = s$. This is done as follows. Let $u_{\mathbf{k}} = u_{k_1} \wedge u_{k_2} \wedge \dots$ be a semi-infinite ordered $q$-wedge. For $i = 1, 2, \dots$, we put $k_i = a_i + e(l - b_i) - elm_i$, where $a_i \in \{1, \dots, e\}$, $b_i \in \{1, \dots, l\}$ and $m_i \in \mathbb{Z}$. For $b = 1, \dots, l$, let $k_1^{(b)} > k_2^{(b)} > \dots$ be the semi-infinite sequence obtained by ordering the elements of the set $\{a_i - em_i \mid b_i = b\}$ in strictly decreasing order. Then the set

$$\{(k_i^{(b)}, b) \mid i \geqslant 1, 1 \leqslant b \leqslant l\}$$

defines an $l$-abacus which corresponds to an $l$-partition $\lambda$ and an $l$-tuple $\mathbf{s}$ under the above bijection. Hence, there is a unique $s_b \in \mathbb{Z}$ such that $k_i^{(b)} = s_b - i + 1$ for sufficiently large $i$ and we can set $\tau_l(u_{\mathbf{k}}) = |\lambda, \mathbf{s}\rangle$., where $\lambda = (\lambda^1, \ldots, \lambda^l)$ and where for $i > 0$, $\lambda_i^{(b)} = k_i^{(b)} - s_b + i - 1$ with $\mathbf{s} = (s_1, \ldots, s_l)$. This defines a bijection $\tau_l$ between the set of semi-infinite ordered $q$-wedges and the set of symbols $|\lambda, \mathbf{s}\rangle$, where $\lambda \vdash_l n$ and $\mathbf{s} = (s_1, \ldots, s_l) \in \mathbb{Z}^l$ is such that $\sum_{1 \leqslant i \leqslant l} s_i = s$.

**Fig. 6.5** A 1-runner abacus representing the sequence $\mathbf{k} = (15, 12, 8, 7, 3, 1, -2, -4, -5, -6, \ldots)$

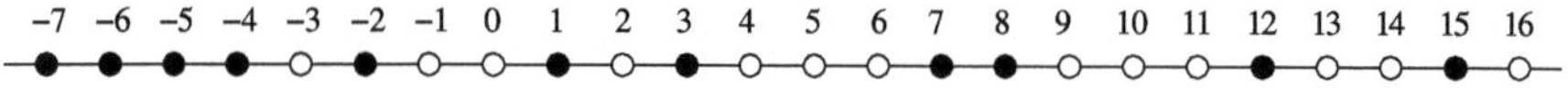


**Fig. 6.6** A 3-runner abacus representing the sequence $\mathbf{k} = (15, 12, 8, 7, 3, 1, -2, -4, -5, -6, \ldots)$ for $e = 4$

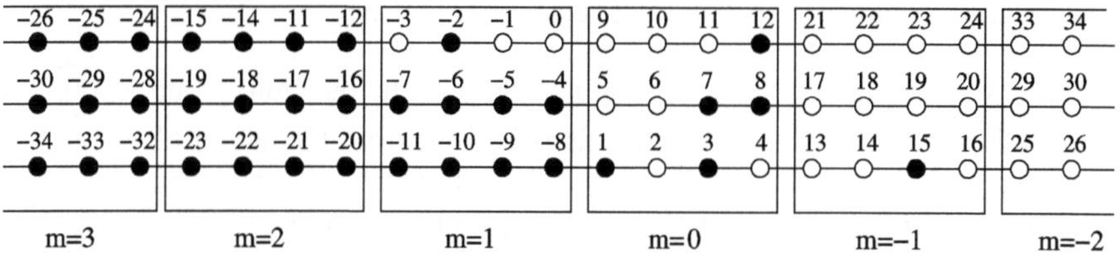


In a less formal way, the infinite decreasing sequence $(k_1, k_2, \ldots)$ can be pictured as a set of coloured beads on an infinite runner. Each element $k_j$ corresponds to the position of a bead on a horizontal abacus which is full of beads on the left and empty on the right (see Figure 6.5). On the other hand, using the decomposition $k_i = a_i + e(l - b_i) - elm_i$, the same sequence can be represented by an $l$-abacus; that is, as a set of coloured beads on $l$ infinite runners (see Figure 6.6).

**Fig. 6.7** A 3-runner abacus representing the sequence $\mathbf{k} = (15, 12, 8, 7, 3, 1, -2, -4, -5, -6, \ldots)$ for $e = 4$ with another indexation of the beads.

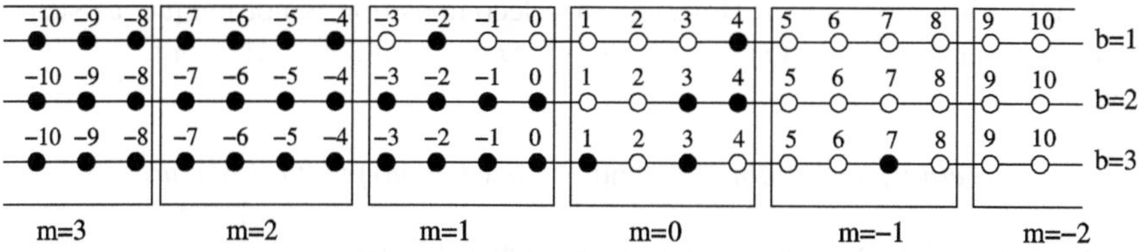

**Fig. 6.8** The 3-runner abacus obtained after moving the beads to obtain $\mathbf{s}$

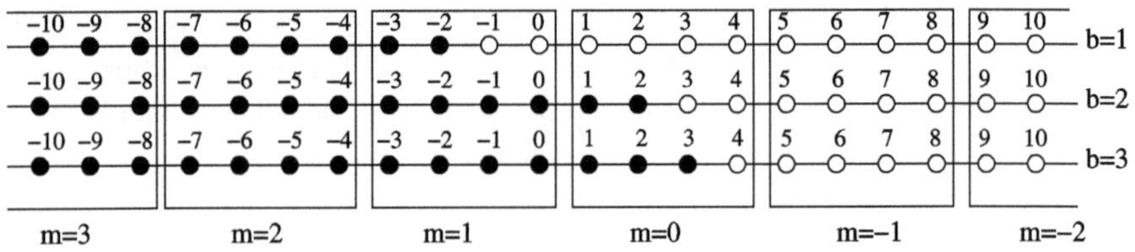


In this representation, the coloured beads are labelled by the integers $k_i$ (with $i = 1,2,\ldots$). We can alternatively label them by the integers $k_i^{(b)}$ (with $i = 1,2,\ldots$ and $b = 1,2,\ldots,l$) as in Figure 6.7.

Let $\lambda$ be the $l$-partition and $\mathbf{s} = (s_1,\ldots,s_l)$ be the $l$-tuple of integers associated to $\mathbf{k}$. For $i = 1,2,\ldots$ and $b = 1,2,\ldots,l$, we have $\lambda_i^{(b)} = k_i^{(b)} - s_b + i - 1$. Thus, one can now easily determine $\lambda$ by counting the number of non coloured beads at the left of each coloured bead on each runner. The multicharge $\mathbf{s}$ can be read in the $l$-abacus by moving the beads to the left as much as possible (see Figure 6.8). Continuing the example, we obtain $\lambda = (6.1, 2.2, 4.1)$ and $\mathbf{s} = (-2,2,3)$.

**6.5.18.** We now show the existence of an action of the quantum group $\mathscr{U}_q(\widehat{\mathfrak{sl}_e})$ on the semi-infinite $q$-wedge product. This is derived from the action of $\mathscr{U}_q(\widehat{\mathfrak{sl}_e})$ on $\Lambda_q^r V$ that we have constructed previously. The action of the Chevalley operator $f_i$ on an arbitrary semi-infinite ordered $q$-wedge $u_{\mathbf{k}}$ is given as follows:

$$f_i(u_{k_1} \wedge u_{k_2} \wedge \ldots) := \sum_{j \geqslant 1} t_i(u_{k_1}) \wedge \ldots \wedge t_i(u_{k_{j-1}}) \wedge f_i(u_{k_j}) \wedge u_{k_{j+1}} \wedge \ldots$$

for $\mathbf{k} \in P^{++}(s)$. Note that, by 6.5.14 and by Proposition 6.5.11, there are only a finite number of $q$-wedges in the above sum. Now, we want to give the action of the $e_i$ and the $t_i$. They are a bit more involved to describe and we will need to introduce more notations to present them. In [291, Section 4.2], this requires in particular the notion of weight of a $q$-wedge. As noted in [298, (3.2.1)], we can bypass this by arguing as follows. First, for $m \in \mathbb{Z}$, we set

$$|elm\rangle := u_{elm} \wedge u_{elm-1} \wedge u_{elm-2} \wedge \ldots$$

Note that each semi-infinite ordered $q$-wedge $u_{\mathbf{k}}$ can be written as $v \wedge |elm\rangle$ for $v \in \Lambda^r V$ with $r = s - elm$. However, this decomposition is not unique, as such a $q$-wedge can also be written as $v' \wedge |elm'\rangle$ for $m' < m$ and $v' \in \Lambda^t V$ for $t = s - elm'$. We now set

$$t_i|elm\rangle := q^{l\delta_{i,0}}|elm\rangle \qquad \text{and} \qquad t_i(v \wedge |elm\rangle) := t_i(v) \wedge |elm\rangle,$$

which is well defined. The action of $e_i$ is given as follows:

$$e_i(v \wedge |elm\rangle) := e_i(v) \wedge t_i^{-1}|elm\rangle.$$

Finally, we set

$$\mathfrak{d}|elm\rangle := -elm\,\mathrm{sgn}(m)\frac{m(m+1)}{2}|elm\rangle,$$

where

$$\mathrm{sgn}(m) = \begin{cases} 0 & \text{if } m = 0, \\ 1 & \text{if } m > 0, \\ -1 & \text{if } m < 0, \end{cases}$$

and we have

$$\mathfrak{d}(v \wedge |elm\rangle) := \mathfrak{d}(v) \wedge |elm\rangle,$$

which is well defined. One can then see that this makes $\Lambda^s$ into a $\mathcal{U}_q(\widehat{\mathfrak{sl}_e})$-module. It is now a natural question to ask what becomes of these formulae when we use the indexation of the basis by the set of $l$-partitions. Recall that the Fock space associated with $\mathbf{s} = (s_1, \ldots, s_l) \in \mathbb{Z}^l$ is the $\mathbb{Q}(q)$-vector space generated by the symbols $|\boldsymbol{\lambda}, \mathbf{s}\rangle$ with $\boldsymbol{\lambda} \vdash_l n$:

$$\mathcal{F}_{\mathbf{s}} := \bigoplus_{n \geqslant 0} \bigoplus_{\boldsymbol{\lambda} \vdash_l n} \mathbb{Q}(q)|\boldsymbol{\lambda}, \mathbf{s}\rangle.$$

Now, we have seen that one can use the indexation by $l$-partitions for the standard basis of $\Lambda^s$. Using this, the Fock space $\mathcal{F}_{\mathbf{s}}$ can be seen as a subset of $\Lambda^s$ where $\sum_{1 \leqslant i \leqslant l} s_i = s$, and from now on we will make this identification.

**Proposition 6.5.19 (Uglov [291, §4.2]).** *For all $\mathbf{s} = (s_1, \ldots, s_l) \in \mathbb{Z}$ such that we have $\sum_{1 \leqslant i \leqslant l} s_i = s$, the space $\mathcal{F}_{\mathbf{s}}$ is a $\mathcal{U}_q(\widehat{\mathfrak{sl}_e})$-submodule of $\Lambda^s$. Moreover, the induced action of $\mathcal{U}_q(\widehat{\mathfrak{sl}_e})$ on $\mathcal{F}_{\mathbf{s}}$ is given in Theorem 6.2.5.*

What the above result says is that the action constructed above generalises the action constructed in Section 6.2. Now, one can now ask if we can give a definition for the canonical basis of the whole space of semi-infinite $q$-wedge product.

## 6.6 Canonical Bases for Fock Spaces

In this section we define an involution on the Fock space which will be used to obtain the definition of the canonical basis. To do this, we use the semi-infinite $q$-wedge product that we introduced in the last section.

**6.6.1.** Let $\mathbf{s} = (s_1, \ldots, s_l) \in \mathbb{Z}^l$. We define an involution on the Fock space. First, recall that we have already introduced an involution on the space $\Lambda_q^r V$ in 6.5.12. If $\mathbf{k} = (k_1, k_2, \ldots) \in P_s^{++}$, we set

$$\deg(u_{\mathbf{k}}) = \min\{i \geqslant 1 \mid k_j = k_i - (j - i) \text{ for all } j > i\}.$$

Then our involution on $\Lambda^s$ is defined as follows.

**Proposition 6.6.2 (Uglov).** *Let* $\mathbf{k} = (k_1, k_2, \ldots) \in P_s^{++}$. *The formula*

$$\overline{u_{\mathbf{k}}} := \overline{u_{k_1} \wedge \ldots \wedge u_{k_r}} \wedge u_{k_{r+1}} \wedge u_{k_{r+2}} \wedge \ldots,$$

*where* $r \geqslant deg(u_{\mathbf{k}})$, *gives a well-defined involution on* $\Lambda^s$.

*Proof.* This comes from the fact that in $\Lambda_q^t V$, if $t > r \geqslant \deg(u_{\mathbf{k}})$, we have

$$\overline{u_{k_1} \wedge \ldots \wedge u_{k_r}} \wedge u_{k_{r+1}} \wedge \ldots \wedge u_{k_t} = \overline{u_{k_1} \wedge \ldots \wedge u_{k_r} \wedge u_{k_{r+1}} \wedge \ldots \wedge u_{k_t}};$$

see [291, Lemma 4.10]. □

An easy, but quite long computation [291, Proposition 3.31] shows us that this definition is compatible with the action of the quantum group and the involution defined in 6.1.14 in the following sense.

**Proposition 6.6.3 (Uglov).** *For all* $w_e \in \mathscr{U}_q(\widehat{\mathfrak{sl}_e})$ *and* $u \in \Lambda^s$, *we have* $\overline{w_e.u} = \overline{w_e}.\overline{u}$. *Moreover, if* $\mathbf{s} = (s_1, \ldots, s_l)$ *is such that* $s = \sum_{1 \leqslant i \leqslant l} s_i$, *then, for all* $u \in \mathscr{F}_{\mathbf{s}}$ *we have* $\overline{u} \in \mathscr{F}_{\mathbf{s}}$.

Let us now fix $\mathbf{s} \in \mathbb{Z}^l$ such that $s = \sum_{1 \leqslant i \leqslant l} s_i$, then the above discussion shows the existence of an involution on the Fock space by restriction:

$$u \in \mathscr{F}_{\mathbf{s}} \quad \mapsto \quad \overline{u} \in \mathscr{F}_{\mathbf{s}}.$$

For all $\lambda \vdash_l n$, there exist coefficients $a_{\lambda,\mu}(q) \in \mathbb{Z}[q]$ such that

$$\overline{|\lambda, \mathbf{s}\rangle} = \sum_{\mu \vdash_l n} a_{\lambda,\mu}(q) |\mu, \mathbf{s}\rangle.$$

We set $A(\mathbf{s}) := (a_{\lambda,\mu}(q))_{\lambda \vdash_l n, \mu \vdash_l n}$. The following proposition gives us a first property of this matrix.

**Lemma 6.6.4.** *For all* $\lambda \vdash_l n$, *we have* $a_{\lambda,\lambda}(q) = 1$.

*Proof.* Let $u_{\mathbf{k}} = u_{k_1} \wedge u_{k_2} \wedge \ldots$ be a semi-infinite ordered $q$-wedge of degree $r$. We define $\mathbf{k}^r := (k_1, \ldots, k_r) \in P_r^{++}$. Then using the notation of 6.5.12, by ($\circ$) in 6.5.12 together with Proposition 6.6.2, we have

$$\overline{u_{\mathbf{k}}} := (-q)^{\upsilon(\mathbf{b}(\mathbf{k}^r))} (q^{-1})^{\upsilon(\mathbf{a}(\mathbf{k}^r))} u_{k_r} \wedge \ldots \wedge u_{k_1} \wedge u_{k_{r+1}} \wedge u_{k_{r+2}} \wedge \ldots$$

Using the relations (R1)–(R4), we can express $u_{k_r} \wedge \ldots \wedge u_{k_1} \wedge u_{k_{r+1}} \wedge u_{k_{r+2}} \wedge \ldots$ as a sum of semi-infinite ordered $q$-wedges. We are only interested in the coefficient of $u_{\mathbf{k}}$ in this expression. This coefficient is precisely $(-q)^{-\upsilon(\mathbf{b}(\mathbf{k}^r))} (q)^{\upsilon(\mathbf{a}(\mathbf{k}^r))}$, as required. □

**6.6.5.** We now wish to study the matrix of the involution $A(\mathbf{s})$. The main aim will be to prove the unitriangularity of this matrix, with the partial order $\ll_{\mathbf{m}}$ that we have defined in 5.7.5. This is a generalisation of the main result of [176, Theorem 4.7].

Recall that we have fixed a sequence $\mathbf{s} = (s_1, \ldots, s_l) \in \mathbb{Z}^l$. Let $\mathbf{v} = (v_1, \ldots, v_l) \in \mathbb{Q}^l$ be such that for all $i < j$ we have $0 < v_j - v_i < e$ and set $\mathbf{m} := (s_1 - v_1, \ldots, s_l - v_l)$. Recall the following partial order. Let $\lambda$ and $\mu$ be both $l$-compositions of rank $n$. As in 5.5.10, we define

$$\lambda \ll_{\mathbf{m}} \mu \qquad \overset{\text{def}}{\Longleftrightarrow} \qquad \lambda = \mu \quad \text{or} \quad \kappa_{\mathbf{m}}(\lambda) \rhd \kappa_{\mathbf{m}}(\mu).$$

In Chapter 5, we used this partial order to study the canonical basic sets for Ariki–Koike algebras in the case where $\mathbf{s} \in \mathscr{S}_e^l$. Here, we consider a general multicharge $\mathbf{s} \in \mathbb{Z}^l$. Using the property of unitriangularity that we prove here, we will be able to deduce the existence of a canonical basis for the Fock space $\mathscr{F}_{\mathbf{s}}$, where $\mathbf{s} \in \mathbb{Z}^l$. Before doing this, we study the form of the elements appearing in the expansion of an element $\overline{|\lambda, \mathbf{s}\rangle}$ with $\lambda \vdash_l n$ in terms of elements $|\mu, \mathbf{s}\rangle$ with $\mu \vdash_l n$. More precisely, we examine the forms of the $l$-partitions appearing in this expansion from the $l$-partition $\lambda$. First, we introduce some more notation. Let $u_{\mathbf{k}}$ be a semi-infinite (possibly non ordered) $q$-wedge. Let $u_{\tilde{\mathbf{k}}}$ be the semi-infinite ordered $q$-wedge obtained from $u_{\mathbf{k}}$ by reordering the $k_i$ in strictly decreasing order. The bijection $\tau_l$ then allows us to associate with $u_{\tilde{\mathbf{k}}}$ a symbol $|\lambda, \mathbf{s}\rangle$ such that $\lambda \vdash_l n$ and $\mathbf{s} = (s_1, \ldots, s_l)$. We denote by $\pi$ the linear map from the vector space generated by all semi-infinite $q$-wedges to $\bigoplus_{\mathbf{s} \in \mathbb{Z}^l, s_1 + \ldots + s_l = s} \mathscr{F}_{\mathbf{s}}$ such that

$$\pi(u_{\mathbf{k}}) = |\lambda, \mathbf{s}\rangle.$$

In particular, $\tau_l$ and $\pi$ coincide on the set of semi-infinite ordered $q$-wedge. Note that the map $\pi$ is not injective.

**6.6.6.** Let $u_{\mathbf{k}} = u_{k_1} \wedge u_{k_2} \wedge \ldots$ be an arbitrary semi-infinite $q$-wedge and assume that this is non-ordered. Then there exists $i \in \mathbb{Z}_{>0}$ such that $k_i < k_{i+1}$. The relations (R1)–(R4) then show how to express $u_{\mathbf{k}}$ in terms of semi-infinite $q$-wedges $u_{\mathbf{k}'}$ with $k'_i > k'_{i+1}$. Let us denote $\pi(u_{\mathbf{k}}) = |\lambda, \mathbf{s}\rangle$ and $\pi(u_{\mathbf{k}'}) = |\lambda', \mathbf{s}\rangle$. Using the representations of $u_{\mathbf{k}}$ and $u_{\mathbf{k}'}$ by $l$-abaci, we can see that $u_{\mathbf{k}'}$ is obtained from $u_{\mathbf{k}}$ by moving two beads representing $k_i$ and $k_{i+1}$ to two beads representing $k'_i$ and $k'_{i+1}$ and lying in the same runners as $k_i$ and $k_{i+1}$. As a consequence, the $l$-partition $\lambda'$ is obtained from $\lambda$ by removing a "ribbon" $R$ of size $N$ (or a rim $N$-hook; see [182, Section 2.7]) from a component $b$ and adding a ribbon $R'$ of size $N$ to a component $b'$. In a more formal way, there exist positive integers $r$, $s$, $p$ and $t$ such that

$$\lambda'^{(b)} = (\lambda_1^{(b)}, \ldots, \lambda_{r-1}^{(b)}, \lambda_{r+1}^{(b)} - 1, \ldots, \lambda_s^{(b)} - 1, \lambda_r^{(b)} - (N - (r - s)), \lambda_{s+1}^{(b)}, \ldots),$$
$$\lambda'^{(b')} = (\lambda_1^{(b')}, \ldots, \lambda_{p-1}^{(b')}, \lambda_t^{(b')} + (N - (p - t)), \lambda_p^{(b')} + 1, \ldots, \lambda_{t-1}^{(b')} + 1, \lambda_{t+1}^{(b')}, \ldots).$$

In fact, the two $l$-partitions $\lambda$ and $\lambda'$ are both obtained by adding the ribbons $R$ and $R'$ to the components $b$ and $b'$ of the same $l$-partition $\mu$ which is characterized as follows. We have $|\mu, \mathbf{s}\rangle = \pi(u_{\mathbf{r}})$, where

$$u_{\mathbf{r}} = u_{k_1} \wedge u_{k_{i-1}} \wedge u_{\hat{k}_i} \wedge u_{\hat{k}_{i+1}} \wedge u_{k_{i+2}} \wedge \ldots$$

Here, $\hat{k}_i$ and $\hat{k}_{i+1}$ are the minimal integers of $\{k_i, k_{i+1}, k'_i, k'_{i+1}\}$ such that the beads associated with $\hat{k}_i$ and $\hat{k}_{i+1}$ lie at the components $b$ and $b'$ of the $l$-abacus of $u_{\mathbf{r}}$ and $\hat{k}_{i+1} < \hat{k}_i$.

**6.6.7.** Let $\mathfrak{B}$ be the shifted $\mathbf{m}$-symbol of $\mu$ of size $k$ (which is assumed to be sufficiently large) and write all the elements appearing in this symbol by $\{\gamma_1, \ldots, \gamma_h\}$. These elements are of the form

$$\mu_j^c - j + s_c - v_c + k, \quad \text{where} \quad c \in \{1, \ldots, l\}, \quad j \in \{1, \ldots, [k + s_c - v_c]\}.$$

Note that $\kappa_{\mathbf{m}}(\mu)$ is the sequence of elements of this set written in decreasing order. By definition of $\kappa_{\mathbf{m}}$ and of the semi-infinite $q$-wedge, there exist $x$ and $y$ such that $\kappa_{\mathbf{m}}(\lambda)$ is the sequence obtained by reordering the elements of $\{\gamma_1, \ldots, \gamma_h\} \setminus \{\gamma_x\} \cup \{\gamma_x + N\}$ in decreasing order and $\kappa_{\mathbf{m}}(\lambda')$ is the sequence obtained by reordering the elements $\{\gamma_1, \ldots, \gamma_h\} \setminus \{\gamma_y\} \cup \{\gamma_y + N\}$ in decreasing order, where $N$ is a certain positive integer.

**Lemma 6.6.8.** *Keeping the above notation, assume that $\lambda \neq \lambda'$; then we have $\gamma_x > \gamma_y$.*

*Proof.* We use the same argument as in [174, pp. 581–583]. Keeping the notation of the previous section, we set $\hat{k}_i =: a + e(l-b) - elm$ and $\hat{k}_{i+1} =: a' + e(l-b') - elm'$. Then, by the discussion in 6.5.17, the elements of $\kappa_{\mathbf{m}}(\mu)$ are closely related to the $(k_1, \ldots, k_{i-1}, \hat{k}_i, \hat{k}_{i+1}, k_{i+2}, \ldots)$. Indeed, we have that

$$\gamma_x = a - em - v_b - 1 + k \text{ and } \gamma_y = a' - em' - v_{b'} - 1 + k.$$

Hence, one has to show that

$$(\circ\circ) \qquad a - em - v_b > a' - em' - v_{b'}.$$

For $j = i, i+1$, we set $k_j =: a_j + e(l - b_j) - elm_j$ and we define $\alpha := (a_{i+1} - a_i) \bmod el$ and $\beta := e(b_{i+1} - b_i) \bmod el$. We assume that $\lambda \neq \lambda'$. Then, regarding the relations (R1)–(R4) in 6.5.10, we have several cases to consider:

- There exists an $m > 0$ such that $k'_i := k_{i+1} - elm$ and $k'_{i+1} = k_i + elm$.
- We have $\beta \neq 0$ and there exists an $m \geqslant 0$ such that $k'_i = k_{i+1} - \beta - elm$ and $k'_{i+1} = k_i + \beta + elm$.
- We have $\alpha \neq 0$ and there exists an $m \geqslant 0$ such that $k'_i = k_{i+1} - \alpha - elm$ and $k'_{i+1} = k_i + \alpha + elm$.
- We have $\beta \neq 0$, $\alpha \neq 0$ and there exists an $m \geqslant 0$ such that $k'_i = k_{i+1} - \beta - \alpha - elm$ and $k'_{i+1} = k_i + \beta + \alpha + elm$.

In each case, we easily check that the inequality $(\circ\circ)$ holds using the fact that $b_i < b_{i+1}$ if and only if $0 < v_{b_i} < v_{b_{i+1}} < e$. □

Note that, as a consequence, we obtain that $\lambda \ll_{\mathbf{m}} \lambda'$ by the definition of $\ll_{\mathbf{m}}$.

**6.6.9.** We are now in a position to prove the unitriangularity of the matrix $A(\mathbf{s})$. Let $\lambda \vdash_l n$ and set $u_{\mathbf{p}} := \tau_l^{-1}(|\lambda, \mathbf{s}\rangle)$. We set $u_{\mathbf{k}} = u_{k_1} \wedge u_{k_2} \wedge \ldots := \overline{u_{\mathbf{p}}}$. We have by definition that $\pi(u_{\mathbf{k}}) = |\lambda, \mathbf{s}\rangle$. Now, after several applications of the discussion in 6.6.6 and 6.6.7, we obtain that if $u_{\mathbf{k}'}$ is a semi-infinite ordered $q$-wedge appearing in the expansion of $\overline{u_{\mathbf{p}}}$, then we have $\lambda \ll_{\mathbf{m}} \lambda'$, where $\tau_l(u_{\mathbf{k}'}) = |\lambda', \mathbf{s}\rangle$. Combining this with Lemma 6.6.4, we obtain the following proposition.

**Proposition 6.6.10.** *Let* $\mathbf{s} = (s_1, \ldots, s_l) \in \mathbb{Z}^l$ *and let* $\mathbf{v} := (v_1, \ldots, v_l)$ *be such that* $0 < v_i - v_j < e$ *if* $j > i$. *Set* $\mathbf{m} = (s_1 - v_1, \ldots, s_l - v_l)$ *and let* $\lambda \vdash_l n$. *Then,* $a_{\lambda,\mu}(q) \neq 0$ *implies* $\lambda \ll_{\mathbf{m}} \mu$. *Moreover, we have* $a_{\lambda,\lambda}(q) = 1$.

The above results will allow us to define the canonical basis for the Fock space. To do this, we set

$$\mathscr{L}[\mathbf{s}] := \bigoplus_{\lambda \vdash_l n} \mathbb{Q}[q] |\lambda, \mathbf{s}\rangle.$$

Then we obtain the following theorem.

**Theorem 6.6.11 (Leclerc and Thibon [209, 4.1] ($l = 1$), Uglov [291, §4] ($l \geqslant 1$)).** *Let* $\mathbf{s} = (s_1, \ldots, s_l) \in \mathbb{Z}^l$. *There exists a unique basis* $\{B_q(\mu, \mathbf{s}) \mid \mu \vdash_l n\}$ *of the Fock space* $\mathscr{F}_{\mathbf{s}}$ *such that*

$$\overline{B_q(\mu, \mathbf{s})} = B_q(\mu, \mathbf{s}) \qquad \text{and} \qquad B_q(\mu, \mathbf{s}) = |\mu, \mathbf{s}\rangle \mod q\mathscr{L}[\mathbf{s}].$$

*Proof.* Let $\mathbf{v} := (v_1, \ldots, v_l)$ be such that $0 < v_i - v_j < e$ if $j > i$. Set $\mathbf{m} = (s_1 - v_1, \ldots, s_l - v_l)$. We construct the elements $B(\lambda, \mathbf{s})$ by induction on $\ll_{\mathbf{m}}$. First, assume that $\mu \vdash_l n$ is maximal with respect to $\ll_{\mathbf{m}}$. As the matrix $A(\mathbf{s})$ is unitriangular by Proposition 6.6.10, we have $\overline{|\mu, \mathbf{s}\rangle} = |\mu, \mathbf{s}\rangle$ and we can set $B_q(\mu, \mathbf{s}) = |\mu, \mathbf{s}\rangle$. Let $\lambda \vdash_l n$. By induction, we assume that we have already constructed the canonical basis elements $B_q(\lambda[1], \mathbf{s})$, $\ldots$, $B_q(\lambda[i-1], \mathbf{s})$ for all the $l$-partitions $\lambda[j]$ satisfying $\lambda \ll_{\mathbf{m}} \lambda[j]$ and $\lambda \neq \lambda[j]$ with $j = 1, \ldots, i-1$. In addition, we assume that they satisfy

$$B_q(\lambda[j], \mathbf{s}) = |\lambda[j], \mathbf{s}\rangle + \sum_{\lambda[j] \ll_{\mathbf{m}} \nu, \lambda[j] \neq \nu} a_{\nu,j}(q) |\nu, \mathbf{s}\rangle$$

for certain coefficients $a_{\nu,j}(q)$ in $q\mathbb{Z}[q]$ with $j = 1, \ldots, i-1$. We set $\lambda[i] := \lambda$. We want to show that there exist $a_{m,i}(q)$ in $q\mathbb{Z}[q]$ with $1 \leqslant m < i$ such that

$$B_q(\lambda[i], \mathbf{s}) = |\lambda[i], \mathbf{s}\rangle + \sum_{1 \leqslant m < i} a_{m,i}(q) |\lambda[m], \mathbf{s}\rangle.$$

First, as $A(\mathbf{s})$ is unitriangular by Proposition 6.6.10 and by the above induction hypothesis, there exist coefficients $\beta_m(q)$ in $\mathbb{Z}[q, q^{-1}]$ with $1 \leqslant m < j$ such that $\overline{|\lambda[i], \mathbf{s}\rangle} - |\lambda[i], \mathbf{s}\rangle = \sum_{1 \leqslant m < i} \beta_m(q) B(\lambda[m], \mathbf{s})$. Applying the bar involution to this equation we obtain that $\beta_m(q^{-1}) = -\beta_m(q)$ for all $1 \leqslant m < i$. Hence, there exist polynomials $a_{m,i}(q) \in q\mathbb{Z}[q]$ with $1 \leqslant m < i$ such that $\beta_m(q) = a_{m,i}(q) - a_{m,i}(q^{-1})$ for all $1 \leqslant m < i$. It is readily checked that the element $B_q(\lambda[i], \mathbf{s}) = |\lambda[i], \mathbf{s}\rangle + \sum_{1 \leqslant m < i} a_{m,i}(q) B_q(\lambda[m], \mathbf{s})$ satisfies the desired property. So the induction ends. Assume that $B_q(\lambda, \mathbf{s})'$ is such that we have $\overline{B_q(\lambda, \mathbf{s})'} = B_q(\lambda, \mathbf{s})'$ and $B_q(\lambda, \mathbf{s})' \equiv |\lambda, \mathbf{s}\rangle \bmod q\mathscr{L}[\mathbf{s}]$ for $\lambda \vdash_l n$. Then the element $G := B_q(\lambda, \mathbf{s})' - B_q(\lambda, \mathbf{s})$ satisfies $\overline{G} = G$ and $G \in q\mathscr{L}[\mathbf{s}]$, which implies that $G$ is zero. The unicity follows. □

The proof of the above theorem also shows us the following result, which is stated in [291] for another partial order.

**Theorem 6.6.12.** *Assume that* $\mathbf{s} \in \mathbb{Z}^l$. *Then for all* $\mu \vdash_l n$, *we have*

$$B_q(\mu, \mathbf{s}) = |\mu, \mathbf{s}\rangle + \sum_{\mu \ll_{\mathbf{m}} \lambda, \mu \neq \lambda} b_{\lambda,\mu}(q) |\lambda, \mathbf{s}\rangle,$$

*where* $b_{\lambda,\mu}(q) \in q\mathbb{Z}[q]$ *if* $\mu \ll_{\mathbf{m}} \lambda$ *and* $\mu \neq \lambda$.

**6.6.13.** Let $\mathbf{s} \in \mathbb{Z}^l$. Consider the submodule $V(\Lambda_{\mathbf{s}})^{\mathbf{s}}$ of $\mathscr{F}_{\mathbf{s}}$ generated by the symbol $|\varnothing, \mathbf{s}\rangle$. By 6.1.16, it is equipped with a canonical basis

$$\{G_q(\mu, \mathbf{s}) \mid \mu \in \Phi_{(\mathbf{s},e)}\}$$

indexed by the set a Uglov $l$-partitions, which has been explicitly computed in 6.4.9 and 6.4.10. We prove the compatibility of this basis with the one just defined.

**Theorem 6.6.14 (Uglov).** *For all* $\mu \in \Phi_{(\mathbf{s},e)}$, *we have* $G_q(\mu, \mathbf{s}) = B_q(\mu, \mathbf{s})$.

*Proof.* By Proposition 6.6.3, the restriction of the bar involution to the irreducible highest weight module $V(\Lambda_{\mathbf{s}})^{\mathbf{s}}$ corresponds to the involution defined in 6.1.16. In addition, the elements $G_q(\mu, \mathbf{s})$ have the same congruence property with respect to $q\mathscr{L}[\mathbf{s}]$ as do the $B_q(\mu, \mathbf{s})$. By the uniqueness of these elements, we can conclude. □

Hence, for all $\mu \vdash_l n$, there exist elements $b_{\lambda,\mu}(q)$ with $\lambda \vdash_l n$ such that

$$B_q(\mu, \mathbf{s}) = \sum_{\lambda \vdash_l n} b_{\lambda,\mu}(q) |\lambda, \mathbf{s}\rangle$$

and we have

$$b_{\lambda,\mu}(q) = d_{\lambda,\mu}(q) \qquad \text{for all } \mu \in \Phi_{(\mathbf{s},e)}(n) \text{ and } \lambda \vdash_l n.$$

So we may now change the notation and define $d_{\lambda,\mu}(q) := b_{\lambda,\mu}(q)$ for all $\mu \vdash_l n$ and $\lambda \vdash_l n$ such that $\mu \notin \Phi_{(\mathbf{s},e)}(n)$. Consequently, the matrix of the canonical basis $D_{(\mathbf{s},e)}(q) = (d_{\lambda,\mu}(q))_{\lambda \vdash_l n, \mu \in \Phi_{(\mathbf{s},e)}(n)}$ for $V(\Lambda_{\mathbf{s}})^{\mathbf{s}}$ is a submatrix of the matrix of the canonical basis of the Fock space $\Delta_{(\mathbf{s},e)}(q) = (d_{\lambda,\mu}(q))_{\lambda \vdash_l n, \mu \vdash_l n}$. One can obtain the first from the latter by keeping only the rows indexed by the elements in $\Phi_{(\mathbf{s},e)}(n)$.

**Example 6.6.15.** Assume that $\mathbf{s} = (0,0)$, $e = 2$ and $l = 2$. Then the matrix of the canonical basis for the associated Fock space can be obtained first by computing the matrix of the involution $A(\mathbf{s})$ and then by following the proof of the existence of such basis (Theorem 6.6.11). We have

$$\Delta_{(\mathbf{s},e)}(q) = \begin{pmatrix} 1 & . & . & . & . \\ q & 1 & . & . & . \\ q & . & 1 & . & . \\ q^2 & q & q & 1 & . \\ . & . & . & . & 1 \end{pmatrix} \begin{matrix} (2,\varnothing) \\ (1.1,\varnothing) \\ (\varnothing,2) \\ (\varnothing,1.1) \\ (1,1) \end{matrix}$$

and the matrix of the canonical basis for $V(\Lambda_{\mathbf{s}})^{\mathbf{s}}$ is given by

$$D_{(\mathbf{s},e)}(q) = \begin{pmatrix} 1 & . \\ q & . \\ q & . \\ q^2 & . \\ . & 1 \end{pmatrix} \begin{matrix} (2,\varnothing) \\ (1.1,\varnothing) \\ (\varnothing,2) \\ (\varnothing,1.1) \\ (1,1) \end{matrix} .$$

Let $(v_1,v_2) \in \mathbb{Q}^2$ be such that $0 < v_2 - v_1 < 2$. Set, as usual, $\mathbf{m} = (s_1 - v_1, s_2 - v_2) = (-v_1, -v_2)$. Then, one can check that the minimal elements with respect to $\ll_{\mathbf{m}}$ appearing with non-zero coefficient in the each column of $D_{(\mathbf{s},e)}(q)$ are $(2,\varnothing)$ and $(1,1)$. This is in agreement with the above discussion and Theorem 6.6.12 because we have $\Phi_{(\mathbf{s},e)}(2) = \{(2,\varnothing),(1,1)\}$. Note that this set corresponds to the set of FLOTW 2-partitions because $(s_1,s_2) \in \mathscr{S}_2^2$ in this case; see Definition 5.7.8.

**6.6.16.** We have already seen how one can compute the matrices of the canonical basis for the irreducible highest weight modules. It is now natural to ask if this is possible to provide such an algorithm for computing the matrices of the canonical basis for the whole Fock space. Such an algorithm can be obtained by computing the matrix of the involution $A(\mathbf{s})$. The most direct way to do that is to use the relations (R1)–(R4). Unfortunately, it is in practice very difficult (and not efficient!) to use these relations. In particular, if $l \gg 0$ and if $s_1 \gg s_2 \gg \ldots \gg s_l$, the number of $q$-wedges involved in such a computation is too high to be done in practice. Yvonne [300] provides a faster algorithm for computing $A(\mathbf{s})$ which does not require the relations (R1)–(R4). The idea is to compute an invariant basis of $\Lambda^s$ with respect to the involution. Such a basis is obtained by studying the action of three algebras on the vacuum vector: the algebra $\mathscr{U}_q(\widehat{\mathfrak{sl}_e})$, the algebra $\mathscr{U}_q(\widehat{\mathfrak{sl}_l})$ and finally a Heisenberg algebra. This result gives us an invariant basis for each weight subspace of the Fock spaces and a transition matrix $T(q)$ between the standard basis and this invariant basis. The matrix of the involution of this weight space follows. This result generalizes earlier works of Leclerc and Thibon [209], corresponding to the case $l = 1$.

## 6.7 Computation of Canonical Basic Sets for Iwahori–Hecke Algebras

We now come back to the problem of parametrising the canonical basic sets for Iwahori–Hecke algebras of type $B$. First, we generalise our previous results on the canonical basic sets for Ariki–Koike algebras and then focus on Iwahori–Hecke algebras of type $B_n$.

**6.7.1.** As in Section 5.8, let $e \in \mathbb{Z}_{>1}$ and let $\mathbf{s} \in \mathbb{Z}^l$. Assume that $k$ is a field of characteristic 0. Let $\theta : A \to k$ be a specialisation such that $\theta(u) = \eta_e$, a primitive root of unity of order $e > 1$. Assume in addition that, for all $j = 1, \dots, l$, we have $\theta(V_j) = \eta_e^{s_j}$ for $\mathbf{s} = (s_1, \dots, s_l) \in \mathbb{Z}^l$. We get a specialised algebra

$$\mathbf{H}_{k,n} := \mathbf{H}_{k,n}(\eta_e; \eta_e^{s_1}, \dots, \eta_e^{s_l}).$$

Let $\mathbf{v} = (v_1, \dots, v_l) \in \mathbb{Q}^l$ be such that $i < j$ implies $0 < v_j - v_i < e$. As before, we set $m^j := s_j - v_j \in \mathbb{Q}$ for all $j = 1, \dots, l$, $\mathbf{m} := (m^1, \dots, m^l)$ and consider $D_{(\mathbf{s},e)} := D_\theta$ the associated decomposition matrix. We want to study the existence of a canonical basic set with respect to the **a**-function $\mathbf{a}^{(\mathbf{m},r)}$ (where $r$ is an arbitrary positive integer). In Theorem 5.8.2, we have already proved the existence of such a canonical basic set, but only in the case where $\mathbf{s} \in \mathscr{S}_e^l$. It is true that if $\mathbf{s} \in \mathbb{Z}^l$ then there exist $\mathbf{s}' \in \mathscr{S}_e^l$ and $w \in \widehat{\mathfrak{S}}_l$ such that $\mathbf{s} = w.\mathbf{s}'$. We then have $D_{(\mathbf{s},e)} = D_{(\mathbf{s}',e)}$, because the associated specialised algebras are isomorphic. However, the partial orders induced by the **a**-invariants $\mathbf{a}^{(\mathbf{m},r)}$ (where $\mathbf{m} = (s_1 - v_1, \dots, s_l - v_l)$) and $\mathbf{a}^{(\mathbf{m}',r)}$ (where $\mathbf{m}' = (s'_1 - v_1, \dots, s'_l - v_l)$) are really different in the two cases; see Example 6.7.3. Hence, the question of existence of a canonical basic set in the general setting is not trivially deduced from the case where $\mathbf{s} \in \mathscr{S}_e^l$.

**Theorem 6.7.2.** *Assume that we are in setting of 6.7.1. Then, with respect to the order on simple $\mathbf{H}_{K,n}$-modules induced by $\mathbf{a}^{(\mathbf{m},r)}$, the algebra $\mathbf{H}_{k,n}$ admits a canonical basic set $\mathscr{B}_\theta$ in the sense of Definition 5.5.19. We have $\mathscr{B}_\theta = \Phi_{(\mathbf{s},e)}(n)$, the set of Uglov l-partitions.*

*Proof.* By Ariki's theorem, the decomposition matrix $D_{(\mathbf{s},e)}$ is equal to the matrix $D_{(\mathbf{s},e)}(1)$. Now by Theorem 6.6.12 together with Proposition 6.6.14, for all $\mu \in \Phi_{(e,\mathbf{s})}(n)$, we have

$$G_q(\mu, \mathbf{s}) = |\mu, \mathbf{s}\rangle + \sum_{\mu \ll_{\mathbf{m}} \lambda, \mu \neq \lambda} d_{\lambda,\mu}(q) |\lambda, \mathbf{s}\rangle.$$

Now note that the rational numbers $v_j$ for $i < j$ satisfy $0 < v_j - v_i < e$. Thus, by Proposition 5.7.7, $\mu \ll_{\mathbf{m}} \lambda$ and $\mu \neq \lambda$ implies $\mathbf{a}^{(\mathbf{m},r)}(\lambda) > \mathbf{a}^{(\mathbf{m},r)}(\mu)$. We deduce that the decomposition matrix is unitriangular with respect to the **a**-invariant and that the associated canonical basic set is given by the set $\Phi_{(\mathbf{s},e)}(n)$. □

What the above result shows is that the simple modules of an Ariki–Koike algebra may be labelled by several sets depending on the choice of the weight sequence $\mathbf{m}$ we take. This is illustrated in the following example.

**Example 6.7.3.** Let $l = e = n = 2$ and $\mathbf{s} = (0,1)$; then the decomposition matrix of the associated Ariki–Koike algebra is

$$D_{((0,1),2)} := \begin{pmatrix} 1 & . \\ . & 1 \\ 1 & 1 \\ 1 & . \\ . & 1 \end{pmatrix} \begin{matrix} (2,\varnothing) \\ (\varnothing,2) \\ (1,1) \\ (1.1,\varnothing) \\ (\varnothing,1.1) \end{matrix} .$$

We set $\mathbf{v} = (0,1)$ and $r = 1$. Then $\mathbf{m} = (0,0)$. The $\mathbf{a}^{(\mathbf{m},r)}$-invariants of the 2-partitions associated with this choice of multicharge are given as follows:

$$\mathbf{a}^{(\mathbf{m},r)}(2,\varnothing) = \mathbf{a}^{(\mathbf{m},r)}(\varnothing,2) = 0, \quad \mathbf{a}^{(\mathbf{m},r)}(1,1) = 1,$$

$$\mathbf{a}^{(\mathbf{m},r)}(1.1,\varnothing) = \mathbf{a}^{(\mathbf{m},r)}(\varnothing,1.1) = 2.$$

We have an associated canonical basic set given by the set of Uglov 2-partitions:

$$\Phi_{((0,1),2)}(2) = \{(2,\varnothing),(\varnothing,2)\}.$$

Now assume that $\mathbf{s} = (2,1)$ and take as above $\mathbf{v} = (0,1)$, then the decomposition matrix $D_{((2,1),2)}$ is the same as above, but the $\mathbf{a}^{(\mathbf{m},r)}$-invariants are now

$$\mathbf{a}^{(\mathbf{m},r)}(2,\varnothing) = 0, \mathbf{a}^{(\mathbf{m},r)}(\varnothing,2) = 3, \quad \mathbf{a}^{(\mathbf{m},r)}(1,1) = 2,$$

$$\mathbf{a}^{(\mathbf{m},r)}(1.1,\varnothing) = 1, \quad \mathbf{a}^{(\mathbf{m},r)}(\varnothing,1.1) = 6.$$

Consequently, we also have an associated canonical basic set given by the set of Uglov 2-partitions:

$$\Phi_{((2,1),2)}(2) = \{(2,\varnothing),(1,1)\}.$$

Finally, assume that $\mathbf{s} = (0,3)$ and take as above $\mathbf{v} = (0,1)$, then the decomposition matrix $D_{((0,3),2)}$ is the same as above, but the $\mathbf{a}^{(\mathbf{m},r)}$-invariants are now

$$\mathbf{a}^{(\mathbf{m},r)}(2,\varnothing) = 3, \quad \mathbf{a}^{(\mathbf{m},r)}(\varnothing,2) = 0, \quad \mathbf{a}(1,1) = 2,$$

$$\mathbf{a}^{(\mathbf{m},r)}(1.1,\varnothing) = 6, \quad \mathbf{a}^{(\mathbf{m},r)}(\varnothing,1.1) = 1.$$

Consequently, we also have an associated canonical basic set given by the set of Uglov 2-partitions:

$$\Phi_{((0,3),2)}(2) = \{(\varnothing,2),(1,1)\}.$$

We have already seen that for all $w \in \widehat{\mathfrak{S}}_l$ we have $D_{(\mathbf{s},e)}(1) = D_{(w.\mathbf{s},e)}(1)$. As noted by Yvonne [298], the following example shows that, in general, this is not the case for the matrix $\Delta_{(\mathbf{s},e)}$.

**Example 6.7.4.** Assume that $e = 2$ and $l = 2$. The matrices of the canonical basis $\Delta_{((4,-3),2)}(q)$ and $\Delta_{((1,-0),2)}(q)$ for the Fock spaces $\mathscr{F}_{(4,-3)}$ and $\mathscr{F}_{(1,0)}$ can be computed using the algorithm described in [300]. The following tables are taken from [298]. We obtain:

$$\Delta_{((4,-3),2)}(q) = \begin{pmatrix} 1 & \cdot & \cdot & \cdot & \cdot & \cdot & \cdot & \cdot & \cdot & \cdot \\ q & 1 & \cdot & \cdot & \cdot & \cdot & \cdot & \cdot & \cdot & \cdot \\ q^2 & q & 1 & \cdot & \cdot & \cdot & \cdot & \cdot & \cdot & \cdot \\ \cdot & q & \cdot & 1 & \cdot & \cdot & \cdot & \cdot & \cdot & \cdot \\ \cdot & q^2 & q & q & 1 & \cdot & \cdot & \cdot & \cdot & \cdot \\ \cdot & \cdot & q & \cdot & \cdot & 1 & \cdot & \cdot & \cdot & \cdot \\ \cdot & \cdot & q^2 & \cdot & q & q & 1 & \cdot & \cdot & \cdot \\ \cdot & \cdot & \cdot & q & q^2 & \cdot & q & 1 & \cdot & \cdot \\ \cdot & \cdot & \cdot & \cdot & \cdot & \cdot & \cdot & \cdot & 1 & \cdot \\ \cdot & \cdot & \cdot & \cdot & \cdot & \cdot & \cdot & \cdot & \cdot & 1 \end{pmatrix} \begin{matrix} (\varnothing,3) \\ (\varnothing,2.1) \\ (1,2) \\ (\varnothing,1.1.1) \\ (1.1,1) \\ (3,\varnothing) \\ (2.1,\varnothing) \\ (1.1.1,\varnothing) \\ (2,1) \\ (1,1.1) \end{matrix},$$

$$\Delta_{((1,0),2)}(q) = \begin{pmatrix} 1 & \cdot & \cdot & \cdot & \cdot & \cdot & \cdot & \cdot & \cdot & \cdot \\ \cdot & 1 & \cdot & \cdot & \cdot & \cdot & \cdot & \cdot & \cdot & \cdot \\ \cdot & q & 1 & \cdot & \cdot & \cdot & \cdot & \cdot & \cdot & \cdot \\ q & q & \cdot & 1 & \cdot & \cdot & \cdot & \cdot & \cdot & \cdot \\ \cdot & q^2 & q & q & 1 & \cdot & \cdot & \cdot & \cdot & \cdot \\ \cdot & \cdot & q^2 & \cdot & q & 1 & \cdot & \cdot & \cdot & \cdot \\ q^2 & \cdot & \cdot & q & \cdot & \cdot & 1 & \cdot & \cdot & \cdot \\ \cdot & \cdot & \cdot & q^2 & q & \cdot & q & 1 & \cdot & \cdot \\ \cdot & \cdot & \cdot & \cdot & \cdot & \cdot & \cdot & \cdot & 1 & \cdot \\ \cdot & \cdot & \cdot & \cdot & \cdot & \cdot & \cdot & \cdot & \cdot & 1 \end{pmatrix} \begin{matrix} (\varnothing,3) \\ (\varnothing,2.1) \\ (1,2) \\ (\varnothing,1.1.1) \\ (1.1,1) \\ (3,\varnothing) \\ (2.1,\varnothing) \\ (1.1.1,\varnothing) \\ (2,1) \\ (1,1.1) \end{matrix}.$$

Note that $(1,0)$ and $(4,-3)$ are in the same orbit for the action of $\widehat{\mathfrak{S}}_2$ on $\mathbb{Z}^2$. However, $\Delta_{((1,0),2)}(q)$ is not equal to $\Delta_{((4,-3),2)}(q)$. In fact, there are not the same number of zeros in the matrices $\Delta_{((1,0),2)}(q)$ and $\Delta_{((4,-3),2)}(q)$ and thus we have $\Delta_{((1,0),2)}(1) \neq \Delta_{((4,-3),2)}(1)$ even after arbitrary permutation of rows/columns! This phenomenon was first observed by Yvonne. However, we have $D_{((1,0),2)}(1) = D_{((4,-3),2)}(1)$.

**6.7.5.** We finally give the explicit characterisation of the canonical basic sets in type $B_n$ in characteristic 0. We assume that $k$ is a field such that $\mathrm{char}(k) = 0$ and that we have a weight function $L : W \to \mathbb{Z}$ such that $L(t) = b > 0$ and $L(s_i) = a > 0$ for all $i = 1, \ldots, n-1$.

$B_n$

Let $v$ be an indeterminate and set $u := v^2$. We consider a specialisation $\theta : R[v, v^{-1}] \to k$ such that $\theta(u) = \eta \in k^\times$ (where $u = v^2$). We assume that $\eta^a$ is a primitive root of unity of order $e$, where $e \in \mathbb{Z}_{>1}$, and we set $\eta_e := \eta^a$. By Example 5.7.3, one may assume that there exists $d \in \mathbb{Z}$ such that $\eta^b = -\eta^{ad}$. Note that this condition ensures that $b/a$ is different from $d + p.e$ for all $p \in \mathbb{Z}$. Taking the notation of the previous section and using Example 5.7.3, we set $l = 2$, $r = a$ and we choose $d$ such that

$$(\ominus) \qquad d - e < b/a < d.$$

We set $\mathbf{s} = (d,e)$ and $\mathbf{v} = (v_1, v_2) = (d - b/a, e)$. Then note that we have $v_2 - v_1 = b/a - d + e$. Hence, we have $0 < v_2 - v_1 < e$. In addition, note that the cyclotomic Ariki–Koike algebra associated with $\mathbf{m} = (s_1 - v_1, s_2 - v_2) = (b/a, 0)$ and $r = a$ is the Iwahori–Hecke algebra of type $B_n$ associated with $L$. This implies that the $\mathbf{a}$-function $\mathbf{a}^{(\mathbf{m},r)}$ is the usual $\mathbf{a}$-function in type $B$ associated with $L$; see Example 5.5.14. Finally, the Ariki–Koike algebra $\mathbf{H}_{k,n}(\eta_e; \eta_e^{s_1}, \eta_e^{s_2})$ is isomorphic to the specialised Iwahori–Hecke algebra of type $B_n$, where $\eta_e := \eta^a$. We can then apply Theorem 6.7.2 and deduce the following result.

**Theorem 6.7.6 (Cf. [125, Theorem 5.4]).** *Assume that $L : W \to \mathbb{Z}$ is such that $L(t) = b > 0$ and $L(s_i) = a > 0$ for all $i = 1, \ldots, n-1$ and that $\mathrm{char}(k) = 0$. Then, in the setting of 6.7.5, the algebra $\mathbf{H}_{k,n}$ admits a canonical basic set $\mathscr{B}_\theta = \Phi_{((d,e),e)}(n)$, the set of Uglov $l$-partitions; see Definition 6.2.13.*

*Remark 6.7.7.* By Remark 6.2.15, we have $\Phi_{((d,e),e)}(n) = \Phi_{((d-e,0),0)}(n)$.

Let us now describe more precisely the canonical basic sets in some cases.

**6.7.8.** Let $W$ be a Weyl group of type $B_n$ and consider a weight function $L : W \to \mathbb{Z}$ such that $L(t) = b > 0$ and $L(s_i) = 1$ for all $i = 1, \ldots, n-1$.

$B_n$ 

We assume that $k$ is a field such that $\mathrm{char}(k) = 0$. We consider a specialisation $\theta : R[v, v^{-1}] \to k$ such that $\theta(u) = \eta_e$, a primitive root of unity of order $e > 1$. By Example 5.7.3, one can assume that there exist $d \in \mathbb{Z}$ such that $\eta_e^b = -\eta_e^d$. Thus $e$ is even. We have to take $d = b + e/2$ so that condition $(\ominus)$ in 6.7.5 is satisfied. We set $l = 2$, $r = 1$, $\mathbf{s} = (b + e/2, e) \in \mathscr{S}_e^2$, $\mathbf{v} = (e/2, e)$. The weight sequence that we have to consider is thus $\mathbf{m} = \mathbf{s} - \mathbf{v} = (b, 0)$. Note that the cyclotomic Ariki–Koike algebra associated with $\mathbf{m} = (b, 0)$ and $r = 1$ is the Iwahori–Hecke algebra of type $B_n$ associated with $L$ and thus that the $\mathbf{a}$-invariant $\mathbf{a}^{(\mathbf{m},r)}$ is the usual $\mathbf{a}$-invariant in type $B_n$ associated with $L$; see Example 5.5.14. Furthermore, the Ariki–Koike algebra $\mathbf{H}_{k,n}(\eta_e; \eta_e^{s_1}, \eta_e^{s_2})$ is isomorphic to the specialised Iwahori–Hecke algebra of type $B_n$.

**Theorem 6.7.9.** *In the setting of 6.7.8, the algebra $\mathbf{H}_{k,n}$ admits a canonical basic set $\mathscr{B}_\theta = \Phi_{((b+e/2,e),e)}(n)$.*

**6.7.10.** We still assume that $W$ is a Weyl group of type $B_n$. We now consider a weight function $L : W \to \mathbb{Z}$ such that $L(t) = 2c + 1$ (where $c \geqslant 0$) and $L(s_i) = 2$ for all $i = 1, \ldots, n-1$.

$B_n$ 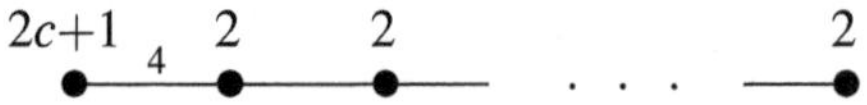


We assume that $k$ is a field such that $\text{char}(k) = 0$. We consider a specialisation $\theta : R[v, v^{-1}] \to k$ such that $\theta(u) = \eta_f$, a primitive root of unity of order $f > 1$. By Example 5.7.3, one can assume that there exist $d \in \mathbb{Z}$ such that $\eta_f^{2c+1} = -\eta_f^{2d}$. Thus, $f$ is even and we can set $f = 2e$ for an integer $e$. Then $\eta_e := \eta_f^2$ is a primitive root of unity of order $e$, where $e$ is odd. We assume that $e > 1$. We have to take $d = (2c+1+e)/2$ so that condition ($\ominus$) in 6.7.5 is satisfied. We set $l = 2$, $r = 2$, $\mathbf{s} = ((2c+1+e)/2, e) \in \mathscr{S}_e^2$, $\mathbf{v} = (e/2, e)$. The weight sequence that we have to consider is thus $\mathbf{m} = \mathbf{s} - \mathbf{v} = ((2c+1)/2, 0)$. Note that the cyclotomic Ariki–Koike algebra associated with $\mathbf{m} = ((2c+1)/2, 0)$ and $r = 2$ is the Iwahori–Hecke algebra of type $B_n$ associated with $L$ and thus that the $\mathbf{a}$-invariant $\mathbf{a}^{(\mathbf{m},r)}$ is the usual $\mathbf{a}$-invariant in type $B_n$ associated with $L$; see Example 5.5.14. In addition, the Ariki–Koike algebra $\mathbf{H}_{k,n}(\eta_e; \eta_e^{s_1}, \eta_e^{s_2})$ is isomorphic to the specialised Iwahori–Hecke algebra of type $B_n$, where $\eta_e := \eta^2$.

**Theorem 6.7.11.** *In the setting of 6.7.10, the algebra $\mathbf{H}_{k,n}$ admits a canonical basic set $\mathscr{B}_\theta = \Phi_{(((2c+1+e)/2, e), e)}(n)$.*

*Remark 6.7.12.* We are now also in position to solve the problem of computing the canonical basic set in type $B_n$ when $L : W \to \mathbb{Z}$ is such that $L(t) = 2$ and $L(s_i) = 1$ for all $i = 1, \ldots, n-1$. The case where $\theta : R[v, v^{-1}] \to k$ is such that $\theta(u) = -1$ was missing in Theorem 5.8.13. Assume that $k$ is a field of characteristic 0. We can here argue exactly as in 5.8.12 and we obtain that, in this situation, the Iwahori–Hecke algebra $\mathbf{H}_{k,n}$ admits a canonical basic set $\mathscr{B}_\theta = \Phi_{((3,2),2)}(n)$. When $\text{char}(k) = p$, this last set is also a canonical basic set, as already remarked in Example 4.4.15.

**6.7.13.** We finally consider the asymptotic case in type $B_n$. In this case, we can just apply Theorem 6.7.6 and obtain the description of the canonical basic sets in characteristic 0. The proof is due to Ariki [9]. However, we can have a stronger statement in this setting.

**Theorem 6.7.14.** *Let $k$ be any field. Let $L : W \to \mathbb{Z}$ be such that $L(t) = b > 0$ and $L(s_i) = a > 0$ for all $i = 1, \ldots, n-1$, and assume that $b > (n-1)a > 0$. Let $e$ be the order of $\theta(u)^a \in k^\times$. Assume that $e \in \mathbb{Z}_{>1}$ and that $\theta(u)^b = -\theta(u)^{ad}$ for some $d \in \{1, \ldots, e\}$ (see Example 5.7.3). Then, $\mathbf{H}_{k,n}$ admits a canonical basic set given by the set of Kleshchev bipartitions (see Example 6.2.16):*

$$\mathscr{B}_\theta = \overline{\Phi}_{((d,0) \bmod e, e)}(n).$$

*Proof.* First assume that $\text{char}(k) = 0$. We can then use Theorem 6.7.6: the algebra $\mathbf{H}_{k,n}$ admits a canonical basic set $\mathscr{B}_\theta = \Phi_{((d,e),e)(n)}$, where $d \in \mathbb{Z}$ is such that $d - e < b/a < d$. Now note that we have $d - e > n - 1 - e$; hence, by Example 6.2.16, the multicharge $(d, e)$ is very dominant. We thus have $\Phi_{((d,e),e)}(n) = \overline{\Phi}_{((d,0) \bmod e, e)}(n)$, the set of Kleshchev bipartitions. This completes the proof in the case where $\text{char}(k) = 0$.

Now assume that $\text{char}(k) = p > 0$. By Example 3.2.10, we already know that a canonical basic set for $\mathbf{H}_{k,n}$ exists. So we can try to use Lemma 3.6.7. Now, since $e > 1$ is the order of $\theta(u)$ in $k^\times$, we have $\Phi_e(\theta(u)) = 0$, where $\Phi_e(u) \in \mathbb{Z}[u]$ denotes the $e$th cyclomotic polynomial. Then the specialisation $\theta : R[v, v^{-1}] \to k$ lies above a suitable specialisation $\theta_0 : R[v, v^{-1}] \to k_0$, where $k_0$ has characteristic 0 and $\Phi_e(\theta_0(u)) = 0$; that is, $\theta_0(u) \in k_0^\times$ is a primitive root of unity of order $e$. Furthermore, Theorem 6.2.22 shows that $|\text{Irr}(\mathbf{H}_{k,n})| = |\text{Irr}(\mathbf{H}_{k_0,n})|$. Thus, all the assumptions of Lemma 3.6.7 are satisfied. So we can deduce that $\overline{\Phi}_{((d,0) \bmod e,e)}(n)$ also is a canonical basic set for $\mathbf{H}_{k,n}$, as claimed. □

## 6.8 Recent Developments and Conjectures

In this final section we discuss several new developments connected to the topics of this book.

**6.8.1.** Both finite Coxeter groups and the groups of type $G(l, p, n)$ are particular cases of finite complex reflection groups. Let $\mathbb{K}$ be a finite abelian extension of $\mathbb{Q}$ and let $V$ be a finite-dimensional $\mathbb{K}$-vector space. We say that $s \in \text{GL}(V)$ is a *pseudo-reflection* if it is a non-trivial element which acts trivially on a hyperplane. A finite subgroup $W$ of $\text{GL}(V)$ generated by pseudo-reflections is called a *complex reflection group*. As noted before, a classification for the irreducible complex reflection groups (i.e., complex reflection groups which act irreductibly on $V$) has been obtained by Shephard and Todd [278]. They are given by:

- the family of groups $G(l, p, n)$ (with $l \geqslant 1$, $p|l$ and $n \geqslant 1$) that we have already introduced in 5.1.9,
- a list of 34 exceptional groups.

Complex reflection groups generalise both the main groups that we are interesting in in this book: the finite Coxeter groups and the groups of type $G(l, 1, n)$. One can also associate with these groups some Hecke algebras. These were introduced and studied by Broué, Malle, Michel, Rouquier; see [35], [36], [238], [241]. It is natural to consider the problem of existence of canonical basic sets for the wider class of complex reflection groups and their Hecke algebras.

**6.8.2.** Let $R$ be an integral domain, $q$ be a unit in $R$ and let $x = (x_1, \ldots, x_d)$ be a sequence of elements in $R$. Following Ariki [6], the cyclotomic Hecke algebra of type $G(r, p, n)$ over $R$ with parameters $q$ and $x$ has a presentation by generators $a_0, \ldots, a_n$ and defining relations

$$\begin{aligned}
&(a_0 - x_1)\dots(a_0 - x_d) = 0,\\
&(a_i - q)(a_i + 1) = 0,\ 1 \le i \le n,\\
&a_1 a_3 a_1 = a_3 a_1 a_3,\\
&a_i a_{i+1} a_i = a_{i+1} a_i a_{i+1},\ 2 \le i \le n-1,\\
&(a_1 a_2 a_3)^2 = (a_3 a_1 a_2)^2,\\
&a_1 a_i = a_i a_1,\ 4 \le i \le n,\\
&a_i a_j = a_j a_i,\ 2 \le i < j \le n,\ j \ge i+2,\\
&a_0 a_1 a_2 = (q^{-1} a_1 a_2)^{2-r} a_2 a_0 a_1 + (q-1) \sum_{i=1}^{r-2} (q^{-1} a_1 a_2)^{1-k} a_0 a_1 = a_1 a_2 a_0,\\
&a_0 a_i = a_i a_0,\ 3 \le i \le n.
\end{aligned}$$

This class of algebras has been defined in [6]. Several aspects of the representation theory of these algebras can be deduced using Clifford theory as in type $D_n$ (see 5.1.9). In the semisimple case, the simple modules are in one-to-one correspondence with the simple modules of the group algebra. These modules have been constructed in [6]. The modular case has been studied by Hu [165], [166], [167] and Genet and Jacon [136]. Morita equivalences for this class of algebras have been obtained by Hu and Mathas [168], in the same spirit as Theorem 5.4.3. Finally, it is possible to prove the existence of basic sets and the unitriangularity of the decomposition matrices using Clifford theory and the existence of basic sets for Ariki–Koike algebras; see Genet and Jacon [136]. The parametrisation of the basic sets is derived from the definition of FLOTW $l$-partitions.

**6.8.3.** Let us come back to the general case of a complex reflection group $W$. Let $\mathscr{S}$ be the set of pseudo-reflections in $W$ and let $c$ be a map from $\mathscr{S}$ to $\mathbb{C}$ which is constant on conjugacy classes. Following Etingof and Ginzburg [77], the rational Cherednik algebra $\mathscr{H}_{t,c}$ attached to $W$ with parameters $c$ and $t \in \mathbb{C}$ is defined as the quotient of the smash product of $\mathbb{C}W$ and the tensor algebra of $V \oplus V^*$ by the relations

$$[x,x'] = 0, \quad [y,y'] = 0, \quad [y,x] = t\langle y,x\rangle - \sum_{s\in\mathscr{S}} c_s \langle \alpha_s, y\rangle \langle x, \alpha_s^\vee\rangle s$$

for all $x, x' \in V^*$, $y, y' \in V$. Here, $\langle y,x\rangle$ is the canonical pairing between $V$ and $V^*$, the element $\alpha_s$ is a generator of $\mathrm{Im}(s|_{V^*} - 1)$ and $\alpha_s^\vee$ is a generator of $\mathrm{Im}(s|_V - 1)$ such that $\langle \alpha_s, \alpha_s^\vee\rangle = 2$. The representation theory of these algebras turns out to be very different if $t = 0$ or $t \ne 0$, but, quite remarkably, both are closely related to the modular representation theory of Hecke algebras. First, let us consider the case $t = 0$. Then the representations of the associated algebras have been studied by Gordon [139] and a conjecture by Gordon and Martino [141] gives connections between these representations when $W$ is a Weyl group and the theory of two-sided cells for finite Weyl groups. A further conjecture by the same authors gives a relation between blocks of the Cherednik algebras and Rouquier families for cyclotomic Hecke algebras. This conjecture has been proved for $W = G(l,1,n)$ (see Martino [242]) and $W = G(l,p,n)$ (see Bellamy [16] and Martino [243]).

**6.8.4.** Assume that $t \neq 0$. Then, by Ginzburg et al. [137], there is a functor, called the KZ functor, from a certain category of $\mathcal{H}_{t,c}$-modules, the category $\mathcal{O}$, to the category of finitely generated modules over the Hecke algebra of $W$. To each irreducible representation of $W$, one can attach a certain "standard module" in this category $\mathcal{O}$ with a simple head. When $W = G(l,1,n)$, we denote these simple modules by $L(\lambda)$ with $\lambda \vdash_l n$. In the associated Grothendieck group, the image of $L(\lambda)$ through the KZ functor corresponds to the class of the associated Specht module (see Shan [276, Lemma 3.1]). This functor also provides a proof for the existence of canonical basic sets for Ariki–Koike algebras in characteristic 0, for certain choices of $\mathbf{m}$ as in Remark 5.7.4; see Ginzburg et al. [137] and Chlouveraki, Gordon and Griffeth [49]. In addition, one can also define some parabolic restriction and induction functors for these algebras; see Bezrukavnikov and Etingof [19]. Using this, one can construct a crystal structure on the set of simple modules $L(\lambda)$ which turns out to be isomorphic to the one of the Fock space; see Shan [276, Theorem 6.3]. Furthermore, connections with the representation theory of $q$-Schur algebras, and even category $\mathcal{O}$, have been given by Rouquier [274]. Finally, conjectures by Yvonne [299] and Rouquier [274] assert that the decomposition matrices for these $q$-Schur algebras should be given by the canonical basis of Fock spaces for a particular choice of the multicharge.

**6.8.5.** Ariki's theorem gives us a deep connection between the representation theory of Ariki–Koike algebras and the theory of canonical bases for quantum groups by saying that the decomposition matrices are equal to the evaluation at $q = 1$ of the matrices for the canonical bases of irreducible $\mathcal{U}_q(\widehat{\mathfrak{sl}_e})$-modules $D_{(\mathbf{s},e)}(1)$. It is natural to ask if the matrices $D_{(\mathbf{s},e)}(q)$ themselves have an interpretation in the representation theory of Hecke algebras. An answer for this problem has been recently given by Brundan, Kleshchev and Wang [39], [40], [41]. We refer to Kleshchev [203] for a complete survey on these results.

The main point is that the Ariki–Koike algebras are isomorphic to (sign-modified) "cyclotomic Khovanov–Lauda–Rouquier" algebra defined in [198], [199], [275]. As these algebras are naturally $\mathbb{Z}$-graded, this shows the existence of a natural grading on Ariki–Koike algebras. Such results have been previously predicted by Turner [290] and Rouquier [273]. It is then possible to show that the Specht modules in Remark 5.3.12 are graded with respect to this gradation. As a consequence, one can define some "$q$-decomposition numbers" for these Specht modules, which are defined with respect to this grading and which correpond to the usual decomposition numbers after evaluation at $q = 1$.

The main result of Brundan and Kleshchev [40] shows that the graded decomposition numbers correspond precisely to the coefficient of the matrix of the canonical basis for the $\mathcal{U}_q(\widehat{\mathfrak{sl}_e})$-modules for a particular choice of the multicharge. This choice corresponds to the case where the Fock space can be seen as a tensor product of Fock spaces with level 1. It thus gives a generalisation and a $q$-analogue of Ariki's theorem. An interpretation for general multicharge is still missing.

In another direction, an interpretation of the "$q$-decomposition numbers" has been given by Shan [277] when $l = 1$ in terms of the Jantzen filtrations. This gives a proof of a conjecture by Lascoux, Leclerc and Thibon [208].

# Chapter 7
# Decomposition Numbers for Exceptional Types

One of the declared aims of this book has been to determine the decomposition matrices and canonical basic sets associated with (principal) specialisations of Iwahori–Hecke algebras. In this chapter we present these results for algebras of exceptional type. These are the culmination of work by various authors over a number of years:

| | |
|---|---|
| $F_4$ (equal parameters) | Geck and Lux [126]; |
| $E_6$ | Geck [94]; |
| $F_4$ (some unequal parameters) | Bremke [30]; |
| $E_7$ | Geck [96], Müller [253]; |
| $E_8$ (partial results) | Müller [253]; |
| $I_2(m)$, $H_3$, $H_4$ | Müller [254]; |
| $F_4$ (any parameters) | McDonough and Pallikaros [249], [250]; |
| $E_8$ (completed) | Geck and Müller [129]. |

In most cases, the proofs rely on computer calculations based on the ideas in Parker's MEATAXE [263], [270] and the GAP part of the CHEVIE system; see [118], [252].

We begin by explaining some fundamental computational methods in Section 7.1. We will then mainly be interested in the case where we consider a $\Phi_e$-modular specialisation as in Example 3.1.15(b); that is, $A$ is the ring of Laurent polynomials in one variable and this variable is specialised to a root of unity in $\mathbb{C}$.

In Section 7.2, we briefly discuss the case where $W$ is of dihedral type; here, everything is easily computed "by hand". In Section 7.3, we consider various choices of parameters for $W$ of type $F_4$. Section 7.4 contains tables for $W$ of type $H_3$, $H_4$, $E_6$, $E_7$ and $E_8$. The decomposition matrices for $W$ of type $E_7$ and $E_8$ appear here for the first time in print. In order to keep the volume of data to be printed within a reasonable size, we shall also rely on the information contained in the tables in Appendix F of [132], concerning the distribution of the irreducible representations into $\Phi_e$-blocks and the Brauer trees of $\Phi_e$-blocks of defect 1. So, here, we will only consider the decomposition matrices of blocks of defect at least 2.

M. Geck, N. Jacon, *Representations of Hecke Algebras at Roots of Unity*, Algebra and Applications 15, DOI 10.1007/978-0-85729-716-7_7,

## 7.1 Algorithmic Methods

Let $W$ be a finite Coxeter group, $\mathbf{H} = \mathbf{H}_A(W,S,L)$ a corresponding generic Iwahori–Hecke algebra and $\theta\colon A \to k$ be a specialisation, where $k$ is a field. Assume that $R$ is $L_0$-good so that we have a decomposition map $d_\theta\colon R_0(\mathbf{H}_K) \to R_0(\mathbf{H}_k)$. As in Section 3.2, let us write

$$d_\theta([E_\varepsilon^\lambda]) = \sum_{M\in\mathrm{Irr}(\mathbf{H}_k)} d_{\lambda,M}\,[M] \qquad (\lambda \in \Lambda).$$

As explained in Remark 3.2.3, if we can manage to compute the decomposition numbers $d_{\lambda,M}$, then we can also determine a corresponding canonical basic set as in Definition 3.2.1. In principle, the numbers $d_{\lambda,M}$ are computed as follows. By Lemma 3.1.13, there exists an $\mathbf{H}$-module $N^\lambda$ which is finitely generated and free over $A$, such that $d_\theta([E_\varepsilon^\lambda]) = [k \otimes_A N^\lambda]$. Hence, $d_{\lambda,M}$ is the multiplicity of $M$ in a composition series of $k\otimes_A N^\lambda$. In order to turn this theoretical result into some efficient practical method, the first thing we need are realisations of the modules $N^\lambda$ in terms of explicitly given representing matrices on the generators $T_s$ ($s \in S$). Suitable candidates for such realisations are provided by $W$-*graph* representations; see Definition 1.4.11. For $W$ of exceptional type, such $W$-graphs are explicitly known by Examples 1.4.12 and 1.4.13. (They are available in electronic form through Michel's development version of the CHEVIE system; see [118], [252].) By specialisation, we obtain corresponding matrix representations of $\mathbf{H}_k$. Our next task is to decompose these representations into irreducible ones.

**7.1.1.** We recall the basic principles of Parker's MEATAXE algorithm [263]. This is a general computational method for deciding whether a finite-dimensional representation of an associative algebra over a field is irreducible, and for finding an explicit non-trivial submodule when it is not.

So let $\underline{H}$ be a finite-dimensional associative algebra over a field $k$. Let $\rho\colon \underline{H} \to M_d(k)$ be a matrix representation of $\underline{H}$. Let $\underline{H}^\circ$ be the opposite algebra of $\underline{H}$. Then we also have a representation $\rho^\circ\colon \underline{H}^\circ \to M_d(k)$ such that $a \mapsto \rho(a)^{\mathrm{tr}}$ ($a \in \underline{H}$). Thus, we have an $\underline{H}$-module $V = k^d$ (via $\rho$) and an $\underline{H}^\circ$-module $V^\circ = k^d$ (via $\rho^\circ$). Concretely, if $\underline{H}$ is generated (as an algebra) by, say $h_1,\ldots,h_n \in \underline{H}$, then $\rho$ and $\rho^\circ$ are of course determined by specifying the representing matrices for $h_1,\ldots,h_n$.

For any matrix $B \in M_d(k)$, we denote by $\ker(B) := \{v \in k^d \mid B.v = 0\}$ the "nullspace" of $B$. For any $0 \neq v \in k^d$, we denote by $\langle v\rangle_{\underline{H}} \subseteq V$ the smallest $\underline{H}$-submodule of $V$ containing $v$. Similarly, $\langle v\rangle_{\underline{H}^\circ} \subseteq V^\circ$ will denote the smallest $\underline{H}^\circ$-submodule of $V^\circ$ containing $v$. Given algebra generators for $\underline{H}$, these subspaces are effectively computed using the "spinning algorithm"; see [234, §1.3] for technical details. Now the following result is the theoretical basis of the MEATAXE:

**Norton's Irreducibility Criterion.** Let $h \in \underline{H}$ be fixed. Then $V$ is irreducible if $k^d = \langle v\rangle_{\underline{H}} = \langle v'\rangle_{\underline{H}^\circ}$ for every $0 \neq v \in \ker(\rho(h))$ and every $0 \neq v' \in \ker(\rho^\circ(h))$.

(See [234, 1.3.3] for a proof; note that the converse is trivially true.) Algorithmically, the critical point is the choice of the element $h \in \underline{H}$ to be used in the above criterion. The ideal case is when one succeeds in finding an element $h \in \underline{H}$ such that $\dim_k \ker(\rho(h)) = 1$. For, in this case, it will be enough to determine only two subspaces under the "spinning algorithm" (corresponding to non-zero vectors in the one-dimensional nullspaces) in order to decide whether $\rho$ is irreducible or not. However, in general, this will not be possible, and then one has to run through all vectors in the nullspaces. Hence, this only provides a practical method when $k$ is a **finite field**. By testing various "standard words" in the generators of $\underline{H}$, the MEATAXE algorithm tries to find an element $h \in \underline{H}$ such that $\ker(\rho(h))$ has, at least, small dimension; some statistics about this can already be found in [263].

In any case, this algorithm has turned out to be extremely efficient in applications. It runs for representations over a finite field $k$ of dimension up to a few thousand; see the remarks at the end of [234, §1.3] for further details and references. A sophisticated computer implementation (together with detailed descriptions) is due to Ringe [270]. In this system, the CHOP function takes as input the representing matrices $\rho(h_1),\ldots,\rho(h_n)$ (where $h_1,\ldots,h_n$ are algebra generators) and combines the above ideas into a recursive procedure to determine a composition series for $\rho$. The output consists of explicit lists of matrices specifying the action of the generators $h_1,\ldots,h_n$ on the various composition factors. It is this function CHOP that we use in our applications to Iwahori–Hecke algebras.

**7.1.2.** We return to the problem of computing decomposition numbers for a generic Iwahori–Hecke algebra $\mathbf{H} = \mathbf{H}_A(W,S,L)$. Let us now place ourselves in the setting of 3.3.1 where we have factorisations of the elements $\mathbf{c}_\lambda$ in terms of a certain subset $\mathscr{P} \subseteq A$. Furthermore, let $\leqslant$ be a monomial order on $\Gamma$ such that $L(s) \geqslant 0$ for all $s \in S$ and assume that $\mathbf{H}$ admits a cellular basis $\{\mathbf{C}^\lambda_{\mathfrak{st}}\}$ as in Theorem 2.6.12, where (♠) and (♣) hold. Let $\Phi \in \mathscr{P}$ and consider the corresponding principal specialisation $\theta_\Phi \colon A \to k_\Phi$, where $k_\Phi$ is the field of fractions of $A/(\Phi)$. Suppose that we can find a specialisation $\theta\colon A \to k$ such that

$$\boxed{\theta(\Phi) = 0, \;\; \Phi \text{ is strongly } \theta\text{-isolated, and } k \text{ is a finite field.}}$$

Using the methods described in 7.1.1, we can find the irreducible representations of $\mathbf{H}_k$ and the corresponding decomposition numbers. The remaining task, then, is to see how results about decomposition numbers related to specialisations into finite fields can be "lifted" to principal specialisations, whose target fields will be of characteristic 0, in most cases. As in 3.7.4, we have

$$\begin{aligned} \operatorname{Irr}(\mathbf{H}_{k_\Phi}) &= \{L^\mu_\Phi \mid \mu \in \Lambda^\circ_\Phi\}, \quad \text{where } \Lambda^\circ_\Phi \subseteq \Lambda \text{ is a canonical basic set for } \mathbf{H}_{k_\Phi}, \\ \operatorname{Irr}(\mathbf{H}_k) &= \{L^\mu_k \mid \mu \in \Lambda^\circ_k\}, \quad \text{where } \Lambda^\circ_k \subseteq \Lambda \text{ is a canonical basic set for } \mathbf{H}_k. \end{aligned}$$

Let $D_\Phi = \big(d^\Phi_{\lambda,\mu}\big)$ and $D_\theta = \big(d^\theta_{\lambda,\mu}\big)$ be the decomposition matrices associated with the specialisations $\theta_\Phi$ and $\theta$ respectively. These are given by the equations

$$d_\Phi([E^\lambda_\varepsilon]) = \sum_{\mu \in \Lambda^\circ_\Phi} d^\Phi_{\lambda,\mu}\,[L^\mu_\Phi] \qquad \text{and} \qquad d_\theta([E^\lambda_\varepsilon]) = \sum_{\mu \in \Lambda^\circ_k} d^\theta_{\lambda,\mu}\,[L^\mu_k].$$

By Example 3.6.4 and Theorem 3.7.16, we have

$$\boxed{\Lambda_\Phi^\circ = \Lambda_k^\circ \text{ and } D_\theta = D_\Phi.D_\theta^\Phi, \text{ where } D_\theta^\Phi \text{ is an "adjustment matrix".}}$$

Now if the general version of James' conjecture was known to hold, then we could conclude that $D_\Phi^\theta$ is the identity matrix and so $D_\Phi = D_\theta$. At the present state of knowledge, we need to use some ad-hoc arguments to obtain this conclusion. We shall illustrate some of these arguments in the examples below, but refer the reader to [96], [129], [253] for further details. The final result is as follows.

**Theorem 7.1.3 (Cf. Geck and Müller [129]).** *Let $(W,S)$ be of exceptional type $H_3$, $H_4$, $F_4$, $E_6$, $E_7$ or $E_8$; let $\Gamma = \mathbb{Z}$ and $L(s) = 1$ for all $s \in S$ (equal parameter case). Assume we are in the setting of Example 3.3.15, where we consider a $\Phi_e$-modular specialisation $\theta_e \colon A \to \mathbb{K}$ $(e \geqslant 2)$. Thus, we have a natural parametrisation*

$$\operatorname{Irr}(\mathbf{H}_{(e)}) = \{L_{(e)}^\mu \mid \mu \in \Lambda_{(e)}^\circ\}, \qquad \textit{where } \Lambda_{(e)}^\circ \subseteq \Lambda \textit{ is a canonical basic set.}$$

*For blocks of defect at least 2, the sets $\Lambda_{(e)}$ and the dimensions of $L_{(e)}^\mu$ for $\mu \in \Lambda_{(e)}$ are given in Table 7.1. (The remaining blocks are covered by the tables in [132, Appendix F], as discussed in Example 3.3.15(b).)*

The basic strategy of the proof is outlined as follows.

**7.1.4.** Assume we are in the general setting of 7.1.2. Now let $\Gamma = \mathbb{Z}$ and $R \subseteq \mathbb{C}$ be the subring generated by $\mathbb{Z}_W$ and all fractions $f_\lambda^{-1}$, where $\lambda \in \Lambda$. Then $R$ is $L$-good and $A = R[v, v^{-1}]$ is the ring of Laurent polynomials in $v = \varepsilon$. Let $e \geqslant 2$ and consider the $\Phi_e$-modular specialisation $\theta_e \colon A \to \mathbb{C}$, where $\theta_e(v) = \exp(\pi\sqrt{-1}/e)$. Let $D_{(e)}$ be the corresponding decomposition matrix and write

$$D_{(e)} = \left(d_{\lambda,\mu}^{(e)}\right)_{\lambda \in \Lambda, \mu \in \Lambda_{(e)}^\circ}.$$

Our task is to find $\Lambda_{(e)}^\circ$ and the entries of $D_{(e)}$. Choosing a suitable prime number $\ell > 0$, we can certainly find a specialisation $\theta \colon A \to \mathbb{F}_\ell$ such that $\theta(\Phi_{2e}) = 0$, where $\Phi_{2e} \in \mathbb{Z}[v]$ denotes the $2e$th cyclotomic polynomial. For example, let $\ell$ be any prime such that $\ell \notin R^\times$ and $2e$ divides $\ell - 1$; then define $\theta$ by sending $v$ to an element of order $2e$ in $\mathbb{F}_\ell^\times$. Choosing $\ell$ large enough, we can also assume that $\Phi_e$ is strongly $\theta$-isolated; see Lemma 3.7.12. (For example, if $L(s) = 1$ for all $s \in S$, then this is the case if $\ell$ does not divide $|W|$.) Thus, by Theorem 3.7.16, we have

$$\Lambda_{(e)}^\circ = \Lambda_k^\circ \qquad \text{and} \qquad D_\theta = D_{(e)}.D_\theta^{(e)}.$$

In particular, this shows that we obtain $\Lambda_{(e)}^\circ$ once we have determined $D_\theta$ and $\Lambda_k^\circ$. To solve the latter problem, we use the methods described in 7.1.1. The sets $\Lambda_{(e)}^\circ = \Lambda_\theta$

**Table 7.1** The canonical basic sets $\Lambda^{\circ}_{\zeta_e}$ for type $H_3, H_4, F_4, E_6, E_7, E_8$

| | | |
|---|---|---|
| $H_3$ | $e=2$ | $\{1_r(1), 3'_s(2), \overline{3}'_s(2)\}$ |
| $H_4$ | $e=2$ | $\{1_r(1), 4_t(4), \overline{4}_t(4), 9_s(5), \overline{9}_s(5), 25_r(17)\}$ |
| | $e=3$ | $\{1_r(1), 16_{rr}(16), 25_r(8), 10_r(1), 40_r(16)\}$, |
| | | $\{4_t(4), \overline{4}_t(4), 16_r(8), 16_t(4), \overline{16}_t(4)\}$ |
| | $e=4$ | $\{1_r(1), 9_s(8), \overline{9}_s(8), 25_r(8), 24_t(8), \overline{24}_t(8), 25'_r(1)\}$ |
| | $e=5$ | $\{1_r(1), 9_s(9), 16_{rr}(6), 8_r(1), \overline{24}_s(9)\}$, |
| | | $\{4_t(4), 16_r(16), 36_{rr}(16), \overline{24}_t(4), 48_{rr}(16)\}$ |
| | $e=6$ | $\{1_r(1), 4_t(4), \overline{4}_t(4), 25_r(16), 18_r(1), 24_s(4), \overline{24}_s(4)\}$ |
| | $e=10$ | $\{1_r(1), \overline{4}_t(4), \overline{9}_s(9), 36_{rr}(22), \overline{24}_s(1), \overline{30}_s(4), 40_r(9)\}$ |
| $F_4$ | $e=2$ | $\{1_1(1), 2_1(2), 2_3(2), 9_1(5)\}$, $\{4_2(4)\}$ |
| | $e=3$ | $\{1_1(1), 2_1(1), 2_3(1), 4_1(1)\}$, $\{4_2(4), 8_1(4), 8_3(4), 16_1(4)\}$ |
| | $e=4$ | $\{1_1(1), 4_2(4), 9_1(4), 6_1(1), 12_1(4)\}$ |
| | $e=6$ | $\{1_1(1), 2_1(2), 2_3(2), 8_1(5), 8_3(5)\}$ |
| $E_6$ | $e=2$ | $\{1_p(1), 6_p(6), 20_p(14), 15_q(14), 30_p(10), 60_p(46\}$ |
| | $e=3$ | $\{1_p(1), 6_p(5), 20_p(14), 15_p(10), 15_q(1), 30_p(25), 64_p(10), 60_p(5)$, |
| | | $60_s(14), 80_s(25)\}$ |
| | $e=4$ | $\{1_p(1), 6_p(6), 15_p(15), 15_q(8), 81_p(60), 10_s(1), 80_s(6), 90_s(15)\}$ |
| | $e=6$ | $\{1_p(1), 6_p(6), 20_p(13), 15_p(14), 30_p(11), 60_p(32), 24_p(11), 60_s(14)$, |
| | | $80_s(13), 60'_p(1), 30'_p(6)\}$ |
| $E_7$ | $e=2$ | $\{1_a(1), 7'_a(6), 27_a(14), 35_b(14), 105'_a(78), 189'_b(56)$, |
| | | $315'_a(126)\}, \{56'_a(56), 120_a(64), 280_b(216)\}$ |
| | $e=3$ | $\{1_a(1), 21_a(21), 35_b(34), 120_a(98), 105_b(7), 168_a(35), 210_a(91), 280_b(14)$, |
| | | $210_b(49), 420_a(196)\}$, |
| | | $\{7'_a(7), 21'_b(14), 56'_a(49), 15'_a(1), 105'_a(35), 70'_a(21), 280'_a(196), 336'_a(91)$, |
| | | $512_a(98), 84'_a(34)\}$ |
| | $e=4$ | $\{1_a(1), 56'_a(56), 105_b(48), 210_a(154), 189_a(35), 405_a(147)$, |
| | | $70_a(21), 315_a(120)\}$, |
| | | $\{7'_a(7), 15'_a(8), 105'_a(105), 189'_c(84), 280_b(168), 378'_a(21), 210'_b(27)\}$, |
| | | $\{27_a(27), 21_a(21), 35_b(8), 216'_a(168), 210_b(7), 105_c(84), 378_a(105)\}$, |
| | | $\{21'_b(21), 120_a(120), 189'_b(48), 35'_a(35), 70'_a(1), 315'_a(147)$, |
| | | $336_a(154), 405'_a(56)\}$ |
| | $e=6$ | $\{1_a(1), 7'_a(7), 21'_b(13), 21_a(21), 35_b(27), 15'_a(14), 105'_a(77), 105_b(43)$, |
| | | $168_a(43), 210_a(92), 70'_a(42), 280'_a(90), 315'_a(13), 84_a(14), 210_b(27)$, |
| | | $420_a(92), 210'_b(1), 420'_a(77), 280_a(21), 315_a(7)\}$ |

**Table 7.1** The canonical basic sets $\Lambda^{\circ}_{\zeta_e}$ for type $H_3$, $H_4$, $F_4$, $E_6$, $E_7$, $E_8$ (cont'd)

| | | |
|---|---|---|
| $E_8$ | $e=2$ | $\{1_x(1), 8_z(8), 35_x(27), 84_x(48), 50_x(42), 210_x(202), 560_z(246), 700_x(126),$<br>$1400_z(792), 1050_x(651), 1400_x(378), 4200_x(1863)\}$,<br>$\{112_z(112), 160_z(160), 400_z(288), 1344_x(1184), 2240_x(1056), 3360_z(2016)\}$ |
| | $e=3$ | $\{1_x(1), 35_x(35), 28_x(28), 84_x(48), 50_x(1), 210_x(147), 300_x(70), 700_x(518),$<br>$1344_x(497), 175_x(28), 350_x(322), 1050_x(35), 1400_x(1225), 2240_x(322),$<br>$4096_z(1036), 4200_x(147), 700_{xx}(48), 3200_x(497), 4200_y(518), 4480_y(1225)\}$,<br>$\{8_z(8), 112_z(104), 160_z(56), 560_z(384), 400_z(8), 448_z(56), 1400_z(848),$<br>$840_z(448), 1400_{zz}(104), 4096_x(1896), 4200_z(384), 5600_z(1896), 7168_w(848)\}$ |
| | $e=4$ | $\{1_x(1), 35_x(34), 112_z(77), 50_x(16), 210_x(176), 567_x(280), 400_z(96), 175_x(1),$<br>$350_x(70), 1050_x(336), 1575_x(946), 525_x(168), 3360_z(1654), 2800_z(1302),$<br>$2835_x(34), 6075_x(280), 3150_y(77), 4480_y(176), 5670_y(946)\}$,<br>$\{8_z(8), 560_z(560), 1344_x(784), 840_z(56), 1400_{zz}(832), 4536_z(2360),$<br>$4200'_z(1008), 2240'_x(1400)\}$,<br>$\{28_x(28), 160_z(160), 300_x(300), 972_x(512), 840_x(28), 700_{xx}(512), 1344_w(160),$<br>$840'_x(300)\}$,<br>$\{56_z(56), 1008_z(1008), 1400_z(1400), 3240_z(832), 2240_x(8), 4200_z(2360),$<br>$3200'_x(784), 4536'_z(560)\}$,<br>$\{84_x(84), 700_x(616), 2268_x(1652), 4200_x(1848), 2100_x(448), 2016_w(84),$<br>$5600_w(1652), 4200'_x(616)\}$ |
| | $e=5$ | $\{1_x(1), 28_x(28), 84_x(83), 567_x(539), 1344_x(722), 972_x(166), 2268_x(1729),$<br>$4096_z(1078), 168_y(1), 1134_y(28), 2688_y(722), 4536_y(1729),$<br>$4096'_z(539), 972'_x(83)\}$ |
| | $e=6$ | $\{1_x(1), 8_z(8), 35_x(35), 28_x(28), 84_x(40), 50_x(41), 210_x(210), 560_z(279),$<br>$300_x(225), 700_x(86), 56_z(56), 448_z(85), 1400_z(489), 175_x(85), 350_x(266),$<br>$1050_x(660), 1400_x(259), 840_z(259), 1400_{zz}(40), 840_x(41), 4200_x(1906),$<br>$2100_x(1036), 2400_z(266), 4200_z(279), 5600_z(489), 420_y(1), 1680_y(56),$<br>$4200_y(660), 4480_y(8), 4536_y(225), 5670_y(28), 4200'_x(35), 1400'_x(210)\}$,<br>$\{112_z(112), 160_z(160), 400_z(288), 1344_x(1072), 2240_x(768), 3360_z(2128),$<br>$3200_x(2128), 1344_w(288), 7168_w(1072), 3360'_z(160), 2240'_x(112)\}$ |
| | $e=8$ | $\{1_x(1), 35_x(34), 160_z(160), 567_x(373), 175_x(174), 1400_x(992), 1575_x(1042),$<br>$525_x(152), 2835_x(1668), 6075_x(3516), 2016_w(174), 5600_w(1042),$<br>$7168_w(992), 2835'_x(1), 6075'_x(373), 1400'_x(34), 1575'_x(160)\}$ |
| | $e=10$ | $\{1_x(1), 8_z(8), 28_x(28), 84_x(75), 567_x(531), 448_z(372), 1008_z(449), 1400_z(786),$<br>$972_x(897), 2268_x(502), 4536_z(2406), 1400_y(449), 3150_y(372), 4200_y(897),$<br>$4480_y(786), 4536'_z(75), 2268'_x(531), 448'_z(1), 1008'_z(28), 1400'_z(8)\}$ |
| | $e=12$ | $\{1_x(1), 35_x(35), 112_z(76), 50_x(50), 210_x(99), 400_z(349), 1050_x(651), 1400_x$<br>$(974), 525_x(449), 3360_z(1386), 2800_z(1202), 1400_y(99), 2688_y(651), 4536_y$<br>$(974), 2100_y(449), 3360'_z(349), 2800'_z(76), 1050'_x(50), 1400'_x(1), 210'_x(35)\}$ |

Each list corresponds to a block of defect $\geqslant 2$; the numbers in brackets give $\dim L^{\mu}_{(e)}$ for $\mu \in \Lambda^{\circ}_{(e)}$.

are printed in Table 7.1 and the matrices $D_\theta$ are the ones printed in the tables in the subsequent sections. It turns out that, in each case, we actually have $D_{(e)} = D_\theta$, but the proof of this statement requires considerably more work and will not be discussed in detail here. We shall just give a few examples in order to illustrate some of the difficulties involved in the proof.

**Example 7.1.5.** Assume we are in the setting of 7.1.4. Then the factorisation $D_\theta = D_{(e)}.D_\theta^{(e)}$ means that the columns of $D_\theta$ are obtained by possibly adding together some columns of $D_{(e)}$. The triangular shape of $D_\theta$ and $D_{(e)}$ then immediately yields the following criterion:

(a) Write $D_\theta = (d_{\lambda,\mu})$ and assume that, for any $\mu \neq \mu'$ in $\Lambda_k^\circ$ such that $\mathbf{a}_\mu < \mathbf{a}_{\mu'}$, there exists some $\lambda \in \Lambda$ such that $d_{\lambda,\mu} < d_{\lambda,\mu'}$. Then $D_{(e)} = D_\theta$.

(The proof is an easy exercise about matrices whose proof is left to the reader.) If (a) is satisfied, we say that the decomposition matrix $D_\theta$ is *strongly triangular*. One can of course also formulate a block-wise version of this criterion. Hence, in this case, we know that $D_{(e)} = D_\theta$, without any further computation!

For example, in type $F_4$ with equal parameters, all the decomposition matrices have this property; see Section 7.3. Unfortunately, this is not always satisfied but the exceptions are relatively rare for $(W,S)$ of exceptional type.

**Example 7.1.6.** Let $W$ be of type $H_3$. Let us apply the procedure in 7.1.4, where $e = 2$. So, choosing a suitable prime number $\ell$, we find the following decomposition matrix $D_\theta$ by the method described in 7.1.1:

| | $\mathbf{a}_\lambda$ | $D_\theta$ | | | | |
|---|---|---|---|---|---|---|
| • $1_r$ | 0 | 1 | . | . | . | . |
| • $3'_s$ | 1 | 1 | 1 | . | . | . |
| • $\overline{3}'_s$ | 1 | 1 | . | 1 | . | . |
| $5_r$ | 2 | 1 | 1 | 1 | . | . |
| • $4_r$ | 3 | . | . | . | 1 | . |
| • $4'_r$ | 3 | . | . | . | . | 1 |
| $5'_r$ | 5 | 1 | 1 | 1 | . | . |
| $3_s$ | 6 | 1 | 1 | . | . | . |
| $\overline{3}_s$ | 6 | 1 | . | 1 | . | . |
| $1'_r$ | 15 | 1 | . | . | . | . |

| $D_{(2)}$? | | | | |
|---|---|---|---|---|
| 1 | . | . | . | . |
| . | 1 | . | . | . |
| 1 | . | 1 | . | . |
| . | 1 | 1 | . | . |
| . | . | . | 1 | . |
| . | . | . | . | 1 |
| . | 1 | 1 | . | . |
| . | 1 | . | . | . |
| 1 | . | 1 | . | . |
| 1 | . | . | . | . |

| $D_{(2)}$? | | | | |
|---|---|---|---|---|
| 1 | . | . | . | . |
| 1 | 1 | . | . | . |
| . | . | 1 | . | . |
| . | 1 | 1 | . | . |
| . | . | . | 1 | . |
| . | . | . | . | 1 |
| . | 1 | 1 | . | . |
| 1 | 1 | . | . | . |
| . | . | 1 | . | . |
| 1 | . | . | . | . |

We see that this matrix is not "strongly triangular" in the sense of Example 7.1.5: Either the second or the third column could be subtracted from the first column. Thus, there are potentially two candidates for a non-trivial adjustment matrix; the corresponding candidates for $D_{(2)}$ are printed above. How can we exclude these possibilities? Let us assume, if possible, that $D_{(2)}$ is the second of the above matrices. This would show that the $\Phi_2$-modular reduction of $3'_s$ is irreducible. Then one has to show by some direct methods that this is not true. (A similar argument applies if $D_{(2)}$ was the third of the above matrices.)

Thus, even though $D_\theta$ is not "strongly triangular", the knowledge of $D_\theta$ considerably restricts the possibilities for the adjustment matrix and it shows precisely to which representations of $\mathbf{H}_K$ some special arguments need to be applied.

As in the above example, a special argument is typically needed to show that the $\Phi_e$-modular reduction of some irreducible representation of $\mathbf{H}_K$ is *not* irreducible. Here, the general techniques of the MEATAXE algorithm can still be helpful. The following is another simple method to find non-trivial submodules.

**Lemma 7.1.7.** *Let $\rho\colon \mathbf{H}_K \to M_d(K)$ be a representation such that $\rho(T_w) \in M_d(A)$ for all $w \in W$. Then there exists a matrix $B \in M_d(A)$ such that*

$$\det(B) \neq 0 \qquad \textit{and} \qquad B.\rho(T_{w^{-1}}) = \rho(T_w)^{\mathrm{tr}}.B \quad \textit{for all } w \in W.$$

*Let $\theta\colon A \to k$ be any specialisation into a field $k$. Let $B_k$ be the matrix obtained by applying $\theta$ to all entries of $B$. If $B_k \neq 0$ and $\det(B_k) = 0$, then the specialised representation $\rho_k\colon \mathbf{H}_k \to M_d(k)$ is not irreducible.*

*Proof.* Consider the contragredient representation $\hat{\rho}\colon \mathbf{H}_K \to M_d(K)$ by $\hat{\rho}(T_w) := \rho(T_{w^{-1}})^{\mathrm{tr}}$ for $w \in W$. By Example 1.2.5, $\rho$ and $\hat{\rho}$ are equivalent. So there exists an invertible matrix $B_1 \in M_d(K)$ such that $B_1.\rho(T_{w^{-1}}) = \rho(T_w)^{\mathrm{tr}}.B_1$ for all $w \in W$. Let $0 \neq d \in A$ be such that $B := dB_1$ has all entries in $A$. Then $B$ satisfies the above conditions. Assume now that $\rho_k$ is irreducible. Then $\hat{\rho}_k$ will also be irreducible. So Schur's Lemma implies that $B_k = 0$ or $B_k$ is invertible. □

**Example 7.1.8.** Let us return to Example 7.1.6, where $e = 2$ and $W$ is of type $H_3$. The representation of $\mathbf{H}_K$ labelled by $3_s'$ is given by

$$T_{s_1} \mapsto \begin{bmatrix} v & 0 & 0 \\ 0 & v & 0 \\ 0 & -\bar{\alpha} & -v^{-1} \end{bmatrix}, \quad T_{s_2} \mapsto \begin{bmatrix} v & 0 & 0 \\ 1 & -v^{-1} & -\bar{\alpha} \\ 0 & 0 & v \end{bmatrix}, \quad T_{s_3} \mapsto \begin{bmatrix} -v^{-1} & 1 & 0 \\ 0 & v & 0 \\ 0 & 0 & v \end{bmatrix},$$

where $\alpha = \frac{1}{2}(-1+\sqrt{5})$ and $\bar{\alpha} = \frac{1}{2}(-1-\sqrt{5})$. We find that a matrix $B$ satisfying the condition in Lemma 7.1.7 is given by

$$B = \begin{bmatrix} v+v^{-1} & -1 & 0 \\ -1 & v+v^{-1} & \bar{\alpha} \\ 0 & \bar{\alpha} & v+v^{-1} \end{bmatrix}.$$

Now apply the specialisation $\theta_2\colon A \to \mathbb{C}$. We obtain

$$B_{\theta_e} = \begin{bmatrix} 0 & -1 & 0 \\ -1 & 0 & \bar{\alpha} \\ 0 & \bar{\alpha} & 0 \end{bmatrix} \qquad \text{and so} \qquad \operatorname{rank}(B_{\theta_e}) = 2.$$

Hence, the $\Phi_2$-modular reduction of $3_s'$ is not irreducible. A similar computation shows that the $\Phi_2$-modular reduction of $\overline{3}_3'$ is not irreducible. Thus, we conclude that, in the setting of Example 7.1.6, the adjustment matrix is the identity.

## 7.2 Decomposition Matrices for $W$ of Dihedral Type

In this section, let $W$ be of type $I_2(m)$ ($m \geqslant 3$), with generators $s_1, s_2$ such that $(s_1 s_2)^m = 1$. Let $\zeta \in \mathbb{C}$ be a root of unity of order $m$ such that $\zeta + \zeta^{-1} = 2\cos(2\pi/m)$. Let $L\colon W \to \Gamma$ be a weight function and $A = R[\Gamma]$, where $m \in R^\times$ and $\zeta + \zeta^{-1} \in R \subseteq \mathbb{C}$. (Thus, $R$ is $L$-good.) Then, given any specialisation $\theta\colon A \to k$, where $k$ is a field, the algebra $\mathbf{H}_k$ is split and we have a corresponding decomposition map $d_\theta\colon R_0(\mathbf{H}_K) \to R_0(\mathbf{H}_k)$. In 7.2.1, we briefly describe how $d_\theta$ can be computed. Then we present a number of particular cases which are relevant in the representation theory of finite groups of Lie type.

**7.2.1.** By Example 1.3.7, we have an explicit description of the irreducible representations of $\mathbf{H}_K$, as follows:

$$
\begin{array}{lllll}
1_W^\varepsilon: & T_{s_1} \mapsto & v_{s_1}, & T_{s_2} \mapsto & v_{s_2},\\
\mathrm{sgn}^\varepsilon: & T_{s_1} \mapsto & -v_{s_1}^{-1}, & T_{s_2} \mapsto & -v_{s_2}^{-1},\\
\mathrm{sgn}_1^\varepsilon: & T_{s_1} \mapsto & v_{s_1}, & T_{s_2} \mapsto & -v_{s_2}^{-1},\\
\mathrm{sgn}_2^\varepsilon: & T_{s_1} \mapsto & -v_{s_1}^{-1}, & T_{s_2} \mapsto & v_{s_2},\\
\sigma_j^\varepsilon: & T_{s_1} \mapsto & \begin{pmatrix} -v_{s_1}^{-1} & 0 \\ \mu_j & v_{s_1} \end{pmatrix}, & T_{s_2} \mapsto & \begin{pmatrix} v_{s_2} & 1 \\ 0 & -v_{s_2}^{-1} \end{pmatrix},
\end{array}
$$

where $\mu_j = v_{s_1} v_{s_2}^{-1} + \zeta^j + \zeta^{-j} + v_{s_1}^{-1} v_{s_2}$; here, $1 \leqslant j \leqslant (m-2)/2$ (if $m$ is even) and $1 \leqslant j \leqslant (m-1)/2$ (if $m$ is odd). We notice that all these representations are realised over $A$. Given a matrix representation $\sigma^\varepsilon$ of $\mathbf{H}_K$ such that all coefficients of $\sigma^\varepsilon(T_w)$ ($w \in W$) lie in $A$, we denote by $\bar{\sigma}$ the representation of $\mathbf{H}_k$ obtained by extension of scalars from $A$ to $k$ (via $\theta$). Thus, we have $d_\theta([\sigma^\varepsilon]) = [\bar{\sigma}] \in R_0(\mathbf{H}_k)$. In the following, we shall write $\xi_{s_i} = \theta(v_{s_i})$ for $i = 1, 2$ and $\bar{\zeta} = \theta(\zeta)$.

(a) If $\sigma^\varepsilon$ is one-dimensional, then $\bar{\sigma}$ clearly is irreducible. The only question is whether some of the one-dimensional representations become equal or not. We have:

If $\xi_{s_1}^2 = -1$ and $\xi_{s_2}^2 = -1$, then $\bar{1}_W = \bar{\mathrm{sgn}} = \bar{\mathrm{sgn}}_1 = \bar{\mathrm{sgn}}_2$.
If $\xi_{s_1}^2 = -1$ and $\xi_{s_2}^2 \neq -1$, then $\bar{1}_W = \bar{\mathrm{sgn}}_2 \neq \bar{\mathrm{sgn}} = \bar{\mathrm{sgn}}_1$.
If $\xi_{s_1}^2 \neq -1$ and $\xi_{s_2}^2 = -1$, then $\bar{1}_W = \bar{\mathrm{sgn}}_1 \neq \bar{\mathrm{sgn}} = \bar{\mathrm{sgn}}_2$.
If $\xi_{s_1}^2 \neq -1$ and $\xi_{s_2}^2 \neq -1$, then $\bar{1}_W$, $\bar{\mathrm{sgn}}$, $\bar{\mathrm{sgn}}_1$, $\bar{\mathrm{sgn}}_2$ are pairwise different.

(Here, in each case, $\mathrm{sgn}_1$ and $\mathrm{sgn}_2$ only appear if $m$ is even; also, the second and the third cases only occur if $m$ is even.)

(b) Let $\sigma_j^\varepsilon$ be one of the two-dimensional representations, where $1 \leqslant j \leqslant (m-2)/2$ (if $m$ is even) or $1 \leqslant j \leqslant (m-1)/2$ (if $m$ is odd). Then $\bar{\sigma}_j$ is reducible if and only if $\bar{\sigma}_j(T_{s_1})$ and $\bar{\sigma}_j(T_{s_2})$ have a common eigenvector. A straightforward computation shows that $\bar{\sigma}_j$ is reducible if and only if $\xi_{s_1} = +\xi_{s_2}^{-1}\bar{\zeta}^{\pm j}$ or $\xi_{s_1} = -\xi_{s_2}\bar{\zeta}^{\pm j}$. (Note that $\xi_{s_1} = -\xi_{s_2}\bar{\zeta}^{\pm j}$ can only occur if $m$ is even and $\xi_{s_1} \neq \xi_{s_2}$.) We have:

If $\xi_{s_1} = +\xi_{s_2}^{-1}\bar{\zeta}^{\pm j}$, then $[\bar{\sigma}_j] = [\bar{1}_W] + [\bar{\text{sgn}}]$.
If $\xi_{s_1} = -\xi_{s_2}\,\bar{\zeta}^{\pm j}$, then $[\bar{\sigma}_j] = [\bar{\text{sgn}}_1] + [\bar{\text{sgn}}_2]$ (assuming $m$ is even).

(c) Finally, let $j, j'$ be such that $\bar{\sigma}_j$ and $\bar{\sigma}_{j'}$ are irreducible. Then we claim that $[\bar{\sigma}_j] \neq [\bar{\sigma}_{j'}]$ if $j \neq j'$. To see this, we compute that $\text{trace}(\sigma_j^{\varepsilon}(T_{s_1s_2})) = \zeta^j + \zeta^{-j}$. Since $R$ is assumed to be $L$-good, the characteristic of $k$ is zero or a prime which does not divide $m$. Hence, the roots of unity $\{\bar{\zeta}, \bar{\zeta}^2, \ldots, \bar{\zeta}^{m-1}\}$ are all distinct in $k$ and so we will have $\text{trace}(\bar{\sigma}_j(T_{s_1s_2})) = \bar{\zeta}^j + \bar{\zeta}^{-j} \neq \bar{\zeta}^{j'} + \bar{\zeta}^{-j'} = \text{trace}(\bar{\sigma}_{j'}(T_{s_1s_2}))$ if $j \neq j'$. Consequently, we also have $[\bar{\sigma}_j] \neq [\bar{\sigma}_{j'}]$ if $j \neq j'$.

The above information completely determines the blocks and their decomposition matrices in all cases.

**Example 7.2.2.** Let $\Gamma = \mathbb{Z}$ (with its natural order); then $A = R[v, v^{-1}]$ is the ring of Laurent polynomials in $v = \varepsilon$. Let $L\colon W \to \mathbb{Z}$ be a weight function such that

$$L(s_1) = c \in \{1, 2, 3, \ldots\} \qquad \text{and} \qquad L(s_2) = 1.$$

Let $e \geqslant 2$ and assume that $R \subseteq \mathbb{C}$ contains all roots of unity of order $e$. We consider the $\Phi_e$-modular specialisation

$$\theta_e\colon A \to \mathbb{K}, \qquad v \mapsto \zeta_{2e} := \exp(\pi\sqrt{-1}/e) \in \mathbb{K};$$

see Example 3.1.15(b). Using 7.2.1, we can compute the corresponding decomposition matrix. See Table 7.2 for the case where $c = 1$ (equal parameter case).

Here, and in the tables below, we use the following conventions. First, there is a table describing the $\Phi_e$-blocks (where blocks of defect 0 are not printed). In each block, the representations are ordered according to increasing value of $\mathbf{a}_\lambda$. For a block of defect 1, this information is sufficient to determine the decomposition matrix; see Theorem 3.3.13. Hence, we only need to further print the decomposition matrices of blocks of defect at least 2. In each case, one sees (independently of the theoretical results in Chapter 2) that there is a canonical basic set (marked by "•").

**Table 7.2** $\Phi_e$-blocks and decomposition numbers for type $I_2(m)$ (equal parameters)

| $e$ | $d$ | $\Phi_e$-blocks of defect $d$ | |
|---|---|---|---|
| 2 | 1 | $\{1_W, \text{sgn}\}$ | ($m$ odd) |
| 2 | 2 | $\{1_W, \text{sgn}_1, \text{sgn}_2, \text{sgn}\}$ | ($m$ even) |
| $j = 3, 4, \ldots, m$ and $j$ divides $m$ | 1 | $\{1_W, \sigma_{m/j}, \text{sgn}\}$ | |

| $e = 2$ | | $d_{\lambda,M}$ |
|---|---|---|
| • $1_W$ | 0 | 1 |
| $(\text{sgn}_1)$ | 1 | 1 |
| $(\text{sgn}_2)$ | 1 | 1 |
| sgn | $m$ | 1 |

Tables 7.3–7.6 show analogous results for $W$ of type $G_2$ and $I_2(8)$ and those weight functions which appear in the context of the representation theory of finite groups of Lie type (see 4.3.7 and Example 4.5.17).

**Table 7.3** $\Phi_e$-blocks and decomposition numbers for type $G_2$ with equal parameters

| $e$ | $d$ | $\Phi_e$-blocks of defect $d$ |
|---|---|---|
| 2 | 2 | $\{1_W, \mathrm{sgn}_1, \mathrm{sgn}_2, \mathrm{sgn}\}$ |
| 3 | 1 | $\{1_W, \sigma_2, \mathrm{sgn}\}$ |
| 6 | 1 | $\{1_W, \sigma_1, \mathrm{sgn}\}$ |

| $e=2$ | | $d_{\lambda,M}$ |
|---|---|---|
| • $1_W$ | 0 | 1 |
| $\mathrm{sgn}_1$ | 1 | 1 |
| $\mathrm{sgn}_2$ | 1 | 1 |
| $\mathrm{sgn}$ | 6 | 1 |

**Table 7.4** $\Phi_e$-blocks and decomposition numbers for type $G_2$ with parameters $3,1$

| $e$ | $d$ | $\Phi_e$-blocks of positive defect $d$ |
|---|---|---|
| 2 | 2 | $\{1_W, \mathrm{sgn}_1, \mathrm{sgn}_2, \mathrm{sgn}\}$ |
| 3 | 1 | $\{1_W, \sigma_2, \mathrm{sgn}\}$, $\{\mathrm{sgn}_1, \sigma_1, \mathrm{sgn}_2\}$ |
| 6 | 2 | $\{1_W, \mathrm{sgn}_1, \sigma_2, \mathrm{sgn}_2, \mathrm{sgn}\}$ |
| 12 | 1 | $\{1_W, \sigma_1, \mathrm{sgn}\}$ |

| $e=2$ | | $d_{\lambda,M}$ |
|---|---|---|
| • $1_W$ | 0 | 1 |
| $\mathrm{sgn}_1$ | 1 | 1 |
| $\mathrm{sgn}_2$ | 7 | 1 |
| $\mathrm{sgn}$ | 12 | 1 |

| $e=6$ | | $d_{\lambda,M}$ | |
|---|---|---|---|
| • $1_W$ | 0 | 1 | . |
| • $\mathrm{sgn}_1$ | 1 | . | 1 |
| $\sigma_2$ | 3 | 1 | 1 |
| $\mathrm{sgn}_2$ | 7 | 1 | . |
| $\mathrm{sgn}$ | 12 | . | 1 |

**Table 7.5** $\Phi_e$-blocks and decomposition numbers for type $G_2$ with parameters $9,1$

| $e$ | $d$ | $\Phi_e$-blocks of defect $d$ |
|---|---|---|
| 2 | 2 | $\{1_W, \mathrm{sgn}_1, \mathrm{sgn}_2, \mathrm{sgn}\}$ |
| 3 | 1 | $\{1_W, \sigma_2, \mathrm{sgn}\}$, $\{\mathrm{sgn}_1, \sigma_1, \mathrm{sgn}_2\}$ |
| 6 | 2 | $\{1_W, \mathrm{sgn}_1, \sigma_1, \mathrm{sgn}_2, \mathrm{sgn}\}$ |
| 12 | 1 | $\{\mathrm{sgn}_1, \sigma_1, \mathrm{sgn}_2\}$ |
| 15 | 1 | $\{1_W, \sigma_2, \mathrm{sgn}\}$ |
| 18 | 1 | $\{1_W, \mathrm{sgn}_2\}$, $\{\mathrm{sgn}_1, \mathrm{sgn}\}$ |
| 24 | 1 | $\{\mathrm{sgn}_1, \sigma_2, \mathrm{sgn}_2\}$ |
| 30 | 1 | $\{1_W, \sigma_1, \mathrm{sgn}\}$ |

| $e=2$ | | $d_{\lambda,M}$ |
|---|---|---|
| • $1_W$ | 0 | 1 |
| $\mathrm{sgn}_1$ | 1 | 1 |
| $\mathrm{sgn}_2$ | 25 | 1 |
| $\mathrm{sgn}$ | 30 | 1 |

| $e=6$ | | $d_{\lambda,M}$ | |
|---|---|---|---|
| • $1_W$ | 0 | 1 | . |
| • $\mathrm{sgn}_1$ | 1 | . | 1 |
| $\sigma_1$ | 9 | 1 | 1 |
| $\mathrm{sgn}_2$ | 25 | 1 | . |
| $\mathrm{sgn}$ | 30 | . | 1 |

**Table 7.6** $\Phi_e$-blocks and decomposition numbers for type $I_2(8)$ with parameters $4,2$

| $e$ | $d$ | $\Phi_e$-blocks of defect $d$ |
|---|---|---|
| 4 | 2 | $\{1_W, \mathrm{sgn}_1, \sigma_2, \mathrm{sgn}_2, \mathrm{sgn}\}$ |
| 8 | 2 | $\{1_W, \mathrm{sgn}_1, \sigma_3, \mathrm{sgn}_2, \mathrm{sgn}\}$ |
| 12 | 1 | $\{1_W, \sigma_2, \mathrm{sgn}\}$ |
| 24 | 1 | $\{1_W, \sigma_1, \mathrm{sgn}\}$ |

| $e=4$ | | $d_{\lambda,M}$ | |
|---|---|---|---|
| • $1_W$ | 0 | 1 | . |
| $\mathrm{sgn}_1$ | 1 | 1 | . |
| • $\sigma_2$ | 2 | 1 | 1 |
| $\mathrm{sgn}_2$ | 5 | . | 1 |
| $\mathrm{sgn}$ | 12 | . | 1 |

| $e=8$ | | $d_{\lambda,M}$ | |
|---|---|---|---|
| • $1_W$ | 0 | 1 | . |
| • $\mathrm{sgn}_1$ | 1 | . | 1 |
| $\sigma_3$ | 2 | 1 | 1 |
| $\mathrm{sgn}_2$ | 5 | 1 | . |
| $\mathrm{sgn}$ | 12 | . | 1 |

## 7.3 Decomposition Matrices for $W$ of Type $F_4$

In this section, let $(W,S)$ be of type $F_4$, with generators and diagram given by

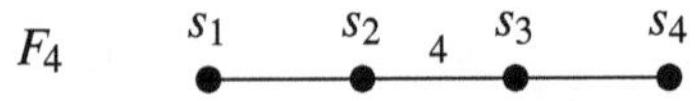


We will consider various principal specialisations $\theta\colon A \to k$, where $k$ is a field of characteristic 0, and describe the corresponding blocks and decomposition numbers. There are two essentially different cases: the one where $\Gamma = \mathbb{Z}$ and the one where $\Gamma = \mathbb{Z}^2$. In each case, one sees (independently of the theoretical results in Chapter 2) that there is a canonical basic set (marked by "•" in the tables). The notation for $\mathrm{Irr}(W)$ is taken from [132, Table C.3 (p. 413)].

**Example 7.3.1.** Let $\Gamma = \mathbb{Z}$ (with its natural order) so that $A = R[v, v^{-1}]$ is the ring of Laurent polynomials in $v = \varepsilon$. Let $L\colon W \to \mathbb{Z}$ be given by

$$L(s_1) = L(s_2) = 1 \quad \text{and} \quad L(s_3) = L(s_4) = c, \quad \text{where} \quad c \in \{1,2,4\}.$$

(These are the ones that arise in the context of representations of finite groups of Lie type; see 4.3.7 and Example 4.5.17). Thus,

$$T_{s_i}^2 = T_1 + (v - v^{-1})T_{s_i} \qquad (i = 1,2),$$
$$T_{s_i}^2 = T_1 + (v^c - v^{-c})T_{s_i} \qquad (i = 3,4).$$

Let $e \geqslant 2$ and assume that $R = \mathbb{K} \subseteq \mathbb{C}$ is a subfield which contains all roots of unity of order $e$. We consider the $\Phi_e$-modular specialisation

$$\theta_e\colon A \to \mathbb{K}, \qquad v \mapsto \zeta_{2e} := \exp(\pi\sqrt{-1}/e) \in \mathbb{K};$$

see Example 3.1.15(b). Note that $\theta_e$ is a principal specialisation where $\Phi = v - \zeta_{2e}$. The $\Phi_e$-blocks and the corresponding decomposition matrices are contained in Table 7.7 ($c = 1$), Table 7.8 ($c = 2$) and Table 7.9 ($c = 4$). These results are due to Geck and Lux [126] and Bremke [30].

**Example 7.3.2.** Let $\Gamma = \mathbb{Z}^2$ and $R = \mathbb{Q}$, so that $A = \mathbb{Q}[V^{\pm 1}, v^{\pm 1}]$ is the ring of Laurent polynomials in two independent indeterminates $V := \varepsilon^{(1,0)}$ and $v := \varepsilon^{(0,1)}$. Let $L\colon W \to \Gamma$ be given by $L(s_1) = L(s_2) = (0,1)$ and $L(s_3) = L(s_4) = (1,0)$, Thus,

$$T_{s_i}^2 = T_1 + (v - v^{-1})T_{s_i} \qquad (i = 1,2),$$
$$T_{s_i}^2 = T_1 + (V - V^{-1})T_{s_i} \qquad (i = 3,4).$$

We choose the usual lexicographic order as a monomial order on $\Gamma$. (Thus, we are in the asymptotic case; see Example 1.1.11.) From the list of the Laurent polynomials $\mathbf{c}_\lambda$ in [132, Table 11.1 (p. 379)], we find that the three conditions in 3.3.1 hold, where

$$\mathscr{P} = \{\Phi_3(V), \Phi_3(v), \Phi_4(V), \Phi_4(v), \Phi_6(V), \Phi_6(v), \Phi_{12}(V), \Phi_{12}(v),$$
$$\Phi_4(Vv^{\pm 1}), \Phi_4(V^2v^{\pm 1}), \Phi_4(V^{\pm 1}v^2), \Phi_8(Vv^{\pm 1}), \Phi_{12}(Vv^{\pm 1})\}.$$

**Table 7.7** $\Phi_e$-blocks and decomposition numbers for type $F_4$ with equal parameters

| $e$ | $d$ | $\Phi_e$-blocks of defect $d$ |
|---|---|---|
| 2 | 2 | $\{4_2, 4_3, 4_4, 4_5\}$ |
| | 4 | $\{1_1, 2_3, 2_1, 9_1, 8_1, 8_3, 1_2, 9_2, 6_2, 1_3, 9_3, 6_1, 8_2, 8_4, 9_4, 2_4, 2_2, 1_4\}$ |
| 3 | 2 | $\{1_1, 2_1, 2_3, 1_2, 1_3, 4_1, 2_2, 2_4, 1_4\}$, $\{4_2, 8_1, 8_3, 4_3, 4_4, 16, 8_2, 8_4, 4_5\}$ |
| 4 | 1 | $\{2_1, 4_3, 2_4\}$, $\{2_3, 4_4, 2_2\}$ |
| | 2 | $\{1_1, 4_2, 9_1, 4_1, 6_1, 12, 9_4, 4_5, 1_4\}$ |
| 6 | 2 | $\{1_1, 2_1, 2_3, 8_1, 8_3, 9_2, 9_3, 12, 8_2, 8_4, 2_2, 2_4, 1_4\}$ |
| 8 | 1 | $\{1_1, 9_1, 16, 9_4, 1_4\}$ |
| 12 | 1 | $\{1_1, 4_2, 6_2, 4_5, 1_4\}$ |

| $e=2$ | | $d_{\lambda,M}$ | | | |
|---|---|---|---|---|---|
| • $1_1$ | 0 | 1 | . | . | . |
| • $2_3$ | 1 | . | 1 | . | . |
| • $2_1$ | 1 | . | . | 1 | . |
| • $9_1$ | 2 | . | 1 | 1 | 1 |
| $8_1$ | 3 | 1 | . | 1 | 1 |
| $8_3$ | 3 | 1 | 1 | . | 1 |
| $1_2$ | 4 | 1 | . | . | . |
| $9_2$ | 4 | . | 1 | 1 | 1 |
| $6_2$ | 4 | 1 | . | . | 1 |
| $1_3$ | 4 | 1 | . | . | . |
| $9_3$ | 4 | . | 1 | 1 | 1 |
| $6_1$ | 4 | 1 | . | . | 1 |
| $8_2$ | 9 | 1 | . | 1 | 1 |
| $8_4$ | 9 | 1 | 1 | . | 1 |
| $9_4$ | 10 | . | 1 | 1 | 1 |
| $2_4$ | 13 | . | 1 | . | . |
| $2_2$ | 13 | . | . | 1 | . |
| $1_4$ | 24 | 1 | . | . | . |

| $e=2$ | | $d_{\lambda,M}$ |
|---|---|---|
| • $4_2$ | 1 | 1 |
| $4_3$ | 4 | 1 |
| $4_4$ | 4 | 1 |
| $4_5$ | 13 | 1 |

| $e=4$ | | $d_{\lambda,M}$ | | | | |
|---|---|---|---|---|---|---|
| • $1_1$ | 0 | 1 | . | . | . | . |
| • $4_2$ | 1 | . | 1 | . | . | . |
| • $9_1$ | 2 | 1 | 1 | 1 | . | . |
| $4_1$ | 4 | . | . | 1 | . | . |
| • $6_1$ | 4 | 1 | . | 1 | 1 | . |
| • $12$ | 4 | . | 1 | 1 | . | 1 |
| $9_4$ | 10 | . | . | 1 | 1 | 1 |
| $4_5$ | 13 | . | . | . | . | 1 |
| $1_4$ | 24 | . | . | . | 1 | . |

| $e=3$ | | $d_{\lambda,M}$ | | | |
|---|---|---|---|---|---|
| • $1_1$ | 0 | 1 | . | . | . |
| • $2_1$ | 1 | 1 | 1 | . | . |
| • $2_3$ | 1 | 1 | . | 1 | . |
| $1_2$ | 4 | . | 1 | . | . |
| $1_3$ | 4 | . | . | 1 | . |
| • $4_1$ | 4 | 1 | 1 | 1 | 1 |
| $2_2$ | 13 | . | . | 1 | 1 |
| $2_4$ | 13 | . | 1 | . | 1 |
| $1_4$ | 24 | . | . | . | 1 |

| $e=3$ | | $d_{\lambda,M}$ | | | |
|---|---|---|---|---|---|
| • $4_2$ | 1 | 1 | 0 | . | . |
| • $8_1$ | 3 | 1 | 1 | . | . |
| • $8_3$ | 3 | 1 | . | 1 | . |
| $4_3$ | 4 | . | 1 | . | . |
| $4_4$ | 4 | . | . | 1 | . |
| • $16$ | 4 | 1 | 1 | 1 | 1 |
| $8_2$ | 9 | . | . | 1 | 1 |
| $8_4$ | 9 | . | 1 | . | 1 |
| $4_5$ | 13 | . | . | . | 1 |

| $e=6$ | | $d_{\lambda,M}$ | | | | | | | |
|---|---|---|---|---|---|---|---|---|---|
| • $1_1$ | 0 | 1 | . | . | . | . | . | . | . |
| • $2_1$ | 1 | . | 1 | . | . | . | . | . | . |
| • $2_3$ | 1 | . | . | 1 | . | . | . | . | . |
| • $8_1$ | 3 | 1 | 1 | . | 1 | . | . | . | . |
| • $8_3$ | 3 | 1 | . | 1 | . | 1 | . | . | . |
| • $9_2$ | 4 | . | 1 | . | 1 | . | 1 | . | . |
| • $9_3$ | 4 | . | . | 1 | . | 1 | . | 1 | . |
| • $12$ | 4 | 1 | . | . | 1 | 1 | . | . | 1 |
| $8_2$ | 9 | . | . | . | . | 1 | . | 1 | 1 |
| $8_4$ | 9 | . | . | . | 1 | . | 1 | . | 1 |
| $2_2$ | 13 | . | . | . | . | . | . | 1 | . |
| $2_4$ | 13 | . | . | . | . | . | 1 | . | . |
| $1_4$ | 24 | . | . | . | . | . | . | . | 1 |

(As in Example 3.3.4, one sees that in each case the ring $A/(\Phi)$ is integrally closed in its field of fractions.) Let $\Phi \in \mathscr{P}$ and consider the corresponding principal specialisation $\theta\colon A \to k_\Phi$; see Section 3.3. The $\Phi$-blocks and corresponding decomposition matrices are given in Table 7.10. These tables are extracted from the results obtained by McDonough and Pallikaros [249], [250].

**Table 7.8** $\Phi_e$-blocks and decomposition numbers for type $F_4$ with parameters $1,1,2,2$

| $e$ | $d$ | $\Phi_e$-blocks of defect $d$ |
|---|---|---|
| 2 | 4 | $\{1_1,2_3,4_2,9_1,1_3,2_1,4_4,9_3,8_1,6_1,12,6_2,4_1,9_2,4_3,8_2,1_2,9_4,2_2,4_5,2_4,1_4\}$ |
| 3 | 2 | $\{1_1,2_3,1_3,2_1,4_1,1_2,2_2,2_4,1_4\}$, $\{4_2,8_3,4_4,8_1,16,4_3,8_2,8_4,4_5\}$ |
| 4 | 1 | $\{4_2,4_3\}$, $\{4_4,4_5\}$ |
|  | 2 | $\{1_1,1_3,9_1,9_3,16,9_2,1_2,9_4,1_4\}$, $\{2_3,8_3,6_1,6_2,8_4,2_4\}$ |
| 6 | 1 | $\{8_3,16,8_4\}$ |
|  | 3 | $\{1_1,2_3,4_2,2_1,9_1,1_3,4_4,8_1,6_1,4_1,12,4_3,8_2,1_2,2_2,9_4,4_5,2_4,1_4\}$ |
| 8 | 1 | $\{2_1,9_2,8_4,1_4\}$, $\{1_1,8_3,9_3,2_2\}$ |
| 10 | 1 | $\{2_3,9_3,8_2,1_4\}$, $\{1_1,8_1,9_2,2_4\}$ |
| 12 | 1 | $\{1_1,9_1,16,9_4,1_4\}$, $\{2_3,8_3,12,8_4,2_4\}$ |
| 18 | 1 | $\{1_1,4_2,6_2,4_5,1_4\}$ |

| $e=2$ |  | $d_{\lambda,M}$ |
|---|---|---|
| $\bullet\, 1_1$ | 0 | 1 . . . . . |
| $\bullet\, 2_3$ | 1 | . 1 . . . . |
| $\bullet\, 4_2$ | 2 | . 1 1 . . . |
| $\bullet\, 9_1$ | 3 | 1 1 1 1 . . |
| $1_3$ | 3 | 1 . . . . . |
| $2_1$ | 3 | . . 1 . . . |
| $4_4$ | 5 | . 1 1 . . . |
| $9_3$ | 6 | 1 1 1 1 . . |
| $\bullet\, 8_1$ | 6 | 1 . 1 1 1 . |
| $6_1$ | 7 | 1 . . 1 1 . |
| $\bullet\, 12$ | 7 | . 1 2 1 . 1 |
| $6_2$ | 7 | 1 . . 1 1 . |
| $4_1$ | 7 | . . . 1 . . |
| $9_2$ | 10 | . . 1 1 1 1 |
| $4_3$ | 11 | . . 1 . . 1 |
| $8_2$ | 12 | 1 . 1 1 1 . |
| $1_2$ | 15 | . . . . 1 . |
| $9_4$ | 15 | . . 1 1 1 1 |
| $2_2$ | 15 | . . 1 . . . |
| $4_5$ | 20 | . . 1 . . 1 |
| $2_4$ | 25 | . . . . . 1 |
| $1_4$ | 36 | . . . . 1 . |

| $e=3$ |  | $d_{\lambda,M}$ |
|---|---|---|
| $\bullet\, 1_1$ | 0 | 1 . . . |
| $\bullet\, 2_3$ | 1 | 1 1 . . |
| $1_3$ | 3 | . 1 . . |
| $\bullet\, 2_1$ | 3 | 1 . 1 . |
| $\bullet\, 4_1$ | 7 | 1 1 1 1 |
| $1_2$ | 15 | . . 1 . |
| $2_2$ | 15 | . 1 . 1 |
| $2_4$ | 25 | . . 1 1 |
| $1_4$ | 36 | . . . 1 |

| $e=3$ |  | $d_{\lambda,M}$ |
|---|---|---|
| $\bullet\, 4_2$ | 2 | 1 . . . |
| $\bullet\, 8_3$ | 3 | 1 1 . . |
| $4_4$ | 5 | . 1 . . |
| $\bullet\, 8_1$ | 6 | 1 . 1 . |
| $\bullet\, 16$ | 7 | 1 1 1 1 |
| $4_3$ | 11 | . . 1 . |
| $8_2$ | 12 | . 1 . 1 |
| $8_4$ | 15 | . . 1 1 |
| $4_5$ | 20 | . . . 1 |

| $e=4$ |  | $d_{\lambda,M}$ |
|---|---|---|
| $\bullet\, 1_1$ | 0 | 1 . . . |
| $\bullet\, 1_3$ | 3 | . 1 . . |
| $\bullet\, 9_1$ | 3 | 1 . 1 . |
| $\bullet\, 9_3$ | 6 | . 1 . 1 |
| $16$ | 7 | . . 1 1 |
| $9_2$ | 10 | 1 . 1 . |
| $1_2$ | 15 | 1 . . . |
| $9_4$ | 15 | . 1 . 1 |
| $1_4$ | 36 | . 1 . . |

| $e=4$ |  | $d_{\lambda,M}$ |
|---|---|---|
| $\bullet\, 2_3$ | 1 | 1 . |
| $\bullet\, 8_3$ | 3 | 1 1 |
| $6_1$ | 7 | . 1 |
| $6_2$ | 7 | . 1 |
| $8_4$ | 15 | 1 1 |
| $2_4$ | 25 | 1 . |

| $e=6$ |  | $d_{\lambda,M}$ |
|---|---|---|
| $\bullet\, 1_1$ | 0 | 1 . . . . . . . |
| $\bullet\, 2_3$ | 1 | . 1 . . . . . . |
| $\bullet\, 4_2$ | 2 | . . 1 . . . . . |
| $\bullet\, 2_1$ | 3 | 1 . . 1 . . . . |
| $\bullet\, 9_1$ | 3 | 1 1 1 . 1 . . . |
| $\bullet\, 1_3$ | 3 | . . . . . 1 . . |
| $\bullet\, 4_4$ | 5 | . 1 . . . 1 1 . |
| $8_1$ | 6 | 1 . 1 1 1 . . . |
| $6_1$ | 7 | 1 1 . . 1 . 1 . |
| $4_1$ | 7 | . 1 . . 1 . . . |
| $\bullet\, 12$ | 7 | . 1 1 . 1 . . 1 |
| $4_3$ | 11 | 1 . . 1 1 . . . |
| $8_2$ | 12 | . 1 . . . 1 1 1 |
| $1_2$ | 15 | . . . 1 . . . . |
| $2_2$ | 15 | . . . . . 1 1 . |
| $9_4$ | 15 | . 1 . . 1 . 1 1 |
| $4_5$ | 20 | . . . . . . . 1 |
| $2_4$ | 25 | . . . . 1 . . . |
| $1_4$ | 36 | . . . . . . 1 . |

**Table 7.9** $\Phi_e$-blocks and decomposition numbers for type $F_4$ with parameters $1,1,4,4$

| $e$ | $d$ | $\Phi_e$-blocks of defect $d$ |
|---|---|---|
| 2 | 4 | $\{1_1,2_3,1_3,4_2,9_1,4_4,2_1,9_3,8_1,4_1,6_1,6_2,12,8_2,2_2,9_2,4_3,9_4,4_5,1_2,2_4,1_4\}$ |
| 3 | 2 | $\{1_1,2_3,1_3,2_1,4_1,2_2,1_2,2_4,1_4\}$, $\{4_2,8_3,4_4,8_1,16,8_2,4_3,8_4,4_5\}$ |
| 4 | 2 | $\{1_1,1_3,8_3,9_1,2_1,9_3,16,2_2,9_2,9_4,8_4,1_2,1_4\}$ |
| 6 | 1 | $\{8_3,16,8_4\}$ |
| | 3 | $\{1_1,2_3,1_3,4_2,4_4,2_1,9_3,8_1,4_1,6_2,12,8_2,2_2,9_2,4_3,4_5,1_2,2_4,1_4\}$ |
| 8 | 1 | $\{1_3,1_4\}$, $\{9_3,9_4\}$, $\{9_1,9_2\}$, $\{1_1,1_2\}$, $\{4_4,4_5\}$, $\{4_2,4_3\}$ |
| | 2 | $\{2_3,8_3,6_1,6_2,8_4,2_4\}$ |
| 10 | 1 | $\{2_3,4_4,2_2\}$, $\{2_1,4_3,2_4\}$ |
| | 2 | $\{1_1,4_2,9_1,4_1,6_1,12,9_4,4_5,1_4\}$ |
| 12 | 1 | $\{4_2,8_1,4_3\}$, $\{4_4,8_2,4_5\}$, $\{2_3,4_1,2_4\}$ |
| | 2 | $\{1_1,1_3,8_3,2_1,9_3,16,2_2,9_2,8_4,1_2,1_4\}$ |
| 14 | 1 | $\{1_3,8_2,9_4,2_4\}$, $\{2_3,9_1,8_1,1_2\}$ |
| 18 | 1 | $\{1_1,8_1,9_2,2_4\}$, $\{2_3,9_3,8_2,1_4\}$, $\{1_3,4_4,6_1,4_3,1_2\}$ |
| 20 | 1 | $\{1_1,9_1,16,9_4,1_4\}$ |
| 24 | 1 | $\{2_3,8_3,12,8_4,2_4\}$ |
| 30 | 1 | $\{1_1,4_2,6_2,4_5,1_4\}$ |

| $e=2$ | | $d_{\lambda,M}$ |
|---|---|---|
| • $1_1$ | 0 | 1 . . . . . |
| • $2_3$ | 1 | . 1 . . . . |
| $1_3$ | 3 | 1 . . . . . |
| • $4_2$ | 4 | . 1 1 . . . |
| • $9_1$ | 7 | 1 1 1 1 . . |
| $4_4$ | 7 | . 1 1 . . . |
| $2_1$ | 9 | . . 1 . . . |
| $9_3$ | 10 | 1 1 1 1 . . |
| • $8_1$ | 12 | 1 . 1 1 1 . |
| $4_1$ | 13 | . . . 1 . . |
| $6_1$ | 13 | 1 . . 1 1 . |
| $6_2$ | 13 | 1 . . 1 1 . |
| • 12 | 13 | . 1 2 1 . 1 |
| $8_2$ | 18 | 1 . 1 1 1 . |
| $2_2$ | 21 | . . 1 . . . |
| $9_2$ | 22 | . . 1 1 1 1 |
| $4_3$ | 25 | . . 1 . . 1 |
| $9_4$ | 27 | . . 1 1 1 1 |
| $4_5$ | 34 | . . 1 . . 1 |
| $1_2$ | 39 | . . . . 1 . |
| $2_4$ | 49 | . . . . . 1 |
| $1_4$ | 60 | . . . . 1 . |

| $e=3$ | | $d_{\lambda,M}$ |
|---|---|---|
| • $1_1$ | 0 | 1 . . . |
| • $2_3$ | 1 | 1 1 . . |
| $1_3$ | 3 | . 1 . . |
| • $2_1$ | 9 | 1 . 1 . |
| • $4_1$ | 13 | 1 1 1 1 |
| $2_2$ | 21 | . 1 . 1 |
| $1_2$ | 39 | . . 1 . |
| $2_4$ | 49 | . . 1 1 |
| $1_4$ | 60 | . . . 1 |

| $e=3$ | | $d_{\lambda,M}$ |
|---|---|---|
| • $4_2$ | 4 | 1 . . . |
| • $8_3$ | 5 | 1 1 . . |
| $4_4$ | 7 | . 1 . . |
| • $8_1$ | 12 | 1 . 1 . |
| • 16 | 13 | 1 1 1 1 |
| $8_2$ | 18 | . 1 . 1 |
| $4_3$ | 25 | . . 1 . |
| $8_4$ | 29 | . . 1 1 |
| $4_5$ | 34 | . . . 1 |

| $e=10$ | | $d_{\lambda,M}$ |
|---|---|---|
| • $1_1$ | 0 | 1 . . . . |
| • $4_2$ | 4 | . 1 . . . |
| • $9_1$ | 7 | 1 1 1 . . |
| $4_1$ | 13 | . . 1 . . |
| • $6_1$ | 13 | 1 . 1 1 . |
| • 12 | 13 | . 1 1 . 1 |
| $9_4$ | 27 | . . 1 1 1 |
| $4_5$ | 34 | . . . . 1 |
| $1_4$ | 60 | . . . 1 . |

| $e=6$ | | $d_{\lambda,M}$ |
|---|---|---|
| • $1_1$ | 0 | 1 . . . . . . . |
| • $2_3$ | 1 | . 1 . . . . . . |
| • $1_3$ | 3 | . . 1 . . . . . |
| • $4_2$ | 4 | 1 1 . 1 . . . . |
| • $4_4$ | 7 | . . . . 1 . . . |
| $2_1$ | 9 | 1 . . 1 . . . . |
| • $9_3$ | 10 | . 1 1 . 1 1 . . |
| • $8_1$ | 12 | 1 1 . 1 . . 1 . |
| $4_1$ | 13 | . 1 . . . 1 . . |
| $6_2$ | 13 | . 1 1 1 . 1 . . |
| 12 | 13 | . 1 . . 1 1 1 . |
| • $8_2$ | 18 | . . 1 . 1 1 . 1 |
| $2_2$ | 21 | . . 1 . . . . 1 |
| $9_2$ | 22 | . 1 . 1 . 1 1 . |
| $4_3$ | 25 | . . . . . . 1 . |
| $4_5$ | 34 | . . 1 . . 1 . 1 |
| $1_2$ | 39 | . . . 1 . . . . |
| $2_4$ | 49 | . . . . . 1 . . |
| $1_4$ | 60 | . . . . . . . 1 |

| $e=4$ | | $d_{\lambda,M}$ |
|---|---|---|
| • $1_1$ | 0 | 1 . . . . . . . |
| • $1_3$ | 3 | . 1 . . . . . . |
| • $8_3$ | 5 | 1 1 1 . . . . . |
| • $9_1$ | 7 | 1 . 1 1 . . . . |
| $2_1$ | 9 | . . . 1 . . . . |
| • $9_3$ | 10 | . 1 1 . 1 . . . |
| • 16 | 13 | . . 1 1 1 1 . . |
| $2_2$ | 21 | . . . . 1 . . . |
| • $9_2$ | 22 | . . . 1 . 1 1 . |
| • $9_4$ | 27 | . . . . 1 1 . 1 |
| $8_4$ | 29 | . . . . . 1 1 1 |
| $1_2$ | 39 | . . . . . . 1 . |
| $1_4$ | 60 | . . . . . . . 1 |

| $e=12$ | | $d_{\lambda,M}$ |
|---|---|---|
| • $1_1$ | 0 | 1 . . . . . |
| • $1_3$ | 3 | . 1 . . . . |
| • $8_3$ | 5 | 1 . 1 . . . |
| • $2_1$ | 9 | 1 . . 1 . . |
| • $9_3$ | 10 | . 1 1 . 1 . |
| • 16 | 13 | 1 . 1 . 1 1 |
| $2_2$ | 21 | . 1 . . 1 . |
| $9_2$ | 22 | 1 . . 1 . 1 |
| $8_4$ | 29 | . . . . 1 1 |
| $1_2$ | 39 | . . . 1 . . |
| $1_4$ | 60 | . . . . 1 . |

| $e=8$ | | $d_{\lambda,M}$ |
|---|---|---|
| • $2_3$ | 1 | 1 . |
| • $8_3$ | 5 | 1 1 |
| $6_1$ | 13 | . 1 |
| $6_2$ | 13 | . 1 |
| $8_4$ | 29 | 1 1 |
| $2_4$ | 49 | 1 . |

**Table 7.10** $\Phi$-blocks and decomposition numbers for type $F_4$ with parameters $v, v, V, V$

| $\Phi$ | $d$ | $\Phi$-blocks of defect $d$ |
|---|---|---|
| $\Phi_3(v), \Phi_6(v)$ | 1 | $\{1_2, 2_4, 1_4\}, \{4_3, 8_4, 4_5\}, \{8_1, 16, 8_2\}, \{2_1, 4_1, 2_2\}, \{4_2, 8_3, 4_4\}, \{1_1, 2_3, 1_3\}$ |
| $\Phi_3(V), \Phi_6(V)$ | 1 | $\{1_3, 2_2, 1_4\}, \{4_4, 8_2, 4_5\}, \{8_3, 16, 8_4\}, \{2_3, 4_1, 2_4\}, \{4_2, 8_1, 4_3\}, \{1_1, 2_1, 1_2\}$ |
| $\Phi_4(v)$ | 1 | $\{1_2, 1_4\}, \{4_3, 4_5\}, \{9_2, 9_4\}, \{9_1, 9_3\}, \{4_2, 4_4\}, \{1_1, 1_3\}$ |
| | 2 | $\{2_1, 8_1, 6_1, 6_2, 8_2, 2_2\}$ |
| $\Phi_4(V)$ | 1 | $\{1_3, 1_4\}, \{4_4, 4_5\}, \{9_3, 9_4\}, \{9_1, 9_2\}, \{4_2, 4_3\}, \{1_1, 1_2\}$ |
| | 2 | $\{2_3, 8_3, 6_1, 6_2, 8_4, 2_4\}$ |
| $\Phi_{12}(v)$ | 1 | $\{2_1, 8_1, 12, 8_2, 2_2\}$ |
| $\Phi_{12}(V)$ | 1 | $\{2_3, 8_3, 12, 8_4, 2_4\}$ |
| $\Phi_4(Vv)$ | 1 | $\{2_3, 4_4, 2_2\}, \{2_1, 4_3, 2_4\}$ |
| | 2 | $\{1_1, 4_2, 9_1, 4_1, 6_1, 12, 9_4, 4_5, 1_4\}$ |
| $\Phi_4(Vv^{-1})$ | 1 | $\{2_2, 4_5, 2_4\}, \{2_3, 4_2, 2_1\}$ |
| | 2 | $\{1_3, 4_4, 9_3, 4_1, 6_2, 12, 9_2, 4_3, 1_2\}$ |
| $\Phi_4(V^2v)$ | 1 | $\{1_1, 8_1, 9_2, 2_4\}, \{2_3, 9_3, 8_2, 1_4\}$ |
| $\Phi_4(V^2v^{-1})$ | 1 | $\{1_3, 8_2, 9_4, 2_4\}, \{2_3, 9_1, 8_1, 1_2\}$ |
| $\Phi_4(Vv^2)$ | 1 | $\{1_1, 8_3, 9_3, 2_2\}, \{2_1, 9_2, 8_4, 1_4\}$ |
| $\Phi_4(V^{-1}v^2)$ | 1 | $\{1_3, 8_3, 9_1, 2_1\}, \{2_2, 9_4, 8_4, 1_2\}$ |
| $\Phi_8(Vv)$ | 1 | $\{1_1, 9_1, 16, 9_4, 1_4\}$ |
| $\Phi_8(Vv^{-1})$ | 1 | $\{1_3, 9_3, 16, 9_2, 1_2\}$ |
| $\Phi_{12}(Vv)$ | 1 | $\{1_1, 4_2, 6_2, 4_5, 1_4\}$ |
| $\Phi_{12}(Vv^{-1})$ | 1 | $\{1_3, 4_4, 6_1, 4_3, 1_2\}$ |

| $\Phi_4(v)$ | $\mathbf{a}_\lambda$ | $d_{\lambda,M}$ | |
|---|---|---|---|
| • $2_1$ | (3,−3) | 1 | . |
| • $8_1$ | (3,0) | 1 | 1 |
| $6_1$ | (3,1) | . | 1 |
| $6_2$ | (3,1) | . | 1 |
| $8_2$ | (3,6) | 1 | 1 |
| $2_2$ | (3,9) | 1 | . |

| $\Phi_4(V)$ | $\mathbf{a}_\lambda$ | $d_{\lambda,M}$ | |
|---|---|---|---|
| • $2_3$ | (0,1) | 1 | . |
| • $8_3$ | (1,1) | 1 | 1 |
| $6_1$ | (3,1) | . | 1 |
| $6_2$ | (3,1) | . | 1 |
| $8_4$ | (7,1) | 1 | 1 |
| $2_4$ | (12,1) | 1 | . |

| $\Phi_4(Vv)$ | $\mathbf{a}_\lambda$ | $d_{\lambda,M}$ | | | | |
|---|---|---|---|---|---|---|
| • $1_1$ | (0,0) | 1 | . | . | . | . |
| • $4_2$ | (1,0) | . | 1 | . | . | . |
| • $9_1$ | (2,−1) | 1 | 1 | 1 | . | . |
| $4_1$ | (3,1) | . | . | 1 | . | . |
| • $6_1$ | (3,1) | 1 | . | 1 | 1 | . |
| • 12 | (3,1) | . | 1 | 1 | . | 1 |
| $9_4$ | (6,3) | . | . | 1 | 1 | 1 |
| $4_5$ | (7,6) | . | . | . | . | 1 |
| $1_4$ | (12,12) | . | . | . | 1 | . |

| $\Phi_4(Vv^{-1})$ | $\mathbf{a}_\lambda$ | $d_{\lambda,M}$ | | | | |
|---|---|---|---|---|---|---|
| • $1_3$ | (0,3) | 1 | . | . | . | . |
| • $4_4$ | (1,3) | . | 1 | . | . | . |
| • $9_3$ | (2,2) | 1 | 1 | 1 | . | . |
| $4_1$ | (3,1) | . | . | 1 | . | . |
| • $6_2$ | (3,1) | 1 | . | 1 | 1 | . |
| • 12 | (3,1) | . | 1 | 1 | . | 1 |
| $9_2$ | (6,−2) | . | . | 1 | 1 | 1 |
| $4_3$ | (7,−3) | . | . | . | . | 1 |
| $1_2$ | (12,−9) | . | . | . | 1 | . |

## 7.4 Decomposition Matrices for Types $H_3, H_4, E_6, E_7, E_8$

In this section, we present the $\Phi_e$-modular decomposition matrices of blocks of defect at least 2 for the exceptional type $H_3, H_4, E_6, E_7, E_8$. Information regarding the block distribution, including defect 0 representations and Brauer trees for blocks of defect 1, already appears in the tables in [132, Appendix F].

Throughout, let $\Gamma = \mathbb{Z}$ (with its natural order) and $L(s) = 1$ for all $s \in S$ (equal parameter case). In particular, $A = R[v, v^{-1}]$ is the ring of Laurent polynomials in $v = \varepsilon$. Let $e \geqslant 2$ and assume that $R = \mathbb{K} \subseteq \mathbb{C}$ is a subfield which contains all roots of unity of order $e$. We consider the $\Phi_e$-modular specialisation

$$\theta_e : A \to \mathbb{K}, \qquad v \mapsto \zeta_{2e} := \exp(\pi\sqrt{-1}/e) \in \mathbb{K};$$

see Example 3.1.15(b). Note that $\theta_e$ is a principal specialisation where $\Phi = v - \zeta_{2e}$. We shall denote extension of scalars from $A$ to $\mathbb{K}$ (via $\theta_e$) simply by a subscript "$(e)$". Thus, for example, $\mathbf{H}_{(e)} := \mathbb{K} \otimes_A \mathbf{H}$ is the corresponding specialised algebra. The assumptions of Proposition 3.2.7 are satisfied. Hence, we have a cell datum for $\mathbf{H}$ (see Theorem 2.6.12) and there is a natural parametrisation

$$\operatorname{Irr}(\mathbf{H}_{(e)}) = \{L^{\mu}_{(e)} \mid \mu \in \Lambda^{\circ}_{(e)}\}, \qquad \text{where } \Lambda^{\circ}_{(e)} \subseteq \Lambda \text{ is a canonical basic set.}$$

As in Section 3.2, the map $\theta_e : A \to \mathbb{K}$ gives rise to a decomposition matrix

$$D = \big(d_{\lambda\mu}\big)_{\lambda\in\Lambda,\, \mu\in\Lambda^{\circ}_{(e)}}.$$

By Proposition 3.2.7 (see also Remark 3.2.2), $d_{\lambda\mu}$ equals the multiplicity of $L^{\mu}_{(e)}$ as a composition factor in the Graham–Lehrer cell module of $\mathbf{H}_{(e)}$ labelled by $\lambda$.

In Example 3.3.15, we have already discussed the example where $W$ is of type $H_3$; we recall the results in Table 7.11. Here, the rows of the decomposition matrix marked with "•" correspond to the canonical basic set, the first column contains standard labels for $\operatorname{Irr}(W)$ and the second column contains the invariants $\mathbf{a}_\lambda$.

**Table 7.11** $\Phi_e$-blocks and decomposition numbers for type $H_3$

| $e$ | $d$ | $\Phi_e$-blocks of defect $d$ |
|---|---|---|
| 2 | 3 | $\{1_r, \overline{3}'_s, 3'_s, 5_r, 5'_r, \overline{3}_s, 3_s, 1'_r\}$ |
| 3 | 1 | $\{1_r, 5_r, 4'_r\}$ + dual |
| 5 | 1 | $\{1_r, 4'_r, \overline{3}_s\}$ + dual |
| 6 | 1 | $\{1_r, 5_r, 5'_r, 1'_r\}$ |
| 10 | 1 | $\{1_r, 3'_s, 3_s, 1'_r\}$ |

| $e=2$ | | $d_{\lambda,M}$ | | |
|---|---|---|---|---|
| • $1_r$ | 0 | 1 | . | . |
| • $3'_s$ | 1 | 1 | 1 | . |
| • $\overline{3}'_s$ | 1 | 1 | . | 1 |
| $5_r$ | 2 | 1 | 1 | 1 |
| $5'_r$ | 5 | 1 | 1 | 1 |
| $3_s$ | 6 | 1 | 1 | . |
| $\overline{3}_s$ | 6 | 1 | . | 1 |
| $1'_r$ | 15 | 1 | . | . |

Tables 7.12–7.15 show similar results for $W$ of type $H_4, E_6, E_7$ and $E_8$, where we omit the tables concerning the block distribution, and where we only print decomposition matrices for blocks of defect at least 2. These results are all due to Müller and the first-named author; see [94], [96], [129], [253], [254].

**Table 7.12** $\Phi_e$-blocks of defect $\geqslant 2$ and decomposition numbers for type $H_4$

| $e=2$ | | $d_{\lambda,M}$ |
|---|---|---|
| $\bullet\ 1_r$ | 0 | 1 . . . . . |
| $\bullet\ 4_t$ | 1 | . 1 . . . . |
| $\bullet\ \overline{4}_t$ | 1 | . . 1 . . . |
| $\bullet\ 9_s$ | 2 | . . 1 1 . . |
| $\bullet\ \overline{9}_s$ | 2 | . 1 . . 1 . |
| $\bullet\ 25_r$ | 4 | . 1 1 . . 1 |
| $36_{rr}$ | 5 | 1 1 1 1 1 1 |
| $6_s$ | 6 | 1 . . 1 . . |
| $\overline{6}_s$ | 6 | 1 . . . 1 . |
| $10_r$ | 6 | . . . 1 1 . |
| $18_r$ | 6 | 1 . . . . 1 |
| $30_s$ | 6 | . . 2 1 . 1 |
| $\overline{30}_s$ | 6 | . 2 . . 1 1 |
| $36'_{rr}$ | 15 | 1 1 1 1 1 1 |
| $25'_r$ | 16 | . 1 1 . . 1 |
| $9'_s$ | 22 | . . 1 1 . . |
| $\overline{9}'_s$ | 22 | . 1 . . 1 . |
| $4'_t$ | 31 | . 1 . . . . |
| $\overline{4}'_t$ | 31 | . . 1 . . . |
| $1'_r$ | 60 | 1 . . . . . |

| $e=4$ | | $d_{\lambda,M}$ |
|---|---|---|
| $\bullet\ 1_r$ | 0 | 1 . . . . . . |
| $\bullet\ 9_s$ | 2 | 1 1 . . . . . |
| $\bullet\ \overline{9}_s$ | 2 | 1 . 1 . . . . |
| $\bullet\ 25_r$ | 4 | 1 1 1 1 . . . |
| $8_{rr}$ | 6 | . . . 1 . . . |
| $\bullet\ 24_t$ | 6 | . . 1 1 1 . . |
| $\bullet\ \overline{24}_t$ | 6 | . 1 . 1 . 1 . |
| $\bullet\ 25'_r$ | 16 | . . . 1 1 1 1 |
| $9'_s$ | 22 | . . . . . 1 1 |
| $\overline{9}'_s$ | 22 | . . . . 1 . 1 |
| $1'_r$ | 60 | . . . . . . 1 |

| $e=6$ | | $d_{\lambda,M}$ |
|---|---|---|
| $\bullet\ 1_r$ | 0 | 1 . . . . . . |
| $\bullet\ 4_t$ | 1 | . 1 . . . . . |
| $\bullet\ \overline{4}_t$ | 1 | . . 1 . . . . |
| $\bullet\ 25_r$ | 4 | 1 1 1 1 . . . |
| $\bullet\ 18_r$ | 6 | 1 . . 1 1 . . |
| $\bullet\ 24_s$ | 6 | . . 1 1 . 1 . |
| $\bullet\ \overline{24}_s$ | 6 | . 1 . 1 . . 1 |
| $25'_r$ | 16 | . . . 1 1 1 1 |
| $4'_t$ | 31 | . . . . . . 1 |
| $\overline{4}'_t$ | 31 | . . . . . 1 . |
| $1'_r$ | 60 | . . . . 1 . . |

| $e=10$ | | $d_{\lambda,M}$ |
|---|---|---|
| $\bullet\ 1_r$ | 0 | 1 . . . . . . |
| $\bullet\ \overline{4}_t$ | 1 | . 1 . . . . . |
| $\bullet\ \overline{9}_s$ | 2 | . . 1 . . . . |
| $\bullet\ 36_{rr}$ | 5 | 1 1 1 1 . . . |
| $\bullet\ \overline{24}_s$ | 6 | 1 . . 1 1 . . |
| $\bullet\ \overline{30}_s$ | 6 | . 1 . 1 . 1 . |
| $\bullet\ 40_r$ | 6 | . . 1 1 . . 1 |
| $36'_{rr}$ | 15 | . . . 1 1 1 1 |
| $\overline{9}'_s$ | 22 | . . . . . . 1 |
| $\overline{4}'_t$ | 31 | . . . . . 1 . |
| $1'_r$ | 60 | . . . . 1 . . |

| $e=3$ | | $d_{\lambda,M}$ |
|---|---|---|
| $\bullet\ 4_t$ | 1 | 1 . . . . |
| $\bullet\ \overline{4}_t$ | 1 | . 1 . . . |
| $\bullet\ 16_r$ | 3 | 1 1 1 . . |
| $8_{rr}$ | 6 | . . 1 . . |
| $\bullet\ 16_t$ | 6 | 1 . 1 1 . |
| $\bullet\ \overline{16}_t$ | 6 | . 1 1 . 1 |
| $16'_r$ | 18 | . . 1 1 1 |
| $4'_t$ | 31 | . . . 1 . |
| $\overline{4}'_t$ | 31 | . . . . 1 |

| $e=3$ | | $d_{\lambda,M}$ |
|---|---|---|
| $\bullet\ 1_r$ | 0 | 1 . . . . |
| $\bullet\ 16_{rr}$ | 3 | . 1 . . . |
| $\bullet\ 25_r$ | 4 | 1 1 1 . . |
| $8_r$ | 6 | . . 1 . . |
| $\bullet\ 10_r$ | 6 | 1 . 1 1 . |
| $\bullet\ 40_r$ | 6 | . 1 1 . 1 |
| $25'_r$ | 16 | . . 1 1 1 |
| $16'_{rr}$ | 18 | . . . . 1 |
| $1'_r$ | 60 | . . . 1 . |

| $e=5$ | | $d_{\lambda,M}$ |
|---|---|---|
| $\bullet\ 1_r$ | 0 | 1 . . . . |
| $\bullet\ 9_s$ | 2 | . 1 . . . |
| $\bullet\ 16_{rr}$ | 3 | 1 1 1 . . |
| $\overline{6}_s$ | 6 | . . 1 . . |
| $\bullet\ 8_r$ | 6 | 1 . 1 1 . |
| $\bullet\ \overline{24}_s$ | 6 | . 1 1 . 1 |
| $16'_{rr}$ | 18 | . . 1 1 1 |
| $9'_s$ | 22 | . . . . 1 |
| $1'_r$ | 60 | . . . 1 . |

| $e=5$ | | $d_{\lambda,M}$ |
|---|---|---|
| $\bullet\ 4_t$ | 1 | 1 . . . . |
| $\bullet\ 16_r$ | 3 | . 1 . . . |
| $\bullet\ 36_{rr}$ | 5 | 1 1 1 . . |
| $\overline{16}_t$ | 6 | . . 1 . . |
| $\bullet\ \overline{24}_t$ | 6 | 1 . 1 1 . |
| $\bullet\ 48_{rr}$ | 6 | . 1 1 . 1 |
| $36'_{rr}$ | 15 | . . 1 1 1 |
| $16'_r$ | 18 | . . . . 1 |
| $4'_t$ | 31 | . . . 1 . |

**Table 7.13** $\Phi_e$-blocks of defect $\geqslant 2$ and decomposition numbers for type $E_6$

| $e=2$ | | $d_{\lambda,M}$ |
|---|---|---|
| $\bullet\ 1_p$ | . | 1 . . . . . |
| $\bullet\ 6_p$ | 1 | . 1 . . . . |
| $\bullet\ 20_p$ | 2 | . 1 1 . . . |
| $15_p$ | 3 | 1 . 1 . . . |
| $\bullet\ 15_q$ | 3 | 1 . . 1 . . |
| $\bullet\ 30_p$ | 3 | . 1 1 . 1 . |
| $\bullet\ 60_p$ | 5 | . . 1 . . 1 |
| $24_p$ | 6 | . . 1 . 1 . |
| $81_p$ | 6 | 1 1 1 1 . 1 |
| $10_s$ | 7 | . . . . 1 . |
| $20_s$ | 7 | . 1 . 1 . . |
| $60_s$ | 7 | . . . 1 . 1 |
| $90_s$ | 7 | . 1 2 . 1 1 |
| $81'_p$ | 1. | 1 1 1 1 . 1 |
| $60'_p$ | 11 | . . 1 . . 1 |
| $24'_p$ | 12 | . . 1 . 1 . |
| $15'_p$ | 15 | 1 . 1 . . . |
| $15'_q$ | 15 | 1 . . 1 . . |
| $30'_p$ | 15 | . 1 1 . 1 . |
| $20'_p$ | 2. | . 1 1 . . . |
| $6'_p$ | 25 | . 1 . . . . |
| $1'_p$ | 36 | 1 . . . . . |

| $e=3$ | | $d_{\lambda,M}$ |
|---|---|---|
| $\bullet\ 1_p$ | 0 | 1 . . . . . . . . . |
| $\bullet\ 6_p$ | 1 | 1 1 . . . . . . . . |
| $\bullet\ 20_p$ | 2 | 1 1 1 . . . . . . . |
| $\bullet\ 15_p$ | 3 | . 1 . 1 . . . . . . |
| $\bullet\ 15_q$ | 3 | . . 1 . 1 . . . . . |
| $\bullet\ 30_p$ | 3 | . 1 . . . 1 . . . . |
| $\bullet\ 64_p$ | 4 | . 1 1 1 . 1 1 . . . |
| $\bullet\ 60_p$ | 5 | . 1 1 . 1 1 1 1 . . |
| $24_p$ | 6 | . . 1 . . . 1 . . . |
| $10_s$ | 7 | . 1 . . . . . 1 . . |
| $20_s$ | 7 | . . . 1 . . 1 . . . |
| $\bullet\ 60_s$ | 7 | 1 1 1 1 1 . 1 1 1 . |
| $\bullet\ 80_s$ | 7 | . 1 . 1 . 1 1 1 . 1 |
| $60'_p$ | 11 | 1 1 . 1 . . . 1 1 1 |
| $24'_p$ | 12 | . . . 1 . . . . 1 . |
| $64'_p$ | 13 | . . . 1 . . 1 1 1 1 |
| $15'_p$ | 15 | . . . . . . 1 1 . . |
| $15'_q$ | 15 | 1 . . . . . . . 1 . |
| $30'_p$ | 15 | . . . . . . . 1 . 1 |
| $20'_p$ | 2. | . . . . 1 . . 1 1 . |
| $6'_p$ | 25 | . . . . 1 . . 1 . . |
| $1'_p$ | 36 | . . . . 1 . . . . . |

| $e=4$ | | $d_{\lambda,M}$ |
|---|---|---|
| $\bullet\ 1_p$ | 0 | 1 . . . . . . . |
| $\bullet\ 6_p$ | 1 | . 1 . . . . . . |
| $\bullet\ 15_p$ | 3 | . . 1 . . . . . |
| $\bullet\ 15_q$ | 3 | 1 1 . 1 . . . . |
| $\bullet\ 81_p$ | 6 | . 1 1 . 1 . . . |
| $\bullet\ 10_s$ | 7 | 1 . . 1 . 1 . . |
| $\bullet\ 80_s$ | 7 | . 1 . 1 1 . 1 . |
| $\bullet\ 90_s$ | 7 | . . 1 . 1 . . 1 |
| $81'_p$ | 1. | . . . . 1 . 1 1 |
| $15'_p$ | 15 | . . . . . . . 1 |
| $15'_q$ | 15 | . . . 1 . 1 1 . |
| $6'_p$ | 25 | . . . . . . 1 . |
| $1'_p$ | 36 | . . . . . 1 . . |

| $e=6$ | | $d_{\lambda,M}$ |
|---|---|---|
| $\bullet\ 1_p$ | 0 | 1 . . . . . . . . . . |
| $\bullet\ 6_p$ | 1 | . 1 . . . . . . . . . |
| $\bullet\ 20_p$ | 2 | 1 1 1 . . . . . . . . |
| $\bullet\ 15_q$ | 3 | 1 . . 1 . . . . . . . |
| $\bullet\ 30_p$ | 3 | . 1 1 . 1 . . . . . . |
| $\bullet\ 60_p$ | 5 | 1 . 1 1 . 1 . . . . . |
| $\bullet\ 24_p$ | 6 | . . 1 . . . 1 . . . . |
| $\bullet\ 60_s$ | 7 | . . . 1 . 1 . 1 . . . |
| $\bullet\ 80_s$ | 7 | . . 1 . 1 1 1 . 1 . . |
| $\bullet\ 60'_p$ | 11 | . . . . . 1 . 1 1 1 . |
| $24'_p$ | 12 | . . . . 1 . . . 1 . . |
| $15'_q$ | 15 | . . . . . . . 1 . 1 . |
| $\bullet\ 30'_p$ | 15 | . . . . . . 1 . 1 . 1 |
| $20'_p$ | 2. | . . . . . . . . 1 1 1 |
| $6'_p$ | 25 | . . . . . . . . . . 1 |
| $1'_p$ | 36 | . . . . . . . . . 1 . |

**Table 7.14** $\Phi_e$-blocks of defect $\geqslant 2$ and decomposition numbers for type $E_7$

| $e=2$ | | $d_{\lambda,M}$ |
|---|---|---|
| $\bullet\, 1_a$ | 0 | 1 . . . . . . |
| $\bullet\, 7'_a$ | 1 | 1 1 . . . . . |
| $\bullet\, 27_a$ | 2 | 1 2 1 . . . . |
| $21'_b$ | 3 | 1 1 1 . . . . |
| $21_a$ | 3 | 1 1 1 . . . . |
| $\bullet\, 35_b$ | 3 | 1 1 1 1 . . . |
| $15'_a$ | 4 | 1 . . 1 . . . |
| $\bullet\, 105'_a$ | 4 | 1 2 1 . 1 . . |
| $\bullet\, 189'_b$ | 5 | 1 2 2 1 1 1 . |
| $105_b$ | 6 | 1 1 2 1 . 1 . |
| $168_a$ | 6 | . 1 2 . 1 1 . |
| $210_a$ | 6 | 2 3 3 1 1 1 . |
| $189'_c$ | 7 | 1 2 2 1 1 1 . |
| $35'_a$ | 7 | 1 1 1 1 . . . |
| $70'_a$ | 7 | . . 1 . . 1 . |
| $280'_a$ | 7 | 2 3 4 1 1 2 . |
| $\bullet\, 315'_a$ | 7 | 1 2 2 1 1 1 1 |
| $189_a$ | 8 | 1 2 2 1 1 1 . |
| $405_a$ | 8 | 1 3 4 1 1 2 1 |
| $378'_a$ | 9 | . 1 3 1 1 2 1 |
| $84_a$ | 10 | . . 1 1 . 1 . |
| $210_b$ | 10 | . . 1 1 . 1 1 |
| $420_a$ | 10 | 2 3 4 2 1 2 1 |
| $105_c$ | 12 | 1 1 2 1 . 1 . |
| $84'_a$ | 13 | . . 1 1 . 1 . |
| $210'_b$ | 13 | . . 1 1 . 1 1 |
| $420'_a$ | 13 | 2 3 4 2 1 2 1 |
| $378_a$ | 14 | . 1 3 1 1 2 1 |
| $105'_c$ | 15 | 1 1 2 1 . 1 . |
| $189'_a$ | 15 | 1 2 2 1 1 1 . |
| $405'_a$ | 15 | 1 3 4 1 1 2 1 |
| $35_a$ | 16 | 1 1 1 1 . . . |
| $70_a$ | 16 | . . 1 . . 1 . |
| $280_a$ | 16 | 2 3 4 1 1 2 . |
| $315_a$ | 16 | 1 2 2 1 1 1 1 |
| $189_c$ | 20 | 1 2 2 1 1 1 . |
| $105'_b$ | 21 | 1 1 2 1 . 1 . |
| $168'_a$ | 21 | . 1 2 . 1 1 . |
| $210'_a$ | 21 | 2 3 3 1 1 1 . |
| $189_b$ | 22 | 1 2 2 1 1 1 . |
| $15_a$ | 25 | 1 . . 1 . . . |
| $105_a$ | 25 | 1 2 1 . 1 . . |
| $21'_a$ | 30 | 1 1 1 . . . . |
| $35'_b$ | 30 | 1 1 1 1 . . . |
| $21_b$ | 36 | 1 1 1 . . . . |
| $27'_a$ | 37 | 1 2 1 . . . . |
| $7_a$ | 46 | 1 1 . . . . . |
| $1'_a$ | 63 | 1 . . . . . . |

| $e=2$ | | $d_{\lambda,M}$ |
|---|---|---|
| $\bullet\, 56'_a$ | 3 | 1 . . |
| $\bullet\, 120_a$ | 4 | 1 1 . |
| $\bullet\, 280_b$ | 7 | . 1 1 |
| $216'_a$ | 8 | . . 1 |
| $336'_a$ | 10 | 1 1 1 |
| $336_a$ | 13 | 1 1 1 |
| $216_a$ | 15 | . . 1 |
| $280'_b$ | 16 | . 1 1 |
| $120'_a$ | 25 | 1 1 . |
| $56_a$ | 30 | 1 . . |

| $e=6$ | | $d_{\lambda,M}$ |
|---|---|---|
| $\bullet\, 1_a$ | 0 | 1 . . . . . . . . . . . . . . . . . . . |
| $\bullet\, 7'_a$ | 1 | . 1 . . . . . . . . . . . . . . . . . . |
| $\bullet\, 21'_b$ | 3 | 1 1 1 . . . . . . . . . . . . . . . . . |
| $\bullet\, 21_a$ | 3 | . . . 1 . . . . . . . . . . . . . . . . |
| $\bullet\, 35_b$ | 3 | 1 1 . . 1 . . . . . . . . . . . . . . . |
| $\bullet\, 15'_a$ | 4 | 1 . . . . 1 . . . . . . . . . . . . . . |
| $\bullet\, 105'_a$ | 4 | . 1 . 1 . . 1 . . . . . . . . . . . . . |
| $\bullet\, 105_b$ | 6 | 1 1 1 . 1 1 . 1 . . . . . . . . . . . . |
| $\bullet\, 168_a$ | 6 | 1 1 1 . 1 . 1 . 1 . . . . . . . . . . . |
| $\bullet\, 210_a$ | 6 | . 1 1 1 . . 1 . . 1 . . . . . . . . . . |
| $\bullet\, 70'_a$ | 7 | 1 . . . 1 . . . . . 1 . . . . . . . . . |
| $\bullet\, 280'_a$ | 7 | . . . 1 . . 1 . . 1 . 1 . . . . . . . . |
| $\bullet\, 315'_a$ | 7 | . 1 1 . 1 . 1 1 1 1 . . 1 . . . . . . . |
| $\bullet\, 84_a$ | 10 | . . . . 1 . . . 1 . . . . 1 . . . . . . |
| $\bullet\, 210_b$ | 10 | 1 . 1 . 1 1 . 1 1 . 1 . . . 1 . . . . . |
| $\bullet\, 420_a$ | 10 | . . 1 . . . 1 . 1 1 . 1 1 . . 1 . . . . |
| $105_c$ | 12 | . . . . . . . . . 1 . . 1 . . . . . . . |
| $84'_a$ | 13 | . . . . . 1 . 1 . . . . . . 1 . . . . . |
| $\bullet\, 210'_b$ | 13 | . . . . 1 . . 1 1 . 1 . 1 1 1 . 1 . . . |
| $\bullet\, 420'_a$ | 13 | . . 1 . . . . 1 . 1 . 1 1 . . 1 . 1 . . |
| $105'_c$ | 15 | . . 1 . . . . . . . . . . . . 1 . . . . |
| $70_a$ | 16 | . . . . . . . . . . 1 . . . 1 . 1 . . . |
| $\bullet\, 280_a$ | 16 | . . . . . . . . . . . 1 . . . 1 . 1 1 . |
| $\bullet\, 315_a$ | 16 | . . 1 . . . . 1 1 . . . 1 . 1 1 . 1 . 1 |
| $105'_b$ | 21 | . . . . . . . . 1 . . . 1 1 1 . 1 . . 1 |
| $168'_a$ | 21 | . . . . . . . 1 . . . . 1 . 1 . 1 1 . 1 |
| $210'_a$ | 21 | . . . . . . . . . . . . 1 . . 1 . 1 1 1 |
| $15_a$ | 25 | . . . . . . . . . . . . . 1 . . 1 . . . |
| $105_a$ | 25 | . . . . . . . . . . . . . . . . . 1 1 1 |
| $21'_a$ | 30 | . . . . . . . . . . . . . . . . . . 1 . |
| $35'_b$ | 30 | . . . . . . . . . . . . . . 1 . 1 . . 1 |
| $21_b$ | 36 | . . . . . . . . . . . . 1 . . . 1 . . 1 |
| $7_a$ | 46 | . . . . . . . . . . . . . . . . . . . 1 |
| $1'_a$ | 63 | . . . . . . . . . . . . . . . . 1 . . . |

| $e=3$ | | $d_{\lambda,M}$ |
|---|---|---|
| $\bullet\, 1_a$ | 0 | 1 . . . . . . . . . |
| $\bullet\, 21_a$ | 3 | . 1 . . . . . . . . |
| $\bullet\, 35_b$ | 3 | 1 . 1 . . . . . . . |
| $\bullet\, 120_a$ | 4 | 1 1 . 1 . . . . . . |
| $\bullet\, 105_b$ | 6 | . . . 1 1 . . . . . |
| $\bullet\, 168_a$ | 6 | 1 . 1 1 . 1 . . . . |
| $\bullet\, 210_a$ | 6 | . 1 . 1 . . 1 . . . |
| $\bullet\, 280_b$ | 7 | 1 . 1 1 1 1 1 1 . . |
| $84_a$ | 10 | 1 . 1 . . 1 . 1 . . |
| $\bullet\, 210_b$ | 10 | . 1 . 1 1 1 . . 1 . |
| $\bullet\, 420_a$ | 10 | . . . 1 . 1 1 . . 1 |
| $512'_a$ | 11 | 1 1 . 1 1 1 1 1 1 1 |
| $105_c$ | 12 | . . . . . . 1 1 . . |
| $336_a$ | 13 | . . . . . 1 1 1 . 1 |
| $35_a$ | 16 | . . . . . 1 . . . . |
| $70_a$ | 16 | . 1 . . . . . . 1 . |
| $280_a$ | 16 | . . . . . 1 . . 1 1 |
| $15_a$ | 25 | 1 . . . . . . . 1 . |
| $105_a$ | 25 | . . . . 1 1 . 1 1 . |
| $56_a$ | 30 | . . . . 1 . . . 1 . |
| $21_b$ | 36 | . . . . 1 . . 1 . . |
| $7_a$ | 46 | . . . . 1 . . . . . |

| $e=3$ | | $d_{\lambda,M}$ |
|---|---|---|
| $\bullet\, 7'_a$ | 1 | 1 . . . . . . . . . |
| $\bullet\, 21'_b$ | 3 | 1 1 . . . . . . . . |
| $\bullet\, 56'_a$ | 3 | 1 . 1 . . . . . . . |
| $\bullet\, 15'_a$ | 4 | . 1 . 1 . . . . . . |
| $\bullet\, 105'_a$ | 4 | 1 1 1 . 1 . . . . . |
| $35'_a$ | 7 | . . . . 1 . . . . . |
| $\bullet\, 70'_a$ | 7 | . . 1 . . 1 . . . . |
| $\bullet\, 280'_a$ | 7 | . . 1 . 1 . 1 . . . |
| $\bullet\, 336'_a$ | 10 | . 1 . . 1 . 1 1 . . |
| $\bullet\, 512_a$ | 11 | 1 1 1 1 1 1 1 1 1 . |
| $\bullet\, 84'_a$ | 13 | . 1 . 1 1 . . . . 1 |
| $210'_b$ | 13 | 1 . 1 . 1 1 . . 1 . |
| $420'_a$ | 13 | . . . . 1 . 1 1 1 . |
| $105'_c$ | 15 | . 1 . . . . . 1 . . |
| $280'_b$ | 16 | 1 1 . 1 1 . . 1 1 1 |
| $105'_b$ | 21 | 1 . . . . . . . 1 . |
| $168'_a$ | 21 | . . . 1 1 . . . 1 1 |
| $210'_a$ | 21 | . . . . . 1 . 1 1 . |
| $120'_a$ | 25 | . . . 1 . 1 . . 1 . |
| $21'_a$ | 30 | . . . . . 1 . . . . |
| $35'_b$ | 30 | . . . 1 . . . . . 1 |
| $1'_a$ | 63 | . . . 1 . . . . . . |

**Table 7.14** $\Phi_e$-blocks of defect $\geqslant 2$ and decomposition numbers for type $E_7$ (cont'd)

| $e=4$ | | $d_{\lambda,M}$ |
|---|---|---|
| $\bullet\, 1_a$ | 0 | 1 . . . . . . . |
| $\bullet\, 56'_a$ | 3 | . 1 . . . . . . |
| $\bullet\, 105_b$ | 6 | 1 1 1 . . . . . |
| $\bullet\, 210_a$ | 6 | . 1 . 1 . . . . |
| $\bullet\, 189_a$ | 8 | . . . 1 1 . . . |
| $\bullet\, 405_a$ | 8 | . 1 1 1 . 1 . . |
| $336'_a$ | 10 | . . . 1 1 1 . . |
| $35_a$ | 16 | . . . . 1 . . . |
| $\bullet\, 70_a$ | 16 | 1 . 1 . . . 1 . |
| $\bullet\, 315_a$ | 16 | . . 1 . . 1 . 1 |
| $189_b$ | 22 | . . 1 . . . 1 1 |
| $120'_a$ | 25 | . . . . . . . 1 |
| $21_b$ | 36 | . . . . . . 1 . |

| $e=4$ | | $d_{\lambda,M}$ |
|---|---|---|
| $\bullet\, 21'_b$ | 3 | 1 . . . . . . . |
| $\bullet\, 120_a$ | 4 | . 1 . . . . . . |
| $\bullet\, 189'_b$ | 5 | 1 1 1 . . . . . |
| $\bullet\, 35'_a$ | 7 | . . . 1 . . . . |
| $\bullet\, 70'_a$ | 7 | 1 . 1 . 1 . . . |
| $\bullet\, 315'_a$ | 7 | . 1 1 . . 1 . . |
| $\bullet\, 336_a$ | 13 | . . . 1 . 1 1 . |
| $189'_a$ | 15 | . . . 1 . . 1 . |
| $\bullet\, 405'_a$ | 15 | . . 1 . . 1 1 1 |
| $105'_b$ | 21 | . . 1 . 1 . . 1 |
| $210'_a$ | 21 | . . . . . . 1 1 |
| $56_a$ | 30 | . . . . . . . 1 |
| $1'_a$ | 63 | . . . . 1 . . . |

| $e=4$ | | $d_{\lambda,M}$ |
|---|---|---|
| $\bullet\, 7'_a$ | 1 | 1 . . . . . . |
| $\bullet\, 15'_a$ | 4 | 1 1 . . . . . |
| $\bullet\, 105'_a$ | 4 | . . 1 . . . . |
| $\bullet\, 189'_c$ | 7 | . . 1 1 . . . |
| $\bullet\, 280_b$ | 7 | 1 . 1 . 1 . . |
| $\bullet\, 378'_a$ | 9 | . . 1 1 1 1 . |
| $\bullet\, 210'_b$ | 13 | 1 1 . . 1 . 1 |
| $105'_c$ | 15 | . . . 1 . 1 . |
| $216_a$ | 15 | . . . . 1 1 1 |
| $21'_a$ | 30 | . . . . . 1 . |
| $35'_b$ | 30 | . 1 . . . . 1 |
| $27'_a$ | 37 | . . . . . . 1 |

| $e=4$ | | $d_{\lambda,M}$ |
|---|---|---|
| $\bullet\, 27_a$ | 2 | 1 . . . . . . |
| $\bullet\, 21_a$ | 3 | . 1 . . . . . |
| $\bullet\, 35_b$ | 3 | 1 . 1 . . . . |
| $\bullet\, 216'_a$ | 8 | 1 1 . 1 . . . |
| $\bullet\, 210_b$ | 10 | 1 . 1 1 1 . . |
| $\bullet\, 105_c$ | 12 | . 1 . . . 1 . |
| $\bullet\, 378_a$ | 14 | . 1 . 1 . 1 1 |
| $280'_b$ | 16 | . . . 1 1 . 1 |
| $189_c$ | 20 | . . . . . 1 1 |
| $15_a$ | 25 | . . 1 . 1 . . |
| $105_a$ | 25 | . . . . . . 1 |
| $7_a$ | 46 | . . . . 1 . . |

**Table 7.15** $\Phi_e$-blocks of defect $\geqslant 2$ and decomposition numbers for type $E_8$

| $e=10$ | | $d_{\lambda,M}$ |
|---|---|---|
| $\bullet\, 1_x$ | 0 | 1 . . . . . . . . . . . . . . . . . . . |
| $\bullet\, 8_z$ | 1 | . 1 . . . . . . . . . . . . . . . . . . |
| $\bullet\, 28_x$ | 3 | . . 1 . . . . . . . . . . . . . . . . . |
| $\bullet\, 84_x$ | 3 | 1 1 . 1 . . . . . . . . . . . . . . . . |
| $\bullet\, 567_x$ | 6 | . 1 1 . 1 . . . . . . . . . . . . . . . |
| $\bullet\, 448_z$ | 7 | 1 . . 1 . 1 . . . . . . . . . . . . . . |
| $\bullet\, 1008_z$ | 7 | . . 1 . 1 . 1 . . . . . . . . . . . . . |
| $\bullet\, 1400_z$ | 7 | . 1 . 1 1 . . 1 . . . . . . . . . . . . |
| $\bullet\, 972_x$ | 10 | . . . 1 . . . . 1 . . . . . . . . . . . |
| $\bullet\, 2268_x$ | 10 | . . . . 1 . 1 1 . 1 . . . . . . . . . . |
| $\bullet\, 4536_z$ | 13 | . . . 1 . 1 . 1 1 . 1 . . . . . . . . . |
| $\bullet\, 1400_y$ | 16 | . . . . . . 1 . . 1 . 1 . . . . . . . . |
| $\bullet\, 3150_y$ | 16 | . . . . . 1 . . . . 1 . 1 . . . . . . . |
| $\bullet\, 4200_y$ | 16 | . . . . . . . . 1 . 1 . . 1 . . . . . . |
| $\bullet\, 4480_y$ | 16 | . . . . . . . 1 . 1 1 . . . 1 . . . . . |
| $\bullet\, 4536'_z$ | 23 | . . . . . . . . . . 1 . 1 1 1 1 . . . . |
| $972'_x$ | 30 | . . . . . . . . . . . . . 1 . 1 . . . . |
| $\bullet\, 2268'_x$ | 30 | . . . . . . . . . 1 . 1 . . 1 . 1 . . . |
| $\bullet\, 448'_z$ | 37 | . . . . . . . . . . . . 1 . . 1 . 1 . . |
| $\bullet\, 1008'_z$ | 37 | . . . . . . . . . . . 1 . . . . 1 . 1 . |
| $\bullet\, 1400'_z$ | 37 | . . . . . . . . . . . . . . 1 1 1 . . 1 |
| $567'_x$ | 46 | . . . . . . . . . . . . . . . . 1 . 1 1 |
| $28'_x$ | 63 | . . . . . . . . . . . . . . . . . . 1 . |
| $84'_x$ | 63 | . . . . . . . . . . . . . . . 1 . 1 . 1 |
| $8'_z$ | 91 | . . . . . . . . . . . . . . . . . . . 1 |
| $1'_x$ | 120 | . . . . . . . . . . . . . . . . . 1 . . |

| $e=12$ | | $d_{\lambda,M}$ |
|---|---|---|
| $\bullet\, 1_x$ | 0 | 1 . . . . . . . . . . . . . . . . . . . |
| $\bullet\, 35_x$ | 2 | . 1 . . . . . . . . . . . . . . . . . . |
| $\bullet\, 112_z$ | 3 | 1 1 1 . . . . . . . . . . . . . . . . . |
| $\bullet\, 50_x$ | 4 | . . . 1 . . . . . . . . . . . . . . . . |
| $\bullet\, 210_x$ | 4 | . 1 1 . 1 . . . . . . . . . . . . . . . |
| $\bullet\, 400_z$ | 6 | 1 . . 1 . 1 . . . . . . . . . . . . . . |
| $\bullet\, 1050_x$ | 8 | . . . 1 . 1 1 . . . . . . . . . . . . . |
| $\bullet\, 1400_x$ | 8 | 1 . 1 . . 1 . 1 . . . . . . . . . . . . |
| $\bullet\, 525_x$ | 12 | . . 1 . . . . . 1 . . . . . . . . . . . |
| $\bullet\, 3360_z$ | 12 | . . . . . 1 1 1 . 1 . . . . . . . . . . |
| $\bullet\, 2800_z$ | 13 | . . 1 . 1 . . 1 1 . 1 . . . . . . . . . |
| $\bullet\, 1400_y$ | 16 | . . . . 1 . . . . . 1 1 . . . . . . . . |
| $\bullet\, 2688_y$ | 16 | . . . . . . 1 . . 1 . . 1 . . . . . . . |
| $\bullet\, 4536_y$ | 16 | . . . . . . . 1 . 1 1 . . 1 . . . . . . |
| $\bullet\, 2100_y$ | 20 | . . . . . . . . 1 . 1 . . . 1 . . . . . |
| $\bullet\, 3360'_z$ | 24 | . . . . . . . . . 1 . . 1 1 . 1 . . . . |
| $\bullet\, 2800'_z$ | 25 | . . . . . . . . . . 1 1 . 1 1 . 1 . . . |
| $\bullet\, 1050'_x$ | 32 | . . . . . . . . . . . . 1 . . 1 . 1 . . |
| $\bullet\, 1400'_x$ | 32 | . . . . . . . . . . . . . 1 . 1 1 . 1 . |
| $525'_x$ | 36 | . . . . . . . . . . . . . . 1 . 1 . . . |
| $400'_z$ | 42 | . . . . . . . . . . . . . . . 1 . 1 1 . |
| $50'_x$ | 52 | . . . . . . . . . . . . . . . . . 1 . . |
| $\bullet\, 210'_x$ | 52 | . . . . . . . . . . . 1 . . . . 1 . . 1 |
| $112'_z$ | 63 | . . . . . . . . . . . . . . . . 1 . 1 1 |
| $35'_x$ | 74 | . . . . . . . . . . . . . . . . . . . 1 |
| $1'_x$ | 120 | . . . . . . . . . . . . . . . . . . 1 . |

**Table 7.15** $\Phi_e$-blocks of defect $\geqslant 2$ and decomposition numbers for type $E_8$ (cont'd)

| $e=2$ | | $d_{\lambda,M}$ |
|---|---|---|
| $\bullet\, 1_x$ | 0 | 1........... |
| $\bullet\, 8_z$ | 1 | .1.......... |
| $\bullet\, 35_x$ | 2 | .11......... |
| $28_x$ | 3 | 1.1......... |
| $\bullet\, 84_x$ | 3 | 1111........ |
| $\bullet\, 50_x$ | 4 | .1..1....... |
| $\bullet\, 210_x$ | 4 | .1...1...... |
| $\bullet\, 560_z$ | 5 | .22.111..... |
| $300_x$ | 6 | ..2...1..... |
| $567_x$ | 6 | 1221.11..... |
| $\bullet\, 700_x$ | 6 | 11111111.... |
| $56_z$ | 7 | .1.1........ |
| $448_z$ | 7 | 1.11..11.... |
| $1008_z$ | 7 | 12311121.... |
| $\bullet\, 1400_z$ | 7 | .221111.1... |
| $175_x$ | 8 | 1..1...1.... |
| $350_x$ | 8 | .12.1.1..... |
| $\bullet\, 1050_x$ | 8 | 11211.1..1.. |
| $\bullet\, 1400_x$ | 8 | .2212121..1. |
| $1575_x$ | 8 | 122211111... |
| $3240_z$ | 9 | 12432131111. |
| $972_x$ | 10 | ..11..1..1.. |
| $2268_x$ | 10 | 123221211.1. |
| $840_z$ | 10 | ...1....1... |
| $1400_zz$ | 10 | .1111.1..11. |
| $525_x$ | 12 | 11211.11.... |
| $840_x$ | 12 | ...11.11..1. |
| $\bullet\, 4200_x$ | 12 | 11331.2111.1 |
| $700_xx$ | 13 | 1..1.....1.. |
| $2100_x$ | 13 | 1222111.11.. |
| $4536_z$ | 13 | ..222.211111 |
| $2835_x$ | 14 | ..11..1..1.1 |
| $6075_x$ | 14 | 125421411211 |
| $2400_z$ | 15 | 12422131.11. |
| $4200_z$ | 15 | 11332.32.111 |
| $5600_z$ | 15 | 135431421111 |
| $70_y$ | 16 | 1.1.1....... |
| $168_y$ | 16 | ....1..1.... |
| $420_y$ | 16 | ....1.....1. |
| $1134_y$ | 16 | ..2.2.21..1. |
| $1400_y$ | 16 | 12322.22..1. |
| $1680_y$ | 16 | 12221121.1.. |
| $3150_y$ | 16 | ..121.22..11 |
| $4200_y$ | 16 | ..121.2..211 |
| $4536_y$ | 16 | 12522141.111 |
| $5670_y$ | 16 | 123421212111 |
| $2100_y$ | 20 | 12222121.11. |
| $2400'_z$ | 21 | 12422131.11. |
| $4200'_z$ | 21 | 11332.32.111 |
| $5600'_z$ | 21 | 135431421111 |
| $2835'_x$ | 22 | ..11..1..1.1 |
| $6075'_x$ | 22 | 125421411211 |
| $4536'_z$ | 23 | ..222.211111 |
| $840'_x$ | 24 | ...11.11..1. |
| $4200'_x$ | 24 | 11331.2111.1 |
| $700_xx'$ | 25 | 1..1.....1.. |
| $2100'_x$ | 25 | 1222111.11.. |
| $840'_z$ | 28 | ...1....1... |
| $1400_zz'$ | 28 | .1111.1..11. |
| $972'_x$ | 30 | ..11..1..1.. |
| $2268'_x$ | 30 | 123221211.1. |
| $3240'_z$ | 31 | 12432131111. |
| $175'_x$ | 32 | 1..1...1.... |
| $350'_x$ | 32 | .12.1.1..... |
| $1050'_x$ | 32 | 11211.1..1.. |
| $1400'_x$ | 32 | .2212121..1. |
| $1575'_x$ | 32 | 122211111... |
| $525'_x$ | 36 | 11211.11.... |
| $56'_z$ | 37 | .1.1........ |
| $448'_z$ | 37 | 1.11..11.... |
| $1008'_z$ | 37 | 12311121.... |
| $1400'_z$ | 37 | .221111.1... |
| $300'_x$ | 42 | ..2...1..... |
| $700'_x$ | 42 | 11111111.... |
| $567'_x$ | 46 | 1221.11..... |
| $560'_z$ | 47 | .22.111..... |
| $50'_x$ | 52 | .1..1....... |
| $210'_x$ | 52 | .1...1...... |
| $28'_x$ | 63 | 1.1......... |
| $84'_x$ | 63 | 1111........ |
| $35'_x$ | 74 | .11......... |
| $8'_z$ | 91 | .1.......... |
| $1'_x$ | 120 | 1........... |

| $e=2$ | | $d_{\lambda,M}$ |
|---|---|---|
| $\bullet\, 112_z$ | 3 | 1..... |
| $\bullet\, 160_z$ | 4 | .1.... |
| $\bullet\, 400_z$ | 6 | 1.1... |
| $\bullet\, 1344_x$ | 7 | .1.1.. |
| $\bullet\, 2240_x$ | 10 | ...11. |
| $1296_z$ | 10 | 1..1.. |
| $\bullet\, 3360_z$ | 12 | .1.1.1 |
| $2800_z$ | 13 | 11111. |
| $3200_x$ | 15 | ...1.1 |
| $448_w$ | 16 | .11... |
| $1344_w$ | 16 | ..1.1. |
| $2016_w$ | 16 | .....1 |
| $5600_w$ | 16 | .1.211 |
| $3200'_x$ | 21 | ...1.1 |
| $3360'_z$ | 24 | .1.1.1 |
| $2800'_z$ | 25 | 11111. |
| $2240'_x$ | 28 | ...11. |
| $1296'_z$ | 30 | 1..1.. |
| $1344'_x$ | 37 | .1.1.. |
| $400'_z$ | 42 | 1.1... |
| $160'_z$ | 52 | .1.... |
| $112'_z$ | 63 | 1..... |

| $e=5$ | | $d_{\lambda,M}$ |
|---|---|---|
| $\bullet\, 1_x$ | 0 | 1............. |
| $\bullet\, 28_x$ | 3 | .1............ |
| $\bullet\, 84_x$ | 3 | 1.1........... |
| $\bullet\, 567_x$ | 6 | .1.1.......... |
| $\bullet\, 1344_x$ | 7 | ..111......... |
| $\bullet\, 972_x$ | 10 | 1.1.11........ |
| $\bullet\, 2268_x$ | 10 | ...1..1....... |
| $\bullet\, 4096_z$ | 11 | .1.11.11...... |
| $\bullet\, 168_y$ | 16 | 1....1..1..... |
| $\bullet\, 1134_y$ | 16 | .1.....1.1.... |
| $\bullet\, 2688_y$ | 16 | ....11.1..1... |
| $\bullet\, 4536_y$ | 16 | ......11...1.. |
| $\bullet\, 4096'_z$ | 26 | .......1.1111. |
| $\bullet\, 972'_x$ | 30 | .....1..1.1..1 |
| $2268'_x$ | 30 | ...........11. |
| $1344'_x$ | 37 | ..........1.11 |
| $567'_x$ | 46 | .........1..1. |
| $28'_x$ | 63 | .........1.... |
| $84'_x$ | 63 | ........1....1 |
| $1'_x$ | 120 | ........1..... |

**Table 7.15** $\Phi_e$-blocks of defect $\geqslant 2$ and decomposition numbers for type $E_8$ (cont'd)

| $e=3$ | | $d_{\lambda,M}$ |
|---|---|---|
| • $1_x$ | 0 | 1 . . . . . . . . . . . . . . . . . . . |
| • $35_x$ | 2 | . 1 . . . . . . . . . . . . . . . . . . |
| • $28_x$ | 3 | . . 1 . . . . . . . . . . . . . . . . . |
| • $84_x$ | 3 | 1 1 . 1 . . . . . . . . . . . . . . . . |
| • $50_x$ | 4 | 1 . . 1 1 . . . . . . . . . . . . . . . |
| • $210_x$ | 4 | . 1 1 . . 1 . . . . . . . . . . . . . . |
| • $300_x$ | 6 | . 1 . 1 . 1 1 . . . . . . . . . . . . . |
| • $700_x$ | 6 | . 1 . . . 1 . 1 . . . . . . . . . . . . |
| • $1344_x$ | 7 | 1 1 1 1 . 1 1 1 1 . . . . . . . . . . . |
| • $175_x$ | 8 | . . . . . 1 . . . 1 . . . . . . . . . . |
| • $350_x$ | 8 | . . 1 . . . . . . . 1 . . . . . . . . . |
| • $1050_x$ | 8 | . . . . . . . 1 1 . . 1 . . . . . . . . |
| • $1400_x$ | 8 | . . 1 . . 1 . . . . . . 1 . . . . . . . |
| • $2240_x$ | 10 | . . . . . 1 . 1 . 1 . . 1 1 . . . . . . |
| • $4096_z$ | 11 | 1 . 1 . . 1 . 1 1 . 1 . 1 1 1 . . . . . |
| $525_x$ | 12 | . . 1 . . . . . 1 . . . . . . . . . . . |
| $840_x$ | 12 | . . . . . . . 1 . . . . . 1 . . . . . . |
| • $4200_x$ | 12 | . . 1 . . 2 1 1 1 1 . 1 1 1 1 1 . . . . |
| • $700_xx$ | 13 | 1 . . 1 1 . 1 . 1 . . 1 . . . . 1 . . . |
| $2100_x$ | 13 | . . 1 . . 1 1 . 1 . 1 . . . 1 . . . . . |
| • $3200_x$ | 15 | 1 1 . 1 1 1 1 1 1 1 . . . 1 1 . . 1 . . |
| $70_y$ | 16 | . . . . . . 1 . . . . . . . . . . . . . |
| $168_y$ | 16 | 1 . . 1 1 . 1 . . . . . . . . . 1 . . . |
| $420_y$ | 16 | . . 1 . . 1 1 . . 1 . . . . . 1 . . . . |
| $1400_y$ | 16 | . . . . . 1 1 . . . . . . . 1 1 . . . . |
| $1680_y$ | 16 | . . . . . . . . . . 1 . . 1 1 . . . . . |
| $2688_y$ | 16 | 1 1 1 1 1 1 2 . 1 1 . 1 . . 1 1 1 1 . . |
| • $4200_y$ | 16 | . 1 1 . . 1 1 1 1 1 1 1 . 1 1 1 . 1 1 . |
| • $4480_y$ | 16 | . . 1 . . 1 . . . 1 1 . 1 1 1 1 . . . 1 |
| $2100_y$ | 20 | . . . . . . 1 . 1 . . . . . 1 . . 1 . . |
| $3200'_x$ | 21 | 1 . 1 . 1 . 1 . 1 . 1 1 . . 1 1 1 1 1 . |
| $840'_x$ | 24 | . . . . . . . . . . 1 . . . . . . . 1 . |
| $4200'_x$ | 24 | . 1 1 . . 1 1 . . 1 1 . . . 1 2 . 1 1 1 |
| $700_xx'$ | 25 | 1 1 . 1 1 . 1 . . . . . . . . . 1 1 . . |
| $2100'_x$ | 25 | . . . . . . 1 . . 1 . . . 1 1 1 . 1 . . |
| $4096'_z$ | 26 | . . . . 1 . . . . 1 1 . . 1 1 1 . 1 1 1 |
| $2240'_x$ | 28 | . . 1 . . . . . . . 1 . . . . 1 . . 1 1 |
| $175'_x$ | 32 | . . 1 . . . . . . . . . . . . 1 . . . . |
| $350'_x$ | 32 | . . . . . . . . . 1 . . . 1 . . . . . . |
| $1050'_x$ | 32 | . 1 . . . . . . . . . . . . . . . 1 1 . |
| $1400'_x$ | 32 | . . . . . . . . . 1 . . . . . 1 . . . 1 |
| $525'_x$ | 36 | . . . . . . . . . 1 . . . . . . . 1 . . |
| $1344'_x$ | 37 | . . . . 1 . 1 . . 1 . 1 . . . 1 1 1 1 . |
| $300'_x$ | 42 | . . . . . . 1 . . . . 1 . . . 1 1 . . . |
| $700'_x$ | 42 | . . . . . . . . . . . 1 . . . 1 . . 1 . |
| $50'_x$ | 52 | 1 . . . 1 . . . . . . . . . . . 1 . . . |
| $210'_x$ | 52 | . . . . . . . . . 1 . 1 . . . 1 . . . . |
| $28'_x$ | 63 | . . . . . . . . . 1 . . . . . . . . . . |
| $84'_x$ | 63 | . . . . 1 . . . . . . 1 . . . . 1 . . . |
| $35'_x$ | 74 | . . . . . . . . . . . 1 . . . . . . . . |
| $1'_x$ | 120 | . . . . 1 . . . . . . . . . . . . . . . |

| $e=3$ | | $d_{\lambda,M}$ |
|---|---|---|
| • $8_z$ | 1 | 1 . . . . . . . . . . . . |
| • $112_z$ | 3 | 1 1 . . . . . . . . . . . |
| • $160_z$ | 4 | . 1 1 . . . . . . . . . . |
| • $560_z$ | 5 | 2 1 1 1 . . . . . . . . . |
| • $400_z$ | 6 | 1 . . 1 1 . . . . . . . . |
| $56_z$ | 7 | . . 1 . . . . . . . . . . |
| • $448_z$ | 7 | 1 . . 1 . 1 . . . . . . . |
| • $1400_z$ | 7 | 1 1 1 1 . . 1 . . . . . . |
| • $840_z$ | 10 | 1 . . 1 . . . 1 . . . . . |
| • $1400_zz$ | 10 | . . . 1 1 1 1 . 1 . . . . |
| • $4096_x$ | 11 | 2 . 1 2 1 1 1 1 . 1 . . . |
| $3360_z$ | 12 | 1 . 1 1 1 1 1 . 1 1 . . . |
| $2800_z$ | 13 | . . 1 . . . 1 . . 1 . . . |
| $2400_z$ | 15 | . . 1 . . . . 1 . 1 . . . |
| • $4200_z$ | 15 | 1 . 1 1 2 1 1 1 1 1 1 . . |
| • $5600_z$ | 15 | 1 . 1 1 1 1 1 1 . 1 . 1 . |
| $448_w$ | 16 | . . . . . . . 1 . . . . . |
| $1344_w$ | 16 | 1 . 1 1 1 1 . 1 . . 1 . . |
| $5600_w$ | 16 | 2 . 1 1 2 1 . 2 . 1 1 1 . |
| • $7168_w$ | 16 | 2 1 2 1 2 2 1 1 1 1 1 1 1 |
| $2400'_z$ | 21 | . . . . . 1 . 1 . . . 1 . |
| $4200'_z$ | 21 | 2 1 1 1 1 1 . 1 . . 1 1 1 |
| $5600'_z$ | 21 | 1 . 1 . 1 1 . 1 . 1 1 1 1 |
| $3360'_z$ | 24 | 1 1 1 . 1 1 . . . . 1 1 1 |
| $2800'_z$ | 25 | . . . . . 1 . . . . . 1 1 |
| $4096'_x$ | 26 | 1 . 1 . 2 1 . 1 . . 2 1 1 |
| $840'_z$ | 28 | . . . . 1 . . 1 . . 1 . . |
| $1400_zz'$ | 28 | 1 1 1 . . . . . . . 1 . 1 |
| $56'_z$ | 37 | . . . . . 1 . . . . . . . |
| $448'_z$ | 37 | . . 1 . 1 . . . . . 1 . . |
| $1400'_z$ | 37 | . . . . 1 1 . . 1 . 1 . 1 |
| $400'_z$ | 42 | 1 . . . 1 . . . . . 1 . . |
| $560'_z$ | 47 | . . . . 2 1 . . 1 . 1 . . |
| $160'_z$ | 52 | . . . . . 1 . . 1 . . . . |
| $112'_z$ | 63 | . . . . 1 . . . 1 . . . . |
| $8'_z$ | 91 | . . . . 1 . . . . . . . . |

**Table 7.15** $\Phi_e$-blocks of defect $\geqslant 2$ and decomposition numbers for type $E_8$ (cont'd)

| $e=4$ | | $d_{\lambda,M}$ |
|---|---|---|
| $\bullet\ 1_x$ | 0 | 1.................. |
| $\bullet\ 35_x$ | 2 | 11................. |
| $\bullet\ 112_z$ | 3 | 111................ |
| $\bullet\ 50_x$ | 4 | .1.1............... |
| $\bullet\ 210_x$ | 4 | .1..1.............. |
| $\bullet\ 567_x$ | 6 | .11.11............. |
| $\bullet\ 400_z$ | 6 | 11111.1............ |
| $\bullet\ 175_x$ | 8 | 1.1...11........... |
| $\bullet\ 350_x$ | 8 | .....1..1.......... |
| $\bullet\ 1050_x$ | 8 | 1211111..1......... |
| $\bullet\ 1575_x$ | 8 | ..1.111...1........ |
| $1296_z$ | 10 | .....1..1.1........ |
| $\bullet\ 525_x$ | 12 | ..1..1.....1....... |
| $\bullet\ 3360_z$ | 12 | 111.112...1.1...... |
| $\bullet\ 2800_z$ | 13 | ....111...1..1..... |
| $\bullet\ 2835_x$ | 14 | 12111121.1..1.1.... |
| $\bullet\ 6075_x$ | 14 | .11.132.111111.1... |
| $70_y$ | 16 | ........1.......... |
| $420_y$ | 16 | .1.1.....1....1.... |
| $1134_y$ | 16 | .....1..11.1...1... |
| $2688_y$ | 16 | .1...1..11..1.11... |
| $\bullet\ 3150_y$ | 16 | 1111.121.1.11.111.. |
| $\bullet\ 4480_y$ | 16 | .1.1112..1..1111.1. |
| $\bullet\ 5670_y$ | 16 | .....12.1.1.11.1..1 |
| $2100_y$ | 20 | .....1..1..1.1.1... |
| $2835'_x$ | 22 | 11.1..21.1..1.2111. |
| $6075'_x$ | 22 | .....12.11.11113111 |
| $3360'_z$ | 24 | ......21....1.11111 |
| $2800'_z$ | 25 | ......1......1.1.11 |
| $1296'_z$ | 30 | ........1......1..1 |
| $175'_x$ | 32 | 1.....11........1.. |
| $350'_x$ | 32 | ........1......1... |
| $1050'_x$ | 32 | ...1..11.1....2111. |
| $1575'_x$ | 32 | ......1........1111 |
| $525'_x$ | 36 | ...........1...11.. |
| $400'_z$ | 42 | ...1..11......1.11. |
| $567'_x$ | 46 | ..............1111. |
| $50'_x$ | 52 | ...1..........1.... |
| $210'_x$ | 52 | ..............1..1. |
| $112'_z$ | 63 | .......1......1.1.. |
| $35'_x$ | 74 | .......1......1.... |
| $1'_x$ | 120 | .......1........... |

| $e=4$ | | $d_{\lambda,M}$ |
|---|---|---|
| $\bullet\ 56_z$ | 7 | 1....... |
| $\bullet\ 1008_z$ | 7 | .1...... |
| $\bullet\ 1400_z$ | 7 | ..1..... |
| $\bullet\ 3240_z$ | 9 | .111.... |
| $\bullet\ 2240_x$ | 10 | ..111... |
| $\bullet\ 4200_z$ | 15 | .1.1.1.. |
| $\bullet\ 3200'_x$ | 21 | 1....11. |
| $\bullet\ 4536'_z$ | 23 | ...1.111 |
| $840'_z$ | 28 | 1.....1. |
| $1400_zz'$ | 28 | ...11..1 |
| $1344'_x$ | 37 | ......11 |
| $560'_z$ | 47 | .......1 |
| $8'_z$ | 91 | ....1... |

| $e=4$ | | $d_{\lambda,M}$ |
|---|---|---|
| $\bullet\ 28_x$ | 3 | 1....... |
| $\bullet\ 160_z$ | 4 | .1...... |
| $\bullet\ 300_x$ | 6 | ..1..... |
| $\bullet\ 972_x$ | 10 | .111.... |
| $\bullet\ 840_x$ | 12 | ..111... |
| $\bullet\ 700_xx$ | 13 | 11...1.. |
| $\bullet\ 1344_w$ | 16 | .1.1.11. |
| $\bullet\ 840'_x$ | 24 | 1....1.1 |
| $700_xx'$ | 25 | ...11.1. |
| $972'_x$ | 30 | .....111 |
| $300'_x$ | 42 | .......1 |
| $160'_z$ | 52 | ......1. |
| $28'_x$ | 63 | ....1... |

| $e=4$ | | $d_{\lambda,M}$ |
|---|---|---|
| $\bullet\ 8_z$ | 1 | 1....... |
| $\bullet\ 560_z$ | 5 | .1...... |
| $\bullet\ 1344_x$ | 7 | .11..... |
| $\bullet\ 840_z$ | 10 | ..11.... |
| $\bullet\ 1400_zz$ | 10 | 11..1... |
| $\bullet\ 4536_z$ | 13 | .11.11.. |
| $3200_x$ | 15 | ..11.1.. |
| $\bullet\ 4200'_z$ | 21 | ....111. |
| $\bullet\ 2240'_x$ | 28 | 1...1..1 |
| $3240'_z$ | 31 | ....1.11 |
| $56'_z$ | 37 | ...1.... |
| $1008'_z$ | 37 | ......1. |
| $1400'_z$ | 37 | .......1 |

| $e=4$ | | $d_{\lambda,M}$ |
|---|---|---|
| $\bullet\ 84_x$ | 3 | 1....... |
| $\bullet\ 700_x$ | 6 | 11...... |
| $\bullet\ 2268_x$ | 10 | .11..... |
| $\bullet\ 4200_x$ | 12 | 1111.... |
| $\bullet\ 2100_x$ | 13 | ..1.1... |
| $448_w$ | 16 | ....1... |
| $\bullet\ 2016_w$ | 16 | 1..1.1.. |
| $\bullet\ 5600_w$ | 16 | ..111.1. |
| $\bullet\ 4200'_x$ | 24 | ...1.111 |
| $2100'_x$ | 25 | ....1.1. |
| $2268'_x$ | 30 | ......11 |
| $700'_x$ | 42 | .....1.1 |
| $84'_x$ | 63 | .....1.. |

| $e=8$ | | $d_{\lambda,M}$ |
|---|---|---|
| $\bullet\ 1_x$ | 0 | 1................ |
| $\bullet\ 35_x$ | 2 | 11............... |
| $\bullet\ 160_z$ | 4 | ..1.............. |
| $\bullet\ 567_x$ | 6 | .111............. |
| $\bullet\ 175_x$ | 8 | 1...1............ |
| $\bullet\ 1400_x$ | 8 | 11.1.1........... |
| $\bullet\ 1575_x$ | 8 | ..11..1.......... |
| $\bullet\ 525_x$ | 12 | ...1...1......... |
| $\bullet\ 2835_x$ | 14 | 1...11..1........ |
| $\bullet\ 6075_x$ | 14 | ...1.111.1....... |
| $\bullet\ 2016_w$ | 16 | ....1...1.1...... |
| $\bullet\ 5600_w$ | 16 | ......1..1.1..... |
| $\bullet\ 7168_w$ | 16 | .....1..11..1.... |
| $\bullet\ 2835'_x$ | 22 | ........1.1.11... |
| $\bullet\ 6075'_x$ | 22 | .......1.1.11.1.. |
| $175'_x$ | 32 | ..........1..1... |
| $\bullet\ 1400'_x$ | 32 | ............1111. |
| $\bullet\ 1575'_x$ | 32 | ...........1..1.1 |
| $525'_x$ | 36 | .......1......1.. |
| $567'_x$ | 46 | ..............111 |
| $160'_z$ | 52 | ................1 |
| $35'_x$ | 74 | .............1.1. |
| $1'_x$ | 120 | .............1... |

**Table 7.15** $\Phi_e$-blocks of defect $\geqslant 2$ and decomposition numbers for type $E_8$ (cont'd)

| $e=6$ | | $d_{\lambda,M}$ |
|---|---|---|
| • $1_x$ | 0 | 1................................ |
| • $8_z$ | 1 | .1............................... |
| • $35_x$ | 2 | ..1.............................. |
| • $28_x$ | 3 | ...1............................. |
| • $84_x$ | 3 | 111.1............................ |
| • $50_x$ | 4 | 11...1........................... |
| • $210_x$ | 4 | ......1.......................... |
| • $560_z$ | 5 | .111..11......................... |
| • $300_x$ | 6 | ..1.1...1........................ |
| • $700_x$ | 6 | 111.1111.1....................... |
| • $56_z$ | 7 | ..........1...................... |
| • $448_z$ | 7 | 111.1..1...1..................... |
| • $1400_z$ | 7 | .1111.1111..1.................... |
| • $175_x$ | 8 | 11..11.......1................... |
| • $350_x$ | 8 | ...1......1...1.................. |
| • $1050_x$ | 8 | .1111..1.......1................. |
| • $1400_x$ | 8 | .1.1.111.1..1...1................ |
| • $840_z$ | 10 | .1.1...1......1..1............... |
| • $1400_zz$ | 10 | 111.21.1.1.1.1.1..1.............. |
| $525_x$ | 12 | .1.1........1.................... |
| • $840_x$ | 12 | 11..11.1.1.1.....1.1............. |
| • $4200_x$ | 12 | .1112..211.11..1..1.1............ |
| $700_xx$ | 13 | ...............1..1.............. |
| • $2100_x$ | 13 | ...1....1.1.1.1......1........... |
| • $2400_z$ | 15 | ...1......1.1.1.1....11.......... |
| • $4200_z$ | 15 | .1.111.1.1..11.11.1.1..1......... |
| • $5600_z$ | 15 | .1.11..111..2.1..1..11..1........ |
| $168_y$ | 16 | .....1...1.........1............. |
| • $420_y$ | 16 | 1...11...1.1.1....11.....1....... |
| • $1680_y$ | 16 | ..........1...1......11...1...... |
| • $4200_y$ | 16 | ....2..1.1.1.1.1..2.1..1...1..... |
| • $4480_y$ | 16 | .1..11.1.2.111..11111..11...1.... |
| • $4536_y$ | 16 | ....1...11..1.....1.11..1....1... |
| • $5670_y$ | 16 | ...1...1.1..1.1.11..11111.....1.. |
| $2100_y$ | 20 | .........1..1........1..1........ |
| $2400'_z$ | 21 | ..............1..1...11.1.1...1.. |
| $4200'_z$ | 21 | ....1..1.1.1.....1111..11..11.1.. |
| $5600'_z$ | 21 | .........1..1...1.1.11112...111.. |
| $840'_x$ | 24 | .....1...1...1..1.11...1.1..1.... |
| • $4200'_x$ | 24 | ....1....1...1....2.1..21..11111. |
| $700_xx'$ | 25 | ....1......................1..... |
| $2100'_x$ | 25 | .....................11.1.1..11.. |
| $840'_z$ | 28 | ................1.....11....1.1.. |
| $1400_zz'$ | 28 | ....1....1.1.1....21...1.1.11..1. |
| $175'_x$ | 32 | ...........1......11.....1..1.... |
| $350'_x$ | 32 | ......................1...1...1.. |
| $1050'_x$ | 32 | ..................1....1...11.11. |
| • $1400'_x$ | 32 | .........1.......1.1...11...1.1.1 |

| | | |
|---|---|---|
| $525'_x$ | 36 | ........................1...1.1.. |
| $56'_z$ | 37 | ..........................1...... |
| $448'_z$ | 37 | ...............1...1...1.1..1..1. |
| $1400'_z$ | 37 | ..........1........1...11...11111 |
| $300'_x$ | 42 | ...................1.........1.1. |
| $700'_x$ | 42 | ..........1........11..1.1..1..11 |
| $560'_z$ | 47 | .......................1....1.111 |
| $50'_x$ | 52 | ....................1....1..1.... |
| $210'_x$ | 52 | ................................1 |
| $28'_x$ | 63 | ..............................1.. |
| $84'_x$ | 63 | ...................1.....1..1..1. |
| $35'_x$ | 74 | ...............................1. |
| $8'_z$ | 91 | ............................1.... |
| $1'_x$ | 120 | .........................1....... |

| $e=6$ | | $d_{\lambda,M}$ |
|---|---|---|
| • $112_z$ | 3 | 1.......... |
| • $160_z$ | 4 | .1......... |
| • $400_z$ | 6 | 1.1........ |
| • $1344_x$ | 7 | 11.1....... |
| • $2240_x$ | 10 | 1.111...... |
| • $3360_z$ | 12 | .1.1.1..... |
| • $3200_x$ | 15 | ...1..1.... |
| • $1344_w$ | 16 | ..1.1..1... |
| • $7168_w$ | 16 | ...1111.1.. |
| $3200'_x$ | 21 | .....1..1.. |
| • $3360'_z$ | 24 | ......1.11. |
| • $2240'_x$ | 28 | ....1..11.1 |
| $1344'_x$ | 37 | ........111 |
| $400'_z$ | 42 | .......1..1 |
| $160'_z$ | 52 | .........1. |
| $112'_z$ | 63 | ..........1 |

# References


1. Ackermann, B.: A short note on Howlett–Lehrer Theory. Archiv Math. **79**, 161–166 (2002).
2. Alvis, D.: The left cells of the Coxeter group of type $H_4$. J. Algebra **107**, 160–168 (1987); see also http://mypage.iusb.edu/~dalvis/h4data.
3. Alvis, D., Lusztig, G.: The representations and generic degrees of the Hecke algebra of type $H_4$. J. Reine Angew. Math. **336**, 201–212 (1982). Corrections: *ibid* **449**, 217–218 (1994).
4. Andrews, G.E.: The theory of partitions. Reprint of the 1976 original. Cambridge Mathematical Library. Cambridge University Press, Cambridge (1998).
5. Ariki, S.: On the semi-simplicity of the Hecke algebra of $(\mathbb{Z}/r\mathbb{Z}) \wr \mathfrak{S}_n$. J. Algebra **169**, 216–225 (1994).
6. Ariki, S.: Representation theory of a Hecke algebra of $G(r,p,n)$. J. Algebra **177**, 164–185 (1995).
7. Ariki, S.: On the decomposition numbers of the Hecke algebra of $G(m,1,n)$. J. Math. Kyoto Univ. **36**, 789–808 (1996).
8. Ariki, S.: Robinson–Schensted correspondence and left cells. In: Combinatorial methods in representation theory (Kyoto, 1998), Adv. Stud. Pure Math., **28**, pp. 1–20. Kinokuniya, Tokyo (2000).
9. Ariki, S.: On the classification of simple modules for cyclotomic Hecke algebras of type $G(m,1,n)$ and Kleshchev multipartitions. Osaka J. Math. **38**, 827–837 (2001).
10. Ariki, S.: Representations of quantum algebras and combinatorics of Young tableaux. University Lecture Series **26**, Amer. Math. Soc., Providence, RI (2002).
11. Ariki, S., Jacon, N.: Dipper–James–Murphy's conjecture for Hecke algebras of type $B_n$. In: Representation theory of algebraic groups and quantum groups (Nagoya, 2006; eds. A. Gyoja et al.), pp. 17–32. Progress in Math. **284**, Birkhäuser, Boston (2010).
12. Ariki, S., Jacon, N., Lecouvey, C.: Factorization of the canonical bases for higher level Fock spaces. Proc. Edinburgh Math. Soc. (to appear).
13. Ariki, S., Koike, K.: A Hecke algebra of $(Z/rZ) \wr \mathfrak{S}_n$ and construction of its irreducible representations. Adv. Math. **106**, 216–243 (1994).
14. Ariki, S., Kreiman, V., Tsuchioka, S.: On the tensor product of two basic representations of $U_v(\widehat{\mathfrak{sl}}_e)$. Adv. Math. **218** (2008), 28–86.
15. Ariki, S., Mathas, A.: The number of simple modules of the Hecke algebras of type $G(r,1,n)$. Math. Z. **233**, 601–623 (2000).
16. Bellamy, G.: The Calogero–Moser partition for $G(m,d,n)$. Nagoya J. Math. (to appear).
17. Benard, M.: Schur indices and splitting fields of the unitary reflection groups. J. Algebra **38**, 318–342 (1976).
18. Bessis, S.: Sur le corps de définition d'un groupe de réflexions complexe. Comm. Algebra **25**, 2703–2716 (1997).
19. Bezrukavnikov, R., Etingof, P.: Parabolic induction and restruction functors for rational Cherednik algebras. Selecta Math. (N.S.) **14**, 397–425 (2009).



M. Geck, N. Jacon, *Representations of Hecke Algebras at Roots of Unity*, Algebra and Applications 15, DOI 10.1007/978-0-85729-716-7,

20. Bleher, B., Geck, M., Kimmerle, W.: Automorphisms of integral group rings of finite Coxeter groups and Iwahori–Hecke algebras. J. Algebra **197**, 615–655 (1997).
21. Bonnafé, C.: Two-sided cells in type $B$ in the asymptotic case. J. Algebra **304**, 216–236 (2006).
22. Bonnafé, C.: Semicontinuity properties of Kazhdan–Lusztig cells. New Zealand J. Math. **39**, 171–192 (2009).
23. Bonnafé, C.: On Kazhdan-Lusztig cells in type $B$. Journal of Alg. Comb. **31**, 53–82 (2010).
24. Bonnafé, C., Dyer, M.: Semidirect product decomposition of Coxeter groups. Comm. Algebra **38**, 1549–1574 (2010).
25. Bonnafé, C., Geck, M., Iancu, L., Lam, T.: On domino insertion and Kazhdan–Lusztig cells in type $B_n$. In: Representation theory of algebraic groups and quantum groups (Nagoya, 2006; eds. A. Gyoja et al.), pp. 33–54. Progress in Math. **284**, Birkhäuser, Boston (2010).
26. Bonnafé, C., Iancu, L.: Left cells in type $B_n$ with unequal parameters. Represent. Theory **7**, 587–609 (2003).
27. Bonnafé, C., Jacon, N.: Cellular structures on Hecke algebras of type $B$. Special issue in honour of Gus Lehrer, J. Algebra **321**, 3089–3111 (2009).
28. Bonnafé, C., Rouquier, R.: Catégories dérivées et variétés de Deligne–Lusztig. Publ. Math. Inst. Hautes Études Sci. **97**, 1–59 (2003).
29. Bourbaki, N.: Groupes et algèbres de Lie, Chap. 4, 5 et 6. Hermann, Paris (1968).
30. Bremke, K.: The decomposition numbers of Hecke algebras of type $F_4$ with unequal parameters. Manuscripta Math. **83**, 331–346 (1994).
31. Bremke, K., Malle, G.: Reduced words and a length function for $G(e,1,n)$. Indag. Math. (N.S.) **8**, 453–469 (1997).
32. Broué, M.: Introduction to complex reflection groups and their braid groups. Lecture Notes in Mathematics 1988. Springer-Verlag, Berlin (2010).
33. Broué, M., Kim, S.: Familles de caractères des algèbres de Hecke cyclotomiques. Adv. Math. **172**, 53–136 (2002).
34. Broué, M., Malle, G.: Zyklotomische Heckealgebren. Astérisque **212**, 119–189 (1993).
35. Broué, M., Malle, G., Michel, J.: Towards Spetses I. Trans. Groups **4**, 157–218 (1999).
36. Broué, M., Malle, G., Rouquier, R.: Complex reflection groups, braid groups, Hecke algebras. J. Reine Angew. Math. **500**, 127–190 (1998).
37. Broué, M., Michel, J.: Blocs et séries de Lusztig dans un groupe réductif fini. J. Reine Angew. Math. **395**, 56–67 (1989).
38. Brundan, J., Dipper, R., Kleshchev, A.: Quantum linear groups and representations of $\mathrm{GL}_n(\mathbb{F}_q)$. Mem. Amer. Math. Soc. **149** (2001), no. 706, viii+112 pp.
39. Brundan, J., Kleshchev, A.: Blocks of cyclotomic Hecke algebras and Khovanov–Lauda algebras. Invent. Math. **178**, 451–484 (2009).
40. Brundan, J., Kleshchev, A.: Graded decomposition numbers for cyclotomic Hecke algebras. Adv. Math. **222** 1883–1942 (2009).
41. Brundan, J., Kleshchev, A., Wang, W.: Graded Specht modules. J. Reine Angew. Math. (to appear).
42. Burkhardt, R.: Über die Zerlegungszahlen der Suzukigruppen $\mathrm{Sz}(q)$. J. Algebra **59**, 421–433 (1979).
43. Cabanes, M., Enguehard, M.: Representation theory of finite reductive groups. New Mathematical Monographs, 1. Cambridge University Press, Cambridge (2004).
44. Carter, R.W.: Simple groups of Lie type. Wiley, New York (1972). Reprinted 1989 as Wiley Classics Library Edition.
45. Carter, R.W.: Finite groups of Lie type: conjugacy classes and complex characters. Wiley, New York (1985). Reprinted 1993 as Wiley Classics Library Edition.
46. Chlouveraki, M.: On the cyclotomic Hecke algebras of complex reflection groups. Ph. D. thesis, Université Paris 7 (2007); available at arXiv:math/0710.0776.
47. Chlouveraki, M.: Rouquier blocks of the cyclotomic Ariki–Koike algebras. Algebra Number Theory **2**, 689–720 (2008).
48. Chlouveraki, M.: Blocks and families for cyclotomic Hecke algebras. Lecture Notes in Mathematics 1981, Springer, Berlin (2009).

49. Chlouveraki, M., Gordon, I., Griffeth, S.: Cell modules and canonical basic sets for Hecke algebras via Cherednik algebras. Preprint (2011).
50. Chriss, N., Ginzburg, V.: Representation theory and complex geometry. Birkhäuser, Boston (1997).
51. Cline, E., Parshall, B., Scott, L.: Generic and $q$-rational representation theory. Publ. RIMS Kyoto Univ. **35**, 31–90 (1999).
52. Curtis, C.W., Iwahori, N., Kilmoyer, R.W.: Hecke algebras and characters of parabolic type of finite groups with BN-pairs. Inst. Hautes Études Sci. Publ. Math. **40**, 81–116 (1971).
53. Curtis, C.W., Reiner, I.: Methods of representation theory Vols. I and II. Wiley, New York (1981, 1987). Reprinted 1990 and 1994 as Wiley Classics Library Edition.
54. Deligne, P., Lusztig, G.: Representations of reductive groups over finite fields. Annals of Math. **103**, 103–161 (1976).
55. Digne, F., Michel, J.: Representations of finite groups of Lie type. London Math. Soc. Students Texts **21**, Cambridge University Press, Cambridge (1991).
56. Dipper, R.: On the decomposition numbers of the finite general linear groups. Trans. Amer. Math. Soc. **290**, 315–343 (1985).
57. Dipper, R.: On the decomposition numbers of the finite general linear groups II. Trans. Amer. Math. Soc. **292**, 123–133 (1985).
58. Dipper, R.: On quotients of Hom-functors and representations of finite general linear groups I. J. Algebra **130**, 235–259 (1990).
59. Dipper, R.: On quotients of Hom-functors and representations of finite general linear groups II. J. Algebra **209**, 199–269 (1998).
60. Dipper, R., Du, J.: Harish-Chandra vertices. J. Reine Angew. Math. **437**, 101–130 (1993).
61. Dipper, R., Fleischmann, P.: Modular Harish-Chandra theory I. Math. Z. **211**, 49–71 (1992).
62. Dipper, R., James, G.D.: Representations of Hecke algebras of general linear groups. Proc. London Math. Soc. **52**, 20–52 (1986).
63. Dipper, R., James, G.D.: Blocks and idempotents of Hecke algebras of general linear groups. Proc. London Math. Soc. **54**, 57–82 (1987).
64. Dipper, R., James, G.D.: The $q$-Schur algebra. Proc. London Math. Soc. **59**, 23–50 (1989).
65. Dipper, R., James, G.D.: $q$-Tensor space and $q$-Weyl modules. Trans. Amer. Math. Soc. **327**, 251–282 (1991).
66. Dipper, R., James, G.D.: Representations of Hecke algebras of type $B_n$. J. Algebra **146**, 454–481 (1992).
67. Dipper, R., James, G.D., Mathas, A.: Cyclotomic $q$-Schur algebras. Math. Z. **229**, 385–416 (1998).
68. Dipper, R., James, G.D., Murphy, G.E.: Hecke algebras of type $B_n$ at roots of unity. Proc. London Math. Soc. **70**, 505–528 (1995).
69. Dipper, R., James, G.D., Murphy, G.E.: Gram determinants of type $B_n$. J. Algebra **189**, 481–505 (1997).
70. Dipper, R., Mathas, A.: Morita equivalences for Ariki–Koike algebras. Math. Z. **240**, 579–610 (2002).
71. Donkin, S.: An introduction to the Lusztig conjecture. In: Representations of reductive groups (eds. M. Geck and R.W. Carter), Publ. Newton Inst., pp. 173–87. Cambridge Univeristy Press (1998).
72. Donkin, S.: The $q$-Schur algebra. London Math. Soc. Lecture Notes Series **253**. Cambridge University Press, Cambridge (1998).
73. Du, J., Parshall, B., Scott, L.: Stratifying endomorphism algebras associated to Hecke algebras. J. Algebra **203**, 169–210 (1998).
74. Du, J., Rui, H.: Ariki–Koike algebras with semisimple bottoms. Math. Z. **229**, 807–830 (2000).
75. DuCloux, F.: Computing Kazhdan–Lusztig polynomials for arbitrary Coxeter groups. Experiment. Math. **11** (2002), 371–381.
76. DuCloux, F.: Positivity results for the Hecke algebras of noncrystallographic finite Coxeter group. J. Algebra **303** (2006), 731–741.

77. Etingof, P., Ginzburg, V.: Symplectic reflection algebras, Calogero–Moser space and deformed Harish-Chandra homomorphism. Invent. Math. **147**, 243–348 (2002).
78. Fayers, M.: Weight two blocks of Iwahori–Hecke algebras of type $B$. J. Algebra **301**, 154–201 (2006).
79. Fayers, M.: Decomposition numbers for weight three blocks of symmetric groups and Iwahori–Hecke algebras. Trans. Amer. Math. Soc. **360**, 1341–1376 (2008).
80. Fayers, M.: An extension of James's conjecture. Int. Math. Res. Not. IMRN, no. 10, Art. ID rnm032 (2007).
81. Fayers, M.: James's conjecture holds for weight four blocks of Iwahori–Hecke algebras. J. Algebra **317**, 593–633 (2007).
82. Fayers, M.: An LLT-type algorithm for computing higher-level canonical bases. J. Pure Appl. Algebra **214**, 2186–2198 (2010).
83. Feit, W.: The representation theory of finite groups. North-Holland, Amsterdam (1982).
84. Fiebig, P.: Sheaves on affine Schubert varieties, modular representations and Lusztig's conjecture. J. Amer. Math. Soc. **24**, 133–181 (2011).
85. Fiebig, P.: An upper bound on the exceptional characteristics for Lusztig's character formula. J. Reine Angew. Math. (to appear).
86. Fleischmann, P.: On Lusztig's isomorphism theorems for Hecke algebras and representation theory. J. London Math. Soc. **47**, 450–472 (1993).
87. Fong, P., Srinivasan, B.: Brauer trees in classical groups, J. Algebra **131**, 179–225 (1990).
88. Foda, O., Leclerc, B., Okado, M., Thibon, J-Y., Welsh, T.: Branching functions of $A_{n-1}^{(1)}$ and Jantzen–Seitz problem for Ariki–Koike algebras. Adv. Math. **141**, 322–365 (1999).
89. Geck, M.: Irreducible Brauer characters of the 3-dimensional special unitary groups in non-defining characteristic. Comm. Algebra **18**, 563–584 (1990).
90. Geck, M.: Verallgemeinerte Gelfand–Graev Charaktere und Zerlegungszahlen endlicher Gruppen vom Lie-Typ. Ph. D. thesis, RWTH Aachen (1990).
91. Geck, M.: On the classification of $l$-blocks of finite groups of Lie type. J. Algebra **151**, 180–191 (1992).
92. Geck, M.: Brauer trees of Hecke algebras. Comm. Algebra **20**, 2937–2973 (1992).
93. Geck, M.: Basic sets of Brauer characters of finite groups of Lie type, II. J. London Math. Soc. (2) **47**, 255–268 (1993).
94. Geck, M.: The decomposition numbers of the Hecke algebra of type $E_6$. Math. Comp. **61**, 889–899 (1993).
95. Geck, M.: A note on Harish-Chandra induction. Manuscripta Math. **80**, 393–401 (1993).
96. Geck, M.: Beiträge zur Darstellungstheorie von Iwahori–Hecke-Algebren. Habilitationsschrift, Aachener Beiträge zur Mathematik **11**, Verlag der Augustinus Buchhandlung, Aachen (1995).
97. Geck, M.: Kazhdan–Lusztig cells and decomposition numbers. Represent. Theory **2**, 264–277 (1998).
98. Geck, M.: Representations of Hecke algebras at roots of unity. Séminaire Bourbaki, 50ème année, 1997–98, Astérisque No. 252, Exp. 836, pp. 33–55 (1998).
99. Geck, M.: The character table of the Iwahori–Hecke algebra of the symmetric group: Starkey's rule. C. R. Acad. Sci. Paris **329**, 361–366 (1999).
100. Geck, M.: On the number of simple modules of Iwahori–Hecke algebras of finite Weyl groups. Bul. Stiit. Univ. Baia Mare, Ser. B **16**, 235–246 (2000).
101. Geck, M.: On the representation theory of Iwahori–Hecke algebras of extended finite Weyl groups. Represent. Theory **4**, 370–397 (2000).
102. Geck, M.: Modular Harish-Chandra series, Hecke algebras and (generalized) $q$-Schur algebras. In: Modular representation theory of finite groups (Charlottesville, VA, 1998; eds. M.J. Collins, B.J. Parshall and L.L. Scott), p. 1–66. Walter de Gruyter, Berlin (2001).
103. Geck, M.: Constructible characters, leading coefficients and left cells for finite Coxeter groups with unequal parameters. Represent. Theory **6**, 1–30 (2002).
104. Geck, M.: An introduction to algebraic geometry and algebraic groups. Oxford Graduate Texts in Mathematics **10**. Oxford University Press, New York (2003).

105. Geck, M.: Computing Kazhdan–Lusztig cells for unequal parameters. J. Algebra **281**, 342–365 (2004).
106. Geck, M.: Left cells and constructible representations. Represent. Theory **9**, 385–416 (2005); Erratum, ibid. **11**, 172–173 (2007).
107. Geck, M.: Kazhdan–Lusztig cells and the Murphy basis. Proc. London Math. Soc. **93**, 635–665 (2006).
108. Geck, M.: Relative Kazhdan–Lusztig cells. Represent. Theory **10**, 481–524 (2006).
109. Geck, M.: Modular principal series representations. Int. Math. Res. Notices, Article ID 41957, pp. 1–20 (2006).
110. Geck, M.: Modular representations of Hecke algebras, In: Group representation theory (EPFL, 2005; eds. M. Geck, D. Testerman and J. Thévenaz), pp. 301–353. EPFL Press, Lausanne (2007).
111. Geck, M.: Hecke algebras of finite type are cellular. Invent. Math. **169**, 501–517 (2007).
112. Geck, M.: Leading coefficients and cellular bases of Hecke algebras. Proc. Edinburgh Math. Soc. **52**, 653–677 (2009).
113. Geck, M.: On the Kazhdan–Lusztig order on cells and families. Comm. Math. Helv. (to appear).
114. Geck, M.: On Iwahori–Hecke algebras with unequal parameters and Lusztig's isomorphism theorem. Jacques Tits special issue, Pure Appl. Math. Q. **7**, 587–620 (2011).
115. Geck, M.: Some applications of CHEVIE to the theory of algebraic groups. Carpath. J. Math. **27**, 64–94 (2011).
116. Geck, M., Hiss, G.: Basic sets of Brauer characters of finite groups of Lie type. J. Reine Angew. Math. **418**, 173–188 (1991).
117. Geck, M., Hiss, G.: Modular representations of finite groups of Lie type in non-defining characteristic. In: Finite reductive groups: related structures and representations (ed. M. Cabanes), pp. 195–249. Birkhäuser, Basel (1997).
118. Geck, M., Hiß, G., Lübeck, F., Malle, G., Pfeiffer, G.: CHEVIE – A system for computing and processing generic character tables for finite groups of Lie type, Weyl groups and Hecke algebras. Appl. Algebra Engrg. Comm. Comput. **7**, 175–210 (1996).
119. Geck, M., Hiss, G., Malle, M.: Cuspidal unipotent Brauer characters. J. Algebra **168**, 182–220 (1994).
120. Geck, M., Hiss, G., Malle, G.: Towards a classification of the irreducible representations in non-defining characteristic of a finite group of Lie type. Math. Z. **221**, 353–386 (1996).
121. Geck, M., Iancu, L.: Lusztig's $a$-function in type $B_n$ in the asymptotic case. Special issue celebrating the 60th birthday of George Lusztig, Nagoya J. Math. **182**, 199–240 (2006).
122. Geck, M., Iancu, L.: On the Kazhdan–Lusztig order on families in type $B_n$. Preprint (2011).
123. Geck, M., Iancu, L., Malle, G.: Weights of Markov traces and generic degrees. Indag. Math. (N.S.) **11**, 379–397 (2000).
124. Geck, M., Iancu, L., Pallikaros, C.A.: Specht modules and Kazhdan–Lusztig cells in type $B_n$. J. Pure Appl. Algebra **212**, 1310–1320 (2008).
125. Geck, M., Jacon, N.: Canonical basic sets in type $B$. Special issue in honour of Gordon Douglas James, J. Algebra **306**, 104–127 (2006).
126. Geck, M., Lux, K.: The decomposition numbers of the Hecke algebra of type $F_4$. Manuscripta Math. **70**, 285–306 (1991).
127. Geck, M., Malle, G.: Cuspidal unipotent classes and cuspidal Brauer characters. J. London Math. Soc. **53**, 63–78 (1996).
128. Geck, M., Malle, G.: On the existence of a unipotent support for the irreducible characters of finite groups of Lie type. Trans. Amer. Math. Soc. **352**, 429–456 (2000).
129. Geck, M., Müller, J.: James' conjecture for Hecke algebras of exceptional type I. Special issue in honour of Gus Lehrer, J. Algebra **321**, 3274–3298 (2009).
130. Geck, M., Pfeiffer, G.: The unipotent characters of the Chevalley groups $D_4(q)$, $q$ odd. Manuscripta Math. **76**, 281–304 (1992).
131. Geck, M., Pfeiffer, G.: On the irreducible characters of Hecke algebras. Adv. Math. **102**, 79–94 (1993).

132. Geck, M., Pfeiffer, G.: Characters of finite Coxeter groups and Iwahori–Hecke algebras. London Math. Soc. Monographs, New Series **21**. Oxford University Press, New York (2000).
133. Geck, M., Rouquier, R.: Centers and simple modules for Iwahori–Hecke algebras. In: Finite reductive groups, related structures and representations (M. Cabanes, ed.), Progress in Math. **141**, pp. 251–272. Birkhäuser, Boston (1997).
134. Geck, M., Rouquier, R.: Filtrations on projective modules for Iwahori–Hecke algebras. In: Modular representation theory of finite groups (Charlottesville, VA, 1998; eds. M.J. Collins, B.J. Parshall and L.L. Scott), pp. 211–221. Walter de Gruyter, Berlin (2001).
135. Genet, G.: On decomposition matrices for graded algebras. J. Algebra **274**, 523–542 (2004).
136. Genet, G., Jacon, N.: Modular representations of cyclotomic Hecke algebras of type $G(r,p,n)$. Int. Math. Res. Not. (2006), Art. ID 93049, 18 pp.
137. Ginzburg, V., Guay, N., Opdam, E., Rouquier, R.: On the category $\mathcal{O}$ for rational Cherednik algebras. Invent. Math. **154**, 617–651 (2003).
138. Goldschmidt, D.M.: Lectures on character theory. Publish or Perish, Berkeley (1980).
139. Gordon, I.: Baby Verma modules for rational Cherednik algebras. Bull. London Math. Soc. **35**, 321–336 (2003).
140. Gordon, I.: Quiver varieties, category $\mathcal{O}$ for rational Cherednik algebras, and Hecke algebras. Int. Math. Res. Pap. IMRP, no. 3, Art. ID rpn006 (2008).
141. Gordon, I., Martino, M.: Calogero–Moser space, reduced rational Cherednik algebras, and two-sided cells. Math. Res. Lett. **16**, 255–262 (2009).
142. Gorenstein, D., Lyons, R., Solomon, R.: The classification of the finite simple groups. Math. Surveys and Monographs, vol. 40.1. Amer. Math. Soc., Providence, RI (1994).
143. Graham, J.J.: Modular representations of Hecke algebras and related algebras. Ph. D. thesis, University of Sydney (1995).
144. Graham, J.J., Lehrer, G.I.: Cellular algebras, Invent. Math. **123**, 1–34 (1996).
145. Green, J.A.: Polynomial representations of $\mathrm{GL}_n$. Lecture Notes in Math. **830**. Springer-Verlag, Berlin (1980).
146. Grojnowski, I.: Affine $\mathfrak{sl}_p$ controls the representation theory of the symmetric group and related Hecke algebras. Preprint (1999); see arXiv:math/9907129.
147. Grove, L.C.: Classical groups and geometric algebra. Graduate Studies in Math., vol. 39. Amer. Math. Soc., Providence, RI (2002).
148. Gruber, J.: Green vertex theory, Green correspondence and Harish-Chandra induction. J. Algebra **186**, 476–521 (1997).
149. Gruber, J., Hiss, G.: Decomposition numbers of finite classical groups for linear primes. J. Reine Angew. Math. **485**, 55–91 (1997).
150. Gyoja, A.: On the existence of a $W$-graph for an irreducible representation of a finite Coxeter group. J. Algebra **86**, 422–438 (1984).
151. Gyoja, A.: Cells and modular representations of Hecke algebras. Osaka J. Math. **33**, 307–341 (1996).
152. Hiss, G.: On the decomposition numbers of $G_2(q)$. J. Algebra **120**, 339–360 (1989).
153. Hiss, G.: Zerlegungszahlen endlicher Gruppen vom Lie-Typ in nicht-definierender Charakteristik. Habilitationsschrift, RWTH Aachen, (1989).
154. Hiss, G.: Harish-Chandra series of Brauer characters in a finite group with a split $BN$-pair. J. London Math. Soc. **48**, 219–228 (1993).
155. Hiss, G.: Supercuspidal representations of finite reductive groups. J. Algebra **184**, 839–851 (1996).
156. Hiss, G.: Hermitian function fields, classical unitals, and representations of 3-dimensional unitary groups. Indag. Math. (N.S.) **15**, 223–243 (2004).
157. Hoefsmit, P.N.: Representations of Hecke algebras of finite groups with BN-pairs of classical type. Ph. D. thesis, University of British Columbia, Vancouver (1974).
158. Hong, J., Kang, S.-J.: Introduction to quantum groups and crystal bases. Graduate Studies in Mathematics, vol. 42. Amer. Math. Soc., Providence, RI (2002).
159. Hotta, R., Takeuchi, K., Tanisaki, T.: $D$-modules, perverse sheaves, and representation theory. Progress in Math. **236**. Birkhäuser, Boston (2008).

160. Howlett, R.B., Kilmoyer, R.W.: Principal series representations of finite groups with split $BN$-pairs. Comm. Algebra **8**, 543–583 (1980).
161. Howlett, R.B., Lehrer, G.I.: Induced cuspidal representations and generalized Hecke rings. Invent. Math. **58**, 37–64 (1980).
162. Howlett, R.B., Lehrer, G.I.: Representations of generic algebras and finite groups of Lie type. Trans. Amer. Math. Soc. **280**, 753–779 (1983).
163. Howlett, R.B., Lehrer, G.I.: On Harish-Chandra induction for modules of Levi subgroups. J. Algebra **165**, 172–183 (1994).
164. Howlett, R.B., Yin, Y.: Computational construction of irreducible $W$-graphs for types $E_6$ and $E_7$. J. Algebra **321**, 2055–2067 (2009).
165. Hu, J.: Crystal bases and simple modules for Hecke algebra of type $D_n$. J. Algebra **267**, 7–20 (2003).
166. Hu, J.: Crystal bases and simple modules for Hecke algebras of type $G(p,p,n)$. Represent. Theory **11**, 16–44 (2007).
167. Hu, J.: The number of simple modules for the Hecke algebras of type $G(r,p,n)$. With an appendix by Xiaoyi Cui. J. Algebra **321**, 3375–3396 (2009).
168. Hu, J., Mathas, A.: Morita equivalences of cyclotomic Hecke algebras of type $G(r,p,n)$. J. Reine Angew. Math. **628**, 169–194 (2009).
169. Iancu, L.: Markov traces and generic degrees in type $B_n$. J. Algebra **236**, 731–744 (2001).
170. Iancu, L.: On the centre of Iwahori–Hecke algebras and the construction of Jucys–Murphy elements. Preprint (2011).
171. Iwahori, N.: On the structure of a Hecke ring of a Chevalley group over a finite field. J. Fac. Sci. Univ. Tokyo **10**, 215–236 (1964).
172. Jacon, N.: Représentations modulaires des algèbres de Hecke et des algèbres de Ariki–Koike. Ph. D. thesis, Université Claude Bernard Lyon 1 (2004); available at "theses-On-line" http://tel.archives-ouvertes.fr/tel-00006383.
173. Jacon, N.: Sur les représentations modulaires des algèbres de Hecke de type $D_n$. J. Algebra **274**, 607–628 (2004).
174. Jacon, N.: On the parametrization of the simple modules for Ariki–Koike algebras at roots of unity. J. Math. Kyoto Univ. **44**, 729–767 (2004).
175. Jacon, N.: An algorithm for the computation of the decomposition matrices for Ariki–Koike algebras. J. Algebra **292**, 100–109 (2005).
176. Jacon, N.: Crystal graphs of higher level $q$-deformed Fock spaces, Lusztig $a$-values and Ariki–Koike algebras. Algebr. Represent. Theory **10**, 565–591 (2007).
177. Jacon, N.: Crystal graphs of irreducible highest weight $\mathcal{U}_q(\widehat{\mathfrak{sl}_e})$-modules of level two and Uglov $l$-partitions. J. Algebraic Combin. **27**, 143–162 (2008).
178. Jacon, N., Lecouvey, C.: Crystal isomorphisms for irreducible highest weight $\mathcal{U}_q(\widehat{\mathfrak{sl}_e})$-modules of higher level. Algebras Represent. Theory **13**, 467–489 (2010).
179. James, G.D.: The representation theory of the symmetric group. Springer Lecture Notes in Mathematics **682**. Springer-Verlag, Berlin (1978).
180. James, G.D.: The irreducible representations of the finite general linear groups. Proc. London Math. Soc. **52**, 236–268 (1986).
181. James, G.D.: The decomposition matrices of $\mathrm{GL}_n(q)$ for $n \leqslant 10$. Proc. London Math. Soc. **60**, 225–265 (1990).
182. James, G.D., Kerber, A.: The representation theory of the symmetric group. Encyclopedia of Mathematics, vol. 16, Addison-Wesley (1981).
183. James, G.D., Mathas, A.: A $q$-analogue of the Jantzen–Schaper theorem. Proc. London Math. Soc. **74**, 241–274 (1997).
184. James, G.D., Mathas, A.: The Jantzen sum formula for cyclotomic $q$-Schur algebras. Trans. Amer. Math. Soc. **352**, 5381–5404 (2000).
185. Jantzen, J.C.: Lectures on quantum groups. Graduate Studies in Mathematics, vol. 6. Amer. Math. Soc., Providence, RI (1996).
186. Jimbo, M., Misra, K.C., Miwa, T., Okado, M.: Combinatorics of representations of $U_q(\hat{\mathfrak{sl}}(n))$ at $q = 0$. Comm. Math. Phys. **136**, 543–566 (1991).

187. Joseph, A.: On the classification of primitive ideals in the enveloping algebra of a semisimple Lie algebra. In: Lie group representations, I (eds. R.L.R. Herb and J. Rosenberg), Lecture Notes in Mathematics **1024**, pp. 30–76. Springer, Berlin (1983).
188. Kac, V.G.: Infinite dimensional Lie algebras. Cambridge University Press (1985).
189. Kashiwara, M.: On crystal bases of the $q$-analogue of universal enveloping algebras, Duke J. Math. **63**, 465–516 (1991).
190. Kashiwara, M.: On crystal bases of quantum groups. Duke Math. J. **69**, 455–485 (1993).
191. Kashiwara, M.: Bases cristallines des groupes quantiques. Cours Spécialisés **9**. Soc. Math. France, Paris (2002).
192. Kashiwara, M., Miwa, T., Stern, E.: Decomposition of $q$-deformed Fock spaces. Selecta Math. (N.S.) **1**, 787–805 (1995).
193. Kawanaka, N.: Generalized Gelfand–Graev representations and Ennola duality. In: Algebraic groups and related topics, Advanced Studies in Pure Math. **6**, pp. 175–206. Kinokuniya, Tokyo, and North-Holland, Amsterdam (1985).
194. Kawanaka, N.: Shintani lifting and Gelfand–Graev representations. In: The Arcata Conference on Representations of Finite Groups (Arcata, Calif., 1986), Proc. Symp. Pure Math., vol. 47, Part 1, pp. 147–163. Amer. Math. Soc., Providence, RI (1987).
195. Kazhdan, D.A., Lusztig, G.: Representations of Coxeter groups and Hecke algebras. Invent. Math. **53**, 165–184 (1979).
196. Kazhdan, D.A., Lusztig, G.: Schubert varieties and Poincaré duality. Proc. Symp. Pure Math. **34**, 185–203 (1980).
197. Kiel, R., Weissauer, R.: Weil conjectures, perverse sheaves and $l$-adic Fourier transform. Results in Mathematics and Related Areas, 3rd Series, vol. 42. Springer-Verlag, Berlin (2001).
198. Khovanov, M., Lauda, A.: A diagrammatic approach to categorification of quantum groups I. Represent. Theory **13**, 872–933 (2009).
199. Khovanov, M., Lauda, A.: A diagrammatic approach to categorification of quantum groups II. Trans. Amer. Math. Soc. **363**, 2685–2700 (2011).
200. Kirillov, A.A., Jr.: Lectures on affine Hecke algebras and Macdonald's conjectures. Bull. Amer. Math. Soc. **34**, 251–292 (1997).
201. Kirwan, F.: An introduction to intersection homology theory. Pitman Research Notes in Mathematics Series, vol. 187. Longman (1988).
202. Kleshchev, A.S.: Linear and projective representations of symmetric groups. Cambridge Tracts in Mathematics, vol. 163. Cambridge University Press, Cambridge (2005).
203. Kleshchev, A.S.: Representation theory of symmetic groups and related Hecke algebras, Bull. Amer. Math. Soc. **47**, 419–474 (2010).
204. Kleshchev, A.S.: Branching rules for the modular representations of symmetric groups III: some corollaries and a problem of Mullineux. J. London Math. Soc. **54**, 25–38 (1996).
205. Knuth, D.E.: The art of computer programming, volume 3: sorting and searching. Addison-Wesley, Second Edition (1998).
206. Kreuzer, M., Robbiano, L.: Computational commutative algebra 1. Springer-Verlag, Berlin (2000).
207. Landrock, P., Michler, G.O.: Principal 2-blocks of the simple groups of Ree type. Trans. Amer. Math. Soc. **260**, 83–111 (1980).
208. Lascoux, A., Leclerc, B., Thibon, J-Y.: Hecke algebras at roots of unity and crystal bases of quantum affine algebras. Comm. Math. Phys. **181**, 205–263 (1996).
209. Leclerc, B., Thibon, J-Y.: Canonical bases of $q$-deformed Fock spaces, Int. Res. Math Notices **9**, 447–456 (1996).
210. Leclerc, B., Miyachi, H.: Constructible characters and canonical bases. J. Algebra **277**, 298–317 (2004).
211. Lusztig, G.: Coxeter orbits and eigenspaces of Frobenius. Invent. Math. **38**, 101–159 (1976).
212. Lusztig, G.: Irreducible representations of finite classical groups, Invent. Math. **43**, 125–175 (1977).
213. Lusztig, G.: Representations of finite Chevalley groups. C.B.M.S. Regional Conference Series in Mathematics, vol. 39. Amer. Math. Soc., Providence, RI (1977).

214. Lusztig, G.: Unipotent representations of a finite Chevalley group of type $E_8$. Quart. J. Math. Oxford **30**, 315–338 (1979).
215. Lusztig, G.: A class of irreducible representations of a finite Weyl group. Indag. Math. **41**, 323–335 (1979).
216. Lusztig, G.: On a theorem of Benson and Curtis. J. Algebra **71**, 490–498 (1981).
217. Lusztig, G.: Unipotent characters of the symplectic and odd orthogonal groups over a finite field. Invent. Math. **64**, 263–296 (1981).
218. Lusztig, G.: A class of irreducible representations of a finite Weyl group II. Nederl. Akad. Wetensch. Indag. Math. **44**, 219–226 (1982).
219. Lusztig, G.: Left cells in Weyl groups. In: Lie group representations, I (eds. R.L.R. Herb and J. Rosenberg), Lecture Notes in Mathematics **1024**, pp. 99–111. Springer, Berlin (1983).
220. Lusztig, G.: Characters of reductive groups over a finite field. Annals Math. Studies, vol. 107. Princeton University Press (1984).
221. Lusztig, G.: Intersection cohomology complexes on a reductive group. Invent. Math. **75**, 205–272 (1984).
222. Lusztig, G.: Cells in affine Weyl groups. Advanced Studies in Pure Math. **6**, Algebraic groups and related topics, pp. 255–287. Kinokuniya and North-Holland (1985).
223. Lusztig, G.: Cells in affine Weyl groups II. J. Algebra **109**, 536–548 (1987).
224. Lusztig, G.: Cells in affine Weyl groups III. J. Fac. Sci. Univ. Tokyo **34**, 223–243 (1987).
225. Lusztig, G.: Leading coefficients of character values of Hecke algebras. Proc. Symp. Pure Math. **47**, pp. 235–262. Amer. Math. Soc., Providence, RI (1987).
226. Lusztig, G.: On the representations of reductive groups with disconnected centre. Astérisque **168**, 157–166 (1988).
227. Lusztig, G.: Modular representations and quantum groups. Contemp. Math. **82**, 59–77 (1989).
228. Lusztig, G.: Canonical bases arising from quantized enveloping algebras. J. Amer. Math. Soc. **3**, 447–498 (1990).
229. Lusztig, G.: A unipotent support for irreducible representations. Adv. Math. **94**, 139–179 (1992).
230. Lusztig, G.: Introduction to quantum groups. Progress in Math. **110**. Birkhäuser, Boston (1993).
231. Lusztig, G.: Hecke algebras with unequal parameters. CRM Monographs Ser. **18**. Amer. Math. Soc., Providence, RI (2003).
232. Lusztig, G.: Comments to my papers, http://math.mit/edu/~gyuri.
233. Lusztig, G., Vogan, D.: Singularities of closures of $K$-orbits on a flag manifold. Invent. Math. **71**, 365–379 (1983).
234. Lux, K., Pahlings, H.: Representations of groups. A computational approach. Cambridge Stud. Adv. Math., vol. 124. Cambridge University Press, Cambridge (2010).
235. Lyle, S., Mathas, A.: Blocks of cyclotomic Hecke algebras. Adv. Math. **216**, 854–878 (2007).
236. Macdonald, I.G.: Symmetric functions and Hall polynomials, Second edition. Oxford Math. Monographs, Clarendon Press, Oxford (1995).
237. Malle, G.: Unipotente Grade imprimitiver komplexer Spiegelungsgruppen. J. Algebra **177**, 768–826 (1995).
238. Malle, G.: Spetses. In: Proceedings of the International Congress of Mathematicians, Vol. II (Berlin, 1998). Doc. Math., Extra Vol. II, 87–96 (1998).
239. Malle, G.: On the rationality and fake degrees of characters of cyclotomic algebras. J. Math. Sci. Univ. Tokyo **6**, 647–677 (1999).
240. Malle, G., Mathas, A.: Symmetric cyclotomic Hecke algebras. J. Algebra **205**, 275–293 (1998).
241. Malle, G., Rouquier, R.: Familles de caractères de groupes de réflexions complexes. Represent. Theory **7**, 610–640 (2003).
242. Martino, M.: The Calogero–Moser partition and Rouquier families for complex reflection groups. J. Algebra **323**, 193–205 (2010).
243. Martino, M.: Blocks of restricted rational Cherednik algebras for $G(m,d,n)$. Preprint (2010); see arXiv:1009.3200.

244. Mathas, A.: Simple modules of Ariki–Koike algebras. In: Group representations: cohomology, group actions and topology (Seattle, WA, 1996), Proc. Pure Symp. Math., vol. 63, pp. 383–396. Amer. Math. Soc., Providence, RI (1998).
245. Mathas, A.: Iwahori–Hecke algebras and Schur algebras of the symmetric group. Univ. Lecture Ser., vol. 15. Amer. Math. Soc., Providence, RI (1999).
246. Mathas, A.: Matrix units and generic degrees for the Ariki–Koike algebras. J. Algebra **281**, 695–730 (2004).
247. Mathas, A.: The representation theory of the Ariki–Koike and cyclotomic $q$-Schur algebras. In: Representation theory of algebraic groups and quantum groups, Adv. Stud. Pure Math., vol. 40, pp. 261–320. Math. Soc. Japan, Tokyo (2004).
248. Matsumura, H.: Commutative ring theory. Cambridge Stud. Adv. Math., vol. 8. Cambridge University Press, Cambridge (1986).
249. McDonough, T.P., Pallikaros, C.A.: On the irreducible representations of the specialisations of the generic Hecke algebra of type $F_4$. J. Algebra **218**, 654–671 (1999).
250. McDonough, T.P., Pallikaros, C.A.: On the irreducible representationsof the specialisations in characteristic 2 and 3 of the generic Hecke algebra of type $F_4$. J. Algebra **226**, 857–864 (2000).
251. McDonough, T.P., Pallikaros, C.A.: On relations between the classical and the Kazhdan–Lusztig representations of symmetric groups and associated Hecke algebras. J. Pure Appl. Algebra **203**, 133–144 (2005).
252. Michel, J.: Homepage of the development version of the GAP part of CHEVIE; see http://www.institut.math.jussieu.fr/~jmichel/chevie/chevie.html.
253. Müller, J.: Zerlegungszahlen für generische Iwahori–Hecke-Algebren von exzeptionellem Typ. Dissertation, RWTH Aachen (1995).
254. Müller, J.: Decomposition numbers for generic Iwahori–Hecke algebras of non crystallographic type. J. Algebra **189**, 125–149 (1997).
255. Murphy, G.E.: On the composition factors of Specht modules. J. Algebra **90**, 259–264 (1984).
256. Murphy, G.E.: On the representation theory of the symmetric groups and associated Hecke algebras. J. Algebra **152**, 492–513 (1992).
257. Murphy, G.E.: The representations of Hecke algebras of type $A_n$, J. Algebra **173**, 97–121 (1995).
258. Neunhöffer, M.: Arithmetical properties of idempotents in group algebras. C. R. Acad. Sci. Paris Sér. I Math. **330**, 535–539 (2000).
259. Neunhöffer, M.: Untersuchungen zu James' Vermutung über Iwahori–Hecke-Algebren vom Typ $A$. Ph. D. Thesis, RWTH Aachen (2003).
260. Neunhöffer, M., Scherotzke, S.: Formulas for primitive idempotents in Frobenius algebras and an application to decomposition maps. Represent. Theory **12**, 170–185 (2008).
261. Okuyama, T., Waki, K.: Decomposition numbers of $\mathrm{Sp}(4,q)$. J. Algebra **199**, 544–555 (1998).
262. Okuyama, T., Waki, K.: Decomposition numbers of $\mathrm{SU}(3,q^2)$. J. Algebra **255**, 258–270 (2002).
263. Parker, R.A.: The computer calculation of modular characters (the MEATAXE). In: Computational group theory (ed. M.D. Atkinson), pp. 267–274. Academic Press, London (1984).
264. Pietraho, T.: Knuth relations for the hyperoctahedral groups. J. Algebraic Combin. **29**, 509–535 (2009).
265. Pietraho, T.: Module structure of cells in unequal parameter Hecke algebras. Nagoya Math. J. **198**, 23–45 (2010).
266. Policzew, D.: Jantzen filtrations for Iwahori–Hecke algebras of exceptional type. Ph. D. thesis, University of Aberdeen, in preparation.
267. Puig, L.: Algèbres de source de certaines blocs de groupes de Chevalley. In: Représentations linéaires des groupes finis (Luminy, 16–21 mai 1988), Astérisque **181–182**, 221–236 (1990).
268. Ram, A.: Seminormal representations of Weyl groups and Iwahori–Hecke algebras. Proc. London Math. Soc. **75** (1997), 99–133.

269. Ram, A., Ramagge, J.: Affine Hecke algebras, cyclotomic Hecke algebras and Clifford theory. In: A tribute to C.S. Seshadri (Chennai, 2002), pp. 428–466. Trends Math., Birkhäuser, Basel (2003).
270. Ringe, M.: The C MEATAXE Release 2.4. Lehrstuhl D für Mathematik, RWTH Aachen (2004); electronically available at http://www.math.rwth-aachen.de/~MTX.
271. Robbiano, L.: Term orderings on the polynomial ring. EUROCAL '85, Vol. 2 (Linz, 1985), pp. 513–517. Lecture Notes in Comput. Sci., 204. Springer, Berlin (1985).
272. Rouquier, R.: Familles et blocs d'algèbres de Hecke. C. R. Acad. Sci. Paris, Sér. I, **329**, 1037–1042 (1999).
273. Rouquier, R.: Derived equivalences and finite dimensional algebras. In: Proceedings of the International Congress of Mathematicians (Madrid 2006), vol. II, p. 191–221. Eur. Math. Soc., Zürich (2006).
274. Rouquier, R.: $q$-Schur algebras and complex reflection groups. Mosc. Math. J. **8**, 119–158 (2008).
275. Rouquier, R.: 2-Kac–Moody algebras. Preprint (2008); see arXiv:0812.5023.
276. Shan, P.: Crystals of Fock spaces and cyclotomic rational double affine Hecke algebras. Ann. Sci. Ecole Norm. Sup. **44** (2010).
277. Shan, P.: Graded decomposition matrices of $v$-Schur algebras via Jantzen filtration. Preprint (2010); see arXiv:1006.1545.
278. Shephard, G.C., Todd, J.A.: Finite unitary reflection groups. Canad. J. Math. **6**, 274–304 (1954).
279. Schubert, V.: Durch den kovarianten Hom-Funktor induzierte Äquivalenzen. Dissertation Universität Stuttgart, Shaker-Verlag, Aachen 1999.
280. Spaltenstein, N.: Classes unipotentes et sous-groupes de Borel. Lecture Notes in Mathematics 946, Springer, 1982.
281. Spaltenstein, N.: A property of special representations of Weyl groups. J. Reine Angew. Math. **343**, 212–220 (1983).
282. Springer, T.A.: A construction of representations of Weyl groups. Invent. Math. **44** (1978), 279–293.
283. Springer, T.A.: Quelques applications de la cohomologie d'intersection, Séminaire Bourbaki (1981/82), exp. 589, Astérisque **92–93** (1982).
284. Srinivasan, B.: The characters of the finite symplectic group $\mathrm{Sp}(4,q)$. Trans. Amer. Math. Soc. **131**, 488–525 (168).
285. Starkey, A.J.: Characters of the generic Hecke algebra of a system of $BN$-pairs. Ph. D. thesis, University of Warwick (1975).
286. Steinberg, R.: Prime power representations of finite linear groups, II. Canad. J. Math. **9**, 347–351 (1957).
287. Steinberg, R.: Lectures on Chevalley groups. Mimeographed notes, Department of Mathematics, Yale University (1967).
288. Takemura, K., Uglov, D.: Representations of the quantum toroidal algebra on highest weight modules of the quantum affine algebra of type $\mathfrak{GL}_n$. Publ. Res. Inst. Math. Sci. **35**, 407–450 (1999).
289. Taskin, M.: Plactic relations for $r$-domino tableaux. Preprint (2008); see arXiv:0803.1148.
290. Turner, W.: Rock blocks. Mem. Amer. Math. Soc. **202**, (2009), no. 947, viii+102 pp.
291. Uglov, D.: Canonical bases of higher-level $q$-deformed Fock spaces and Kazhdan–Lusztig polynomials. In: Physical combinatorics (Kyoto, 1999), Progress in Math. **191**, pp. 249–299. Birkhäuser, Boston (2000).
292. Varagnolo, M., Vasserot, E.: On the decomposition matrices of the quantized Schur algebra. Duke Math. J. **100**, 267–297 (1999).
293. White, D.L.: Decomposition numbers of $\mathrm{Sp}(4,q)$ for primes dividing $q \pm 1$. J. Algebra **132**, 488–500 (1990).
294. White, D.L.: Decomposition numbers of $\mathrm{Sp}_4(2^a)$ in odd characteristics. J. Algebra **177**, 264–276 (1995).
295. Williamson, G.: Why the Kazhdan-Lusztig basis of the Hecke algebra is a cellular basis. Honours essay, University of Sydney (2003).

296. Xi, N.: A new proof for the classification of irreducible modules of a Hecke algebra of type $A_{n-1}$. Special issue in honour of Gus Lehrer, J. Algebra **321**, 3594–3600 (2009).
297. Yin, Y.: $W$-Graph representations for Coxeter groups and Hecke algebras. Ph. D. thesis, University of Sydney (2004).
298. Yvonne, X.: Base canonique d'espaces de Fock de niveau supérieur. Ph. D. thesis, Université de Caen (2005); available at "theses-On-line" http://tel.archives-ouvertes.fr/tel-00137705.
299. Yvonne, X.: A conjecture for $q$-decomposition matrices of cyclotomic $v$-Schur algebras. J. Algebra **304**, 419–456 (2006).
300. Yvonne, X.: An algorithm for computing the canonical bases of higher-level $q$-deformed Fock spaces. J. Algebra **309**, 760–785 (2007).

# Index

**a**-invariants, 13, 277
addable node, 264
adjustment matrix, 182, 183
affine Hecke algebra of type $A$, 334
affine symmetric group, 319
Ariki–Koike algebra, 259, 266
asymptotic case, 6, 149

balanced representation, 21
block, 200
block idempotents, 199
border nodes, 172, 264
Brauer graph, 137
Bruhat–Chevalley order, 60

canonical $\lambda$-tableau, 271
canonical basic set, 136, 144, 232, 284
Cartan matrix, 310
cell datum, 105
cell module, 105
cell representation, 105
cellular basis, 105
central character, 137
central idempotent, 199
centrally primitive idempotent, 199
character, 12
character table, 8
CHOP function, 365
Clifford system, 90
Clifford's theorem, 93, 151
column standard, 270
complex reflection group, 359
composition, 177
constructible representations, 197
content, 172
contragredient module, 9
Coxeter group, 2

crystal, 314
crystal basis, 314
crystal graph, 314
crystallographic type, 3
cuspidal, 217
cyclic module, 312
cyclotomic Ariki–Koike algebras, 276
cylindrical multipartition, 291

decomposition map, 135
decomposition matrix, 107, 135, 213
decomposition numbers, 213
defect, 156
degrees, 142
discrete valuation ring, 135
distinguished left coset representatives, 272
divided powers, 312
dominance order, 75, 77, 177, 271
dominant weights, 310

$e$-regular, 176
$e$-residue, 177
$e$-residue sequence, 175
equal-parameter case, 3
exceptional representations, 187
extended affine symmetric group, 319
extended Iwahori–Hecke algebra, 90

FLOTW $l$-partition, 291
Fock datum, 162
Fock space, 316
fundamental weights, 310

generalised Gelfand–Graev representation, 245
generic degree, 12
generic Iwahori–Hecke algebra, 4
good addable $i$-node, 320

M. Geck, N. Jacon, *Representations of Hecke Algebras at Roots of Unity*,
Algebra and Applications 15, DOI 10.1007/978-0-85729-716-7,

good removable $i$-node, 321

Harish-Chandra induction, 216
Harish-Chandra restriction, 216
Harish-Chandra series, 218
Hecke algebra, 219
height, 177, 269
highest weight vector, 312
Hom functor, 208, 220

$i$-word, 320
inertia group, 221
integrable module, 313
irreducible components, 2
isomorphism of crystal graphs, 315
Iwahori–Hecke algebra, 3

$\tilde{\mathbf{J}}$-cells, 38
James's conjecture, 185, 190
Jucys–Murphy elements, 156, 166, 267

Kashiwara operators, 313
Kashiwara–Lusztig canonical basis, 315
Kazhdan–Lusztig basis, 60
Kazhdan–Lusztig cells, 67
Kleshchev $l$-partitions, 322
Kostka number, 126

$L$-bad, 32
$l$-charge, 316
$l$-composition, 262
$L$-good, 32
$l$-partition, 263
ladder method, 176
leading matrix coefficients, 21
left cells, 36
level, 312
level of the Fock space, 316
Levi complement, 216
lies above, 182
LLT-graph, 174

Mackey formula, 217
monomial order, 5
multicharge, 316
multicomposition, 262
multipartition, 263

nodes, 264
normal addable $i$-nodes, 320
normal removable $i$-nodes, 320

opposite cell datum, 107
oppositive cell datum, 148
ordered $q$-wedge, 339
ordinary $\mathbf{m}$-symbol, 277
ordinary symbol, 269
orthogonality relations, 12

partition, 18
$\Phi_e$-modular decomposition map, 140
$\Phi_e$-modular specialisation, 160
pre-order, 36
principal series representations, 13, 220
principal specialisation, 155
pseudo-reflection, 359

$q$-wedge, 339
$q$-wedge product, 337
quantum affine algebra, 311
quasi-split case, 87, 228

reduced $i$-word, 320
reflections, 156
removable node, 264
residue, 287
residue sequence, 296
right cells, 36
root lattice, 310
row-standard, 270

Schur elements, 12
Schur functor, 208
Schur relations, 12
semi-infinite $q$-wedge, 343
semi-infinite $q$-wedge product, 343
shape, 177
shifted $\mathbf{m}$-symbol, 277
simple roots, 310
Specht modules, 273
special representation, 74
special unipotent classes, 74, 231
specialisation, 4
spetsial Ariki–Koike algebra, 277
split $BN$-pair, 216
Springer correspondence, 74
standard, 270
Starkey's rule, 169
Steinberg representation, 225
strongly triangular, 369
supercuspidal, 246
Suzuki group, 227
symmetric algebra, 12, 107

tableau, 270
$\theta$-blocks, 137
$\theta$-isolated, 194
Tits's deformation theorem, 8, 142

two-sided cells, 37

Uglov $l$-partitions, 321
unipotent principal series representations, 220
unipotent representations, 231
unipotent support, 231

very dominant multicharge, 322

$W$-graph, 27, 364
weight, 312
weight function, 4
weight lattice, 310
weight sequence, 276
weight space, 312
weight vectors, 312
Weyl group, 3, 216

Young diagram, 171, 177
Young tableau, 264
Young's rule, 126

MIX
Papier aus verantwortungsvollen Quellen
Paper from responsible sources
FSC® C105338

If you have any concerns about our products,
you can contact us on
**ProductSafety@springernature.com**

In case Publisher is established outside the EU,
the EU authorized representative is:
**Springer Nature Customer Service Center GmbH**
**Europaplatz 3, 69115 Heidelberg, Germany**

Printed by Libri Plureos GmbH
in Hamburg, Germany